The Practice of
Sheep Veterinary Medicine

This book is available as a free fully searchable ebook from
www.adelaide.edu.au/press

The Practice of
Sheep Veterinary Medicine

Kym Abbott

With contributions from

Philip Hynd (Chapter 6)

Simon de Graaf and Tamara Leahy (Chapter 8)

John Larsen (Chapter 9)

THE UNIVERSITY
of ADELAIDE

UNIVERSITY OF
ADELAIDE PRESS

A NOTE ABOUT CURRENCY

All dollar figures in this manuscript are given in Australian dollars unless otherwise stated.

Published in Adelaide by
University of Adelaide Press
Barr Smith Library
The University of Adelaide
South Australia 5005
press@adelaide.edu.au
www.adelaide.edu.au/press

The University of Adelaide Press publishes peer-reviewed scholarly books. It aims to maximise access to the best research by publishing works through the internet as free downloads and for sale as high-quality printed volumes.

For the full Cataloguing-in-Publication data please contact the National Library of Australia: cip@nla.gov.au

ISBN (paperback) 978-1-925261-77-6
ISBN (ebook: pdf) 978-1-925261-78-3

Senior editor: Rebecca Burton
Book design: Midland Typesetters
Cover: Emma Spoehr
Cover image: Lesley Abbott

CONTENTS

PREFACE

This book is intended to be a reference text for veterinarians who provide clinical services to sheep producers. The book is directed primarily at Australian sheep-raising systems, but we hope that the approaches described herein have wide application in countries where sheep are raised under extensive grazing conditions.

Australia has a unique history of involvement in the sheep industry. From humble beginnings in the late 18th century, the wool industry emerged in the mid-19th century to dominate the Australian economy for over 100 years. In the 1960s wool was still the largest single contributor to Australian export income but much has changed since then. Other commodities and services and other agricultural products have grown substantially while the wool industry has contracted to less than half the size it was in the early to mid-20th century.

There are some important legacies of the country's early dependence on wool. Australian sheep production is still based on the Australian Merino sheep — a breed (or group of breeds) derived from European Merinos and other breeds in the 19th century and developed into a remarkable new breed — a specialist producer of large quantities of high-quality fine wool. The dominance of the Merino and Merino crossbred in the national sheep flock is a unique characteristic of Australian sheep production and one which strongly influences the nature of sheep veterinary medicine in this country.

The production and export of fine wool remains one of Australia's most valuable sources of foreign income amongst agricultural commodities — worth around $3 billion annually to the country. Australian fine wool dominates the global market — accounting for a quarter of all wool traded in the world and contributing a much higher proportion to high-quality apparel production. The income from wool makes a very large contribution to the revenue earned on nearly all of the 30 000 sheep farms in Australia.

A major development in the Australian sheep industry since the late 20th century is the massive expansion of the sheep meat industry. While over 40% of lamb production is consumed in the domestic market, Australia is now also the largest exporter of sheep meats in the world — an export industry which has grown to be of similar magnitude to that of wool. New Zealand exports similar quantities of lamb and, together, these two countries dominate the world market in sheep meat.

Thanks to the strengths of these two industries, Australian sheep production remains a profitable and fulfilling agricultural pursuit for a large number of farm owners. This book is intended to assist those who work in the industry to add to the profitability and efficiency of sheep production systems, the quality of sheep products and the welfare of the sheep in those systems.

The book provides details about the way disease processes develop and manifest in sheep flocks and contains numerous references for those who wish to read further. Most of the important conditions of sheep in Australia are relatively straightforward to diagnose but the establishment of effective and economically sound control strategies is often the most difficult part of health management, particularly for those who are less familiar with sheep production systems. The first six chapters are intended to provide a basic understanding of some of the business and science underpinning sheep production and it is hoped that the reader is familiar with these before exploring any of the other chapters, which deal with reproduction and disease conditions, ordered largely on a systems basis. An underlying assumption of much of the text is that, given a sound understanding of how and why particular disease conditions occur, most well-informed producers and sheep health advisers will be able to develop effective control programs specific to each individual flock and its unique physical and financial environment. The knowledge that is critical to good control programs includes an understanding of the factors which lead to the development of poor health and poor productivity and the factors that predispose sheep to disease. This text aims first and foremost to provide that information.

ABOUT THE AUTHORS

Dr Kym Abbott is a sheep veterinary specialist and adjunct professor of sheep medicine at the University of Adelaide. Dr Abbott was a farm animal practitioner then sheep veterinary consultant in South Australia and western Victoria before taking up academic appointments at the University of Sydney and the Royal Veterinary College, London. He was the founding head of the Veterinary School at Charles Sturt University, Wagga Wagga, New South Wales, then head and professor of sheep medicine at the University of Adelaide. Dr Abbott completed his MVS at the Mackinnon Project of the University of Melbourne in 1986 and PhD in ovine footrot at the University of Sydney in 2000.

Associate Professor Simon de Graaf, co-author of Chapter 8, is an associate professor of animal reproduction and director of the Animal Reproduction Unit in the Faculty of Science at the University of Sydney. He has also held academic positions at the Royal Veterinary College, London and has been a visiting scholar at INRA, France. Dr de Graaf is a world expert in sheep reproduction, seminal plasma, sperm sexing and artificial insemination. He consults to the Australian artificial breeding industry, providing instructional courses on controlled breeding, standardisation of semen assessment, processing and freezing for semen processing centres. He is currently secretary general of the International Congress on Animal Reproduction, vice-president of the Association of Applied Animal Andrology, a member of the scientific advisory boards of Enterprise Ireland and an Editorial Board member of the international journal *Animal Reproduction Science*.

Professor Philip Hynd, the author of Chapter 6, is emeritus professor of animal production at the University of Adelaide. His teaching and research interests have centred on the application of animal physiology to animal production issues. Dr Hynd's research activities have included wool and hair biology of sheep and rumen function and nutrient yield in grazing ruminants. More recently, his work has centred on the interaction of nutrition and genes in developing embryos as foetal programming is becoming increasingly recognised as a critical determinant of the lifetime health and productivity of animals. Professor Hynd was elected a fellow of the Australian Society of Animal Production in 2010.

Associate Professor John Larsen, author of Chapter 9, is a senior researcher with the Mackinnon Project at the University of Melbourne Veterinary School, Werribee. He was director of the Mackinnon Group from 2001 until early 2018. After graduation, Dr Larsen worked in field and pathology positions with Agriculture Victoria until 1997. He initiated the 'Wormplan Focus Farms' extension program to promote better worm control practices on Victorian sheep farms in the 1990s and completed a PhD in immunoparasitology in 1998. Since then he has been involved in major industry-funded research on disease and internal parasites of sheep and beef cattle, and sheep blowflies, including the AWI Sustainable Control of Internal Parasites (SCIPS) and Integrated Parasite Management (IPM-s) projects, MLA

'Lifting the Limits' project and, most recently, field studies to validate models of worm infections funded by MLA.

Dr Tamara Leahy, co-author of Chapter 8, is a research fellow of reproductive biology in the Faculty of Science at the University of Sydney. Her current research focus is the investigation of the interaction between ram sperm, seminal plasma and the ewe's reproductive tract, with the aim of improving fertility following the cervical deposition of frozen-thawed ram sperm. Dr Leahy received her PhD from the University of Sydney in 2010 by demonstrating how seminal plasma proteins could be used to improve ram sperm function during processing for sex sorting or storage, and she has held a research fellowship position at the University of Queensland that detailed the proteome of bull sperm. Her research programs have been awarded over $1 million in funding since 2011 and the results have been published in over 20 refereed articles in international journals and presented at several national and international conferences.

1

VETERINARY SERVICES TO SHEEP FARMS

THE ROLE OF THE VETERINARY PRACTITIONER IN THE AUSTRALIAN SHEEP INDUSTRY

There is a general perception amongst veterinary students and sheep producers that there are limited opportunities for a 'sheep vet' because individual sheep are generally of low value and the cost of veterinary involvement is too high. It is quite true that the value of individual sheep in commercial flocks is generally too low for sheep diseases with a low incidence to attract veterinary intervention. A whole flock of sheep, however, may consist of 100 to 1000 or more individuals and the cost of disease in such a large group of animals can justify significant levels of veterinary intervention. The practice of sheep veterinary medicine is usually concerned with the diagnosis of disease in a portion of the flock, perhaps the first few cases of an epidemic or a flare-up of an endemic disease, and the institution of preventive plans to protect the rest of the flock. The large number of animals at risk and the large productive value of the flock can justify significant expense on veterinary investigation and provide the veterinary practitioner with ample financial scope to display his or her diagnostic skills. It does require that the veterinarian approaches cases in a manner quite different from that used for individual sick animals. A good sheep veterinarian has excellent diagnostic skills, and a solid understanding of the epidemiology of sheep diseases, and can apply that knowledge in the context of often complex farm business operations.[1]

One would have to say that Australian sheep flocks are still under-serviced by private veterinary practitioners. The reasons for this are numerous. One major factor has been the emphasis on individual animal medicine in veterinary education and in most facets of clinical veterinary work. Sheep growers have perceived this, usually correctly, and used their veterinary practitioner for services to individual animals of value — rams, for example, and farm animals of other species such as cattle. Occasionally, sheep are presented at clinics for examination or necropsy but the determination of action required on the farm in the light of the diagnosis has been very much in the hands of the client rather than the veterinarian.[2] Many opportunities to make significant improvements in farm productivity are then lost because the veterinarian is insufficiently familiar with the details of the farm operation or perhaps lacks the confidence to follow through and review the outcome of any remedial action.

There are three important components of the approach taken by successful sheep veterinarians. All of these components require either experience gained in the field, postgraduate training or both. These components are listed below.

1. Disease management in sheep flocks requires an epidemiological approach to both the investigation and the recommendations for prevention or treatment. Sheep flocks are populations and the approaches to disease control should be those of population medicine.

2. The economic consequences of both the disease and the steps necessary to reduce the disease prevalence must be considered. For many disease conditions, the law of diminishing returns applies to time and money spent on control. Resources should be allocated to control of a disease condition only while prevention costs less than the disease.[3] This topic is discussed further in Chapter 3.

3. The complexity and interconnectedness of management of a farm business requires that the farm be treated as a system.[4] When one aspect of a farm system is altered, there are consequences on other parts of the system. Farm managers often have a sense of the structure of their own farming system but have not usually formalised it in a way that can be presented to advisors such as veterinarians.[5] It is essential, therefore, that advisors seek to understand the farm system before creating plans that could have unforeseen and unintended consequences on the business as a whole. This topic is also further discussed in Chapter 3.

This book aims to encourage an interest in the practice of sheep veterinary medicine which is compatible with sound sheep management systems. The veterinarian must remain a sheep health expert but his or her knowledge of sheep management and sheep production systems and strategies must be developed to a moderate degree at least. This presents difficulties for many, particularly those who have not been exposed to rural life significantly before graduation. The problem, however, is far from insurmountable and the rewards are large. Sheep producers react quickly to the presence in their community of a veterinarian who, in their words, 'knows what sheep farming is all about!' They seek opinions on a wide range of sheep health matters and, if the advice is considered practicable, will implement the recommendations. This offers great satisfaction to the veterinarian, who will be able to witness the confirmation of the diagnosis and judge the effectiveness of the recommendations in the improvement of profits for the client and the health and welfare of the animals.

First, however, the veterinarian must develop knowledge of sheep-grazing systems both in general and specifically for the district and the client's property. A primary rule for sheep veterinarians emerges — *you must attend the farm*. Much becomes obvious when sheep and their environment are viewed first-hand — provided that the veterinarian knows what to look at, what to look for and what to ask. While high levels of skill only come with experience, the following description of some veterinary activities might help develop a basic approach.

THE VETERINARY ROLE ON SHEEP FARMS

The three roles which veterinarians in rural practice commonly have on sheep farms are

to investigate the occurrence of a disease at the request of a flock manager, and to make appropriate recommendations for treatment, control or elimination of the disease. The investigation occurs after a disease outbreak has occurred and the veterinarian visits the farm, collects a history, examines the environment and the affected portion of the flock, and collects such specimens as necessary to make or confirm a diagnosis. Specimens may include blood and faeces from a sample of the flock and possibly tissues collected at necropsy

- to make more general flock management and preventive medicine plans which will enable the producer to avoid serious disease problems, and to enact such plans. The plans might be designed to control (or eliminate) problems associated with, for example, clostridial diseases, internal parasites, ovine Johne's disease, footrot, lice, improper feeding, nutritional deficiencies and supplementary feeding or poor reproductive performance

- to carry out preprogrammed production-improving plans, such as assisted reproduction procedures like artificial insemination (AI) and multiple ovulation and embryo transfer (MOET).

To be effective in any of these roles, particularly the first two, veterinarians need to know how sheep farm businesses work and how farm decision making occurs. This requires a sound working knowledge of the sheep flock productive cycle, climatic seasonality, pasture growth, farm calendars and flock structures. Farm decisions are ultimately driven by a need to maintain a robust and profitable business. They are made to offer the greatest *sustainable* return to the producer, within a framework of limitations imposed by personal objectives or external regulation. Examples of personal objectives are the desire to have a low risk of business failure and the desire to avoid employing any additional staff; and examples of limiting regulations are the restrictions on the use of certain agricultural or veterinary chemicals, and constraints around the land use of some areas of the farm.

Farm decisions are not made to maximise condition score, animal health, wool production per head or the lamb marking percentage, but to make profits from the farm as a whole. This important principle will be developed in Chapters 2 and 3 in discussing farm economics and farm systems. To be most effective, veterinarians should understand that a desire for business success drives farm decision making. Ultimately, for there to be a long-term relationship between the veterinarian and the client, veterinary advice must increase the farm profitability and the financial security of the client.

There are some veterinarians who work as consultants on sheep farms. Effective veterinary sheep consultants or specialists must be able to carry out all of the above roles and, in addition, be able to give sound advice about other important flock management strategies, such as setting an appropriate stocking rate and flock structure, lambing time and shearing time, selecting the most appropriate genotype of sheep to run and, possibly, means of genetic improvement. Some consultants are also sufficiently familiar with pasture production, wool clip preparation, marketing and financial management to be able to integrate advice on those issues into their consultancy services. A full training in these latter fields is beyond the scope of an undergraduate course; various forms of postgraduate training are necessary for graduates who choose to develop their careers in this direction. Nevertheless, the generalist rural veterinarian does require some familiarity with these topics in order to develop sound recommendations and plans.

KEY ELEMENTS OF A SHEEP PRODUCTION SYSTEM

A sheep production system can be well described by defining the following:

(1) the breed and genotype of sheep in the flock

(2) the production objective of the flock

(3) the flock size and composition

(4) the farm's management calendar.

Breed and genotype[a]

The various types and breeds of sheep present in Australia are well described elsewhere (see Cottle (2010) in Recommended Reading below). In short, purebred Merinos dominate the national sheep flock, making up over 75% of the sheep shorn in 2014.[6] Merinos are considered a wool-producing breed with limited suitability as a meat sheep, but the larger-framed medium and strong-wool Merinos are increasingly being selected for characteristics which enhance both meat and wool production. There is a significant difference in the productivity of Merino sheep of different genotypes, and information about these differences is becoming increasingly available to Australian sheep producers through the national genetic evaluation programme MERINOSELECT. The important role of genetic evaluation for wool producers is discussed further in Chapter 5.

Approximately 10% of the national flock are Border Leicester x Merino ewes — often just called first-cross ewes — and these are the preferred type used as dams in flocks breeding second-cross lambs for meat. Second-cross lambs are those sired by a ram of a meat breed, and their high growth rate, excellent muscling and relative leanness make them highly suitable for meat production. Second-cross lambs are often called *prime lambs*, reflecting their advantage over first-cross lambs for meat.

The breeds which are used as prime lamb sires include Poll Dorset, White Suffolk, Texel and Suffolk. Sheep of these breeds are selected for their meat-production and wool is of no economic importance. The genetic evaluation programme used by breeders of these sheep is called LAMBPLAN.

There are some purebreeds considered dual-purpose (meat and wool) and the Corriedale is the most populous of these in Australia. The Dohne — a breed derived from a wool-type Merino and the German Meat Merino in the 1940s in South Africa — has been increasing in number in Australia since its introduction in 1988. The SAMM breed (South African Mutton Merino) was introduced into Australia in 1996.

The Dorper and White Dorper were two other breeds developed in South Africa — this time from the Black Headed Persian and the Dorset Horn breeds — which were introduced to Australia in 1996. Sheep of this breed shed their fleeces naturally, so shearing is not required and no income, therefore, is derived from wool. These breeds are examples of *clean-skin* sheep breeds now in Australia — sheep which shed their fleeces naturally each year — and they include Damara, Wiltshire Horn, Wiltipolls and breeds derived from a range of crosses.

A detailed discussion about the breed characteristics and the factors which make some breeds and genotypes more suitable for particular environments is beyond the scope of this text but is essential knowledge for veterinarians working with sheep.

---

a The word *genotype* usually describes a subpopulation of a breed, the individuals of which share distinctive genetic characteristics; it may therefore be used to denote *a strain* or *a bloodline* of a breed.

Production objective

Production objectives vary between farms within districts and between districts but, in commercial Merino flocks[b], production of high-value wool is usually the dominant objective, with income from the sale of surplus sheep being of secondary importance. Some Merino flock managers join a portion of the ewe flock to Border Leicester rams so that income from wool is supplemented by the sale of first-cross lambs or hoggets. In prime lamb flocks the chief objective is to concentrate on income from the sale of lambs, but significant income is still derived from the sale of wool and cast-for-age ewes. In ram-breeding flocks, including stud flocks, the production objectives differ in emphasis from commercial flocks, reflecting the value to the business of income from the sale of rams to other producers.

Flock structure and stocking rate

Flock structure and stocking rate will be examined in later chapters (Chapters 3 and 6).

Farm management calendars

The timing of major sheep husbandry and management events on farms (the farm management calendar) is important information to veterinarians for three reasons. First, the timing of events may be an important predisposing factor to outbreaks of disease. The clearest examples of this are the relationship between the time of lambing and the incidence of pregnancy toxaemia in ewes and the incidence of nutrition-related diseases in recently weaned lambs. Second, preventive medicine strategies, like drenching, vaccinating or footrot control, should be integrated with other management events which require mustering, to save time and labour for the farm operator. Veterinarians should be prepared to take the usual management calendar into account when recommending the timing of preventive therapies. Third, the timing of particular management events can have implications for total farm productivity unrelated to occurrences of disease. Examples include the time of lambing or time of shearing — two events for which the timing is critical to the success of the farm operation. Advice about timing of such activities is generally not considered to be part of the role of the general practitioner but it does form a significant part of the work of sheep specialist veterinarians.

The optimisation of the management calendar for a particular farm depends on the production objective and is complex, being influenced by environmental, health management and economic considerations. Sheep flocks may be non-breeding or breeding enterprises and the management calendar of a non-breeding enterprise generally has much more flexibility than that of a breeding flock. On non-breeding farms, the key decision is when to shear. On breeding properties, the key decision is when to join, followed by when to shear. The timing of most other husbandry practices will be related to these key decisions. Bell (2010) discusses this in some depth (see Recommended Reading).

Non-breeding flocks

Throughout the 20th century it was common for some Merino flocks to consist of wethers only or to be composed of a breeding flock and a wether flock in which most wethers were

---

b   The term *commercial flocks* refers to those flocks where the growing and selling of wool or lambs is the primary objective, in contrast to ram-breeding flocks, where ram sales are the primary source of income.

retained to adult ages. While this is unusual now, it simplifies an examination of management calendars to examine one for a non-breeding flock first. Sheep husbandry practices include some or all of the following:

(a) shearing and wool classing

(b) dipping or the use of *pour-ons* to control lice

(c) crutching

(d) jetting

(e) drenching

(f) foot paring and foot bathing

(g) vaccination

(h) disposal of cast-for-age sheep and purchase of young replacement sheep.

A sample calendar for a farm running mediumwool Merino wethers only is shown in Table 1.1.

Breeding flocks

In breeding flocks there are additional husbandry practices which relate to the reproductive cycle and the management of pregnant and lactating ewes, lambs and weaners. These include some or all of the following:

(i) joining

(j) pregnancy diagnosis

(k) lambing

(l) lamb marking/mulesing

(m) weaning

(n) culling breeders

(o) classing ewe hoggets.

A sample calendar for a Merino farm in southern Australia with a winter-dominant rainfall pattern and an autumn lambing is shown in Table 1.2. An example for a Merino flock

Table 1.1: Hypothetical management calendar for a non-breeding flock.

Practice	Time	Comments
Shearing	May	Winter shearing may expose sheep to risk of cold exposure. Off-shears sale prices may be high in winter.
Dipping	Two weeks after shearing	Or *pour-on* immediately off-shears. Are lice present and is dipping necessary?
Crutching	September and March, depending on when shearing occurs	How much crutching do wethers require? What is the duration of the blowfly season? Is a pre-shearing crutch necessary?
Jetting	September	If flystrike is occurring or likely to occur in spring.
Drenching	December and February	Tactical treatments at other times.
Pizzle rot prevention	September	Only necessary on improved pastures.

Table 1.2: Hypothetical management calendar for an autumn lambing Merino flock in southern Australia (winter rainfall zone).

Practice	Time	Comments
Joining	Dec–January	For 6 weeks from 1 Dec
Crutching	Early February	Shear rams
Vaccinating all ewes	April	Pre-lambing booster
Lambing	May–June	Lamb over 7 weeks
Marking mulesing and vaccinating lambs	Late June	Lambs 1 to 8 weeks old
Weaning lambs	Early September	At 3 to 4 months of age
Shearing	September	All sheep including rams
Classing ewe hoggets	September	Before or at shearing
Purchasing rams	September	Ready to join in December
Dipping all sheep	September	2 weeks off-shears
Selling cull maidens, CFA ewes, CFA rams	October	
Isolating rams from ewes	October	6 to 8 weeks before joining
Selling wether weaners	November	These weaners may also be retained.

in northern NSW is shown in Table 1.3. These calendars are incomplete. Not considered are such topics as:

(1) nutritional management of ewes to regulate condition score at joining and lambing

(2) management of the previous year's drop of young sheep

(3) worm control, blowfly control, and other essential husbandry activities.

The calendars in Tables 1.1 to 1.3, although fairly typical, are not necessarily the most suitable for all Merino properties. The optimisation of individual calendars will be examined in subsequent chapters but, in brief, it includes further examination of topics such as stocking rate, seasonality in pasture quantity and quality, reproductive performance, and the availability of markets for lambs, weaners and other surplus sheep.

INVESTIGATIONS OF DISEASE OR POOR PERFORMANCE IN A SHEEP FLOCK

Sometimes, veterinarians are asked to investigate a specific problem by the flock owner. The most common conditions which give rise to these requests are

- poor reproductive rate
- outbreaks of disease with significant mortality
- diarrhoea
- lameness
- fleece derangement.

Table 1.3: Hypothetical management calendar for a spring lambing Merino flock in northern NSW (summer rainfall zone).

Practice	Time	Comments
Joining	March-April	For 6 weeks from 1 March (inside the breeding season)
Shearing	June	All sheep including rams
Dipping all sheep	July	2 weeks off-shears
Lambing	August-September	Lamb over 5-6 weeks
Marking mulesing and vaccinating lambs	Late September	Lambs 1 to 7 weeks old
Weaning lambs	Early December	At 3 to 4 months of age
Purchasing rams	December	To use in March
Crutching	January	Shear rams
Classing maidens	January	
Selling cull maidens, CFA ewes, CFA rams	February	

At other times the request may be more general, such as a request to investigate

- poor growth or unexpectedly poor body condition in adults or young sheep
- weaner ill-thrift.

Occasionally a producer may make a request specifically for a preventive medicine programme. The most likely trigger for this is concern about control of internal parasites and anthelmintic resistance management, but it could also include a producer's concern about a vaccination programme, or nutritional supplementation, including trace element and vitamin nutrition.

Step 1

The veterinarian should arrange a time to visit the farm and have an agreement in advance with the client about what is to be examined and how long the visit might take. The client should be made aware that there is an hourly charge for the visit and should be advised of its likely cost. In general, clients appreciate that it is necessary to spend two to four hours to become sufficiently familiar with the farm and the sheep and to gather a good history. It is necessary, however, that the length and thoroughness of the farm visit is aligned with the client's expectations. For example, the producer may be expecting a visit and quick necropsy of two sheep, and there may be some resistance to the veterinarian unilaterally deciding to extend the visit and the cost without apparent reason or agreement in advance. The extended visit might have to wait until a short initial visit is completed; much depends on the client and his or her confidence in the sheep expertise of the veterinarian.

Step 2: The history

During the farm visit the veterinarian should gather both a history and a sense of the owner's understanding or prior experience of the problem. If the scope of the request is broad, the history gathering should be comprehensive. If the request is specific, then the questions should clearly relate to the problem at hand. For example, if the veterinarian is requested to investigate an outbreak of lameness, then the history collection should relate to that condition in the first instance at least. After some animals have been examined and the diagnostic possibilities have been narrowed, further history collection will be important, but the owner should be made aware of the relevance of the questions.

In contrast, if the investigation is to address a problem which is less well defined and more obviously a complex issue, such as the investigation of a poor reproductive rate, it is wise to spend an hour or so collecting history and reviewing records before inspecting or examining any sheep.

So, depending on the particular situation, the history collection could include the following:

- *Signalment of the affected animals*: Determine which sheep are affected and which are not, with groups being defined by their age group, sex, management group, reproductive status and breed.

- *The timing of events, relating to the time of year, climatic events, any husbandry procedures*: Determine when the condition first became noticeable and attempt to relate that to shearing, crutching, joining, drenching, moving from one pasture to another, or any other management events. Determine whether the problem has happened on previous occasions.

- *Relationship to introductions*: Is the affected group a purchased or a home-bred mob? If introduced, the time and source of the introduction should be noted.

- *Gathering some basic epidemiology*: The problem should be quantified, if possible, based on the number of sheep in the flock and in affected groups, their sex, age and other identifying factors. The number affected, number of sheep dying and the time sequence and pattern of disease occurrences should be recorded.

- *Gathering the general management history*: This should include joining and lambing dates, length of joining, shearing and crutching dates, weaning dates, normal drenching and vaccination dates.

- *Gathering the specific management information*: The details of anthelmintics, vaccines, supplementary trace element and vitamin nutrition should be collected, including products, dosages and frequency of treatment.

- *Gathering the nutritional history*: In addition to trace element and vitamin supplementation, other supplementary feeding programmes should be noted. If hay, grain or silage is used, the amount provided and the frequency of feeding should be noted. Supplementary feed should be calculated to a daily rate per head, in grams.

Other information relating to specific problems which should be collected with the history is discussed in more detail in the chapters dealing with particular syndromes. For example, you may need to know much more about fertiliser treatments, the source and health status of introduced sheep, stocking rates, nutritional supplements, parasite control approaches and other management interventions, depending on the problem being investigated.

The information gathered should be recorded for future reference. Usually, taking contemporaneous notes is best. By the time the history is collected, much more should be known about the *context* of the problem — the flock size and type, the management system, the animals affected, the broad nature of the problem — but also a sense of the owner's skill and experience and an appreciation of the *persistence* of the problem. It is important for the veterinarian to decide, for example, if the producer is a very able and experienced person who has battled the problem for years, finally calling in the veterinarian for assistance, or if this is a first-time occurrence of a relatively minor or straightforward problem. Elucidation of the intransigence of the problem and the ability of the producer will provide an indication of the expected scope of the investigation.

Step 3: The environment

If and when appropriate, the veterinarian should ask to be taken to see the sheep at pasture. This may involve an inspection of all of the different mobs on the farm or just a sample of the mobs. Inspecting the sheep at pasture also provides an opportunity to review the infrastructure of the farm. The veterinarian should note the quality of the sheep-handling facilities and the existence or not of a laneway system for moving sheep to and from the handling yards. Pastures should be inspected to determine whether they are improved, well fertilised and relatively weed-free. The current state of the pastures in terms of proportion of green or dry and the availability should be noted. Pasture availability may also provide some insight into the stocking rate of the pastures, particularly if pastures are set-stocked rather than rotationally grazed. Watering points should also be inspected to determine the ease of access of sheep to good-quality drinking water.

Step 4: The sheep at pasture

The sheep should be examined as a flock undisturbed at pasture. Examination should include both the affected groups and at least some of the non-affected groups. If possible, the mobs should be inspected first with relatively little disturbance and with any sheepdogs remaining in the farm vehicle. Some lameness conditions are best assessed when the sheep are walking quietly or grazing rather than when they are moving quickly. The mob should be observed as a whole in order to decide whether some animals have isolated themselves and are behaving differently from their flock mates.

After inspection without disturbance the mob could be gathered, provided it is safe to do so. Ewes with young lambs should not be disturbed, if possible. Moving the sheep to close-by yards or to another part of the paddock will enable an assessment of exercise tolerance and lameness. Coughing may not be apparent in some cases until the flock is made to move quickly for a short distance.

Presenting signs such as diarrhoea (based on presence of dags), lameness and fleece derangement are readily evident from an inspection of the mob at pasture, and a rough estimate of prevalence can be made. The general health and condition of the sheep can also be assessed by observing the fullness of the abdomen, or an obvious excessive range in size or condition. In the case of lambs, healthy, well-fed lambs are strong, and they run quickly and play, while unhealthy, underfed lambs adopt a more sedate or plodding form of walking. The *bloom* on lambs which are still on their mothers provides an indication of the quality of the lactation of the ewes.

It is possible to catch individual sheep in the field. Usually a producer with a good dog can hold a mob in a corner of a paddock while one or more sheep are caught for closer examination.

If this is done it is essential that the corner has sound fencing and does not include a boundary fence. If the sheep press onto the fence and break through, the veterinarian may be deemed at least partly responsible.

It is possible to complete this section of the flock work-up by inspecting the flock or a mob in the sheep yards. This is generally not as satisfactory as examination at pasture because behaviours can be dramatically changed by the stress of mustering. Lameness, for example, may be much less evident in the yards than when the sheep are inspected without disturbance in the field. Nevertheless, when a flock or mob is confined in a yard it is much easier to catch multiple individual sheep for more detailed examination.

Step 5: The sheep close up

In most instances, after the sheep have been examined as a group, individual sheep need to be examined in order to either confirm a diagnosis or to provide further clinical evidence, or specimens, so as to help arrive at a definitive diagnosis.

The technique for clinical examination of the individual sheep is described elsewhere (for example, Jackson and Cockcroft (2002) — see Recommended Reading below).

Condition scoring and body weighing

Sheep are condition scored on a 1 to 5 scale.[7] Scoring is best done with a group of sheep standing in a race. It is often useful to condition score a random sample of the flock (20 to 30 animals), to record the scores and to calculate an average. Condition scoring provides an instant assessment of the nutritional status of the sheep and can be used as a basis for continuing monitoring. The condition score should be related to the current physiological state of the animals. For example, a mean condition score below 2.0 of a group of late-lactation, prolific ewes may be acceptable and normal. The same condition score in late pregnancy could well be associated with a high risk of pregnancy toxaemia, high lamb mortality or long-term negative effects on the productivity of the progeny.

Weighing a sample of the flock can also be useful as a basis for monitoring, but without a benchmark or a second weight for comparison it is not possible to use bodyweight alone as a diagnostic clue for adult sheep. For young sheep body weighing can be very useful, particularly when compared to the reference weight for adult sheep of the same genotype. For example, the mean body weight of a group of weaned Merino lambs can be very informative about the risk of malnutrition and death as a result of poor feed quality.

Sampling the flock

Specimens are frequently collected from live sheep to aid the diagnostic process. Tissues typically include blood (for biochemistry such as trace element nutrition, and for immunology such as the detection of rising titres to infectious disease as well as for proving disease freedom) and faeces for parasitological diagnosis.

It is important that sufficient numbers of animals are sampled but, frequently, too few are tested to produce dependable results. The more animals that are tested from the one management group, the more reliable is the estimate of the mean value (Figure 1.1). For very variable parameters, such as faecal egg counts, 10 is the lowest number of animals that should be sampled to usefully estimate the mean value for the group from which the sample of animals was derived.[8,9] The higher the number tested, the narrower the confidence interval around the estimate. Note that the confidence interval around the estimate of a mean is, for large populations and small sample sizes, independent of the population size. It is misleading to suggest that an appropriate sample size can be based on a percentage of the population — it is the absolute size of the sample that matters, not the size of the population from which the sample is drawn nor the relative size of the sample compared to the whole population. One should choose an appropriate sample size based on the desired level of accuracy and the degree of variability within the population (the standard deviation) using standard and straightforward statistical formulae.

Remember what confidence intervals tell us — in the case of 95% confidence intervals, we are 95% confident that the true population mean lies within the range specified. Another way to consider this is that, if we took 100 samples of a particular size and estimated the mean and 95% confidence limits each time, we would expect that the true mean would lie within that range on all but five occasions.

Necropsy

Necropsy of sheep is a very valuable diagnostic tool and the opportunities it presents should not be wasted by poor techniques or lack of specimen collection for expert review in a diagnostic laboratory. The sheep chosen for necropsy should be recently dead or sacrificed on the basis of advanced clinical signs. Multiple necropsies are advisable — three animals with consistent evidence of similar syndromes give much more compelling evidence than just one animal. Every opportunity to collect tissues for further examination should be made — including the gastrointestinal tract for total worm count (faecal egg counts are *not* useful in one animal or animals in ill health), liver sections for trace element assays and a full set of tissues for microbiological and histopathological testing — including brain.

RECOMMENDED READING

Bell KJ (2010) Sheep management. In: International sheep and wool handbook, ed DJ Cottle. Nottingham University Press: UK, pp 407-24.
Cottle DJ (2010) World Sheep and Wool Production. In: International sheep and wool handbook, ed DJ Cottle. Nottingham University Press: UK, pp 7-36.
Court J, Webb Ware J and Hides S (2010) Sheep farming for meat and wool. CSIRO Publishing: Collingwood, Victoria, pp 16-18 and Appendix 1.
Jackson P and Cockcroft P (2002) Clinical examination of farm animals. Blackwell Science: UK.
West DM, Bruere AN and Ridler AL (2009) The sheep: Health, disease & production. 3rd ed. VetLearn Foundation: New Zealand, pp 1-9.

REFERENCES

1   Sackett DM (1997) A specialist sheep veterinarian's view of the needs of the sheep industry. In: Proceedings of the Fourth International Congress for Sheep Veterinarians, 2-6 February, Armidale, ed MB Allworth. Australian Sheep Veterinary Society: Indooroopilly, Qld, pp. 31-5.
2   Kaler J and Green LE (2013) Sheep farmer opinions on the current and future role of veterinarians

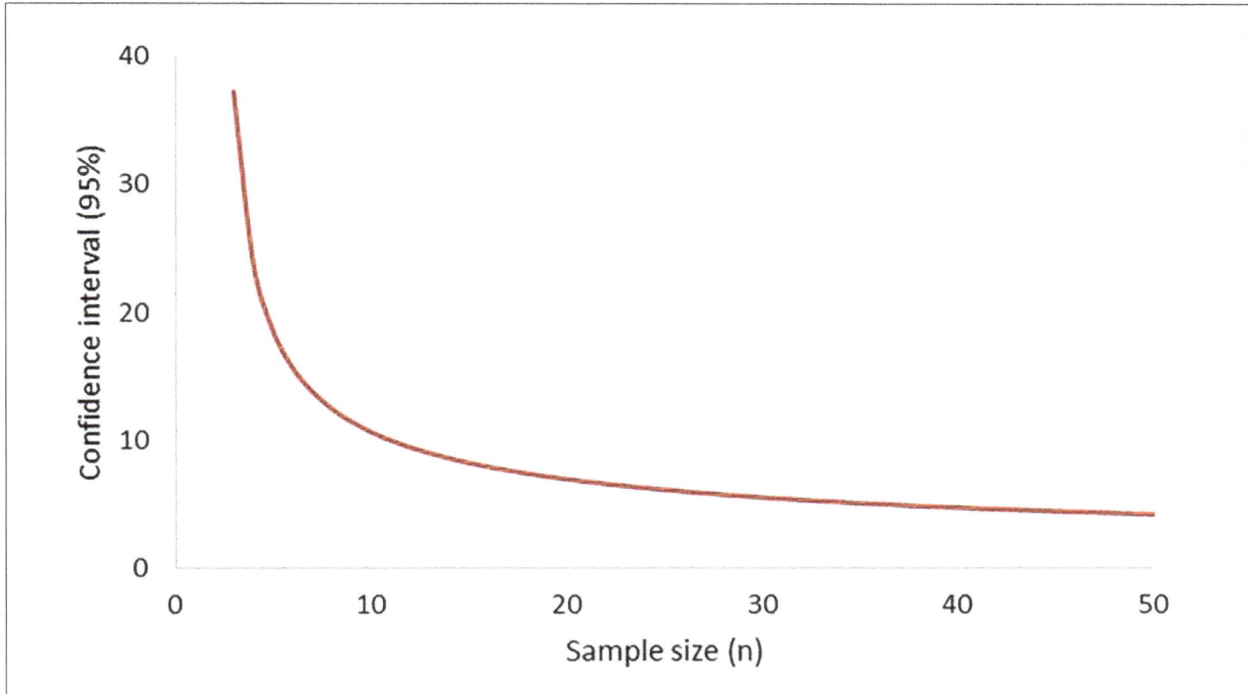

Figure 1.1: An illustration of the relationship between sample size and the confidence intervals (ci) for an example of glutathione peroxidise (GSHPx) measurements in lambs, used as a guide to selenium nutrition status. Paynter et al. (1979)[10] found that groups of lambs from different flocks had mean GSHPx levels ranging from 10 to 50 units per gram of haemoglobin but that only flocks of lambs with mean levels below 30 responded with increased growth rates if supplemented with selenium (i.e. 30 could be considered to be the threshold for diagnosing selenium deficiency). In their study, the standard deviation of values within each group of lambs tested was variable but around 15. This graph tells us that, for a standard deviation of 15, 95% confidence intervals could not be considered 'acceptable' unless 10 lambs or more have been tested. If we test only 10 lambs then we are 95% confident that the true mean lies within a 21.4 unit range (95% ci for n=10 is $\pm$ 10.7). With fewer than 10 lambs tested we may not be sufficiently confident that the result will be useful to the client unless the sample mean is approaching an extreme value. The figure is a graph of n against $t_{n-1(.95)}.s/\sqrt{n}$ where s is the sample standard deviation and the value for $t_{n-1(.95)}$ can be found from tables of the t distribution or using the Microsoft Excel® function T.INV.2T (0.05,n-1). Source: KA Abbott.

in flock health management on sheep farms: A qualitative study. Prev Vet Med **112** 370-7. https://doi.org/10.1016/j.prevetmed.2013.09.009.

3   Perry BD and Randolph TF (1999) Improving the assessment of the economic impact of parasitic diseases and of their control in production animals. Vet Parasitol **84** 145-68. https://doi.org/10.1016/S0304-4017(99)00040-0.

4   Wilson J (1995) Changing agriculture. An introduction to systems thinking. 2nd ed. Kangaroo Press: Sydney.

5   Fairweather JR and Hunt LM (2011) Can farmers map their farm system? Causal mapping and the sustainability of sheep/beef farms in New Zealand, Agric Human Values **28** 55-66.

6   Australian Wool Innovation. Available from: http://www.wool.com/market-intelligence/wool-and-sheepmeat-survey/. Accessed 9 February 2016.

7   Jefferies BC (1961) Body condition scoring and its use in management. Tasmanian J Agric **39** 19-21.

8   Brunsdon RV (1970) Within-flock variations in strongyle worm infections in sheep: The need for adequate diagnostic samples. NZ Vet J **19** 185-8. https://doi.org/10.1080/00480169.1970.33896.

9   Morgan ER, Cavill L, Curry GE et al. (2005) Effects of aggregation and sample size on composite faecal egg counts in sheep. Vet Parasitol **131** 79-87. https://doi.org/10.1016/j.vetpar.2005.04.021.

10   Paynter DI, Anderson JW and McDonald JW (1979) Glutathione peroxidise and selenium in sheep, II. The relationship between glutathione peroxidise and selenium-responsive unthriftiness in Merino lambs. Aust J Agric Res **30** 703-9. https://doi.org/10.1071/AR9790703.

2

THE SHEEP FARM AS A BUSINESS

INTRODUCTION

This chapter is intended for veterinarians working as rural practitioners and servicing the grazing industries who wish to better understand the financial environment in which their clients operate their businesses. The objective of this chapter is to explain some of the economic tools used to describe and plan farm businesses and to provide some examples which illustrate their size, financial structure and economic constraints. In general, veterinarians do not have access to the details of their clients' business arrangements but, with some knowledge of industry norms, veterinarians can provide advice which includes consideration of the likely economic impact on the farm business. When veterinarians make recommendations which include accurate economic assessments, their credibility is enhanced and the client benefits from the sound and appropriate advice.

Advice is *economically sound* if the net financial benefit which arises from implementing the advice increases profit or decreases the risk of loss for the farm business. Such advice is not always easy to give, for it demands a proper appreciation of the whole farm system, not just the isolated component currently under investigation. For example, consider a farm in southern Australia on which a client's spring-born Merino weaners are dying during the summer from malnutrition as a consequence of their low body weights. The client could be advised to move the time of lambing to autumn or winter to ensure that the weaners are better grown before summer. While this advice is technically correct — earlier-born weaners will be bigger by summer — it is not the best advice that a veterinarian can give. An earlier lambing will also have the effect of decreasing the productivity of the ewes. The client, if he or she took the advice, might lose more money from the poorer productivity of the ewes than he or she gains in better weaner health. An economically sounder solution could be to institute handfeeding with high-energy supplements earlier before the weaners lose weight. Veterinary advice should take into consideration the effect of any changes on the economic structure of the business as a whole, not just a single component. The need to adopt a *systems approach* when advising changes in farm activities is discussed further in Chapter 3.

When making recommendations which involve substantial up-front costs, veterinarians often need to make judgements about the financial strength of a farm business, even though the details are usually not available to them. Not all farm businesses are sufficiently robust to sustain extra expenditure unless the financial benefits materialise quickly. When veterinarians do not know the financial position of the client's business, an awareness of the financial position of the 'average' farm can assist the development of recommendations. It may be wise to offer a range of options, so that the client can choose one which is *appropriate* for his or her current financial position.

For example, following the diagnosis of footrot on a property, the client may be correctly advised that the disease is sufficiently serious to warrant eradication. It may be that the same client is in a tenuous financial position with limited ability to fund any further capital expenditure or employ further labour. The lack of funds will reduce the chance of a successful eradication campaign; the extra expenditure may push the farm business over the brink of viability. It is of little solace to a producer to have achieved eradication of footrot after three years if the farm business is no longer financially viable. In such a case, appropriate advice would be to establish effective, low-cost control measures which limit the effect of the disease while an experienced financial adviser addresses the immediate financial problems which the producer is facing.

Advice which is economically inappropriate is unlikely to be adopted or, worse, may be adopted and contribute to a deteriorating financial situation. Producers are accustomed to receiving veterinary advice which they choose not to implement for economic reasons. In the case of dairy farmers, it has been shown that producers generally do not perceive their veterinarian as competent in farm finance or business management[1,2], with implications, therefore, for compliance with veterinary recommendations. The same attitude is likely to be common amongst sheep producers. Producers continue to have trust in their veterinarian and to use veterinary services to make diagnoses, but the perceived lack of financial wisdom of the veterinarian often means that the veterinarian's suggestions for managing a problem are ignored or highly modified by the producer. Producers do not necessarily doubt the science behind veterinary advice, but they may question its practicality for farm businesses.[3]

In order for veterinarians to give appropriate and effective advice, with an expectation of a high level of producer compliance, several elements are necessary. First, it is important that they understand the basic financial structure of farm businesses. This means having an appreciation of the economic operation of farms in general, even if information about the specific client is unavailable. Second, veterinarians need to understand the effects that particular management strategies have on the profitability of those businesses.[4] Third, it is often best to offer a range of options when advising clients, particularly if doing so without privileged insights into the client's financial affairs.

SOME BASIC ACCOUNTING — FINANCIAL STATEMENTS

It is possible to gain useful insights into the soundness and profitability of a business by examining its financial statements. Two of the most useful and familiar statements for this purpose are the *Balance Sheet* and the *Profit and Loss Statement*. While most veterinarians in general practice will not have access to a client's financial statements it is still useful for veterinarians to be familiar with accounting terminology and the financial structure of farm businesses.

Balance sheet (also known as the *Statement of financial position*)

The Balance Sheet is a list of the *assets* and the *liabilities* of the business. An asset is an item which is of value to the business — which could be sold to realise cash. Assets usually enable the business to produce income or, in the case of cash, allow the purchase of income-producing items. Assets include buildings, tractors, livestock and cash in the bank. Assets can also include debts owed to the business. In veterinary practice, for example, clients who do not pay for veterinary services immediately but delay payment for some days or weeks are known as *debtors* of the business and constitute an asset of the business. Similarly, farmers waiting for their wool cheque from their wool-selling agency after the wool is sold have that agency as a debtor to the farm business.

In contrast, debts owed by the business are *liabilities*. In most businesses, loans to the business constitute the major liabilities, and these loans are usually made by banks or private individuals. These lenders are known as *creditors* of the business. In veterinary practice, suppliers of pharmaceuticals to the practice may wait up to 30 days or more for payment. After they have supplied the goods and while they are awaiting payment, they are *creditors* of the business.

In accountancy terms, the *business* rather than the *proprietor* owns the assets. The proprietor is one of the creditors or suppliers of funds to the business — along with the bank, the stock firm, relatives or anyone who has lent money to the business. The liability of the business to the proprietor has a special name — *equity*, or *owner's equity*. It represents the capital funds introduced to the business when the proprietor started it and so is often referred to on Balance Sheets as the owner's *capital account*. *Equity* is the farm owner's investment in the farm business.

By definition, on a Balance Sheet,

$$\text{Assets} = \text{Liabilities} + \text{Equity}$$

The Balance Sheet, therefore, is always in balance. The top part of the Balance Sheet (Table 2.1) lists and totals the assets of the business; the second part lists and totals the liabilities and the owner's equity. The two totals will always be equal.

The Balance Sheet demonstrates how much money the owners could realise if they sold all the assets and paid all the debts. It shows, therefore, the *financial position* of the owners in relation to the business *at any one point in time*. It does not reflect the profitability of the business, although inferences could be made (with knowledge of the likely productivity of their particular enterprise) about the *viability* of the business. The farm business is *viable* if it has the ability to service (pay interest on) its debts and still have some money left over to provide the farm family with living expenses. If it does not have that ability, its debts will increase and the owner's equity will decline.

A simple Balance Sheet is illustrated in Table 2.1. The farm business had, at July 2017, assets worth a little over $2.8m. Another way to say this is that 'the farm is worth $2.8m'. The farm business has liabilities of $120 000 — a loan from a relative, perhaps one whose share in the farm was purchased by the present proprietors but who was not paid in full, preferring to be paid interest by the farm business. The other 'loan' is a bank account that has an overdraft facility but is currently not overdrawn and that has, therefore, a zero balance.

By definition, the remainder of the value of the farm business belongs to the two proprietors of the business — the farm owners. If the farm assets are truly valued at current market values, then the farm could be sold and all debts ($120 000) paid; and the proprietors would keep the rest of the proceeds. This amount ($2 688 500) is their *equity* in the business. Equity is

Table 2.1: Balance sheet for a sheep farm business.

Balance Sheet
July 1 2009
J Smith and Son

ASSETS

Farm land and improvements	2,500,000
Plant and equipment	148,000
Livestock	160,000
Bank account	500
TOTAL ASSETS	**2,808,500**

LIABILITIES

Loan from SA Smith	120,000
Overdraft at bank	0
TOTAL LIABILITIES	**120,000**

EQUITY

Capital account J Smith	1,344,250
Capital account J Smith Jnr	1,344,250
TOTAL EQUITY	**2,688,500**
Liabilities plus Equity	**2,808,500**

sometimes expressed as a percentage of the total assets. These proprietors have 2 688 500 ÷ 2 808 500 × 100 = 95.7% equity in their farm business.

Although it is not shown on the Balance Sheet, we can expect that the business has to pay interest on its debts. The level of indebtedness will vary through the year but, if $140 000 were the average debt, the interest bill might be around $7000 per annum, depending on the interest rate. This amount of interest will show on the Profit and Loss Statement, because interest is one of the expenses of the business. On some farms, equity is very low and farm debts are very high. In such cases the interest bill may be so high as to cripple the business — because it cannot generate enough income to pay the interest. A business which cannot pay the interest on its loans without further borrowings is not viable.

Profit and Loss Statement (also known as the *Income statement*)

The Profit and Loss Statement reveals the revenue earned and the expenses incurred in the operation of the farm business over a defined period — usually a financial year but sometimes a shorter period. Revenue earned on-farm normally comes from the sale of produce (wool, meat, livestock, milk, grain, etc.) and expenses are those monies spent on products completely consumed in the production process (shearing, drench, fuel, etc.) rather than on assets with a much longer life (buildings, tractors, etc.) or family expenses unrelated to the farm production (clothes, groceries for the immediate family, school fees, fuel for the family car, etc.). In

Table 2.2 the Profit and Loss (P & L) Statement shows that the farm business made a profit in the 12 months to 30 June 2018 of $241 502. There are several important points to notice in this statement.

Different types of expense account

Each line in the statements relate to one item or *account*. The expense accounts in the P & L Statement can be classified into several categories. Payment of interest, for example, is a *financial* expense. A major distinction exists between *variable* and *fixed* expenses. The difference between these two categories is important in some forms of farm financial analyses, which will be discussed below. *Variable expenses* are those which vary with the size or intensity of the enterprise. Note the first seven expense accounts in the example shown in Table 2.2. If there were no sheep on the farm, these would have zero balances — that is, nothing would be spent on shearing, animal health, etc. These are examples of variable expenses because they would become bigger if the enterprise were to run more sheep. Other costs would not vary if there were more sheep; administrative costs, rates and taxes, insurance for fences and sheds, for

Table 2.2: Profit and Loss Statement for a sheep farm business.

Profit and Loss Statement
June 30 2018
J Smith and Son

REVENUE	
Income from wool	243,339
Income from livestock sales	322,962
Bank interest	1,320
Total	567,621
EXPENSES	
Shearing	67,000
Animal health	36,000
Supplementary feed	32,500
Flock rams	21,000
Wool and livestock freight & costs	27,500
Livestock insurance	800
Miscellaneous stock requisites	2,500
Fertilizer	18,819
Interest	7,000
Rates & taxes	15,000
Administration	12,000
Farm insurance	9,000
Repairs and maintenance	35,000
Miscellaneous	18,000
Depreciation	24,000
Total	326,119
PROFIT	241,502

example, would remain the same. These costs are *fixed* — they exist whether the paddocks are stocked, cropped or empty.

Owner's labour

On this farm, the owners have supplied the main portion of the labour and management of the farm. The sum of $241 502, therefore, represents the reward for their management and labour and provides a return on their equity (their financial investment in the farm business). If the two owners had paid themselves a proper 'wage' for their work on the farm the surplus would have been substantially smaller. Being owners rather than employees, funds which they withdraw from the business to spend on personal items are termed *drawings*, rather than wages.

Return on capital

The total capital value of the farm is $2 808 500. In order to estimate the value of their farm and farm business as an investment, the owners could allocate a value to their labour and management of, for example, $100 000, leaving a surplus of $141 502. That surplus, expressed as a percentage of the total capital value, represents a 5% rate of return.

Return on equity

If the surplus is expressed as a percentage of their equity ($2 688 500), the owners could consider that they have received a 5.2% return for their investment (their equity) in the farm business.

Allocation of the profit

Assuming the family do not withdraw all of the profit for personal expenses, there is a surplus produced in the 2017-18 year which can be applied either to reinvestment in the farm infrastructure (a new piece of equipment, for example, or a new farm building) or to debt reduction. Either way, if spent wisely, the expenditure should increase the likelihood of future profits in the business. Alternatively, the family may choose to withdraw some of the profit for investment off-farm.

Depreciation

Depreciation of the value of some assets is an *expense* of the farm business but is not a *cash expense*. It is an expense on paper only. It is good business practice to account for the decline in value of farm assets like tractors, shearing plant, etc., and this allowance for depreciation is deductible from taxable income. Nevertheless, in the example in Table 2.2, the net *farm cash income* is actually $24 000 higher than the profit shown because the effect of the depreciation of assets will not be realised until the depreciated assets are sold.

Cash flow

The business received interest as income but also paid interest as an expense. This fact reveals an important characteristic of *cash flow* on sheep farms. Income in sheep enterprises is largely derived from a few major events each year, such as shearing and sheep sales, which often occur within just a few weeks or months of that year. When wool or sheep are sold, some short-term debts (such as bank overdrafts) are paid, while surplus cash is invested in term deposits (for three to six months, perhaps), and the farm business receives some interest from the invested funds. Core debt, such as farm mortgages or other long-term loans, is not repaid because some cash will be required for operating funds during the year. By the time shearing and sheep sales come around again, cash could be in short supply and the business may be operating with bank overdrafts again. One can see that, if farm debt were high, interest expenses could be so high that the interest bill would exceed the profit. When that happens, unless remedial action is taken, additional funds must be borrowed every year. The farm business can then fall with increasing speed into overwhelming indebtedness.

FINANCIAL PERFORMANCE OF FARMS PRODUCING SHEEP

The types of farms running sheep in Australia cover a broad range, from those which are specialist wool or lamb producers to those with a mixture of several enterprises with sheep forming a small part of the whole farm business. The size of flocks also varies considerably. A commercial flock may consist of just a few hundred ewes to 20 000 or more.

A family-operated farm — still the most common type of farm business in Australia — might need to run 3000 or more ewes to provide a satisfactory income for the family, if sheep were the only farming enterprise on the farm.

The statements in Table 2.2 are an example of the financial statements for a family-operated sheep farm of 7800 dry sheep equivalents (DSE[a]), running 3000 adult ewes on 800 hectares. Most ewes are mated to Merino rams but some are mated to meat-breed rams, producing first-cross slaughter lambs. Farm income is of the order of $189 per ewe present, which includes wool from the ewes, ewe hoggets and lambs, as well as the sale of crossbred lambs, Merino wether weaners, cull ewe weaners and cast-for-age sheep. Fleece values of adult Merino sheep have generally ranged between $30 and $60 in the decade to 2018 and the sale price of lambs at slaughter age has ranged between $80 and $150. Many factors influence these values, but some knowledge of the approximate productive value of sheep is essential to the process of advising producers about health and management strategies. Note also that profit is, in this example, about $30 per DSE but might be significantly less than that if the true value of family labour was included as a cost.

It is of interest to examine some financial statistics collected from lamb-producing farms across Australia. The following statistics for these farms, as discussed below and illustrated in Figure 2.1, are supplied by the Australian Bureau of Agricultural Resource Economics (ABARE).[5] Farm businesses reported here are those which derive a significant portion of their income from lamb sales by selling 200 or more lambs per year. Most such farms operate with a mixture of enterprises, including cropping and beef cattle.

In 2016-17, around 18 000 Australian farms sold more than 200 lambs for slaughter and nearly half of those sold more than 500 annually. About half of the total number of slaughter

---

a   A dry sheep equivalent (DSE) refers to the nutritional requirements of one head of livestock compared to that of a non-reproductive adult sheep. A ewe is typically rated as 2.0 to 2.5 DSE. See Chapter 6 for further discussion.

lambs produced came from farms selling 500 to 2000 lambs annually. These farms also receive income from wool and, usually, other farming enterprises.

- The average *farm cash income* for such farms over the past decade has ranged between $120 000 and $265 000 (Figure 2.1). Farm cash income is defined as total cash receipts less total cash costs. Cash costs include interest payments, employed labour, fertiliser, repairs and maintenance, etc. Cash income has been highly variable over the past 20 years as a consequence of variation in seasonal conditions and in the price received at sale for lambs, adult sheep, wool, beef and crops.

- Average *farm business profit* — defined as farm cash income adjusted for changes in the livestock and fodder inventories over the year, depreciation of assets and an imputed cost for unpaid family labour — was $141 000 in 2016-17.

- Farm business profit is strongly influenced by the level of indebtedness and therefore by the size of the annual interest bill. For example, a farm with a debt of $500 000 borrowed at 6% will pay $30 000 per year in interest.

- The average farm debt on lamb-producing farms in 2016-17 was estimated to be $736 000. Most farms operators (around 60%) have 90% or greater equity in the farm business, but 12% are operating with equity below 70%.

- Interest payments have consumed between 6% and 10% of farm cash income on lamb-producing farms over the decade to 2017.

- The average rate of return on total capital invested in farms has ranged between −1% and 5% over the past 20 years, but it fluctuates markedly between farms in response to the level of indebtedness, commodity prices, the mix of enterprises on each farm and the quality of management.

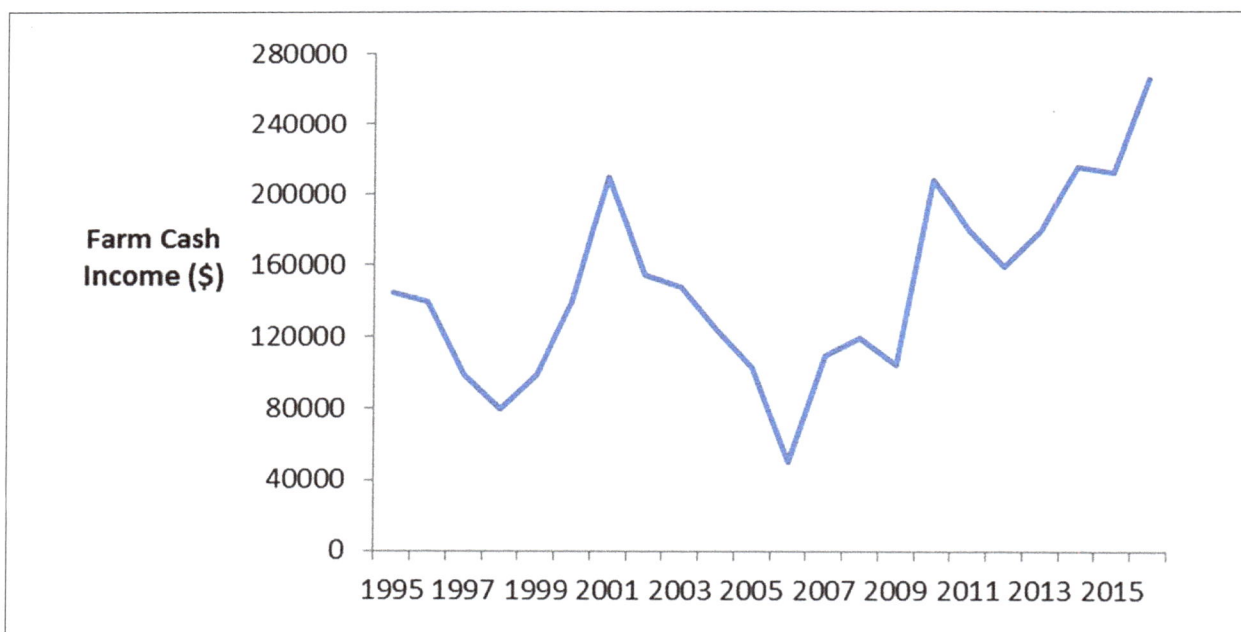

Figure 2.1: Average farm cash income from 1995-96 to 2016-17 (in 2017 dollars) for farm businesses selling more than 200 lambs per year lamb. Drawn by KA Abbott. Based on data from ABARE surveys, Department of Agriculture and Water Resources; see van Dijk, Frilay and Ashton (2018).[5]

On farms deriving a large proportion of income from wool, the levels of income and profit have largely followed the fluctuations in the wool price (Figure 2.2). Through the latter part of the 1980s the market price was held artificially high by the Reserve Price Scheme, but the floor price was dropped dramatically to 700 cents per kg clean in June 1990 before the price stabilisation scheme was abandoned completely in February 1991. Prices commenced recovery in May 1993, only to fall again from mid-1995.

Rapidly rising meat prices since around 2003 led to a substantial shift in emphasis in the Australian sheep industry, away from wool production and into meat production. There has, however, been a marked increase in the wool price since 2010, with prices for wool in 2018 higher than at any time since 1988 and, in US dollar terms, the highest ever received.

Conclusions from financial measurements

Two points arise from these statistics. First, the figures produced from farm surveys reflect average farm performances. The performance of the best farms is substantially better than the worst. Some of the differences between farms can be attributed to sheep management strategies which are associated, directly or indirectly, with health and production plans. Veterinarians have a role in developing and implementing these plans and can, therefore, assist in turning loss-making farms into profit-making farms.

Second, one has to wonder why farmers would tolerate low incomes and low returns on equity during periods of sustained low commodity prices. Some part of the answer lies in the choice of lifestyle and in the fulfilment of family tradition but, from a purely financial point of view, farms have represented a good investment over the long term because of *capital*

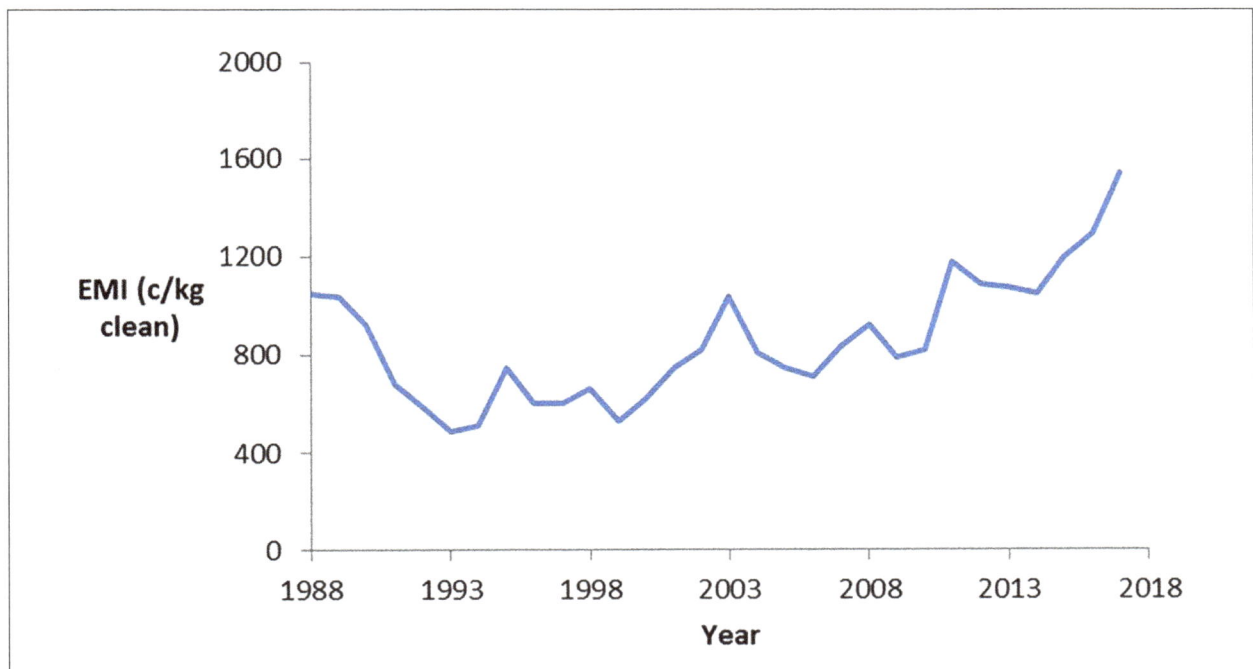

Figure 2.2: Wool price reflected by the eastern market indicator (EMI), in nominal terms, from 1988 to 2017. The EMI is a commonly used index of wool price, related to the price received per kg for an 'average' fleece, on a clean-wool basis. Source: KA Abbott.

growth — that is, increase in the value of farms over time has proceeded faster than inflation. Provided that equity is high and farm business profits are on average positive, farms do represent a good investment in the long term. Consistently achieving positive business profit in the sheep and beef industries, however, increasingly demands high standards of management.

ECONOMIC ANALYSIS OF ENTERPRISES AND STRATEGIES

The previously discussed statements, the Balance Sheet and the P & L Statement, are useful documents in that they are readily available and they give a broad overview of the state of health of a business. They suffer from the drawback that they are prepared primarily for taxation purposes rather than as aids to identification of problem areas in the business structure or in its performance.

Consequently, other economic tools have been developed to provide insights into farm businesses and as an aid to farm planning. The two most commonly used tools are *gross margin analysis* and *partial budgeting* but there is a range of other, more sophisticated techniques available, including *linear programming* and whole-farm economic modelling. A recent example of the latter technique is provided by van der Voort et al. (2017).[6]

Partial budgeting

A *budget* is a statement of expected income and expenses for a period of time in the future. In a *partial budget*, two or more alternative plans are compared with budgets which show only the extra expenses and extra income associated with each alternative. The budget is *partial* because only items which are relevant to the proposal are shown. It is a common procedure for evaluating veterinary intervention in grazing enterprises, such as improvements in worm control which might be achieved by altering the frequency of anthelmintic treatment, for example, or by examining the costs and benefits of a proposed vaccination programme.

Gross margin analysis (GMA)

A *gross margin* is the difference (the *margin*) between *gross income* and *variable costs*. In a *gross margin analysis*, the gross margins of particular farm activities (for example, wool from a wether flock, wool and lamb production from a prime lamb flock, vealer production from a beef herd, wheat grown from a cropped paddock) are calculated and compared in order to assist in farm-planning decisions. GMA ignores fixed and financial costs because these are unique to each individual farm and each farm business and do not usually affect the financial merits of the enterprise under review.

Gross income is the value of the total production from the enterprise for the period of time under analysis. It is not the same as total income for the period, for it does not include products sold in the analysis period which were produced outside the analysis period, but it does include products arising in the analysis period which remain unsold at the end of the period.

Variable costs are the expenses incurred for resources consumed during the analysis period *which vary with the intensity of the activity*. They are also called *direct costs*, because they are costs which can be attributed directly to the operation of the enterprise, rather than the costs which occur whatever enterprise is run.

The P & L Statement from Table 2.2 can be used to calculate a gross margin for the farm of J Smith & Son (Table 2.3). This farm runs sheep on 800 ha and has 7800 dry sheep equivalents

Table 2.3: A gross margin analysis for the 7800 DSE Merino sheep flock of J Smith and Son, running on 800 hectares. The flock includes 3000 medium-wool Merino ewes, some of which are joined to Merino rams and some joined to a terminal sire for crossbred lamb production.

Gross Margin Analysis

J Smith and Son

GROSS INCOME	
Income from wool	243,339
Income from livestock sales	322,962
Total	566,301
VARIABLE EXPENSES	
Shearing	67,000
Animal health	36,000
Supplementary feed	32,500
Flock rams	21,000
Wool and livestock freight & costs	27,500
Livestock insurance	800
Miscellaneous stock requisites	2,500
Total	187,300
GROSS MARGIN	379,001
Gross margin per hectare	$ 474
Gross margin per DSE	$ 49

(DSE). In comparing farm activities, it is usually necessary to compare gross margins relative to another resource. Gross margins are frequently quoted on a *per hectare* (GM/ha) basis, so the gross margin for the sheep enterprise is $474 per hectare. Occasionally, gross margins are quoted on a *per-head* or *per-DSE* basis. These references are usually less useful because stock numbers are rarely the *limiting resource* for graziers; but land area and capital funds are. Gross margin per head is *not* constant with stocking rate, nor is gross margin per hectare. Producers are much more likely to wish to maximise gross margin per hectare than gross margin per DSE, and the two maximum points rarely occur at the same stocking rate.

Note that the gross margin is not the same as profit. Profit in the above example could be calculated by subtracting fixed and financial costs from the gross margin.

The application of gross margin analysis

Gross margins are used particularly for two purposes.

1. For farm planning:
    - This process involves comparing alternative management strategies for the operation of an enterprise — for example, to compare the gross margin likely to be achieved

from a Merino flock keeping wethers to two years of age, rather than selling them as weaners.

- This process involves comparing alternative enterprises within the one farm business; for example, to compare the gross margin achievable from a Merino flock to that from a beef herd. Frequently, such comparisons lead to changes in the relative scale of enterprises on the farm. While wool prices remained low in the first decade of this century, many graziers have scaled down their Merino flocks and scaled up their prime lamb flocks, their beef herds and their cropping operations, all of which had similar or higher gross margins than Merino flocks in many districts.

2. For farm analysis:
   - Gross margin analysis can be used to highlight inefficient practices in the operation of farm activities, particularly by comparing details of the analysis to a *district standard* or to other farms in the district (an inter-farm comparison).

There are a number of limitations to the application of GMA for these purposes but, provided care is exercised, the process can be particularly useful. The advantage of GMA is that fixed, financial, personal and tax expenses can be ignored in the analysis. While these expenses can vary markedly between farms, they are difficult to allocate to parts of farms or specific enterprises within farms. In any case, those expenses are unaffected by the operation of any of the possible enterprises and are therefore irrelevant to the comparisons.

One of the frequent criticisms of the GMA technique is that it fails to properly account for differences in the requirements for capital resources, particularly capital funds and labour, between different enterprises. Other criticisms include the technique's failure to account for longer-run effects of different enterprises, complementarity between enterprises on farm, and taxation and cash-flow implications involved in different enterprises.

Some expenses are difficult to describe as completely fixed or completely variable. For example, should fertiliser and pasture renovation be considered variable expenses? Labour is also difficult to allocate, partly because it behaves economically in a *lumpy* or *step-wise* fashion. If labour is a slack resource (under-utilised), a more labour-intensive enterprise may not increase expenditure on labour. At some level of labour utilisation, however, increases in intensity will add a quantum amount to labour expenses.

Modified gross margin analysis

It is possible to overcome some of the principal weaknesses in GMA for evaluation of alternative sheep or cattle management plans by making allowance for changes in requirements for capital and labour. Whereas it would not be correct to include changes in required capital as either an expense or source of income, we could allow for the *opportunity cost* of the capital required.

Opportunity cost

This is effectively the income forgone by investment of capital funds into a particular activity. In many cases, opportunity cost is equivalent to the interest foregone by not investing in an alternative money-making scheme, or simply the cost of bank overdraft interest if additional funds are provided by a bank loan. The major changes in capital often relate to the size of the

Table 2.4: A modified gross margin analysis of a 2600 DSE Merino ewe enterprise, for comparison to Table 2.5.

Gross Margin Budget
1000 ewe sheep enterprise

GROSS INCOME		
	Income from wool	80,000
	Income from livestock sales	110,000
Total		190,000
VARIABLE EXPENSES		
	Shearing	22,000
	Animal health	12,000
	Supplementary feed	11,000
	Flock rams	7,000
	Wool and livestock freight & costs	9,000
	Livestock insurance	250
	Miscellaneous stock requisites	900
	Opportunity cost, livestock	7,200
	Labour (imputed value)	30,000
Total		99,350
GROSS MARGIN		90,650
Gross margin per hectare	$	363
Gross margin per DSE	$	35

flock or herd, so we can introduce the opportunity cost of the value of the stock as an extra expense when comparing plans with different numbers or types of livestock. The capital can be recouped by selling the livestock at some time in the future, if desired, but in the meantime there is income foregone from alternative uses of the money invested in the livestock.

Accounting for labour expenses in GMA can be done by allocating a particular value to the cost of labour per head or per DSE. For example, in the modified GMA of the J Smith and Son flock, the allocation of labour costs has been $30 per ewe (Table 2.4). This is equivalent to saying that one labour unit costing $90 000 per year can manage a 3000-ewe flock single-handedly.

Using a modified gross margin analysis to compare enterprises

Let's assume the Smith family have purchased an additional 250 ha property adjacent to their existing farm. They are considering either an expansion of their current sheep enterprise or a beef cattle enterprise. In either case, the extra livestock will need to be purchased. They can buy 1000 Merino ewes for $120 each or 175 cows at $1500 per head. Once stocked, and with young stock on the property as well, those numbers represent the same grazing pressure — about 2600 dry sheep equivalents. One of the factors which attracts the Smiths to a cattle enterprise is their lesser requirement for labour per DSE compared to breeding ewes.

A normal gross margin analysis does not account for the differences in capital cost of livestock or the labour requirement of each enterprise, but a modification of the gross margin analysis technique can include those expenses (Tables 2.4 and 2.5).

The comparison of the gross margins for the two enterprises suggests that the Smiths would not be wise to replace their sheep flock with a cattle enterprise. The gross margins, however, are reasonably close, so the difference may be sensitive to assumptions which have been made about the prices received for produce, or the costs of livestock. A *sensitivity analysis*, exploring the effect of varying the assumptions, is likely to be worthwhile.

Note that the difference in the funds required to purchase the ewes ($120 000) or the cows ($255 000) has been included by allocating the opportunity cost of the capital at a value of 6% of the capital cost.

Table 2.5: Gross margin analysis of a 2600 DSE beef cow herd selling vealers, using a modification to account for the opportunity cost of capital expenditure and the imputed cost of labour. The gross margin is directly comparable to that in Table 2.4, because both are scaled to the same number of DSE, so both enterprises are expected to require the same amount of pasture area. Differences in labour requirement and capital investment required to buy cows have been accommodated as described for modified gross margin analysis.

Gross Margin Budget
170 cow beef enterprise

GROSS INCOME

Income from sale of steers & heifers	106,000
Income from sale of cfa bulls and cows	12,000
Income from sale of culls	21,000
Total	139,000

VARIABLE EXPENSES

Animal health	1,750
Supplementary feed	3,938
Replacement bulls	10,500
Livestock freight and selling costs	10,150
Livestock insurance	400
Miscellaneous stock requisites	200
Opportunity cost, livestock	15,750
Labour (imputed value)	20,000
Total	62,688

GROSS MARGIN	76,313

Gross margin per hectare	$	305
Gross margin per DSE	$	29

Even if the cattle enterprise appeared to be a better option, additional factors deserve consideration. These include any additional capital costs required for a switch to cattle production (for example, strengthened fencing and water troughs, cattle yards) and their own expertise in managing cattle. However, there may be complementarity between the two enterprises which favours the cattle enterprise — such as differences in the timing of major labour-intensive husbandry events, such as calving, calf marking, shearing, lambing, lamb marking, etc.

Gross margin calculations often contain more detail than given here. For example, income might detail the expected wool cut and wool price from ewes, wethers, hoggets and weaners. Similarly, income from cattle might show the relative contribution from sale of cows, finished vealers, store vealers and bulls. This enables some assessment of the degree of optimism or pessimism employed in the comparison. Details have been omitted here to avoid confusing the general application of GMA with the details, which vary from year to year.

Gross margin analysis is used at several points in later chapters of this book, particularly to evaluate procedures which involve major changes in the operation of the flock, such as alternative genotypes, changing reproductive rate and optimising stocking rate. One strategy we will discuss is the optimisation of flock structure, and we can use a GMA approach to come to general conclusions about the optimisation of flock composition and the *sensitivity* of the gross margin to alternative compositions. Computer models are frequently used for this purpose because the number of mathematical relationships requiring recalculation for a range of strategies can be tiresome if done by hand and calculator, particularly where the results are sensitive to small changes in a large number of variables, such as the decline in productivity of ewes and wethers with age. White[3] used a comparison of gross margins to evaluate change in the reproductive rate of sheep flocks and the gross margins were calculated by a computer model which considered a very large number of effects and interactions for a large set of environmental conditions.

GMA has also been used in veterinary fields to justify an increased level of veterinary services to dairy farms. A four-year controlled study of a dairy herd health programme conducted from the Melbourne University Veterinary School from 1973 to 1977 was evaluated in economic terms using a GMA.[4] Gross margins per hectare on farms using the herd health programme were increased by $23, $1, $43 and $68 in each of the four years, relative to surveillance farms which did not have herd health programmes. The method and the conclusions, however, were challenged by economists who argued, inter alia, that GMA was not the best method of analysing the results.[5] Johnstone et al.[6] used GMA in analysing a field experiment to show that suppressive treatment (11 per year) with anthelmintics was financially superior to preventive, curative and salvage treatment. While the result was correct, it has since been demonstrated that the suppressive treatment was not optimal, either — other untested strategies are now known to be better.

This report is a straightforward application of GMA for those readers interested in a simple illustration of the technique. It should be noted both that only a few income and expense items are used and that the results could equally well have been analysed with a partial budget. In other more recent studies, the impact of ovine Johne's disease on the gross margin of sheep flocks in Australia and the losses associated with bovine viral diarrhoea virus in a dairy herd in Europe were assessed using the technique of gross margin analysis.[7,8]

REFERENCES

1   Kristensen E and Enevoldsen C (2008) A mixed methods inquiry: How dairy farmers perceive the value(s) of their involvement in an intensive dairy herd health management program. Acta Vet Scand **50** 50. https://doi.org/10.1186/1751-0147-50-50.

2   Noordhuizen JPTM, van Egmond MJ, Jorritsma R et al. (2008) Veterinary advice for entrepreneurial Dutch dairy farmers. From curative practice to coach-consultant: What needs to be changed? Tijdschr Diergeneesk **133** 4-8.

3   Garforth CJ, Bailey AP and Tranter RB (2013) Farmers' attitudes to disease risk management in England: A comparative analysis of sheep and pig farmers. Prev Vet Med **110** 456-66. https://doi.org/10.1016/j.prevetmed.

4   Morris RS (1969) Assessing the economic value of veterinary services to primary industries. Aust Vet J **45** 295-300. https://doi.org/10.1111/j.1751-0813.1969.tb01955.x.

5   van Dijk J, Frilay J and Ashton D (2018) Lamb farms: Farm financial performance. Available from: http://www.agriculture.gov.au/abares/research-topics/surveys/lamb. Accessed 6 June 2018.

6   van der Voort M, Van Meensel J, Lauwers L et al. (2017) Economic modelling of grazing management against gastrointestinal nematodes in dairy cattle. Vet Parasitol **236** 68-75. https://doi.org/10.1016/j.vetpar.2017.02.004.

7   White DH (1984) Economic values of changing reproductive rates. In: Reproduction in sheep, eds DR Lindsay and DT Pearce. Australian Academy of Science: Canberra, p 371.

8   Williamson NB (1980) The economic efficiency of a veterinary preventive medicine and management program in Victorian dairy herds. Aust Vet J **56** 1-9. https://doi.org/10.1111/j.1751-0813.1980.tb02529.x.

9   Alston JM and Ryan TJ (1981) The economic efficiency of a veterinary preventive medicine and management program in Victorian dairy herds: An alternative viewpoint. Aust Vet J **57** 572-3. https://doi.org/10.1111/j.1751-0813.1981.tb00442.x.

10 Johnstone IL, Darvill FM, Bowen FL et al. (1976) Gross margins in a Merino weaner sheep enterprise with different levels of parasite control. Proc Aust Soc Anim Prod **11** 369.

11 Bush RD, Windsor PA and Toribio J-A (2006) Losses of adult sheep due to ovine Johne's disease in 12 infected flocks over a 3-year period. Aust Vet J **84** 246-53. https://doi.org/10.1111/j.1751-0813.2006.00001.x.

12 Fourichon C, Beaudeau F, Bareille N et al. (2005) Quantification of economic losses consecutive to infection of a dairy herd with bovine viral diarrhoea virus. Prev Vet Med **72** 177-81. https://doi.org/10.1111/j.1751-0813.2006.00001.x.

3

THE SHEEP FARM AS A PRODUCTION SYSTEM

SHEEP FARMING IS A PRODUCTION SYSTEM

Successful sheep producers enjoy the benefits of a profitable farm business and also receive personal satisfaction from the physical appearance of their animals, pastures and farm environment. In many cases they would like to do more to enhance the health and welfare of their livestock, or to improve their infrastructure, but they are constrained in their ability to do so profitably. These constraints can be described almost completely by two economic principles. The first is that, when a certain level has been reached, it is no longer worthwhile to spend more in order to increase the health, welfare or productivity of their flock. This fact is encapsulated in the principle referred to in the study of economics as the *law of diminishing returns*.

The second is that most of the management strategies that producers put in place have effects on a wide range of farm and flock events, such that many husbandry decisions reflect a compromise between multiple outcomes, none of which is as good as one might hope but all of which, when combined, represent a result which is the best overall outcome. This is a consequence of the *systems behaviour* of farm businesses.

These two principles are critically important in determining how a sheep farm (or any complex production system) operates; and a thorough appreciation of these will help the farm advisor in two ways — first, to understand the drivers behind the decisions that successful producers make, and second, to ensure that veterinary advice that is given to producers is economically and practically sound.

Sheep production and the law of diminishing returns

Sheep producers make a number of decisions about the management of their farm and flock in order to have healthy, productive animals and a profitable farm business. It is not possible, however, to have animals which are perfectly healthy or producing to their maximum genetic potential and, at the same time, to be making a profit.[1] The cost of ensuring perfect health or providing very high levels of nutrition exceeds the value of the production of the animals. From experience, producers learn that inputs into an agricultural system obey *the law of diminishing returns* and the relationship between inputs and production (outputs) is curvilinear (Figure 3.1). For a sheep farmer, inputs are items like veterinary treatments or additional feed. There comes a point where the extra production gained by increases in inputs is not worth as much as the cost of the extra input.

In economic terms, small changes in input or production (say, from point A to B or from point C to D in Figure 3.1) are called *marginal* changes. It is rational, therefore, to continue to

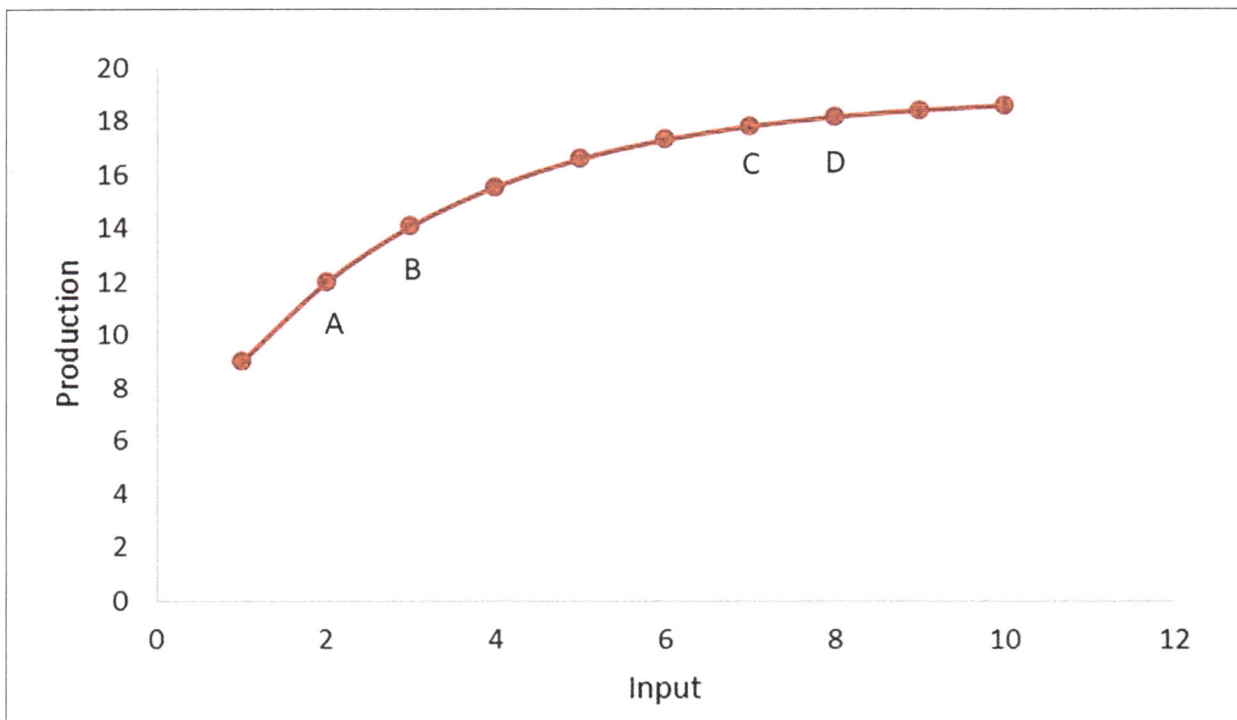

Figure 3.1: The law of diminishing returns. As the level of inputs increases, production also increases but it does so at a declining rate. Eventually, at high levels of input, further increases in inputs produce zero, or near-zero increases in production. An increase of one unit of input from 2 to 3 (point A to point B) increases production by 2.1 units. An increase of one unit from 7 to 8 (point C to point D) results in only 0.4 units of extra production. Source: KA Abbott.

increase the level of input as long as the cost of the marginal increase in input is less than the value of the marginal increase in production. For example, if the units on both axes in Figure 3.1 were dollars, then an input of $5 would be rational (production of $16.60), but an input of $6 would not (production of $17.30). The extra dollar spent on inputs would return only $0.70.

The law of diminishing returns is virtually universal in biological systems such as sheep production. It applies to the nutrition of sheep (an input) and the number of lambs they produce, or the amount of wool they produce (outputs). It applies in a less obvious way to their health. As an example, treating young sheep with anthelmintics three times per year may be economically wise on farms in some regions of Australia, because it prevents severe losses of production. Treating the sheep six times per year will reduce their worm burdens even further, but the extra production they might make because of the lower worm burdens will not justify the cost of the extra three treatments.

Producers take a whole-farm view

Producers also know that many of the management strategies that they employ on their farms are interrelated, and that if they change one component of their management, there will be impacts on multiple parts of the farm operation.

For example, a producer may decide to increase the survival rate of lambs by lambing the ewes in the warmer month of October rather than the cold weather of August. Survival rate of lambs may improve but the lambs will be younger and lighter when pastures become dry in summer. This may lead to an increase in weaner mortality in autumn and a decline in the productivity and profitability of the whole farm business.

Experienced producers know that virtually all of the major management strategies that they employ interact with one another. The whole farm, its climate, pastures and livestock, are interrelated parts of a *system*. It is not possible to disturb one part of the system without disturbing other parts. 'Improving' one part may cause another part to perform very badly.

The skill of the farm manager is to combine resources — sheep, pastures, labour, financial capital — in such a way as to maximise the return from the business as a whole. The return may be simply the financial reward of a good profit, or it may be tempered with the sense of achievement from establishing a farm of high ethical standards, with good animal welfare and responsible management of the physical environment, or meeting a range of other personal and business goals.

Science-trained advisers need a systems approach

A scientist's education is usually broken up into disciplines, such as nutrition, agronomy and animal health. Consequently, the tendency for scientists when advising producers is to emphasise their own discipline area and neglect others. The challenge we have as veterinarians servicing the needs of sheep producers is to know the extent to which our field of expertise influences the productivity of the whole system. This does not require that we be expert in all fields. It does require, however, that we are prepared to take a *systems approach* to formulating advice for producer clients.

What defines a good production system?

The success of a production system is usually measured in economic terms. This does not always perfectly predict the producer's objectives, but it usually is close. The financial success of the system cannot be described by simply recording the success of parts of the system, such as the value of lambs sold, or the total income from lines of fleece wool. The financial success must be measured in terms of the whole system and with consideration of the costs involved in producing all income. For reasons discussed in the previous chapter, gross margins are often used as tools to measure the productivity of farming systems. Because farm area is usually the most limiting resource for graziers, gross margins are usually quoted as dollars of gross margin per hectare (GM/ha). The objective of the farm system is, therefore, to maximise the GM/ha. To achieve this, the components of the farm system will be *optimised*, rather than *maximised*. For example, in one particular flock, the GM/ha may be $200/ha when the weaning rate of lambs is, on average over a few years, around 85%. If additional funds are expended in an attempt to increase the weaning rate, unless the increased return from a higher weaning rate equals or exceeds the extra costs, GM/ha will fall. Thus, for that particular system, weaning rate is optimal at 85% (assuming that a lower weaning rate is also less profitable).

Land area is not always the most limiting resource. Sometimes labour or capital funds are the most limiting. In such cases, a farmer may wish to maximise GM/labour unit, or GM/$100 of capital. In Australia now, it is uncommon for the availability of labour to be limiting, so GM/ha and GM/$100 of capital are the more common indices of farming success. GM/animal is not a useful index of success of the system as a whole because maximal values of GM/sheep or GM/DSE do not usually coincide with maximal values of GM/ha.

What are the boundaries of a farm system?

It is true that a farm is not a full system in itself but is part of a bigger system. For example, an increase in meat production from one farm has an effect on the price received by all producers. The effect is usually immeasurably small when only one farm is considered, so it is often valid to consider that a farm system stops at the farm boundary. Regional policy makers cannot simplify farm systems to this degree when fixing price schedules. Thus milk production from dairy farms under restrictive licensing arrangements may operate with a quite different set of farm objectives from those of wool production based on a wool price determined on a deregulated world market. When working with individual sheep producers it is usually an acceptable simplification to limit the system under consideration to the one farm.

Computers and farm systems

All agricultural systems are complex, and it is natural that computers are applied to describing and understanding farm systems. Computer models in many cases replace simpler models of the system which are drawn on paper or even exist only in the mind of the farm operator. Computer models are unlikely to be perfectly accurate, but they should be at least as good as any model we can devise without computer assistance and much easier to use. In many cases, computer models have a useful educational role for advisers, alerting them to possible consequences of new farm management strategies which previously had been ignored or had not been quantified. Examples of sheep farm models include the decision support tool GrassGro, from Horizon Agriculture Pty Ltd, which was developed by the CSIRO Division of Plant Industry.

DESIGNING A SHEEP PRODUCTION SYSTEM

Once the decision has been made to operate a sheep production enterprise, there are a number of key decisions which have to be made. These include

- the *production objective*: is meat, wool (or milk) the principal farm product?
- the *breed and genotype* of the sheep
- the *flock structure*: what is the best age to sell ewes and wethers?
- the *stocking rate*
- the *time of lambing within the year*
- the *time of shearing*.

Although these key decisions can be seen as something of a hierarchy, in that the first two are long-term strategies which cannot be easily changed in one or two years, all of them are interrelated and none can be considered in isolation. For example, to compare specialist sheep meat production to Merino wool production, it is necessary to compare them on an equivalent stocking rate basis and to ensure that the optimal implementation of each form of production has been considered. A meat-lamb production operation may be more profitable than a Merino sheep enterprise which keeps all wethers to 4 years of age, but less profitable than one which sells all the wether lambs at 6 months of age. Table 3.1 illustrates a comparison of the likely profitability of a range of sheep production systems and flock structures, using gross margin per hectare as the comparator. Such comparisons are strongly dependent on relative prices of wool of varying quality (fibre diameter), as well as on meat and sheep prices; and the ranking of the various systems can vary significantly both between years and within the same year.

The first two decision areas listed above (that is, the production objective and the breed and genotype of the sheep) are often reviewed by producers and their advisers. Veterinarians who develop a special interest in sheep production may be involved in decision making on these two points through consultations with a sheep-farming client. Veterinarians who are in general rural practice, rather than consultancy practice, are more likely to be providing their clients with advice limited to animal health, reproduction and nutritional management. Nevertheless, these issues are interrelated with the four major profit-drivers in the farm business — flock structure, stocking rate, time of lambing and time of shearing — and it is important to understand something of the relationship that these four factors have with whole-farm production before making recommendations to improve particular aspects of flock performance.

PRODUCTION OBJECTIVES

Broadly, commercial sheep producers choose to focus on wool production, sheep meat production or a combination of the two. Commercial producers require a supply of purebred rams which are produced in specialist ram-breeding flocks — usually referred to as stud flocks.

Specialist wool production is based on the Merino breed of sheep. The classical form of Merino wool production in Australia is based on the self-replacing flock in which rams are the only sheep introduced onto the farm. All male lambs are castrated and the wethers are sold as lambs or hoggets or at any age up to 5 years old or, uncommonly, at an age beyond this. Ewe lambs are shorn as lambs, then classed as hoggets. Some of the ewe hoggets are retained as replacements for the breeding flock and the remainder are culled. At 19 months of age ewes enter the breeding flock at joining, and they are cast for age after four (usually) or more lambings.

The specialist sheep meat producer has a flock of ewes which, in Australia, most commonly comprises first-cross (Border Leicester x Merino) ewes. The first-cross ewes are mated to a terminal sire and all lambs (referred to as *second-cross lambs*) are sold, usually before the age of 8 months. Ewes enter the breeding flock at 19 months of age or, if very well grown, 7 months of age, and are cast for age after five (or more) lambings. Replacement ewes are purchased at around 15 months of age and all rams are introduced. Second-cross lambs are also called *prime lambs*.

First-cross ewes are produced from Merino flocks in which maternal meat rams (such as Border Leicester) are mated to a proportion of the Merino ewes. The male (first-cross) lambs are sold for meat and the first-cross females are sold to prime lamb producers as lambs or hoggets.

Dual-purpose flocks based on the Merino ewe are becoming increasingly common as a result of both increases in the price received for lambs and older sheep for meat, and an easing in the relatively high value for fine wool since 2003.

Producers may choose to breed only Merinos, selling wether lambs for meat, or they may mate some or all of the ewes to a terminal sire breed, selling all first-cross lambs for meat. Depending on the proportion of the breeding flock mated to terminal sires, it may be necessary to purchase some or all replacement ewes.

The relative profitability of each of these enterprises is reviewed from time to time, as economic conditions in the sheep industry change. For example, a comparison of the following production systems was performed in 2006[2]:

- superfine and fine-wool Merino wethers (17.5 µm or 19 µm)
- self-replacing Merino ewes (19 µm or 21 µm) selling Merino lambs at 4 or 12 months of age
- dual-purpose Merino ewes (19 µm or 21 µm) selling first-cross lambs at 4 or 6 months of age
- prime lamb; first-cross ewe flock selling second-cross lambs at 4 months or 6 months of age.

Using prices for wool and meat over a range of years (1996-2003) it was concluded that the dual-purpose Merino and the prime lamb systems were the most profitable, with the dual-purpose Merino proving relatively resilient to price changes. This example is provided for illustration only — the relative merits of different systems will change over time and producers must evaluate how they can best alter their systems to benefit from current and future market conditions. Government departments of agriculture in each state generally provide updated estimates of the relative profitability of each of several different production systems, such as that used to compile Table 3.1.

There are many alternatives to the systems and, particularly, the breeds of sheep described above. For example, there have been for decades a number of breeds of sheep in Australia which have been considered dual-purpose and which can be maintained in self-replacing or partially self-replacing flocks. Examples include the Corriedale, Polwarth and, particularly in New Zealand, the Romney. In Australia there has been growing interest in dual-purpose Merino breeds from South Africa and in sheep which shed their fleeces annually and require no shearing.

It is common to find a mixture of these enterprises on sheep farms in Australia. For example, a producer may breed first-cross ewes from a Merino flock, and retain some or all of the females to form a prime-lamb-producing flock. Another example is for owners of self-replacing Merino flocks to retain ewes to 6 years of age, mating them to a maternal meat ram or terminal sire for their last lamb. Many studs operate commercial flocks in addition to their ram-breeding enterprise.

There is a very small sheep dairy industry in Australia and the appropriate breeds and management strategies for sheep dairy producers are quite different from those focusing on wool or meat. The topic is discussed further in Chapter 4.

FLOCK STRUCTURE

The *flock structure* or *flock composition* refers to the age and sex profile of the flock — that is, the relative numbers of sheep of each age and sex. The flock owner determines the flock composition on the basis of economic and management considerations, and implements the plan by adopting appropriate buying and selling policies. The composition is also influenced by reproductive and mortality rates. It is usual for flock owners to adopt a broad aim for a particular flock structure but then to make minor adjustments in response to market fluctuations and year-to-year variation in reproductive rates. Optimisation of flock structure is one of the more powerful strategies which producers can use to maximise profitability.

Table 3.1: A comparison of the predicted profitability of a range of sheep production systems, summarised from NSW Department of Primary Industry (2010), for commodity prices current in October 2015.[5]

Flock type	GM $/ha	Assumptions	Comments
First-cross ewes terminal meat sire	$341	2.8 DSE/ewe 118% weaned	22% of the ewe flock purchased each year
Merino ewes 20 μm terminal meat sire	$333	2.5 DSE/ewe 90% weaned	22% of the ewe flock purchased each year
Dorper ewes Dorper rams	$297	2.8 DSE/ewe 118% weaned	No wool income No shearing costs Self-replacing flock
Merino ewes 18 μm Merino rams	$274	2.1 DSE/ewe 83% weaned Wether weaners sold at $50	Self-replacing flock Wether weaners sold at 4 months
Merino ewes 20 μm Maternal meat rams	$359	2.6 DSE/ewe 90% weaned	22% of the ewe flock purchased each year.
Merino ewes 20 μm 75% Merino ram 25% terminal sire	$317	2.6 DSE/ewe 87% weaned	Self-replacing flock
Merino ewes 20 μm Merino rams Wether lambs finished	$302	2.6 DSE/ewe 86% weaned Wether weaners sold at $124	Self-replacing flock Wether weaners sold at 10 months
Merino wethers 18 μm	$246	0.1 DSE/wether Kept for 5 shearings	18% of the wether flock purchased each year.
Merino ewes 20 μm Merino rams	$258	2.3 DSE/ewe 86% weaned Wether weaners sold at $50	Self-replacing flock Wether weaners sold at 4 months
Merino wethers 20 μm	$219	0.2 DSE/wether Kept for 5 shearings	18% of the wether flock purchased each year.

Determination of the best flock structure is strongly influenced by the production objective. For example, prime lamb producers with crossbred ewes and terminal sires have restricted options. Variations in the sex balance of flocks are of most interest in self-replacing flocks.

When planning, flock owners will attempt to predict the profitability of alternative flock structures. To do this correctly, it is necessary to ensure that the total size of each proposed flock remains the same and this is usually achieved by comparing flock structures with the same total *dry sheep equivalents* (DSEs).

Figure 3.2 illustrates the composition of five different flocks, all of the same size in terms of grazing intensity or DSEs. Note that, if wethers are retained, fewer ewes can be run. Note also that the larger size and prolificacy of crossbred ewes restrict their numbers, compared to Merino ewes.

Optimum age to sell wethers

Merino-wool-producing flocks are broadly of three types: all-wether flocks, flocks purchasing replacement ewes or self-replacing flocks. All-wether flocks are resilient in tough conditions and

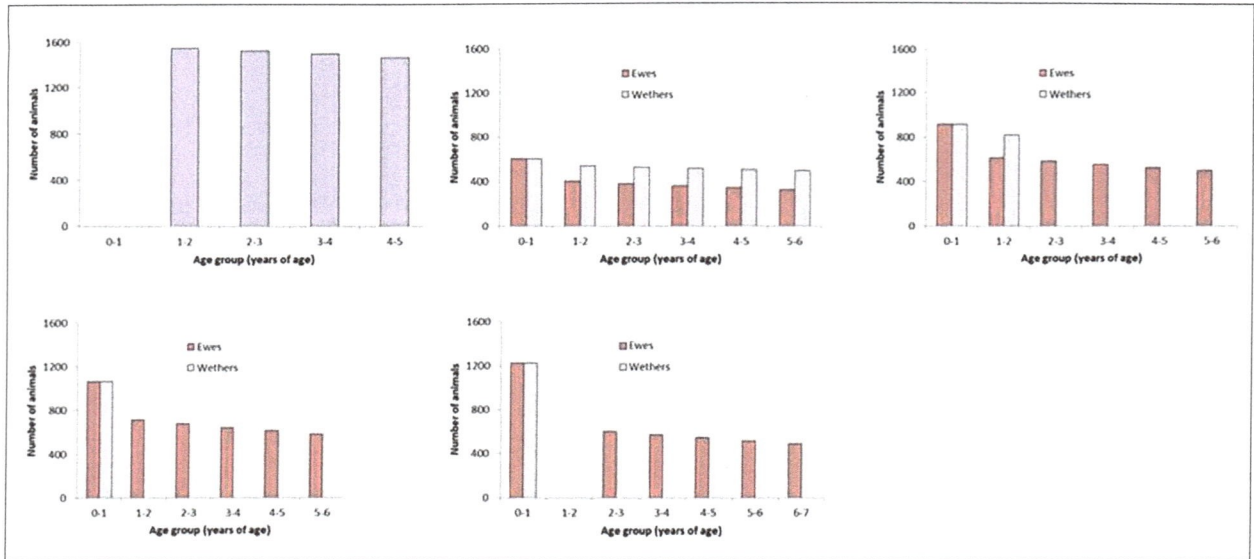

Figure 3.2: The comparative composition of some 6000 DSE flocks. From left to right (top): all wether, self-replacing with maximum wether numbers, self-replacing selling wether hoggets; (bottom) self-replacing selling wether lambs, prime lamb flock. Note that a 6000 DSE all-wether flock contains about twice as many adult animals as a 6000 DSE prime lamb flock. Source: KA Abbott.

have low labour inputs. In recent times there has been strong demand for wethers for the live export market which has underpinned high sale prices. Consequently, wethers are now usually sold as lambs to the domestic meat market or as young adults to the live export market. All-wether flocks or breeding flocks retaining a high proportion of wethers are therefore relatively uncommon.[4]

Production systems which depend on the purchase of replacement stock (ewes or wethers) suffer the disadvantage that the genotype available to buy may not be as productive as the genotype which could be maintained in a self-replacing flock. This is particularly important in Merino wool production, where there are large differences in profitability between flocks of different genetic background. Additionally, there is a constant risk of introducing diseases like Johne's disease, footrot and lice.

When wethers are sold as young sheep, rather than retained, more breeding ewes are maintained to stock the farm at appropriate levels. The relative merits of any particular structure are largely determined by the following characteristics:

- Flocks consisting of a high proportion of wethers produce more wool but fewer lambs than flocks which sell wethers at young ages.

- The presence of mobs of adult wethers facilitates worm control in young sheep and breeding ewes.

- Flocks which maintain large wether flocks have less demand for labour (per hectare) or less demand for *timeliness* of labour inputs, or both.

- Flocks which maintain large wether flocks are more resilient in the face of poor seasons[5] and are, therefore, often run at higher stocking rates (higher DSE/ha) than flocks in which ewes predominate.

- Flocks which sell wethers at young ages have a relatively higher proportion of their gross income from stock sales.
- Fleece weight, fibre diameter, fertility, fecundity, survival rates and sale price are all age-related variables which directly affect flock productivity and profitability (Table 3.2).

The optimum size of a wether flock varies with the wool type

The adoption of a particular flock structure by a producer with a self-replacing Merino flock is strongly influenced by both the price of wool relative to the price of sale sheep, and the rate of *depreciation* of wether sheep with age. Wether flocks produce more wool per hectare than breeding flocks so superfine wool producers are more likely to maintain large wether flocks because their wool is relatively valuable but their sale value (fine wool wethers are of small frame size) is often relatively low. Influenced by the same market conditions, producers with medium- and strong-wool Merinos are more likely to sell their wethers at young ages and have relatively larger breeding flocks.[2]

Optimum age to sell ewes

Ewes are often retained in the breeding flock for four seasons, lambing first at the age of 2 years then being sold at the age of 5½ years. Merino flocks generally have relatively low reproductive rates, so three lambings are insufficient to maintain breeder numbers without purchasing at least some replacement ewes. It is likely to be economically wise to keep ewes for five or more lambings when the return from livestock sales is high (thus favouring the higher reproductive ability of

Table 3.2: Age- and sex-related variables which affect optimality of flock structures.

Character[a]	Sheep of either sex	Breeding ewes
Fleece weight	Fleece weights in ewes and wethers peak between 3 and 4 years of age, then decline. Wethers produce more wool than dry ewes but this is related to body size (skin area) and, consequently, feed intake.	Pregnancy and, more particularly, lactation depress fleece weights. Lactation for a single Merino (Mo) lamb depresses fleece weight by 12% to 15%.
Fibre diameter (FD)	In all sheep, FD increases with age, so fleece values tend to decline from 2 or 3 years onwards.	Pregnancy and lactation depress FD, which increases wool value per kg provided tensile strength is not also reduced significantly.
Reproductive rate		Prolificacy and rearing ability increase with age, peaking at 5 to 7 years.
Purchase price	Purchase price is highest for well-grown weaners or, more usually, hoggets (1½ years). Purchase prices usually decline beyond that age.	
Sale price	Sale price follows the same pattern as purchase price. The rate of decline with age varies with market demand.	Sale price of older ewes is variable — sometimes there is a significant drop between 4½ and 5½ years of age.
Mortality rates	High in weaners and in sheep over 5 to 6 years of age. Older wethers have high mortalities on phyto-oestrogenic pastures.	Similar to wethers but added stress of reproduction increases mortality rates in ewes over 5 years of age.

a   The effects of age on reproductive rates are discussed further in Chapter 7.

older ewes) and the return from wool is low (thus discounting the deterioration in fleece value as ewes age). Prime lamb flocks are an example of flocks where ewes have low fleece values and high initial purchase cost but produce valuable lambs for sale. It is usual in such flocks to keep ewes to 6½ or more years of age.

The 'best' decision will depend on the particular economic environment in which each flock operates. If the economic relativities between wool and livestock values change and the producer expects the new economic environment to persist, it is rational to review the flock structure. Vizard (1990)[6], in a computerised linear programming exercise, found that higher wool prices favoured retaining more wethers; spring lambing favoured selling wethers at younger ages than was best for an autumn lambing flock; and, although there is no single flock structure which can be recommended for all flocks, there are some structures which perform well under a wide range of economic conditions.

STOCKING RATE

The *stocking rate* of a sheep farm is the number of animals run per unit area. It is usually expressed in terms of *sheep per hectare* or *DSE per hectare*. Generally, discussions about changes in stocking rate assume that no change in inputs has occurred which would alter the pasture productivity of each hectare of land. Thus, changing *sheep per hectare* is equivalent to changing *sheep per tonne of dry matter grown*. Occasionally, producers talk of increasing stocking rate following pasture improvements — such as additional fertiliser or a change to more productive pasture species. If the additional inputs result in extra pasture production, the higher stocking rate (SR) may not represent a change in *sheep per tonne of dry matter grown*. Our discussions here, however, are concerned solely with changes in SR where it is the only variable changed. Thus, unless increasing SR itself increases pasture production (and it usually does not), increasing SR means that more sheep have to share each tonne of dry matter grown.

At stocking rates which are commonly in the range adopted by commercial wool producers, the effect of increasing stocking rates is to reduce the pasture availability. Unless feed availability is very high (1500 to 1800 kg of pasture dry matter per hectare), daily intake of pasture is a function of availability.[b] Thus, in most seasons of the year, increasing SR reduces daily pasture intake of each grazing animal. Increases in the total amount of pasture consumed per hectare (because a higher proportion of the pasture is consumed, not because more is grown) may, however, lead to substantial improvements in the efficiency of conversion of pasture to animal product.

For the veterinarian involved in rural practice, the effect of SR on farm profitability is generally a peripheral issue. Nevertheless, it is important for the practitioner to be aware of the profound influence of SR on profitability because some of the most common sheep diseases involve nutrition — and nutrition in grazing animal management is almost entirely a function of stocking rate. For example, the clinical expression of worm parasites, pregnancy toxaemia, weaner ill-thrift and uncomplicated undernutrition all have some relationship with pasture availability and, hence, stocking rate. The astute practitioner, realising the association but also realising the importance of SR to the client's income, will not advise a reduction in SR to

---

b   Pasture intake is most strongly influenced by pasture availability and pasture quality. This relationship will be discussed further in Chapter 6.

limit the effects of disease, but will attempt to institute control measures which allow the maintenance of profitably high SR.

For the veterinarian who is a sheep specialist or who provides a consultancy service to sheep producers, analysis of the sensitivity of farm profit to changes in the current SR is an essential activity on every client's farm. Some key review papers are listed in the references at the end of this chapter and the topic is discussed further in Chapter 6.

Effects of SR on soils and pastures

At high SR, lower levels of soil cover can lead to an increased risk of soil and nutrient loss through wind and the scouring effect of running water. In many of the low rainfall areas of Australia, experience indicates that stocking rates have been excessive and damage to the soil-pasture ecosystem has been severe. In medium and high rainfall areas, the deleterious effects of high stocking rates may be avoided by taking steps to prevent erosion by means other than reduced stocking rates. Nevertheless, it may be necessary to substantially destock some pastures in some years — particularly in years when pasture growth has been much lower than expected.[7]

SR also has an effect on the amount of pasture present, the proportion of particular pasture species which make up the pasture sward and, usually, an inverse relationship with total annual pasture productivity.

Effects on animal health

Increased stocking rates lead to lower liveweights at most times of the year and the consequences of this for ewes depend largely on when the low liveweights occur in relation to the annual reproductive cycle. For example, if undernutrition on short autumn pastures coincides with late pregnancy, pregnancy toxaemia may be expected to occur at a higher incidence at high SR than at a low SR. If, however, the same increase in SR has reduced the liveweight of the ewes at conception and, subsequently, reduced their conception rates, the expected outbreak of pregnancy toxaemia may not occur. Probably the most consistent effect of higher SR on breeding flocks is a deterioration in lamb survival, particularly of twin lambs.[8]

For all stock, but particularly for weaners, higher stocking rates generally lead to increased need of supplementation for two reasons. First, liveweights are often lower at the start of prolonged *feed trough* periods, so supplementation needs to start earlier and be continued for longer. Second, the profundity of pasture shortfalls is greater at high SR, so the level of supplementation may need to be greater. Diseases associated with undernutrition and with grain feeding are expected to occur more commonly at high SR.

The association between SR and helminthosis is less clear.[9] Whereas an increase in contamination of pastures with nematode eggs is expected at higher SR and sheep are expected to be more susceptible to parasites when underfed, shorter pastures in fact may lead to higher mortality rates of free-living nematode parasite larvae. This effect may be more useful with parasites which are particularly susceptible to environmental exposure, like *Haemonchus contortus*, than resistant parasites, such as *Nematodirus* spp.[10,11] One could conclude from the work of Brown et al.[12] that a good nematode control programme is still sufficient to maintain productivity at high SR, but the penalty of an inadequate programme is increased.

Effects on animal productivity

As already stated, increasing SR leads to reductions in pasture intake per sheep at most times of the year and, consequently, lower levels of individual animal productivity. Thus, at higher SR, sheep grow lighter fleeces of lower fibre diameter with shorter staple length and lower staple strength.[8] Additionally, sheep have lower liveweights, at least for part of the year, and ewes have lower reproductive rates; survival rates may decline and the requirement for supplementary feed may be increased. Sheep which have been run at higher stocking rates will often have a lower average sale price.

Despite these negative effects on individual productivity, a change from a low to a medium stocking rate generally leads to increases in wool produced per hectare, as well as to the number of lambs produced per hectare and the number of sale animals available per hectare. The relationship between production and SR is curvilinear; that is, total production rises to a maximum with increasing SR, then declines. (See Chapter 6 for a more detailed description of the relationship between production and SR.)

Most variable costs rise, on a *per hectare* basis, linearly with SR. The major exception is the cost of supplementary feed, which usually rises on a *per head* basis with increasing SR. Most other variable costs rise on a *per hectare* basis but remain constant on a *per head* basis (shearing costs, for example).

TIME OF LAMBING

The time of lambing chosen by producers varies from district to district, but the most common choices are the five months between and including May and September. April and May lambing is usually referred to as *autumn lambing*; June and July lambing as *winter lambing*; August, September and October lambing as *spring lambing* (despite the fact that August is late winter).

Autumn lambing is favoured in districts where the normal pasture-growing season is short and the summer-autumn dry period is prolonged, and in flocks where lambs are sold at the end of the growing season in their first year of life for a price based on their liveweight. Spring lambing is favoured in districts with a pasture-growing season extending into early summer, which facilitates the growth of weaners, and in flocks for which the main production objective is wool, rather than young sale sheep. Winter lambing is something of a compromise — often favoured by flock owners whose principal objective is wool production but who operate in districts with prolonged summer-autumn dry periods. It is usually only considered viable in areas with mild climates in winter which are therefore not severely adverse to lamb survival. It is also common practice in prime lamb flocks in districts where lambs can achieve market weights before pastures become dry in summer.

In many wool-growing districts, late winter is one of the most critical times for maintaining the nutritional level of sheep because the climatic conditions place extra nutrient demands on sheep and the pastures are often short and growing slowly, if at all. More ewes can be carried on the farm all year round if their peak demands for reproduction (lactation) are delayed until spring, when feed surpluses start to appear. In some districts this is possible, but in other districts the rapid deterioration of pasture quality in summer makes the management of spring born weaners very difficult. Consequently, it is common for producers to lamb in late winter and sacrifice some carrying capacity (that is, they run fewer ewes) for the sake of easier weaner management. There are a number of important factors which influence the merit of different times of lambing. Consider, for example, the factors involved in selecting a best time to lamb on a farm in the winter rainfall zone of southern NSW:

- *photoperiod effects on ovulation rate* — ovulation rate tends to increase to a maximum in March to April, for a given body weight[c]

- *body weight effects on ovulation rate* — body weights of ewes are highest in early summer and decline until pasture growth recommences after the autumn break (March, April or May). Body weight at joining has a very strong effect on the number of multiple births[13]

- *climate and lamb survival rates* — for the same birth weight and ewe nutritional status, lamb survival rates are highest when born in warm, dry weather and lowest when born in wet, cold, windy weather. Lamb mortality rates associated with exposure and chilling are, therefore, likely to increase from April (best weather) to August (worst weather) or September[14]

- *ewe nutrition and lamb survival rates* — survival rates of lambs are also strongly influenced by lamb birth weight which, in turn, is influenced by ewe nutritional state during pregnancy and at parturition.[15] Ewes require substantial quantities of supplementary feed if late pregnancy coincides with periods of poor pasture quality or low pasture availability. Feed quality improves after the autumn break and pasture availability improves from April or May onwards, sometimes with a *trough* in July and August, until October or November

- *ewe nutrition and wool quality* — autumn-lambing ewes produce wool with lower tensile strength than spring-lambing ewes, because the nutritional demands of late pregnancy and lactation coincide with periods of low pasture availability and poor pasture quality[16]

- *pasture availability and lamb growth rate* — lamb growth rates are higher when their dams have better feed intakes.[17] Pasture quality and availability increase to a peak in October and November. The maximum dietary energy demands of a ewe flock occur in the second and third months after lambing starts — when all ewes have lambed and when lambs are grazing with their dams

- *the length of the growing season and lamb final weight* — lambs and weaners will continue growing through spring and early summer until the digestibility of the pasture diet is no longer sufficiently high to support growth — usually November or December — although growth rates will be declining significantly over the last month of the growing season. Early-born lambs have more opportunity to grow to a satisfactory weight by the end of the growing season, but later-born lambs enjoy better nutrition in their first few weeks of life and show higher early growth rates[18]

- *coinciding seasonal pasture availability with ewe flock nutritional demands* — the greatest stocking rate of ewes which can be adequately maintained all year is determined by the stocking rate which can be sustained without uneconomic levels of supplementary feeding in the periods of the year when the gap between livestock nutritional requirements and pasture availability is greatest. The advantage of a later lambing is that ewes can be run at a higher stocking rate without a need for high rates of supplementary feeding to prevent

---

c   Many of the aspects of time of joining which affect reproductive rates are discussed further in Chapter 7. These two sections should be read conjointly.

uneconomic production and health losses from undernutrition during late pregnancy and lactation. The limiting period of the year is a function of both pasture growth and time of lambing — often it occurs in July and August.

All of these effects and a few others[19,20] must be considered when selecting a time of lambing. The higher optimum stocking rate is a major advantage of a later lambing, but the full benefit of the high stocking rate facilitated by a late lambing is constrained by the subsequent difficulties in lamb marketing (for first-cross or second-cross lambs) or in weaner management over summer and autumn (for Merino and wool-producing breeds).

Field experiments to compare different times of lambing[13,18,21] have been performed but are expensive and complex and cannot be repeated for every region; nor can they be performed for the full range of the variation in climate that occurs from year to year. Consequently, computer modelling is frequently used to compare strategies such as time of lambing, and high-quality biological models linked to historic weather databases provide some of the best available sources of advice for reviewing complex grazing systems.

The interrelationship between the production objective, climatic region and the optimum time of lambing was demonstrated in a modelling exercise (using GrassGro) by Warn et al. (2006).[22] Simulations were run using weather data for 37 years at Mortlake, Rutherglen, Naracoorte and Cowra, four regions in southern Australia which differ in average rainfall, rainfall pattern and the length of the growing season (Table 3.3, Figure 3.3). The optimum time of lambing for store-lamb production was latest in Mortlake, with the longest pasture-growing season, and earliest in Cowra, which had the equal-shortest growing season. The optimum time of lambing was about one month earlier if lambs were kept for longer and fed grain to achieve higher sale weights.

It should be noted that the stocking rate in this exercise was not altered and the authors observed that even further economic benefit would come from optimising the stocking rate to match the optimum time of lambing. The full economic benefit of choosing the optimum time of lambing depends on choosing the stocking rate which takes full advantage of that time of lambing. Unless the number of ewes run per hectare is increased, a change of the time of lambing from autumn to winter-spring may produce only a small economic advantage. It should also be noted that increasing ewe numbers has a capital cost and the increase in gross margin per hectare from a higher stocking rate should be compared to the opportunity cost of the additional financial investment in a larger flock (see 'Modified gross margin analysis', Chapter 2 p. 26).

In summary, time of lambing and stocking rate are interrelated and should be co-optimised with regard to production objective and climatic region. If gross margin analysis is used to

Table 3.3: Climatic data of four regions used for simulations illustrated in Figure 3.3.

	Mortlake	Rutherglen	Naracoorte	Cowra
Mean annual rainfall (mm)	663	619	567	633
Rainfall pattern	Winter-dominant	Winter-dominant	Winter-dominant	Uniform
Length of pasture-growing season (months)	9	8	7	7

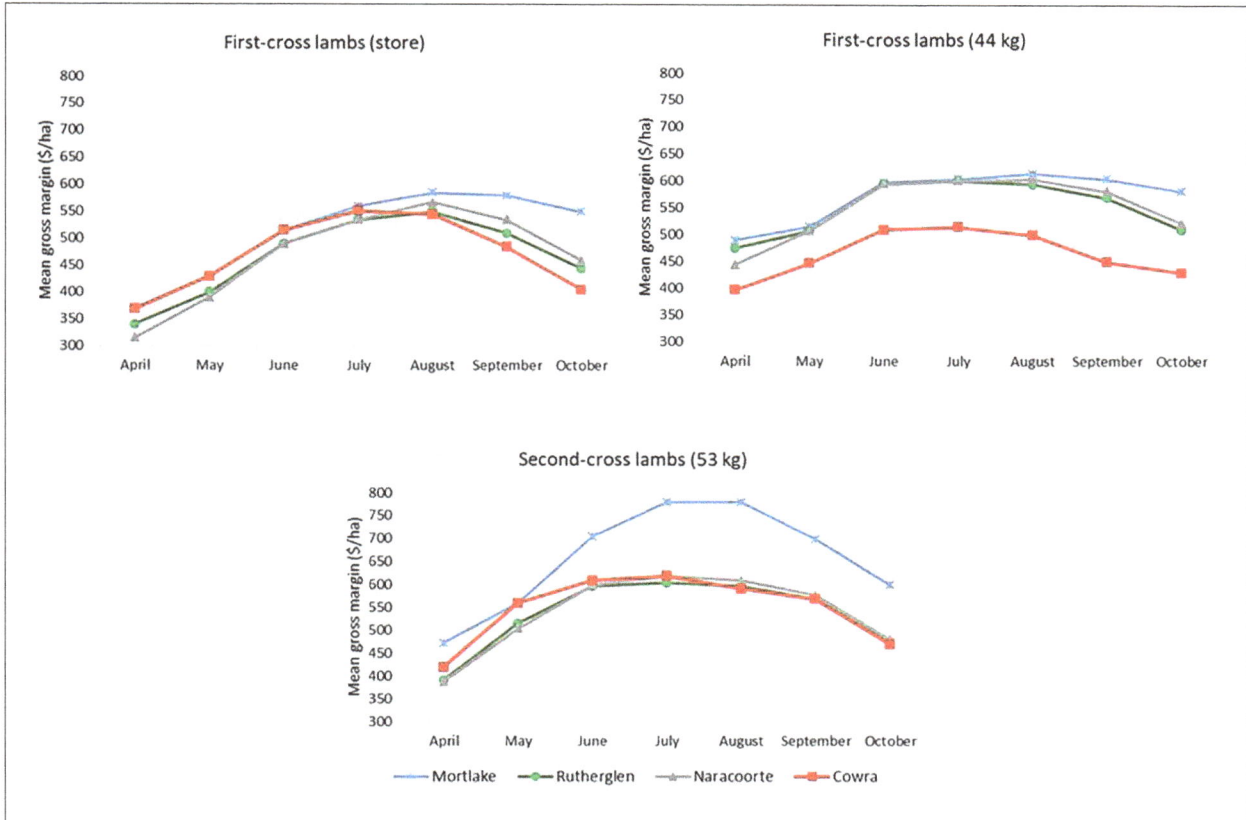

Figure 3.3: Results of a computer simulation examining the effect of time of lambing on mean gross margin per hectare for four locations at a stocking rate of 10 ewes/hectare. Top left: fine-wool Merino ewes producing first-cross lambs (terminal sire) at store weight, top right: fine-wool Merino ewes producing first-cross lambs (terminal sire) grown to 44 kg liveweight; bottom: first-cross ewes producing second-cross lambs grown to 53 kg liveweight. Gross margins are averaged for runs with 37 years of climatic data. Drawn by KA Abbott. Based on data from Warn, Geenty and McEachern (2006).[2]

model and compare strategies, then the cost of the capital investment in livestock should be included in the comparison.

In self-replacing Merino flocks, ewe lambs are not sold but are kept over the following summer. In some cases, wether lambs are also retained over the summer. Chapter 12 discusses weaner management in more detail, but it is relevant to briefly mention the subject here. In general, the later in the year that lambing occurs, the lighter the lambs are at the end of the growing season and the more the lambs will be below critical liveweights and at risk of death or debilitating inter-current disease because of undernutrition and low body weight.

There is the opportunity on many properties to improve the quality of weaner management to the extent that lambing can be postponed — from early winter to late winter, or from late winter to early spring. The benefits which arise from such a change in management come from the ability to run more ewes but still achieve similar levels of productivity and health. Effectively, by coinciding the winter feed trough with an earlier, and less nutritionally demanding, phase of the reproductive cycle, the *dry sheep equivalent* value of each ewe in winter can be lowered and the number of ewes raised to compensate. Consequently, more ewes can be run per hectare, more ewes can be shorn and, provided the reproductive rate does not fall, more lambs can be weaned per hectare. There are, however, some drawbacks to a later lambing. The three major economic ones are as follows:

- The later-born lambs have less wool and shorter wool at shearing, unless the time of shearing is changed as well.

- Later-born lambs are smaller at any given month after birth (despite usually having higher growth rates), and this will reduce the opportunity to sell lambs at marketable liveweights before pastures become dry.

- Later-born lambs which are retained over summer will probably require extra supplementation in the subsequent summer-autumn because they will be lower in liveweight when pasture quality can no longer sustain animal growth.

Plant (1983)[20] lists 13 factors which need to be considered when producers select a time for lambing. Morley (1983)[19] discusses five factors in more detail, particularly the relationship between optimum stocking rate and time of lambing.

TIME OF SHEARING

Shearing times are chosen by producers after consideration of a number of factors, including

- the availability of shearers, which is often determined by normal district practice

- the best time to sell surplus sheep, which may be timed to meet a particular market or to reduce stock numbers before pasture feed levels decline too much

- the need to manage grass-seed infestation in wool[23]

- the need to avoid flystrike conditions in woolly sheep

- the need to complement joining or lambing events

- the need to influence the position of break which is determined when the wool is tested for staple strength

- the need to avoid cold stress off-shears.

Many producers choose to shear in spring or early summer for some of the reasons listed above, but shearing later may increase the utilisation of spring feed surpluses by maintaining higher sheep numbers and thereby growing more harvestable wool from the animals destined for sale as cull or cast-for-age. In southern Australia, sheep shorn annually in autumn may produce more wool than sheep shorn annually in spring.[24,25] An additional advantage of an autumn shearing is that it may increase the attractiveness of later lambing (spring rather than autumn or winter) by ensuring that lambs have sufficient fleece weight and fleece length to make their shearing a profitable activity.

In some districts, the presence of annual grasses with damaging awns makes summer or autumn shearing inadvisable. Changes to pasture management may reduce the prevalence of grass seeds and provide opportunities to choose a time of shearing without the constraint imposed by annual grasses and weeds.

The position of break when wool is tested for staple strength (see Chapter 4) usually occurs at the point where the fibre diameter along the fibres is lowest, and that point is usually determined by the level of nutrition available to the sheep at the time when the wool fibre is produced in the skin follicle. Moving the weakest point of the staple to the tip or the base of the fibres has benefits for both tensile strength and position of break. On the assumption that the finest wool is produced at the time of the autumn break, one could assume that autumn shearing would produce wool of higher staple strength than spring or summer shearing, but this is not always the case.[25] With spring-lambing ewes in particular, it is difficult to predict whether an autumn feed deficit or the physiological competition for nutrients during late pregnancy will have the greater effect on staple strength.

In cold climates, particularly with winter-dominant rainfall patterns, winter shearing is inadvisable because of the increased feed requirements of sheep with short fleeces exposed to cold climates. The increased energy requirements may be managed by feeding supplements or running sheep in high body condition at low stocking rates, but both options are expensive and sheep may be severely affected by unaccustomed low temperatures immediately after shearing despite the availability of feed.

REFERENCES

1   Morris RS (1969) Assessing the economic value of veterinary services to primary industries. Aust Vet J **45** 295-300.

2   Warn LK, Geenty KG and McEachern S (2006) What is the optimum wool-meat enterprise mix? Wool meets meat. In: Proceedings of the 2006 Australian Sheep Industry CRC Conference, eds PB Cronje and D Maxwell. Australian Sheep Industry CRC: Armidale, NSW, pp. 60-9.

3   NSW Department of Primary Industries (2010) Livestock gross margin budgets. Available at: http://www.dpi.nsw.gov.au/agriculture/budgets. Accessed 7 June 2018.

4   Curtis K (2009) Recent changes in the Australian sheep flock. Available at: https://archive.sheepcrc.org.au/files/pages/information/publications/australias-declining-sheep-flock/Recent_changes_in_the_Australian_sheep_industy.pdf. Accessed 30 August 2018.

5   White DH, McConchie BJ, Curnow BC et al. (1980) A comparison of levels of production and profit from grazing Merino ewes and wethers at various stocking rates in northern Victoria. Aust J Exp Agric Anim Husb **20** 296-307. https://doi.org/10.1071/EA9800296.

6   Vizard AL (1990) Optimum structure of Merino and Corriedale flocks. Aust Adv Vet Sci **1990** 70.

7   Daniel G (1991) Stocking rate & soil erosion. Proceedings of the Australian Sheep Veterinary Society Conference. 13-17 May, Darling Harbour, Australia, pp. 19-30.

8   Abbott KA (1991) The effect of stocking rate on farm profitability. Proceedings of the Australian Sheep Veterinary Society Conference. 13-17 May, Darling Harbour, Australia, pp. 1-18.

9   Thamsborg SM, Jorgensen RJ, Waller PJ et al. (1996) The influence of stocking rate on gastrointestinal nematode infections of sheep over a 2-year grazing period. Vet Parasitol **67** 207-24.

10 Morley FHW (1983) Stocking rates. In: Sheep production and preventive medicine. Proceedings No 67. University of Sydney Postgraduate Committee in Veterinary Science: Sydney, pp. 95-102.

11 Abbott KA (1988) Stocking rate and flock structure — their effect on profitability. In: Sheep health and production. Proceedings No 110. University of Sydney Postgraduate Committee in Veterinary Science: Sydney, pp. 399-405.

12 Brown TH, Ford GE, Miller DV et al. (1985) Effect of anthelmintic dosing and stocking rate on the productivity of weaner sheep in a Mediterranean climate environment. Aust J Agric Res **36** 845-55. https://doi.org/10.1071/AR9850845.

13 Egan JK, Thompson RL and McIntyre JS (1977) Stocking rate, joining time, fodder conservation and the productivity of Merino ewes. 1. Liveweights, joining and lambs born. Aust J Exp Agric Anim Husb **17** 566-73. https://doi.org/10.1071/EA9770566.

14 Donnelly J (1984) The productivity of breeding ewes grazing on lucerne or grass and clover pastures on the tablelands of southern Australia. III. Lamb mortality and weaning percentage.

Aust J Agric Res **35** 709-21. https://doi.org/10.1071/AR9840709.

15 Oldham CM, Thompson AN, Ferguson MB et al. (2011) The birthweight and survival of Merino lambs can be predicted from the profile of liveweight change of their mothers during pregnancy. An Prod Sci **51** 776-83.

16 Foot JZ and Vizard AL (1993) Current sheep management: South Australia and Victoria. In: Proceedings of a National Workshop on Management for Wool Quality in Mediterranean Environments, eds PT Doyle, JA Fortune and NR Adams. Department of Agriculture: Perth, Western Australia, p. 67.

17 Kenney PA and Davis IF (1973) Effect of time of joining and rate of stocking on the production of Corriedale ewes in southern Victoria. 2. Survival and growth of lambs. Aust J Exp Agric Anim Husb **13** 496-501. https://doi.org/10.1071/EA9730496.

18 Fitzgerald RD (1976) Effect of stocking rate, lambing time and pasture management on wool and lamb production on annual subterranean clover pasture. Aust J Agric Res **27** 261-75. https://doi.org/10.1071/AR9760261.

19 Morley FHW (1983) Date of joining. In: Sheep production and preventive medicine. Proceedings No 67. University of Sydney Postgraduate Committee in Veterinary Science: Sydney, pp. 83-7.

20 Plant JW (1981) Infertility in the ewe. In: Sheep. Proceedings No 58. University of Sydney Postgraduate Committee in Veterinary Science: Sydney, pp. 675-705.

21 Reeve JL and Sharkey MJ (1980) Effect of stocking rate, time of lambing and inclusion of lucerne on prime lamb production in north-east Victoria. Aust J Exp Agric Anim Husb **20** 637-53. https://doi.org/10.1071/EA9800637.

22 Warn L, Webb Ware J, Salmon L et al. (2006) Analysis of the profitability of sheep wool and meat enterprises in southern Australia. Final Report for Project 1.2.6. Australian Sheep Industry CRC: Armidale, NSW.

23 Warr GJ, Gilmour AR and Wilson NK (1979) Effect of shearing time and location on vegetable matter components in the New South Wales woolclip. Aust J Exp Agric Anim Husb **19** 684-8. https://doi.org/10.1071/EA9790684.

24 Arnold GW, Charlick AJ and Eley JR (1984) Effects of shearing time and time of lambing on wool growth and processing characteristics. Aust J Exp Agric Anim Husb **24** 337-43. https://doi.org/10.1071/EA9840337.

25 Campbell AJD, Larsen JWA and Vizard AL (2011) The effect of annual shearing time on wool production by a spring-lambing Merino flock in south-eastern Australia. Anim Prod Sci **51** 939-51. https://doi.org/10.1071/AN10270.

4

SHEEP FARM PRODUCTS: WOOL, MEAT, SKINS AND MILK

INTRODUCTION

Since the mid-19th century in Australia, sheep have been raised primarily for their wool, and sheep meat was a secondary but important product. Since the 1990s, the production of sheep meat has become of increasing economic importance to sheep graziers to the extent now that, in some cases, sheep are raised purely for meat production and their wool is not even harvested. Nevertheless, on most sheep farms income from both wool and meat production is important. When sheep are sold for slaughter, producers also receive payment for the sheep skin, which makes an additional contribution to the value of a slaughtered animal. The fourth potential product of sheep farming is milk. Internationally, some 20 million sheep are used for dairying but, in Australia, dairying is a very small and specialised sector of the sheep industry.

The health management of the sheep flock has an influence on the quantity and quality of all four sheep products. In this chapter the important characteristics of these products will be presented so that the impact of disease or disease control strategies on the output of the sheep flock and, therefore, the income of the producer can be appreciated.

WOOL

Importance of wool to the sheep farm's gross income

Most Merino flocks are self-replacing, which means that a proportion of the young female sheep bred in the flock are retained as replacements for aged ewes culled from the flock at 5 or 6 years of age. Flocks are described as self-replacing even if rams are introduced from outside sources, which is usually the case. In Merino flocks, income from wool usually makes up around 60% of total gross income and the remaining 40% is derived from the sale of cast-for-age adult sheep, young wethers and cull ewe hoggets. The actual proportion derived from wool varies with the flock composition, the relative value of wool and sheep, and the reproductive success of the flock ewes. To some extent, the emphasis on wool income relative to income from sheep sales is determined by the flock owner, but some districts are better suited to breeding and growing young sheep, while others are better suited to an emphasis on wool production.

Towards the other end of the spectrum, prime lamb flocks based on crossbred ewes mated to terminal sires typically receive 20% of their gross income from wool sales and the balance from sales of prime lambs and cast-for-age adults. If, however, the cost of purchased replacement ewes is deducted from the sheep sale income, wool income becomes a higher proportion of total income relative to *sheep trading profit*. So, even in specialist lamb-producing flocks, wool is usually an important source of income.

In recent years, some flock owners have adopted breeds of sheep which shed their fleeces and, therefore, do not require shearing or produce wool income. These breeds include Damara, Wiltipoll, Dorper and White Dorper.

Harvesting and marketing of wool

Frequency of shearing

In virtually all Australian sheep flocks where shearing is practised, sheep are shorn once annually. More frequent shearing leads to increased harvesting (shearing) costs and a reduction in the price received for wool on a per kilogram basis, because of low staple length.

Time of shearing

The time (month) of shearing is a critical decision for sheep producers because it has ramifications for so many other management strategies. Time of shearing is the second most important decision, after time of lambing, in determining the flock's management calendar. The decision has implications for wool quality attributes (staple strength, vegetable matter contamination) and for the methods which are appropriate to use for flystrike control.

Age of lambs when first shorn

The time of shearing must be determined with time of lambing in mind, for several reasons. First, ewes should not lamb when they are carrying 10 months' wool or more because they will have an increased risk of becoming cast during lambing and because shearing will clash with the early part of lactation. Second, the age and therefore wool length of the lambs at their first shearing have a large effect on the value of each kilogram of wool they produce. The age at which lambs are shorn cannot be decided in isolation because, at their second shearing, they are usually shorn with the adult flock.

Clip preparation in the woolshed

In Australia, sheep are shorn in woolsheds by professional shearers who each shear between 100 and 200 sheep per day. Shearers generally work in teams of two to twelve, depending on how many shearing stands (shearing machines) are in the woolshed. Typically, shearing occurs for two to four weeks each year, with 5% to 10% of the flock being shorn on each working day.

The fleece is removed from the sheep and thrown onto a skirting table, where the sweaty fribs are removed from the points and edges, corresponding to the belly, crutch, neck and mid-leg regions of the fleece, in a process called *skirting*. The fleece is then rolled into a loose ball and placed on the classer's table. Wool classers[a] are trained and licensed professionals whose job it is to examine a sample of the fleece, assess (subjectively) a number of characteristics of the fleece,

---

a   Wool classers are not the same as sheep classers. Sheep classers class sheep, usually when they are carrying a full or nearly full fleece, and visually assess a number of characteristics of the sheep, including wool characters, before deciding whether the sheep should be retained or culled.

and allocate it physically to the appropriate wool bin. A wool presser removes wool from the bins and, using a hydraulic or (rarely now) a hand press, presses the wool into bales of approximately 195 kg. The presser brands the wool under direction of the wool classer (Table 4.1). Wool bales from one farm at one shearing event which bear the same brand constitute a *line* of wool and, generally, all the bales in one line are tested and offered for sale as one lot.

The auction system for selling wool

The shorn wool clip is transported from the farm to a woolstore where it is prepared for sale by auction. At the woolstore, lines of wool are tested by core sampling for fibre diameter and yield and, if additional measurement is requested, for staple length, staple strength and position of break. These attributes are then made available to potential buyers when the wool is catalogued for sale at auction. Wool can be sold in as little as two weeks after it arrives in the woolstore, or the producer may prefer to delay the sale of some or all of the clip.

Table 4.1: A typical adult Merino sheep flock wool clip consists of a number of lines of wool, identified by a brand on each bale. The example below shows how a 60-bale clip might be prepared for sale.

Bale Brand	Number of bales	Description of each line
AAAM	31	Main fleece line
AAM	2	Fleece wool of significantly shorter staple length
BBB	8	Slightly broader fibre diameter than main fleece line
FLC	2	Tender fleece
DGY	1	Doggy fleece
COL	1	Discoloured fleece
PCS	7	Pieces — skirtings free from stain
BLS	4	Bellies free from stain
STN	1	Stained wool — well dried
LKS	2	Locks are the short pieces of wool falling from the fleece
BC	1	Bulk class — non-matching wool

FACTORS DETERMINING THE VALUE OF A WOOL CLIP

Key attributes determining price

Fleece weight and yield

Assuming that all wool quality characters are held constant, increasing the weight of wool produced (the annual wool *clip*) will increase the income received. The amount of wool produced from a property is determined by the amount of wool produced per head of sheep (the fleece weight) and the number of sheep shorn. There is an important interaction between fleece weight and nutrition (usually controlled through stocking density), but it should be noted that sheep are not, under commercial conditions, fed at such high levels as to maximise their individual fleece weights.

Wool sold at the woolstores consists of about 30% by weight of grease and other non-wool constituents. One of the pre-sale tests applied to samples of lines of wool is to estimate the yield (specifically the *Schlumberger dry combing yield*) of clean wool. The presence of grease in wool has virtually no bearing on its price when calculated on a clean basis; in other words, buyers calculate their offers to purchase wool on a *cents per kg clean* basis, then multiply that price by the yield stated for the line in the sale catalogue before making their offer on a greasy basis.

Consequently, the first key determinant of the value of a wool clip is its clean weight or, to state this in another way, its greasy weight and its yield.

Fibre diameter

The most important determinant of the price paid per kg of clean wool is the mean fibre diameter (FD) of the line of wool. Non-fleece wools — the shorter wools in pieces, bellies and locks — are responsible for considerable variation in price not accounted for by fibre diameter; but for fleece wools, fibre diameter accounts for over 80% of the variation in price between sale lots on any one selling day.[1]

Fibre diameter is measured in micrometres (μm), frequently abbreviated to micron or μ. The finer the wool, the higher the price received. In industry parlance, wools below 20.4 μm are fine, wools between 20.5 μm and 22.4 μm are medium and wools over 22.4 μm are broad. Broad wools are also called *strong* wools, which is a confusing term because strength is also used in describing the tensile strength of the fibres. Fine wools under 18.5 μm are further categorised as superfine, ultrafine and extrafine. During the five years to 2013, the greatest production was in the 18.5 to 20.4 μm range (Figure 4.1). Virtually all wool under 26 μm is

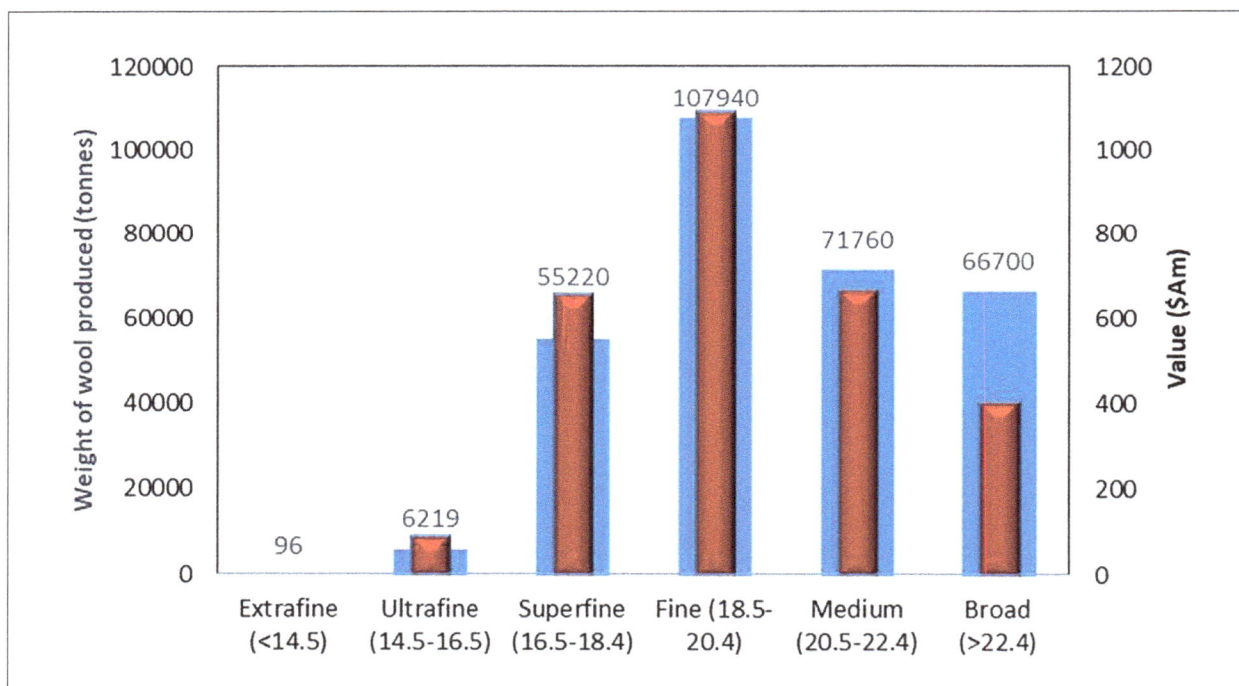

Figure 4.1: The size of the Australian wool clip 2008 to 2013. The weight of wool produced is shown in the broad, blue columns, with data labels; its value is shown in narrow red columns. The largest sector in terms of wool weight and income is the 18.5 to 20.4 μm range, with an average of nearly 108 000 tonnes produced per year, worth $1.1b. Drawn by KA Abbott. Based on data from Nolan (2014)[1] prepared for Australian Wool Innovation Ltd.

produced by Merino sheep, and most wool in the 26 to 32 μm range is produced by Merino crosses such as the Border Leicester x Merino crossbred ewe.

The relationship between price and fibre diameter has varied considerably in the past 20 years and, currently, over the range from 16.5 μm to 32 μm is approximately linear and worth about $0.46 per μm (Figure 4.2). The near-linear relationship in this graph contrasts markedly with the relationship which existed before 2000, in which there was a very steep increase in price as fibre diameter fell below 19 μm.

For some discussions, the relationship between price and fibre diameter is expressed as *micron premium*, which is the percentage increase in price for a 1 μm fall in fibre diameter. The micron premium of 19 μm wool over 20 μm wool in Figure 4.2 is around 3%, while the micron premium of 24 μm over 25 μm is 13%. The low micron premium for fine wool is a reversal of the relationship which existed before 2000, when the micron premium for wool under 19 μm was often greater than 50%.

The fibre diameter which is reported for a line of wool is an average of thousands of fibres and the range in diameter between individual fibres is large. Variation in fibre diameter within a line occurs at several levels — within a fibre (along its length), between fibres within a staple[b], between staples from different parts of the sheep's fleece and between sheep.[2]

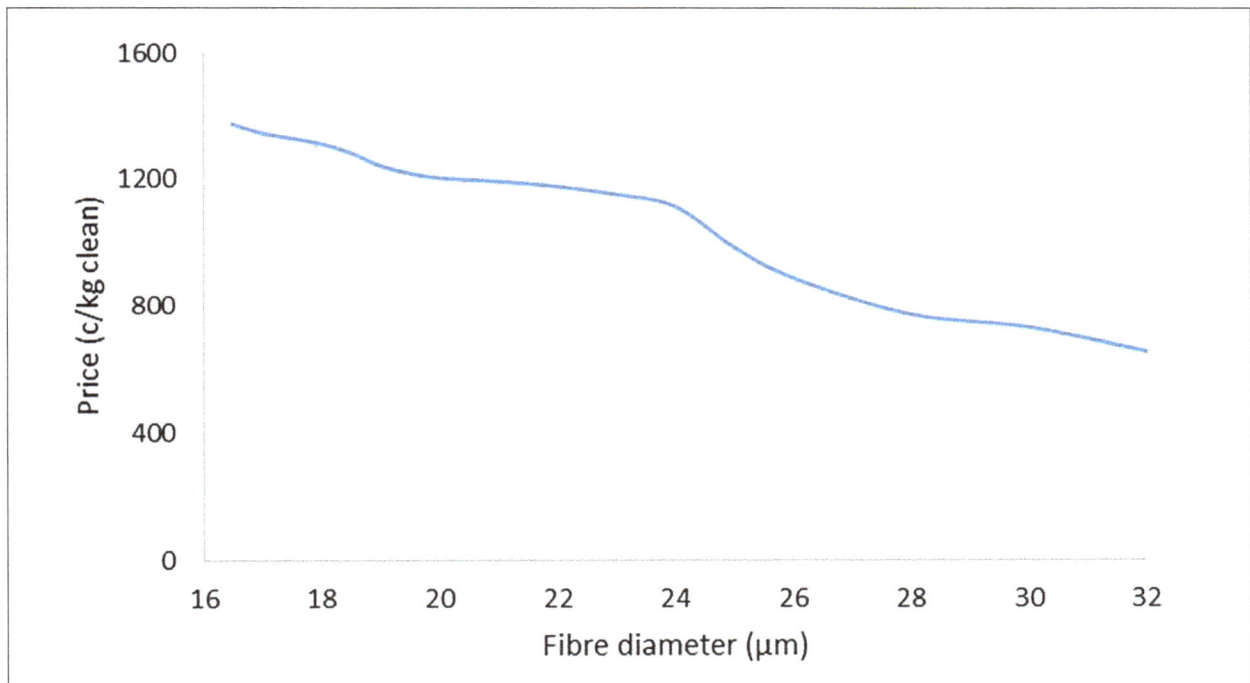

Figure 4.2: The category indicator prices for wool sold at auction in Australia, averaged for the 2014-15 selling season. Drawn by KA Abbott. Based on data from Australian Wool Exchange Limited Annual Report (2015) Annual Report Australian Wool Exchange 2015. Available from: http://www.awex.com.au/media/1500/awex-2015-annual-report-final.pdf. Accessed 17 July 2018.

---

b   A *staple* of wool is a collection of many fibres which grow from a group of adjacent wool follicles and adhere together, forming a lock; they are most evident when a fleece is loosely opened by hand. The fibres are held loosely together by the scales on the individual fibres, which tend to interlock.

One of the most important causes of variation in diameter along fibres is variation in the level of nutrition available to the sheep during the season. Wool fibres are being produced from wool follicles in the sheep skin constantly, and sheep which receive good nutrition produce a greater mass of wool each day compared to sheep subject to poor nutrition. The increase in mass arises from a higher rate of fibre production (the fibre gets longer faster) and an increase in fibre diameter (the fibre is thicker). Thus, for sheep grazing low-quality feed, such as occurs in southern Australia in summer, the wool fibres which are being produced in the wool follicles will have a low diameter while the period of nutrient deprivation continues. When the availability of higher quality feed increases, such as occurs after autumn rains, the availability of nutrients to each wool follicle increases and the part of the fibre produced then is thicker. Thus the variation in fibre diameter along fibres reflects the variation in levels of nutrition available to the sheep over the preceding months. More specifically, the variation in diameter is influenced by the level of nutrition available to the wool follicles, so, during periods when there is increased demand for nutrients within the sheep for other productive functions, such as occurs in late pregnancy, there will be an impact on fibre production similar to that caused by deprivation of feed intake.

The degree of variation in fibre diameter between fibres within a staple, or between staples on different parts of the sheep, is largely under genetic control. Ram breeders now routinely measure the coefficient of variation of fibre diameter (CV_{FD}) in wool samples taken from sheep (samples which are likely to include several staples from one site on the body — usually the mid-side) — and estimates of the heritability of CV_{FD} are typically around 0.5[3], indicating that a large proportion of the variation in fibre diameter between fibres from one site in the body is due to genetic influences.

Staple length

The staple length (SL) of wool in a 12-month fleece varies with the breed and the strain within the breed; the genetic characteristics of the individual sheep; its age; its nutritional state; and its reproductive activities since the last shearing. Typically, in Merinos, adult fine wool sheep grow 85 mm to 95 mm of wool each year, while strong wool sheep grow 95 mm to 105 mm. Short wool has always been penalised at wool sales, relative to normal-length wools, but, over the past two decades, wool buyers have increasingly penalised long wools as well (Figure 4.3). Wools below 65 mm in staple length are heavily discounted because they are more likely to be directed into a lower value sector of the manufacturing industry — into *woollen yarn* manufacture, rather than into *worsted* goods. For worsted manufacture, wools are *combed* to produce a *wool top*, so raw wools which are processed this way are said to be *combing wools* or *of combing length*.

Staple strength

After wool is received at a woollen mill, it is cleaned (a process called scouring), then processed (carding, gilling and combing) into a *top* — a continuous, thick strand of aligned wool fibres which is then drawn, to reduce its thickness, and spun into yarn. The top consists of a series of overlapping and interlocking wool fibres. The average length of fibres in the top is called the *Hauteur*. Wool of high Hauteur can be spun into lower weight yarns and at higher speeds and

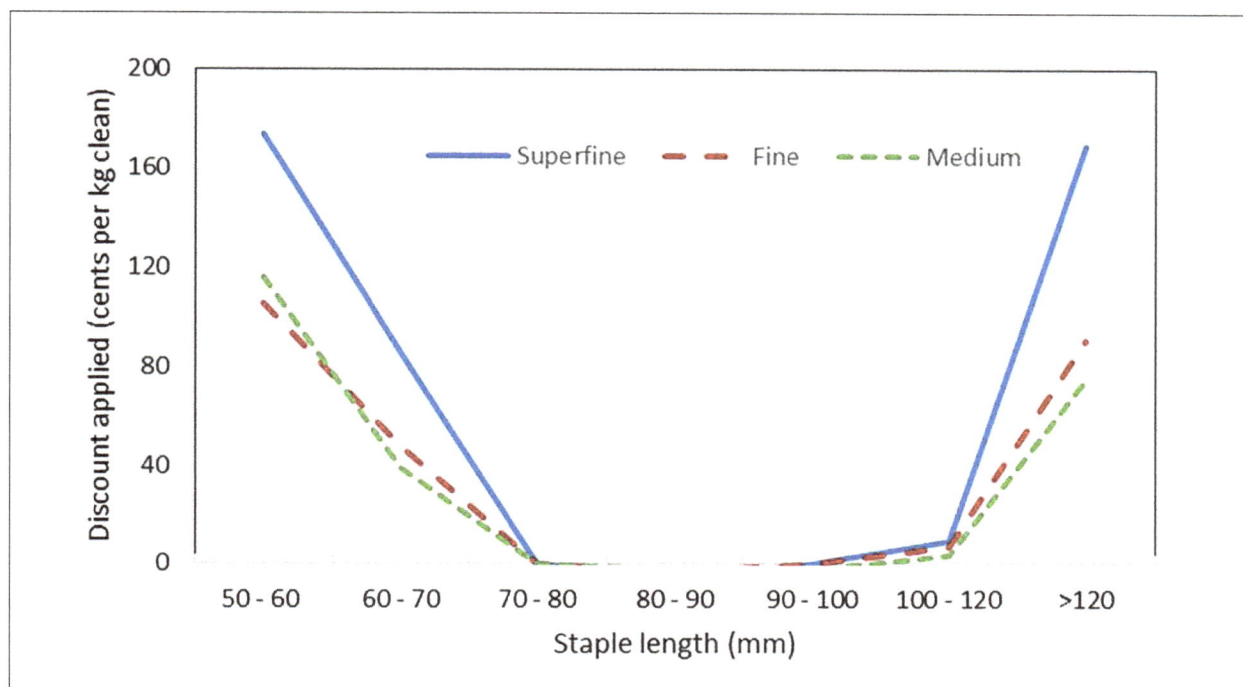

Figure 4.3: Both short and long staple lengths are penalised, with the heaviest penalties applied to finer wools. Data from 2008 to 2013. Drawn by KA Abbott. Based on data from Nolan (2014)[1], prepared for Australian Wool Innovation Ltd.

suffer fewer breakages during spinning compared to low Hauteur wools. Hauteur is strongly influenced by average fibre diameter, staple length and staple strength, and, consequently, these are attributes which are valued by wool buyers.

Staple strength (SS) (also called tensile strength) is the force, in *newtons*, required to break a staple of wool of defined mass and length, or *linear density*. Linear density is measured in units of *tex*, where one *kilotex* (*ktex*) equals 1 gram per metre. SS, therefore, is reported in units of *Nktex*[1] (*newtons per kilotex*).

To measure SS, samples of wool are collected from lines of wool held in the woolstore before sale and submitted for testing. Staple strength is measured in wool testing laboratories with one of two types of machines which, basically, clamp each end of a staple of wool of known length and mass, and pull it until it breaks.

In these machines, the staple of wool is held at the tip (the end on the outer part of the fleece) and the base (the end cut at shearing). The point at which it breaks is called the *position of break* (POB). Because in any one sample a large number of staples is tested, the POB for a line of wool is reported as the percentage breaking in the tip, middle or base part of the staple.

Both the SS and the POB are important in wool manufacturing. In particular, wools with low SS and a high percentage of breaks in the middle are not desirable. If wool fibres break in the middle during processing, both pieces of the fibre are of similar length and short. Consequently, discounts are applied to wools of low SS and higher discounts applied if the percentage of mid-fibre POB is high (Figure 4.4). In recent seasons, the point at which discounts commence is about 38 Nktex[1] in fine wools. The discount varies depending on

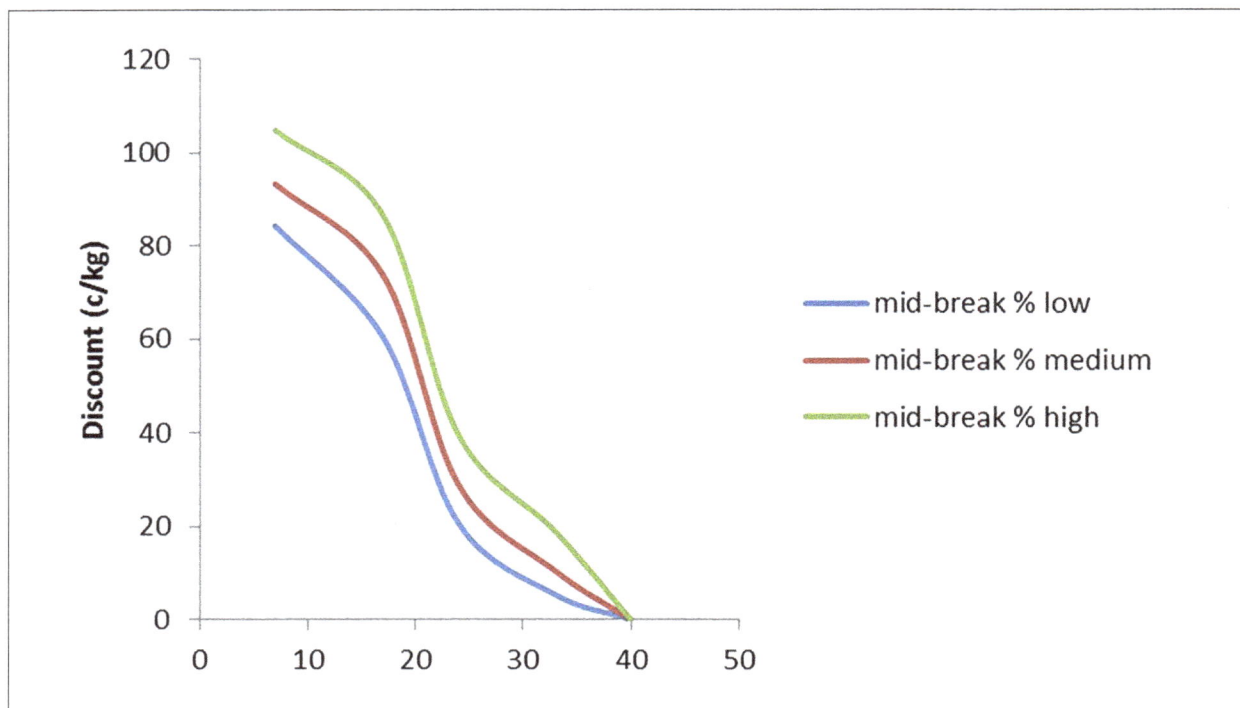

Figure 4.4: Wools with staple strengths below about 38 Nktex[1] are subject to discounts. Illustrated is the discount for fine fleece wool (18.5 to 20.4 µm) for the five years up to and including 2013. Wools with a higher percentage (>75%) of breaks at the mid-point of the staple are discounted more heavily than those with a low percentage (<46%) of mid-breaks. Drawn by KA Abbott. Based on data from Nolan (2014)[1], prepared for Australian Wool Innovation Ltd.

demand but, generally, a decrease of 10 Nktex[1] reduces the value of the wool by about the same as an increase of 1 µm in FD. The mean SS of the Australian fleece-wool clip is typically around 35 Nktex[1], so much of it receives some penalty every year.

Wools of high tensile strength are called *sound*. Wools of low tensile strength are called *tender*. A *wool break* results from the development of a point in the staple of such low staple strength that the staple breaks spontaneously and the outer part of the fleece is shed.

Vegetable matter

As sheep graze they come in contact with bushes, weeds, trees and other forms of plant material, some of which can adhere to the fleece. If the material is smooth, it will generally fall out, but spiny or hairy burrs, seeds or stalks may become entangled in the wool. This material has to be removed from wool in the manufacturing process and different forms of plant material vary in the ease with which they can be removed. The vegetable matter (VM) content of the core sample of wool is reported as a percentage, on a dry weight basis, and type, whether it be *burr*, *shive* or *hardhead*. Both the amount and type of vegetable fault in a line of wool affect its value. VM levels up to 1% usually have no or little effect on the price of wool at auction, and levels between 1% and 2% have only a small impact. VM above 2%, however, will attract significant discounts which become greater as the VM level increases.

Colour

Colours can arise in fleeces from inherent pigmentation; staining from urine or chemicals applied to sheep; diseases like fleece rot; or skin irritation from lice infestations. Putting these attributes aside, there is still variation in the colour of fleeces which can be attributed to either greasy wool colour, which will therefore be removed when the wool is scoured, or colour in the clean wool, which is more or less permanent. Both forms of variation in the colour of wool are a minor source of variation in price.

Style

Wool style is appraised subjectively on a grab sample of wool removed from each line of wool at the woolstore. While style is highly correlated with yield, staple length and greasy wool colour, it also takes into account a number of other features of wool relating to its general appearance. These include crimp frequency and definition, tippyness, dust penetration and weathering.

The degree of variation in price paid for different style grades from MF1 (Choice) to MF7 (Inferior Topmaking) varies with FD. For wool between 19.5 and 20.5 μm there is no extra benefit, in price received, for improving the style of wool beyond MF4 (best topmakers). However, for 19 μm and finer wool, any improvement in style will result in an increase in price.

Factors which affect key wool attributes

Many of the disease conditions discussed in this book add to their economic impact through their effect on wool quantity and quality. Nutrition in general is the major effect but internal parasites, external parasites (lice, blowflies), fleece diseases such as dermatophilosis, trace element deficiencies and diseases causing lameness also affect wool quality and quantity.

Fibre diameter

The principal determinant of a flock's mean fibre diameter is the genotype of the sheep. Further controls can be exercised by the level of nutrition of the sheep (better fed, higher FD) and age profile of the flock (older sheep have higher FD than young sheep).

Staple length

Within a breed of sheep, staple length is largely determined by the length of the inter-shearing interval. Provided there have been 12 months between shearing events, Merinos and Merino crossbred sheep will produce staples about 80 mm to 100 mm long. Low staple length is generally only a serious issue for producers, in terms of any effect on the wool price, when lambs are shorn with less than 9 or 10 months' wool and when shearing times are changed such that adult sheep are shorn with only 10 months' growth of wool, or less.

Staple length is also affected by nutrition. Sheep fed high levels of nutrition have higher fleece weights than sheep fed on a low plane; and the increase in weight arises from an increase in the diameter of wool fibres and an increase in the length of fibres, rather than from a change in the number of fibres growing. Other activities which compete with wool growth for nutrients, particularly late pregnancy and lactation, will reduce fleece weight by reducing mean fibre diameter and staple length.

Similarly, genetic selection for increased fleece weight generally results in increases in staple length. In genetic selection programmes, fibre diameter is generally controlled by direct measurement, so the only biological avenues for increasing fleece weight are to increase fibre numbers (selecting bigger sheep with more skin area, or sheep with more densely packed wool follicles) or to increase staple length.

Staple strength

The diameter of a wool fibre varies as it grows in response to changes in the level of nutrients available to the wool follicle. The variation in nutrients may be simply a result of seasonal variation in feed available to the sheep, or it may be complicated by competition from reproductive demands — particularly those of late pregnancy[4,5] — in the case of ewes. When wool is tested for staple strength, it breaks at the point where the diameter of the fibres is lowest. If there is a big variation in the diameter of the fibres of a wool staple along their length, the reported SS will be low.

If sheep are fed a constant low level of nutrition such that the diameter of fibres along a staple is relatively constant, the reported SS will be high. In fact, the intrinsic strength of the wool fibre material does not change at times of low nutrition. The fibres become thinner, but the wool proteins and structures which compose the fibres retain the same strength.

The explanation for the apparent anomaly lies in the way SS is measured. SS is measured in newtons per kilotex, where kilotex is a measure of linear density and is akin to volume. Fibres can achieve a given volume (linear density) by being consistent in diameter along their length or by being thick in some parts and thin in others. If diameter varies along the fibre, the strength will be determined by the strength of the thinnest point. As that point will be thinner than any point in an even-thickness fibre of the same volume, it will break with a lower force. In reporting the SS, the applied force to break the fibres is divided by the linear density, so the fibres with the most variable diameters will have the lowest reported SS. To illustrate this in another way, any one fibre with an average thickness of 20 μm but a thin point of 16 μm will require the same force to break it as a fibre which is an even 16 μm along its whole length. The SS of each fibre, however, is different, because the fibre which averages 20 μm has a higher linear density and, therefore, a lower SS in terms of Nktex[1] (Figure 4.5).

It appears, therefore, that one way for sheep producers to influence staple strength is to reduce the variation in diameter of the fibres the sheep are growing. This may require taking steps to

• reduce the severity of the most severe nutritional check applied to the sheep during the year (which will have the effect of increasing the minimum fibre diameter)

• reduce the amount of feed offered to the sheep during the time of the year when feed is most available, usually the spring time (which will reduce the linear density of a staple of the fleece and probably also reduce the mean fibre diameter).

The understanding of staple strength described here is relatively new and management strategies to use these approaches have not yet been fully tested. In Western Australia, where

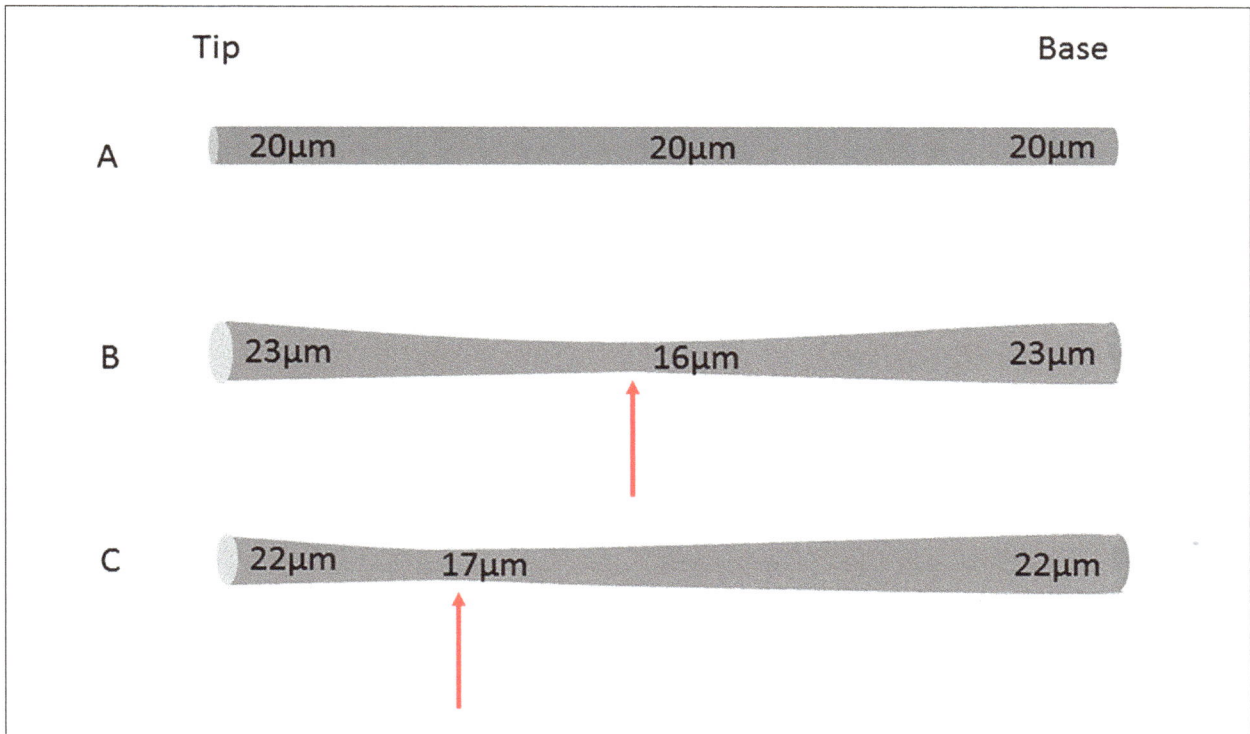

Figure 4.5: The diagram illustrates three fibres of the same average fibre diameter (20 μm) and of the same tex (length and weight). Fibre A is of constant diameter and will break at a random point along its length. Fibre B will break in the middle at the point of lowest fibre diameter. Fibre C will break near the tip, at its point of lowest fibre diameter. The force required to break the fibre will be highest for fibre A — because its minimum diameter is 20 μm — and lowest for fibre B, because its minimum diameter is 16 μm. Source: KA Abbott.

low SS is a major problem for sheep producers, some early results have indicated that the restriction of feed offered to young sheep in spring will lead to increases in SS, although the coincident reduction in fibre diameter has a greater beneficial effect on wool price than does the improvement in SS.

Staple strength is influenced by other attributes of wool fibres within a wool staple. These include the variation in fibre diameter between fibres (independent of variation along fibres), the correlated variation in fibre length within staples and the proportion of shed fibres in the staple. Variation in diameter and length between fibres appears to be under genetic control.

Fibre shedding in Merino sheep occurs when wool follicles become inactive in response to stress.[6] Follicle shutdown leading to the shedding of fibres can be reproduced by the administration of cortisol.[7] Fibre shedding is likely to have a role in the occurrence of fleeces with low staple strengths, but only when SS is below 30 Nktex[1]. In cases where sheep have wool of such low staple strength, nutritionally induced fibre shedding may simply be contributing to the low staple strength induced by the low fibre diameter produced by the same nutritional check.[6]

At greater staple strengths, variations in staple strength appear to be most strongly influenced by variations in fibre diameter. The complete fleece loss (a wool break) which follows severe nutritional or health insults is, however, a consequence of generalised follicle shutdown.

Vegetable matter

The amount of vegetable matter in fleeces and, therefore, in lines of wool is generally a consequence of the timing of shearing in relation to the late-spring/early-summer appearance of grass seeds and burrs. If sheep are shorn just before this time, they pick up much less vegetable material, and have longer to lose what they do pick up before the next shearing. If sheep are shorn soon after this time, such as in summer, there may be very heavy levels of vegetable fault in the fleece. The seriousness of the issue depends on the type of pasture and weeds present on the farm and the grazing system used. In higher rainfall areas and on improved pastures with good weed control, summer or early autumn shearing may not necessarily lead to levels of VM in fleece wools above 2%, but the pieces may be quite high in VM.

Grass seeds can cause problems beyond those associated with wool price. Grass seeds of some plants can penetrate the skin and eyes of sheep and lead to mortalities or losses of weight gain and wool productivity. The effect of grass seeds on carcase quality is discussed below in the section on meat production.

Fleece value per sheep

Reports of the wool market are made frequently in the rural press and usually express the prices received for wool by quoting a *category indicator* price for wools of a particular fibre diameter. The indicator price is an index and is calculated from the average price received for a range of wool types in a particular fibre diameter range which were sold at auction in the previous time period. The quoted price is an approximate estimate of the value of a sound fleece of wool from a sheep, taking into account that some wool from each fleece is of relatively low value (PCS, BLS, LKS — see Table 4.1) and some is of high value — the wool in the *main line*. For example, from Figure 4.1, the average category indicator for 20 μm wool in 2014-15 was 1200 cents per kg of clean wool. For a sheep which produces 3.5 kg of clean wool each year worth, on average 1200 cents, the gross return therefore is around $42. After selling costs and levies, the producer should receive over 95% of that amount.

MEAT

Sheep meat production

Sheep meat is produced from lambs and from adult sheep. The highest value meat is from sheep with no permanent incisors erupted, which are described as lambs in the meat trade, even though they may be up to 13 or 14 months of age. Meat produced from hoggets (sheep with two permanent incisors) and older sheep is usually of lower value and returns 20% to 30% per kg less to the producer than lamb.

In some flocks, sheep meat is a secondary product after wool, and the only animals sold for slaughter may be cast-for-age sheep or young sheep which are culled from the flock due to low productivity or poor wool quality. In specialist meat-producing flocks, the principal product is lamb and these flocks are usually described as prime lamb flocks. All lambs of both sexes are sold — replacement ewes are purchased from outside sources.

Lambs produced in prime lamb flocks are usually crossbred and sired by a ram of a breed which has been selected for its strong meat-producing characteristics. In Australia the most commonly used meat breeds are Poll Dorset, White Suffolk, Texel and Suffolk. Because all progeny produced within the prime lamb flock are sold, the rams are described as terminal sires.

The dams of prime lambs are most commonly, in Australia, a Border Leicester x Merino cross. The lambs in such systems are termed second-cross lambs and the ewes, their dams, are called first-cross ewes.

Terminal sires may also be joined to Merino ewes, producing first-cross lambs for slaughter. First-cross and second-cross lambs usually receive a higher price per kilogram than purebred Merino lambs.

Sheep may be sold from farms direct to an abattoir; the producer then receives payment based on a previously agreed price per kilogram of carcase weight. This form of sale is called *over the hooks*. Alternatively, sheep for slaughter may be consigned to a saleyard where buyers compete for the sheep at auction. The offered price is based on the buyer's estimate of what the sheep's carcase weight is likely to yield.

In Australia, the preferred carcase weight for lambs is between 18 and 34 kg. Dressing percentage (the carcase weight as a proportion of the liveweight) is variable but in the range of 45% to 48%. A lamb of 50 kg liveweight is expected to yield a 22.5 kg carcase. Heavy carcases tend to have greater proportions of fat, and lighter carcases may be too lean. The value of a carcase and, therefore, the price paid to the producer are influenced by both size and fatness. The price is usually determined on a price grid (Figure 4.6). There is considerable variation in the price paid within and between seasons, depending on the availability of suitable lambs and the demand for particular types of carcase or sheep meat products from exporters, domestic butcher shops and consumers.

Disease conditions which affect meat quality and price

A number of disease conditions directly affect the carcase quality of animals which are slaughtered at abattoirs. In some cases, the cost of carcase condemnation — partial or total — is borne by the owner of the abattoir. On other occasions, particularly when animals are consigned directly from farm to abattoir, the cost is passed back to the producer. In many cases the disease conditions are preventable.

In Australia, the National Sheep Health Monitoring Project commenced in 2007. The programme seeks to record and monitor disease conditions which are detectable when animals

Fat depth (mm)	Carcase weight (kg)										
	14.1-16.0	16.1-18	18.1-20	20.1-22	22.1-24	24.1-26	26.1-28	28.1-30	30.1-32	32.1-34	>34
0-5	1.50	3.50	4.50	4.50	4.50	4.50	4.50	4.50	4.50	4.50	4.00
6-10	1.80	3.80	4.80	4.80	4.80	4.80	4.80	4.80	4.80	4.80	4.30
11-15	1.80	3.80	4.80	4.80	4.80	4.80	4.80	4.80	4.80	4.80	4.30
16-20	1.80	3.80	4.80	4.80	4.80	4.80	4.80	4.80	4.80	4.80	4.30
>20	1.80	3.60	4.50	4.50	4.50	4.50	4.50	4.80	4.80	4.80	4.30

Figure 4.6: An example of a price grid for crossbred lambs indicating the price paid ($ per kg carcase weight). Lambs that are too light, too heavy, too fat or too lean are penalised. Source: KA Abbott.

Table 4.2: Disease conditions of sheep identified from abattoir monitoring, NSW, 2006-11. Adapted from National Sheep Monitoring Project (https://www.animalhealthaustralia.com.au/nshmp/). Only the eight most prevalent conditions are shown.

Disease agent or condition	Common name	Proportion of consignments positive	Proportion of animals positive	Annual cost to Australian industry
Taenia hydatigena	Bladder worm	68%	18%	
Fasciola hepatica	Liver fluke	27%	8%	
Taenia ovis	Sheep measles	51%	4.0%	
Pleurisy-pneumonia		47%	4.3%	
Caseous lymphadenitis	Cheesy gland	33%	3.9%	
Echinococcus granulosus	Hydatids	1.6%	0.15%	
Oes columbianum	Nodule worm	1.8%	0.4%	
Sarcocystis spp	Sarco	1.8%	0.2%	$0.89m

are slaughtered at abattoirs. For some conditions there is a direct cost in terms of condemnation of organs, carcases or part-carcases. In other cases, the abattoir simply provides a convenient way to monitor the level of a disease in the sheep population.

Metacestode infections of sheep

Sheep are the intermediate hosts of three cestode tapeworms of canids: *Echinococcus granulosus*, *Taenia ovis* and *Taenia hydatigena*. Each of these parasites exists as a tapeworm in the definitive host — dogs and other canids — and in a metacestode form in the intermediate host — the sheep or other hosts. The metacestode form of *T ovis* and *T hydatigena* is a cysticercus, while for *E granulosus* the intermediate stage is a hydatid cyst. *T ovis* and, to a lesser extent, *T hydatigena* have a negative impact on the Australian meat industry, but neither presents any risk to the health of humans. *E granulosus* infection in dogs is the primary source of hydatid infection in humans. Sheep are the principal source of *E granulosus* infection for domestic dogs.

Echinococcus granulosus (hydatids)

Echinococcus granulosus is a tapeworm of dogs, dingoes and foxes. Intermediate hosts are infected by ingestion of eggs passed by the definitive host. The hydatid cyst develops on the viscera, particularly the liver and lungs. Intermediate hosts in Australia include humans, sheep, goats, cattle, feral pigs, macropods and wombats. Approximately 100 human patients are diagnosed with cystic hydatidosis in Australia each year. Most human cases are treated surgically and there is a significant case fatality rate.[8]

Two transmission cycles exist in Australia, with crossover between them on the mainland. One cycle exists between domestic dogs and sheep, cattle or goats. There is also a sylvatic cycle between wild dogs, dingoes and foxes (as definitive hosts) and kangaroos, wallabies and feral pigs (as intermediate hosts). Crossover occurs when wild dogs and dingoes prey on domestic livestock or when wild canids contaminate sheep pastures with eggs.[9]

Following a programme aimed at elimination of hydatidosis from Tasmania, the state was declared provisionally free in 1996. The programme was based on public education, abattoir monitoring, banning of offal feeding to dogs and six-weekly cestocidal treatment of dogs. The absence of a sylvatic cycle in Tasmania made successful statewide elimination feasible.[10] *E granulosus* still exists in Tasmania at a very low level and is apparently being transmitted between dogs and cattle in a few restricted areas. New infections in humans have not been reported[8] and sheep infections are rare.[11]

To prevent infection in sheep, all dogs should be prevented from eating the offal of domestic animals, macropods or feral pigs, and they should be dosed six-weekly with praziquantel — or both. Transmission from dingoes and wild dogs is much more difficult to prevent and will involve further research and development before control programmes can be attempted.[10]

The risk of transmission to humans, particularly children, from farm dogs should be emphasised by veterinarians in conversations with farm owners. The eggs passed by dogs are sticky and can adhere to the dog's coat and transfer to human hands when patting the dog. Washing hands with soap and water reduces the risk of infection.

On farms where sheep measles (*Taenia ovis*) has been diagnosed, the risk of hydatid transmission between sheep and dogs should also be considered, because both parasites have similar life cycles.

Taenia hydatigena (bladder worm)

The metacestode form of *T hydatigena* was formerly known as *Cysticercus tenuicollis*. Cysts form as fluid-filled sacs on abdominal organs, particularly the liver. Infection is found very commonly at slaughter, but the impact is limited to condemnation of the organ (the liver, usually) or trimming of the cysts. Condemnation of the liver can be due to the presence of multiple cysts or of tracts through the liver from migrating larvae or fibrous tissue on the liver surface where cysts existed previously.

T hydatigena tapeworms occur in the small intestine of canids, including domestic dogs, wild dogs, dingoes and foxes[12], and pine martens, weasels, stoats and polecats.[13] Cats are not suitable hosts.[14] After a pre-patent period of 8 to 12 weeks[15], proglottids containing, on average, 28 000 eggs are passed in the faeces, although most eggs are released in the intestinal tract. An infected dog will typically harbour one or two tapeworms, passing up to 100 000 eggs daily, and will remain infected for about one year.[14]

Eggs ingested by the intermediate host hatch in the small intestine and the larvae migrate through the intestinal wall, enter the bloodstream and travel to the liver, entering the liver parenchyma from the portal vessels. The larval cysticerci burrow through the liver for 18 to 30 days before emerging into the peritoneal cavity and then, usually, attaching to the abdominal viscera, mesentery or greater omentum. The cystercerci complete their development and become infective to the definitive host as early as 34 days post-infection. Mature cysticerci are 1 to 6 cm long and filled with clear fluid and a single scolex.

The migrating larvae produce necrotic, haemorrhagic tracks, but, in small numbers, their activity causes no clinical illness. More severe damage to the liver can occur if large numbers of larvae invade the organ more or less simultaneously, as can occur when a sheep ingests an entire tapeworm segment containing thousands of eggs. Under such circumstances the sheep may develop clinical signs of traumatic hepatitis and peritonitis and, if these conditions are sufficiently severe, may die acutely. Infectious necrotic hepatitis (black disease), caused by *Clostridium novyi*, may be precipitated by the migrating cysticerci[16] but this appears to be rare.

Sheep, goats, cattle, pigs and wild ruminants can all serve as intermediate hosts.[13]

Control of *T hydatigena* is similar to that for *T ovis* (discussed below).

Taenia ovis (sheep measles)

Canids contract infections with *Taenia ovis* tapeworms by ingesting sheep meat containing viable cysts of the cysticercus form of the parasite. The tapeworms establish in the small intestine and the infection becomes patent seven weeks or more after ingestion of the cyst.[15] Each infected dog usually has only one tapeworm which may remain for nine months or more. As with *T hydatigena*, eggs are passed in the faeces in very large numbers either free or contained within proglottids and can remain viable on pasture for up to 300 days.[14]

Both sheep and goats can serve as intermediate hosts and ingest the eggs while grazing. The eggs hatch in the small intestine and the activated oncospheres cross the intestinal wall and migrate to muscle tissue via the bloodstream. The larvae encyst in muscle forming the cysticercus. The cysts are 4 to 6 mm in diameter and each contains one cestode scolex. The cysts become potentially infective to canids after approximately 46 days and remain viable for several months, after which time cysticerci die as a result of the immune reaction of the host. As the cysts degenerate they become at first pus-filled and then mineralised.

Cysts form most commonly in the cardiac, diaphragmatic and masseter muscles but can be widespread throughout the skeletal musculature. The presence of cysts — viable or calcified — in the muscle leads to condemnation of a part, or the entirety, of the carcase at the abattoirs. Neither the cysts in the sheep nor the tapeworms in canids pose any threat to humans. The reason for the condemnation is the undesirable appearance of the cysts in the meat and the risk of rejection of large shipments of meat by importing countries. The presence of one cyst will exclude a carcase from export markets. The presence of more than a few cysts will lead to rejection of the carcase from domestic markets as well.

Losses due to condemnation of carcases and offal (hearts in particular) are suffered by both abattoirs and the producers consigning sheep to the abattoirs. The greatest losses are experienced with older sheep, rather than with lambs. A high prevalence of infection can occur unexpectedly — from farms with no previous history of infection — and losses can be severe. Heavy contamination of a pasture can be caused by one or a few canids and, perhaps as a result of favourable climatic conditions, can lead to high levels of infection in sheep which graze the pasture, leading to high levels of carcase rejection when the sheep are slaughtered. Such outbreaks appear to be a worldwide phenomenon.[17,18]

The definitive canid hosts include domestic dogs, dingoes, dingo-dog hybrids, foxes and, in some other countries, wolves. In Australia, the prevalence of infection in domestic dogs is low[11], presumably because dogs are usually fed with commercial dog foods rather than raw sheep meat and are treated with highly effective anthelmintic preparations containing praziquantel. Farm dogs fed raw meat from sheep but not subject to effective and timely anthelmintic treatment are a potential source of *T ovis* eggs to grazing sheep. The feeding of raw sheep hearts to dogs in rural environments may be a particularly common and, for *T ovis* transmission, high-risk practice.[11] *T ovis* tapeworms have been recovered at a low frequency from foxes in NSW and WA (two out of 499 examined), but out of 52 wild dogs examined in NSW and ACT no tapeworms were recovered.[12]

To prevent sheep measles from occurring in flocks, farm dogs should be fed sheep meat or offal only if it has been cooked or frozen before feeding. If offal is to be cooked for dog food it must be boiled for at least 40 minutes. Freezing for 15 days at -5 °C is sufficient to kill cysticerci of *T saginata*[19] and so is likely to be sufficient for *T ovis* cysts.

Dead sheep and discarded sheep offal should be burnt or buried out of reach of dogs and foxes. In addition, dogs should be wormed with a product containing praziquantel at intervals of six weeks or less. In New Zealand, the industry-funded sheep measles control programme (Ovis Management Ltd) recommends treatment at 28-day intervals.[20] Access of other dogs (belonging to neighbours or farm visitors, for example) to sheep pastures should be prevented unless they, too, have been adequately treated.

Sarcosporidiosis

Lesions of sarcosporidiosis ('*Sarco*') in sheep appear as small white cysts, resembling rice grains, in the oesophagus, tongue, diaphragm and striated muscles. They rarely have any effect on the health of the sheep but are commonly encountered when sheep are slaughtered. Heavy levels of infection lead to partial or total condemnation of carcases. Estimates of the prevalence of detected sarcosporidiosis in Australian carcases range from 0.4% to 0.9%, with much higher prevalence detected in southern regions, especially Tasmania. There is a marked increase in the prevalence of macroscopic sarcocysts with increasing age of sheep.[21]

Sarcosporidiosis is caused by a protozoon parasite of the genus *Sarcocystis*. All *Sarcocystis* spp have indirect life cycles with a carnivore as the definitive host and another species, typically a prey animal, as the intermediate host. There are over 120 species of *Sarcocystis* and each is host-specific with respect to the species of intermediate host and the family of definitive host.[22] Four *Sarcocystis* species occur in sheep, two with canids as the definitive host (*S tenella*, *S arieticanis*) and two with cats as the definitive host (*S gigantean* and *S medusiformis*).[23] The red fox may also be a definitive host for *S gigantea*.[22]

The species of *Sarcocystis* recorded in goats (with definitive hosts) are *S capracanis* (canid), *S hircicanis* (dog), *S moulei* (dog) and, possibly, *S caprifelis* (cat).[22,24]

Sheep are infected by eating sporocysts passed in dog or cat faeces. Excystation occurs in the intestine, and sporozoites are released and penetrate the intestinal wall. Asexual multiplication occurs within endothelial cells of capillaries and arterioles releasing merozoites into the bloodstream. These enter muscle cells and form sarcocysts which contain many cystozoites (bradyzoites). If dogs or cats ingest sheep meat containing mature sarcocysts of the appropriate species, then a sexual cycle of development occurs in the intestine, leading to formation of oocysts which sporulate, forming sporocysts.[25] Infections in cats are patent 11 to 21 days after ingestion of sporocysts.[23]

Only the cat-derived sarcocysts of sheep are macroscopically visible and therefore cause a problem for meat inspection.[26] *S gigantean* has a predilection for the oesophagus and forms cysts of a particular shape (three to five times longer than they are wide) described colloquially as *fat*. *S medusiformis* does not exhibit the same muscle site predilection and forms *thin* cysts in muscles, including laryngeal and abdominal muscles and the diaphragm. Mature cysts of *S medusiformis* are fusiform in shape, around 10 times longer than wide, measuring up to 5 mm long and 0.5 mm across.[27] Neither of these *Sarcocystis* species is pathogenic for sheep.

The cysts of *S tenella* and *S arieticanis* are small (<1 mm)[24] and are not detected by gross inspection, and they are not, therefore, a problem of meat quality. Both species produce cysts in the striated muscle and *S tenella* has also been recorded in the CNS and Purkinje fibres of the heart. Both have some pathogenicity for sheep, *S tenella* being the more pathogenic. Clinical sarcosporidiosis due to *S tenella* has been associated with outbreaks of myeloencephalitis, pneumonia and abortion.[28,29] (*S tenella* is also known as *S ovicanis*.)

After ingestion of sporocysts of *S medusiformis*, sheep develop sarcocysts which are detectable by gross examination of affected muscles within 11 months, although the sarcocysts are not mature and, therefore, they are not infective for cats for several more months.[27] There is no apparent effect of *Sarcocystis* on the health of infected cats. Control of cat-derived sarcosporidiosis is based on minimising the access of cats to raw sheep meat — by burning or burying all sheep carcases, as well as by avoiding deliberate feeding of raw sheep meat to cats — and by prevention of contamination of sheep feed sources by cats. There is no treatment available for cats. Effective feral cat control is difficult, but attempts to reduce the feral cat population have begun in Kangaroo Island, SA, with a view to eliminating cats from the island by 2030. Sarcosporidiosis and toxoplasmosis are both factors considered in justification of the plan.

Pleurisy and pneumonia

Respiratory infections of sheep are discussed in Chapter 20.

Arthritis

Arthritis is discussed in Chapter 13.

Caseous lymphadenitis (Cheesy gland)

CLA is discussed in Chapter 19.

Grass seeds

The seeds of a number of grasses and weeds can lodge in the fleece of sheep and lambs and damage and penetrate the skin. Any seeds which do penetrate the skin may rest in the subcutaneous tissues and thus lie on the carcase, where they are detected during slaughter and processing. Reaction around the grass seed may be minimal, or an abscess may develop. Both carcase and skin values may be heavily penalised by processors if grass seeds have damaged the skin or lodged on or in the carcase.

The effect of grass seeds is not restricted to post-slaughter costs. Grass seed irritation can markedly reduce lamb growth rate, and grass seeds may also lodge in the eye, leading to corneal abrasion and kerato-conjunctivitis.[30,31] Grass seeds also contribute to vegetable matter contamination of fleeces, with subsequent effects on fleece value, although there are many more plants which cause fleece contamination without skin penetration.

The species of grasses which cause carcase contamination include those which are common in pastures and possess awns which readily attach to wool and are capable of penetrating the skin. The most likely to cause carcase contamination include *Erodium* spp (storksbill, geranium), *Aristida* spp (wiregrass), *Austrostipa* spp (spear grass) and *Vulpia* spp (silver grass), but *Hordeum* spp (barley grass) and *Bromus* spp (brome grass) can occasionally penetrate the skin and also cause significant damage to eyes and skin.

Grass seed accumulation in the fleece may be prevented by adjusting shearing time or the time at which sheep are sold in order to avoid grazing sheep with fleeces of significant length in paddocks containing problem grasses. Barley grass seed accumulation in the fleece can be reduced by removing sheep from barley-grass-infested pastures for one to two months while seeds fall from the plants to the ground. Slashing the pasture may also be effective. However, with regard to wiregrass, the extended period over which this grass produces seeds makes such an approach less useful.[32]

Shearing lambs before exposing them to pastures containing barley grass at the time of seed shedding (late spring) leads to increased carcase weights and improved quality of the tanned skin.[31] Longer-term management options include

- the provision of low-risk pastures or fodder crops to be used for lambs when seed set is occurring in paddocks of annual grasses
- feedlotting lambs during the periods when seed set is occurring
- chemical treatment of pastures to reduce problem grasses or weeds[30] or to alter seed development.

SHEEP SKINS

The pelt of the sheep is removed at slaughter and processed to produce either nappa leather or skins still bearing wool (*wool-on skins*). The value of the skin at the time of slaughter is determined by supply and demand factors current at any one time but is also influenced by the breed of sheep and the quality of the product which can be produced from it. The removal of wool or hair from the skin is called fellmongering and, in the case of fine wool such as that from Merino and Merino-derived breeds of sheep, the wool removed at the fellmongery is called *slipe wool* and can add significantly to the value of the skin at the point of slaughter. In Australia, lambskins — from sheep under 1 year of age — are typically used as wool-on leather products such as seat covers, bed underlays and clothing.[33]

Merino sheep tend to have more skin wrinkles than other breeds, and this characteristic is retained into the processed leather, where the wrinkles give an appearance of ribs in the skin, which is then referred to as *ribby. Ribbyness* in a pelt reduces its value as a leather product.[34] Consequently, Merino skins often attract lower prices than those of smooth-skinned breeds.

Cuts of the skin such as those caused by the shearing handpiece leave scars on the skin which affect skin values. Shearing scars reduce the value of skins for leather production and are most likely to occur in Merino sheep — which are relatively wrinkly and prone to shearing cuts — and in older sheep which have been shorn more often than young sheep.

In Australia, the value to the producer of a sheep skin when a sheep is sold for slaughter varies with the breed, age and wool length of the sheep in addition to the influence of market conditions for skins. For a crossbred lamb with a fleece length of 50 to 75 mm and little vegetable matter contamination, the skin may add 10% to the value of the slaughtered animal, while the value of the skin of an adult, cast-for-age Merino with <20 mm of wool may be zero.

For lambs, crutching pre-sale is recommended to minimise the risk of faecal contamination of carcases during processing. To increase the value of the skin, the area crutched should be no more than required for hygienic processing of the carcase.

The importance of minimising grass seed contamination of the fleece was discussed above in relation to its role in affecting carcase quality. Grass seed damage to the skin can also occur with or without complete penetration of the skin and provides another reason to manage the exposure of sheep, particularly young sheep destined for slaughter, to pastures which present a risk of grass seed accumulation in the fleece.

Several factors further influence the value of sheep skins to the industry, but the costs of these are not necessarily passed back to the producer directly. They include vaccination lesions, dark fibres in the fleece, marking paints or brands applied to areas on the body rather than the head, and several conditions known by the terms used in the leather industry — that is, *whitespot*, *cockle*, *pinhole* and *grain strain*.

Whitespot is caused by dermatophilosis and *cockle* is caused by infestations of the sheep body louse *Bovicola ovis*. Both dermatophilosis and lice are discussed further in Chapter 10. *Grain strain* can be caused during the slaughter process when the skin is removed from the carcase by pulling. It can also be caused pre-mortem by sheep handlers who catch or hold sheep by the wool.

MILK

The milking of sheep has a very long history in parts of central and eastern Europe, France, Italy, Spain, Greece, Turkey and Iran. Several breeds have been developed for their ability to sustain long lactations at relatively high levels and to adapt to the relatively infrequent removal of milk from the udder, which contrasts with the very frequent suckling of lambs under natural conditions. These breeds include the Awassi and the East Friesian, which were both introduced into Australia in the 1990s but in very small numbers. Most milking sheep in Australia are crosses between Awassi, East Friesian and other breeds more common in Australia, such as the Poll Dorset and Border Leicester. Because of their wide availability in Australia, Border-Leicester x Merino crossbreds are the foundation source of many dairy flocks.[35]

Worldwide and in Australia, almost all sheep milk is used for the production of cheese or yoghurt. Australia imports over $8 million of sheep milk products annually (cheeses such as pecorino, roquefort, feta[c], haloumi and ricotta) and there may be an opportunity for the establishment of a profitable sheep dairying industry in Australia. Despite interest over several decades in establishing an Australian industry, along with the importation of suitable breeds of

---

c   *Feta* is traditionally of Greek origin and is made from sheep's milk or a mixture of sheep's milk and up to 30% goat milk. *Haloumi* is made from a mixture of sheep and goat milk. *Ricotta* may be produced from sheep, cow, goat or buffalo milk.

sheep, there are still very few commercial sheep dairies in Australia; altogether, they produce around 500 000 L of milk annually from about 5000 ewes.[36] The largest sheep dairy in Australia is a mixed sheep and goat enterprise in Victoria, milking over 1000 ewes and over 5000 does.

Several issues of veterinary importance confront the Australian sheep dairy industry. These include the commercial need to have ewes lambing year-round, preferably every nine months, and the dilemma associated with lamb raising. Allowing ewes to nurse lambs for three to four weeks before machine milking reduces the productivity of ewes over the lactation by 25%. This is in contrast to the highly selected milking breeds of Europe and the Middle East, where both anatomical (larger cisterns) and endocrine differences in dairy sheep allow ewes to nurse lambs and to adapt to twice-daily milking without incurring any significant penalty in milk production later in the lactation.[37,38]

In Europe, clinical mastitis is reported to be an important cause of involuntary culling in dairy ewes and the incidence of subclinical mastitis is up to 50% in some flocks.[39] In Australia, while data are scarce, the incidence of subclinical mastitis appears to be low and somatic cell counts are not considered useful.[37]

REFERENCES

1   Nolan E (2014) The economic value of wool attributes Phase 2. Report prepared for Australian Wool Innovation Limited. AWI: Melbourne, Australia.

2   Dunlop AA and McMahon PR (1974) The relative importance of sources of variation in fibre diameter for Australian Merino sheep. Aust J Agric Res **25** 167-81. https://doi.org/10.1071/AR9740167.

3   Safari A and Fogarty NM (2003) Genetic parameters for sheep production traits: Estimates from the literature. Technical Bulletin **49**. NSW Agriculture: Orange, Australia.

4   Masters DG, Stewart CA and Connell PJ (1990) Changes in plasma amino acid patterns and wool growth during late pregnancy and early lactation on the ewe. Aust J Agric Res **44** 945-57. https://doi.org/10.1071/AR9930945.

5   Robertson SM, Robards GE and Wolfe EC (2000) The timing of nutritional restriction during reproduction influences staple strength. Aust J Agric Res **51** 125-32. https://doi.org/10.1071/AR98150.

6   Thompson AN, Schlink AC, Peterson AD et al. (1998) Follicle abnormalities and fibre shedding in Merino weaners fed different levels of nutrition. Aust J Agric Res **49** 1173-9. https://doi.org/10.1071/A98011.

7   Ansari-Renani HR and Hynd PI (2001) Cortisol-induced follicle shutdown is related to staple strength in Merino sheep. Livest Prod Sci **69** 279-89. https://doi.org/10.1016/S0301-6226(00)00253-0.

8   O'Hern JA and Cooley L (2013) A description of human hydatid disease in Tasmania in the post-eradication era. Med J Aust **199** 117-20. https://doi.org/10.5694/mja12.11745.

9   Grainger HJ and Jenkins DJ (1996) Transmission of hydatid disease to sheep from wild dogs in Victoria, Australia. Int J Parasitol **26** 1263-70.

10 Jenkins DJ (2005) Hydatid control in Australia: Where it began, what we have achieved and where to from here. Int J Parasitol **35** 733-40.

11 Jenkins DJ, Lievaart JJ, Boufana B et al. (2014) *Echinococcus granulosus* and other intestinal helminths: Current status of prevalence and management in rural dogs of eastern Australia. Aust Vet J **92** 292-8. https://doi.org/10.1111/avj.12218.

12 Jenkins DJ, Urwin NAR, Williams TM et al. (2014) Red foxes (*Vulpes vulpes*) and wild dogs (dingoes (*Canis lupus dingo*) and dingo/domestic dog hybrids), as sylvatic hosts for Australian *Taenia hydatigena* and *Taenia ovis*. Int J Parasitol **3** 75-80.

13 Soulsby EJL (1974) Helminths, arthropods and protozoa of domesticated animals. Publ Bailliere, Tindall and Cassell: London.

14 Arundel JH (1972) A review of cystercoses of sheep and cattle in Australia. Aust Vet J **48** 140-55.

15 Ransom BH (1913) *Cysticercus ovis*, the cause of tapeworm cysts in mutton. J Agric Res **1** 15-57.

16 Hreczko I (1959) Infectious necrotic hepatitis in sheep in South Australia, possibly associated with *Cysticercus tenuicollis*. Aust Vet J **35** 462-3. https://doi.org/10.1111/j.1751-0813.1959.tb08373.x.

17 Soehl H (1984) An outbreak of *Cysticercus ovis* in Nova Scotia. Can Vet J **25** 424-5.

18 Eichenberger RM, Karvountzis S, Ziadinov I et al. (2011) Severe *Taenia ovis* outbreak in a sheep flock in south-west England. Vet Rec **168** 619. https://doi.org/10.1136/vr.d887.

19 Hilwig RW, Cramer JD and Forsyth KS (1978) Freezing times and temperatures required to kill cysticerci of *Taenia saginata* in beef. Vet Parasitol **4** 215-19.

20 Simpson B (2010) Role for veterinarians in management of sheep measles. Vetscript **23** (Nov 2010) 22.

21 Munday BL (1975) The prevalence of sarcosporidiosis in Australian meat animals. Aust Vet J **51** 478-80. https://doi.org/10.1111/j.1751-0813.1975.tb02384.x.

22 Levine ND (1986) The taxonomy of *Sarcocystis*. J Parasitol **72** 372-82.

23 Collins GH, Atkinson E and Charleston WAG (1979) Studies on *Sarcocystis* species III: The macrocystic species of sheep. NZ Vet J **27** 204-6. https://doi.org/10.1080/00480169.1979.34651.

24 Tenter AM (1995) Current research on *Sarcocystis* species of domestic animals. Int J Parasitol **25** 1311-30.

25 Buxton D (1998) Protozoan infections (*Toxoplasma gondii*, *Neospora caninum* and *Sarcocystis* spp) In: Sheep and goats: Recent advances. Veterinary Research BioMed Central **29** 289-310.

26 Collins GH and Charleston WAG (1979) Studies on *Sarcocystis* species. 1. Feral cats as definitive hosts for sporozoa. NZ Vet J **27** 80-4. https://doi.org/10.1080/00480169.1979.34605.

27 Obendorf DL and Munday BL (1987) Experimental infection with *Sarcocystis medusiformis* in sheep. Vet Parasitol **24** 59-65.

28 Caldow GL, Gidlow JR and Schock A (2000) Clinical, pathological and epidemiological findings in three outbreaks of ovine protozoan myeloencephalitis. Vet Rec **146** 7-10. https://doi.org/10.1136/vr.146.1.7.

29 Agerholm JS and Dubey JP (2014) Sarcocystosis in a stillborn lamb. Reprod Dom Anim **49** e60-3.

30 Little DL, Carter ED and Ewers AL (1993) Liveweight change, wool production and wool quality of Merino lambs grazing barley grass pastures sprayed to control grass or unsprayed. Wool Tech Sheep Breed **41** 369-78.

31 Holst PJ, Hall DG and Stanley DF (1996) Barley grass seed and shearing effects on summer lamb growth and pelt quality. Aust J Exp Agric **36** 777-80. https://doi.org/10.1071/EA9960777.

32 Lodge GM and Hamilton BA (1987) Grass seed contamination of the wool and carcases of sheep grazing natural pasture on the north-western slopes of New South Wales. Aust J Exp Agric Anim Husb **21** 382-6.

33 Holst PJ, Hegarty RS, Fogarty NM et al. (1997) Fibre metrology and physical characteristics of lambskins from large Merino and crossbred lambs. Aust J Exp Agric **37** 509-14. https://doi.org/10.1071/EA96098.

34 Scobie DR, Young SR, O'Connell D et al. (2005) Skin wrinkles of the sire adversely affect Merino and halfbred pelt characteristics and other production traits. Aust J Exp Agric **45** 1551-7. https://doi.org/10.1071/EA03202.

35 Stubbs A, Abud G and Bencini R (2009) Dairy sheep manual: Farm management guidelines. RIRDC Publication No 08/205. Rural Industries Research and Development Corporation: Barton, ACT.

36 Foster M and the Agricultural Commodities Section (ABARES) (2014) Emerging animal and plant industries: Their value to Australia. RIRDC Publication No 14/069. Rural Industries Research and Development Corporation: Barton, ACT.

37 Cameron A (2014) Optimising generics, reproduction and nutrition of dairy sheep and goats. RIRDC Publication No 10/070. Rural Industries Research and Development Corporation: Barton, ACT.

38 Bencini R, Knight TW and Hartman PE (2003) Secretion of milk and milk components in sheep. Aust J Exp Agric **43** 529-34. https://doi.org/10.1071/EA02092.

39 Carta A, Casu S and Salaris S (2009) Invited review: Current state of genetic improvement in dairy sheep. J Dairy Sci **92** 5814-33. https://doi.org/10.3168/jds.2009-2479.

5

GENETICS ON THE SHEEP FARM

INTRODUCTION

The aim in this chapter is to describe some of the genetic principles which underpin commercial sheep production. Veterinarians who service commercial sheep producers are often involved in artificial insemination programmes, examinations of rams for breeding soundness, strategies to lift reproductive rate and other strategies which may have implications for the genetic merit of a commercial sheep flock. It is important therefore that they have not only a strong understanding of the contribution that good genetic decisions can make to the productivity of a commercial sheep flock but also an appreciation of the dominant role in sheep genetics played by the ram-breeding sector of the industry.

Veterinarians are not usually involved in the design of selective breeding programmes in ram-breeding flocks — at least, not without postgraduate training in the field — and that topic is not discussed in this chapter.

Production of sheep meat and wool in the Australian sheep industry occurs principally in the 15 000-20 000 commercial sheep flocks in the country — flocks in which breeding occurs almost exclusively by natural mating and which require the acquisition of new rams every year or every second year. Across Australia, to mate the roughly 45 million ewes joined every year, about 1 million rams are required. Given that the average working life of a ram is less than four years, one could estimate that commercial flocks acquire around 300 000 new rams every year. These rams are largely produced in specialist ram-breeding flocks and sold to the operators of commercial flocks. These specialist flocks are expected to produce rams of high genetic merit and to be making steady genetic improvement so that the rams they sell each year are genetically better than those produced in the previous year.

The transfer of ewes makes virtually no contribution to the movement of genetic material between flocks in the Australian sheep industry. Notwithstanding the availability of embryo transfer technologies, the limited reproductive rate of ewes (which typically have only three to six lambs per lifetime) compared to rams (which may produce 100 or more lambs each) means that the fastest and most efficient way to make genetic change in the Australian sheep flock is by the perpetual introduction of rams from ram-breeding flocks into commercial flocks. Consequently, it is through the genetic selection practices in the specialist ram-breeding flocks and the transfer of rams from them to commercial flocks that genetic improvement of the Australian sheep flock is achieved.

WHAT CONSTITUTES GENETIC IMPROVEMENT IN THE AUSTRALIAN SHEEP INDUSTRY

Wool industry

The Merino (including the Poll Merino) is the dominant breed in the Australian wool industry and there has been a very large body of research addressing the genetic improvement of the breed conducted in Australia over many decades. The discussion which follows in this section refers specifically to this breed.

A few of the important traits contributing to the efficient production of high-value wool which have a significant genetic basis are listed in Table 5.1. Some of these are directly related to the quantity and quality of wool produced — traits like clean fleece weight (the amount of clean wool produced annually by a sheep) and average fibre diameter — and some traits, like reproductive rate or disease resistance, influence the efficiency of production.

Since the mid-19th century, breeders of Merino sheep in Australia have sought to increase the quantity and quality of wool produced by their sheep, although the emphasis given to each of the important traits has varied — sometimes in response to market conditions and sometimes in response to the climatic conditions in different sheep-raising areas of Australia.

Since the latter part of the 20th century there has been an increasing adoption of objective measurement of production traits and multi-trait selection indices in ram-breeding flocks.[a] In such flocks, young rams are run together as one management group and their performance measured, and they are ranked on an index which may include estimates of their breeding values for fleece weight, fibre diameter, body size, reproductive rate, faecal egg count and a range of other wool and carcase attributes. Ram breeders using this approach are likely to achieve annual improvements in the genetic merit of their flocks for a number of traits simultaneously. Genetic improvement is not usually limited to wool-producing traits, for the Merino breed makes a significant contribution to sheep meat production as well as wool production on many farms.

In Australia, sectors of the sheep industry have worked together to adopt a common approach to measuring, recording, analysing and reporting the genetic merit of sheep in participating ram-breeding flocks. The umbrella organisation is called Sheep Genetics and includes three distinct sheep databases — MERINOSELECT for Merino and Poll Merino breeders, LAMBPLAN for breeders of meat sheep and dual-purpose maternal sires, and DOHNE MERINO for the Dohne Merino breed. A common approach to measurement and reporting is used across all breeds. Estimated breeding values (EBVs) used in the Australian sheep industry are referred to as ASBVs (Australian Sheep Breeding Values) and the ASBVs are comparable across flocks and across years.

Sheep meat industry

A wide range of sheep breeds contribute to sheep meat production in Australia but the most important production systems are those based on the Border Leicester — as a maternal sire

---

a   A selection index is a numerical expression of the estimated genetic merit of an animal combining a number of traits (an animal's estimated breeding value or EBV for each trait) weighted by the value to the producer (usually monetary value) of a one-unit change of each trait and then added together. See Appendix 5.1 for more details about selection indices.

Table 5.1: The estimated heritabilities of some important genetic traits of Merino sheep, based on data from Safari et al. (2005).[1] The estimates are based on a review of multiple studies. Standard errors are published in the cited paper.

Trait	Heritability (h^2)	Trait	Heritability (h^2)
Wool traits		**Reproductive traits**	
Greasy fleece weight (GFW)	0.37	Number of lambs born per ewe	
Clean fleece weight (CFW)	0.36	joined (NLB/EJ)	0.10
Mean fibre diameter (FD)	0.59	Scrotal circumference	0.21
Coefficient of variation of FD	0.52	**Disease resistance traits**	
Staple length	0.46	Faecal egg count (FEC)	0.27
Staple strength	0.34	Fleece rot incidence	0.17
Yield	0.56	Fleece rot severity	0.23
Crimp frequency	0.41		
Growth traits			
Post-weaning weight	0.30		
Adult body weight	0.42		

breed — and the terminal sire breeds such as the Poll Dorset, Suffolk, White Suffolk and Texel. The Merino breed, despite being primarily a wool-producing breed, makes the largest contribution to sheep meat production through the role of Merino ewes, either directly as the dams of first-cross lambs or as the dams of the crossbred ewes (Border Leicester x Merino) used in prime lamb production. In addition, many pure Merino ewes and wethers are ultimately slaughtered for sheep meat.

As a consequence of the different production systems, improvement in the efficiency of sheep meat production through genetic selection has had different drivers in the different breed types. In Merinos, growth rate, carcase and reproductive traits have received only limited attention, with most emphasis applied to wool productivity. In the terminal sire breeds, there has been very strong selection for traits affecting the growth rate and carcase quality of lambs (Table 5.2). In the Border Leicester, because the breed is used to sire the dams of prime lambs, more attention is paid to reproductive rate.

Again, as with Merinos, most breeders of meat and dual-purpose breeds select for a range of traits using either selection indices or more subjective measures. Whichever approach they

Table 5.2: Estimated heritabilities of some important genetic traits of meat-producing sheep, based on data from Safari et al. (2005).[1]

Trait	Heritability (h^2)
Growth traits	
Birth weight	0.15
Post-weaning weight	0.22
Adult body weight	0.29
Carcase traits	
Fat depth	0.30
Eye muscle depth	0.24
Lean meat yield	0.35

take, their goal should be to select for traits which match the production objectives of their ram-buying clients.

WHO IS RESPONSIBLE FOR GENETIC IMPROVEMENT?

The term *commercial* is used here to describe flocks which do not produce rams for sale but produce wool and sheep for slaughter. Rams used for breeding are purchased from other flocks, rather than bred within the flock. One distinguishing feature of a commercial flock therefore is that all male lambs are castrated or sold before they reach breeding age. Breeding rams, when they are required, are purchased from specialist ram producers.[b] The rams that are purchased for use in commercial flocks are called *flock rams* to distinguish them from *stud rams*, which are used in ram-breeding flocks (stud flocks) to sire the next generation of rams. In commercial flocks, replacement ewes may be selected from the ewe lambs born in the flock (as occurs in self-replacing flocks) or they may be purchased when required.

To qualify as a stud flock, the individual stud sheep must be registered with a breed society. For Australian Merinos and Poll Merinos, the relevant society is the Australian Association of Stud Merino Breeders Limited. Not all suppliers of rams are studs, however. Registered studs produce about 80% of the rams used in the Australian wool industry and the remainder are supplied from non-registered flocks. The term *ram-breeding flock* is used here to include both stud and non-registered flocks which supply rams to other breeders.

Genetic improvement in the national sheep flock occurs as a result of selective breeding practices in the specialist ram-breeding flocks. Traditionally, in the Merino and Poll Merino breeds, genetic improvement has been in the hands of relatively few elite ram-breeding flocks, known as *parent* studs. Parent studs generally became closed to outside introductions soon after they were formed which, for many of the Merino studs, of Australia was late in the 19th or early in the 20th century.

There are generally considered to be one or two layers of ram breeding interposed between the elite ram-breeding flocks and the commercial producer. These flocks are *multiplier* flocks[2] (also called *daughter studs* if they are linked exclusively to one parent stud or *general studs* if they acquire rams from multiple parent or daughter studs) because they acquire rams (or semen from rams) which are considered to be of high genetic merit from a parent stud, mate those rams to selected ewes and offer the ram progeny for sale to commercial producers. This multiplication step, while diluting some of the genetic superiority of the elite sire, provides access to improved genetic material at an affordable cost for commercial producers. Multiplier flocks do not add to the genetic gains made at the parent stud level but, if the rams purchased from a parent stud are of high genetic merit, they significantly increase the number of high-quality flock rams available to the commercial producer, compared to the number which could be produced by parent studs.

Each year, if improvement is occurring in the elite flocks, it is expected that the average merit of the rams sold to other flocks is better than the previous year. In this way, improved genetic material passes down in a hierarchical fashion from the ram-breeding flocks to the commercial flocks (Figure 5.1).

---

b   There are exceptions. Some commercial flock owners breed their own flock rams, as is discussed further below. Those that do, however, do not sell flock rams to others.

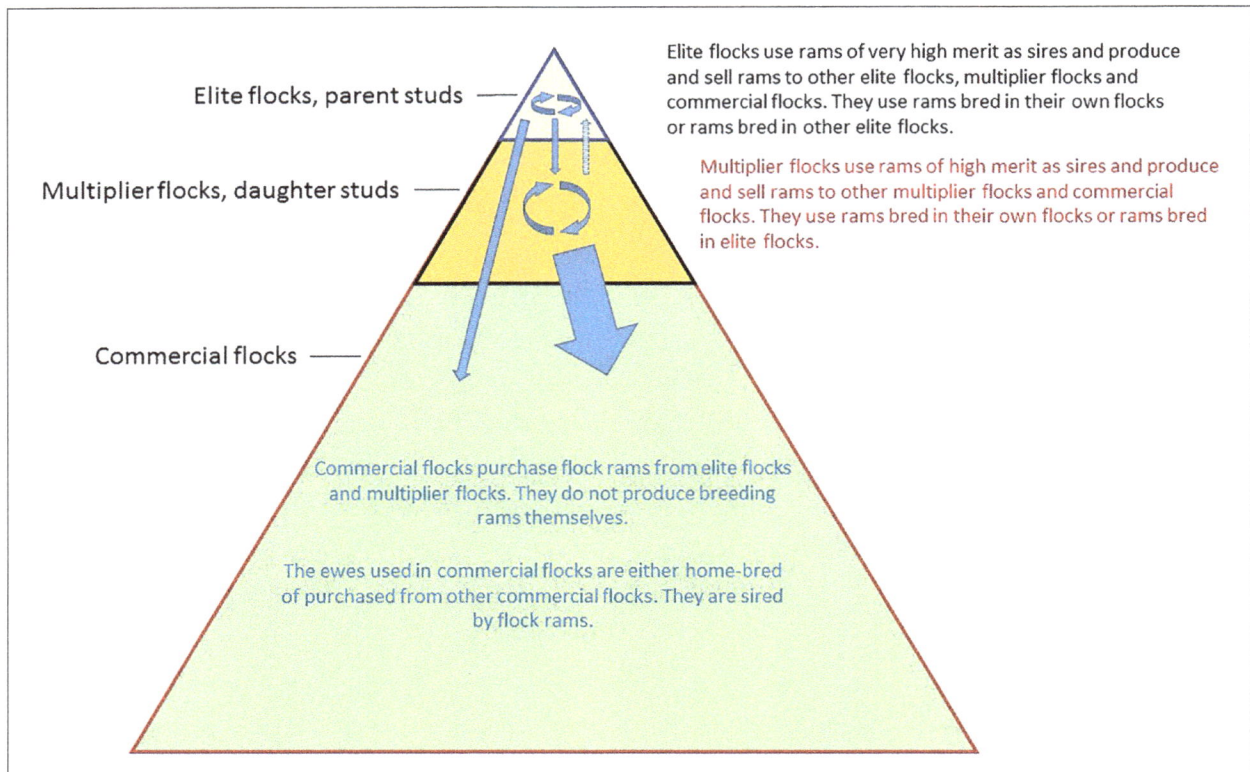

Elite flocks, parent studs

Multiplier flocks, daughter studs

Commercial flocks

Elite flocks use rams of very high merit as sires and produce and sell rams to other elite flocks, multiplier flocks and commercial flocks. They use rams bred in their own flocks or rams bred in other elite flocks.

Multiplier flocks use rams of high merit as sires and produce and sell rams to other multiplier flocks and commercial flocks. They use rams bred in their own flocks or rams bred in elite flocks.

Commercial flocks purchase flock rams from elite flocks and multiplier flocks. They do not produce breeding rams themselves.

The ewes used in commercial flocks are either home-bred of purchased from other commercial flocks. They are sired by flock rams.

Figure 5.1: The breeding pyramid. The genetic development of the Australian Merino and Poll Merino is largely controlled by the genetic selection programs of a small number of elite flocks which are at the apex of the breed. In the past, the elite flocks have been effectively closed to introductions of rams or ewes. Rams and their genetic material have moved from the elite flocks down to commercial flocks, either directly or through multiplier flocks. With the advent of across-flock sire evaluation programs since the 1990s and the development of AI programs using frozen semen, increasingly the elite flocks are open to introductions of rams from other elite flocks and from multiplier flocks. Source: KA Abbott.

Parent studs have in the past been effectively closed to outside introductions, possibly to maintain a commitment to producing a particular 'type' of sheep and possibly also because of uncertainty that a ram from another flock could usefully contribute to genetic improvement. Much has changed in this regard since the latter part of last century. Many ram-breeding flocks submit rams (or semen from rams) to progeny-testing schemes which include rams from other flocks (see later) and receive reliable genetic information about their own rams by comparison to those bred in other flocks. Consequently, there has been an increasing trend for progressive ram breeders to use proven rams of high genetic merit regardless of their flock of origin. The parent flocks and elite ram-breeding flocks which now drive genetic improvement in the Australian sheep industry are therefore no longer necessarily closed.

HOW THE BENEFITS FLOW TO COMMERCIAL PRODUCERS

Rams available to buyers vary in genetic merit and price

Ram-breeding flocks supply rams to other ram-breeding flocks and to commercial flock owners. The greatest number of rams go to commercial flocks, but the best rams are either retained in the ram-breeding flock where they were bred or they are sold to other ram breeders. The rams that commercial growers buy (flock rams) are usually offered for sale in two or more grades. Top-grade flock rams often sell in the range of $800 to $2000, while lower-grade rams may sell for $500 to $800 (2018 prices).

Index scores are commonly used to rank rams for sale

Where rams are measured objectively and ranked using an index, ram breeders commonly divide each year's drop of rams into grades based on their ranking on index value. Index scores, based on continuously variable traits like fleece weight, fibre diameter and body weight, follow a normal distribution, so it is expected that the range of index scores of a group of animals raised as a management group will be distributed in a fashion similar to that illustrated in Figure 5.2.

The top 2% of rams are more than two standard deviations of the index above average. Rams in the top decile are at least 1.3 standard deviations above average and are likely to be retained within the ram-breeding flock for continuing evaluation and use in the ram-breeding nucleus or its associated commercial flock. Rams in the next one or two deciles are likely to be offered for sale at auction. These rams are between 0.5 and 1.3 standard deviations above average. Rams in the lower deciles (fifth, sixth, seventh decile) are likely to be offered for sale in grades corresponding to the decile in which their index value falls and priced accordingly.

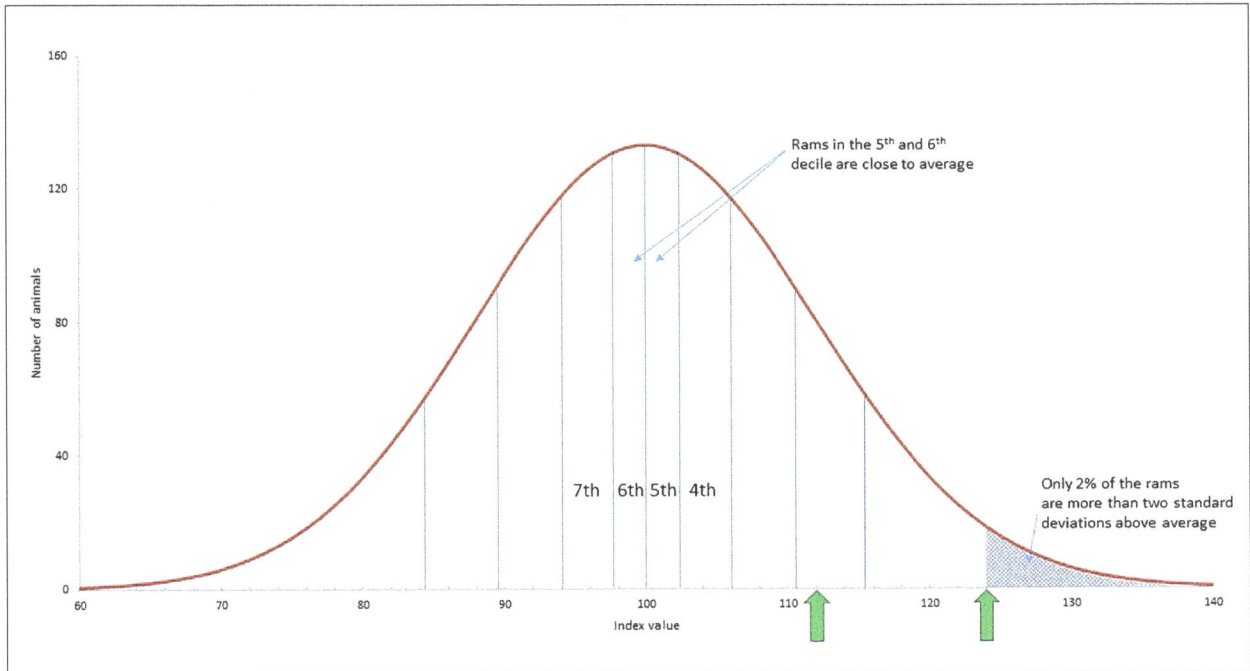

Figure 5.2: The distribution of index scores for 4000 young rams in a ram-breeding flock is expected to be normal. In this example, the average index score is 100. Each decile, consisting of 400 rams, is delineated by a vertical blue line. The 4th, 5th, 6th and 7th deciles are labelled. The 800 rams ranked in the 5th and 6th decile are within a few index points of average. The green arrows indicate points which are one and two standard deviations above average. Around 16% of rams are at least one standard deviation above average. Source: KA Abbott.

Rams which fall in the fifth or sixth decile are close to the average in genetic merit. Rams in the lowest three or four deciles are likely to be culled for slaughter.

It is possible to put a monetary value on each ram's genetic merit when index scores are based on measurable traits of economic value, using the information in Tables 5.3 and 5.4. Even approximate calculations provide a sense of the magnitude of the differences in the productive value of rams with different index scores. To understand how index scores relate to productive values, it is necessary to review the way in which a selection index is developed.

The ram breeder first identifies the genetic traits which are to be improved by the breeding programme. These traits form the basis of the *breeding objective*. Next, *relative economic values* must be allocated to the traits in the breeding objective. The economic value for each trait is an estimate of the marginal profit arising from a one-unit change in the trait. When deciding on the economic values, ram breeders must consider the effect of genetic change in the production

Table 5.3: Average prices for medium and fine Merino wool in 2017. The Australian Wool Exchange Ltd (AWEX) publishes weekly price indicators intended to reflect the value of a typical 12-month-grown fleece of the specified mean fibre diameter. This table shows the average values for 2017 sales and the micron premium associated with each micron category — the extra value of wool per kg clean for a 1 μm reduction in fibre diameter.

Fibre diameter (μm)	AWEX micron price guide 2017 $ per kg clean wool	Micron premium ($)	Micron premium (%)
17	23.42		
18	21.88	1.53	7
19	19.30	2.58	13
20	17.01	2.29	13
21	16.17	0.84	5
22	15.29	0.88	6
23	14.14	1.15	8

Table 5.4: Two of the most economically important, genetically controlled traits in Merino sheep are clean fleece weight and mean fibre diameter of the fleece. Typical values for young sheep of a medium-fine wool type are tabulated. The fleece value is based on the production of 3.5 kg of clean wool worth $17 per kg ($60 per fleece). Standard deviation estimates are derived from Atkins (1997).[3]

Trait	Mean individual value	Standard deviation (units of trait)	Coefficient of variation (SD/mean)	Change in fleece value per phenotypic standard deviation	
				In $	As a %
Hogget clean fleece weight (12 months)	3.5 kg	0.5 kg	15%	$8.90	15%
Mean fibre diameter	20 μm	1.4 μm	7%	$11.20	19%

systems of their ram-buying clients, because it is in the commercial flocks that the benefit of genetic improvement will be realised. Decisions around the economic value must also include considerations of any additional costs associated with a one-unit increase in productivity as well as any likely changes in the value in the future — because the desired genetic change will not be realised in the commercial flock until several years after the ram breeder defines the objective.

In relation to changes in clean fleece weight and fibre diameter, reference to Table 5.3 shows that, using 2017 commodity prices, a producer whose adult sheep grow 3.5 kg of clean wool per year, with an average fibre diameter of 20 μm, could expect a 1 kg increase in clean wool production to increase wool income by $17 per fleece, assuming no change in fibre diameter. Similarly, a 1 μm fall in mean fibre diameter is expected to increase fleece values by $2.29 per kg, or $8.15 per fleece. Using those economic values, the breeding objective (H) can be written thus:

$$H = 17 \times ACFW - 8.15 \times AFD$$

where ACFW is the adult annual clean wool production (kg) and AFD is the mean fibre diameter of the adult fleece, in microns.

In choosing economic values to apply in the development of selection indices, it is the *relativity* between the values which influences the emphasis applied to each component of any genetic change, rather than the quantum of the values. The statement made by the breeding objective above is: 'A 1 kg increase in clean wool production is as valuable to the producer as a 0.48 micron decline in fibre diameter. It does not matter which trait changes the most, just as long as the aggregate (H) increases as much as possible'.

The ram breeder wishes to increase the value of the adult fleece but must make the selection decisions before the rams reach adult age. Consequently, the rams will be measured at hogget age (13 to 18 months) by fleece weighing and side sampling. The traits which will be used to calculate index scores will be hogget clean fleece weight (HCFW) and hogget fibre diameter (HFD), because these are traits which are highly correlated with the adult expressions. The selection index which is derived from this breeding objective, using these *selection criteria*, is

$$I = 5.5 \times HCFW - 4.1 \times HFD.$$

See Appendix 5.1 for further information about the development of selection indices.

Depending on the way in which they are reared, a flock of young Merino rams could have a mean clean fleece weight of 3.5 kg (for 12 months' growth of wool) with a standard deviation of 0.5 kg or 15%, and a standard deviation of mean fibre diameter of 1.4 μm.[3,4]

Using that index, EBVs for a ram which is one standard deviation better than average could be around +0.03 kg for adult clean fleece weight (+0.9%) and -0.7 μm for fibre diameter. These EBVs have economic values which, in total, add to $6. Some rams in the group with similar index scores will have higher clean fleece weight EBVs and higher (less valuable) fibre diameter EBVs, while some will be genetically finer woolled but with lower fleece weight EBVs. The progeny of a ram of that merit will, on average, inherit half of his genetic superiority and are therefore expected to produce fleeces at least $3 per head more valuable at each adult shearing, compared to the progeny of an average-grade ram.

For traits which are expressed every year, like wool value and reproductive traits (in the case of ewes), the extra value will be realised repeatedly through the progeny's life. Consequently, in some genetic evaluation programmes, the allocation of relative economic values is based on lifetime productivity, in an attempt to recognise the difference between traits which have multiple expressions in the lifetime of a ewe or wether, and those like body weight at sale, which is only expressed once.

In ram-breeding flocks using MERINOSELECT or LAMBPLAN, buyers are presented with lists of rams with EBVs for a wide range of traits and pre-prepared index scores based on breeding objectives developed for several different types of commercial production systems.

Evaluating rams on EBVs directly

Rather than using the presented index scores, a buyer can directly apply relative economic values of his or her own choosing to the EBVs for each trait in order to compare two or more rams.

For example, Table 5.5 compares the EBVs for two flock rams offered for sale in a 2017 sale catalogue:

Ram 4059 has an EBV for fibre diameter 0.7 µm finer than the second ram, and an EBV for clean fleece weight 9% higher — equivalent to 0.3 kg of clean wool when applied to the commercial flock average of 3.5 kg. If the commercial producer interested in buying one of these rams chooses to apply relative economic values of +$17 per kg and -$8.15 per micron directly to the EBVs, then ram 4059 will be found to have EBVs for wool production at least $10 more than ram 4124 — a difference which is expected to add $5 more fleece value to the progeny of ram 4059 compared to ram 4129. Effectively, the producer is using EBVs as the *selection criteria*, creating the following index:

$$I = 17 \times EBV_{CFW} - 8.15 \times EBV_{FD}$$

The two rams in this example were offered for sale in different sale grades — the first is priced by the ram breeder at $200 more than the second. Depending on the type of production system operated by the ram buyer in his or her commercial flock, the extra $200 extra may be considered a small price to pay for a ram which will produce progeny with $5 of extra fleece value every year.

In most sale catalogues there are other traits reported for each ram which the astute buyer will examine, but this simple example is used to illustrate the way in which EBVs can be used to evaluate one or more rams.

Table 5.5: Two rams offered for sale by a ram breeder with their EBVs for fibre diameter (FD) and clean fleece weight (CFW). The publication of EBVs at point of sale allows prospective buyers to compare rams on the basis of the likely productivity of their offspring.

Ram ID	FD EBV µm	CFW EBV %
4059	-1.745	28.6
4124	-1.062	19.8

Not all ram breeders present the results of comparative objective measurement to the commercial producer; some instead offer flock rams for sale at varying prices based on evaluations based largely (or entirely) on visual assessment. This method of sale, which has been very widely practised in the past and is still commonplace, demands a high level of trust between the commercial ram-buying client and the operator of the ram-breeding flock. In these more traditional ram-breeding flocks, the use of objective measurement is often limited and combined with visual appraisal, but rams are still graded into elite, reserve and sale categories (Figures 5.3 and 5.4).

Rates of genetic improvement in the Australian sheep industry

Theoretically, genetic improvement programmes in large, closed parent flocks could increase fleece weight by 1% per year while simultaneously reducing fibre diameter by 0.15 to 0.2 μm per year, using a selection index based on a 10% micron premium.[3] In practice, however, most ram-breeding flocks achieve less than 50% of the potential gains in fleece quantity and quality, due to inefficiencies in selection, inclusion of visually assessed traits and extended generation lengths of both males and females in many ram-breeding flocks.[3]

With the increasing adoption of nationwide, across-flock genetic evaluation programmes for the Australian sheep industry, it is now possible to monitor the rate of genetic improvement in the major breeds and breed types. An estimate in 2008 suggested that terminal sire breeds using the LAMBPLAN database had improved by almost three standard deviations of their selection index, worth about $17 per ewe, over a 15-year period to 2005, while the Merino had improved by 1.3 standard deviations (6% to 7% greater fleece weight, 1.0 to 1.2 μm finer) over the same period.[5] The New Zealand sheep industry estimates annual rates of gain of 0.1 to 0.2 genetic standard deviations in meat breeds[6], similar to those in Australia.

Figure 5.3: Flock rams are often graded, priced and presented by the owner of the ram-breeding flock, and buyers can choose one or more rams from each grade following a visual inspection. Photograph courtesy of LA Abbott.

Figure 5.4: Rams identified by the owner of the ram-breeding flock as elite may be offered for sale by auction or retained within the flock for potential use as elite sires. In such cases they may be shown to ram-buying clients as examples of the best the flock can produce. Photograph courtesy of LA Abbott.

STRATEGIES COMMERCIAL PRODUCERS ADOPT TO INCREASE THE GENETIC MERIT OF THEIR FLOCKS

The rate of genetic improvement in commercial flocks

Genetic improvement occurs as a result of the selective breeding practices of the parent studs and other elite ram-breeding flocks. These genetic gains are passed on to commercial producers who buy flock rams — either directly from the parent stud or via a multiplier flock. Each year, it is expected that the average genetic merit of the ram-breeding flock improves, so the rams sourced from that flock each year will be genetically superior to those purchased in previous years. Commercial producers with self-replacing flocks produce their own ewe replacements which are sired by the purchased rams. The average genetic merit of the ewe flock improves each year as a consequence of the improvement in the rams which are purchased each year. For producers who buy rams of a similar grade from the same ram breeder each year, the genetic improvement in the ewes and wethers of the commercial flock will proceed at the same rate as that achieved by the ram supplier, but the genetic merit of the commercial flock will lag behind that of the ram-breeding flock.

The improvement lag between the ram breeder and the commercial flocks

The time lag separating the genetic merit of the two flocks is termed the *improvement lag*. In the case of commercial flocks which have sourced their rams from the same supplier for a long period, the improvement lag is the amount of genetic improvement in the ram-breeding flock which occurs in the number of years taken for two *generation lengths* of the commercial flock. The generation length in a commercial flock is defined as the average age of the parents when the lambs are born, and this is typically 3 to 4 years. Thus the genetic merit of a commercial flock usually equals that of the ram-breeding flock six to eight years previously. See Appendix 5.2 for a more detailed explanation.

If the commercial breeder purchases rams from a multiplier flock, there is an additional improvement lag placed between the parent stud and the commercial producer's flock. As the generation interval in the multiplier flock is likely to be of similar magnitude to that in the commercial flock, an additional six to eight years is interposed between the parent stud and the commercial producer, leading to an improvement lag which could be as great as 16 years.

Fortunately, the operators of most multiplier flocks do not purchase average-grade sires to breed more flock rams, but purchase rams from the parent stud which are substantially above the 'average' grade. If, for example, a multiplier flock manager purchases rams which are genetically one standard deviation above average from the parent stud, the extra degree of superiority (one standard deviation) may compensate for much of the six to eight years of improvement lag incurred before the flock rams are offered for sale to the commercial producer.

Identifying a ram breeder with a flock of high genetic merit

Traditionally, there have been very few reliable comparisons of the genetic merit of each of the parent flocks, so commercial producers have found it very difficult to decide objectively which are the best sources of rams for them. In many cases, the relationship between ram breeder and commercial client has been based on a sense of loyalty, trust and other interpersonal relationships. Over the past 20 to 30 years this situation has changed as more objective comparisons between ram-breeding flocks and individual sires have become available. Two of the important sources of information are described below.

Merino bloodline performance

The comparison of some aspects of the productivity of Merino sheep from different genetic sources (different bloodlines, representing different ram-breeding flocks) can be performed by conducting *wether trials*. Groups of at least 10 wethers are submitted from each of the participating flocks and the groups are run together under the same conditions for two or more years. Their productivity, particularly in wool traits, is measured annually. Because the wethers are run together for extended periods of time, the differences between the productivity of each group is predominantly due to genetic differences. Provided that a sufficiently large and random sample of the source flock wethers are entered into the trial, the results can provide reliable information about the genetic merit of the source flock, and this information can assist commercial flock owners when deciding on a supplier of rams for their commercial flocks. National organisations which now operate in Australia (such as Merino Bloodline Performance[7]) collect and collate the results of trials that have been conducted under an agreed set of rules. By combining results for the same bloodline from multiple trials, the reliability of the comparisons between bloodlines is dramatically increased. The completed results are published periodically, providing information on multiple bloodlines for traits including clean fleece weight, fibre diameter, liveweight, staple length, staple strength and a number of visually assessed traits. When the characteristics of each bloodline are combined in a computer model, a value for profit ($/DSE) can be estimated. Typically, these values show a $6 range between the best few bloodlines and the lowest-performing bloodline. Such information is a reliable and practical guide for commercial producers wishing to compare bloodlines and ram-breeding flocks.

Central test and on-farm sire evaluation

Another source of comparative information available to producers is derived from sire evaluation schemes, which involve the comparison of individual rams on the basis of their progeny's performance. Some progeny tests are conducted in privately owned flocks, particularly ram-breeding flocks, while others are conducted at a central site and overseen by personnel from a government department or a university. The schemes are 'linked' together by the use of common sires, usually by artificial insemination (AI). Once a linkage is established, it is possible to compare *all* the sires used in *all* linked flocks.

The results are published periodically in a report — *Merino Superior Sires* — which is readily available online and free of charge.[c] Both stud and commercial breeders are able to evaluate the relative merits of a number of rams and, in most cases, purchase semen from the best-performing sires.

It should be noted that information about an individual ram does not necessarily reflect on the average genetic merit of the ram-breeding flock in which the ram was produced, although it is more likely that outstanding sires will be produced in ram-breeding flocks which used leading sires in previous generations and which practise sound genetic selection strategies.

Shortening the generation interval

Once a commercial producer has identified the ram-breeding flock which will supply flock rams, there are a number of other flock management decisions which can influence the genetic quality of the commercial flock ewes. One of these is to take steps to shorten the generation interval in the commercial flock in order to reduce the improvement lag between the ram breeder's flock and the commercial flock. This can be done by reducing the average age of either the flock rams or the flock ewes, or both.

Reducing the average age of the flock rams

Flock rams are generally used for three, four or five years, after which time they are cast for age. The average age of the flock rams in a commercial flock when their progeny are born is normally between 3 and 4 years. If a commercial producer reduces the age at which rams are culled, the average age of the ram flock declines but a greater number of young rams must be

c   Merino Superior Sires: http://www.merinosuperiorsires.com.au.

purchased each year to maintain the size of the ram team. Given the usually large difference between the purchase price of young rams and sale price of cast-for-age rams, this practice increases the annual cost of maintaining the flock ram team. This cost is generally greater than the increase in productive value of the flock which results from the move to a younger ram team. A worked example is given in Chapter 7, but it should be noted that the financial merits are influenced strongly by the purchase price of flock rams. While the strategy is genetically sound, it is not usually economically wise.

Reducing the average age of the flock ewes

In self-replacing Merino flocks, ewes are commonly retained in the breeding flock until they are 5 or 6 years of age, then cast for age. The reasons for retaining ewes for at least four lambings (at 5 years of age) include the high productivity of adult ewes up to that age and the relatively low reproductive rate of Merinos, making it difficult to maintain ewe flock numbers if ewes are removed from the flock after three lambings only. The choice for producers is often one of applying a high selection intensity to young ewes entering the flock and keeping breeding ewes to a greater age, or of reducing the selection intensity on ewe replacements and reducing the age at which breeding ewes are cast for age. For most commercial Merino flocks, with reproductive rates between 80% and 90%, keeping breeding ewes until 5 years of age (after their fourth lambing) is the most acceptable compromise. Each year, about 75% of the ewe hoggets must enter the breeding flock to maintain its size, and therefore 25% can be culled. The small gain in flock genetic merit which may be achieved by reducing the generation length of the ewe flock by one year (a roughly 15% reduction in generation length of the ewes from around 3.5 years to around three years) is unlikely to compensate for the difference in productivity between the cast-for-age 4-year-old ewes and the extra ewe hoggets which must be retained to replace them. Nevertheless, if both the reproductive rate of the flock and the sale price of 4-year-old ewes are relatively high, the benefits of a lower ewe generation length may add to the benefits of selling at least some ewes after their third lambing.

Buying rams of higher grade than average

Most of the discussion up to this point has assumed that a commercial producer is purchasing flock rams which are close to the average in genetic merit of their age group in the ram-breeding flock. These rams are usually sold at moderate prices which meet the expectations of most commercial flock owners. A commercial producer could, however, choose to purchase one or more rams from a higher grade and be prepared to pay more to access animals of a higher-than-average genetic merit.

Assuming that the commercial producer intends to use the ram simply as a flock ram, it is relatively straightforward to compare the extra cost of an above-average ram to the likely increased productivity of his progeny. (The special case where a commercial producer purchases an above-average ram to breed more flock rams, thereby acting as a *multiplier*, is discussed further below.)

In an earlier calculation, it was shown that a ram in the second decile of an index-ranked group of young rams is likely to be around one standard deviation above average and to pass on to his progeny a 0.5 standard deviation advantage, compared to an average-grade ram. Furthermore, the ewe progeny of the superior ram will pass on one half of their superiority to the next generation of progeny born in the flock. The genetic influence of the superior ram flows through the flock for a number of generations, gradually being diluted by each generation of ram introductions. The impact diminishes over time such that, after 20 years, less than 2% of the genetic superiority from that single ram introduction remains in the flock.

The increased financial return from the superior ram commences about one year after the ram purchase, when the lamb progeny are shorn. In subsequent years, more progeny are shorn, increasing the return from the purchase of the superior ram, with the greatest impact occurring about six years after the purchase.

Because the extra cash flows occur several years after the purchase, it is necessary to discount the future return to a net present value (NPV). Using a discount rate of 5%, the NPV of a superior ram used for four years as a sire in a self-replacing Merino flock can be approximated by the expression

$$NPV = k \times V \times (n_1 + n_2/1.05 + n_3/1.05^2 + n_4/1.05^3)$$

where V is the value, in dollars, of the ram's superior breeding value and n1, n2, etc. are the number of direct progeny born in the first, second and subsequent years following the ram's introduction.[8] The value of k varies with the flock structure, which influences how many progeny of the superior ram are retained in the flock, and to what ages. For a flock in which all wether lambs are sold after the lamb shearing, the value of k is around 1.6. Thus, if n1 is 35, n2, n3, n4 are 55 and V is $6, the NPV of that ram is $1770. Under these conditions, it is sensible spending $1770 extra to purchase a ram which has a breeding value (index score) which is worth $6 more than an alternative ram. This approach can be used to evaluate the financial merits of purchasing a superior ram to use for natural mating for a range of flock management scenarios and discount rates.[8]

Some caution must be expressed, however, about the risks of purchasing a single expensive ram on the basis of estimated breeding values. There is always some uncertainty attached to the prediction of the ram's genetic superiority. If the accuracy of his index score is low (such as 0.4), the confidence limits around the estimate of his breeding value are quite wide. There is a significant chance that his true breeding value is substantially different (better or worse) than the predicted value. The uncertainty is easier to manage for owners of large commercial flocks who buy several rams at a time. The uncertainty around the *average* estimated breeding value of a group of rams, each with low to moderate accuracies, is much less than the uncertainty around any one of the individuals. If five or more rams of similar index score and accuracy are purchased, the uncertainty can largely be ignored because the *average* of the group's estimated breeding values will be close to the average of their true breeding values (see Appendix 5.1).

Producing homebred rams within the commercial flock

Some commercial flock owners, rather than buying flock rams, breed their own by selecting some of their best ewes and mating them to a high-merit ram, either through natural mating or with purchased frozen semen. The high-merit ram used as the sire of the flock rams is almost always produced in a dedicated ram-breeding flock and not home-bred in the commercial flock.

Despite the fact that this procedure introduces an additional improvement lag while the flock rams are produced, the access to rams of high merit can make this option financially attractive. By using a similar approach to that described above for evaluating the purchase of a superior ram for use directly in the flock, it can be shown that breeding flock rams from one purchased superior sire can justify the additional expense of around $10 000 for a ram that is one standard deviation above average, or over $15 000 for one which is two standard deviations above average.[8]

Potential drawbacks of this approach include the need to have a flock exceeding 2500 ewes to justify the purchase of one superior ram (because of the number of home-bred flock rams which are necessary to justify the purchase), and the uncertainty which results from using rams with EBVs of low or moderate accuracy.

To avoid both of these constraints, the commercial producer could consider breeding flock rams using artificial insemination with semen from either one proven sire (a ram with highly accurate EBVs) or from a number of sires with moderate or higher accuracies. Provided that care is taken in choosing the AI sires, current prices for semen and insemination procedures make this approach readily justifiable for the commercial producer with a Merino flock of substantial size.

Selection of replacement females

In self-replacing commercial flocks where breeding is practised at least partly to maintain the female breeding flock, young ewes are selected to replace aged ewes which are culled and ewes which have died. If there are more young ewes available for selection than are required to maintain the breeding flock, then selection can be practised on the young ewe flock to ensure that the best ewes are retained and the worst ewes are culled or, at least, not used for breeding future replacements.

Selection of the best ewe hoggets may be based on visual appraisal (often called *classing* the ewes), or on an objective evaluation of one or more characteristics of each young ewe, or a combination of both. Commonly, this is performed when the ewe hoggets are aged between 12 and 18 months. The selection may be as simple as choosing the biggest, or as complicated as weighing all or most fleeces at the first adult shearing and measuring the average fibre diameter of a sample of the fleece around the same time (Figure 5.5). If the desired flock structure requires that some young ewes are to be culled, it makes sense to retain the most productive sheep, particularly if the traits under selection have a genetic basis. These sheep will be more productive for the rest of their lives in the flock compared to the culled ones, and their progeny can be expected to inherit half of the genetic superiority of their dams. The potential benefits can be estimated by examining them in two parts, as follows.

Current generation gain

The higher average productivity of the selected ewes (and wethers if they, too, are selected on measured performance) in each subsequent year is called the *current generation gain*. The current generation gain is only realised if the less productive sheep are culled from the flock. (If they are retained as dry sheep or for mating to terminal sires, there is no gain in the current

Figure 5.5: Side-sampling for estimation of the fleece mean fibre diameter in Merino sheep requires the clipping of a small sample (about 50 g) of wool from an area of the body midway between the shoulder and hip and midway between the back and the belly. The wool should be clipped at skin level and from the same region of the body of all sheep in the group being measured. Young adult sheep are often tested at 15 to 18 months of age — before breeding age — to test for fleece fineness. While mostly confined to ram-breeding flocks, it may be economically justified to select ewes to enter the breeding flock of self-replacing commercial wool-producing flocks. Photograph courtesy of LA Abbott.

generation because the less productive animals are still present on the farm and producing wool, even though they are not contributing to the breeding of future replacements.)

The difference between the average phenotype of the selected sheep and the average phenotype of the group before selection is called the *selection differential* (s). As an illustration of the potential magnitude of the selection differential, consider a flock in which the reproductive rate is 85% and 15% of the ewe hoggets are *classed out* on visual traits such as poor conformation. Of the remainder, 80% must be selected to enter the breeding flock.

The selection intensity (i) associated with the best 80% of the population is 0.35 (see Table 5.9 in Appendix 5.3). Using greasy fleece weight as an example, we can predict the magnitude of the selection differential. One standard deviation of greasy fleece weight in a flock of young ewes is approximately 0.5 kg. The selection differential is the product of the two factors i and s — 0.18 kg. In other words, the average greasy fleece weight of the best 80% is likely to be 0.18 kg greater than the average greasy fleece weight of the whole group of ewes which were tested.

If the selection were based simply on fibre diameter, the best 80% would be, on average, 0.5 µm finer than the group as a whole (0.35 $\times$ 1.4 µm). If the two traits were combined into an index weighted to maintain fleece weight and decrease fibre diameter, the best 80% would be around 0.4 µm finer than the group as a whole, but with the same average fleece weights.

Having been selected on the basis of their productivity as hoggets, these sheep will continue to be more productive than a randomly selected group at each subsequent adult shearing. The degree of their superiority at subsequent shearing events depends on the repeatability of the trait. Generally, the repeatability of a trait measured at hogget age lies between the values for heritability and one.[9] It would be reasonable to assume values of 0.6 to 0.8 for the repeatability of fleece weight and fibre diameter after measurement and selection at 1 to 1.5 years of age. The current generation gain is effectively the phenotypic response to selection (R_p), which can be predicted from the expression

$$R_p = r \times s$$

where r is the repeatability and s is the selection differential.[10]

Selection to achieve the current generation gains has a cost which is incurred at the time of measurement. The economic benefits start to appear after the next shearing event (one year later, usually) and continue to recur annually for the life of the selected animals. When evaluating the benefits, it is necessary to discount the future additional cash flows back to a *net present value* matching the time at which the cost of measurement is incurred.

Future generation genetic gain

The selected ewes have, on average, a genetic superiority which can be predicted from the estimate for heritability (h^2) of the trait or traits and the selection differential, such that the genetic response to selection is

$$R_g = h^2 \times s.$$

For selection based on multiple traits using an index, the expression for the predicted response is derived from the selection intensity (i) and the covariance of each selection criterion with the breeding objective, rather than heritability, as in the case of the single trait. See Appendix 5.1 for an explanation.

The selected ewes will pass on to their progeny one half of their genetic superiority. If those progeny are themselves subject to measurement and selection, they will in turn pass on to their progeny one half of their own genetic superiority, plus one quarter of the genetic superiority which existed in their dams. With each generation of breeding and selection of the ewes, the contribution to genetic superiority of the previous generation is halved by their mating to unrelated rams. Gradually, over a number of generations, if selection is continued, the genetic superiority of the ewes before selection approaches R_g asymptotically and, after selection, the extra productivity of the selected sheep is ~R_g + R_p greater than would be the case had selection never taken place in this way in the flock.

Because heritability is usually less than the repeatability of the trait, it can be seen that the current generation gains are likely to be greater than the future generation genetic gains. They are certainly quicker to arrive. The genetic benefits passed on to offspring are not significant for several years after selection commences. Unlike genetic selection in ram-breeding flocks, this form of selection has an upper limit and the higher productivity of the flock is only sustained while selection is continued (Figure 5.6). Should selection based on measured performance cease, the accumulated gains will dissipate over several generations.[10]

Despite its limitations, objective measurement and selection of ewes in commercial flocks can be of value when the reproductive rate of the flock is sufficiently high to allow moderately high selection intensity of the young ewes. In fact, when plans are under development to increase the reproductive rates in a self-replacing flock through changes in husbandry, the benefit of increased selection intensity of ewe hoggets is an additional factor to include in a

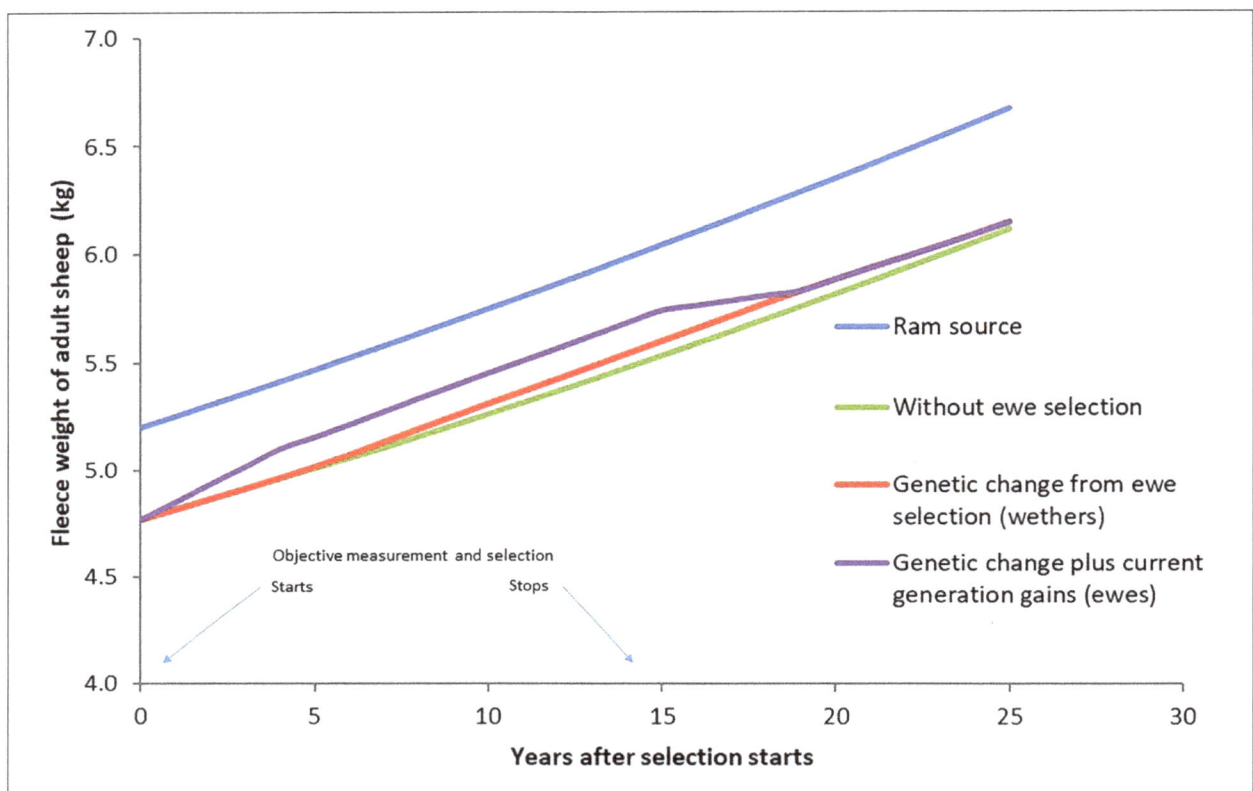

Figure 5.6: Continuing selection of ewes in a commercial flock produces gains in productivity which can be significant in magnitude but are not significantly cumulative. Using greasy fleece weight as an illustration, gains are rapid in the early years but asymptotically approach a maximum value of Rp + Rg. (see text). The ram-breeding flock supplying rams to the commercial flock is improving at 1% per year. With no selection, the merit of the adult ewe flock proceeds at the same rate but with a 7- to 8-year lag. Selection of ewes at hogget age is performed by culling the bottom 20% of ewe hoggets on greasy fleece weight, commencing in year 0, which results in improvements in the average flock productivity. Current generation gains are responsible for most of the early improvement, with 65% of the total gains achieved in the first 5 years. Genetic change, here illustrated by the average merit of the non-measured wether flock, takes longer, achieving 66% after 11 years. Selection ceases in year 16 after which the productivity of the flock gradually returns to the baseline determined by the merit of the purchased rams. Source: KA Abbott.

cost-benefit analysis. It is important to appreciate, however, that the genetic gain is generally small and slow to develop and is not significantly cumulative beyond two generations. This is an important difference between ram-breeding flocks (where female selection is important) and self-replacing commercial flocks.

The economic value of ewe hogget selection

The information provided above illustrates the magnitude of the potential increases in productivity and fleece value which can be obtained by objective measurement of ewe hoggets prior to selection into the breeding flock. The combination of current generation gains and future generation genetic improvement of 0.5 to 1.0 μm in fibre fineness, for example, would be readily achievable in flocks in which 20% of the ewe hoggets are culled after testing.

With premiums in the wool market of $1 to $2 per micron (Table 5.3) and adult clean fleece weights of 4 kg or more, a reduction in fibre diameter of 0.75 μm across the flock could increase the average value of a fleece by at least $3 to $6 per head. This premium is received in each year that the selected sheep are shorn, even though the cost is incurred only once. Note, however, that the cost of testing applies to 100% of the young ewes, while the benefit is received only from the proportion which are retained.

Fibre diameter testing on farm, a service commonly provided by contractors, is now available at a cost well below $2 per head and can be used to measure young ewes at shearing. The benefits of basing ewe hogget selection on objective measurements such as greasy fleece weight and fibre diameter are greatest when reproductive rates are high, when the identified inferior sheep are culled from the flock (and replaced by selected superior sheep) and, even more so, when wethers are kept as wool-growing sheep. The wethers may not be subject to individual selection like the ewe hoggets but they will still express R_g — the genetic lift achieved by several generations of selection of the flock ewes.

If measurement of ewe hoggets is accompanied by identification and recording of individual performance, additional use of the information can be made later in the sheep's life. The data can be used to improve clip preparation[11,12] and can be used also to permit discriminatory culling of older ewes in the flock, such that the less productive ewes are cast for age one or more years earlier than the more productive ewes. This, in turn, may allow for increased selection intensity of ewe hoggets.

In summary, the gains to be made in the first five years of selection compare favourably to the gains passed on to the commercial flock by genetic improvement programmes in the ram-breeding flock supplying rams to it, and the gains are additional to those passed on from the rams.[13] Unlike the improvement in the ram-breeding flock, however, the gains made by ewe selection effectively plateau after two to three generations and are only maintained at that level by continuing selection in the commercial flock.

REFERENCES

1   Safari E, Fogarty NM and Gilmour AR (2005) A review of genetic parameter estimates for wool, growth, meat and reproduction traits in sheep. Livest Prod Sci **92** 271-89. https://doi.org/10.1016/j.livprodsci.2004.09.003.

2   Nicholas FW (2010) Introduction to Veterinary Genetics. 3rd ed. Wiley-Blackwell: UK.

3   Atkins KD (1997) Genetic Improvement of wool production. In: The genetics of sheep, eds L Piper and A Ruvinsky. CAB International: UK, pp. 471-504.

4   Mortimer SI and Atkins KD (2003) Genetic parameters for clean fleece weight and fibre diameter in young Merino sheep. Proc Assoc Advmt Anim Breed Genet **15** 143-6.

5   Swan AA, Brown DJ and Banks RG (2008) Genetic progress in the Australian sheep industry. Proc Assoc Advmt Anim Breed Genet **18** 326-9.

6   Blair HT and Garrick DJ (2007) Application of new technologies in sheep breeding. NZ J Agric Res **50** 89-102. https://doi.org/10.1080/00288230709510285.

7   NSW Department of Primary Industries (2018) Merino bloodline performance and analysis. Available from: https://www.dpi.nsw.gov.au/animals-and-livestock/sheep/management/merino-bloodline-performance/merino-bloodline-performance. Accessed 22 March 2018.

8   Abbott KA (1994) Cost-benefit evaluation of artificial insemination for genetic improvement of wool-producing sheep. Aust Vet J **71** 353-60. https://doi.org/10.1111/j.1751-0813.1994.tb00926.x.

9   Falconer DS (1981) Introduction to quantitative genetics. 2nd ed. Longman Group Ltd: England, p. 127.

10 Abbott KA (2001) The (limited) genetic effects of selection of females in commercial flocks and herds. In: Proceedings of the Australian Sheep Veterinary Society, vol. 11, 2001, eds J Larsen and J Marshall. Australian Veterinary Association: St Leonards, Australia, pp. 92-7.

11 Vizard AL and Williams SH (1993) A model to estimate the economic value of using individual fleece fibre diameter measurements to class wool. Agric Syst **41** 475-86. https://doi.org/10.1016/0308-521X(93)90046-5.

12 Kelly MJ, Swan AA and Atkins KD (2007) Optimal use of on-farm fibre diameter measurement and its impact on reproduction in commercial Merino flocks. Aust J Exp Agric **47** 525-34. https://doi.org/10.1071/EA06222.

13 McGuirk BJ (1976) Estimating genetic progress in the Merino. Proc Aust Soc Anim Prod **11** 13-16.

14 Kinghorn BP (1997) Genetic improvement of sheep. In: The genetics of sheep, eds L Piper and A Ruvinsky. CAB International: UK, pp. 565-91.

15 Brown DJ, Huisman AE, Swan AA et al. (2007) Genetic evaluation for the Australian sheep industry. Proc Assoc Advmt Anim Breed Genet **17** 187-94.

16 Bichard M (1971) Dissemination of genetic improvement through a livestock industry. Animal Production (British Society of Animal Production) **13** 401-11.

Appendix 5.1

WHAT IS AN EBV?

EBVs measure breeding values relative to a base figure

An estimated breeding value (EBV) is just that — an estimate of the amount of genetic merit that an animal is likely to pass on to its offspring. EBVs are always relative to some base level. In its simplest form, an EBV is quoted for an animal based on a comparison to its peers — those animals born in the same flock at roughly the same time and, as much as possible, exposed to the same environment from birth to the age at which measurement is made.

If flocks can be genetically linked, the average merit of the flocks can be ranked such that two or more animals from different flocks can be compared — effectively by comparing each animal to the average of its own flock and then comparing the averages of the linked flocks. Similarly, genetic links across years can be created. Linkages are made by using common sires in multiple flocks and in multiple years, usually by AI, and subsequent measurement of their progeny compared to the progeny of new sires.

Calculating a simple EBV for a ram

As an example of a simple case, consider a ram-breeding flock in which 1000 ram lambs are born and raised together to 1½ years. At shearing at that age, their fleeces are weighed and, for each animal, a greasy fleece weight (GFW) is recorded. The distribution of GFW for the 1000 animals forms a bell curve — representing a normal distribution. The mean value in this example is 4 kg; the standard deviation (SD) is 0.6 kg. The best ram has a fleece weight of 5.8 kg, three standard deviations above average. The selection differential (s) for that animal is 1.8 kg, which represents the phenotypic deviation of his GFW from the group mean (Figure 5.7). A part of his phenotypic superiority is likely to be due to environmental effects — such as good luck in his health and nutrition in his early life — and some of it will be due to his genetic make-up. That part of his superiority which is due to additive genetic effects is his *breeding value*, because he will pass that part of his superiority onto his progeny.

We can predict the effect of using this ram as a sire compared to using a ram with an average value for GFW (a ram for whom s = 0). We expect the progeny of the superior ram to be better, but how much better? Two factors must be considered in estimating the value of his genes to his progeny.

First, we know that he will pass on only half of his genes to his offspring, the other half coming from the dam.

Second, he will only pass on the genetic component of his superiority. His phenotypic superiority was +1.8 kg but heritability estimates tell us that only about 35% of that superiority

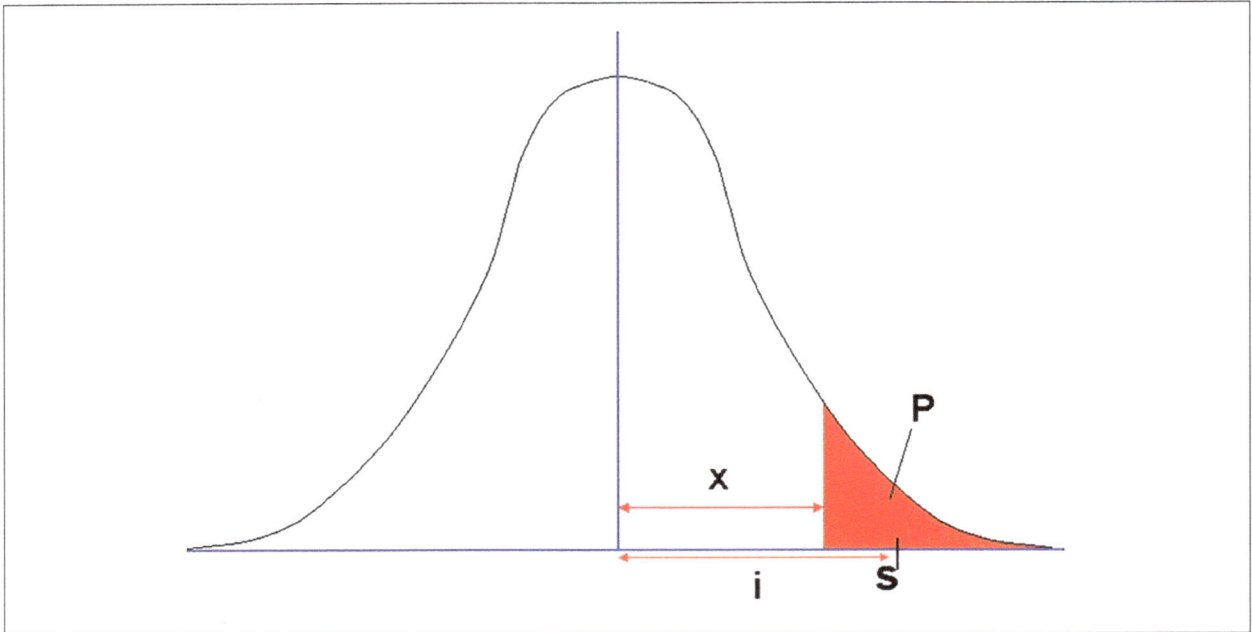

Figure 5.7: A normal distribution. The horizontal axis contains the values for a trait, such as fleece weight. The vertical axis contains values for the numbers of animals with each value for the trait. The animals with the highest value for the trait are at the right-hand end of the distribution, in the red-coloured area P which refers to the proportion selected. The mean value (e.g. fleece weight) of the selected proportion is s, the selection differential, and represents the average superiority of the selected group over the entire unselected population. The selection differential equals the product of i, the selection intensity and the standard deviation of the trait. The selection intensity (i) can be predicted from tables for a given value of P (see Table 5.9). The cut-off point for the selected population is x, which can also be predicted from tables. Source: KA Abbott.

is likely to be due to additive genetic effects. Thus, we predict his breeding value to be 0.35 × 1.8 = 0.63 kg. This is his EBV for GFW.

His progeny will receive on average half that amount and therefore will have fleeces 0.315 kg heavier than those sired by the average ram, if each ram were mated to similar ewes.

Making an index of one or more traits using EBVs

A selection index is effectively a multi-trait EBV in which a number of desirable genetic traits are included and aggregated to produce a single value for overall genetic merit. The units of overall value are often, but not always, currency (for example, dollars). To discuss further the development of a selection index, we will proceed with the example described previously in this chapter.

The breeding objective is

$$H = 17 \times ACFW - 8.15 \times AFD$$

and the selection criteria to be used are

- hogget clean fleece weight (HCFW) (the product of greasy fleece weight and yield)
- hogget fibre diameter (HFD).

Unfortunately, hogget fleece weight is not a perfect predictor of adult fleece weight, nor is hogget fibre diameter a perfect predictor of adult fibre diameter. The traits are, however, highly correlated and, moreover, fibre diameter is correlated with fleece weight, although to a lesser extent. The genetic correlations tell us that some of the genes which contribute to one trait also contribute to another trait of interest, so we can use that information to make a more accurate prediction of each animal's true breeding value for each trait. Moreover, the phenotypic correlation between the two selection criteria in this example can also tell us something about the common effect of the environmental influences on the expression of those traits in the animal. These correlations are known and can be used to improve the accuracy of the index and make it a better predictor of each ram's true breeding value.

The parameters required to calculate the selection index (Table 5.6) are

- the heritability of the selection criteria and the traits in the objective
- the phenotypic variances of the same traits
- the phenotypic correlations between each of the traits used as selection criteria
- the genetic correlations between each of the traits in the objective and the selection criteria
- the relative economic values to be applied to each trait in the objective (in this case $17 and -$8.15).

Construction of the index requires some mathematical calculations (described in detail in other texts[14]), the object of which is to derive weighting factors for each of the selection criteria which make the index as accurate as possible as a predictor of each animal's breeding value — which is now expressed in dollars of fleece value because the breeding objective is expressed in dollars.

An index calculated to rank animals on their breeding value, as described in the breeding objective above, is

$$I = 5.5 \times HCFW - 4.1 \times HFD.$$

Typically, the values inserted into the index to calculate an index value for each animal are deviations from the group's phenotypic mean. Thus, an animal which has a fleece weight 0.2 kg greater than average, and a fibre diameter 0.5 μm finer than average, will have an index score of

$$5.5 \times 0.2 - 4.1 \times -0.5 = 3.15.$$

Indices are frequently scaled for ease of calculation, but the advantage of using the index in its base form is that the differences between individuals are directly related to the economic values assigned to the traits in the objective.

Table 5.6: Heritabilities (on diagonal) and genetic correlations (below diagonal) of traits used in the breeding objective and as selection criteria, and the phenotypic correlation (above diagonal) between the two selection criteria, for calculation of the selection index in the example (see text).

	ACFW	HCFW	AFD	HFD
ACFW	**0.4**			
HCFW	0.8	**0.4**		0.2
AFD	0.3	0.2	**0.59**	
HFD	0.2	0.2	0.9	**0.59**

The calculations of EBVs and selection indices used in Australian industry-wide genetic evaluation programmes (MERINOSELECT, LAMBPLAN) for participating flocks are performed using information from multiple flocks and from relatives (including sires, progeny, dams, if available) of individual animals. Linkages across years and across flocks have allowed the evaluation programmes to create databases in which EBVs and index scores are comparable between breeds, between flocks and across years.[15] When across-flock evaluations began, index scores were given mean values of 100 by the addition of a constant to the index expression, providing a base from which continuing improvement of flocks and breeds can be monitored.

Many traits are now measured and recorded for participating ram breeders and ram buyers can assess the merit of individual rams sold from participating flocks by examining the EBVs for the traits of interest, or by viewing the index scores. Because the EBVs (known as ASBVs in MERINOSELECT and LAMBPLAN) have been calculated using all known correlations between traits and related animals, commercial flock owners can evaluate rams using their own preferred economic values for each trait, using EBVs as selection criteria, weighted directly by their economic value. An example of this method is given earlier in the text of this chapter.

The accuracy of an EBV and an index score

How likely is it that any one progeny of a superior ram will be as good as predicted? There is always some uncertainty and it arises chiefly from two sources.

First, while all of the ram's progeny receive half of the sire's genes, not every one of the progeny will get exactly half of the *good* ones. On average over a large number of progeny, they will get half of his good genes but, individually, some will get more than half and some less. While this variability in outcome is beneficial in breeding programmes where measurement and selection are repeated in every generation, it does mean that one cannot assume that every son or daughter of a pair of champions is a champion!

Second, the original estimate of the ram's breeding value may have been inaccurate. The explanation for this is that, for that one individual in the GFW example, the assumption that 35% of his phenotypic superiority was genetic may have been incorrect. The figure of 35% for the heritability (h^2) of the trait is calculated from experimental observations on large numbers of animals, but it is only a 'best guess' for any one individual. That ram may have been blessed by a particularly outstanding environment as a young sheep, leading to a phenotypic superiority which was all due to his good luck! It may be that his true breeding value is close to zero. On the other hand, his true breeding value may be substantially better than predicted by his EBV.

Because it is known that there is uncertainty associated with the estimate of an individual's breeding value, the *accuracy* of the estimate is also often published. This provides the user with an indication of the reliability of the EBV and, for those who wish to, allows the calculation of confidence limits around the prediction of the progeny's merit. Accuracies are reported on a 0-1 scale and are effectively the correlation between the animal's estimated breeding value (EBV or index score) and its true, but unknown, breeding value. In the simple, one-trait case we have discussed here, the accuracy is h, the square root of the heritability — 0.6 or 60% in the case of hogget GFW. In more complex cases where multiple traits comprise the breeding objective, different but correlated traits serve as selection criteria and information from relatives is included, the accuracy of an index depends on the heritability of each trait, the correlations between traits and the degree of relationship between the sheep under selection and the relatives for which performance records exist.[2] Accuracies over 0.75 are considered to be high, those between 0.4 and 0.75 are moderate and those below 0.4 are low.

The accuracy of the index calculated on the previous page, using HCFW and HFD to rank animals on ACFW and AFD, is 0.67. For a group of animals ranked on that index, this is the predicted correlation between their index scores and their true breeding values. Information about accuracies can be used in a number of ways and Figure 5.8 illustrates one useful inference about accuracies and ram selection which is particularly relevant to ram buyers.

The example in Figure 5.8 shows a scatter plot for a group of 100 rams produced by a computer model allocating at random non-additive genetic and environmental variation to known true breeding values. (In reality, true breeding values are never known). By 'creating

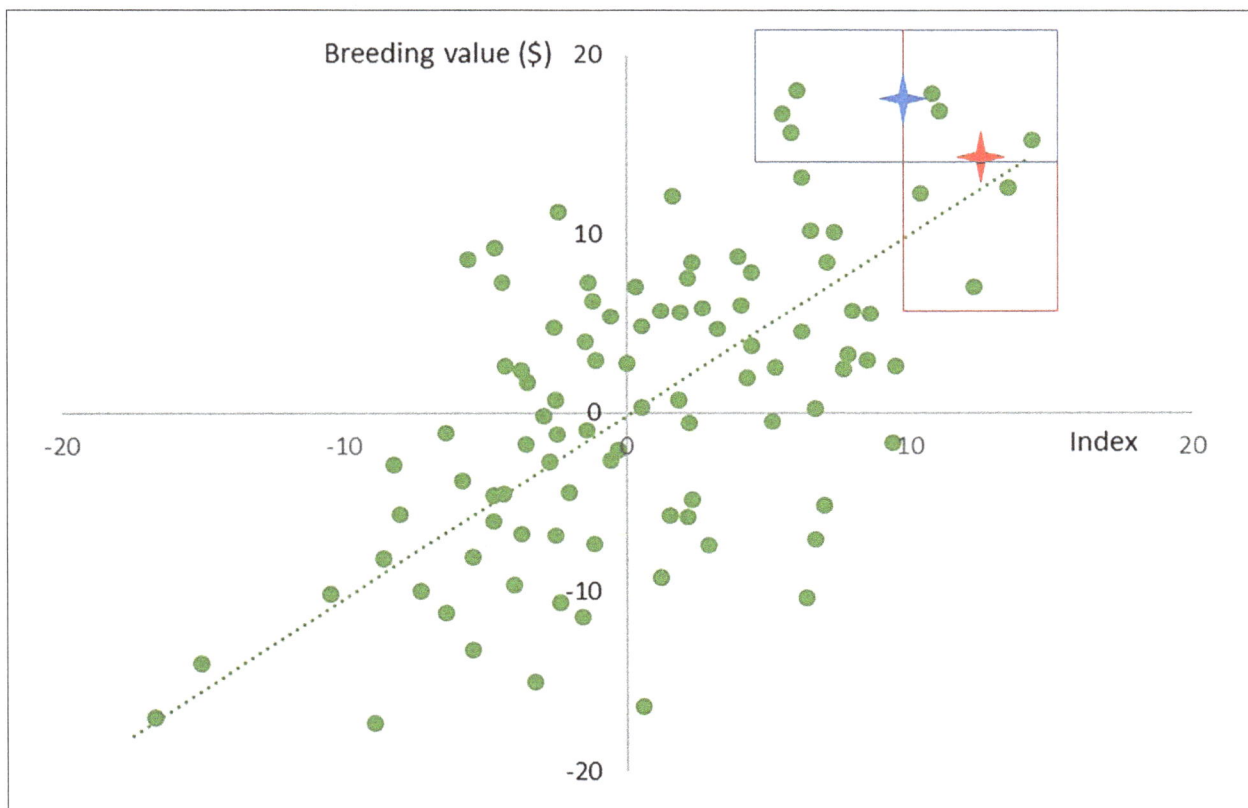

Figure 5.8: This hypothetical example has been created using a stochastic computer model to illustrate the practical outcome of using an index with a moderate level of accuracy — in this case 0.67. The index scores of 100 rams have been plotted against their true breeding values. The best six rams on index scores (in the red rectangle) have mean index scores of 12.0. The average of their true breeding values is 13.7 (shown as a red star). The best six rams on true breeding value (in the blue rectangle) have a mean true breeding value of 17.7 (blue star). If the index were used to select six rams for breeding (those in the red rectangle) it would lead to substantial genetic improvement in the next generation but not as efficiently as a more accurate index. The ram with the highest index score is the sixth highest in true breeding value. In reality, true breeding values are unknown, although, for some sires with hundreds of progeny in sire evaluation schemes, accuracies are very close to 100%. Source: KA Abbott.

and knowing' their true breeding values we can plot them against the index score, which attempts to predict their true breeding values. The correlation in this model is the same as that estimated for the index calculated in the example above (0.67). We can observe from the figure that, for that level of accuracy, there is a good relationship between the predicted and the true value for most rams, but some rams have a significantly overestimated breeding value while others are much better than predicted. Choosing one ram and hoping that it would breed true to prediction might be hazardous but, if six rams were selected (such as those in the red rectangle), the average of their true breeding values would be close to the average of their estimated breeding values (index scores).

With a more accurate estimate of their breeding value, the correlation between index and true breeding value would be higher, the scatter would be less (all dots closer to the line) and the best six rams on index scores would

(a)   have higher average true breeding values

(b)   be individually closer to the predicted average of their index scores (less scatter again).

Appendix 5.2

A COMMERCIAL FLOCK IS TWO GENERATIONS BEHIND ITS RAM SUPPLIER

Consider a commercial self-replacing Merino flock which has been supplied with rams from stud X for 10 years or more. Rams are purchased from around the average for phenotypic merit in the stud flock. Let us refer to the improvement in genetic merit in Stud X as b units per year, which may refer to fleece weight, fibre diameter, growth rate or a combination of traits, or the value in dollars of their genetic differences.

The rate of improvement in the commercial flock cannot exceed that of the stud from which rams are purchased and, when the association between the two flocks is long term, the rates of improvement become the same. There is, however, an *improvement lag*, with the commercial flock lagging behind the stud flock in average genetic merit, despite the fact that both are improving at the same rate.

For simplicity, we will assume that the commercial flock has a stable flock composition, and that the same number of new rams are purchased from the ram-breeding flock each year.

In the commercial flock, the lambs receive half their genetic merit from the rams. It is therefore possible to compare the genetic merit of the commercial flock lambs to those in the ram-breeding flock. If, for example, the average age of the rams is 3 years, then the half of the commercial flock lambs' genes which came from the rams is equivalent in genetic merit to the average of the ram-breeding flock's lambs three years before.

The commercial flock ewes, however, did not come from the stud but were bred in the commercial flock from rams bought from the stud several years earlier. To further complicate the description of the genetic merit of the ewes, remember that their dams were also bred in the flock, from rams purchased from the stud several years earlier again. To assess the contribution of the commercial flock ewes to the genetic merit of the commercial flock lambs, we need to find a way to estimate their genetic merit as well.

Generation length

The easiest way to relate the merit of the ewes in the commercial flock to that of the stud flock is to consider the genetic merit of subsequent *generations* of ewes. In a commercial self-replacing flock, a generation on the ewes' side (GL_{ewes}) is defined as the average age of the dams at the time of birth of the progeny which will later enter the breeding flock.

As an example, consider a self-replacing Merino flock in which ewes lamb at 2, 3, 4 and 5 years of age and then leave the Merino-breeding flock. They may be sold or mated to terminal sires — the important point is they no longer produce potential flock replacements after their fourth lambing (in this example). There are more 2-year-old sheep than 3-, 4- and 5-year-old sheep because a few ewes die each year. The ewe flock structure is shown in Table 5.7. Note that the oldest ewes have a higher reproductive rate than the youngest ewes.

In this example, the average age of the ewes in the flock when the progeny are born is 3.55 years. If we further assume that the probability of the young ewe lambs surviving to breeding age and entering the breeding flock themselves is the same for lambs born to all age groups of ewes, then the generation length in the ewe flock is, as calculated, 3.55 years. However, if the lambs from the older ewes were more likely to enter the breeding flock, the weighted-average age of the dams of future breeders (and therefore the generation length) would be higher. Alternatively, if the lambs from the youngest ewes were selected preferentially, the generation length would be lower.

For the rams that sire these lambs, a similar calculation can be done. Assuming an age structure of the ram flock as shown in Table 5.8, and further assuming that all rams have an equal opportunity of siring a ewe lamb which later enters the breeding flock, the generation length of the rams (GL_{rams}) = 2.83 years.

Table 5.7: An example of a self-replacing Merino flock, consisting of four age groups of ewes. It is assumed that 3% of ewes die each year. The reproductive rate of the ewes increases every year such that, in this example, more ewe lambs are born to 4-year-old ewes than to other age groups. For the ewe lambs born, the average age of their mothers is the sum of the products of lambs born in each age group and their mothers' age, divided by the number of lambs.

Age group of ewes at lambing	Number of ewes lambing	Number of lambs born	Number of ewe lambs born which are eligible to enter the breeding flock	Product of number of ewe lambs and dam age
2 years	500	400	200	400
3 years	485	450	225	675
4 years	470	460	230	920
5 years	455	450	225	1125
Total	1910	1760	880	3120

Sum of products (last column) divided by number of ewe lambs = 3.55 years

Table 5.8: In the example of the self-replacing Merino flock shown in Table 5.7, there are three age groups of rams used each year. Mortalities and culling account for two rams from each age group each year.

Age group of rams used for joining	Number of rams used	Product of number and age
2 years	10	20
3 years	8	24
4 years	6	24
Total	24	68
Average age of ram team		2.83 years

The generation length in the commercial flock is the average age of the parents when the lambs are born: $GL_{flock} = (GL_{ewes} + GL_{rams})/2$. In this example, $GL_{flock} = 3.19$ years

The genetic merit of the ewes is derived from multiple generations of rams

Having calculated the generation length of the ewes and the rams in the commercial flock, let us consider the genetic connection between the commercial flock and the stud from which the rams are purchased each year.

For the commercial flock ewes, half of their genes came from the rams in use in the commercial flock one ewe generation ago. The other half came from their dams, for whom half of their genes came from the sires in use two generations ago. The other quarter came from their grand-dams, half of whose genes came from the sires in use three generations ago. In summary, the genes in the present ewe flock (in Generation zero or *Gen0*) came from

$$\frac{1}{2} \text{ from Gen1} + \frac{1}{4} \text{ from Gen2} + \frac{1}{8} \text{ from Gen3} + \frac{1}{16} \text{ from Gen 4} + \frac{1}{32} \text{ from Gen 5...}$$

where *Gen1* to *Gen5* represents the sires in use 1, 2, 3, 4 and 5 generations ago.

If the difference in genetic merit between each generation is the same (assuming a constant rate of genetic improvement in the stud flock), we can sum the sources of genetic merit for the ewes. Let the merit in the sires in use one ewe generation ago be M and the improvement in each ewe generation be a, then, the average genetic merit of the ewes is

$$\frac{1}{2}.M + \frac{1}{4}.(M - a) + \frac{1}{8}.(M - 2a) + \frac{1}{16}.(M - 3a) + \frac{1}{32}.(M - 4a) + ... + \frac{1}{2^{(n+1)}}.(M - na)$$

which, summed over a large number of generations, approaches M − a. This is the same as the merit of the sires that were used two generations ago. (The contribution to the genetic make-up of the current ewes made by ewes more than four generations ago is very small — less than 3%. It is valid therefore, in flocks which have had a long association (three or four ewe generations) with one stud or a stud of equivalent standing, to ignore the effect — positive or negative — of 'foundation ewes' in the flock more than four generations previously.) For many flocks, with generation lengths of ewes similar to the example used in Table 5.7, one could consider it a 'rule of thumb' that the genetic merit of the ewes is equivalent to that of the rams used in the commercial flock around seven years previously. This is a reasonable assumption for flocks which are in a steady state and whose managers have been buying flock rams from the same ram-breeding flock (or one with sale rams of similar genetic merit) for several generations of ewes.

Linking the commercial flock to the stud

We have related the genetic merit of the commercial flock ewes to that of the flock rams. The next step is to relate the genetic merit of those rams to that of the stud from which they came.

The average age of the flock rams is the generation length of the flock rams (GL_{rams}). If the flock rams are selected from an average grade in the stud then, in any one year, their genetic merit equals the average genetic merit of the ram breeding flock, GL_{rams} years previously. (If, as usual, the ram team consists of multiple age groups, it makes more sense to say that their genetic merit is, on average, the weighted average of the number of rams in each age group

(2, 3 and 4 years in the example used) and the average genetic merit of the stud, 2, 3, 4 years ago.)

We showed above that the ewes in the current flock are two ewe generations ($2 \times GL_{ewes}$ years) behind the rams in merit.

The rams to which they are mated are one ram generation (GL_{rams} years) behind the merit of the lambs born in the current year in the stud.

The merit of the ewes is $2 \times GL_{ewes}$ years behind the rams and therefore $2 \times GL_{ewes} + GL_{rams}$ years behind the lambs born in the stud.

The progeny born in the commercial flock are behind the stud in merit by an amount equal to the average of their parents:

$$= 0.5 \times (2 \times GL_{ewes} + GL_{rams} + GL_{rams}) = GL_{ewes} + GL_{rams} = 2 \times GL_{flock}$$

As the generation length of the flock is the average of the generation length of the ewes and rams, it can be seen that the merit of the lambs born in the commercial flock are two commercial-flock generations behind that of the lambs born in the stud in the same year. (See also Bichard, 1971.[16])

Factors which can reduce or increase the improvement lag

Several assumptions have been made in coming to the conclusion that the commercial flock lags two generations behind its ram supplier flock. It is worthwhile reviewing these assumptions.

First, we assumed that the commercial flock has been derived from the one ram-breeding flock for many years. If this were not the case, then the relationship between the genetic merit of the commercial flock and of the ram-breeding flock may still be approaching the steady state of the two-generation lag. It may be that the commercial flock started further behind and is catching up progressively with every passing generation, or it may be that the commercial flock was previously based on a superior ram-breeding flock and is now moving backwards towards the merit of the new but inferior ram supplier.

Second, we have assumed that the ram-breeding flock is improving genetically and doing so at a reasonably constant rate. If the ram-breeding flock is improving only slowly or not at all, the commercial flock will be very close in genetic quality to that of the ram-breeding flock (the improvement lag still exists but constitutes a very small or zero amount of genetic merit).

Third, the commercial breeder may purchase rams which are from a higher grade than average in the ram-breeding flock. If so, the commercial flock will, ultimately, still lag behind the ram-breeding flock but the relationship will be based on a two-generation lag behind the merit of the *higher-grade animals* in the ram-breeding flock. Depending on the price of rams in the higher grade, this may be a financially wise action on the part of the commercial producer.

Appendix 5.3

Table 5.9: Truncated normal distribution with a mean of zero and standard deviation of 1. P = percentage of the population selected; x = the closest value to the mean for the selected percentage; i = the mean value for the selected percentage. See also Figure 5.7.

P(%)	x	i	P(%)	x	i
99.9%	-3.090	0.003	30.0%	0.524	1.159
99.0%	-2.326	0.027	20.0%	0.842	1.400
90.0%	-1.282	0.195	10.0%	1.282	1.755
80.0%	-0.842	0.350	5.0%	1.645	2.063
70.0%	-0.524	0.497	2.0%	2.054	2.421
60.0%	-0.253	0.644	1.0%	2.326	2.665
50.0%	0.000	0.798	0.5%	2.576	2.892
40.0%	0.253	0.966	0.1%	3.090	3.367

Values for x and i can be found in tables or, in Microsoft Excel®, the following expressions can be used:

x = -1*NORM.S.INV(P) and i = EXP(-0.5* x ^2)/2.5066283/P, where P has values between 0.00001 and 0.99999.

In reference to animal performance traits which are normally distributed within a population, values in Table 5.9 can be interpreted as follows:

- An animal which is the best in 100 is 2.33 standard deviations above average. Animals in the best 1% are, on average, 2.67 standard deviations above the average of the entire population.

- The best 70% of the population have a mean which is 0.5 standard deviations above average, while the worst 30% have a mean which is 1.16 standard deviations below average.

6

THE ENERGY AND PROTEIN NUTRITION OF GRAZING SHEEP

INTRODUCTION

Australia has the largest grazing land area in the world (4.4 million km^2) followed by China (2.4 million km^2), Brazil (1.7 million km^2) and Argentina (1.4 million km^2).[1] Grazed pastures are the cheapest source of nutrients for ruminants. By making some assumptions about land costs, pasture establishment and maintenance costs, and digestible dry matter production, the cost of pasture can be estimated at approximately $0.12/kg dry matter (DM) consumed ($120/ tonne DM) or $0.011/MJ of metabolisable energy[2], in comparison to grain, for example, at $280/tonne DM and $0.023/MJ. Feed represents the highest cost of production in animal enterprises, so grazing pastures and crop residues will continue to play an important role in sheep production systems, and increasingly so as competition for human-digestible feeds increases. Sheep in mixed cropping/sheep enterprises also play an important role in utilising poor-quality, low-nutritive-value crop residues, recycling of nutrients, management of weeds, diversification of income and risk management for producers.

Despite their benefits, the quantity and quality of pastures is highly seasonal and often poorly aligned to the nutritional requirements of the sheep grazing them. In Mediterranean environments, plant growth is limited in summer by low rainfall and in winter by low temperatures, leaving a period in spring of rapid pasture growth. The quality (metabolisable energy, protein and mineral content) of the pasture plants also varies throughout the year. Together this means that sheep are faced with effectively four pasture scenarios: spring (high quality, high quantity), summer (low quality, high quantity), autumn (low quality, low quantity) and winter (high quality, low quantity). In cold-temperate climates (for example, the tablelands of NSW) similar variations occur but with a strong influence of cold temperatures on plant growth. Poor matching of the nutrients available to the demands of grazing sheep is a major contributor to poor productivity and profitability of grazing enterprises.

Poor nutrition is also responsible for many of the health and production problems in sheep enterprises, including pregnancy toxaemia, reduced fertility and fecundity of ewes, neonatal lamb deaths, poor weaner growth, susceptibility to gastrointestinal parasites and presentation of mineral deficiencies (see Chapters 7, 9 and 11). In this chapter, the energy and protein requirements of grazing sheep, the factors influencing the supply of nutrients post-absorptively and the consequences of mismatches between the two are outlined. In particular, the impact of animals on pasture growth and of pastures on animal growth and health are considered. Means of manipulating these interactions are also considered, including management of stocking rate, lambing time, shearing time and supplementation to optimise sheep-grazing systems. These rely on an understanding of ruminant nutrition, feed intake regulation in sheep, nutrient supply from pastures throughout the year, the impact of pregnancy, lactation, activity and climate on nutrient requirements, selective grazing by sheep, and the effects of supplements and stocking rate on sheep production and health. These topics are considered below, but the reader is encouraged to explore the reference list at the end of this chapter if more detail is required on any one topic.

ENERGY AND PROTEIN REQUIREMENTS OF GRAZING SHEEP
Ruminant digestion

In an evolutionary sense, ruminants have been among the most successful groups of mammals in the world, largely as a consequence of a symbiotic relationship between the host animal and its resident anaerobic microbes. The host provides an environment conducive to anaerobic fermentation (that is, an environment which is anaerobic, pH-buffered, isotonic and temperature-controlled, and from which waste products are removed continuously, and with a wide diversity of microenvironments). The microbes, in turn, digest feed, producing volatile fatty acids (VFA) as end-products that the host uses as an energy source, a supply of microbial protein of moderate quality, a supply of microbial lipids and a supply of the vitamins B and K. The VFA are mainly acetic (C2), propionic (C3) and butyric (C4) acids with small amounts of longer-chain acids such as iso-butyric (C4), valeric (C5) and iso-valeric (C5) acids. These acids are the end products of the fermentation of simple sugars and complex carbohydrates like cellulose, hemicellulose and starches by bacteria, fungi and ciliated protozoans. The type of microbes present (for example, cellulolytic, amylolytic, proteolytic, lactate-utilisers, methanogens) and the relative proportions of the VFA produced are determined by the composition of the diet consumed, and the prevailing physico-chemical environment. The fermentation acids are absorbed across the ruminal epithelium by two main mechanisms — direct exchange of dissociated acids with bicarbonate ions (about 50%) and passive absorption of undissociated VFA (about 40%). Both mechanisms tend to buffer the ruminal fluid from large pH drops, because the undissociated acids take their hydrogen ion with them out of the fluid (increasing pH), and the dissociated ions are swapped with a bicarbonate buffering ion.[3] Butyrate is converted to β-OH butyrate on passage through the ruminal epithelium, but the acetate and propionate are unchanged. All three acids travel to the liver through the portal vein and enter the biochemical pathways that generate energy, glucose, ketone bodies, lipids and proteins.

In the process of fermenting the dietary carbohydrates, the microbes obtain ATP which they can then use for metabolic processes, including synthesis of proteins essential to life. These proteins are largely synthesised from simple nitrogenous compounds like ammonia and small peptides (Figure 6.1). This unique ability of the anaerobic microbes to produce protein from non-protein nitrogen accounts for the remarkable success of the ruminant animals in a wide range of environments around the world.

We can take advantage of this utilisation of non-protein nitrogen to form proteins by maintaining sheep on low-nitrogen diets, or very low-quality diets supplemented with urea or other inexpensive non-protein sources. The resulting microbial protein then joins with dietary proteins that are not degraded in the rumen.

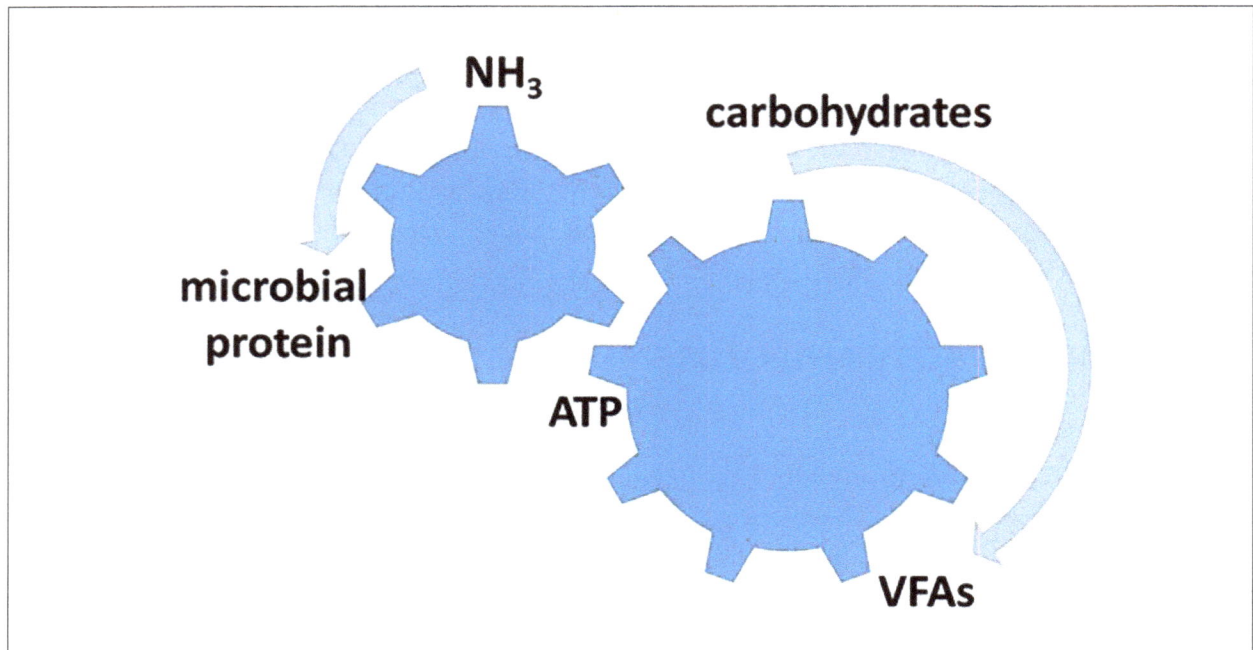

Figure 6.1: Carbohydrates like cellulose, starches and sugars are fermented to simple volatile fatty acids, creating energy-rich ATP which drives the synthesis of microbial protein from simple nitrogen compounds like ammonia. The simpler the carbohydrates, the faster the gear spins and the more rapid the microbial protein synthesis. Source: P Hynd.

Carbohydrates differ widely in the rate at which they are fermented. Simple sugars and starches are the most rapidly fermented, followed by hemicelluloses and pectins, and finally celluloses which are slowly fermented. The rate of fermentation is important because it largely dictates the amount of feed the sheep can consume: voluntary intake by ruminants is directly related to the rate at which the feed material is digested in, or leaves, the rumen. High-fibre diets containing a large amount of slowly digested cellulose are slowly degraded, so the feed particles remain in the rumen for many days, contributing to rumen 'fill' and thereby restricting intake. High-fibre diets are also digested to a lower extent than lower-fibre diets. Digestibility is the main determinant of the energy density of the diet (Figure 6.2).

The digestibility of a feed therefore determines the two most important components of energy available to the sheep: its energy density and how much of it the sheep can consume. Together these determine the total intake of available energy. The digestibility of a feed depends on the content of indigestible (lignin, silica) and poorly digestible (cellulose) components, but other factors including its nitrogen content, mineral content, particle size and structural features can also play a role. Any factor that limits microbial activity will reduce digestibility, intake and performance.

The microbial protein produced during fermentation joins with dietary protein that is not degraded in the rumen and becomes available to the host as amino acids after digestion in the abomasum and small intestines (Figure 6.3). Figure 6.3 summarises the digestive processes taking place in the reticulorumen, and the supply of nutrients available to the sheep for metabolism and production.

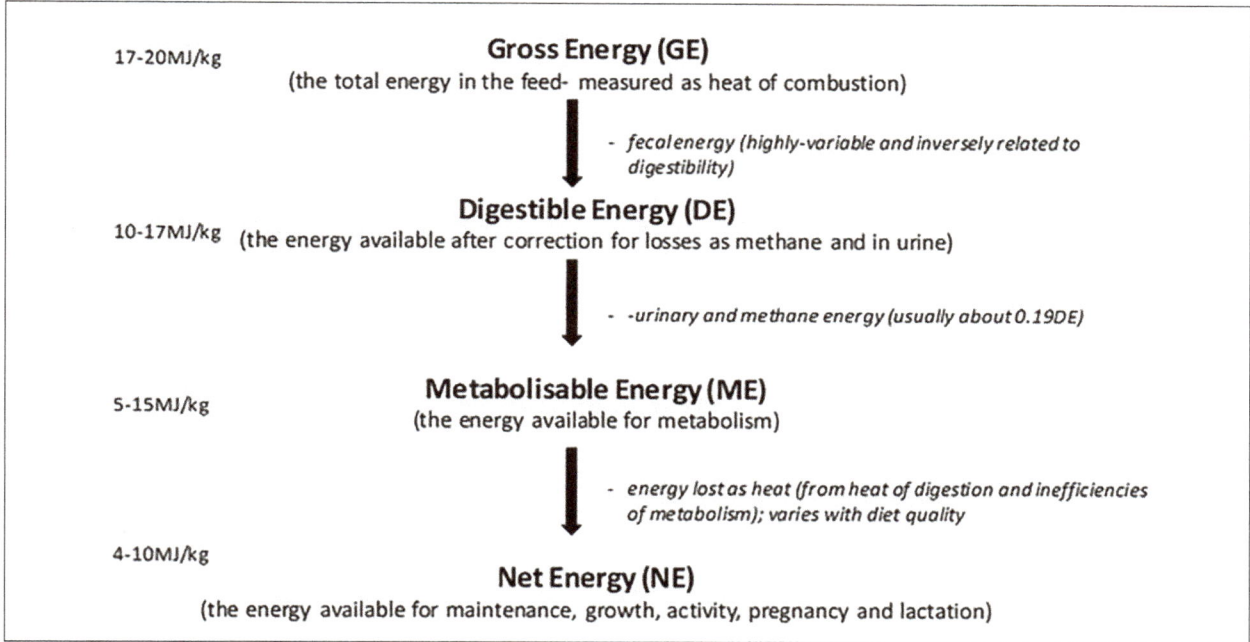

Gross Energy (GE)
(the total energy in the feed- measured as heat of combustion)

17-20MJ/kg

- *fecal energy (highly-variable and inversely related to digestibility)*

Digestible Energy (DE)
10-17MJ/kg (the energy available after correction for losses as methane and in urine)

- *-urinary and methane energy (usually about 0.19DE)*

Metabolisable Energy (ME)
5-15MJ/kg (the energy available for metabolism)

- *energy lost as heat (from heat of digestion and inefficiencies of metabolism); varies with diet quality*

4-10MJ/kg

Net Energy (NE)
(the energy available for maintenance, growth, activity, pregnancy and lactation)

Figure 6.2: Partitioning of feed energy in sheep. Typical energy values in megajoules/kg of dry matter are shown to the left of the diagram. Note the main determinant of energy available to the animal is digestibility of the feed. Source: P Hynd.

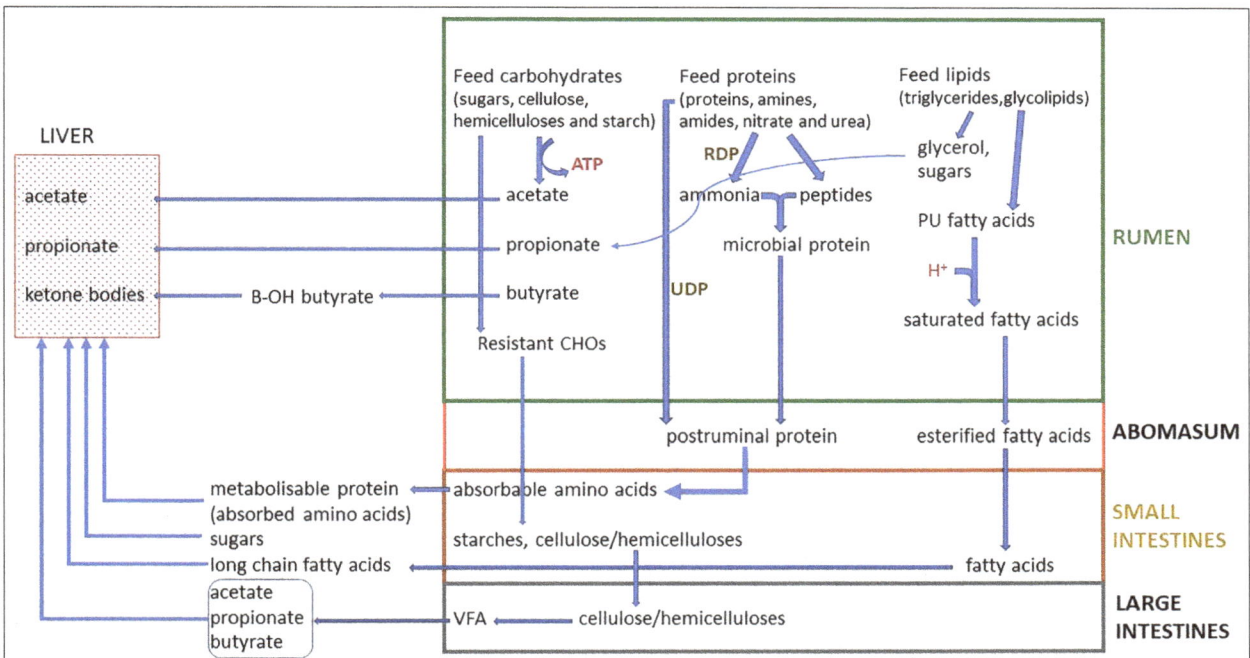

Figure 6.3: Flows of nitrogen and energy in the grazing sheep. Source: P Hynd.

The major absorbed nutrients are the VFA (acetate, propionate), the ketone body β-OH butyrate, amino acids, long-chain saturated fatty acids, simple sugars, minerals and vitamins. Importantly, of these, only certain amino acids (gluconeogenic amino acids) and propionate are able to produce the glucose that is essential for brain function and foetal growth. This becomes important when pregnant ewes, and particularly those bearing multiple lambs, suffer an energy deficit in the last month of pregnancy, resulting in pregnancy toxaemia. The relative proportions of the VFA change with the diet. Roughage diets and poor-quality, senesced pastures and crop residues initiate fermentations characterised by relatively high acetate levels. High-grain rations and lush, green pastures produce relatively high levels of propionate. High-sugar diets produce relatively high levels of butyrate. Figure 6.4 shows the impact of the ratio of concentrate to roughage on relative VFA proportions, and the minimum pH likely to be generated. The rumen pH is directly and inversely related to the total concentration of VFA.

Estimating the energy requirements of grazing sheep

It is clear from Figures 6.2 and 6.3 that the most important form of energy available to the sheep for maintenance and production is called the *net energy* (NE), which is the energy remaining after fermentation, digestion, absorption and metabolism. Net energy, however, is difficult to measure because it requires feeding the animal and collecting all faeces, urine, gas outputs and heat production for each feed available. Given the infinite variety of forages available to a sheep throughout the year, and even throughout the day, it would be impossible to do this. Instead we use a *metabolisable energy* (ME) system in which the ME content of the feed is estimated from average tabulated values for digestibility, methane output and urine energy output. The efficiency of conversion of ME to NE is then estimated using efficiency factors, called *k factors*, which are calculated from the feed quality and the purpose for which the energy is being used (for example, lactation, maintenance, gestation, activity, growth). The total ME required by the sheep can then be calculated by adding up all of the ME requirements for each purpose as follows:

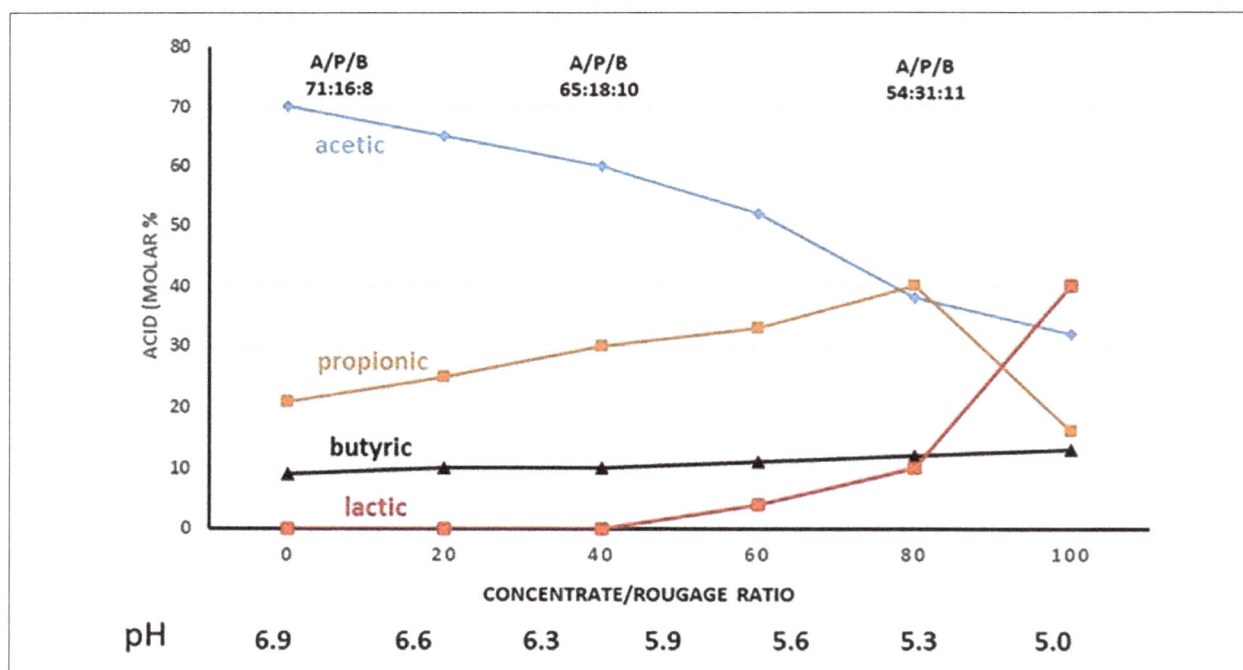

Figure 6.4: Effect of the ratio of concentrate to roughage on relative proportions of VFA and minimal ruminal. Source: P Hynd.

$$\text{ME requirement (MJ/day)} = \text{ME}_m + \text{ME}_l + \text{ME}_p + \text{ME}_g + \text{ME}_a + \text{ME}_c \qquad (1)$$

where

ME_m = ME for maintenance

ME_l = ME for milk production

ME_p = ME for pregnancy

ME_g = ME for growth

ME_a = ME for activity (walking, grazing, ruminating, standing)

ME_c = ME required to maintain core body temperature.

This chapter now considers these in detail, along with the consequences of not providing sufficient ME for each purpose.

Maintenance energy requirement of penned sheep

The energy required to support the basic life-sustaining processes of a sheep (cellular metabolism, osmoregulation, respiration, circulation) can only be measured if the animal is maintained in pens (so that it cannot move other than standing/lying), at a thermoneutral temperature (so that it is not using energy to keep warm or cool down) and fasted (so that it is not using energy to digest food). The animal is also not growing, lactating or maintaining a pregnancy. The heat produced by such an animal is called the *fasting heat production* (FHP) or *basal metabolic rate* (BMR). This FHP is the net energy for maintenance and is closely related to the body weight of the animal. For sheep the equation is as follows:

$$\text{FHP (MJ/d)} = 0.23W^{0.75}$$

The sex of the animal has an effect, however, because males have a higher basic metabolic rate than ewes. Age also has an effect, so the final equation for FHP is as follows:

$$\text{FHP (MJ/d)} = S \times (0.23W^{0.75} \times e^{-0.03\text{Age}})$$

where S = 1.0 for females, 1.15 for males and age is measured in years.

Given that FHP is effectively the NE required for maintenance and we want to use ME as the unit, we need to correct it for the efficiency of utilisation of the ME for NE (k factor). This depends on the feed quality; high-quality feeds are used more efficiently than low-quality feeds. The relationship between efficiency of use of ME for NE (k factor) is described by the following equation:

$$K_m = 0.02\text{MD} + 0.5$$

where MD is the metabolisable density of the diet in units of MJ/kg DM.

Typical values for a wide range of diets lie between 0.62 and 0.74. Most feeds for grazing sheep are used with an efficiency of about 70%, so the ME required for maintenance is the FHP/0.70. While the decision support software systems available for managing grazing sheep take all the factors above into account, for practical sheep feeding the energy required for maintenance can be easily and quite accurately estimated using the following equation:

$$\text{ME}_m \text{ (MJ/d)} = 0.13 \times W + 1.2 \qquad (2)$$

This equation was derived by MAFF (1975)[4] assuming a constant efficiency of utilisation of ME for NE, no sex or age effects and no scale effect of body weight. Despite all these assumptions, the effects on actual ME_m required are very small indeed. A comparison of the values from equation (2) with the more detailed models of NRC (2007)[5] across a range of diets and body weights showed relative agreements of 0.95-1.10, with an average close to 1.0. From a practical feeding viewpoint, equation (2) can be used to quickly estimate the ME requirement for maintenance of grazing sheep.

Maintenance energy requirement of grazing sheep

The energy costs of grazing relate to the body weight of the sheep, the distance it walks, the steepness of the topography and the energy costs of eating and ruminating. These are shown in Table 6.1.

For a 50 kg sheep grazing on hilly terrain for 8 hours/day, ruminating for 8 hours/day, walking, say, 10 km/day and climbing 500 metres, the additional ME required above maintenance would be about 2.4 MJ/day, which is about 30% higher than the maintenance energy it would require if it was housed in a pen. Obviously, sheep grazing very sparse pastures will graze considerably longer than those on high-density pastures and their ME requirement will be about 40% higher than their penned counterparts. A sheep grazing very sparse pastures on very hilly country will require up to 80% more energy than its penned counterpart.

Maintenance energy requirements of cold-stressed sheep

The maintenance energy requirement of sheep grazing in cold environments is increased dramatically, particularly if there is a combination of rain and wind such that the wind chill factor is increased. The temperature below which additional heat is required to achieve homeothermy is called the *lower critical temperature* (LCT). LCT depends on age, fleece cover, body condition (fatness), level of feeding, rainfall and wind speed. The higher the LCT, the more susceptible the sheep is to cold stress, hypothermia and death. Table 6.2 shows a small subset of LCTs for a few scenarios of age, condition and coat cover.

The combinations of wind speed, rainfall, body weight and coat depth are combined in an equation to estimate the additional ME required to produce additional heat below the lower critical temperature. The matrix of wind chill factors, coat cover and body condition are available in NRC (2007)[5] and are estimated in GrazFeed®.[6] Under practical feeding conditions, however, the following considerations are relevant:

Table 6.1: Energy costs of grazing activities in kilojoules (kJ).

Activity	ME cost (kJ/kg body weight)
Standing	10 kJ
Changing body position	0.26 kJ
Walking (horizontal)	2.6 kJ/km
Walking (vertical)	28.0 kJ/km
Eating	2.5 kJ
Ruminating	2.0 kJ

Table 6.2: Lower critical temperatures (°C) of sheep in dry, still air. Adapted from NRC (2007).[5]

Age	Condition	Wool cover	LCT (°C)
Lamb	Newborn		28
Lamb	1 mth		10
Adult	Maintenance	Shorn	25
Adult	Fasted	Shorn	31
Adult	Full-fed	Shorn	18
Adult	Maintenance	50 mm	9
Adult	Maintenance	100 mm	-3

- Sheep on higher levels of feed (for example, lactating ewes) produce more heat from ruminal fermentation and are therefore less susceptible to cold stress (lower LCT) than sheep at maintenance levels of feeding.

- Neonatal lambs are particularly susceptible to hypothermia. Ewes well fed in the last trimester and in good body condition at lambing (condition score 3 to 4) and producing lambs of 4 to 5.5 kg have lower lamb mortality than those producing lighter-weight lambs (Figure 6.5).

- At the same level of energy intake, roughages produce more heat than concentrates (lower km values).

- Sheep off-shears are particularly prone to cold stress and hypothermia and should be protected from wind and rain by shedding or paddock shelters.

Metabolisable energy requirements for gestation and lactation

In the first trimester (days 0-50) little additional energy is required to sustain pregnancy as the placental and foetal growth demands are small. In the mid-trimester (days 50-100) placental growth is rapid but foetal growth is low. In the last trimester placental growth is low but foetal demands are increasing exponentially, particularly for multiple-bearing ewes. The additional ME required for gestation in single- and twin-bearing ewes is shown in Table 6.3.

These figures are derived using an efficiency of utilisation of ME for NE for gestation of a constant 0.133 (K_p). A 50 kg ewe has an ME requirement for maintenance of about 7.7 MJ/day, so these represent significant increases in the ME requirement to maintain the pregnancy. With total ME required of 13 MJ/day (for the single-bearing ewe) and 15.7 MJ/day (for the twin-bearing ewe), it is unlikely that the ewe will be able to consume sufficient feed to achieve these levels of energy intake because the placenta and foetus encroach on the rumen and reduce the total capacity before stretch receptors are stimulated. For low- to medium-quality roughage diets of, say, 7-9 MJ/kg DM would require an intake of 1.9 kg (single) and 2.2 kg (twin) which exceed the likely maximum intake of about 1.8 kg/day. The ewe will make up the shortfall by mobilising body reserves, but too large a shortfall due to low-quality feed or low-pasture biomass will result in rapid mobilisation of reserves and an increased risk of pregnancy toxaemia.

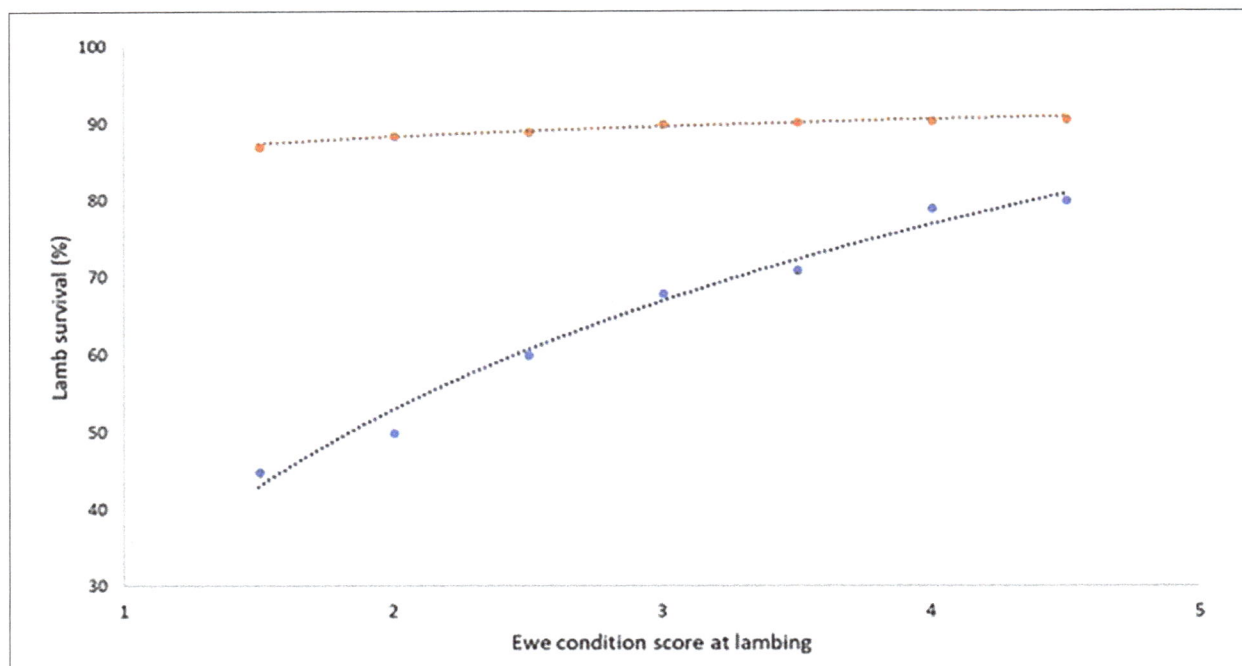

Figure 6.5: Ewes lambing at higher body condition scores have higher lamb survival than those at lower scores. Single lambs — orange line; twin lambs — blue line. The effect of condition score on lamb survival is greater for twin lambs than singles. Reproduced with permission from Lifetime Wool (http://www.lifetimewool.com.au/legal.aspx).[7]

Table 6.3: Additional ME (MJ/day) required for gestation in single- and twin-bearing ewes giving birth to lambs weighing 4.0 kg (singles) and 3.0 kg (twins).

Weeks pre-term	12	8	6	4	2	term
Singles	0.4	1.1	1.7	2.6	3.8	5.3
Twins	0.6	1.7	2.6	3.9	5.7	8.0

Metabolisable energy requirements for lactation

The efficiency of utilisation of ME for lactation depends on diet MD as follows:

$$K_l = 0.02MD + 0.4$$

where MD is the metabolisable density of the diet (MJ/kg DM).

For most sheep diets the K_l value will be between 0.56 (for a medium-quality pasture hay) to 0.64 (for a high-quality concentrate ration). The ME required for each litre of milk depends on its fat content but, assuming a constant fat content of sheep milk of 80 g/kg and an average K_l of 0.60 for most sheep rations, the ME required for milk production is approximately 8.1 MJ/kg milk. Most ewes on standard pastures at 2000 kg feed on offer (FOO) will achieve growth rates of lambs approaching 200 g/d (Figure 6.6).[d]

---

d   The amount of pasture in a paddock is measured and reported using two different systems. *Feed on offer* (FOO) includes all above-ground plant material while *herbage mass* or *pasture mass* assumes that 300 kg of dry matter (the lowest 0.5 cm of herbage) is unavailable to sheep. Estimates of pasture mass are lower than estimates of feed on offer.

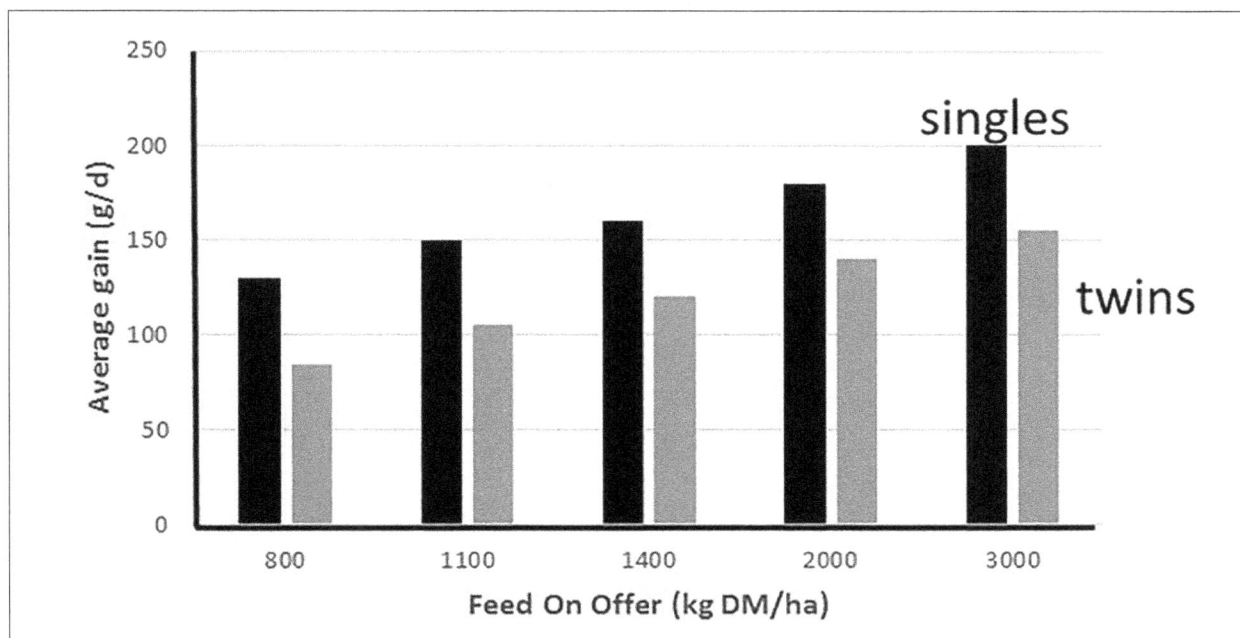

Figure 6.6: Impact of feed on offer on growth rate of single lambs and twin lambs to weaning. Reproduced with permission from Lifetime Wool (http://www.lifetimewool.com.au/legal.aspx).[7]

So for 60 kg ewes feeding single lambs, an additional 8 MJ of ME is required on top of their maintenance requirement of about 9 MJ/d (total = 17 MJ/d). For 60 kg ewes feeding twins, the additional requirement for milk production is about 12 MJ/d (total = 21 MJ/d). Most medium- to high-quality pastures (>10 MJ/kg DM) will provide this amount of energy in the daily intake of about 2.1 kg/d for a 60 kg ewe. Should the quality of pasture fall below 10 MJ/kg DM (a dry matter digestibility of about 68%), the ewe must be able to mobilise tissue reserves to fund the energy deficit for milk production. The ability of the ewe to achieve this fat mobilisation depends on her condition score at lambing.

Guidelines for managing pregnant ewes to ensure high rates of lamb survival and lamb growth to weaning

The following guidelines for ewes approaching lambing are designed to minimise pregnancy toxaemia, to ensure high lamb survival rates and to achieve high growth rates of lambs to weaning:

- Twin-bearing ewes should be at condition score >2.7 and preferably 3.2 by lambing. To achieve this, feed on offer should be >1000 kg DM/ha of good-quality green pasture. If pasture is less than this or if the pasture quality is poor (<9 MJ/kg DM or 60% digestibility), supplements containing moderate to high energy (>10.5 MJ/kg) should be fed to achieve condition scores of about 3.

- Any ewe with condition score <2 should be removed from the mob and managed separately.

- Shelter should be provided for pregnant ewes, particularly twin-bearing ewes, to increase lamb survival, particularly in cold climates (see Chapter 7).

Metabolisable energy requirements for growing weaners

The NE requirement of growing sheep depends on their stage of development (age and maturity) because the energy content of the gain increases as the animal matures and the fat content of the gain increases. At 10% of mature weight (say, a 6 kg lamb), the energy content of the gain is 9 MJ/kg; at 50% of mature weight (a 25 kg weaner, for example) the net energy content of the gain is 20 MJ/kg; and at 80% of mature weight, the energy content of the gain is 25 MJ/kg. The efficiency of utilisation of ME to NE for gain depends on the feed ME density as follows:

$$K_g = 0.043MD.$$

Typical values for K_g are about 30% to 52% for most diets.

This means a typical weaner of 30 kg growing at 200 g/d requires an additional 4 MJ of NE per day to grow at this rate (from the 50% energy content value above). If we assume the weaner is eating a high-quality pasture or feedlot ration of about 11.5 MJ/kg DM, the K_g will be 0.49 and the additional ME required will be 4/0.49 = 8.2 MJ/day in addition to its maintenance requirement of 5.1 MJ/day — a total ME intake of 13.3 MJ/d is therefore required. At 11.5 MJ/kg DM, the weaner would have to eat 1.15 kg DM/day, which is close to its maximum likely intake (predicted to be around 1.05 kg/day, based on 3.5% of body weight).

Protein requirements of grazing sheep

In ruminants, unlike monogastric animals, the protein nutrition must include consideration of the entire nitrogen economy, including the fate of non-protein nitrogen. This is because

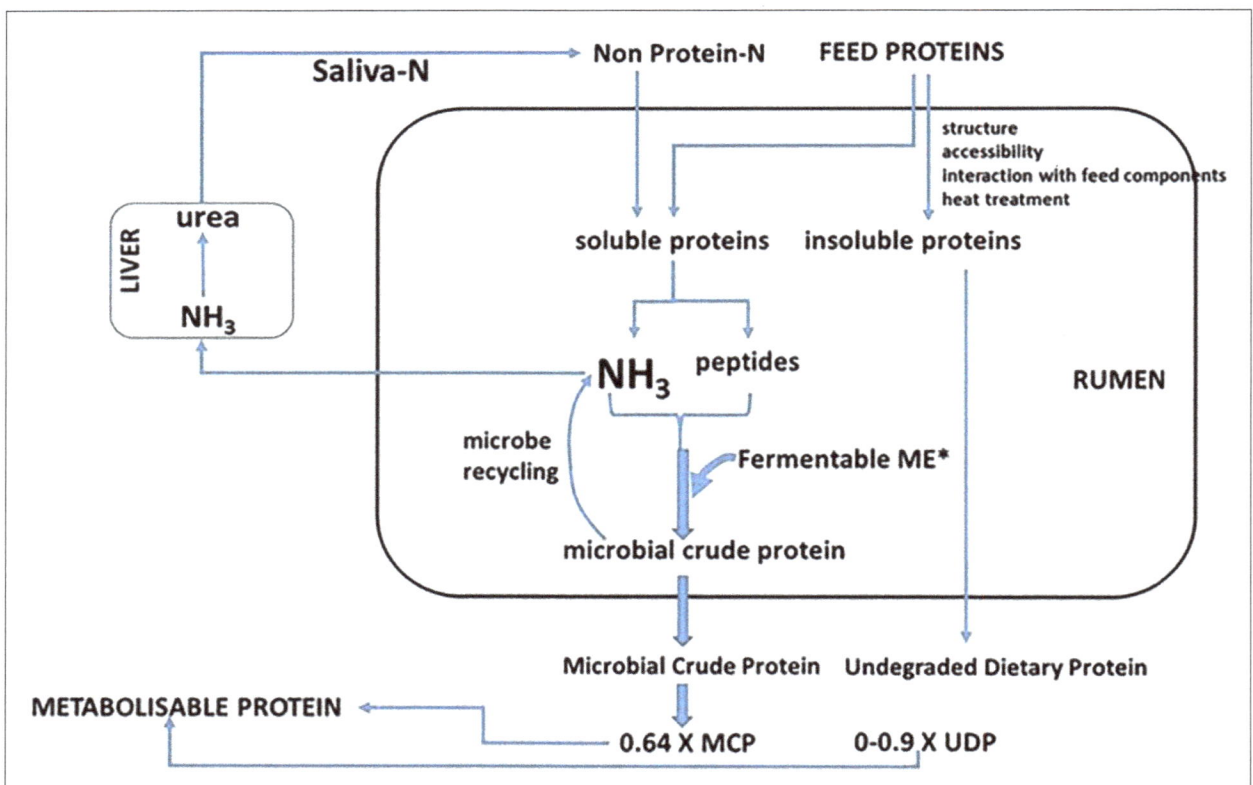

Figure 6.7: Protein digestion and absorption in sheep. *FME = Feed ME – lipids ME – UDP ME. Source: P Hynd.

non-protein nitrogen sources entering the reticulorumen are rapidly degraded to ammonia, which is then either incorporated into microbial proteins, absorbed from the rumen and transported to the liver through the portal vein, or passed on to the omasum. Liver ammonia is converted into urea, which is then either recycled to the rumen via saliva or excreted in the kidney tubules. Feed proteins that are soluble or readily accessible to extracellular microbial proteases are broken down to simple peptides and ammonia, the latter entering the rumen ammonia pool with ammonia from non-protein sources. Peptides are taken up directly by some rumen microbes.[8] The dietary protein that is degraded in the rumen to ammonia and peptides is called the *rumen degradable protein* (RDP). The remaining dietary protein that is not degraded in the rumen usually comprises proteins that are insoluble at ruminal pH and therefore are not accessible by microbial proteases. Other dietary proteins are not degradable because of the tertiary structure of the protein. This undegradable protein is called the *undegraded dietary protein* (UDP). The relative proportion of RDP in dietary proteins varies with the protein source. Green grasses, legumes and hays are typically 70% to 80% RDP; grain legumes are typically 60%; and animal proteins like blood meal and meat meal are 50% to 60% RDP. The flows of nitrogen in ruminants are shown in Figure 6.7.

Microbes take up ammonia and small peptides using energy (ATP) generated by fermentation of carbohydrates. The energy generated by this fermentation is called the *fermentable metabolisable energy* (FME), which is defined as

$$\text{FME} = \text{Dietary ME} - \text{ME}_{fat} - \text{ME}_{ferm} - \text{ME}_{UDP}$$

where ME_{fat} is the ME contained in the fat component of the dietary ME; ME_{ferm} is the ME contained in fermentation acids already present in the feed (from silage or waste products of brewing); and ME_{UDP} is the energy contained in proteins that are not degraded in the rumen.

ME_{fat} and ME_{ferm} are subtracted because the ruminal microbes cannot obtain energy from fats, oils or end-products of a previous fermentation (like VFA and lactate). Consequently, in considering the amount of energy in the diet which is available to rumen microbes, and which therefore can support microbial protein synthesis, the ME content of the dietary fats and oils and the ME content of any fermentation acids in the diet and the ME contained in UDP proteins must be subtracted from the total dietary ME content, to calculate the *fermentable metabolisable energy* (FME) content of the diet. It is necessary to estimate FME of a feed in order to accurately predict the protein value of a feedstuff.

Most of the commonly used sheep feeds have some fat or oil content. The FME of oats, for example, is about 1.4 MJ less than its ME.

The rate of synthesis of microbial protein therefore depends on a supply of RDP, a supply of essential minerals required by the microbes (particularly S, P and cobalt for vitamin B_{12} synthesis), and a supply of energy from fermentation of carbohydrates and proteins. It is also important that the availability of energy and of RDP are synchronous, or close to synchronous. If energy is limiting at any stage, there is a significant loss of uncaptured ammonia across the ruminal epithelium, which is excreted in the urine as urea. If RDP is limiting, there is insufficient nitrogen to synthesise the proteins. For this reason, diets high in ME (for example, a feedlot ration) should be matched by adequate levels of RDP.

Similarly, on poor-quality diets (like crop stubbles or poor-quality hays) a deficit of RDP can limit not only microbial protein synthesis but also digestion of the fibre. This then not only

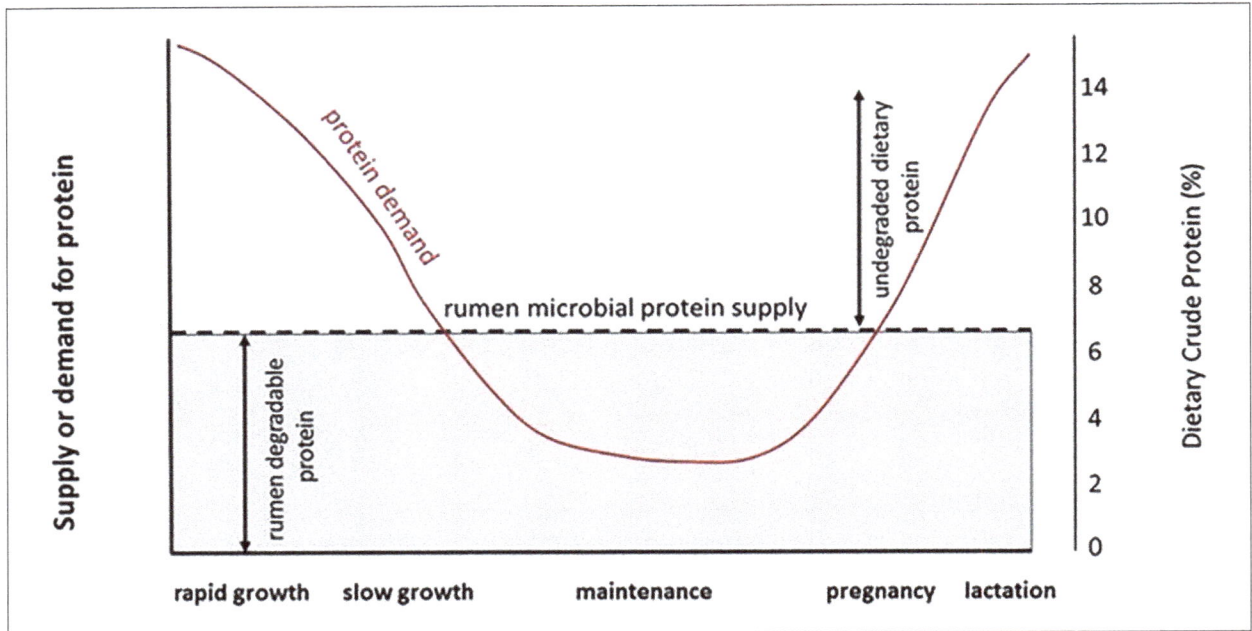

Figure 6.8: Theoretical relationship between protein demand and protein supply in grazing sheep. Microbial protein is synthesised from ruminally degraded proteins or non-protein nitrogen. Microbial protein is adequate for maintenance, slow growth, early pregnancy and low milk production. Rapid growth, late pregnancy and high milk yields (early lactation) require dietary protein that escapes ruminal degradation. Adapted from Kempton, Nolan and Leng (1978).[10]

limits energy availability per kg feed but also reduces feed intake because the rate of passage of the fibrous particles is reduced. In such cases positive responses to supplementation may be obtained with non-protein nitrogen sources such as urea. Unfortunately, supplementation with non-protein nitrogen is ineffective in many cases because microbial protein synthesis appears only to be limited when the ruminal ammonia nitrogen concentration falls below 5 mg NH_3–N/100 mL (2.9 mM) — which equates to a dietary crude protein content of about 12% to 13%.[9] Given the ability of sheep to select a diet as much as 3.5 times higher in crude protein than the average feed on offer (PI Hynd, unpubl), there are few situations when grazing sheep would be deficient in RDP. Urea is also toxic and should not be fed at a rate above 0.5 g/kg body weight; it should also be introduced carefully to diets in order to allow adaptation of the rumen microbes.

Table 6.4: Crude protein requirements of sheep at different life stages. Derived from GrazFeed®.

Live weight	Status	CP (%)
Weaner (<20 kg)	growing	14-16
Weaner (20-25 kg)	growing	12-14
Weaner (>25 kg)	growing	10-12
50 kg	maintenance	6-8
50 kg	mid-pregnancy	8-10
50 kg	peak lactation	12-14

Microbial protein is sufficient to meet the protein requirements of sheep that are maintaining weight or growing slowly, but it is insufficient for sheep experiencing rapid growth, late pregnancy and lactation. Such animals need an additional source of UDP (Figure 6.8).

Together the microbial protein and the UDP enter the small intestines. The digestibility of microbial protein is only 64% in the intestines, and the digestibility of UDP is highly variable depending on the protein source, heat treatment and other components of the feed (like tannins). Digestibility of UDP ranges from 0 to 90% but is usually about 60% to 80%. Together the digestible microbial protein and the digestible undegraded dietary protein (DUP) form the metabolisable protein, which is available for providing the essential amino acids necessary for maintenance, growth, gestation and lactation.

Detailed estimates of the RDP required and supplied by diets can be determined using decision support software like GrazFeed®, but for grazing sheep broad protein requirements are given in Table 6.4.

Nutrition and wool growth

For grazing sheep the rate of wool growth is largely determined by the amount of *digestible protein leaving the stomachs* (DPLS) each day (Figure 6.9). Depending on the genetically determined wool growth efficiency of the sheep, increasing DPLS results in increased wool growth until a plateau is reached, reflecting the maximum wool growth rate.

The relationship between DPLS and wool growth is also affected by energy intake (compare Figure 6.9A with 6.9B). There is an optimum ratio of ME to DPLS such that, when ME is limiting (Figure 6.9B), wool growth is mainly energy-limited, whereas when energy availability is high, wool growth responds mainly to DPLS (Figure 6.9A). The optimum ratio of DPLS to ME is about 12 g protein for each MJ of metabolisable energy.[11,12] So in Figure 6.9B, the lower ME intake means maximum wool growth is met at a significantly lower DPLS value, hence a lower wool growth rate. The main determinant of DPLS is the digestible dry matter intake (or fermentable metabolisable energy (FME) intake) because high FME drives high microbial synthesis and high levels of microbial protein entering the intestines. The energy required for wool growth is small and is usually ignored in calculating energy requirements of the sheep. NRC (2007)[5] estimates the ME cost of wool growth using the following equation:

$$\text{ME wool (MJ/d)} = 0.13 \, (\text{GFW-6})$$

where GFW = greasy fleece growth (g/d).

Wool continues to grow when sheep are losing, maintaining or gaining weight, so sheep producing 3.5 kg greasy fleece/annum would require only 0.5 MJ ME/day — about 5% of the total maintenance energy intake of the animal. For this reason, wool growth is normally considered a component of the maintenance requirement.

Forages that maximise dry matter intake have a high content of protein (legumes, young grasses), some of which escapes ruminal degradation (lupins, cottonseed meal, dried grass/ legume forages), and contain all the essential mineral elements will maximise wool growth rate. Optimising stocking rate is a key target for maximising returns in wool producing enterprises (see below).

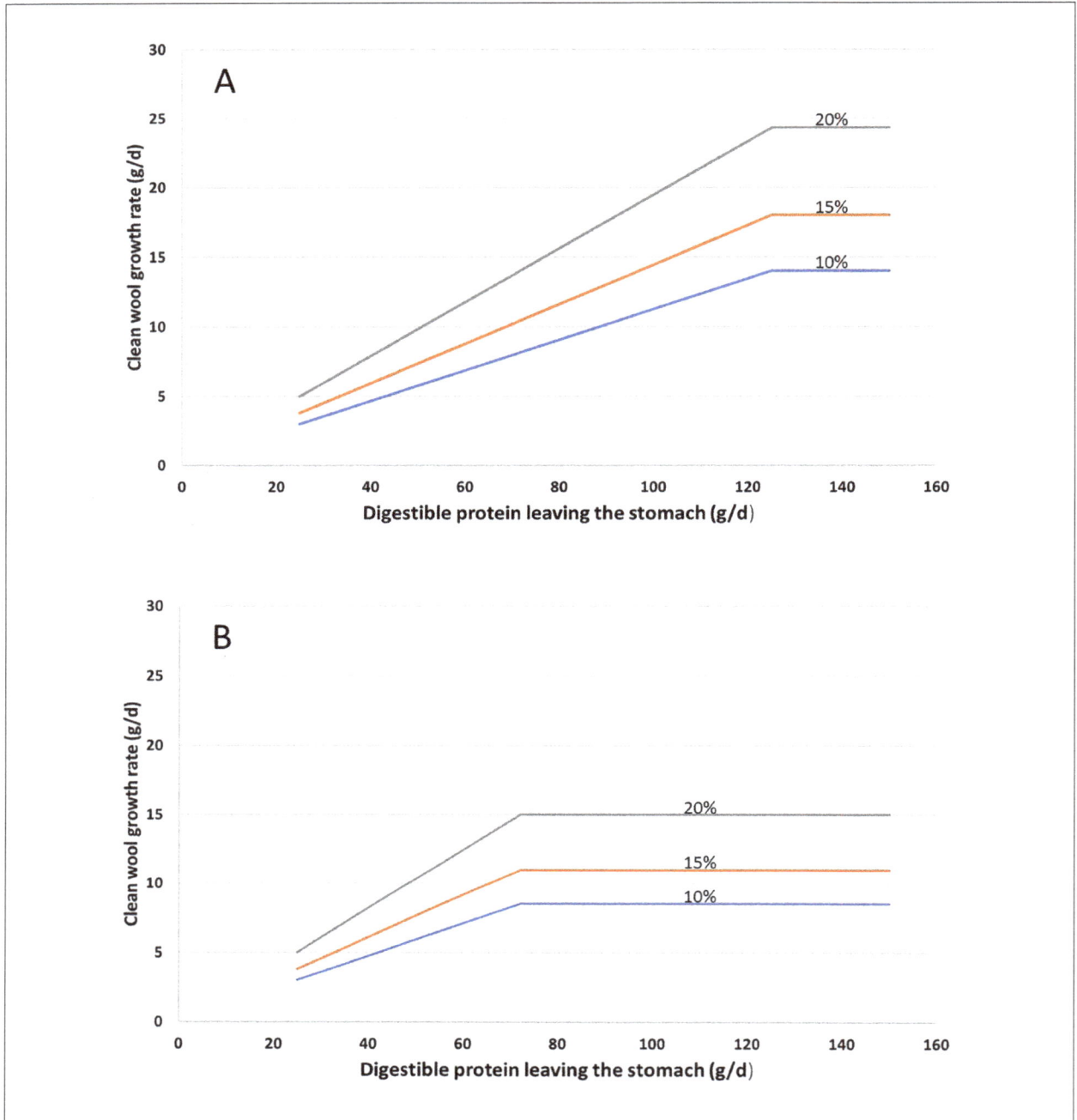

Figure 6.9: Wool growth responses to increasing digestible protein leaving the stomach (g/d) for sheep differing in efficiency of wool growth and fed a diet containing 10 MJ of metabolisable energy/day (A) and 6 MJ/day (B) (adapted from NRC, 2007[5]). Efficiencies are defined as clean fleece weights relative to mature body weight (e.g. 5 kg clean wool/50 kg mature weight = 10% efficiency). Source: P Hynd.

NUTRITION AND MANAGEMENT OF GRAZING SHEEP FOR GOOD HEALTH AND PRODUCTION

The objectives of managing sheep in grazing systems are to maximise profitability, maintain a high-quality, stable pasture feed base, and minimise deleterious health outcomes. Achieving these outcomes requires knowledge of pasture growth patterns throughout the year, the effects

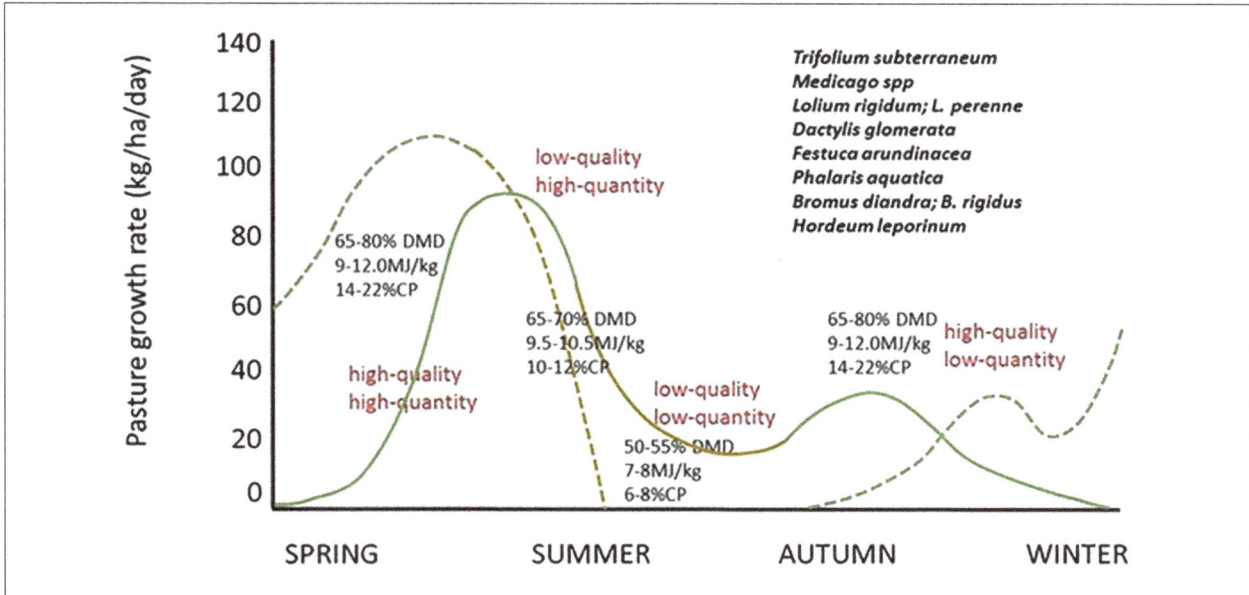

Figure 6.10: Pasture growth profile for pastures in Mediterranean (dotted line) and cool temperate environments (solid line). %DMD = % dry matter digestibility; MJ/kg is megajoules of metabolisable energy/kg dry matter and %CP is crude protein as a % of dry matter. Some typical pasture species found in this zone are indicated. Green lines represent green plants and brown represents mature, senesced plants. Source: P Hynd.

of animals on pasture growth, and the effects of management decisions on feed utilisation and animal production. For example, matching the feed demand with feed supply requires decisions on lambing time, shearing time, culling timing, weaning time and other husbandry activities superimposed on the seasonal variation in pasture nutrient supply. Decision support tools, such as GrazFeed® and GrazPlan®[6], allow producers to achieve these outcomes but rely on good inputs of data on pasture availability and pasture quality and composition. Many sheep health problems are associated with a mismatch of the animal's feed requirements as outlined above, with the availability of nutrients from pastures associated with inadequate feed intake and inadequate nutrient density in the feed. These are considered below.

Nutrient supply from pastures in Mediterranean and cool temperate climates

Mediterranean and cool temperate environments are characterised by cool, wet winters and hot, dry summers, with a period of rapid pasture growth in spring (Figure 6.10). At higher latitudes in the Mediterranean zones, pasture species are dominated by self-seeding annual grasses.

Self-sown seeds from the previous growing season germinate after the opening rains in autumn but grow slowly because of low soil temperature and short day-length. The phases of pasture growth rate and changes in the nutritive value of the pasture are shown in Figure 6.10.

It is clear from Figure 6.10 that sheep during the late summer/early autumn are exposed to a significant nutritional challenge. The little pasture available is low in metabolisable energy and protein and is inadequate for pregnancy, lactation or rapid growth without supplementation.

This period is known as the *autumn feed gap* and can be managed by taking one or more of several measures:

- **Provide supplementary feeds** to meet the energy, protein and mineral needs of the animals to make up for the pasture shortfall.

- **Conserve fodder** (hay, silage) during the period of rapid forage growth (spring/early summer) to feed in the gap.

- **Remove the animals** from the newly germinating pasture to allow good establishment of the young plants without grazing pressure (this is referred to as deferred grazing) and feed them in a confined space with conserved fodder and grain.

- **Introduce new pasture species** that have a longer growing season and can therefore 'fill the gap'.

The main pasture legume grown in southern Australia is subterranean clover (*Trifolium subterraneum*), of which there are several varieties to suit areas with annual rainfall from 400 to 800 mm. The varieties differ in their time of flowering. Early-flowering varieties are best adapted to low rainfall areas because they can set viable seed before the growth season ends. Sub-clover has a unique survival mechanism whereby the plant buries its seed in the spring. These seeds have a hard coat which repels moisture and resists germination during brief periods of rain which may occur during the summer. In drier areas, annual medics are commonly grown in place of sub-clover.

Where there is good summer rainfall or irrigation, white and red clovers (*T repens*, *T pratense*) are sown. White clover is a perennial plant and red clover is a short-lived, summer-growing perennial. Where there are deep, well-drained soils which are neutral to alkaline, lucerne (*Medicago sativa*), which is a perennial plant, is grown. The very deep rooting system of lucerne allows it to survive very dry conditions whilst, with irrigation, it can be very productive, producing over 25 tonnes DM/ha annually.

In the ley-farming zone, also known as the wheat-sheep zone, annual grasses grow with the annual legumes. These grasses include annual ryegrass (*Lolium rigidum*), barley grass (*Hordeum* spp) and silver grass (*Vulpia* spp). On the NSW tablelands, perennial grasses grow with the annual and perennial legumes. In all areas, grasses extend the growing season and increase pasture production potential beyond that achievable by legumes alone. Native grasses, such as *Microlaena* (weeping ricegrass) and *Danthonia* (white top or wallaby grass) can be as nutritious as introduced grasses, with better persistence and drought tolerance but with lower levels of productivity in winter.

Matching energy demand with energy supply from pastures

The major activity in the yearly sheep management calendar that can be manipulated to match the requirements of the sheep with the nutrient supply from pastures is lambing time. By mating five or six months before the period of rapid pasture growth (that is, by mating in February/March for lambing in July/August), the high energy demands of the ewe for lactation can be met with the subsequent effect of high milk yields generating rapid lamb growth and high weaning weights (thus reducing subsequent weaner mortality). Lambs can then be weaned onto high-quality pastures in October and maintain high growth rates through to December or even January, depending very much on the latitude (and therefore the length of the pasture-growing season) of the property. Alternatively, weaners can be placed into an on-farm feedlot and grown out to a carcase weight of 24-26 kg (50-60 kg live weight), or they can be sold as store animals to a specialist feedlot. Similarly, Merino weaners retained for wool production will enter the stressful autumn-winter period at higher body weights if weaned at relatively high weights and supplemented appropriately in summer.

Attempts are being made to identify pasture species (grasses and legumes) that are more persistent and provide longer pasture-growing years and a higher-quality feed-base. Predictions about climate change and altered rainfall patterns present both potential threats and opportunities to the development of the best future pasture base for grazing systems. Greater summer rainfall may allow the wider use of perennials which, with their deeper root systems, extend the growing season and provide opportunistic responses to summer rainfall events. New alternative legumes adapted to deep sandy and acid soils, such as serradella (*Ornithopus compressus*), biserrula (*Biserrrula pelecinus)* and balansa clover (*Trifolium michelianum*), have also been introduced. Legumes provide an attractive option in pasture mixes because they support higher rates of animal production than grasses, as discussed further below.

Relative feeding value of legumes versus grasses

Fulkerson et al. (2007)[13] compared the relative feeding values for dairy cattle of grasses and legumes from the warm temperate regions. Generally, legumes support higher rates of animal production than grasses (Table 6.5).

Clearly, legumes support higher levels of sheep production than grasses do. This was confirmed in studies in Victoria, where the average daily gain of crossbred weaners on legume pastures exceeded 300 g/d (Figure 6.11).

Why do legumes support higher levels of animal production than grasses?

Legumes support faster growth rates of sheep than grasses because they support higher voluntary feed intakes at any digestibility of the feed (Figure 6.12), and they have a higher content of water-soluble carbohydrates, a higher protein content and less structural fibre. Most of the legume protein is digested in the small intestines, which results in a high ratio of protein

Table 6.5: Comparative growth rates of sheep fed various grasses and legumes.

	Relative live weight gain	Reference
Perennial ryegrass	100	
Timothy, common	129	
Lucerne	170	
Lotus pedunculatus	143	Ulyatt (1978)[14]
White clover	186	
Sub clover	121	
Lucerne	146	Freer and Jones (1984)[15]
Phalaris	102	

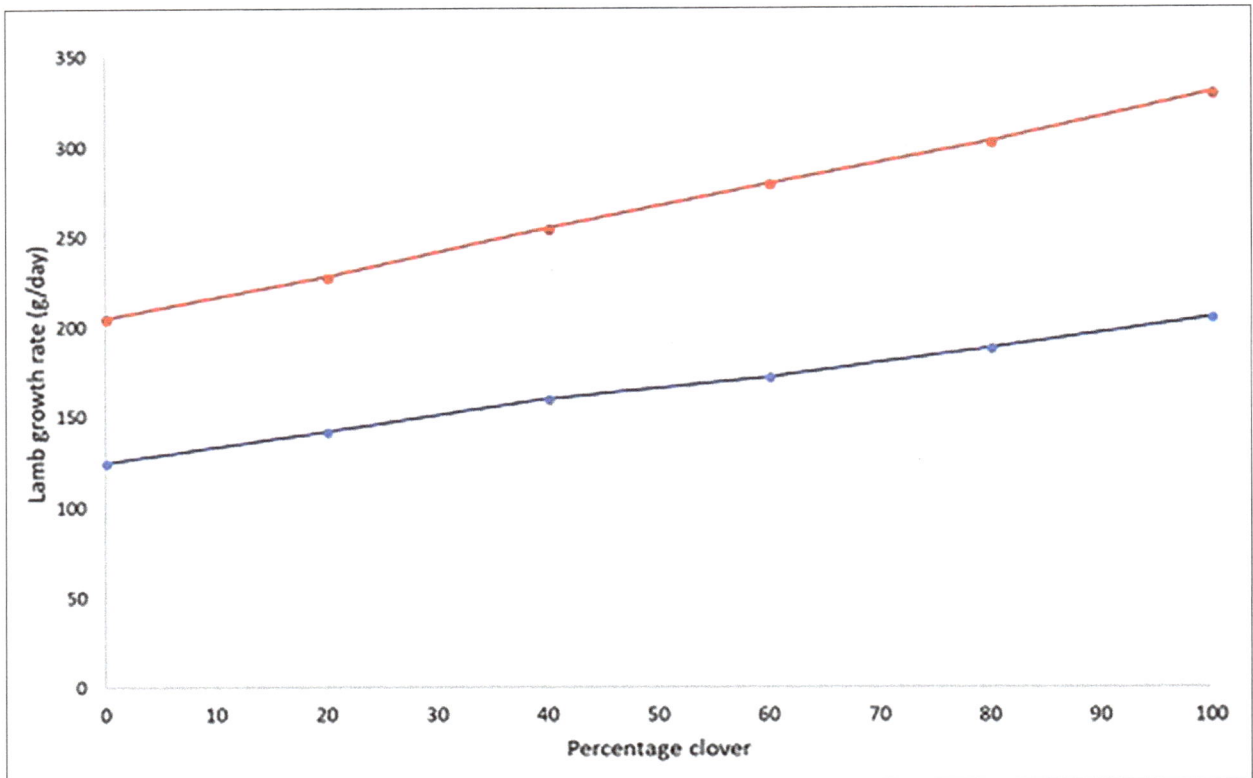

Figure 6.11: Impact of increasing the legume content of pastures for weaner lambs. Reproduced with permission from Lifetime Wool (http://www.lifetimewool.com.au/legal.aspx).[7]

to energy in the absorbed nutrients. Higher feed intakes are a consequence of the faster rate of digestion and faster rumen emptying for legumes, compared to grasses, even at the same digestibility (Figure 6.12).

Nutrient supply from pastures in subtropical climates

Plant species in the tropical and subtropical zones are often characterised by the C4 photorespiration pathway. C4 plants physically separate the carbon compounds produced by photosynthesis from oxygen, thereby reducing the loss of these carbon compounds by respiration. They are therefore very efficient at capturing and storing carbon rather than losing it through respiration pathways. C4 plants are highly productive for this reason and grow very rapidly. However, they also have a high fibre content (particularly as they senesce) and a low protein content, and their digestibility drops rapidly as they mature. Voluntary feed intake by animals is low, and phosphorus (in particular) and minerals (in general) are often deficient in the feed. The pasture-growing season in many low-latitude, summer rainfall regions is short and pasture quality declines rapidly. Sheep in such regions (for example, Queensland, arid and semi-arid NSW, SA, WA) are presented with problems associated with inadequate energy, protein and mineral supply at various times.

Feeding behaviour of grazing animals

Sheep are highly selective in the feed consumed from a highly variable pasture base. Generally, they prefer green material over dead, and leaves over stems, and select for higher protein content,

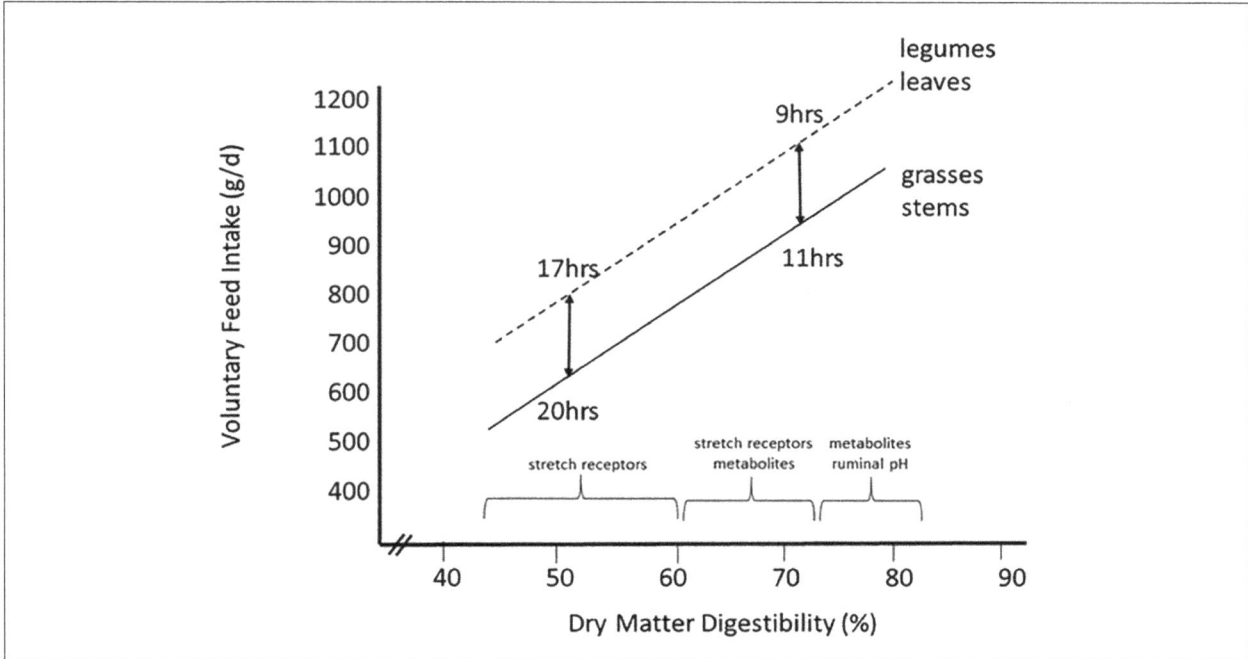

Figure 6.12: Generalised relationship between diet dry matter digestibility (%), ruminal retention time (hrs) and voluntary feed intake (g/d). Source: P Hynd.

lower fibre content and less tough fodders. They actively avoid feeds containing secondary metabolites (Tables 6.6 and 6.7).

Generally, then, sheep select a diet that is significantly higher in protein, slightly lower in fibre and slightly higher in digestibility than the average of the feed on offer. The protein to energy ratio of absorbed nutrients would thus be higher than predicted from an unselected sample of the pasture. Sheep select these characteristics by touch, taste and smell, but the thresholds for these are altered by hunger and previous experience.

This selective ability is important for two reasons:

- The feed consumed is of higher nutritive value than predicted by observation of the pasture and the need for, and responses to, supplements may not be as expected.

- There is pressure on high-value plants and species which may be eliminated from the sward unless grazing management steps are taken to ensure that more uniform pressure is applied to all of the sward. This is discussed further below.

GRAZING AND PASTURE MANAGEMENT FOR SHEEP PRODUCTION

Sheep production involves management of a complex set of interactions between pasture growth rate, pasture growth phase, the total biomass available to the animals (herbage availability), the nutritive value of the pasture, the intake/sheep, stocking rate and intake/hectare (Figure 6.13).

If one adds to this complexity variables such as pasture species, soil types, weather data, animal class and genotype, management calendar and economic data, it is clear that making tactical decisions requires sophisticated dedicated software with all of these inputs and their variances built in. The CSIRO has developed such programmes (GrazPlan® and GrassGro®)[6],

Table 6.6: Diet selection by grazing sheep. Based on data from Kenney and Black (1984).[16]

Select for:	Select against:
Leaf	Stem
Young, green forages	Older, dry forages
High nitrogen, high digestibility	Low nitrogen, low digestibility
Low fibre content	High fibre content
Low mechanical strength materials	High mechanical strength materials
High cell contents	High cell wall contents
Low-physical irritants (prickles)	High physical irritants
Low plant secondary metabolites	Plant secondary metabolites

Table 6.7: Chemical profile of feed on offer and of feed consumed by sheep on two pasture types. Adapted from Coleman and Barth (1973).[17]

Constituent	Available	Selected	Selection (%)	Forage type
Crude protein (%)	15.8	20.5	+30	Fescue/Lespedeza pasture
ADF (%)	37.6	36.0	-5	
NFE + EE (%)	46.5	43.4	-7	
DDM (%)	60.3	62.3	+3	
Crude protein (%)	18.8	22.1	+18	Orchard grass/sub. Clover pasture
ADF (%)	37.2	36.0	-4	
NFE + EE (%)	44.1	41.8	-5	
DDM (%)	64.2	66.3	+3	

ADF = acid detergent fibre; NFE + EE = nitrogen-free extract plus ether extract; DDM = digestible dry matter

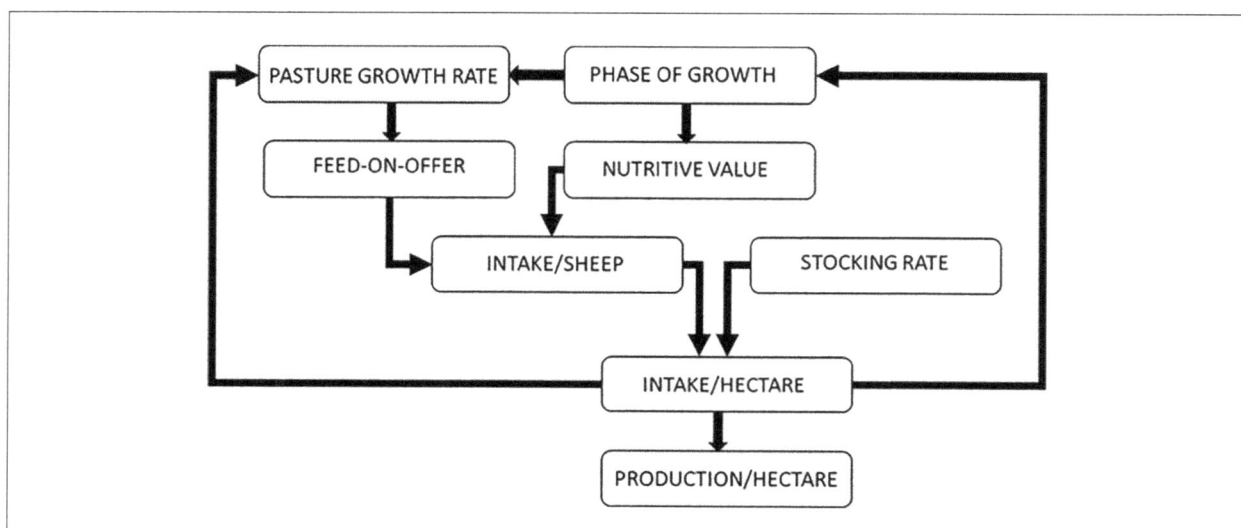

Figure 6.13: Interactions between pastures and animals in a grazing system. Source: P Hynd.

which are available to producers to compare management options for their own properties based on historical weather data and the incorporation of all of the nutrient requirement algorithms discussed above. These are discussed later in this chapter; however, it is useful to dissect the major interactions above because failures within them are responsible for many of the problems that veterinarians, nutritionists and consultants confront on sheep properties.

Pasture growth rate and phase of plant growth

Grasses follow a sigmoidal growth curve over the growing period regardless of the pasture species and its climatic environment (Figure 6.14).

Phase I is characterised by slow growth because the plant has insufficient leaf area to maximise photosynthesis. The nutritive value (digestibility or metabolisable energy content) of the grass declines slowly as the cell wall structural contents increase. In Phase II the rate of growth increases as the leaf area approaches its optimum and photosynthesis is maximised. After this, in Phase III, the rate of growth declines and the feed quality declines rapidly as the plant nutrients are redirected towards flowers rather than tillers and the plant cell walls become increasingly composed of more complex carbohydrates which are of lower digestibility to the ruminant.

In the early stages of growth (Phase I), the plant uses reserves of water-soluble carbohydrates stored in the stems to provide energy for the regrowth of new tillers. As the plant traverses Phase II, photosynthesis increases to a maximum at the optimum leaf area index, and the new photosynthate accumulates in the plant. Management of the rate and severity of defoliation of the plant to ensure rapid recovery of the plant after defoliation is

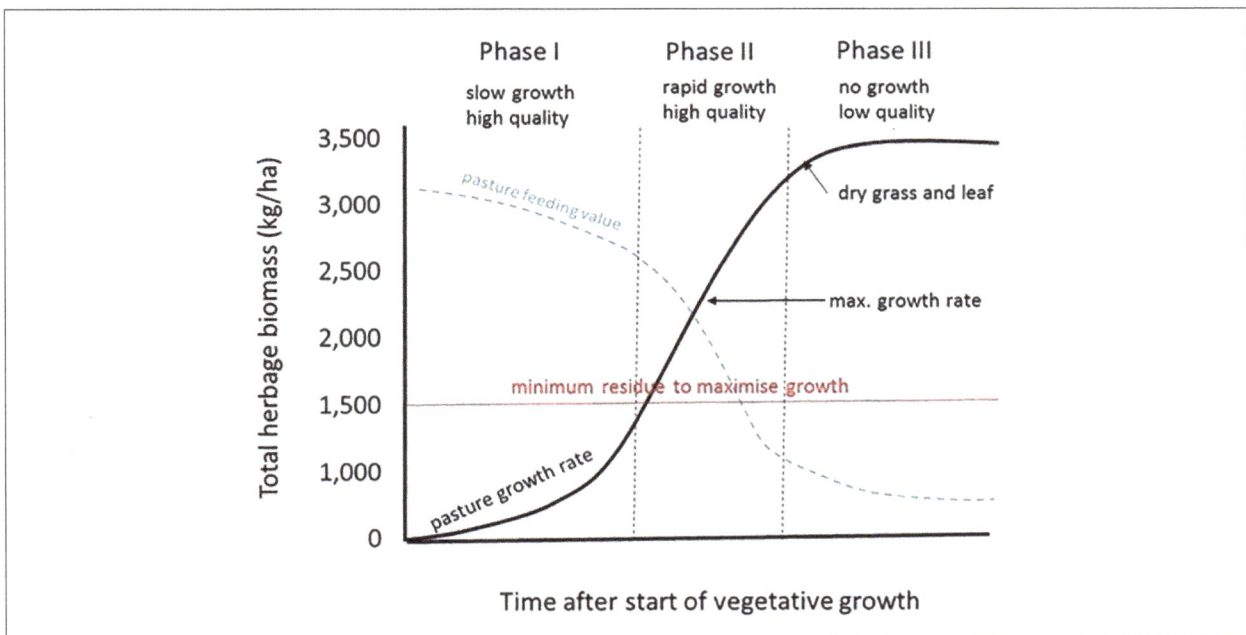

Figure 6.14: Growth curve of pastures over a growing season showing the three phases of growth and the changes in feeding value as the plants transition through each phase. Maximum growth rate occurs midway through Phase II after which the feeding value declines rapidly as the plant flowers and earlier tillers begin to senesce. Source: P Hynd.

the major objective of grazing management, as discussed below. If the defoliation is too severe, the roots are deprived of photosynthate, which is directed towards new leaf growth, and the resilience of the plant is compromised. Continued defoliation will ultimately kill the plant and it may be replaced by a less valuable weed or the ground may be left exposed to erosion and summer weed invasion. Maintaining adequate ground cover is important to discourage undesirable plants. Ensuring good ground coverage, effective weed control and high-quality pastures that are resilient to seasonal weather changes are key drivers of profitable sheep enterprises.

Relationships between herbage availability, nutritive value of pasture and energy intake in sheep

In general, the more herbage available to sheep the more they will eat until they reach a limit which is set by rumen fill. The degree of rumen fill is determined largely by the rate of ruminal digestion of the feed, which is, in turn, largely determined by its dry matter digestibility. Digestibility affects energy intake adversely in two ways:

(1) The lower the digestibility, the slower the rate of digestion and onward passage of feed particles to the omasum, so digesta remains in the reticulorumen for long periods of time delaying further intake of feed.

(2) Digestibility is the main determinant of the energy content of a feed (Figure 6.2).

Consequently, low digestibility feeds greatly reduce metabolisable energy intake and cannot meet the ME needs of sheep other than those maintaining weight or growing slowly. Figure 6.15 shows the impact of pasture digestibility on the ME intake of 50 kg ewes that are dry or pregnant, lactating (with a single at foot) and lactating (with twins at foot).

Note the large impact of pasture digestibility on ME intake of the ewes. Even if a large amount of herbage is available, if its digestibility is below 60% the ewes cannot consume sufficient pasture to maintain their own body weight and condition and, at this digestibility, they are unable to produce sufficient milk for their lamb(s). Lactating ewes require pasture of at least 70% digestibility (MD = 10.3 MJ/kg DM) and an available herbage mass of at least 1000 kg DM/ha (1300 kg DM FOO/ha). If the quantity and quality of pasture fall below these values, the ewes will produce less milk with subsequently lower growth rates of lambs and weaning weights. If the quantity and quality fall too far below the requirement the risk of metabolic diseases such as pregnancy toxaemia increases greatly. In twin-bearing ewes late in pregnancy, the energy demands are similar to those of the lactating ewe, so with pastures of low herbage availability and/or digestibility (say, <70%), the risk of pregnancy toxaemia is greatly increased, particularly if accompanied by sudden stressors like cold snaps, shedding, overzealous dogs and so on.

Effect of diet selection on pasture growth

Selective grazing by sheep puts great pressure on a small subset of the forage available and ultimately will remove that component of the pasture sward. The selection pressure can be intense — as little as 1% of the forage available to the sheep can make up 80% of the animal's diet. One of the objectives of grazing management systems is to reduce or eliminate this

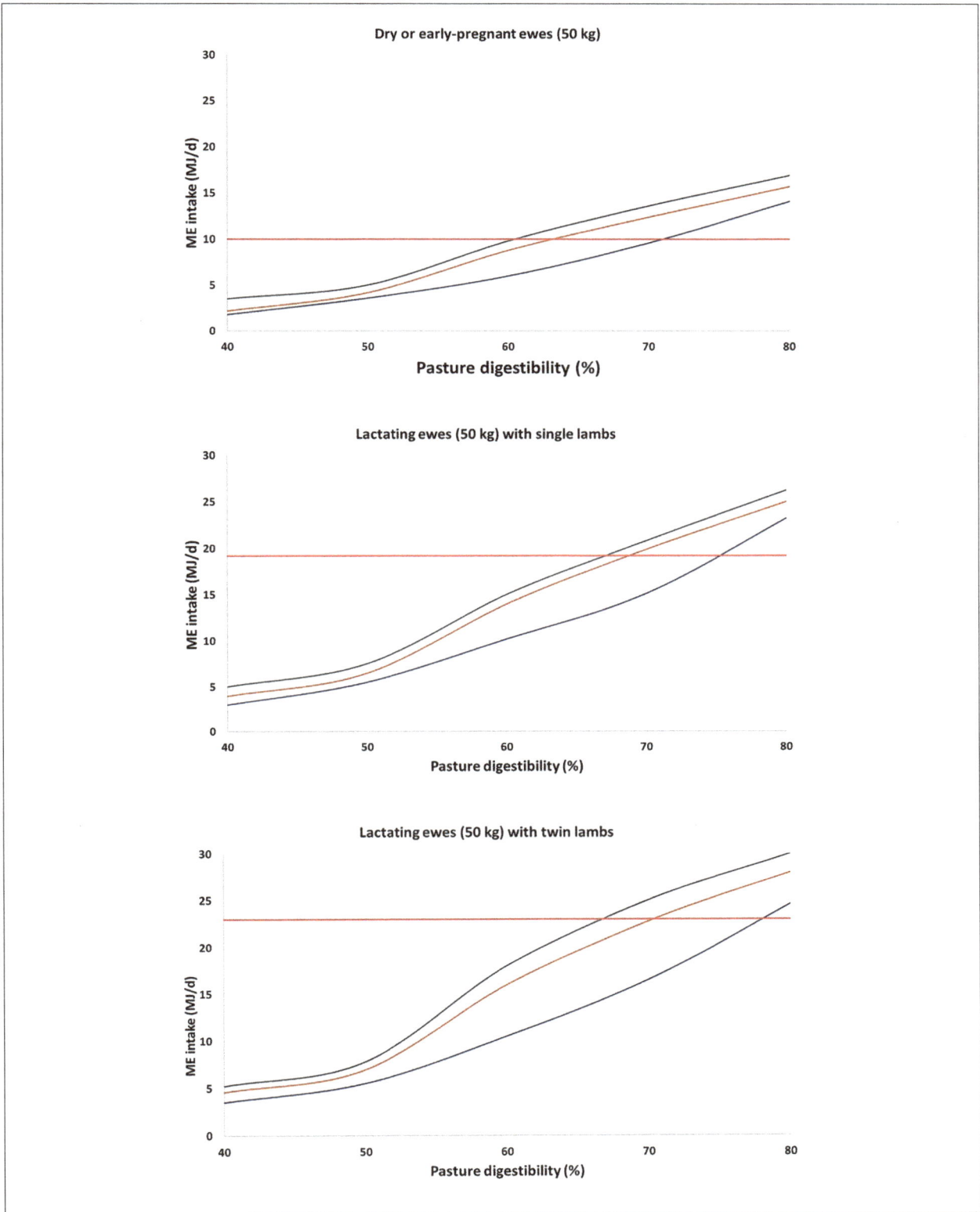

Figure 6.15: Effect of pasture digestibility (%) and herbage mass available to sheep (kg DM/ha) on metabolisable energy intake of 50 kg ewes that are dry (or pregnant), lactating (with singles) or lactating with twins at foot. The three lines represent the herbage mass available (kg DM/ha) at 500 (blue line), 1000 (orange line) and 1500 (grey line). The red line is the ME required/day for the ewes in each class. Note that feed on offer (FOO) is herbage mass + 300 kg DM/ha. Adapted from NSW Department of Primary Industries (2017) (ProGraze® manual).[18]

differential selection pressure by forcing the animals to place the same selection pressure on *all* components of the sward. Another objective of grazing management is to maintain the plants at a height and biomass that maximise the capture of light into plant growth.

Grazing management to optimise sheep production per hectare

Grazing management is aimed at controlling the rate and frequency of plant defoliation in order to achieve high levels of plant dry matter production, maintain high-quality plant species in the pasture, ensure resilience of the plants to dry periods, and optimise the utilisation of pasture.[19,20] These are achieved by controlling the following variables:

- the mix of pasture species in the sward
- the stocking density
- the recovery period between grazing events
- the length of the grazing period
- the residual herbage left after grazing.

The objective is to maximise pasture growth rate, pasture quality, the extent of pasture utilisation, and the persistence of desired pasture species over time.

Continuous grazing

Continuous grazing refers to the practice of maintaining a set number of animals all year on the pasture, at a stocking rate usually dictated by the amount of feed available at the most limiting season for pasture growth and availability (for example, autumn in Mediterranean climates). Such systems operate well below the biological capacity of the system and lead to degradation of the land.[21] Continuous grazing allows sheep to select heavily for preferred species and preferred plant parts such that the quality of their diet is improved but the quality of the remaining herbage becomes progressively poorer. Overgrazing at times of the year when herbage growth is low leaves bare ground which is susceptible to invasion by weed species, some of which thrive in dry seasons or rapidly invade when rain falls. Bare ground is also prone to erosion by wind and water.

Set stocking

Set stocking refers to stocking large areas of land with relatively small numbers of sheep with little movement of animals and no assessment of herbage availability. The rate of pasture utilisation (defined as the proportion of herbage mass consumed relative to herbage mass grown) is inevitably low (<30%), and the total pasture biomass produced is often low because the plants are not in the optimal leaf area index for photosynthesis.

Rotational grazing

Under rotational grazing, sheep are moved from one paddock to another to allow pasture to rest and recover. The paddocks so utilised are often defined by their physical characteristics (topography, soil type and vegetation type). Many producers move the animals on the basis of time, assuming that time reflects pasture herbage mass, but this obviously varies with season and site, and a defined rest period for one paddock will be different for another, resulting in varied herbage availabilities.

Strip grazing

Commonly used for dairy cattle, strip grazing has been used for sheep in high rainfall areas during periods of rapid pasture growth and where the vegetation is relatively uniform (for example, forage crops or sown pastures). The grazed areas are typically very small and the movements frequent. Such systems are labour-intensive.

Cell grazing

Cell grazing refers to the practice of rotating animals through a large number (usually >10) of small areas at high stocking densities. Movements are based on set times, until herbage mass has been so reduced that there is a reduction in animal performance, or until herbage biomass reaches a desired level.

Planned grazing

Planned grazing is based on the principles of plant growth described in Figure 6.14. The objective is to manipulate grazing intensity and time in order to achieve high pasture growth rate, optimum pasture utilisation, maintenance of desired pasture species over time, and maintenance of maximum groundcover (weed control). To achieve this requires knowledge of the herbage mass at any given time, the rate of pasture growth and the requirements of the sheep at any time of the year. Generally, in high rainfall areas a set minimum of 1000 kg dry matter/ha is the target, as at this level the plant reserves of carbohydrate are sufficient for rapid recovery and little loss of roots. The quality of the herbage remains high as it stays in the early Phase II stage (Figure 6.14). The key components of planned grazing are knowledge of the starting quantity of herbage on the paddock, the rate of pasture growth, the number of sheep grazing the area and the target residual herbage quantity. While planned grazing is mainly applicable to higher rainfall, more intensive production systems, planned grazing can also be implemented in low-intensity rangelands situations. Use of smaller paddocks, manipulation of water points to achieve more uniform grazing intensity and more careful attention to stocking rates are key components of increasing sheep productivity and sustainability.

Stocking rate, stocking density, dry sheep equivalent and sheep production

Clearly, matching the number of sheep that can be sustained on the pastures to their carrying capacity is a key requirement of grazing management and a major determinant of profitability of sheep enterprises. Optimising stocking rate requires knowledge of both the energy requirements of the sheep (as described above) and the quantity and quality of the pastures at different times of the year as seasons change (described below). Because stocking rate is effectively a metric of the quantity of feed consumed/hectare we need to express it in terms of intake/animal. The standard way of expressing the intake of pasture dry matter by ruminants is as 'dry sheep equivalents' or DSE. The standard sheep is a non-pregnant, non-lactating ewe or wether weighing 50 kg (although in some regions a 45 kg dry sheep is used as the baseline

DSE). A rough rule of thumb is that such animals will consume 1 kg pasture dry matter/day (equivalent to the maintenance energy requirement of a 50 kg dry sheep (= 8 MJ/day, assuming the pasture MD is about 8 MJ/kg DM or 56% DMD). The intake of other classes of sheep (and cattle) is then expressed relative to this (Table 6.8).

Clearly, the energy requirement and pasture intake/day are greatly influenced by late pregnancy and lactation, particularly if the ewes are raising twin lambs.

For growing sheep, the DSE requirements are shown in Table 6.9.

In order for weaners to grow rapidly they must be offered double the pasture allowance/day which would allow weaners to grow slowly (Table 6.9).

With this information the required pasture intake can be calculated for all the classes of sheep on the farm at different times of the year. The requirements of each class of sheep at each time of the year can be added together to determine the total pasture requirement and its distribution over the year. Together with pasture growth rate, this allows the periods of pasture shortfall to be identified and either tolerated, because the degree of live weight loss is acceptable, or compensated by the use of supplementary feeds (as above). The average stocking rate throughout the year is then the total number of DSE divided by the total grazed hectares, and it is often dictated by the period of lowest pasture availability (for example, late summer or autumn).

It is important here to distinguish between stocking rate and stocking density. Stocking rate is the number of DSE on an area of land, which can be a paddock, group of paddocks, or a whole property, over a 12-month period (for example, 750 DSE/100 ha = 7.5 DSE/ha).

Table 6.8: Table of dry sheep equivalents (DSE) for sheep.

Category	Body weight (kg)							
	40	45	50	55	60	65	70	100
Ewes (dry or <3 mths pregnant)	0.8	0.9	1.0	1.1	1.2	1.2	1.3	
Ewes (last 2 mths pregnancy)	1.1	1.2	1.3	1.4	1.5	1.6	1.7	
Ewes (last 2 mths with twins)	1.2	1.4	1.5	1.6	1.7	1.8	1.9	
Ewes (lactating with singles)	1.6	1.7	1.9	2.1	2.2	2.4	2.5	
Ewes (lactating with twins)	1.9	2.1	2.3	2.5	2.7	2.9	3.0	
Wethers	0.8	0.9	1.0	1.1	1.2	1.2	1.3	
Rams								2.2

Table 6.9: DSE requirements of growing sheep.

Category	Body weight (kg)		
	20-30 kg	30-40 kg	40-50 kg
Weaners (<50 g/d)	0.9	1.0	1.1
Weaners (50-100 g/d)	1.1	1.3	1.4
Weaners (100-150 g/d)	1.3	1.6	1.8
Weaners (150-200 g/d)	1.6	2.0	2.2

Stocking density, in contrast, is the number of DSE grazed on an area on any one day and may be as high as 200 DSE/ha for that day. Given that 1 DSE requires roughly 1 kg of pasture DM per day, the stocking density infers the pasture intake per hectare per day. For example, if the stocking density is 200 DSE/ha, the pasture DM removal will be 200 kg/day. Given a particular pasture growth rate (in kg DM/day), the number of grazing days available on that area can be calculated. The greater the number of paddocks, the greater the control the grazing manager has to match pasture production to pasture intake.

As stocking rate increases, the intake per head begins to decrease because intake/head and pasture biomass available are closely related (Figure 6.16).

Note the large effect of pasture digestibility on voluntary feed intake and the large decline in ME intake as better-quality pastures (60% DMD) decline in herbage biomass.

As intake/head declines, production per animal decreases, but the overall intake/hectare increases as more of the pasture is harvested. If stocking rates are too low, the proportion of pasture that remains after grazing is high and this uneaten pasture either decays or is blown away or trodden into the ground. Such grazing systems are very inefficient at capturing the biomass produced by photosynthesis, with as little as 20% of the biomass being captured by the animals. Low stocking rates also allow the sheep to graze selectively, thereby putting disproportionate pressure on a small fraction of the available biomass. Eventually, these desirable species disappear from the sward and the pasture quality declines. If the stocking rate is too high, there may be insufficient pasture biomass to support the animals because pasture availability and intake/head are closely related (see Figure 6.16). Production and animal health can then be compromised severely, and exposure of soil to erosion and weed invasion may also occur.

However, despite these negative effects of high stocking rates, the efficiency of sheep production increases with stocking rate because the capture of otherwise-wasted biomass means that up to four times more of the total pasture available can be consumed at these high rates. So, as the stocking rate increases from low levels, pasture availability declines and feed intake per head declines, but a higher proportion of the available pasture is consumed and a

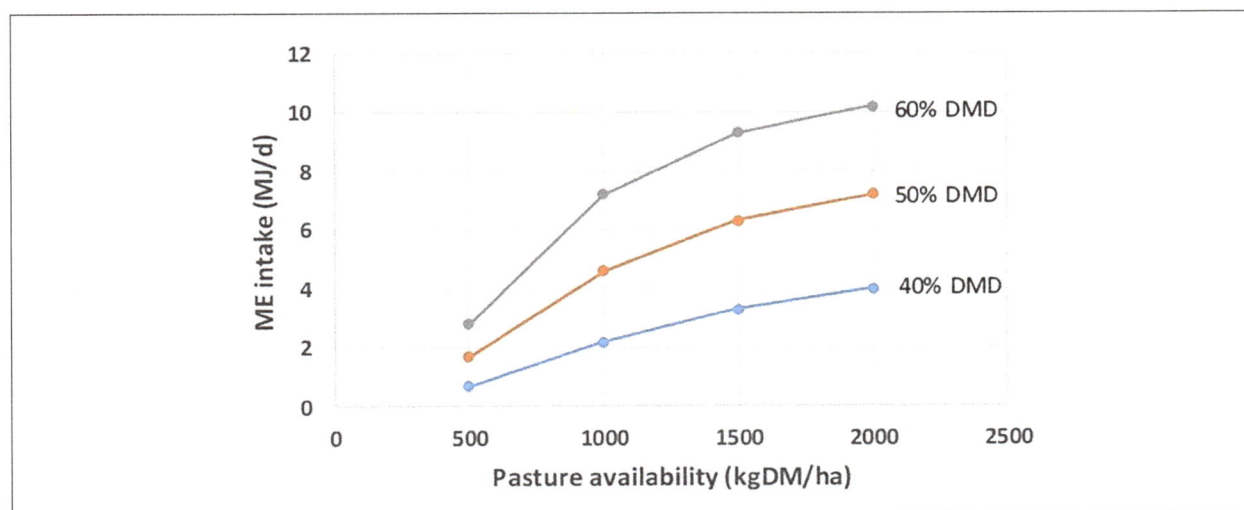

Figure 6.16: The effect of pasture biomass availability on metabolisable energy intake of sheep grazing on pastures differing in DM digestibility. Adapted from NSW Department of Primary Industries (2017) (ProGraze® manual).[22]

greater total mass of pasture is consumed. Animal production responses to stocking rate are illustrated in Figure 6.17.

Note that as stocking rate increases, the production/area increases and net return, or gross margin (income minus variable costs), also increases. Both reach a maximum, with gross margin reaching the maximum before production/area. The point of maximum gross margin is defined as the optimum stocking rate, but often producers choose to stock at a level below the predicted optimum as a risk management decision. To decide on the optimum stocking rate an assessment of risk can be made by using the year-to-year variation in gross margin at each stocking rate and choosing a balance between high gross margin and low risk. Such decisions are best made using decision support software such as GrazPlan® or GrassGro®, which allow multi-year comparisons of gross margins and gross margin variations, using historical data from the property.

Wool production per head decreases as stocking rate increases in response to the decrease in intake/head. Wool/ha increases with increased stocking rate as intake/hectare increases. As intake/head decreases, the efficiency of wool production also increases because the relationship between wool growth/head and intake is not linear (Figure 6.9). Of the two factors increasing wool/hectare (intake/hectare and increased efficiency), it is mainly intake/hectare that drives the extra wool production (>80%), although increasing efficiency contributes up to 20% to the extra wool production. Thus, at higher stocking rates, more wool is produced per hectare. Because intake/head is reduced, the mean fibre diameter of each fleece is also reduced, contributing to higher-value wool, at least up to the point at which staple strength is compromised.[24]

Estimating the digestibility and total quantity of feed on offer in grazed pastures

To implement the planned and strategic grazing systems above and to calculate the appropriate stocking rate, it is essential that producers have the skills to estimate the pasture herbage

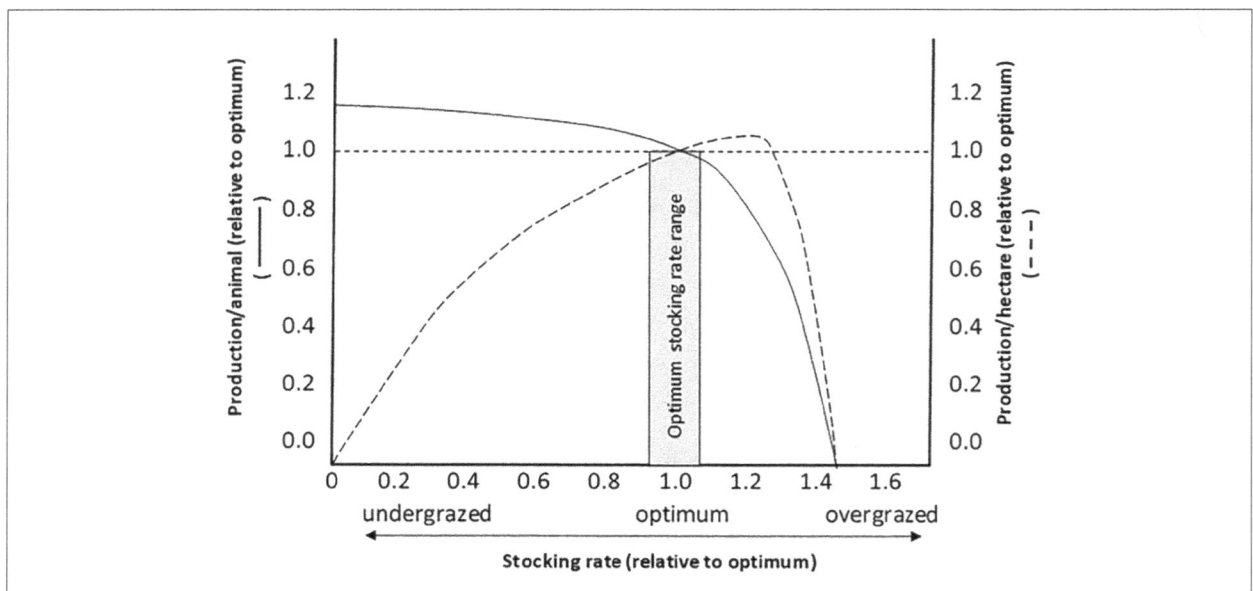

Figure 6.17: Effect of stocking rate on production/animal, production/area. Adapted from Allworth (1993).[23]

matter available, the groundcover percentage, the legume content, the proportion of green/ dead, and the digestibility of the green and dead components.

These data are required inputs into the GrazPlan® model, but can also be used by producers on a day-to-day, decision-making basis to achieve the following:

- to match the stocking rate and grazing interval with available herbage
- to determine how much supplementary feed to provide
- to manipulate the pasture growth rate and stage of plant development to maximise plant growth rate and to capture a high proportion of biomass produced (see below)
- to ensure good ground cover to protect soil from run-off, maximise water infiltration into the soil and reduce weed invasion.

The herbage biomass or feed on offer is estimated on a dry matter basis to correct for the large variations in water content of pasture forages throughout the growing season. The herbage biomass available depends on pasture height, plant density and water content. Figure 6.18 shows the relationship (generic) between pasture height (cm), feed on offer (kg DM/ha) and dry matter intake of pasture by sheep (kg/day).

Note that there is a strong linear relationship between pasture height and feed dry matter on offer per hectare. Note also that the feed intake of sheep plateaus at about 1600 kg DM/ha (FOO). Below that there are rapid increases in total available biomass and intake. There is an added benefit to grazing pastures to approximately 1500 kg DM/ha: at this height, the pasture is operating at maximum photosynthetic rate (light capture) (see grazing management below). Using pasture height to estimate herbage available is practically very

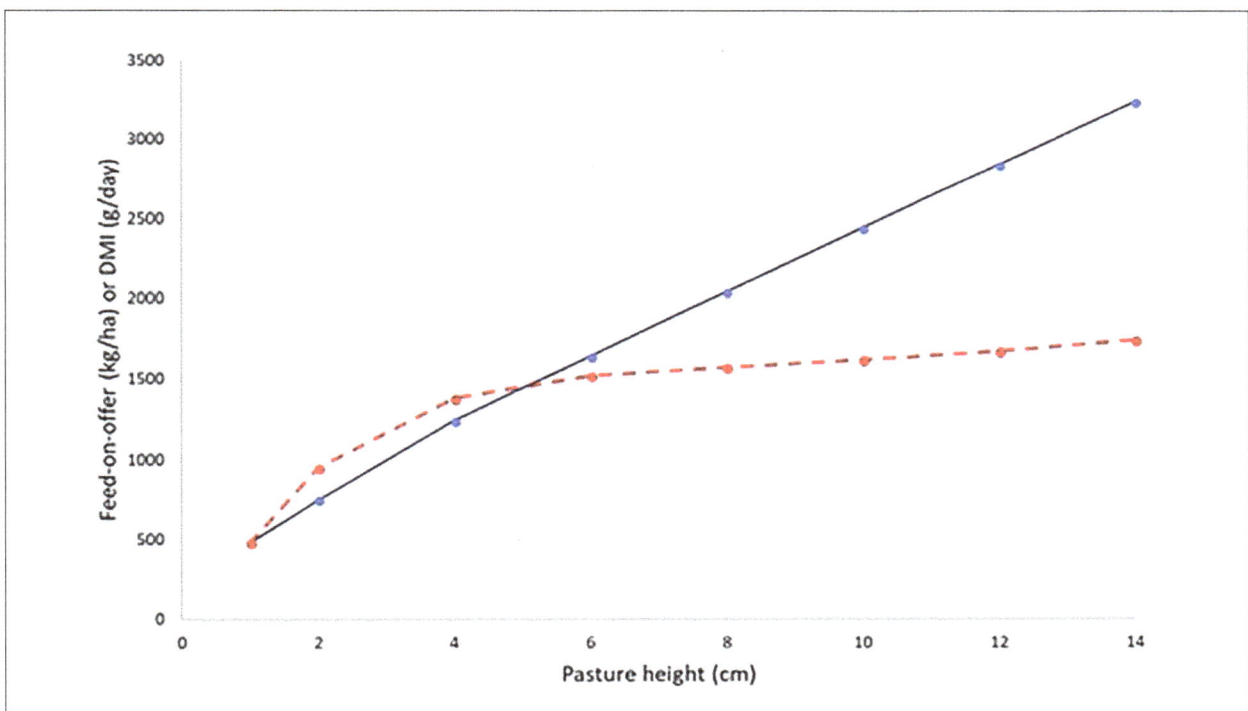

Figure 6.18: Relationships between pasture height (cm), feed on offer (kg DM/ha = solid line) and voluntary feed intake (kg DM/day = dashed line) of a 50 kg sheep. Note: FOO is 300 kg higher than pasture herbage available to the sheep. Drawn by P Hynd using the MLA pasture ruler data (which relates pasture height to FOO) and data of DMI versus FOO from the literature.

MLA Pasture Ruler

Pasture Height to Pasture Quantity Indicator

mla

See other side of ruler and 'Tips & Tools' provided for information about how to use this ruler to predict animal performance and $ returns

Height (cm)	Pasture quantity for a moderately dense pasture (indicative estimates of kg green dry matter/ha)	
14 & OVER	3000 PLUS	Surplus quantity – growth slows, no real additions to intake, quality declines
12	2500	
10	2200	Preferred range of DM/ha for animal and pasture production
8	1900	
7	1700	
6	1600	
5	1400	
4	1200	
3	1000	Too little quantity – retards regrowth, restricts intake and increases erosion hazard
2	700	
1	400	

Figures are indicative only. Very dense, closely grazed pastures will have a higher (up to +25%) kg green DM/ha at the same height. Conversely, with lightly grazed open pastures, differences due to density are greater at heights above 6cm.

Developed by MLA Southern Beef Program

How much feed do I need to get the performance I want?

Estimated **minimum** pasture quantity (kg green DM/ha) required to achieve targeted production levels of livestock

Livestock class	Pasture energy density MJ ME/kg DM (and approximate dry matter digestibility %)		
	11.2 (75%) Active pasture growth, green	10.1 (68%) Late vegetative to early flowering green	9.0 (60%) Mid to late flowering some dead
Dry cow	700	1100	2600
Preg. cow, 7-8 mths, no calf	900	1700	np
Lactating cow calf 2 mths	1100	2200	np
Growing steer 0.61 kg/day	800	1600	np
0.85 kg/day	1200	2600	np
1.12 kg/day	2200	np	np
Dry sheep	400	600	1200
Pregnant ewes - mid	500	700	1700
- last month	700	1200	np
Lactating ewes - single	1000	1700	np
- twins	1500	np	np
Growing sheep 125 g/day	600	1000	np
175 g/day	800	1700	np
225 g/day	1600	np	np

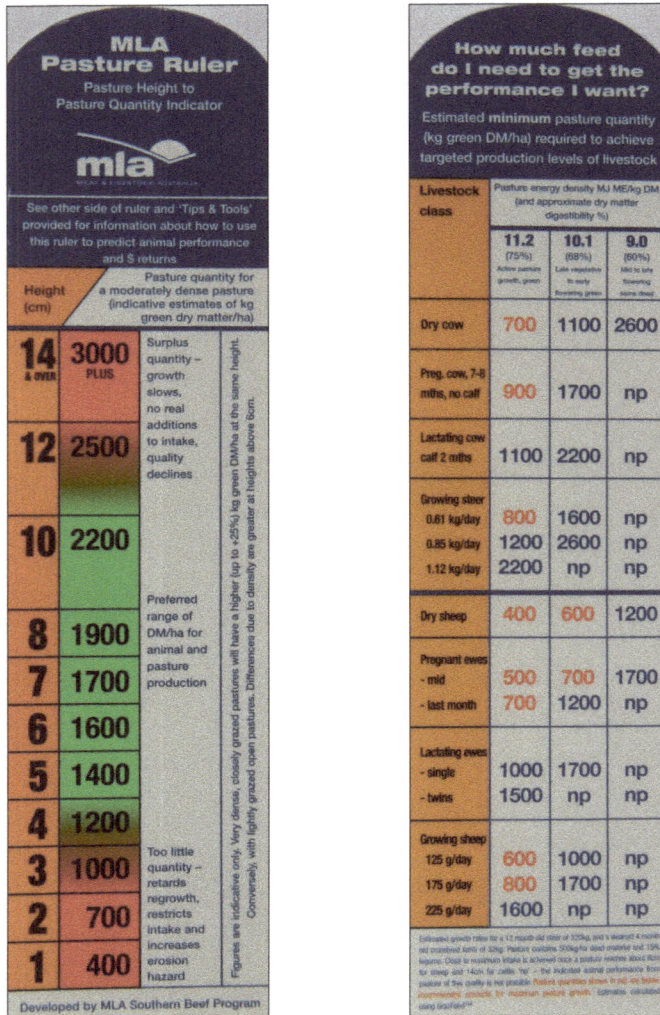

Figure 6.19: MLA pasture ruler for estimating feed on offer based on pasture height in dense, growing perennial pastures. Note: corrections must be made for pasture density if groundcover is <100%. Reproduced with permission from the MLA (http://mbfp.mla.com.au/Pasture-utilisation/Tool-31-Pasture-rulers-sticks-and-meters).

useful because it can be done rapidly using simple visual means like photos and simple meters (see below).

Assessing herbage biomass (or feed on offer) in pastures

Measuring or estimating the feed on offer is an important aspect of managing grazing sheep. Herbage mass can be estimated using several methods, as outlined in the sections below.

1. Cutting quadrats of known area

The most accurate method of estimating herbage mass is to cut pasture quadrats of known area (for example, 0.25 m²) to ground level. Taking a photograph of the area before cutting, or of the immediately surrounding area, allows visual calibration when the cut material is dried and weighed. Multiply the dry weight in grams by 40 to estimate kg dry matter per hectare (1 g/0.25 m² = 40 kg/ha).

2. Pasture height and a calibrated ruler

Figure 6.19 shows the Meat and Livestock Australia (MLA) pasture ruler used to estimate food on offer.

3. Rising plate meter

A probe containing a plate which is pushed up the probe by the pasture provides an estimate of the herbage biomass when calibrated against quadrats cut as above.

4. Electronic probe

This works on the basis of electrical conductivity of the amount of moisture between the legs of the probe, and from this the amount of pasture between the legs. The tool is only reliable for green forages.

5. Use of a photo library to provide a visual cue

A series of photos of various pastures differing in feed on offer are available at the Lifetime Wool website (http://www.lifetimewool.com.au/newpastures.aspx).

As a rule of thumb, producers should graze pastures to a height equal to the top of their boots halfway from the toe to the ankle (about 5 cm or 1500 kg DM/ha), at which point the sheep should be moved to another paddock.

Assessing the digestibility of the feed on offer

Digestibility is an important feature of pasture assessment and management for the following reasons:

Digestibility and metabolisable energy content of pasture plants are closely related

For roughages like pasture forages, the accepted relationship between dry matter digestibility (DMD) and ME (MJ/kg DM) is as follows:

$$\text{ME (MJ/kg DM)} = 0.172 \text{DMD} - 1.71.$$

So a pasture containing on average a DMD (%) of, say, 50% would contain about 6.9 MJ/kg DM; a 65% DMD pasture would contain about 9.5 MJ/kg DM; and a lush, green, young pasture at 80% DMD would contain 12.1 MJ/kg DM. GrazFeed® uses this relationship between digestibility and ME density together with the predicted dry matter intake to calculate stocking rate and time of grazing.

Digestibility and crude protein content of pasture plants are closely related

Digestibility and protein content are closely related in pasture forages (Figure 6.20), although the relationship differs between legumes and grasses. The plot shown here was derived from data obtained over the period of senescence of the plants in a Mediterranean environment.[25] Note that the digestibility of the grasses and the legume declines at a similar rate but at any digestibility, legumes contain more crude protein. The rate of decline in digestibility and crude protein content also varies between species — compare brome grass with barley grass in Figure 6.20.

Digestibility is closely related to voluntary feed intake in grazing sheep

Unlike monogastric animals, ruminants cannot easily regulate their feed intake to meet their energy requirements because the rate of emptying of the reticulo-rumen is rate-limiting to feed

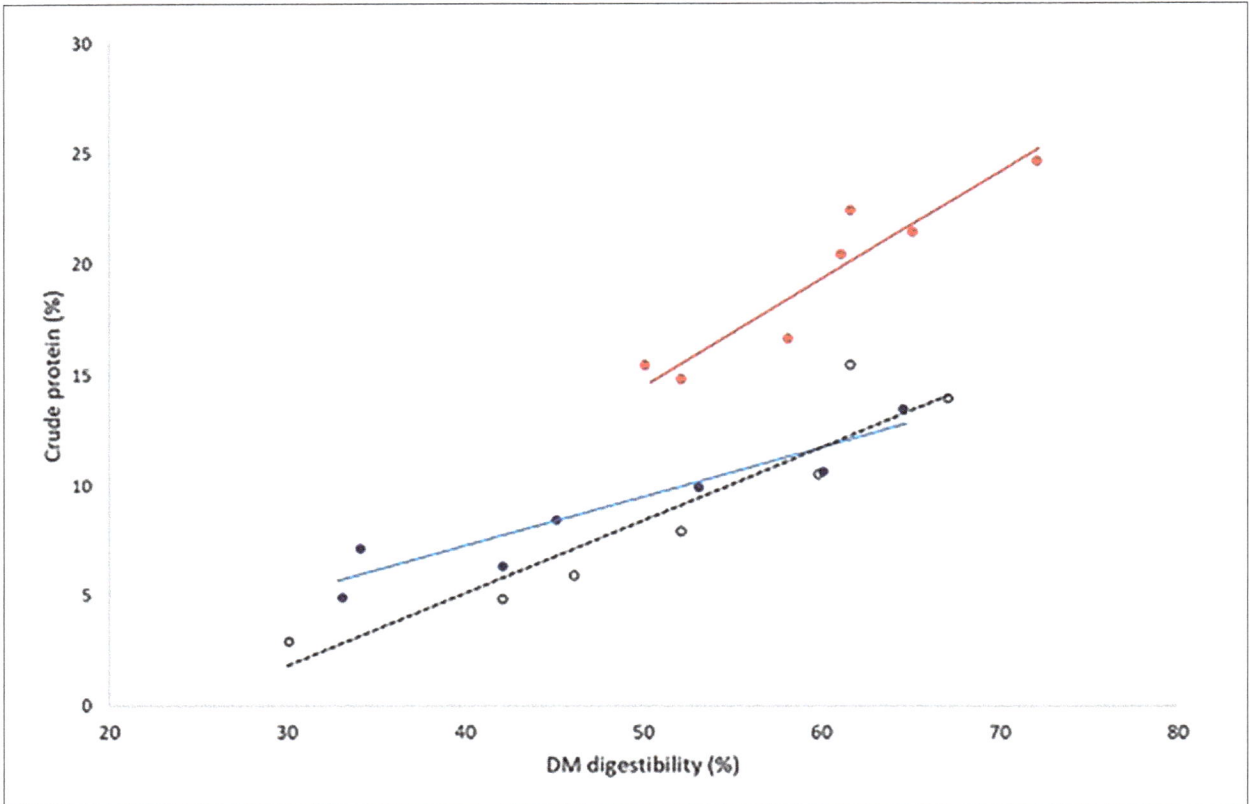

Figure 6.20: Relationship between pasture dry matter digestibility (%) and crude protein content (%DM) for a legume (*Medicago sativa*: black line), and two grasses (*Bromus* spp: dashed line) and *Hordeum leporinum*: blue line) during the period of senescence in a Mediterranean environment. Adapted from Radcliff and Cochrane (1970).[25]

intake. In other words, they cannot consume more until the rumen has digested the feed or moved it on to the omasum and abomasum. The rate of digestion is governed by digestibility as more digestible feeds are digested faster. This then means the greater the digestibility of the feed, the higher the voluntary feed intake, at least until the point when non-physical factors start to limit intake (see Figure 6.12). Physical satiety signals (stretch receptors) govern most of the feeding behaviour until absorbed metabolites and possibly rumen-active metabolites and pH begin to operate. Note that, at any digestibility, the rate of passage, hence voluntary intake, of legumes is greater than that of grasses, and leaves are more rapidly digested than stems. Sheep select for legumes over grasses and leaves over stems presumably because they are perceived in some way (probably ease of prehension and texture) as nutritionally superior. The ability of sheep to select strongly for certain plant species puts disproportionate pressure on these species, and ultimately will remove them from the sward. Given that these are the most desired plants nutritionally, the sward is thereby degraded. Grazing management, as outlined below, is aimed at maintaining sufficiently high grazing pressure such that *all* or *most* plant species in the sward are consumed equally.

The digestibility of pastures depends on several factors:

• the phenology or stage of growth (Figure 6.14). As pastures mature they decline in digestibility from 75% in phase 1, to 65% to 70% in phase II and 60% in phase III. Dead pasture averages about 45% to 50% digestibility, so the greater the proportion of dead/green material the lower the digestibility

- forage species. Legumes are about 10% higher in digestibility (0.5 MJ/kg DM higher) than grasses

- Leaf:stem ratio. Leaves are higher in digestibility than stems — which accounts for some of the decline with maturation of the plant.

Regulation of total feed intake in herbivores

The feeding value of a pasture depends on the combination of its nutritive value (see above) and the voluntary feed intake of it by sheep. The latter depends largely on its digestibility because intake is usually limited by the rate of breakdown of physically resistant cell wall constituents like cellulose and hemicelluloses. Consequently, for most forages and stages of growth, there is a close positive relationship between digestibility and voluntary feed intake. The form of the relationship varies with forage type (legumes versus grasses, leaves versus stems and so on), because the retention time of legumes and leaves is less than that of grasses and stems.

Relationships between intake rate, bite volume, bite rate and grazing time of grazing sheep

The following equation describes the relationships between daily intake of feed, bite size, bite rate and grazing time:

$$\text{Intake (kg/d)} = (\text{bite rate} \times \text{bite volume}) = (\text{kg/min}) \times \text{grazing time (min/d)}.$$

When available forage falls below about 1600-2000 kg dry matter per hectare (or <5 cm high), the intake rate starts to decline because the sheep is unable to compensate for the rapidly declining bite volume by increasing its biting rate. This declining *rate* of herbage intake is compensated for by an increase in grazing time, which can compensate for the lower rate of intake up to the point where the additional energy spent foraging is not compensated by the extra energy available.

Pattern of grazing by herbivores

Sheep tend to graze around dawn and dusk, but as the gap between the two decreases (days shorten) the grazing period becomes almost continuous.

Supplementation of grazing sheep

Supplementary feeding of grazing sheep is used to correct a deficiency of energy, protein or minerals or to take pressure off pastures at critical times in the growth curve.

Energy supplements

Usually, grains or conserved fodders (hay or silage) are used to make up shortfalls in ME intake. Supplements high in energy density, like grains, are easier to handle and feed out than supplements low in energy density, such as hay and silage. They also tend to be more cost-effective when expressed as dollars per MJ ME and per g crude protein (Table 6.11).

To calculate the real energy and protein costs of each supplement the following calculation can be done:

Cost of 1 kg dry matter = $/tonne as fed/DM%

Energy content in 1 kg dry matter = MJ/kg (from table values)

Protein content in 1 kg dry matter = g/kg dry matter (from table values).

Typical values (depending on current feed costs and nutritive values) for common feedstuffs are shown in Table 6.10.

Note the variation in the range of costs/MJ (1.3 to 3.6 c/MJ) and /kg crude protein (80-188 c/kg). Different supplements differ in their relative energy and protein costs, so it is important to be clear about what is limiting production. The value of barley and oat grain as sheep feed energy and protein supplements is apparent. Lupins are a cheap source of protein but an expensive source of energy. Lupins have an advantage over cereal grains as sheep supplements in that they contain virtually no starch and sheep do not require an adaptation period to lupin feeding as they do for cereal grain feeding. The risk of acidosis with lupins feeding is therefore low.

Substitution effects of energy supplements

When energy supplements are given to animals at pasture, the impact on production is difficult to quantify because the supplement can cause a reduction in forage intake, no change in forage intake or an increase in forage intake. These so-called substitution effects are due to associative effects of the supplement on digestion of the base feed. Negative associative effects occur when starchy grains are digested in the rumen and the resulting pH decline reduces the activity of cellulolytic bacteria and the digestibility of cellulose. The increased retention time of the fibre in the rumen reduces intake rate. Positive associative effects occur when the energy supplement provides a nutrient that is limiting either microbial activity or animal metabolism. Commonly this will be nitrogen or sulphur, which will stimulate microbial activity in the rumen, or phosphorus or a trace element for animal metabolism. An example of negative associative effects is shown in Figure 6.21.

As the grain intake increases, the intake of the basal straw diet decreases. The total energy intake increases due to the high energy content of the grain supplement. The extent of this substitution effect varies with supplement and basal forage.

Protein and non-protein nitrogen supplements

The value of supplementing grazing animals with nitrogen depends on the level of protein (nitrogen) in the forage, and the protein requirements of the grazing animals (Figure 6.11).

Mineral supplements

The most cost-effective supplements for grazing animals are those that correct deficiencies of essential minerals. For a few cents/day most minerals can be supplied to meet the microbial requirements and the requirements for the animal's metabolism. Correcting either of these will increase food intake and the efficiency of maintenance and production.

Table 6.10: Comparison of the typical costs of different energy supplements for grazing animals.

Supplement	Cost ($/tonne as fed)	DM (%)	Cost c/kg DM	ME content (MJ/kg DM)	CP content (kg/kg DM)	c/MJ ME	c/kg CP
Barley grain	140	90	16	12	0.11	1.3	145
Lupins	250	89	28	13	0.35	2.2	80
Oats	130	90	14	11	0.09	1.3	155
Field peas	300	90	33	12	0.25	2.8	132
Silage	45	35	13	10	0.08	1.3	163
Meadow hay	135	89	15	8.5	0.08	1.8	188
Lucerne hay	300	88	34	9.5	0.22	3.6	155
Faba beans	280	90	31	12	0.28	2.6	111

Deleterious compounds found in common pasture forages and invasive weeds of pastures

A number of forage chemicals or physical attributes of common forage pasture plants can reduce the intake and/or digestibility of nutrients and thereby reduce performance of the sheep.[27] Others have deleterious effects on sheep health (Table 6.11).

Decision making in grazing sheep enterprises

Decisions about the management of grazing sheep (stocking rate, grazing periods, pasture composition, shearing time, lambing time, supplementary feeding) have traditionally been made empirically, based on the experience and intuition of the producer rather than on the basis of quantitative modelling. Australia (CSIRO) has led the way in the development of decision support software (GrazFeed®, GrassGro®, GrazPlan®) designed specifically to simulate all aspects of the pasture/animal/soil interface.[28] These models allow tactical and strategic

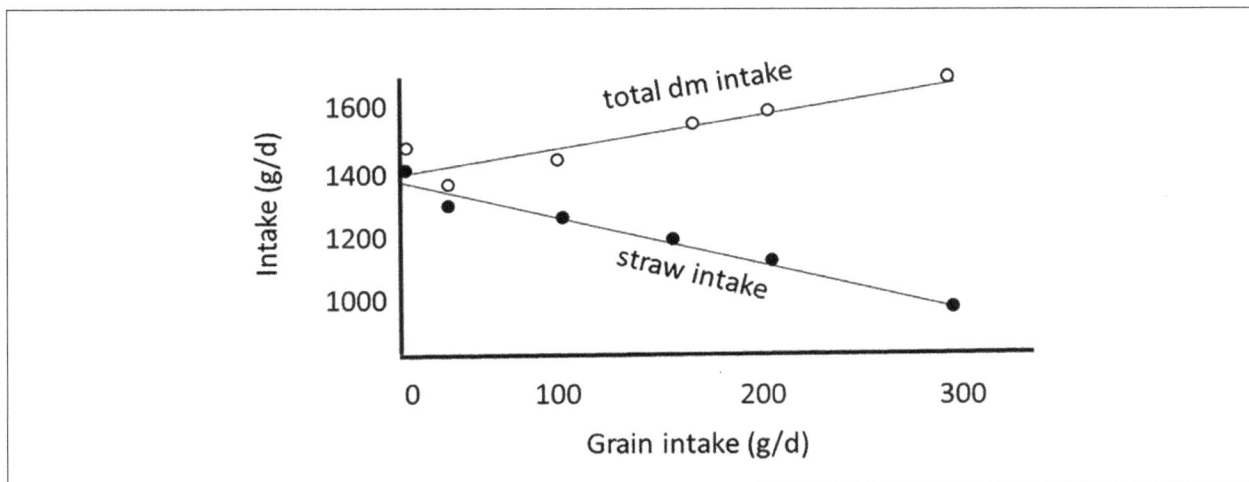

Figure 6.21: Negative associative (substitution) effects of a grain supplement on straw intake in sheep. Adapted from Dixon and Stockdale (1999).[26]

Table 6.11: Common deleterious compounds found in common pasture forages.

Factors	Forage species	Intake	Digestibility	Comments
Lignin		↓	↓	Indigestible
Condensed tannins	Trefoil (*Lotus* spp), Sainfoin (*Onobrychis viciifolia*), crown vetch (*Coronilla coronate*), Sulla (*Hedysarum coronarium*), sorghums (*Sorghum* spp), acacias (*Acacia* spp)	↓	↓	Astringent taste aversion; depressed protein digestibility; inactivate digestive enzymes; anti-microbial effects
Resins		↓	↓	Indigestible
Silica	Most grasses, crop stubbles (e.g. rice straw)	↓	↓	Indigestible
Waxes		↓	↓	Indigestible
Gossypol	Cotton (*Gossypium hirsutum*)		↓	Inactivate digestive enzymes
Essential oils			↓	Anti-microbial effects
Glycosides (cyanogenic, cardiac, saponins, glucosinolates, diterpenoids, bracken glycosides, calcinogenic glycosides, phenolic	*Astragalus* spp, *Lotus* spp, *Coronilla* spp, *Trifolium repens*, *Prunus virginiana*, *Digitalis* spp, *Apocyanaceae* spp, *Lilliaceae* spp, *M sativa*, brassicas, *Pteridium aquilinum*, *Solanum* spp			Haemorrhage; convulsions; death; cyanide poisoning; cardiac pathology
Alkaloids	Potato weed (*Heliotropium europaeum*), Patterson's curse (*Echium plantagineum*), *Senecio* spp, *Crotolaria* spp, *Phalaris aquatica*)	↓		Hepatoxic; photosensitisation; teratogenic; neurotoxic; reproductive failure
Mycotoxins	*Phalaris aquatica*, *Brachiara* spp, *Panicum* spp, *Lolium* spp	↓		Neurological; photosensitisation; diarrhoea; jaundice; death
Nitrates	Oats (*Avena sativa*), maize (*Zea mays*), wheat (*Triticum aestivum*), barley (*Hordeum vulgare*), docks (*Rumex* spp), lucerne (*Medicago sativa*)			Hypoxia and hypoxaemia; convulsions; death
Oxalates	*Oxalis* spp., *Cenchrus ciliaris*, *Atriplex* spp, *Pennsietum* spp			Urolithiasis; death

decision making based on historical rainfall and pasture growth, overlaid with an animal model which simulates the animal's response to the changing grazing environment. The nutrition component of these models is called GrazFeed®. GrazFeed® is based on algorithms derived from the ARC (1980)[29] updated by the SCA (1990).[30] It uses the ME system for energy calculations and the digestible Undegraded Dietary Protein (UDP) and Rumen Degradable Protein (RDP) system for evaluating protein status. GrazFeed® relies on estimating the voluntary intake of feed and diet selection, to allow the total energy and nutrient availability to be partitioned to maintenance and production. There are strong interactions between voluntary feed intake, biomass available and herbage quality. For example, biomass available influences diet quality because at low herbage levels the opportunity for selective grazing and selection of more digestible feeds decreases. Higher-quality feed is also consumed at a higher rate because digestibility drives intake in ruminants. The selective behaviour of sheep is simulated assuming they select from the higher-digestibility classes first and then work their way down the quality spectrum. The amount of material in each class is estimated and multiplied by the relative ingestibility to determine actual intake of that material. The quantity of the next class is then chosen and so on, until the cumulative relative intake is derived and the average quality of the feed determined.[19] The impact of supplements given to grazing sheep is problematic because supplements can increase, have no effect on, or more commonly decrease the intake of base pasture. GrazFeed® assumes that the supplement is treated similarly to the forage component — that is, sheep will select the supplement only when it is equal to, or better than, the available herbage.

REFERENCES

1 Asner GP, Elmore AJ, Loander LP et al. (2004) Grazing systems, ecosystem responses, and global change. Ann Rev Env Res **29** 261-99. https://doi.org/10.1146/annurev.energy.29.062403.102142.

2 Hynd PI (2014) Growing and finishing beef cattle at pasture and in feedlots. In: Beef cattle: Production and trade, eds D Cottle and L Kahn. CSIRO Publ: Australia, pp. 381-99.

3 Penner GB, Aschenbach JR, Gabel G et al. (2009) Epithelial capacity for epithelial uptake of short chain fatty acids is a key determinant for intraruminal pH and susceptibility for subacute ruminal acidosis in sheep. J Nutr **139** 1714-20. https://doi.org/10.3945/jn.109.108506.

4 MAFF (1976) Energy allowances and feeding systems for ruminants. Technical Bulletin **33**. HMSO: London.

5 NRC (2007) Nutrient requirements of domesticated ruminants. CSIRO Publ: East Melbourne, Australia.

6 Donnelly JR, Moore AD and Freer M (1997) Decision support systems for Australian grazing enterprises-I. Overview of the GRAZPLAN project and a description of the MetAccess and LambAlive. DSS Agric Sys **54** 57-76.

7 Lifetime Wool. Lifetime ewe management. Available from: http://www.lifetimewool.com.au/LTEM.aspx. Accessed 27 June 2018.

8 Griswald KE, Hoover WH, Miller TK et al. (1996) Effect of form of nitrogen on growth of ruminal micorbes in continuous culture. J Anim Sci **74** 483-91. https://doi.org/10.2527/1996.742483x.

9 Satter LD and Roffler RE (1975) Nitrogen requirement and utilisation in dairy cattle. J Dairy Sci **58** 1219-37. https://doi.org/10.3168/jds.S0022-0302(75)84698-4.

10 Kempton TJ, Nolan JV and Leng RA (1978) Principles for the use of non-protein nitrogen and bypass proteins in the diets of ruminants. World Animal Review. FAO Animal Production and Health Paper 12.

11 Black JL, Robards GE and Thomas R (1973) Effects of protein and energy intake on wool growth of merino wethers. Aust J Agric Res **24** 399-412. https://doi.org/10.1071/AR9730399.

12 Kempton TJ (1979) Protein to energy ratio of absorbed nutrients in relation to wool growth. In: Physiological and environmental limitations to wool growth, eds JL Black and PJ Reis. UNE Publ Unit: Armidale, NSW, pp. 208-22.

13 Fulkerson WJ, Neal JS, Clark CF et al. (2007) Nutritive value of forage species grown in the warm temperate climate of Australia for dairy cows: Grasses and legumes. Livest Sci **107** 253-64. https://doi.org/10.1016/j.livsci.2006.09.029.

14 Ulyatt MJ (1978) Aspects of the feeding value of pastures. Proc Agron Soc NZ **8** 119-22.

15 Freer M and Jones DB (1984) Feeding value of subterranean clover, lucerne, phalaris and Wimmera ryegrass for lambs. Aust J Exp Agric Anim Husb **24** 156-64. https://doi.org/10.1071/EA9840156.

16 Kenney PA and Black JL (1984) Factors affecting diet selection by sheep 1. Potential intake rate and acceptability of feed. Aust J Agric Res **35** 551-63. https://doi.org/10.1071/AR9840551.

17 Coleman SW and Barth KM (1973) Quality of diets selected by grazing animals and its relation to quality of available forage and species composition of pastures. J Anim Sci **36** 754-61. https://doi.org/10.2527/jas1973.364754x.

18 NSW Department of Primary Industries (2017) Prograze: Profitable, sustainable grazing. 9th ed. Available from: http://www.dpi.nsw.gov.au/__data/assets/pdf_file/0005/700457/Prograze-ninth-edition.pdf. Accessed 17 August 2017.

19 Freer (2014) The nutritional management of grazing sheep. In: Sheep nutrition, eds M Freer and H Dove. CSIRO Publishing and CABI Publishing: Australia, pp. 357-75.

20 Earl J (2014) Grazing and pasture management and utilisation in Australia. In: Beef cattle production and trade, eds D Cottle and L Kahn. CSIRO Publishing: Victoria, Australia, pp. 339-79.

21 Tothill JC and Gillies C (1992) The pasture lands of Northern Australia: Their condition, productivity, and sustainability. Occasional Publication No 5. Tropical Grassland Society of Australia: Brisbane.

22 NSW Department of Primary Industries (2017) Prograze: Profitable, sustainable grazing. 9th ed. Available from: http://www.dpi.nsw.gov.au/__data/assets/pdf_file/0005/700457/Prograze-ninth-edition.pdf. Accessed 17 August 2017.

23 Allworth B (1993) Profitable pasture utilisation — sheep. In: Proceedings of the 8th Annual Conference of the Grassland Society. 7-8 July, Orange, NSW, pp. 67-71.

24 Hynd PI and Masters DG (2002) Nutrition and wool growth. In: Sheep nutrition, eds M Freer and H Dove. CSIRO Publishing and CABI Publishing: Australia, pp. 165-187. https://doi.org/10.1079/9780851995953.0165.

25 Radcliff JC and Cochrane MJ (1970) Digestibility and crude protein changes in ten maturing pasture species. Proc Aust Soc Anim Prod **8** 531-6.

26 Dixon RM and Stockdale CR (1999) Associative effects between forages and grains: Consequences for feed utilisation. Aust J Agric Res **50** 757-73. https://doi.org/10.1071/AR98165.

27 Cheeke PR (2014) Endogenous toxins and mycotoxins in forage grasses and their effects on livestock. J Anim Sci **73** 909-18. https://doi.org/10.2527/1995.733909x.

28 Donnelly JR, Moore AD and Freer M (1997) Decision support systems for Australian grazing enterprises-I. Overview of the GRAZPLAN project and a description of the MetAccess and LambAlive. DSS Agric Sys **54** 57-76.

29 ARC (1980) The nutrient requirements of ruminant livestock. Technical review by an Agricultural Research Council Working Party. Commonwealth Agricultural Bureaux: Slough, England.

30 Corbett JL, Freer M, Hennessy DW et al. (1990) Feeding standards for Australian livestock, ruminants (a report to the Standing Committee on Agriculture from the Ruminants Subcommittee). SCA and CSIRO Publ: East Melbourne, Australia.

7

REPRODUCTIVE MANAGEMENT AND DISEASES IN NATURALLY MATED FLOCKS

INTRODUCTION

This chapter is divided into six sections:

- Section A: The reproductive rate (RR) of sheep in Australia; its contribution to sheep production systems and the major factors which influence it
- Section B: Factors affecting fertilisation, conception rate and pregnancy up to mid-gestation; ewe factors and ram factors
- Section C: Abortion and prenatal diseases of lambs, and disorders of ewes in late pregnancy
- Section D: Perinatal mortality
- Section E: Management and diseases of lactating ewes
- Section F: Investigations of poor reproductive rate in commercial sheep flocks.

SECTION A

THE REPRODUCTIVE RATE (RR) OF SHEEP IN AUSTRALIA

The role of reproduction in the productivity of sheep-grazing systems

The reproductive capacity of sheep serves two essential functions in the industry. The first of these is the production of young, sale animals which are slaughtered for meat. The second is to produce young animals to enter breeding flocks as ewe replacements or ram replacements in order to maintain flock size as older animals die or are cast for age.

In commercial flocks, the emphasis applied to each of these activities varies with the production aims of the flock. At one end of the spectrum are the specialist meat-lamb producers typically based on the first-cross ewe producing second-cross lambs. In these systems, the reproductive rate is a very important contributor to the overall success of the sheep flock business. In general, the higher the reproductive rate, the more lambs are sold per ewe present, although, if reproductive rates are too high, the poor survival rate of lambs from multiple births reduces the economic benefits.

At the other end of the spectrum are the self-replacing flocks, in which the role of reproduction involves both functions — producing young sale animals as well as replacing the older females which are sold from the breeding flock. In some cases, wethers are also retained to adult ages as wool producers. These flocks are predominantly based on the Merino or other breeds which produce wool of high value. In such flocks, reproductive capacity chiefly serves to replace older animals which have been retained as wool growers (wethers) or wool growers and breeders (ewes) — often up to the age of 5 or 6 years. While the sale of surplus young sheep and cast-for-age sheep also makes a significant contribution to farm income, the relative importance of a high reproductive rate to the overall profitability of the enterprise is less than it is in flocks which sell all offspring in their first year of life.

In ram-breeding flocks, reproduction also provides the capacity to select those animals with the best combination of economically important production traits, leading to genetic improvement in the future generations. High rates of reproduction allow more intense selection and more rapid genetic gains.

The different emphasis on the economic importance of reproduction between sheep production systems means that the amount of resources applied to improving reproductive efficiency or maintaining a high reproductive rate varies between flocks. Similarly, the level at which a sheep producer considers that reproductive rate is too low also varies with the production objective. A prime lamb producer may become concerned if a crossbred ewe flock is only producing 90 lambs per 100 ewes, but a Merino flock owner is likely to be very satisfied with that rate of reproduction. Veterinarians who are consulted about poor reproductive rates in ewe flocks need to appreciate the differences in the importance of reproductive capacity between the various sheep production systems.

The major factors influencing reproductive rate

Most sheep flocks in Australia are comprised of Merino or Merino-crossbred ewes. The reproductive performance of Australian sheep flocks is primarily limited by the relatively low genetic potential for reproduction of Merino ewes and the relatively low level of nutrition which is made available to them under normal commercial management conditions. These two factors — nutrition and breed — are undoubtedly the two most important factors limiting the reproductive performance of Australian sheep flocks.

The level of nutrition of ewes under Australian conditions is usually determined by stocking rate — the number of ewes grazing per hectare of pasture. Supplementary feeding (conserved fodder or grain) is generally used only to fill short-term deficiencies in seasonal pasture availability and therefore makes only a small contribution to the annual supply of energy, protein and other nutrients provided to the ewe flock.

As discussed in Chapter 3, stocking rate (the number of ewes grazed per hectare) is one of the critical management strategies determining the financial success of a sheep-grazing system, and it influences many parameters contributing to the economic and biological efficiency of the farm system. The stocking rate chosen for a sheep-grazing enterprise is unlikely to be one which maximises flock reproductive rate — reproductive rate is only one factor contributing to flock profitability — and a reduction in stocking rate in order to increase reproductive rate may have negative consequences for other components of the farm system.

It is essential that veterinarians who are asked to investigate poor reproductive performance are aware of the very strong association between nutrition (stocking rate) and reproductive rate. Without that awareness, and without a strong understanding of the underlying physiological processes affecting reproductive rates in sheep, there is a risk that a veterinary investigation of low reproductive rate could focus too heavily on the possibility of an infectious or toxic cause, to the exclusion of a nutritional or physiological explanation.

In addition to stocking rate, other husbandry strategies may also adversely affect reproductive success, and, for each individual farm, a compromise must be reached which seeks to maximise profit rather than any one index of sheep flock productivity. The potential financial benefits for the producer from increasing reproductive rate are discussed in Section F of this chapter to provide a context for planning rational veterinary intervention. A proper evaluation of reproductive rate and its contribution to the economic performance of a sheep flock requires a *systems approach*, as discussed in Chapter 3. When embarking on a planned investigation of low reproductive rates in a sheep flock, it is important to review all of the ramifications of permanent changes in reproductive performance, for they are unlikely to be limited to a change in numbers of sale sheep only. Increases in reproductive rates have a wide impact on the flock management, usually including a reduction in the number of adult sheep which can be supported per hectare. The complex contribution of reproduction to a sheep production system is also discussed further in Section F.

The components of reproductive rate

Reproductive rate (RR) can be measured and reported in a number of different ways. Producers commonly report *lamb marking rates* because the first accurate count of the number of lambs present is usually made at marking and, under usual circumstances, the death rate of lambs between marking and weaning is low. The rate is reported either as lambs marked per 100 ewes present at marking, or per 100 ewes present at joining. It is important to know which number is used as the denominator, as there will often be up to 5 percentage points' difference between the two figures. The lambing rate — the number of lambs *born* per ewe joined — is rarely known in commercial flocks, although approximations can be made if ultrasound pregnancy diagnosis is used.

A shorthand way to record the method of calculation is to use abbreviations: the rate of lambs marked per ewe joined is written as *LM/EJ*, and the number of lambs born per ewe joined as *LB/EJ*, and so on.

In Australian Merino flocks, marking rates (LM/EJ) of 75% to 85% are commonly achieved and considered satisfactory by most producers. In areas well suited to the health and productivity of Merino sheep, such as the medium to low rainfall areas of southern Australia, marking rates are commonly 85% to 95%. In areas not so well suited, either through nutritional limitations, predation or other environmental effects, marking rates may be less than 70% and flock owners may find it difficult to maintain the size of the ewe flock without retaining ewes into old age. Rates are generally higher with strong-wool strains of Merino than with medium-wool strains, and fine-wool strains typically have even lower rates.

In meat-producing flocks using, for example, Border Leicester x Merino (BL-Mo) crossbred ewes, reproductive rates are usually significantly higher. Marking rates of 110% are common and some flocks achieve higher rates — up to 120% or 130%. There has been much interest in recent years in importing and developing breeds or crossbreeds with high fecundity and there have been experimental studies aimed at identifying more prolific maternal breeds suited for Australian conditions, capable of raising a higher weight of lambs to weaning. The East Friesian-Merino crossbred ewe, for example, outperformed the BL-Mo and other crossbreeds in one study conducted at multiple sites in Australia.[1]

From a mathematical point of view, rather than a physiological one, marking rates are determined by fertility, fecundity and the survival rate of lambs from birth to marking. By definition, *fertility* is the proportion of ewes present at the time of lambing which lamb; *fecundity* is the average number of young born per lambing ewe. Fertility is sometimes reported as the proportion of ewes detected pregnant in mid-pregnancy by ultrasound. This measure of fertility is generally valid but ignores foetal losses beyond mid-pregnancy.

Fertility, fecundity and survival rate of lambs to marking age

The fertility of mature ewes is usually high. Often over 90% of ewes conceive in a five-week joining period, provided no adverse factors, such as oestrogenic pastures, are operating. It does vary, however, with breed, date of joining, duration of mating and other management factors. Fecundity varies markedly with season, nutrition, age, breed and bloodline. Lamb survival rates also vary markedly. For Merinos, 80% to 90% of single and 60% to 80% of twin-born lambs should survive, if environmental conditions are good. Litter size greater than two is unusual in commercial Merino flocks and the survival rate of triplet or higher-order lambs is very low under commercial paddock-lambing conditions.

The association between fecundity and ovulation rate is self-evident; the more ova that are produced at any one ovulation the more that can be fertilised and that can implant. Fertility is also influenced by ovulation rate in that, if two ova are produced at an ovulation, the chance of at least one being fertilised and implanting is higher than if only one ovum is produced. (Fertilisation of ova following multiple ovulations tends to be an 'all-or-none' event, but the survival of each of multiple embryos in a ewe appears to be independent of the survival of the other or others.[2])

As an illustration of how fertility, fecundity and survival contribute to marking rate, Table 7.1 shows two possible ways by which an 85% marking rate can be achieved. Of the two examples, the second involves higher lamb losses and is a less efficient process, both in biological and economic terms.

The fertility of a flock can be estimated by pregnancy testing and by observations at lambing time. When fertility is low, however, it is instructive to further consider the factors which contribute to fertility so that appropriate corrective action can be taken. *Infertility* is defined as a failure to

Table 7.1: Two different combinations of fertility, fecundity and lamb survival rate — an example to illustrate some of the components of reproductive rates, as measured at lamb marking.

Fertility Ewes pregnant (per 100 mated)	Fecundity Ewes with twins (per 100 pregnant ewes)	Survival of single, twins	Calculation	Marking rate
88	10	90%, 80%	$2 \times 88 \times .1 \times .8 +$ $88 \times .9 \times .9 =$	85%
92	20	85%, 60%	$2 \times 92 \times .2 \times .6 +$ $92 \times .8 \times .85 =$	85%

lamb. Because abortion generally occurs at a low frequency in Australian sheep flocks, infertility usually results from failure to mate, failure of fertilisation in ewes which do mate, or embryo mortality. Infertility or prenatal wastage, if measured by pregnancy diagnosis in mid-pregnancy, is usually the result of deficiencies in the management of ewes and rams before and during joining.

The next section reviews these factors — the physiological, management and disease factors which influence the pregnancy rate of ewes up to the point at which pregnancy diagnosis by ultrasound is performed. The subsequent section (Section C) examines causes of abortion.

SECTION B

FACTORS AFFECTING FERTILISATION, CONCEPTION RATE AND PREGNANCY UP TO MID-GESTATION

Fertility and fecundity of ewes

To achieve high fertility, ewes must ovulate and be mated; the rams that mate them must have good semen quality; and the uterine environment must be conducive for fertilisation and embryo development. If rams and ewes are in good health and free of disease conditions which directly affect reproduction, the ovulation rate is the most labile determinant of fertility and the key determinant of fecundity. Some of these factors are physiological or nutritional and are most strongly controlled by flock husbandry decisions; some are specific disease entities which operate independently of husbandry decisions; and some involve aspects of physiology, nutrition and disease.

On sheep-breeding properties a key management decision is when to join. The decision is usually based upon consideration of

(a)   the optimal time for marketing the resulting progeny

(b)   the reliability, quantity and quality of pasture available at different times of the year to support late pregnant and lactating ewes and to permit good lamb growth.

(The interrelationships are discussed, with an example, in Chapter 3.) The decision should also take account of the underlying physiology of reproduction in the ewe and ram and the normal variation in reproductive parameters of sheep during the year. There are two major environmental factors which influence ovulation rate in the ewe: the time of year and therefore the photoperiod, and her nutritional state.

Photoperiodicity in the ewe

Ewes of most breeds have a distinct breeding season. They are *short-day breeders* and, in the breeding season, ewes cycle or return to oestrus every 16-17 days. The first experiments of using artificial light to influence the cyclic activity of ewes[3] led to the conclusion that increasing day-length was the trigger for the onset of cyclic activity, with oestrus activity beginning 13-16 weeks after the summer solstice.[a] In some breeds, however, natural sexual activity commences before the summer solstice while days are still lengthening and the strength of the association between changes in day-length and the beginning of breeding activity has been questioned[4] and the importance of day-length per se has been demonstrated.[5]

---

a   On 21 or 22 December in the southern hemisphere, 20 or 21 June in the northern hemisphere.

Some workers have also suggested that natural triggers work to end sexual activity rather than to begin it, and that, without those triggers, ewes may continue to cycle for extended periods. Despite uncertainty about the control mechanism, it is clear that the peak of breeding activity in all breeds under natural conditions is in autumn when day-length is both short and decreasing.

Despite the overall similarity in the pattern, there are minor breed differences in the timing of the natural breeding period. Merino ewes, when run continuously with vasectomised rams, display a seasonal pattern of sexual activity which, in Australia, begins in November-December and is at its peak in February and March, declining in April or May. Merino ewes are in anoestrus, either completely or nearly so, from July to October[6] (Figure 7.1). The timing of onset of oestrus activity in spring and the depth of anoestrus are influenced by the type of Merino — the strong-wool (South Australian) type tending to be less strongly seasonal than the medium-wool (Peppin) type.

Most British breeds have a sharply defined breeding season, from late February to the end of June.[7] The Dorset Horn and Poll Dorset are notable exceptions, with breeding patterns very similar to Merinos. The Dorper breed — developed in the mid-20th century in Africa from the Dorset Horn and Blackhead Persian and introduced into Australia in the 1990s — is similarly able to breed for extended periods of the year. Border Leicester sheep are strongly seasonal, but Border Leicester x Merino crossbred ewes are intermediate between the two parent breeds, enabling prime lamb producers to join first-cross ewes in early summer, if desired. In the northern hemisphere, the breeding patterns follow the same seasonal pattern as in Australia, with peak breeding activity in the northern autumn (September to November).

Merino ewes differ in another way from the highly seasonal British breeds in that they are more responsive when exposed to rams out of season. The sudden introduction of rams outside the breeding season will cause the majority of Merino ewes to cycle, particularly in the transition period between deep anoestrus and the normal time at which cyclic activity commences spontaneously. British breeds, other than the Dorset breeds, only respond in the period shortly before the natural breeding season commences — for example, after 10 February.[8] Both Merinos and Border Leicester x Merino crossbred ewes will respond and start cycling when exposed to rams as early as November.[9]

For this response to occur reliably it is generally necessary that the ewes are completely isolated from rams for a period of at least 17 days before introduction, with 30 days of isolation being widely recommended.[10] This separation is required because ewes which are run continuously with rams adapt to their presence and show a similar seasonal pattern of anoestrus to ewes which are continuously isolated from rams. While complete separation of rams and ewes when mating is not underway is the usual practice on farms, it may be worthwhile noting that the *ram effect* can still be elicited in ewes which have habituated to rams running with them by the introduction of different rams — rams which are unfamiliar to the ewes.[11,12] The manipulation and application of the *ram effect* is discussed further in Chapter 8.

Ovulation rate (OR) and, consequently, conception rate decline during the breeding season. For example, Border Leicester x Merino crossbred ewes often exhibit a mean OR of about 1.7 in late February-March but only about 1.3 in May-June.

Table 7.2 presents results of spring and autumn joinings of Merino ewes at Trangie, western New South Wales, averaged over several years. The actual dates of joining varied

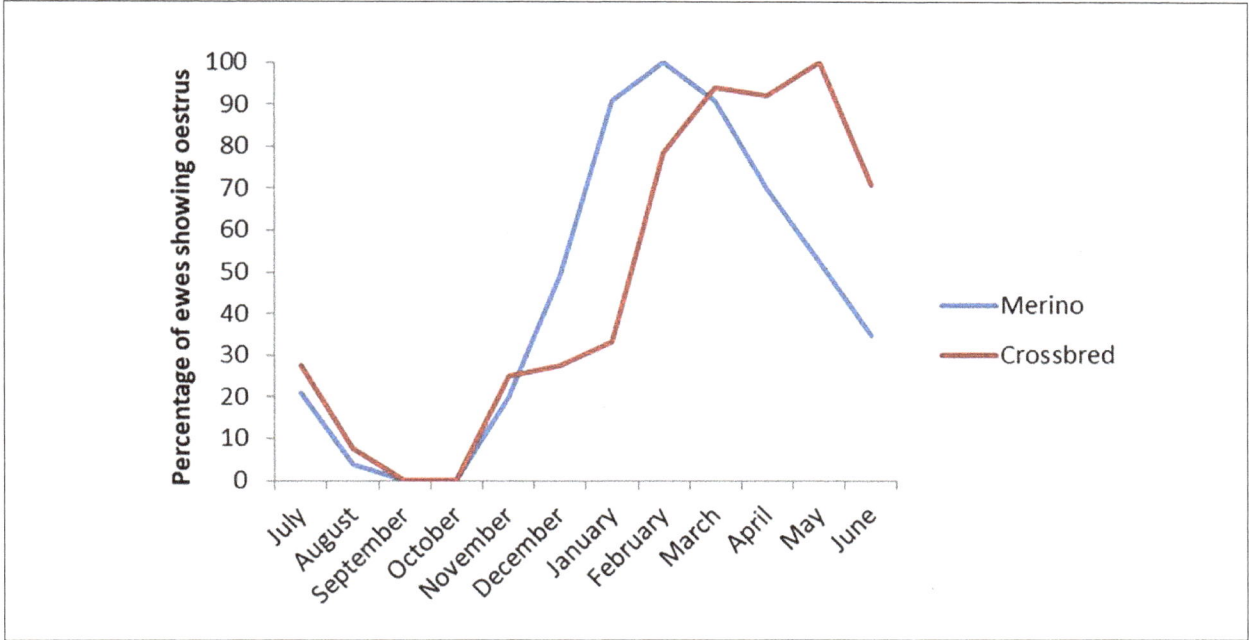

Figure 7.1: The incidence in Australia of oestrus in Merino and Border Leicester x Merino ewes run continuously with vasectomised rams (data averaged over two or three years. Drawn by KA Abbott. Based on data from Underwood, Shier and Davenport (1944).[6]

Table 7.2: Effect of time of mating on reproductive performance of Merino ewes at Trangie, NSW (averages of 4-6 seasons).[13]

Time of joining	Unmated ewes %	Wet ewes %	Twins %	Lambs mothered %	Lambs weaned (of ewes joined) %
Spring	18	72	12	76	67
Autumn	5	88	38	113	99
Advantage to autumn mating	+13	+16	+26	+37	+32

a little between years but spring joining included the period from late September to early November and autumn joining included the months of March and April. The ORs in these ewes were about 1.2 and 1.5 in spring and autumn, respectively. The table shows clearly the reproductive superiority of in-season joinings for Merinos. Nevertheless, for other economically more important reasons often related to pasture availability or the timing of regional sheep-sale events, Merinos are often joined out of season. A summer joining in December-January is expected to produce a result intermediate between spring and autumn joinings.

The proportion of ewes which lamb after spring and summer joinings may be increased by extending the period of joining by two to four weeks and can also be increased by utilising certain controlled breeding techniques (Chapter 8). Generally, flock owners do not attempt to join Merino ewes early in the non-breeding season since lambing in December-February is not desirable because of the effect of summer heat and the usual absence of adequate pasture. In any case, most ewes are in deep anoestrus in July-September and exhibit unsatisfactory reproductive performance when joined at that time.

Effects of body weight and nutrition on ewe fertility and ovulation rate

In mature ewes there is a strong relationship between body weight and ovulation rate. Over a wide range of body weights, ovulation rate can be expected to increase by 2-3% for every additional kilogram of body weight.[14,15] The strength of the relationship is weaker out of season (October to December) and becomes more apparent when joining occurs from January onwards.[16] The effect of season of joining on the response to changes in body weight is greater in Border Leicester x Merino ewes than in pure Merino ewes.[14]

The increase in ovulation rate arises from an increasing proportion of ovulations which are multiple. For example, based on one experiment with Corriedale ewes[15], it was predicted that 10% of ewes at 35 kg would have multiple ovulations but, at 60 kg, over 75% of ewes would produce multiple ova at ovulation. In another[17], the increase in multiple ovulations over a similar body weight range in Merinos of the South Australian strain was from 10% of ewes at 35 kg to 60% at 60 kg.

Not all ova that are produced in an ovulation are fertilised or lead to an implanted embryo. Consequently, the relationship between body weight and lambing rate is usually between 1.5% and 2% for every kilogram of body weight. By this measure, increasing the average body weight of the ewes in a flock by 10 kg should lead to the birth of 15-20 more lambs for every 100 ewes joined. The higher ovulation rate lifts fertility as well as fecundity — the extra lambs are the result of fewer barren ewes as well as more multiple births.[16]

The comparison between heavy ewes and lighter ewes is only valid for ewes of the same age and same breed being joined at the same time of the year. Knowledge of the relationship could be used, for example, by a producer who is considering the provision of additional feed to a flock of ewes pre-joining and is attempting to predict the net economic benefit of doing so.

This relationship is not to be interpreted as an indication that sheep of larger frame size have higher ovulation rates than sheep of small stature. The ovulation response discussed here is one related to the body reserves of energy of the ewe — as evident through her condition score.

Ovulation rates in 18-month-old Merino ewes are lower than in older ewes, but the relationship between body weight and ovulation rate still exists in young ewes, although at a lower level.[16] Because the onset of puberty is related to both age and weight, and because heavier maidens have higher ovulation rates than lighter ones, body weight is the principal factor influencing their fertility.

For medium-wool Merinos a target body weight for joining of 45-50 kg is often recommended. Because of the differences in frame size between strains of Merino, it is more useful to specify target condition scores (such as score 3 on a scale of 1 to 5) rather than to target body weights. Condition score 3 is frequently recommended as the 'ideal' for joining because the ewes will be in sufficiently good condition to have a reasonable frequency of twin births but not so high that there will be a significant number of triplet pregnancies. Furthermore, provided their ongoing nutrition is adequate, they will have sufficient reserves to support the pregnancy and subsequent lactation and to wean lambs of an adequate body weight. In Merino sheep, there is also a relationship between the level of foetal nutrition and the lifetime fleece value (determined by fleece weight and fibre diameter) of the progeny. While the economically optimum level of nutrition and reproductive rate varies from season to season — chiefly related to variations in pasture availability and the cost of supplementary feed — condition score 3 at joining is likely to be close to the optimum in most seasons. Several extension programmes (such as *Lifetime Wool* and *Lifetime Ewe Management*) sponsored by industry funds have been delivered to sheep producers across southern Australia to encourage the adoption of target condition scores for ewes at varying stages of the reproductive cycle as a guide to the optimum nutritional management of the flock.[18,19]

Flushing ewes

Flushing is the practice of increasing the feed available to ewes prior to and during joining, such that they gain weight. It is intended that the ewes will have an increased lambing rate in response to the *change* in liveweight, independent of any effect due to higher liveweight alone. The response to an increase in liveweight is known as the *dynamic* effect and is in contrast to the *static* effect, described above, which is a result of being at a higher liveweight when ovulating.

Experiments aimed at exploring the extent of the ovulatory response to weight change have long been frustrated by the fact that flushing has the obvious effect of ensuring that ewes are actually heavier at the time of ovulation than they were before the increased feeding commenced, so any increase in ovulation rate could be due to either the higher weight achieved (the static effect), the change in weight (the dynamic effect) or a combination of both.

Attempts to separate the two responses have led to a suggestion that flushing for six weeks, such that ewes gain about 1 kg per week commencing three weeks before joining, will lead to an increase in lambing rate of 6-7% independent of, and in addition to, a response of 1.5-2% for every kilogram gained. This predicted level of response is based on results from a New Zealand study with mixed-age Border Leicester x Corriedale ewes, joined from mid-March.[20]

Other studies have failed to detect a consistent response in lambing rate to the dynamic effect of weight change following flushing.[15] While the response to the static effect of higher body weights is predictable, producers should be cautioned against investing resources in attempting to manipulate the timing of any liveweight change of ewes to ensure a period of rapid weight gain at the time of joining. The circumstances under which a response to this strategy occurs are not clear, and the results appear to be inconsistent and unpredictable.[14]

A specific example of increased ovulation rate in response to dietary supplementation occurs when lupin grain is provided to ewes in the period preceding and during ovulation. The feeding of lupins at the rate of 230-450 g per head per day, commencing as few as six days before joining, can result in significant increases in ovulation rate (8-25%) and lambing rate (5-23%).[21,22] The response to lupins occurs even if there is no change in body weight, and it is greater than that which occurs in response to any change in body weight brought about by non-lupin supplementation[22], indicating that lupins have an effect which is different from that described as dynamic or static effects of body weight.

Lupin grain has characteristics which make it highly suitable as a feed source for ruminants. It is relatively high in metabolisable energy (>12 MJ ME/kg DM) and the energy is derived from fermentable non-starch polysaccharide, in contrast to the high starch content of cereal grains and other legume grains such as peas and faba beans.[23] The carbohydrates of lupins are fermented relatively slowly, greatly reducing the risk of lactic acidosis, even if the grain is introduced to the diet at high levels without an introductory period. The fermentation pattern of carbohydrates in lupins is less likely to disturb fibre digestion in the rumen than that of cereal grains. Consequently, supplementation of a low-quality, high-fibre diet with lupins can be very effective because the intake of the base diet is relatively unaffected by the addition of lupins.

Lupin grain is also a rich source of protein, and earlier experiments with lupin supplementation led to the suspicion that their high protein content was responsible for the ovulatory response. It is now recognised that the response in ovulation rate to lupin supplementation is a result of the ability of lupins to markedly increase the availability of glucose for intracellular metabolism[24,25] even when suddenly introduced to the diet in high quantities. Experimentally, infusion of glucose directly into the bloodstream will stimulate an increase in ovulation rate similar to that produced by lupin supplementation[25], and the effect of lupins on the number of follicles recruited in the luteal phase of the oestrous cycle is a consequence of increased blood glucose levels and, possibly, the accompanying increase in insulin concentrations.[26]

In areas where the grain is grown, flushing Merino ewes with lupins has become popular. The ewes typically receive lupin grains at the rate of 500 g per day. Increases in OR occur within seven days of commencing feeding. It is usually recommended that ewe flocks be fed the grain for one week prior to and during the first three weeks of joining. In flocks which respond well, the OR increases by 0.3-0.4 and the percentage of ewes lambing by 10-15 percentage points. Methods for feeding out lupins and the possible occurrence of lupinosis are discussed in Chapters 6 and 17.

Despite its frequent use, responses to lupin supplementation are not consistent and may be more likely to occur in situations where ewes are receiving diets which are inadequate in nutrition during joining. Lupin supplementation has been found on one occasion to increase ovulation rate but also to increase embryonic loss.[27]

Table 7.3 shows some effects of flushing on body weight and twinning rate in spring- and autumn-joined Merinos. Note that flushing had a large impact on OR in autumn-joined ewes. As already noted, OR is higher in autumn and more responsive to nutritional manipulation. Many producers with Merino flocks do not want large increases in the twinning rate and the higher rate of lamb mortality that accompanies multiple births in Merinos. Instead, they seek to utilise some degree of flushing to increase the proportion of ewes which lamb (fertility).

Table 7.3: Effect of flushing on lambing in Merinos.

Time of joining	Treatment	No. of ewes lambing	Increase in body weight (kg) **	Ewes with twins %
Spring	Flushed*	295	3.5	5
	Nil	297	3.0	7
Autumn	Flushed*	324	3.0	15
	Nil	324	1.0	6

* three weeks prior to joining and first three weeks of joining

** increase during period of flushing

Giving additional feed over several weeks to ewes which are in condition score 3 or less can be expected to raise both body weight and OR.

Effect of ewe age on fertility and ovulation rate

Both the ovulation rate of ewes and their maternal ability increase with age. Ovulation rate peaks at 5 to 7 years of age. Mature ewes out-compete maiden ewes for the attention of rams at joining. Compared to mature (2½ years or more) ewes, maiden ewes have a shorter oestrus and less overt oestrous behaviour, are less attractive to rams and are mated on fewer occasions during each oestrus. These effects all contribute to the lower fertility of maiden ewes compared to older ewes. (Adding to the lower fertility of maiden ewes is their poorer performance at and after lambing — their maternal inexperience, rather than their age, contributes to a generally poorer maternal ability and a lower survival rate of their lambs compared to multiparous ewes.[28])

Ovulation without oestrus

Some ovulations in the ewe are not accompanied by oestrus and, therefore, mating will not occur even if a ram is present. Ovulation without oestrus behaviour is called a *silent heat*. Such ovulations occur at the onset of puberty, at the start and possibly the end of each breeding season and, in the case of Merinos, quite commonly during spring and summer. Silent heats occur when ovulation is not immediately preceded by a period of progesterone priming in the brain. Within the breeding season this priming is reliably supplied by the corpus luteum of the previous oestrous cycle. This *luteal progesterone* has other important roles, permitting normal functioning of the new corpus luteum and uterine endometrium after the next ovulation. Whereas ovulation without oestrus is quite common, oestrus without ovulation appears to be very unusual.

Failure of fertilisation due to maternal factors

Fertilisation may fail due to faults in the maternal reproductive tract environment. Pasture oestrogens, for example, interfere with sperm transport through the cervix, as discussed further below. Various controlled breeding techniques (for example, synchronisation of oestrus with progestagen sponges and superovulation of donor ewes for MOET[b]) also alter cervical function and impair sperm transport. Physical abnormalities of the female genital tract may also influence the likelihood of fertilisation — Quinlivan et al. (1966)[29] found some abnormality in 6% of parous 2½-year-old ewes and in 15% of nonparous ewes of the same age group.

In naturally cyclic ewes which ovulate more than one ovum, fertilisation is nearly always 'all or none'. Hence, if a ewe ovulates two ova and is mated, the outcome is usually two zygotes or no zygotes.[2] Little is known about possible causes of defective oocytes in ewes under natural conditions, and their occurrence is probably rare.

Management of joining

Within flocks, ewes are sometimes joined in mobs of a single age group or, more commonly, multiparous ewes are joined in mixed-age mobs and the maiden ewes are joined as a separate

---

b   MOET (multiple ovulation and embryo transfer) programmes are described in Chapter 8.

mob. Maidens tend to be poor competitors with older, more experienced ewes that exhibit a longer and stronger oestrus, and maiden ewes are less active at seeking rams when in oestrus than older ewes. Experience of this in the field varies and some producers obtain satisfactory or good fertility in the maidens when all ewes are joined as a single mob. The risk of reduced fertility of maiden ewes, joined as a single-age group or in a mixed-age flock, can be avoided to some extent with higher ram-joining percentages.[30]

To ensure reasonable fertility, maiden Merinos should have reached a body weight approximately 80% of their adult weight by the time they are first joined. For medium-wool types of Merino ewe, this represents a body weight of 40-45 kg.

Some husbandry procedures can adversely affect reproduction. Ewes may be crutched just prior to joining without any negative effect. Shearing of ewes reduces the attractiveness of ewes to rams[31], although the effect gradually declines over the 10-week post-shearing period. The lesser sexual stimulation of rams from shorn ewes may reduce the number of times an oestrous ewe is mated — with consequent negative effects on fertility.

It is preferable to organise the farm management calendar so that few, if any, procedures need to be carried out on the ewes or rams during the joining period. Repeated mustering, yarding and handling interfere with ram-ewe contact and may reduce flock mating activity and fertility. Some chemicals employed to control parasites and other diseases, if applied just before or during joining, may interfere with fertilisation or embryo development.

Phyto-oestrogenic infertility

Phyto-oestrogenic infertility in ewes occurs as a result of the oestrogenic activity of isoflavone metabolites which are found mainly in the laminae of green legumes, particularly some cultivars of subterranean clover (sub-clover) (*Trifolium subterraneum*) and red clover (*T pratense*). The sub-clover cultivars with the highest isoflavone levels are *Yarloop*, *Dwalganup* and *Dinninup*. Medium levels are found in *Geraldton* and *Tallarook*.

There are two different clinical manifestations of phyto-oestrogenic infertility in ewes. *Temporary infertility* occurs when ewes graze green oestrogenic clover for a short time, including the period of mating. Fertility recovers to normal levels within a few weeks of removal from the pasture. *Permanent infertility* ('clover disease') occurs when the ewes graze such pastures for extended periods, leading to irreversible changes in the reproductive tract. The abnormalities induced by the chronic ingestion of phyto-oestrogens are different from those of temporary infertility and are not an extension of the responses occurring during episodes of temporary infertility.[32]

Over the past few decades, the incidence of the most severe forms of this disease has declined markedly but subclinical permanent infertility probably remains very common. Chronic exposure to phyto-oestrogens also leads to urinary tract problems in wethers, but rams do not appear to be affected.

Pathophysiology of permanent infertility

Two of the ingested isoflavones, formononetin and daidzein, are converted in the rumen to equol. Equol is absorbed and most of it is conjugated with glucuronic acid in the plasma. The small proportion of unconjugated equol is strongly oestrogenic and is the major active oestrogen responsible for clover infertility.[33] The principal lesions causing lower reproductive rates occur in the cervix. These lesions, which present no clinical signs, are the most important abnormalities occurring in subclinical permanent infertility.

The superficial epithelium of the cervix of ewes exposed chronically to oestrogenic clovers contains a lower proportion of stratified squamous and mucus cells and a greater proportion of single-layered columnar cells. The changes in the cervix are, in effect, a trans-differentiation of the endocervix so that it resembles endometrium.[34] As a consequence of both structural changes and a loss of ability to respond to endogenous oestrogen, the cervical mucus in affected ewes is very watery; it has a decreased *spinnbarkheit* (visco-elasticity). Without this normal mucus structure, sperm transport through the cervix is greatly impaired[35] and the number of sperm reaching the oviducts is greatly reduced.

In addition to the cervical changes, chronically affected ewes develop endometrial cysts as a result of cystic hyperplasia of the endometrium. This may be complicated by a bacterial endometritis. The endometrial pathology appears to have less effect on conception rates than does the interference with sperm transport.

Dystocia

Clover-affected ewes often suffer from dystocia. In clover-affected flocks the incidence can be high and can result in the deaths of 30% to 40% of lambs and 15% to 20% of ewes unless intense observations are made and assistance rendered quickly to ewes with difficulties. The dystocia is apparently due to a failure of the vulva to dilate. The lamb is presented into the vagina but the ewe is unable to expel the lamb through the vulva. Strong traction, with or without episiotomy, is required to deliver the lamb.[36] Septic metritis is a common sequel to dystocia. Without assistance, the lamb is not expelled and secondary uterine inertia develops.

Lactation in maiden ewes and wethers

Maiden ewes and wethers may lactate as a result of prolonged exposure to oestrogenic pasture. The mammary secretion varies from a yellow viscous fluid to apparently normal milk. The degree of teat elongation in wethers has been used as an index of pasture oestrogenicity but may be too sensitive to distinguish differing levels of oestrogenic intake.[37]

Other effects on wethers

Uroliths containing crystals of equol metabolites can obstruct the urethra, particularly at the urethral process. The squamous metaplasia of the prostate and bulbourethral glands which also occurs in wethers grazing oestrogenic pastures may contribute to the formation of crystals and may encourage obstruction in the pelvic urethra.[c]

Effects on fertility

While the lesions of 'permanent infertility' are indeed permanent, the infertility is not complete. Affected ewes can still produce lambs. The effect of the phyto-oestrogens is to

---

c   See Chapter 18 for a more detailed discussion of this disease in wethers.

reduce the probability of each ewe conceiving under normal mating conditions and hence to reduce the flock's reproductive capacity. Although the classical 'clover disease' is now rarely seen, on properties where it does occur ewe survival rates are reduced by conditions such as hydrops uteri, dystocia and uterine prolapse. With both the severe form and the subclinical form, phyto-oestrogens have their predominant effect on the number of ewes lambing. As the effects accumulate, ewe conception rates progressively decline with age.

Diagnosis

Phyto-oestrogenic infertility must be differentiated from other common causes of low marking percentages, particularly low body weight of ewes at joining, infertility of rams, particularly that caused by ovine brucellosis, and high perinatal lamb mortalities from a variety of causes. Pregnancy diagnosis of ewes and examination of the rams will assist the diagnosis.

Classical clover disease should be suspected when there are clinical signs in ewes and wethers, a high incidence of dystocia and a history of severely depressed marking rates. Subclinical permanent infertility is more difficult to diagnose because clinical signs are absent and the history is often unremarkable — marking rates may be lowered by only 10% to 20%. Definitive diagnosis will be made on the cervical histopathology of a sample (minimum of 12) of older ewes from the flock and the agronomic identification of a significant content of oestrogenic cultivars of *T subterraneum* in the pastures.

Control

The long-term solution is pasture renovation in order to replace the oestrogenic cultivars. This is neither easy nor cheap; the cost of the problem must be weighed up against the potential loss of often very productive pastures.[38]

Short-term control includes the assessment of the risk associated with each pasture on the farm. The young ewes (weaners and hoggets) are grazed on the least oestrogenic pastures during the times of the year when the pastures are green. Dry ewes (never to be mated), wethers and old ewes (with short breeding futures) are grazed on the high-risk pastures.

Extension of the joining season by leaving the ram in with the ewes for long periods may increase the lambing percentage and is practised on some affected properties to allow the production of sufficient ewe lambs for future breeding ewe replacements.

Temporary infertility

Ewes grazed on green oestrogenic pastures during joining may show reduced conception rates due to temporary clover infertility. The effects of the phyto-oestrogens include a reduction in the incidence of oestrus, interference with ovum transport and a reduction in sperm transport.[39] Some experiments have also shown a negative effect on ovulation rate.[40] Some of the effects of grazing plant oestrogens may persist for three weeks after removal from the pasture but appear not to persist for longer than four weeks.

Coumestans in *Medicago* spp

Coumestans in lucerne (*Medicago sativa*) and some other medic species are also oestrogenic but their effects are always temporary. They reduce ovulation rate during the period that ewes are grazing them but the effects disappear upon removal. Lucerne often develops high coumestan levels late in its growing season[41] — typically March onwards.

Embryo mortality

In the absence of phyto-oestrogens or other substances interfering with sperm transport, the vast majority (85% to 95%) of ovulated ova are fertilised when a ewe is mated by a fertile ram.[42] Losses of embryos in the subsequent 30 days, however, are relatively high and embryo mortality is the major cause of loss of lambs between fertilisation and birth. Infectious abortion can cause high losses, but outbreaks are sporadic and infrequent under the extensive grazing conditions common in Australia. By contrast, embryo mortality occurs to a significant level in most or all flocks, in most or all years.

Estimates of embryo mortality commonly range between 20% and 30%[43] and the majority of these losses occur before day 18 post-fertilisation, when implantation is expected to occur[29,42], and most of these losses occur between day 4 and day 14.[44]

The embryo normally passes through the fallopian tube to the uterus on day 4 or day 5 following fertilisation. The embryo's persistence and survival require the continuing production of progesterone by the corpus luteum. In a ewe with no embryos, PGF2α synthesis in the uterus leads to luteolysis and the subsequent resumption of follicular development and a return to oestrus. In the ewe with one or more healthy embryos, *interferon tau* is secreted by the embryonic trophoblast before day 13 and acts as a signalling mechanism preventing the synthesis of PGF2α. This process is the basis of the maternal recognition of pregnancy.[45,46]

Ewes in which all embryos die before day 13 return to oestrus between days 16-18 (during the normal breeding season) and are therefore not distinguishable in the field from ewes in which fertilisation failed. Ewes which lose all embryos after day 13 have a delayed return to oestrus.[47] Ewes that lose one or more embryos but retain one or more embryos can be expected to have a normal pregnancy. The significance of embryo mortality, from the viewpoint of flock productivity, is that the survival of each embryo within a multiple-ovulating ewe is independent of survival of the other embryo(s), so that most embryonic mortalities in multiple-ovulating ewes do not lead to the ewe losing all conceptuses and returning to oestrus. Considering that most multiple conceptions in commercial sheep production are twins, embryonic mortality leads to a reduction in the number of twin-bearing ewes and an increase in the number of single-bearing ewes. By contrast, when there is a failure of fertilisation, this usually means that no ova are fertilised and the ewe will return to oestrus and, provided the rams are still present, have another chance to conceive. In the case of ewes with a single ovulation, there is no difference in the outcome between fertilisation failure and embryonic loss before day 13. In both cases she will return to oestrus. The greatest impact of embryonic mortality is in reducing the number of lambs born to multiple-ovulating ewes. The next most significant effect is to increase the number of ewes which lamb late in the lambing period.

The chance of an embryo dying is higher following multiple ovulations than in single ovulations. Estimates vary (summarised and analysed by Geisler et al. (1977)[48]) but survival rates of 95% for embryos from a single ovulation, 85% for twin ovulations and 70% for triplet ovulations have been proposed. These estimates seem high, given the data of Quinlivan et al. (1966)[29] (Table 7.4), and lower survival rates (75% to 90% for twin ovulations) have been found in an Australian study with Merino sheep.[49] Three factors — the relatively high rate

Table 7.4: The reproductive performance of 2.5-year-old Romney ewes conceiving to service at one oestrus, roughly half of which were parous in the previous year and half were nonparous, indicates that embryonic mortalities account for around 20% of fertilised ova. Adapted from Quinlivan et al. (1966).[29]

	Parous ewes		Nonparous ewes	
	Number	**Percentage**	**Number**	**Percentage**
Ova produced per 100 ewes	136		132	
Fertilisation rate	129	94%	116	85%
Survival to 18 days post-fertilisation	104	76%	97	71%
Survival to 30 days	97	71%	92	68%
Survival to 140 days	95	70%	86	64%

of embryonic mortality, its randomness (usually affecting only one of twins) and its higher frequency in multiple pregnancies than in single pregnancies — combine to make embryonic mortality a significant obstacle to greater reproductive efficiency in sheep.

Embryo loss in some flocks or flocks in some geographical regions may be associated with delayed returns to oestrus (or apparent failure to return to oestrus, depending on the length of the joining period). This is seen most clearly in the few cases of selenium deficiency which have been associated with ewe infertility in Australia (see Chapter 11).

Aetiology

The reasons for most embryo losses in commercial flocks are unknown. Some losses are attributable to intrinsic faults within the embryo and hence their loss is desirable (rather than the subsequent abortion or birth of abnormal foetuses). While some embryo loss results from the fertilisation of ova by heat-damaged spermatozoa, *in vitro* culture and *in vivo* embryo transfer studies suggest that most losses are due to failures in the uterine environment. Faulty nutrition, abnormal temperatures, endocrine imbalances and asynchronous development of embryo and endometrium have all been suggested as causes. Deficiencies of zinc, iodine and selenium have occasionally been implicated (see Chapter 11).

Undernutrition during the period immediately following mating has been occasionally associated with increased embryonic mortality but the level of undernutrition must be severe and the duration prolonged (at least three weeks) before the effect is significant. The severity of the undernutrition required to cause embryonic mortality is much greater than would occur with normal standards of sheep husbandry.[43] The effect may be greater in young ewes than in mature ewes and in ewes with multiple embryos rather than singles.[42]

Sustained high environmental temperatures, particularly around the time of mating, may increase the incidence of both fertilisation failure and embryonic mortality. In a Western Australian study with Merino ewes, a negative association was found between the number of lambs born and the number of days over 32.5 °C during the mating period and the mean daily maximum in the three weeks after mating, an effect considered to be due to increased embryonic mortality.[50] Experimentally, the effect of increased temperatures has been inconsistent but one factor which appears to reduce the impact of high daytime temperatures is diurnal variation in temperature — mimicking the effect of cool nights between hot days.[42,43]

The relative importance of embryo mortality

In individual flocks experiencing an infertility problem, the relative importance of failure to mate, failure of fertilisation and embryo mortality varies. Normally, with return rates of 20-30%, fewer than a third of returns are due to failure of fertilisation. The presence of pasture oestrogens or some controlled breeding procedures will increase the risk of fertilisation failure but, provided mating with fertile rams occurs, embryo loss is usually more important than failure of fertilisation. Failure to mate is the least common of these three causes of failures to conceive and generally only happens with out-of-season joinings (when ewes fail to cycle) or the joining of ewes to a single, incapacitated ram.

Fertility of rams

Photoperiodicity in the ram

Rams of all breeds show a seasonal variation in testicular size that is influenced by photoperiod and mediated through seasonal changes in gonadotrophin secretion, but the magnitude of the variation within a year varies markedly between breeds. Breeds developed in high latitudes (more northerly breeds in the northern hemisphere) have a short and sharply defined breeding season in ewes and a marked variation in testicular size, testosterone levels, libido and sexual activity in rams. These breeds, such as the Herdwick, Wiltshire and Scottish Blackface behave similarly to wild sheep (the Mouflon) and feral sheep (the Soay), with a peak in testicular size in September-October in the northern hemisphere. Breeds developed in more southerly latitudes of the northern hemisphere, such as the Merino, have a smaller variation in testicular size and peak about one month earlier than the northern breeds.[51] Portland sheep (now a rare breed), Dorset Horn sheep (derived from the Portland breed) and the Poll Dorset[52] (derived from the Dorset Horn) show characteristics similar to the Merino in this regard. The trait was first recorded in Portland sheep in the 18th century[53], possibly reflecting the introduction of sheep of Mediterranean origin in their very early development in southern Britain, although the evidence for that is not strong.

In Australia, the strong seasonal effect on testicular size has been observed in Border Leicester, Suffolk and Romney rams. When the effects of seasonal variation in nutrition are removed, Suffolk rams show an increase in scrotal circumference starting in late spring (November), peaking in early autumn (March) and then declining through late autumn, winter and early spring. The magnitude of the variation in adult rams is of the order of 31 cm (minimum) to 37 cm (maximum). By contrast, scrotal circumference in Merino rams peaks about one month earlier than in Suffolk rams, starts to increase two to three months earlier and shows a smaller seasonal variation — from 34 cm to 37 cm.[54] Romney rams show a variation similar to that of Suffolk sheep, with testicular volume doubling between a low value in November to a seasonal maximum in May.[55]

While testicular volume, libido, sexual interest in ewes and sperm concentration in semen all decline to some extent in all breeds during winter and early spring, rams do remain fertile and able to mate at any time of the year. The manner in which photoperiod controls or influences the seasonal variation is unclear. The key difference between breeds developed in regions of the world differing in latitude may in fact be the degree to which photoperiod effects are overridden by other stimuli, such as changes in nutrition or, possibly, an endogenous sexual cycle which is influenced, but not controlled, by changes in day-length. Even in breeds like the Soay — which has existed in northern Britain for centuries and is genetically close to its wild sheep ancestors — the increase in gonadotrophin secretion commences before the summer solstice, while day-length is still increasing. It appears that the hypothalamic response to increasing day-length during late winter and spring is to cause a delay in the onset of the breeding season, rather than to prevent it altogether, and eventually the pineal gland becomes refractory to the increasing day-length and responds, instead, to the improved nutritional conditions of spring. Rams of breeds with less circumscribed breeding seasons respond differently, in that the delaying effect of the increasing day-length is weaker and more easily overridden by other stimuli, such as nutrition. The difference between breeds is likely to be related to the way the hypothalamus interprets and responds to the melatonin signal produced by the changing day-length.[51,56]

Nutrition also influences testicular mass, so the two factors (season and nutrition) combine to affect testicular size. In breeds like the Suffolk, where the seasonal variation in testicular size is large, nutritional effects are weak and testis mass will increase in summer even if the liveweight of the ram is declining due to poor nutrition. In the Merino, by contrast, the influence of nutrition on testis mass is stronger and will override the effect of photoperiod and season.[57] The negative consequence of this under Australian conditions is that, without supplementation, the testes of Merino rams grazing dry summer-autumn pastures will decline significantly in size — despite the effect of decreasing day-length — at exactly the time when joining occurs and large testes and active spermatogenesis are required.[58] The positive consequence is that the testes of Merino rams will increase markedly in size if the rams are grazed on, or supplemented with, a high-energy, high-protein diet over summer in the period leading up to joining. Both the nutritional response[59] and the photoperiod response in testicular growth are mediated by the frequency of luteinising hormone pulses.

Body weight, nutrition and fertility in the ram

Testicular mass and daily sperm production

For mature rams in good health, the larger the testes, the greater the number of sperm produced per day (Figure 7.2). A ram with 400 g of testes produces about 6000×10^6 sperm per day and every additional 100 g of testicular tissue adds around 2000×10^6 sperm to the daily production.[60]

Notwithstanding the photoperiod effects and breed differences described above, it is in general true that testicular mass responds to changes in nutrition. Increases in body condition score through increased level of feeding result in increases in the size of the testes. The response in sperm production is not, however, immediately measurable in the ejaculates. There is a relatively invariable 49-day period between the initiation of spermatogenesis in the testicular tubules and the appearance of the resultant mature spermatozoa in the distal epididymis. During

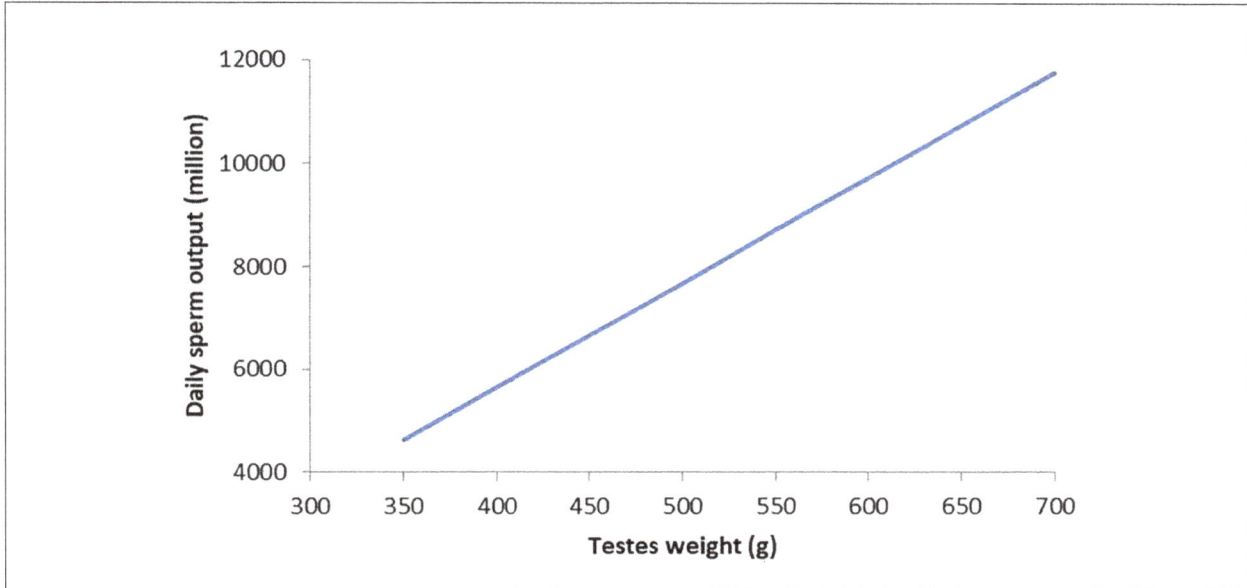

Figure 7.2: The relationship between testicular mass (g) and the daily sperm output from the urogenital tract ($\times 10^6$) is linear over the range 350 g to 700 g, with a slope of 20 and a correlation coefficient of 0.83. Drawn by KA Abbott. Based on data from Lino (1972)[60] and Foster et al. (1989).[61]

a six-week joining period, therefore, the spermatozoa that are produced in ejaculates result from spermatogenesis that commenced in the period from seven weeks pre-joining to one week pre-joining. The condition score and the testicular mass in the two-month period *before* joining commenced are critical in determining the daily sperm output of a ram during joining once epididymal reserves are expended — usually in the first few days, depending on mating load.

A condition score of 3-4 for rams at the commencement of joining is recommended. At that condition score, Merino rams are expected to have at least 400 g or 400 mL of testes and, for mature rams in that condition score, it should be significantly more. (It is assumed that 1 mL of testicular tissue has a mass of 1 g.) Excessive feeding resulting in obesity may reduce libido, whereas inadequate nutrition reduces testis size, libido and semen volume.

Estimates of testicular volume

The testicular volume of rams can be estimated by physical examination of the scrotum and scrotal contents. There is a strong relationship between the circumference of the paired testes, as measured within the scrotum, and testicular mass. Foster et al. (1989)[61] compared the testicular mass and scrotal circumference (SC) of 110 normal Merino rams and found a linear relationship (total testicular mass (g) = 31.4 $\times$ SC (cm) – 564). The measurement of scrotal circumference was highly predictive of testicular mass (r=0.92). The results of that study, and one by Lino (1972)[60] with 10 Merino rams, are summarised in Figure 7.3. From the two studies, one can predict that Merino rams with a scrotal circumference of 32 cm will have testes of around 400 to 440 g and every additional cm adds 23 to 31 g to the testes mass.

A purpose-made, metallic scrotal measuring tape is recommended for measuring the scrotal circumference. The most commonly used tape is made by Ideal Instruments, Lexington, Kentucky and is available for purchase in Australia from animal health product suppliers.

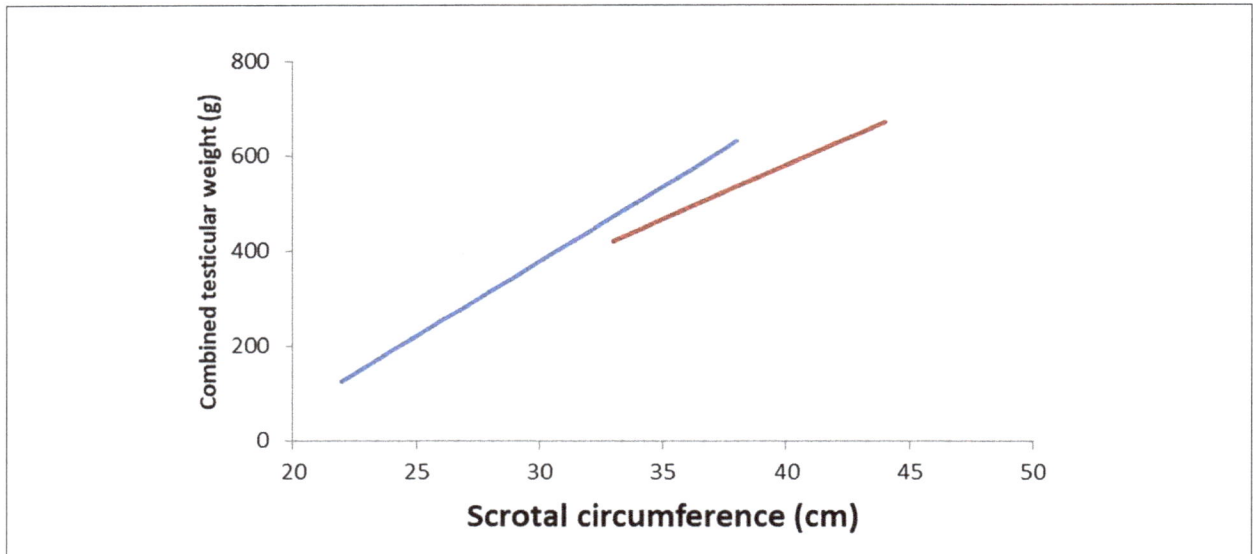

Figure 7.3: There is a strong relationship between scrotal circumference and the mass of the two testes. Drawn by KA Abbott. Blue line — based on data from a study of 110 Merino rams by Foster et al. (1989)[61]; red line — based on data from a study of 10 fine-wool Merino rams by Lino (1972).[60]

The circumference is measured by holding the two testes in the bottom of the scrotum with one hand around the neck of the scrotum. The tape is placed around the scrotum at its widest point and the circumference in centimetres is recorded. The ram can be examined while standing with the operator squatting behind, or with the ram sitting and restrained by a second person, and the operator squatting between his hind limbs (Figure 7.4).

Sperm production during natural joining

In rams that are not joined and are not ejaculating frequently — as they would do if joined to oestrous ewes — spermatozoa accumulate in the tail of the epididymis, vas deferens and ampulla, effectively creating a reservoir of sperm. This reservoir allows rams to include very large numbers of sperm in ejaculates for the first few days of joining. Rams under natural mating conditions may encounter four to six oestrous ewes per day for the first 17 days of joining and are capable of 10 to 20 ejaculates per day. The ejaculates produced on the first day of joining may each contain 2000 to 4000 $\times$ 10^6 spermatozoa. After two to three days of frequent ejaculations, sperm numbers per ejaculate fall dramatically (typically by 90%) and the daily sperm output in ejaculates remains at the lower level — effectively matching daily sperm production — until later in the joining period when the number of ewes in oestrus each day declines (Figure 7.5).[62,63,64]

Spermatozoa numbers required to ensure high conception rates from natural service

The numbers of spermatozoa required for satisfactory conception rates are best known for artificial insemination. Salamon (1962)[65] concluded that 120 to 125 $\times$ 10^6 were necessary with AI with extended fresh semen. For good fertility with natural service, the number of sperm can only be estimated, and it is widely considered that 120 $\times$ 10^6 sperm are adequate.

Figure 7.4: The scrotal circumference is measured with a scrotal measuring tape placed around the scrotum over the two testes, while the testes are held in the scrotum with one hand. Source: KA Abbott.

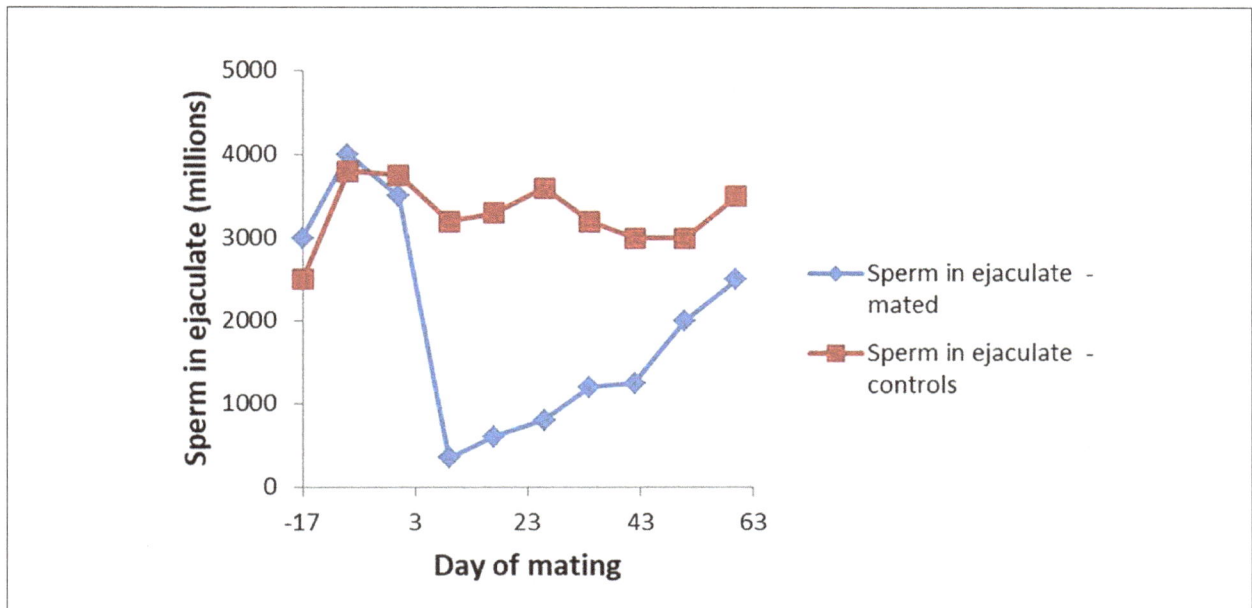

Figure 7.5: The number of sperm per ejaculate declines markedly after the first 16 to 24 ejaculates in the first two or three days of joining, then recovers later in the joining period when fewer ewes are in oestrus each day. Drawn by KA Abbott. Based on data from Raadsma and Edey (1985).[63]

Mating behaviour and flock fertility

Ewes are more likely to conceive if served more than once while in oestrus[62] (Figure 7.6) and generally ewes are served two, three or more times by one or more rams, when joined to multiple-sire groups. Under single-sire mating conditions (where a ram has no competition for access to ewes), rams can be expected to serve three, four or five ewes per day and possibly more.[64,66] They do, however, show preference for some ewes over others, so some ewes may be served repeatedly while some, apparently less attractive, ewes, are served less often or not at all.[67]

Under conditions of multiple-sire joining, rams demonstrate less preferential behaviour and, compared to single-sire mating at equivalent ewe:ram ratios, a higher proportion of the ewes are served in multiple-sire joining groups, although interaction between rams leads to a reduction in the average number of services per ewe.[66,67,68]

There is considerable variation between rams in the maximum number of times each day that they will mount and successfully serve ewes. The number is sufficiently high for most rams (10-20, for example[62,64]) to ensure that, if four or five oestrous ewes are presented in one day, each ewe can be served at least once and, usually, multiple times.

From a mathematical point of view, a healthy, fit, libidinous ram which was producing 6000×10^6 sperm per day in the weeks immediately leading up to joining will be able to sustain 12 to 18 ejaculates per day for the first three weeks of joining, each ejaculate containing over 120×10^6 sperm, and thereby inseminate six ewes, each on two or three occasions, during the hours they are in oestrus. If there is no significant synchronisation of oestrus occurring, it can be expected that, in a flock of 100 ewes, five or six will be in oestrus each day for the first 17 days of joining.

In the field, this conclusion is largely found to be reliable, even given the fact that there is wide variation between rams in terms of their libido, behaviour and fitness and, in group-mating

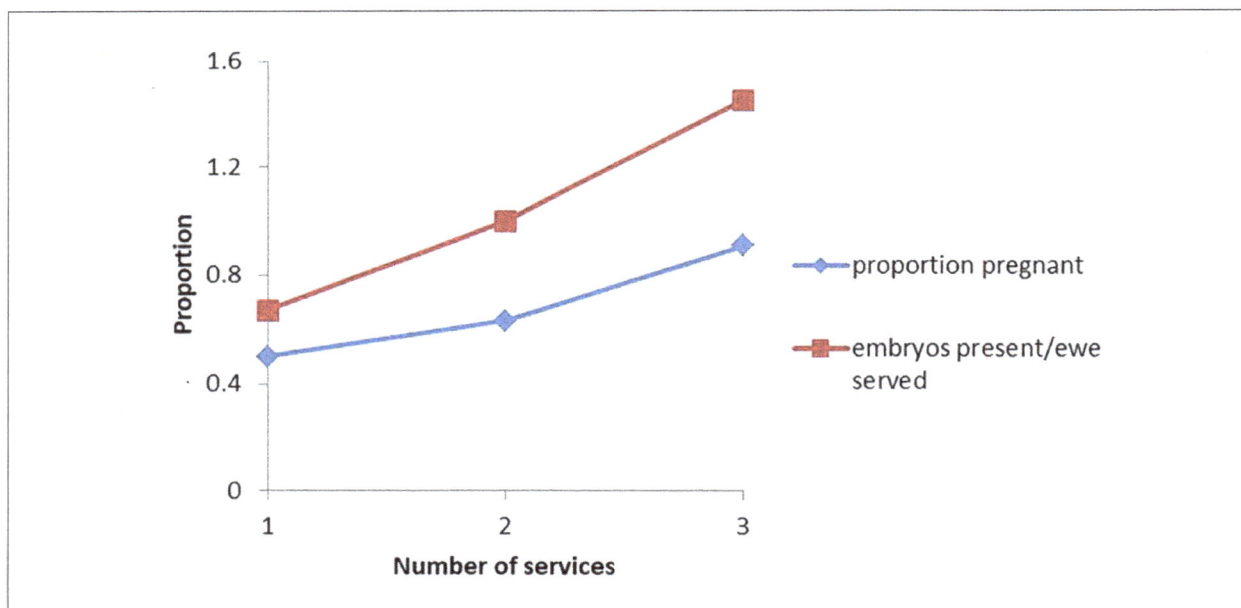

Figure 7.6: Under natural mating conditions, the proportion of ewes becoming pregnant, and the proportion of pregnant ewes with multiple embryos, are higher if the ewes are served two or three times, compared to once. Drawn by KA Abbott. Based on data from Mattner and Braden (1967).[62]

systems, in the interactions and dominance behaviour between rams. This means that, unless other factors (as discussed below) are in effect, it is possible to join rams at the rate of one per 100 ewes and expect good rates of conception.[69]

Puberty and age effects in rams

Merino ram lambs reach puberty and become capable of mounting and impregnating a ewe when aged between 125 and 200 days. Ram lambs which are well nourished and grow quickly reach puberty at ages in the lower part of the age range. Puberty in rams is associated with the appearance of spermatozoa in semen and the loss of adhesions between the penis and prepuce, permitting intromission. Sexual maturity therefore is indicated by a freely moveable penis within the prepuce and growth and firmness of the testes.[70] The development of libido — demonstrated by mounting behaviour, masturbation and the secretion of androgens — precedes the appearance of spermatozoa.[71]

Rams of other breeds tend to reach puberty at younger ages than Merinos, an effect which may be related to the natural prolificacy of the breed as indicated by the relatively high ovulation rate of females.[72]

To avoid unplanned pregnancies, ram lambs should be weaned and removed from access to their dams and other ewes before 5 months of age in the case of Merinos[73] and, to be safe, by 4 months of age in other breeds.

By 12 months of age young rams can produce ejaculates containing a high concentration of highly motile spermatozoa. Rams are usually used for the first time for natural flock matings at 1½ years of age but can be successfully joined to small flocks of ewes at younger ages, such as 7 months of age, if well grown.

The age structure of the ram flock in commercial flocks

Rams can remain in the breeding flock to extended ages provided they remain active without lameness, are capable of maintaining good body condition and, therefore, good testicular volumes, and are otherwise sound for breeding. Generally, however, their physical health deteriorates by the time they are 4 or 5 years of age, having been used for three or four breeding seasons. Producers therefore often anticipate replacement of one-quarter to one-third of the ram flock annually.

Commercial flock owners usually buy rams from specialist ram-breeding flocks or studs, which are expected to be improving genetically each year. Consequently, a 1½-year-old flock ram purchased from a ram-breeding flock is expected to be genetically superior to a flock ram purchased from the same flock four years previously. There is evidence that the value of the annual genetic improvement may be of the order of $0.50 to $1 per ewe progeny per year.[74] The difference in the genetic merit of the progeny of young rams compared to the progeny of older rams will rarely justify the decision to purchase replacement rams at younger ages on genetic grounds alone. An illustration follows in Table 7.5.

Table 7.5 shows that the maintenance of a ram team of 100 rams of four age groups requires the annual purchase of 29 rams but, if only three age groups are maintained, it is necessary to purchase 37 rams annually. The younger ram team can be expected to produce progeny of a

Table 7.5: The approximate age structure of a ram team of 100 rams in a commercial flock varies with the age at which all rams are culled. The example shown is for flocks in which rams are sold after two, three or four years in the breeding flock where each year's purchase of rams is genetically superior to those of the previous year. Using the four-age-group model as a base, the extra productive value of the ewe progeny in flocks with earlier ram-culling practices can be compared to the cost of the extra rams purchased to maintain a younger ram team. Note the following assumptions: Ram team of 100. 10% of rams die or are culled annually. The value of the productivity of the ewe progeny improves by $0.75 each year. Each ram produces 40 ewe progeny per year.

	Number of age groups of rams		
	2	3	4
Number of 1½-year-old rams	53	37	29
Number of 2½-year-old rams	47	33	26
Number of 3½-year-old rams		30	24
Number of 4½-year-old rams			21
Difference in total annual productive value of progeny	$2638	$1281	0

higher productive value (higher fleece value, for example, in the case of Merino sheep) but, if the improvement in value each year is only $0.75 per ewe progeny, the difference in the total annual productivity of the ewes sired by the three-age group and four-age group ram teams is only $1281. That difference in value will rarely justify the purchase of an additional eight flock rams, although, for commercial flocks which breed their own ram replacements, the additional return may contribute to the financial justification of the self-replacement strategy.

Calculations like this can be performed under a range of conditions, including the depreciation rate of rams (the difference between their purchase price and sale value at end of breeding life), the number of progeny likely to be born from each ram each year and the extent to which their productive value might remain in the flock (the number of progeny retained as breeders or for repeated shearing, and to what age). It is likely that, under most circumstances, the genetic difference between a cast-for-age ram and a young ram is likely to be only a minor contributing factor to a decision to sell old rams after three or four years of breeding.

Husbandry procedures and ram fertility

To ensure rams are in good condition with high sperm output when joining starts, extra attention should be given to their management, health and nutrition, beginning 10-12 weeks earlier. Keeping rams in good general health and free of infectious disease is an important component of reproductive management and one of the easiest to accomplish. The ram team is a numerically small component of the sheep flock, so, compared with the cost of managing the whole flock, it is relatively inexpensive to provide them with excellent nutrition and health care.

Major husbandry events (shearing, crutching) should be avoided in the eight weeks before joining. Some producers choose to shear Merino rams every six months but the pre-joining shearing should be planned such that they are carrying three to four months' wool growth when introduced to the ewes. Heat stress decreases semen quality. This occurs commonly in rams housed in sheds where ventilation and temperature control are inadequate. Under paddock conditions, heat stress is uncommon in Merinos but more common in British breeds. The difference is explained in part by the more pendulous nature of the Merino's scrotum, although Merino rams with heavy skin wrinkle have a relatively poor ability to control testicular temperature.

Because of the strong relationship between condition score and testicular volume and, consequently, the number of spermatozoa produced in the seven-week period before joining commences, steps should be taken to ensure that rams are in good condition *before* seven weeks pre-joining and are maintained in good condition (score 3-4) until joining. Lupin supplementation can be used to increase ram condition score, testicular volume and fertilising ability leading up to joining. A supplement fed at the rate of 500 g lupin grain per day for 8 to 10 weeks can increase testicular volume in Merinos from around 400 mL to about 600 mL.[75] Cereal grains can be used instead of lupins, but more care must then be taken to avoid grain poisoning. Supplementation of rams to increase testicular mass may enable the use of fewer rams at joining, and the saving which arises from purchasing fewer rams is usually much greater than the cost of the supplement.

Management of joining

Veterinary inspection pre-joining

Inspection of a team of rams before mating involves an assessment of both their genital and their general health. It is important to consider the general physical condition of the rams, their age, their condition score and the soundness of their feet and limbs because these factors have a bearing on the ability of the rams to remain active and sexually interested throughout the mating period. Protection by vaccination against clostridial wound infection should be encouraged; the timing of husbandry procedures which might temporarily reduce fertility should be discussed with the owner. As joining usually occurs during periods when there is a risk of flystrike, rams should be treated with protective insecticides to ensure that any wounds, such as fighting wounds, do not become flystruck. Ram examination is best carried out 10-12 weeks before joining is due to commence. This allows time to feed rams to ensure that their condition score and testes volume are optimal for joining and, if necessary, to buy in and acclimatise additional rams if it is found that there are too few fit and healthy rams in the ram team.

Inspection of rams is generally confined to physical inspection of body condition and feet and palpation of the external reproductive organs. More involved techniques, like semen collection and evaluation and serving capacity tests[76,77], may have a place when single-sire mating groups are employed (usually only in ram-breeding flocks) or when investigating specific failures of conception. The routine collection of semen is not warranted, both because of the cost of the procedure and because there is not a clear relationship between measured parameters in the ejaculate and the level of ram fertility — although infertile rams can generally be identified.

Palpation of the scrotal contents should include

- measurement of scrotal circumference, as an indirect measurement of testicular volume
- estimation of the firmness and resilience of the epididymal tails and testes[78]

- examination of the head and body of the epididymis to detect any abnormalities
- palpation of the spermatic cords and inguinal regions, to detect hernias, abscesses or varicocoeles
- examination of the scrotum itself, particularly noting the presence of scrotal mange
- examination of the penis by palpation within the prepuce, although it may also be desirable in some cases to exteriorise the penis and examine the penis and everted prepuce for evidence of injury or infection.

Serving capacity tests conducted in pens, under observation, have been used in an attempt to predict the capacity of rams to serve frequently in the field. While rams do vary in their breeding capacity, the ability of pens tests to accurately predict their field performance is only moderate. Some rams which perform poorly in pen tests — possibly as a result of inhibition by the pen environment or by the presence of other rams — do perform satisfactorily in the field.[79] The tests are no longer commonly used under commercial conditions.

Allocation of rams to syndicates based on testicular size

As described above, larger testes allow greater production, storage and ejaculation of sperm. Large testicular size is an evolutionary adaptation to increase mating success of rams because of the promiscuity of ewes — it is usual for ewes to be mated by more than one ram in any oestrous period when multiple rams are available. In natural mating, and where there is intense competition for ewes (high joining percentage or relatively few ewes), social and physical dominance are the important determinants of mating success — the biggest, strongest, most dominant rams prevent lesser rams from mating oestrous ewes.

At low joining percentages or where the ewe flock is large, dominant rams are less able to monopolise all of the oestrous ewes all of the time and smaller, subordinate rams have a better chance of mating oestrous ewes. In these circumstances where multiple rams are mating each ewe, larger testes and greater sperm numbers in each ejaculate provide a competitive advantage because the more numerous sperm from the larger testes will outnumber the sperm produced by a competitive ram with smaller testes.[80]

While natural selection has led to increased testis size to favour rams where multiple sires are mating ewes, in farming systems controlled by non-natural means, increased testis size allows flock managers to reduce competition for ewes by lowering joining percentages, whilst still ensuring that sufficient sperm are deposited in the reproductive tract of every oestrous ewe.

Producers therefore have a choice. They may ignore testis size and join rams at a relatively high percentage (2% or more). The biggest, most dominant rams will sire the most offspring. If the biggest, strongest rams are of low fertility, however, the conception rates may not be as high as could be achieved with highly fertile rams because the low-fertility, dominant rams monopolise the mating opportunities. (One study showed that flock fertility falls if completely infertile rams are dominant to fertile rams. In addition to a higher percentage of non-pregnant ewes, more ewes will conceive later in the joining period, increasing the spread of lambing.[81])

Alternatively, producers may choose to select rams for joining on the basis of testicular size, and to reduce the joining percentage accordingly. Social and physical dominance becomes less important because

(a) the dominant ram is known to have large testes and therefore expected to have good fertility

(b) the subordinate rams will share in the mating more equally and they, too, will be known to have good testis size.

Knowing that all rams used have large testis volume therefore obviates two problems. First, if relatively few rams are responsible for most of the matings, conception rates should still be satisfactory. Second, the lower joining percentage will reduce the effect of dominance behaviour, and all rams should therefore be represented as sires of the offspring.

Lower joining percentages also reduce the number of rams which need to be purchased and maintained. Flock managers can then either spend less on ram purchase by buying fewer rams, or spend a similar amount by purchasing better, more expensive rams, or make use of a combination of both strategies.

Duration of joining

In practice, management decisions about the joining ratio and duration of joining are considered together, since they are somewhat interrelated. It is not easy to determine in advance on a particular property the minimal number of rams and minimum length of joining period which will result in good flock fertility.

When joining occurs during the natural breeding season, ewes should be cycling regularly when the rams are introduced, so five weeks is usually adequate. Most ewes become pregnant in the first 17 days and those that do not should return to oestrus in the second 17-day period.

For out-of-season joinings, where most or all the ewes are in anoestrus at the commencement of joining, few ewes have overt, fertile oestrus in the first two weeks, so a joining period of seven weeks is necessary to allow two cycles after mating begins in earnest. Alternatively, teasers (testosterone-treated wethers or vasectomised rams) can be introduced for 14 days (exploiting the *ram effect*) before the fertile rams, in order to increase the probability that all ewes are cycling.

When joining extends longer than five weeks (in the breeding season) or longer than seven weeks (out of season) the few ewes that fail to conceive in the first two cycles have a third opportunity. Usually, this is a small percentage of the ewe flock (<5%), so the advantage of a small number of extra lambs must be weighed up against the several management penalties associated with prolonged lambing periods. The few late-lambing ewes may cause delay in marking (to avoid having lambing ewes or very young lambs in the flock when it is gathered for marking) and weaning, and the late-born lambs will be smaller and will possibly require extra nutritional care compared to the lambs born in the first six weeks of lambing. Extended joining periods may be justified when flock fertility is low, as is the case when ewes are grazing oestrogenic pastures.

Composition of ram syndicates

A group of rams which is joined to one mob of ewes is termed a *ram syndicate*. Normal practice is for syndicates to consist of some mature, experienced rams and some inexperienced rams (maiden or 1½ years old). Under these conditions, the dominance behaviour of older rams will generally ensure that the young rams are underrepresented in successful matings. Some producers join young rams separately from older rams, but this may lead to slightly reduced fertility in the flock or a slightly extended lambing period unless joining percentages are 4% or higher.[68] In the interests of maximising flock fertility, maiden rams should probably be omitted from syndicates joined to maiden ewes, and it is unwise to mate syndicates composed only of maiden rams to mobs of maiden ewes. Young rams are less experienced at mating behaviour than older rams and are subject to social domination by older rams if they are used in mixed-age ram syndicates.

Joining ratio

As a general rule, healthy, fit rams with testicular volumes >400 mL (scrotal circumference >32 cm) are able to cover ewes adequately at a ratio of 1% rams to ewes.

There are several circumstances where 1% may be insufficient. Examples of such cirumstances are listed below.

- The rams are in poor condition with low testis volume or they are excessively obese.
- The ram syndicate contains a high proportion of 1½-year-old or very old rams.
- The joining period is less than five weeks.
- The ewes are maidens.
- Joining occurs during spring and summer, especially if (highly seasonal) British-breed rams are used.
- Joining takes place in rough and scrubby terrain and/or at very low stocking densities.
- Flocks are very small and the use of one or two rams only may involve excessive risk that the poor performance of one ram severely compromises the flock fertility. To manage this risk, it is commonly advised to join rams at 1% plus 1 — in this way, small joining groups have a higher ram:ewe ratio than larger groups. For example, using this formula, a flock of 100 ewes would be joined to two rams while a flock of 400 ewes would be joined to five rams.
- Some degree of oestrus synchronisation occurs, as a result of the *ram effect* or the use of controlled breeding techniques. Following the successful use of the *ram effect* with teasers for a spring-summer joining, it is possible that a high proportion of the ewe flock is in oestrus at some time between 18 and 25 days after the introduction of teasers (see Chapter 8) or for 4 to 11 days after the fertile rams are introduced. To achieve good conception rates at that first oestrus, it may be necessary to join the ewes to as many as 4% of rams.

Joining ratios should be increased above 1% when one or more of these circumstances are present. In practice, many producers simply run 2% of rams with their ewe flocks as a blanket rule. If some of the circumstances listed above do not apply, this policy could be quite wasteful.

Active rams lose condition during joining, especially during the first three weeks (the first oestrous cycle of the ewes) as a consequence of reduced time spent grazing. Testicular volume and scrotal circumference decrease markedly during joining.[63] The same rams should not be used for successive joinings with substantial numbers of cycling ewes unless they are given a rest period of six to eight weeks with adequate nutrition between the end of one joining period and the commencement of the next.

Failure of fertilisation due to inadequate number of rams

When adequate numbers of rams are employed, it is expected that 95% to 100% of ova will be fertilised and, after some embryo losses, 80% to 85% of the ewes with fertilised ova will conceive. If the joining ratio is very low, conception rates will decline below this level. Table 7.6 shows the percentages of ewes in a flock expected to lamb following joinings of one, two or three cycle lengths. A *conception rate* of 70% (percentage of ewes lambing to one oestrus) is considered satisfactory for Merinos. As conception rate decreases with diminishing ram percentages, so, too, does the percentage of ewes lambing, but the effect of conception rate on the percentage lambing can be moderated by increasing the length of the joining period. Most of the reduction in conception rate is due to oestrous ewes not being mated at all, but some of the reduction may be associated with failure of fertilisation in mated ewes. The data in Table 7.6 are misleading in one respect: if the conception rates to the first cycle is only 60% and 50%, due to too few rams, they should rise to around 70% for the second and third cycles (by which time the 'functional' joining ratio is much increased).

Failure of fertilisation due to other ram factors

When natural mating is well managed, nearly all ova should be fertilised. Failure of the ram to deposit into the vagina adequate, normal, motile sperm is occasionally a cause of low fertility. This could possibly result from the use of too few rams or overworked rams, but it is more often related to abnormalities of the testes or other parts of the male reproductive tract, due to infectious diseases or injury. Note, however, that rams with brucellosis and some other genital infections can often still produce ejaculates with good numbers of normal, motile sperm. Defective spermatozoa commonly result from the excessive heating of rams. Some rams, however, routinely produce ejaculates of low quality for no obvious reason.

Non-specific abnormalities of the epididymis which affect ram fertility

A *spermatocoele* is a cystic dilatation of the epididymal duct with the accumulation of sperm in the cyst. It follows acquired or congenital occlusion of the duct. If extravasation of sperm occurs, the stromal tissue produces a characteristic granulomatous response. Spermatic granulomas which develop secondarily to congenital occlusion are usually in the head of the epididymis. Congenital obstruction in rams and male goats is not uncommon and is usually unilateral.

Table 7.6: Fertilisation and the effect of decreasing ram power and the consequent decrease in conception rate to each cycle.

Percent conception to each cycle	% of ewes lambing		
	1 cycle	2 cycles	3 cycles
70	70	91	97
60	60	84	94
50	50	75	88

Obstruction of the epididymal duct in the head, body or tail leads to testicular degeneration, although the process is relatively slow — the efferent ducts are able to resorb most of the products of the seminiferous epithelium except sperm. Obstruction of the efferent ducts, in contrast, leads rapidly to testicular atrophy.[82]

Spermatic granulomas which develop secondary to bacterial infection are usually found in the tail of the epididymis because the majority of bacterial infections start, and are most severe, at that site. The two most common bacterial infections of epididymides in Australian sheep flocks are those caused by *Brucella ovis* and by *Actinobacillus seminis*. The palpable lesion of the epididymis typical of *B ovis* infection is, in fact, a spermatic granuloma. Spermatic granulomas following obstruction of the duct may also be caused by trauma.[83]

Epididymitis caused by *Brucella ovis* infection (Ovine brucellosis, OB)

Epididymitis caused by *B ovis* is the most common lesion of the genitalia of Merino rams culled from Australian flocks. In one survey, epididymitis was identified in 19% of rams and *B ovis* was associated with 47% of those.[84] There is a lower flock prevalence of infection in Merinos than in British breeds. The prevalence of infection on some properties may exceed 50% of rams but this is uncommon. Generally, fertility is not compromised until the proportion of rams with chronic, palpable lesions exceeds 10% to 20%. Economic wastage occurs from extension of the lambing period[85,86], reduction of lambing percentage and an increased size and rate of turnover of the ram team.[87]

Epidemiology

Rams can become infected at any age over 4 months.[88] Transmission of infection occurs mainly from ram to ram, via the ewe's vagina principally, but also by homosexual activity between rams.[89] Infection can occur by inoculation of mucosal surfaces including the prepuce, conjunctiva and nasal mucosa. Infection in ewes is *usually* short-lived. Experimental infection of ewes at mating[90] and field evidence[91] indicate that infection can persist in the ewe, leading to returns to service, abortion, birth of weak lambs and perinatal mortality. The incidence of infection in ewe flocks is, however, low and the role of persistently infected ewes in the maintenance of infection in the ram flock is insignificant. In chronically infected rams, active excretion of bacteria in semen probably persists indefinitely.

Pathogenesis[d]

Following a bacteraemia there is localisation in the epididymis, usually unilaterally and in the tail, producing degenerative, inflammatory and proliferative changes. The resulting sperm stasis and epithelial damage may result in extravasation of sperm with subsequent spermatic granuloma formation. Histopathological and bacteriological studies suggest that the epididymal tail, ampulla, ductus deferens and seminal vesicles are the most frequently involved sites of infection; the testis and the head of the epididymis are involved less frequently.[92,93]

Seroconversion (as detected by the warm complement fixation test (CFT)) occurs 10 to 66 days after artificial infection, earlier with more sensitive testing procedures.[94] Semen culture is

---

d   The pathogenesis of *B ovis* for ewes will be discussed in the section on abortion.

generally positive five to ten weeks post-infection[93] and lesions caused by the initial infection are usually palpable from nine weeks onwards. (Both of these events can occur sooner than this — positive semen culture and clinically palpable lesions may be detected as soon as four weeks post-infection.[88]) Some challenged rams never develop any evidence of infection; others develop serological evidence only. These rams recover and are said to have had 'abortive' infections.[95] Serological reactions decline in recovered animals over a period of four to five months. In animals which remain chronically infected, serological responses remain relatively constant.[87]

Clinical findings

A deterioration in semen quality occurs early in the disease and the semen contains many leucocytes. In the acute stages of the disease, there is oedema and inflammation of the epididymis and tunics, palpable as a general swelling and a loss of definition of the scrotal contents. There is a systemic reaction which is rarely detected. Regression of the acute syndrome is followed by a latent period of two to three months before chronic lesions with palpable abnormalities develop in one or both sides of the scrotum.

The usual chronic lesion is an enlargement of one or both the epididymides, usually in the tail, which is hard due to fibrosis. The epididymis may be two to three times normal size or even more. There is no orchitis and initially the testes feel normal, but degeneration and atrophy lead to a decline in the size and firmness of the testes. Less commonly, the enlargement and hardening may involve more of the epididymis, or the head only, or it may not involve the epididymis at all, being restricted to one or more of the accessory sex glands.

Diagnosis

The presence in a flock of a prevalence of chronic epididymal lesions greater than 5% is suggestive of brucellosis. Lesions of chronic epididymitis must be differentiated from those caused by trauma and other bacteria, particularly *Actinobacillus seminis*. Lesions caused by *A seminis* generally show a more acute reaction and are located in the head of the epididymis more frequently than lesions caused by *B ovis*. Either semen examination and demonstration of the weakly acid-fast bacilli in smears or semen culture for *B ovis* is necessary for a definitive diagnosis. Neither test is particularly sensitive, primarily because of intermittent shedding of the organism by infected rams, and culture may be the more sensitive technique when laboratory procedures are commenced soon after sample collection.[96,97]

The CFT has long been used in Australia and New Zealand to eradicate the disease from flocks. In 1983, an enzyme-linked immunosorbent assay (ELISA) test was developed with a specificity comparable to the warm CFT (0.5% false positives) but significantly more sensitive.[98] (The increased sensitivity, however, means that some CFT-negative, ELISA-positive rams are detected which will never excrete *B ovis* and will eventually become ELISA-negative.)

If an investigation is carried out soon after *B ovis* is introduced into a flock, a high proportion of rams which will never become excretors may be detected serologically. This fact should be considered when planning eradication programmes. Testing immediately after joining or soon after sexual activity has started in flocks of young rams could lead to the identification and culling of recently infected animals, many of which will ultimately recover and become serologically negative.

Rams with low CFT titres (1:8 or 1:16) are frequently found to be uninfected and are probably recovering from abortive infections. Currently, it is recommended that low-titre-positive animals with no palpable lesions be isolated from other rams and retested after four weeks. Persistent low titres may warrant the slaughter and detailed necropsy examination of the rams in question so that the true status of the flock can be determined.[99]

Eradication by isolation of old rams from young rams

In commercial flocks, brucellosis can be readily eradicated by isolating the existing, infected ram flock, purchasing replacement rams from accredited OB-free studs and keeping them at all times separate from the old rams. Eradication from the older rams can be attempted, by 'test and slaughter', or these rams can be used for mating and cast for age progressively over three to four years. There is a significant danger that the infected rams will gain access to the young rams and cause a breakdown, so the shorter the duration of the 'two-flock' system the safer.

Eradication by 'test and slaughter'

In ram-breeding flocks, a programme of testing (ELISA serology) and culling for slaughter of any reactors (the aforementioned 'test and slaughter' strategy) will successfully eradicate brucellosis, provided that new cases are detected before they commence excretion of B ovis organisms, possibly as early as four weeks post-infection.[88] Serial testing should be performed, therefore, every three weeks and all positive reactors slaughtered.[100] Any older rams with lesions of epididymitis should be culled regardless of the serological result because some false negative results can occur with chronically infected animals.[93] Infection of ewes is potentially a source of breakdown during eradication but this is rarely a problem of any practical significance.

Control by vaccination (but not in Australia)

Vaccination against OB has been used in Australia as a control measure but is no longer permitted. The usual practice was the simultaneous administration of a formalin-killed B ovis saline-in-oil emulsion and B abortus strain 19 vaccine, although the equal efficacy of some B ovis vaccines when administered twice, two weeks apart, was demonstrated.[101] B melitensis Rev1 vaccine, a modified live vaccine developed for the control of B melitensis infection, is now considered a better choice for B ovis control. Vaccination is no longer used as a way of limiting B ovis infection in Australia but is used as a control measure in some countries.[102]

Accredited OB-free flock scheme

Voluntary accreditation schemes for ram breeders operate in all Australian states. In flocks to be accredited, all rams, cryptorchids and teasers over 4 months of age are subject to veterinary examination by palpation of the scrotal contents, and those over 10 months of age are serologically tested. Initial accreditation requires two consecutive negative tests at an interval of 60 to 180 days combined with a process of enquiry and inspection by the certifying veterinarian to ensure that the farm biosecurity is adequate. Subsequent testing, performed annually for three years, then biennially, involves the palpation of all rams, cryptorchids and teasers over 10 months and blood testing of all over 2½ months of age. In large ram flocks, provisions exist for testing a sample of the sale rams, rather than the entire flock, provided the retained stud sires have been tested and found clean. Testing must be accompanied by restrictions on the introduction of further rams into the flock to ensure those animals are also free of OB.

The testing procedure is performed by private practitioners and the register of accredited flocks is maintained by State Departments of Agriculture. The flock owner pays all costs incurred by the practitioner and the laboratory.

Other causes of epididymitis

After *Brucella ovis*, two bacterial species dominate the list of infectious causes of epididymitis in rams. These are *Actinobacillus seminis* and *Histophilus somni*. Both are gram-negative pleomorphic organisms and the epidemiology and pathogenesis of both infectious agents in sheep are so similar that they may, for many purposes, be considered together. The two species are genetically and phenotypically similar and it has been proposed that one may be a variant of the other.[103,104]

Actinobacillus seminis

A seminis was first reported as a cause of epididymitis in rams in 1960 in Australia[105] and there have been numerous reports of *A seminis* epididymitis from other countries since then.[106]

What is currently understood or hypothesised about the epidemiology of infection with *A seminis* in sheep flocks has been summarised by Al-Katib and Dennis (2009).[106] The organism is not persistent in the environment outside sheep but fomites may be a short-term source of infection following contamination by infected semen or discharges from the genital tract of an infected ewe. Transmission between animals can be venereal or non-venereal and transmission from ewes to perinatal lambs seems likely. Such transmission may be postnatal — through umbilical infection, for example — or prenatal. In young rams, organisms can often be isolated from the prepuce in the absence of any clinical signs but the frequency of infection of the prepuce declines with age and is significantly less common in rams over 12 months of age. In some individuals, however, ascending infection from the prepuce, or possibly a bacteraemic episode, may lead to infection of the testes, epididymides or accessory sex organs.[107]

Infection in the epididymis and testis can be asymptomatic or can begin as an acute purulent epididymitis. Those with clinical signs of acute infection tend to have a progressive course, developing chronic epididymitis. Those with less obvious clinical signs may develop chronic lesions or may become free of infection spontaneously. Infected rams usually have low fertility, and chronic cases develop bilateral testicular atrophy. The epididymitis is clinically indistinguishable from brucellosis.[108] Persistent subclinical infection can also occur, in which abnormalities are undetectable on clinical examination but organisms are present in semen.

Orchitis is a more common additional finding to epididymitis with *A seminis* infection than with *B ovis* infection. *A seminis* has also been associated with severe posthitis of a ram and polyarthritis in lambs in Australia[109], metritis, mastitis and abortion in ewes in the UK.[110]

Histophilus somni

H somni is a relatively new species name which includes organisms formerly known as *Histophilus ovis, Haemophilus somnus* and *Haemophilus agni* [111], but many publications still refer to isolates from sheep as *Histophilus ovis*. The names are used interchangeably here, depending on the name given in the cited reference.

H ovis has been reported as causing epididymitis, suppurative polyarthritis in lambs, meningo-encephalitis in adult sheep and mastitis in ewes.[112] The epidemiology of *H ovis* in sheep flocks is uncertain, but the occurrence of polyarthritis caused by this organism in early postnatal lambs (before marking) and the isolation of *H ovis* from the genital tract of ewes and prepuce of young rams suggest that the reservoir of infection in flocks is the genital tract of ewes and, possibly, rams.[113] Infection of the epididymis and testis may be an occasional sequel of a preputial infection of young rams which does not remain localised to the prepuce or resolve spontaneously with age.

Establishing an aetiological diagnosis of epididymitis is often important, although serological tests will generally be sufficient to confirm *Brucella* infections. Serological tests are available for *H somni* and *A seminis* but many normal animals have antibodies to them and cross reactions with each other can confuse the diagnosis. Culture of the organisms from semen is complicated by their slow growth, lack of selective media and likelihood of overgrowth by other semen microflora. Recently, polymerase chain reaction (PCR[e]) tests have been developed which can quickly and reliably identify the two organisms and distinguish between them.[114]

Miscellaneous infections

Trueperella pyogenes, Actinobacillus ligneriesi, Corynebacterium pseudotuberculosis, Yersinia pseudotuberculosis, Escherichia coli (causing abscessation with fistula formation in the scrotum and orchitis[115]) and *B abortus* (strain 19) have been reported in sporadic cases of epididymitis.

Testicular abnormalities

Cryptorchism

Cryptorchism (or cryptorchidism) can be either unilateral or bilateral — the unilateral form is sometimes referred to as monorchism or monorchidism and is more common than the bilateral form.[116] When unilateral, it is usually the right testis which fails to descend. In Merino sheep the foetal testes normally pass through the inguinal canal between days 75 and 80 post-conception and rarely, if ever, will a retained testis descend through the inguinal canal after birth.[117]

The retained testis or testes may be in the abdomen or inguinal canal where the high testicular temperature prevents normal spermatogenesis. Rams with bilateral cryptorchism are sterile, but unilateral cryptorchids are likely to be fertile. While their fertilising capacity is probably reduced to some extent by the functional loss of one testis, the fully descended testis does often show some compensatory hypertrophy.

---

e   PCR testing is a technique commonly used to amplify DNA segments such that the presence of small amounts can be detected by further tests.

In the mid-20th century the incidence of cryptorchism in Australian Merino sheep was estimated to be around 0.4%[118,119], although some flocks reported higher incidences. A later study found an incidence 10-fold higher.[120] An incidence of 0.6% with substantial variation between flocks has been reported in UK sheep flocks.[116] The inheritance of the trait suggests that it is inherited as a recessive gene with a low degree of penetrance or as a threshold polygenic trait.[121] In ram-breeding flocks, the cryptorchid animal should be culled but attempts to further reduce the incidence in the flock are usually only advisable or necessary if the frequency of cryptorchidism is high. In those cases, the parents and all full siblings of the affected animal should be culled.

In commercial flocks, unilateral cryptorchids could feasibly be used to produce offspring for slaughter, although, if the genes contributing to cryptorchidism are already present in the ewe flock to a significant degree, there may be an increased incidence of cryptorchidism in the lambs. If the lambs are sold for slaughter entire, this may not be a problem but, if the custom is to castrate all male lambs, cryptorchids are generally considered a nuisance. Most commercial producers choose not to use unilateral cryptorchid rams as sires.

According to Dolling and Brooker, cited in other sources[122], an association between polledness and cryptorchism has been reported in some studies but is not present in the Australian Merino.

Hypoplasia

Once testes have descended into the scrotum they should increase in size, gradually at first and more quickly at and after puberty. Bilateral small testes in young rams can occur as a result of undernutrition, in which case the size and condition of the animal also reflect the poor growth. Small testes can also be the result of a failure to develop for, presumably, genetic reasons, in which case the condition is one of hypoplasia. That hypoplasia is the reason for the underdevelopment is most obvious when the ram is young, when it is in reasonable or better nutritional condition, and if only one testis is affected. Testicular hypoplasia can be either unilateral or bilateral (Figure 7.7). Both unilateral and bilateral forms may have an inherited basis.

In some cases, the occurrence of one or two abnormally small testes in a young ram may be due to late or delayed development, in which case the hypoplasia may be temporary.[123] In older rams, hypoplasia must be differentiated from atrophy in which a previously normally developed testis becomes smaller and soft, usually following a local or systemic illness. Hypoplasia of the testes of rams has been associated with zinc deficiency experimentally and the possibility exists that natural cases could occur in the field in some regions.[124]

Degeneration

The testicular germinal epithelium is very sensitive to many adverse influences. Degeneration may be unilateral or bilateral, which may help determine whether the cause is systemic or local. The degenerated testis may remain a normal size but be soft and flabby or it may become small and firm. Softness and flabbiness often indicate rapidly progressing degeneration, while fibrosis takes several months to develop. Sperm production is reduced and the semen becomes thin and milky or watery. Regeneration can occur but takes longer than the degeneration.

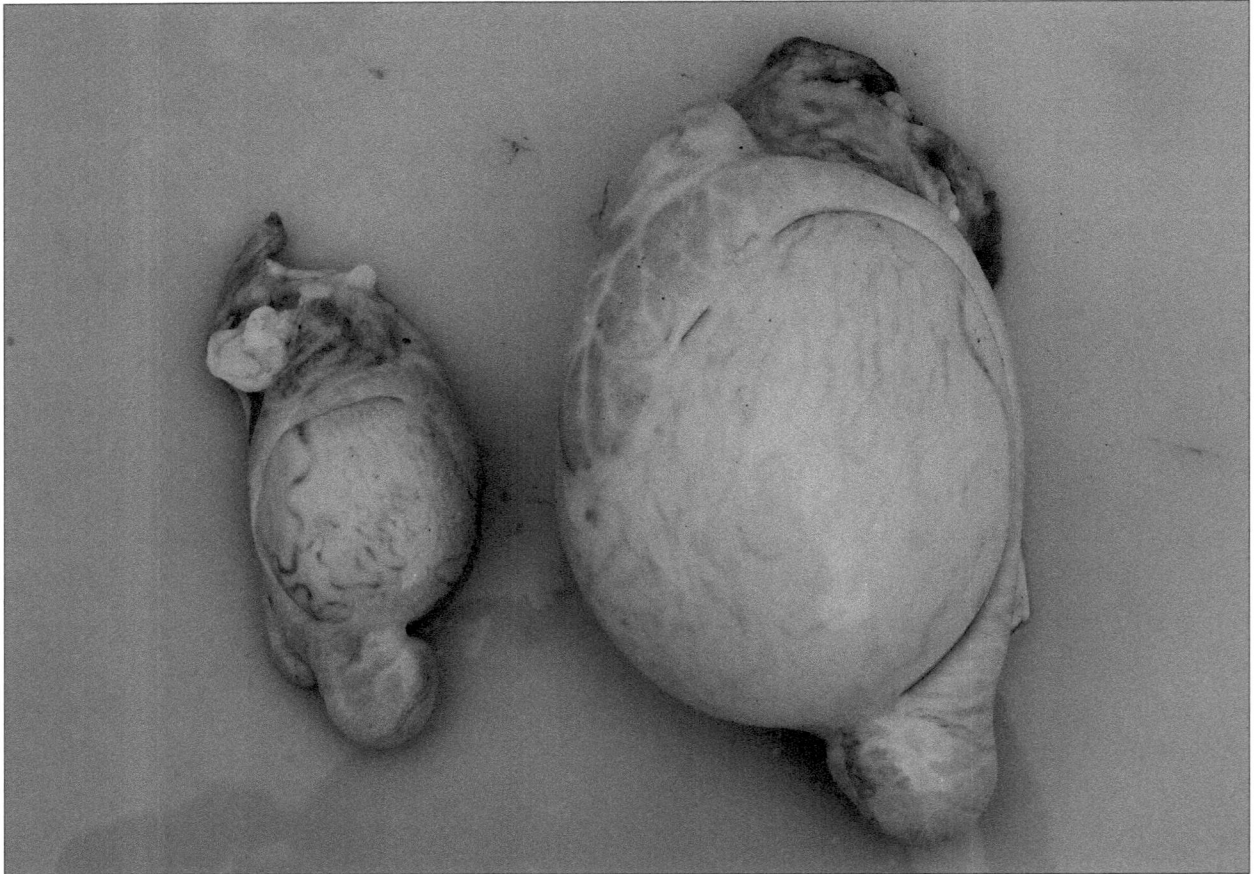

Figure 7.7: Unilateral testicular hypoplasia. The small testis is grossly normal other than being much smaller than expected. The condition is more commonly unilateral than bilateral and if detected in a young ram, may be a temporary condition. Source: KA Abbott.

The causes can be grouped under a number of headings:

- Thermal: Maintaining rams at 35 °C for periods exceeding four days leads to a loss of motility, and increased proportion of abnormal sperm. If the period is extended there is a measurable decrease in the concentration of sperm and ultimately cessation of spermatogenesis
- Local or systemic infection: Fever, toxaemia or local inflammation
- Nutritional deficiencies: Vitamin A, phosphorus, severe deficiencies of protein or energy which lead to very low condition score (1-1½)
- Vascular lesions
- Obstructive lesions of the efferent tubules: The backpressure leads rapidly to degeneration of seminiferous epithelium. The testes degenerate but are enlarged with fluid.

Other lesions of the male genitalia

Balanoposthitis

Balanoposthitis (pizzle rot or sheath rot) is caused by *Corynebacterium renale* under particular dietary conditions and is principally a problem of wethers but it may also occur in rams. That condition is discussed further in Chapter 18. A different condition — a severe, ulcerative balanitis — occurs sporadically in rams, particularly in Border Leicester rams, but also in other British breeds. Signs of the disease are often observed first during joining and may occur concurrently with vulvo-vaginitis in the ewe flock. A high proportion of the ram flock may be involved.[125] The condition is characterised by necrosis and ulceration of the penis often with extensive blood clots in the prepuce and in the ulcers, and paraphimosis. When outbreaks are observed during joining it is necessary to remove the rams from the ewes for individual treatment with systemic and topical antibiotics and antiseptic lavage of the penis and prepuce. Some animals do not recover. The cause of the condition is unknown but *Trueperella pyogenes* is frequently present in the lesions. Several outbreaks have been reviewed by Watt et al. (2016).[126]

Varicocoele

A varicocoele is a dilatation of the spermatic vein and is usually of little or no significance. Bilateral varicocoeles of long duration may adversely affect semen quality (possibly due to anoxia) or they may, if large enough, incapacitate rams so that they are unable to walk normally due to pain. They do, however, increase in size very slowly and rarely reach large proportions.

Scrotal mange

Mange of the scrotum, caused by *Chorioptes bovis*, is associated with seminal degeneration and testicular atrophy.[127] (See also Chapter 10.)

SECTION C

ABORTION AND PRENATAL DISEASES OF LAMBS, AND DISORDERS OF EWES IN LATE PREGNANCY

Abortion and prenatal diseases of lambs

Depending on the stage of gestation at which they occur, abortion and prenatal disease of lambs can be considered to be contributors to infertility or to perinatal mortality. Generally, abortion is an uncommon cause of reproductive wastage in sheep in Australia. Abortion is a more important source of reproductive wastage in countries with more intensive systems of lambing management. When abortion storms do occur in Australia, they are often associated with unusually high stocking densities of the ewe flock. As well as causing sporadic outbreaks of abortion and neonatal lamb deaths, many of the causative agents probably also cause a low level of undetected losses on many farms. One would particularly expect this to be the case with toxoplasmosis. The common infectious causes of abortion in sheep are summarised in Table 7.8.

Toxoplasmosis

Toxoplasma gondii is an obligate intracellular protozoan parasite. In addition to its role in causing abortion in sheep it is an important human pathogen. Humans are infected by ingestion of oocysts or the encysted stages of the parasite in raw or undercooked meat and *T gondii* infection is considered to be responsible for about one-quarter of deaths attributed to foodborne pathogens in the United States.[128] The risk of severe infection is highest in immunologically impaired individuals and pregnant women. Pork and sheep meats are considered as more important sources of infection than poultry or beef. Pigs and sheep are very susceptible to infection with *T gondii* and, unlike cattle, those species do not become clear of infection but remain infected for life.[129]

The parasite's asexual cycle can occur in most warm-blooded animals but the sexual cycle occurs only in cats and other felidae. Wild rodents, with encysted bradyzoites in brain and muscle, act as a reservoir of infection which is spread and enormously amplified by cats. Cat faeces can contain 10^6 oocysts/g. Theoretically, 50 g of cat faeces can provide 250 000 sheep-infective doses.[130] Susceptible sheep become infected and remain so permanently after ingesting oocysts. Infection of the placenta and conceptus occurs when a previously unexposed sheep ingests oocysts when pregnant. The outcome of infection depends on the stage of pregnancy (Table 7.7).

Diagnostic aids

Histopathology of selected foetal and placental tissues reveals characteristic lesions. Serology of ewes is of little use because *Toxoplasma* titres are frequently high in ewes which have not aborted but may be low in ewes aborting due to toxoplasmosis if tested at the time of abortion. Serology of non-autolyzed foetal lambs is, however, useful as a diagnosis of congenital toxoplasmosis, particularly on a flock basis rather than an individual basis.[131] Definitive diagnosis requires the demonstration of *T gondii* in fixed tissue sections or bioassay in mice.

Control

Infection occurs following ingestion of contaminated feedstuffs and pasture — there is no significant sheep-to-sheep transmission. Prevention of infection, therefore, involves the reduction in contamination of the environment and livestock feedstuffs by cats. Young cats pose the greatest threat — oocysts are usually produced only during initial infection which occurs as the young cat starts to catch and eat small prey animals. The vaccine *Toxovax* is available and used in New Zealand[132] and the United Kingdom.

Campylobacteriosis

Campylobacter fetus ss fetus is transmitted by ingestion. The bacteria survive in the gall bladder and gut of infected sheep, and crows and magpies can become infected for several months

Table 7.7: The outcome of infection with *Toxoplasma gondii* depends on the stage of pregnancy of the ewe when infected.

Time of infection	Likely outcome of infection
Infection in early pregnancy	foetal resorption
Infection in mid-pregnancy	birth of stillborn or weak lambs, the cotyledons containing small white foci of necrosis, the intercotyledonary membranes unaffected
Infection in late pregnancy	persistently infected but clinically normal lambs, both ewe and lamb immune to further challenge

and thus become vectors.[133] Outbreaks of abortion are usually preceded by a period of high stocking, particularly in winter and spring, but also after periods of handfeeding in summer and autumn.[134] If the flock has been previously exposed, the older ewes do not abort but the younger ones do. Abortion occurs in the third, fourth or fifth month of gestation. Sometimes an abortion storm follows two to three weeks after the first sporadic abortion. Ewes may retain their membranes and develop metritis. Aborting ewes develop good immunity and are unlikely to abort from this cause again. The disease has occurred repeatedly on some farms.[135]

Diagnostic aids

Some aborted lambs have 'rosette-like' necrotic foci in the liver and this can be a useful differential feature from toxoplasmosis. Definitive diagnosis is based on isolation of organisms from aborted membranes and the foetal stomach.

Control

Aborted ewes should be removed from the lambing flock. Handfeeding should be stopped or changed to a new area each feed. The ewes should be 'spread out' into clean paddocks. Ewe hoggets can be grazed on lambing paddocks after an abortion storm in order to infect them while non-pregnant. The vaccine 'Campylovexin' has been used in New Zealand.[136]

Salmonellosis

The Salmonella Reference Laboratory in Adelaide reports that, in Australia, the salmonella serovars most commonly isolated from sheep are *S typhimurium* and *S bovis-morbificans*.[137] There are a number of other serovars involved in outbreaks from time to time. *S dublin*, a common cattle isolate, occurs much less commonly in sheep. Outbreaks of salmonellosis in pregnant ewes will usually cause abortion in a large proportion of ewes in addition to signs of enteric infection. In one report, four outbreaks of *S typhimurium* infection in WA affecting autumn lambing ewes were characterised by significant mortality of ewes (8% to 18% of the mob), diarrhoea, foetid dark red vaginal discharge, foetal death, abortion, retained foetal membranes, septicaemia and high fever.[138] It is likely that in Australia abortion will usually be accompanied by clinical illness in the ewe. In the United Kingdom, *S abortus-ovis* and *S montevideo* have caused numerous outbreaks of ovine abortion in which other clinical signs in the ewe are largely inapparent.[139] *S abortus-ovis* has declined in importance in the UK since 1975[140] and the serovar is not present in Australia. *S montevideo* is isolated from sheep only rarely in Australia. Outbreaks of *Salmonella* abortion are usually associated with stressful conditions, including overcrowding, handfeeding, undernutrition and pregnancy, as is the case in enteric salmonellosis in non-pregnant sheep (discussed further in Chapter 16).

Diagnosis

The usual presence of clinical signs related to gastroenteritis and septicaemia, including high fever, usually differentiate abortion caused by *Salmonella* spp from that caused by *Toxoplasma* and *Campylobacter*. Confirmation follows isolation of the organisms from the foetal stomach and placenta.

Treatment and control

Separation of affected ewes from unaffected ewes and a reduction in stocking rate of the latter may reduce the incidence of new cases but is unlikely to stop them altogether. Affected sheep can be treated with antibiotics as a life-saving measure. Antibiotic resistance is frequent in *Salmonella* isolates, but least likely for ampicillin, trimethoprin/sulphadiazine, tetracyclines and neomycin. Oral medication with furazolidone, continued for up to seven days, may be attempted to prevent salmonellosis in in-contact sheep but caution must be exercised to avoid overdosage.

Listeriosis

There are two species of *Listeria* which cause disease in animals, including humans: *Listeria ivanovii* and *L monocytogenes*.[141] *L monocytogenes* causes meningoencephalitis, abortion and gastroenteritis in humans and animals. *L ivanovii* is very rarely reported in humans[142] but predominantly affects sheep, goats and cattle, causing abortion and enteritis, but not brain infections. *Listeria ivanovii* was formerly known as *Listeria monocytogenes* serotype 5.

Listeria are often present in the gut of normal sheep and can survive and multiply in faecal material and soil. Compared to many other bacterial species, *Listeria* are tolerant of a wide range of pH, temperature and salt conditions and so are found in a variety of environments, including soil, water, effluents and foods. The feeding of silage, particularly silage with a pH above 5.5, is often associated with outbreaks of listeriosis. The disease occurs commonly under wet, muddy conditions and its occurrence is sporadic and unpredictable.[143] It is probably a cause of widespread losses from abortion and perinatal death but with a generally low flock prevalence.[144] Listeriosis caused by *L ivanovii* should be considered a potential zoonosis, particularly for pregnant women or those who are immuno-compromised.

Diagnosis

Listerial abortion is characterised by foetal loss at 3½-5 months of gestation. The organism can be isolated from the foetus (liver and lungs) and placenta.

Control

Antibiotics are not generally effective but could be considered in the face of an outbreak. Effective control involves changing the predisposing conditions but abortions are likely to continue for some time after intervention.

Enzootic abortion of ewes (EAE)

The causative agent of EAE, *Chlamydia abortus* (formerly called *Chlamydia psittaci* serotype 1)[f], is, like all *Chlamydia* spp, a specialised, antigenically complex, intracellular bacterium. Infections

---

f   In the late 1990s, the family *Chlamidiaceae* was split into two genera, *Chlamydia* and *Chlamydophila*. *Chlamydia psittaci* was reclassified to be of the genus *Chlamydophila* and split into four species: *Chlamydophila psittaci, C abortus, C caviae* and *C felis* (Everett, 2000).[150] More recently, the two genera have been reunited to one — *Chlamydia* — and this genus includes *C trachomatis, C pneumoniae, C psittaci, C muridarum, C pecorum, C suis, C avium, C gallinacea, C abortus, C caviae,* and *C felis* (Sachse et al., 2015).[147]

of sheep by chlamydial spp causing polyarthritis, pneumonia and kerato-conjunctivitis are widespread in Australia, but chlamydial abortion is rare[145,146] and there is no evidence that the reported cases of chlamydial abortions in Australia were caused by the organism now known as *C abortus*.[147] The infrequency of outbreaks of chlamydial abortion, combined with the evidence from serological[148] and molecular[149] surveys of Australian sheep, strongly suggest that *C abortus* is not enzootic within Australia and that EAE does not occur. Enzootic abortion is common overseas, both where *C abortus* has been isolated from sheep, cattle and goats, and in association with abortion in other species, including humans.[150]

Ewes aborting with EAE remain chronically infected but do not abort again. Vast numbers of infectious chlamydial elementary bodies are shed at the time of abortion and these remain a potent source of infection for animals and man.

Brucellosis

B ovis can cause an increase in returns to service, foetal mortality and the birth of lambs with low body weight and a reduced chance of survival when OB-infected rams are mated to uninfected ewes. The infection is more likely to be a cause of perinatal mortality than of abortion, although sporadic abortion does occur.[151] The main result of infection is a placentitis which interferes with foetal nutrition. Lambs born from infected ewes are usually of normal gestational age but of significantly lower birth weight.

Infection does not appear to persist in the ewe flock. Although the placentae from infected ewes are a source of *B ovis* organisms and the vaginal discharge of infected ewes contain brucellae for up to 10 days after parturition, the ewes do not appear to be a significant source of infection to other ewes or to rams.[152] It appears that it is necessary for the ram flock to remain infected for the disease to appear from year to year in the ewe flock.[153]

Ovine pestivirus

This virus causes Border Disease (Hairy Shaker Disease, *hypomyelinogenesis congenita*), embryonic death and abortion. The virus is serologically similar to the pestivirus of bovine viral diarrhoea-mucosal disease (BVD-MD). The disease has been reproduced in sheep with cattle-derived virus. There is a range of strains of which most, if not all, will infect both cattle and sheep. Nevertheless, there are some differences in host affinity, pathogenicity and cross-immunity between strains.[154]

Infection of ewes less than 50 days pregnant commonly results in abortion due to placental degeneration. Infection at 50 to 80 days can result in foetal death and abortion or the birth of live or dead lambs with hairy coats and varying degrees of hypomyelinogenesis, cerebellar and cerebral dysgenesis, arthrogryposis, kyphosis and brachygnathia.

Generally, only a few ewes in the flock abort. Affected lambs born alive fail to thrive and die from complicating illness.

Akabane disease

The virus enters the blood through an insect bite; viraemia is followed by invasion of the placenta and foetus. The principal vector is the biting midge *Culicoides brevitarsis*. Immunity lasts for years, although cross-protection with the other related arbovirus, Aino, may not exist.[155] There is some evidence that the sheep foetus is only susceptible to infection between 30 and 36 days. Lambs are aborted or born with congenital defects, including microencephaly, hydrocephalus, arthrogryposis, kyphosis and cerebellar agenesis.[156]

Abortion caused by *Histophilus somni* (*ovis*)

This organism has been isolated from sporadic ovine abortions in Victoria. It is more usually associated with polysynovitis and septicaemia of lambs and epididymitis/orchitis of rams.

Romulosis

Onion grass (*Romulea rosea* var *australis*, previously *R bulbocodium*, and also known as Guildford grass) occurs in unimproved and low-fertility pastures in some areas of southeastern Australia. Ingestion of the plant, or of a fungus (*Helminthosporium*) infesting the plant, has been associated with infertility and abortion in sheep in Victoria.[157] Infertility is characterised by extremely low (even near zero) lambing rates in one year, with normal rates in the previous and subsequent years. Abortion, if it occurs, does so in mid-pregnancy. Surviving full-term lambs may have long bone deformities.[158] Nervous signs, including posterior paresis, may occur in flocks experiencing onion grass poisoning. Note that onion grass is a different plant from onion weed — a common problem plant for lawn gardeners.

Disorders of ewes in late pregnancy

Gestation length

In Merino ewes, the length of gestation is 147 to 155 days. Gestation is approximately two days shorter in British longwool breeds and a further four days shorter (140 to 148 days) in shortwool breeds. Gestation length is also influenced by the age of the ewe, litter size, and nutrition of the ewes.[159] The genotype of the foetus has a strong influence on the length of gestation — an important fact when MOET (see Chapter 8) involves recipients of a different breed.

Nutrition during pregnancy

Implantation in the ewe is a diffuse, gradual process that involves only the superficial tissues of the endometrium. Initially the trophoblast adheres to the epithelium lining the caruncles. Firm attachment is not apparent until around day 30. The growth of the cotyledonary placenta proceeds rapidly after day 40 and exceeds foetal growth until about days 90-100. After this time the mass of cotyledons tends to decline slowly, until term. At day 90 a typical single Merino foetus weighs about 500 g. Thereafter its growth rate accelerates, to maximum rates of about 70-80 g per day during days 120-140 and at term it weighs 3-5 kg. The maintenance of pregnancy requires an adequate supply of progesterone. Initially progesterone comes from the corpus luteum, but after about day 50 the cotyledons are the principal source of this steroid.

There are probably no nutritional requirements specific to pregnancy until about day 90. The plane of nutrition prior to day 90 may influence placental weight but variation in feeding at this stage has relatively little influence on lamb birth weight. Commencing at about this stage

Table 7.8: Summary of infectious abortion of ewes. The last four conditions, in the shaded text, are not reported from Australia.

Cause	Transmission	Time and incidence	Clinical characteristics	Diagnostic tests	Serology	Prophylaxis
Toxoplasma gondii	Ingestion of oocysts from cat faeces	Last 8 weeks, typically last 10 days, stillborn or weak lambs. Up to 40% abort	Small white necrotic foci in cotyledons, intercotyledonary areas unaffected, mummified foetuses	Histopathology of foetus or placenta	Sabin Feldman dye test, Indirect Fluor. antibody test and others	Expose non-pregnant ewes. Control farm cats. Monensin. Vaccination
Campylobacter fetus	Ingestion	Last 12 weeks, typically 2 to 6 weeks pre-term. 5% to 30% abort	Severe endometritis in ewes; 40% of foetuses with 'rosette' lesions in liver, thickened intercotyledonary area	Isolation of organisms from foetal stomach	Agglutination test	Expose non-pregnant ewes (good flock immunity). Vaccination
Salmonella spp	Ingestion	Last 6 weeks. Up to 40% abort	Metritis, septicaemia, mortality in ewes usually	Isolation of organisms from foetal stomach	Agglutination test	Avoid predisposing conditions
Listeria ivanovii	Ingestion ? inhalation ? coitus	Last 6 weeks and postpartum lamb deaths. 2% to 20% abort	Metritis, septicaemia in some ewes	Isolation of organisms from foetal tissues	None	Avoid feeding (poor-quality) silage
Brucella ovis	Coitus, ingestion	Late abortion but usually low birth-weight lambs. Very low incidence	Epididymitis in rams	Isolation of organisms from foetal stomach, placenta	CF test of ewes and rams	Eliminate infection in rams. **B ovis** vaccine, twice or with **B abortus** strain 19
Pestivirus	Ingestion at 12 to 80 days gestation	Low incidence, variable timing	'Hairy shaker' lambs	Virus isolation, histopathology of foetal nervous system	Not useful	Remove surviving lambs from flock, as they shed virus. Vaccination
Akabane virus	Insects, particularly *C brevitarsis*	Variable timing and premature weak lambs	'Dummy' and arthrogrypotic lambs also born	Histopathology, virus isolation not possible	Agglutination test	
Chlamydia abortus	Ingestion	Last 3 weeks	No premonitory sign, necrotic placentitis, thick, leathery intercotyledonary area, foetus looks normal, none mummified	Chlamydial EBs in cotyledon smears	Serology of little use because antibodies widespread	Usually immune after abortion. Vaccination
B melitensis	Ingestion, etc.	Last 10 weeks	Dull grey cotyledons	Foetal stomach isolation	CF test	Vaccination
Rift Valley fever	Mosquito		Heavy mortality in lambs	Viral antigen in blood	Several tests	Vaccination
Wesselsbron disease	Mosquito					

ewes require additional nutrients to support adequate foetal growth and, at the same time, maintain a satisfactory ewe condition score. During the last three to four weeks of gestation nutrients are also required to support mammogenesis.

Table 7.9 gives some data for the birth weights of single and twin lambs as well as changes in ewe liveweight associated with three different levels of nutrition during only the last five weeks of pregnancy. Note that at this late stage of gestation, adjustments to feeding had a greater influence on the birth weights of twin rather than single lambs. These three levels of feeding are also likely to influence udder development and maternal behaviour in the ewes at lambing. As noted above, the optimisation of lamb birth weights may require that adjustments to feeding commence at least eight weeks before lambing.

Undernutrition of ewes in late pregnancy leads to a spectrum of disease ranging from minor losses of production, to increased lamb mortality, reduced lactation and, at the most obvious, to mortalities of ewes. Many of the syndromes which are subclinical at lambing time have serious consequences in later weeks and months — including death of lambs at a few weeks of age from malnutrition possibly aggravated by intestinal parasitism, poor growth rate of lambs to weaning, difficulties in weaner management in summer and autumn as a consequence of low weaning weights, and poor wool production and subsequent reproductive performance from the ewes. Pregnancy toxaemia, with its high mortality rate of ewes, is the clinical 'tip' of a subclinical 'iceberg' of lost productivity from both ewes and their lambs.

Pregnancy toxaemia

This is a disease of ewes in late pregnancy characterised by dullness, inappetence and recumbency which, unless treated early, progresses to death within a few days of the first appearance of clinical signs. The condition arises when dietary and body reserve sources of glycogenic precursors are unable to meet the glucose requirements of the ewe and the foetus or foetuses. Initially, hypoglycaemia and hyperketonaemia are the predominant physiological changes but, without successful early intervention, a more generalised and irreversible metabolic collapse develops.

Predisposing factors

In general terms, the disease occurs when the nutrition of a ewe with multiple foetuses is inadequate, but the condition is often precipitated by husbandry events which interrupt the feed intake for a period of 12-24 hours. Some ewes are more predisposed to the disease than others.

Table 7.9: Effect of level of nutrition of grazing ewes during the last five weeks of pregnancy on birth weights.

Level of nutrition	Liveweight change of ewes (kg)*	Birth weights (kg)	
		Singles	Twins
High	11.4	4.7	4.0
Medium	5.7	4.6	3.8
Low	0.0	4.4	3.3

* liveweight change includes weight of uterus and contents

Risk factors include

- twin or higher order pregnancy, rather than single
- older ewes, rather than maiden ewes
- low feed quality or quantity in the last four weeks of pregnancy, particularly if the nutritional status of the ewe at joining was high
- sudden interruptions of feed intake for 'at-risk' ewes, such as
  - yarding for crutching or shearing or any other husbandry procedure which leads to the cessation of feed intake for more than 12 hours
  - intercurrent disease such as foot abscess, footrot or hypocalcaemia
  - digestive disturbance, such as acidosis from supplementary feeding with grain
- over-fatness in ewes, which appears to reduce feed intake in late pregnancy.

Pathogenesis

Ruminants absorb virtually no glucose from the digestion of their diet. Ruminal digestion of all carbohydrates, from simple sugars and starches to cellulose and other complex substances, produces volatile fatty acids (VFA) which are absorbed from the digestive tract. VFA are relatively small molecules consisting of two, three or four carbon atoms. The three VFA which are most important are, in their ionised forms, acetate, propionate and butyrate, and these three provide most of the energy supply for the animal. The type of diet influences the relative proportion of acetate or propionate produced in the rumen. With roughage diets, acetic acid predominates, with approximately 3.5-4 moles produced for every single mole of propionic acid. Diets with a higher content of starch and soluble carbohydrate lead to higher levels of propionate production. Consequently, as the grain content of a diet increases, the proportion of propionate produced increases, relative to acetate. The third of the VFA — butyrate — is produced in relatively small amounts in the rumen and is largely converted to the ketone bodies acetoacetate and ß-hydroxybutyrate in the ruminal epithelium[160] before entering the portal circulation.

There are three ketone bodies — acetoacetate, ß-hydroxybutyrate and acetone. The first two are four-carbon (C_4) molecules and acetone (a C_3 molecule) is formed by the loss of a carboxyl group from acetoacetate. Acetoacetate and ß-hydroxybutyrate can be readily converted to acetyl-CoA in many body tissues and, consequently, used for energy production, provided that other substrates of the TCA cycle are available.

Ruminants have adapted to their lack of dietary glucose by meeting most of their energy needs directly by oxidation of VFA and ketone bodies. Acetate, ketone bodies and VFA are the substrates for 60% to 70% of the oxidative energy production of sheep but the energy requirements of the brain, eye, erythrocytes, foeto-placenta and mammary gland (for the production of lactose) must be met by glucose — which must therefore be produced by gluconeogenesis.

Because glucose is not absorbed from the digestive tract of ruminants (in contrast to monogastric digestion) but is required for the function of some tissues, gluconeogenesis is a constant metabolic process in ruminants.

During periods of fasting or other interruptions of feed intake, body tissues are catabolised to provide energy-producing substrates. The principal source of stored energy is fat and so, when dietary sources of VFA are insufficient, lipolysis releases fatty acids for energy production.

Gluconeogenesis. Gluconeogenesis is the process through which the body produces glucose, thus maintaining blood glucose levels and ensuring there is sufficient glucose available for those tissues which depend on it as an energy-producing substrate. The gluconeogenic pathway commences with pyruvate and oxaloacetate — substances which may be derived from a number of sources (glucogenic amino acids, lactate, glycerol) but, in adequately fed ruminants, propionate is the principal substrate for gluconeogenesis.[161] Gluconeogenesis occurs in the liver and a limited number of other body tissues.

Glycogen is a polysaccharide of glucose which serves as a glucose storage mechanism. Breakdown of glycogen is another method by which blood glucose levels can be maintained.

Acetate and propionate metabolism. Acetate is absorbed from the digestive tract and enters the bloodstream. It can be converted to acetyl CoA and used directly to produce energy via the tricarboxylic acid (TCA) cycle in several body tissues including skeletal muscle, heart and kidney, thus sparing glucose as an energy source. Supplies of acetate beyond that which is needed immediately for energy production are used for fat synthesis and storage in adipose tissues — provided glucose is also available for glycerol synthesis.

When feed intake declines acetate concentration in plasma also falls — acetate is not a significant transportable energy resource like glucose, fats or ketone bodies.

Fat stores, glycerol and fatty acids. Adipose tissue is composed principally of triglycerides — esters composed of three fatty acids and one glycerol molecule. When adipose tissue is mobilised to meet energy demands, glycerol and fatty acids are released. The breakdown of the ester molecule releases non-esterified fatty acids (NEFA, also known as free fatty acids or FFA), which are large fat molecules (C_{16}, C_{18}) released into the bloodstream, on an albumin carrier. Both glycerol and NEFA are used for energy production — glycerol for gluconeogenesis, producing glucose for energy production, and NEFA for oxidation, with the production of acetate. The reverse process — the production of triglycerides — will occur when both acetate and glucose are available. When glucose is in short supply, synthesis of adipose tissue ceases and fat stores are broken down to free glycerol for gluconeogenesis.

Ketone bodies. The oxidation of NEFA to form acetate and other C_2 compounds occurs in the hepatic mitochondria. Most acetate formed in this way is condensed into acetoacetate and passed out of the mitochondria; some is then converted to β-hydroxybutyrate and some to acetone in the hepatic cell cytosol. Ketogenesis occurs at two sites in the body — β-hydroxybutyrate is formed in the rumen epithelium and liver from butyrate absorbed from the rumen, adding to the production of the ketone body from the metabolism of NEFA in the liver.

Ketone bodies enter the bloodstream from the liver and are distributed to tissues which can use acetate as an energy source (heart, kidney, skeletal muscle) or a substrate for fat synthesis in adipose tissues and, in lactating animals, the mammary gland, where it is used for the production of milk fat. Ketone bodies are an important energy source in ruminants and the oxidation of acetate through the TCA cycle completes the capture of energy from the breakdown of FFA.

The catabolism of acetate as a source of energy in the TCA cycle requires the availability of the C_4 molecule, oxaloacetate, to form the C_6 molecule citrate. Oxidation of citrate produces energy (captured in NADH) and CO_2 and releases succinate which is then returned to oxaloacetate through the pathway of the cycle (Figure 7.8). When there is high demand for glucose (to prevent hypoglycaemia), cellular levels of oxaloacetate may decline to such an extent that the pathway for oxidation of acetate is reduced and acetate accumulates in the mitochondria and is removed to the cytosol as ketone bodies.

Two steps, therefore, lead to the accumulation of ketone bodies in the hepatic cells. The first is the mobilisation of fatty acids and their transport to the liver and other tissues for oxidation — usually in response to a negative energy balance but also in response to endocrine signals during late pregnancy and early lactation — and the second is a high demand for oxaloacetate for gluconeogenesis. It should be noted that ruminant diets which supply higher levels of propionate compensate to some extent for the shortage of oxaloacetate because propionate enters the TCA cycle by conversion to succinate, via methylmalonate. Diets with a high content of soluble carbohydrates, such as occurs in grains, are therefore important for ruminants with high requirements of glucose.

When the supply of NEFA to the liver exceeds the capacity of the liver cells to produce ketone bodies — the products of NEFA partial oxidation — fat accumulates and is stored in

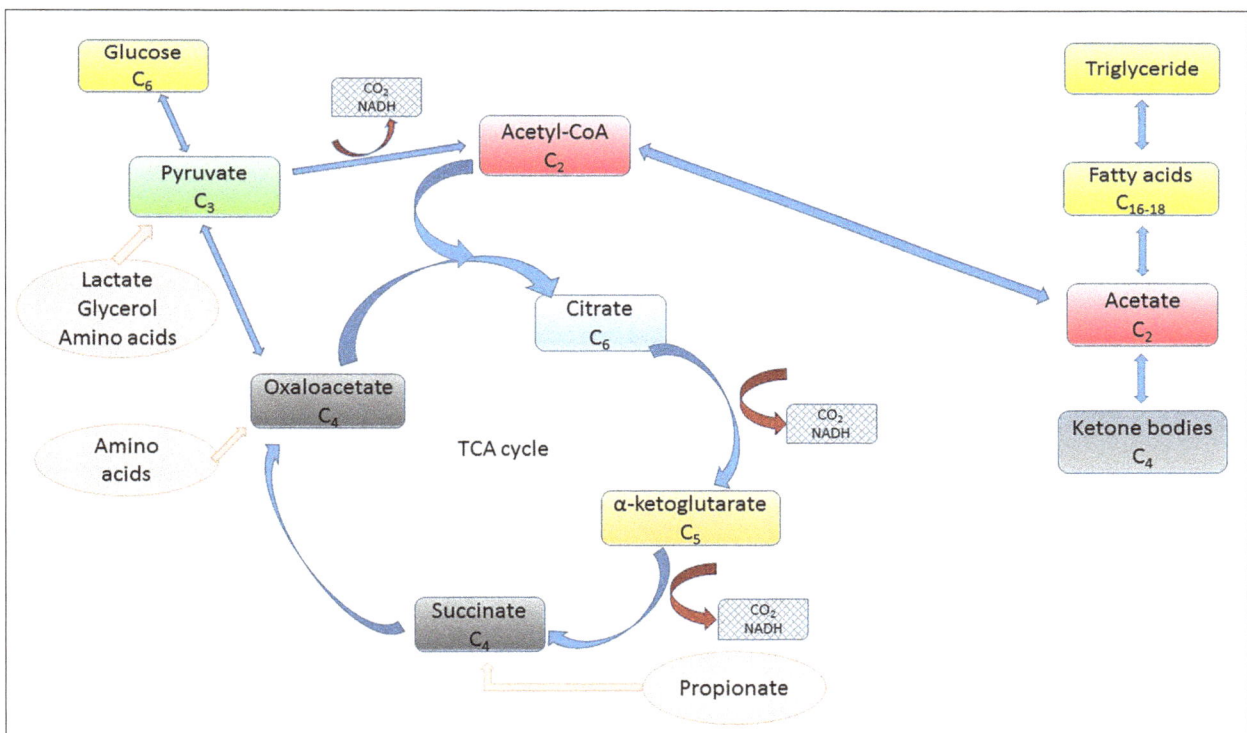

Figure 7.8: The TCA cycle or citric acid cycle provides a metabolic pathway for energy production from glucose or from fats. The common link is the two-carbon intermediary acetyl-CoA, but its incorporation into citrate and subsequent oxidation is dependent on the availability of oxaloacetate. Oxaloacetate is also a substrate for glucose production (gluconeogenesis). Arrows with double heads indicate reversible pathways, those with single heads indicate irreversible pathways. Source: KA Abbott.

the liver as triglycerides. When large amounts of triglycerides accumulate the liver becomes grossly 'fatty' — pale, swollen and greasy to touch. When metabolic conditions permit, triglycerides stored in the liver can be mobilised and transported to other tissues in the body.

Release of NEFA from adipose tissue occurs during periods of hypoglycaemia: when glucose is in short supply, glycerol is used for gluconeogenesis and triglyceride breakdown proceeds faster than synthesis. NEFA are used for energy production by complete oxidation — without the accumulation of ketone bodies — in many of the body tissues but much of the released NEFA is taken up by the liver, producing ketone bodies and, if gluconeogenesis is already consuming all available oxaloacetate, a fatty liver.

Hypoglycaemia, gluconeogenesis and ketogenesis, therefore, tend to occur together. When there is a high demand for glucose or a shortage of glucose precursors, or both, blood glucose levels tend to fall, the stimulus for gluconeogenesis becomes stronger, fats are mobilised but cannot be oxidised beyond the level of acetate in the liver, ketone bodies accumulate and their concentration in the bloodstream rises.

Foetal requirement for glucose. Foetal blood concentrations of acetate and ketone bodies are low relative to maternal concentrations. The foetus, however, has a high requirement for glucose and will take up 8-9 g/kg of foetal weight per day. A single foetus of 5 kg requires approximately 45 g of glucose daily. For comparison, a dietary intake of 750 g of roughage — sufficient to maintain a 50 kg non-pregnant ewe — will provide about 110 g of glucose. Maternal hypoglycaemia is readily induced in twin-bearing ewes in late pregnancy because of the high foetal demand for glucose and the limited amounts produced by gluconeogenesis from propionate and amino acids. The foetus is also highly efficient at capturing maternal glucose — foetal blood levels may remain near normal even when the ewe is severely hypoglycaemic and has virtually no liver glycogen reserves.

Development of pregnancy toxaemia. Provided they are adequately and consistently fed, ewes can cope with the demands for glucose of their placentae and foetuses and can maintain their blood glucose concentrations within a normal range. During late pregnancy, it is normal for a moderate degree of hypoglycaemia to develop but submaintenance feeding can lead to very low blood glucose levels. While non-pregnant ewes normally maintain blood glucose levels above 30 mg/100 mL, in moderately underfed ewes around 115 days pregnant, glucose levels may fall to 25-30 mg/100 mL in single-bearing ewes and 15-20 mg in those carrying twins. More severe underfeeding depresses blood glucose further in multiple-bearing ewes, but has less effect in those carrying singles.

Depression of blood glucose levels is accompanied by a rise in blood ketones, to levels 10 or more times higher than that of non-pregnant sheep. The effect is much greater in ewes with multiple foetuses than in ewes with single pregnancies and is directly related to the degree of undernutrition (Figure 7.9).

Maternal glucose homeostasis. The development of hypoglycaemia and hyperketonaemia in the late gestation, twin-bearing ewe which suffers an interruption to feed supply is not simply a consequence of the high foetal demand for glucose but also a result of the impaired ability of ewes in this physiological state to respond to the low dietary glucose supply with increased

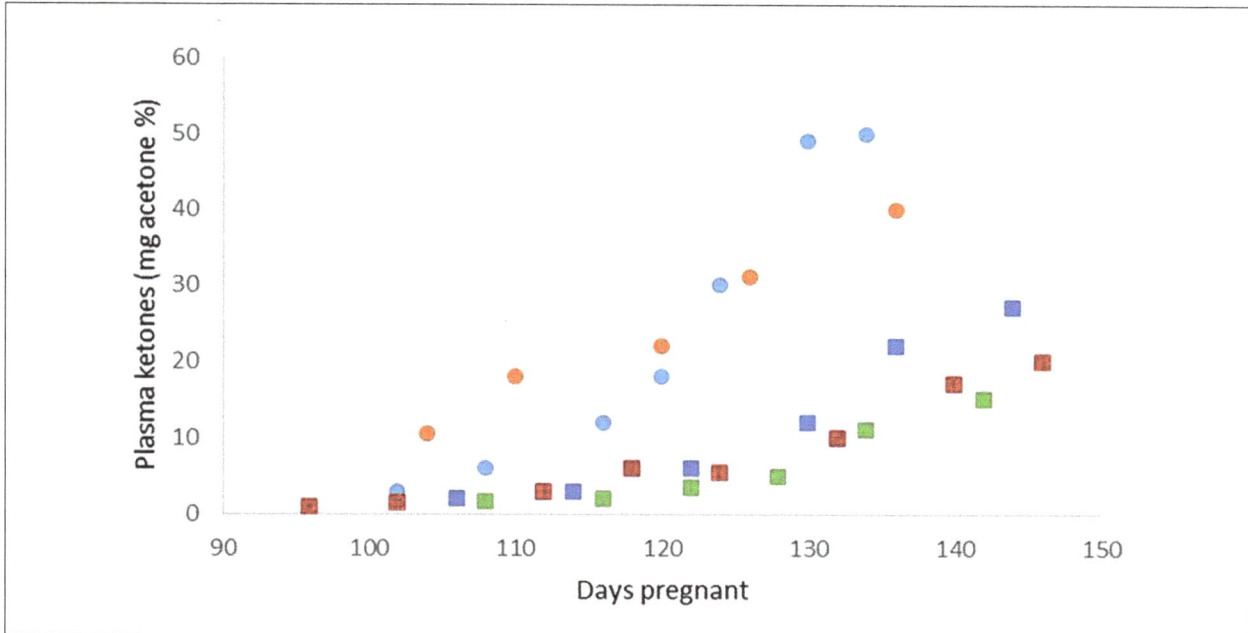

Figure 7.9: The relationship between diet and hyperketonaemia in twin-bearing ewes fed chopped hay. Two ewes fed daily at 0.75 kg feed per 50 kg bodyweight (circles) and three ewes fed at the rate of 1.0 kg per 50 kg (squares). Maintenance for the non-pregnant ewe is 0.75 kg per 50 kg per day. Drawn by KA Abbott. Based on data from Reid (1968).[162]

glucose production.[163] In contrast to the period of late pregnancy, ewes in early lactation — when the demand for glucose for milk production is even higher than for foetal nutrition — can respond effectively to hypoglycaemia with increased gluconeogenesis. The metabolic, hormonal and homeostatic mechanisms which lead to the development of pregnancy toxaemia in ewes are not completely understood but are clearly more complicated than the development of ketosis in the early-lactation dairy cow. The more profound metabolic disturbance of pregnancy toxaemia in ewes is the reason that the condition is much more difficult to correct with appropriate dietary supplementation than is ketosis in cows, and explains why the mortality rate in affected sheep is much higher than in ketotic cattle.

There is considerable variation between individual sheep in their susceptibility to pregnancy toxaemia which appears to be related to their ability to maintain glucose homeostasis in the face of interruptions in feed intake. Sheep which are more susceptible to pregnancy toxaemia have a greater degree of insulin resistance than those which are less susceptible[164], and are less able to make an effective hepatic gluconeogenic response when starved in late pregnancy.

Development of clinical signs. Severe hypoglycaemia, hyperketonaemia and weight loss can occur in ewes without the development of clinical pregnancy toxaemia. The reasons for the development of clinical signs in some cases of hyperketonaemia and not in others are not clear.

Elevated plasma cortisol. Although hypoglycaemia and hyperketonaemia do not consistently lead to the development of clinical disease, one consistent association is the presence of elevated blood cortisol levels in ewes with clinical signs — a feature not present in hyperketonaemic, but clinically normal, ewes. The high level of serum cortisol is a manifestation of the ovine disease which is not present in bovine ketosis and which, among other things, may lead to the wool break in affected but surviving ewes.

Progression to irreversibility. Following persistent hypoglycaemia and hyperketonaemia for one to three days, further metabolic changes occur which are effectively untreatable. Excessive production of ketone bodies produces an acidosis. Prolonged urinary excretion results in loss of sodium and potassium and a lowering of plasma alkali reserve. With sudden cessation of dietary intake the hyperketonaemia worsens, exacerbating the CNS depression. Ewes become comatose, dehydrated, anuric and uraemic.

Clinical signs

In an affected flock, pregnancy toxaemia usually appears as a continuing outbreak over a period of two to three weeks, with a few ewes developing clinical signs each day. The course of the disease is usually four to seven days, although ewes are not always observed in the early stages. Affected ewes separate from the mob and, initially, stand with head held low, appearing depressed. They do not graze and are easily approached. They appear blind but may make some movement to face an approaching dog or human, or to walk away (Figure 7.10). The gait is staggery and weak and they collapse readily, particularly as the disease progresses.

The pupillary light reflex is diminished and the eye preservation reflex is absent. Ruminal movements are normal or reduced. They are usually constipated, disinterested in food and

Figure 7.10: In the early stages of pregnancy toxaemia, ewes cease grazing, separate from the flock and are easily approached. Source: KA Abbott.

become more sleepy as the disease progresses. Recumbency follows two to three days after the ewes are first observed to separate from the flock. Death follows in a further few days.

Signs of nervous derangement may be observed at all stages of the disease. These include muscle tremors of the face, jaw champing and lateral or dorsal head flexion. There may even be tonic-clonic convulsions. Foetal death is a sequel when the disease is prolonged.

Clinical pathology

The presence of ketone bodies in the plasma and urine may be detected using sodium nitroprusside reagents (Acetest® tablets, Ketostix® test strips). Tenfold dilution of urine before testing is recommended to reduce the possibility of false positive reactions. Serum β-hydroxybutyrate levels are elevated. In advanced cases, plasma bicarbonate concentrations are measurably reduced and BUN concentrations elevated. Blood glucose levels are generally not helpful to diagnosis because ewes may become hyperglycaemic as the disease progresses.

Necropsy

The presence of two or more near-term foetuses is the usual finding. The liver is fatty and pale yellow, friable and greasy on cut section. The adrenal glands are enlarged.

Diagnosis

The disease must be differentiated from hypocalcaemia, which is more characterised by paralysis and muscle weakness and, if uncomplicated, responds quickly to treatment with calcium borogluconate. Pregnancy toxaemia is, however, often superimposed on hypocalcaemia unless the latter is treated promptly.

Treatment

The response to treatment depends on the manner in which the disease developed and the duration of time which passes before treatment is instituted. A small proportion of cases will recover spontaneously without treatment, either because dietary intake recommences or because the foetuses are born or die *in utero*. In this latter case, death from septicaemia is also a likely outcome.

In cases which have occurred following sustained undernutrition, treatment is of little value, possibly because acidosis and renal failure are present from the outset. If heroic treatment is commenced, intravenous electrolytes, bicarbonate solutions, glucose and insulin as indicated are rational therapies. Oral administration of glucose precursors (propylene glycol, glycerine) is successful at raising blood glucose concentrations but rarely leads to clinical recovery. In fact, many cases are so advanced when presented that hyperglycaemia may be present before treatment can start.

In those cases where the clinical signs have developed as a result of more sudden deprivation, treatment is more likely to be successful. Treatment, to have any chance of success, must commence early and be vigorous and frequent. Once cases advance to permanent recumbency, treatment is unlikely to succeed.

Concentrated oral rehydration solutions, containing glucose, glycine, NaCl, KH_2PO_4, potassium citrate and citric acid have been shown to be more effective at raising plasma glucose concentrations than the same amount of glucose administered alone. A field trial using the concentrated oral rehydration solution (160 mL every 4 to 8 hours) found that 90% of ewes classified as having mild signs and 55% of those with severe signs recovered completely, producing live lambs.[165] Mild signs were defined as separation from the flock, disinclination to move and anorexia. Severe signs were one or more of the following: blindness, drowsiness or excess salivation. No other treatment was compared in the field trial.

Recommended treatments include

*   rehydration solutions *per os*, for example, *Vy-Trate* (Jurox), 160 mL every 4-6 hours
*   glucose, by injection. 120 mL of 20% dextrose (24 g of glucose) twice daily (the requirements of the ewe and foetuses could approach 200 g daily)
*   glycerol, 120 mL every 6 to 8 hours or propylene glycol, 100 mL every 12 hours
*   anabolic steroids: 30 mg of trenbolone acetate[166]
*   flunixin meglumine, 2.5 mg/kg, by intramuscular injection, daily[167]
*   corticosteroids, such as 10-15 mg of dexamethasone, to induce parturition
*   caesarean section.

Propylene glycol. Propylene glycol (1,2-propanediol — $C_3H_8O_2$) is highly digestible and rapidly cleared from the rumen. Some of it is fermented in the rumen, producing propionate, but most is absorbed and converted to glucose in the liver.[168] By providing gluconeogenic substrates for the TCA cycle, propylene glycol provides increased opportunity for the oxidation of acetyl Co-A, diverting it from further ketogenesis. Propylene glycol increases blood glucose levels, dramatically increases blood insulin levels and reduces concentrations of non-esterified fatty acids and ß-hydroxybutyrate. The stimulation of insulin secretion decreases mobilisation of fatty acids from adipose tissues, further limiting ketogenesis. Some sheep may show signs of toxicity to propylene glycol. In cows, the median toxic dose is reported to be 2.6 g/kg of body weight which, if the same for ewes, would be about 160 g (270 mL) for a 60 kg ewe.[169]

Flunixin meglumine. This potent nonsteroidal anti-inflammatory drug may increase the successful response to treatment of ewes with pregnancy toxaemia when given in addition to dextrose (120 mL of 20% solution) and calcium borogluconate (50 mL), both administered subcutaneously twice daily. In one report flunixin was administered once daily by intramuscular injection for up to three days, the authors warning against intravenous administration.[166]

Prevention

Supplementary feeding with high-energy foodstuffs will reduce the incidence of pregnancy toxaemia. Preferably, the need for supplementation is foreseen before any cases occur, the need being recognised from an assessment of the quantity and quality of the available pasture. Often, however, supplementation is introduced or increased after some ewes have succumbed.

Feeding rates of cereal grains may vary between 400 to 700 g daily, depending on the degree of nutritional deficiency. Hay is unlikely to provide sufficient energy density (megajoules of ME per kg of dry matter or M/D) to prevent the development of pregnancy toxaemia in high risk flocks. Methods of feeding grain and, particularly, the need for care in introduction were discussed in the section on handfeeding.

The existence of pregnancy toxaemia in a flock is suggestive of other (probably more important) subclinical losses of production from high lamb mortalities, poor lamb growth and poor ewe health and production. This subject is discussed further by Foot (1983).[170]

In light of that, producers whose flocks chronically suffer from the disease, or who frequently need to take action to prevent it, should possibly consider changing their management strategies for the lambing of ewes. If the disease is related to poor pasture quality, lambing at a time when feed quality is higher will probably be advantageous. If the disease is related to low pasture availability, decreasing the stocking rate of the ewes is advisable. If pregnancy toxaemia is related to husbandry procedures (crutching, shearing, etc.), changes should be made to ensure that the ewes have shorter periods off-feed and less disturbance close to lambing.

Identification of ewes with multiple pregnancies by ultrasound scanning can be used to form mobs of multiple-bearers which receive differential treatment. This strategy is most cost effective when the lambing rate is around 150% (because about half of the ewes are carrying multiple foetuses.)

Monitoring of body weight and condition score can give useful guides to the need for preventive action. Single-bearing ewes which maintain uterine-free body weight will gain at least 8-9 kg of liveweight over pregnancy (mostly in the last eight weeks) due to the mass of the foetus, foetal fluids and udder development. As discussed previously, such objectives are both difficult to meet in the field and unnecessary in commercial sheep production, from an economic viewpoint. Satisfactory production and health are probably associated with liveweight gains of 2-5 kg in this period, equivalent to the loss of 0.5-0.7 of one condition score. The ability to lose that amount safely depends on bodily condition in early pregnancy — lean ewes should not be allowed to lose that much; fat ewes should be limited to losing no more than that amount of condition. For ewes with twin foetuses, liveweight gains of 5-8 kg are desirable.

Hypocalcaemia

Hypocalcaemia in ewes generally occurs in the last month of pregnancy rather than during lactation. The clinical syndrome is usually precipitated by sudden changes in feed intake or by stress. Moving ewes to another paddock and driving for husbandry procedures, particularly if the weather is cold and wet, can readily precipitate outbreaks. Generally, only a small proportion of the flock is affected. Occasionally the disease occurs in association with exposure following shearing, and large numbers of ewes may be affected. An increase in the incidence of hypocalcaemia of ewes two to six weeks before lambing was observed in the six months following the end of the 1982-83 drought in southern Australia. Most cases occurred spontaneously and some were precipitated by pre-lambing crutching.[171] Older ewes are usually more frequently and more severely affected. Hypocalcaemia also occurs in weaners.

Clinical signs

Initially, affected ewes are ataxic and hyperaesthetic but soon become recumbent, and the paralysis becomes flaccid. They are frequently seen in sternal recumbency with legs stretched behind and head turned to the flank (Figure 7.11). The pupils are dilated. Unless treated, affected ewes usually die within one to two days of collapse.

Treatment

50 mL of 40% calcium borogluconate injected slowly intravenously or subcutaneously is effective, particularly when given early. Failure to respond may be due to intercurrent pregnancy toxaemia. In ewes which do respond, treatment with oral glucose precursors is advisable in an attempt to avoid pregnancy toxaemia.

Vaginal prolapse

Vaginal prolapse occurs in the pregnant ewe up to four weeks before lambing. The incidence in Merino ewes is generally low but may exceed 5% annually in some flocks of British breed sheep. There appears not to be any one aetiological factor but a number of factors which may contribute variably to the development of the syndrome. A major factor is probably the increase in intra-abdominal mass in late pregnancy, particularly in ewes on bulky feed. When the ewe lies down, the intra-abdominal pressure is transmitted to the flaccid pelvic structures, tending to balloon

Figure 7.11: Ewe with hypocalcaemia. Usually hypocalcaemia affects ewes in late pregnancy but this ewe was affected several weeks after lambing. The condition was associated with grazing wheat and the ewe responded quickly to intravenous and subcutaneous calcium borogluconate solutions. Source: KA Abbott.

the relaxed and loosely attached vaginal walls through the lips of the vulva. The exposed mucosa becomes dry and irritated, stimulating tenesmus which exacerbates the prolapse.[172]

Treatment

Treating the prolapse by reduction is useless if it has been present for so long that the mucosa is devitalised or the submucosal tissues bruised or torn. Early cases can be treated by reduction after disinfection with mild antiseptics (suitable for sensitive mucous surfaces) and retention by a variety of means. Epidural anaesthesia, by reducing tenesmus for an hour or so, will facilitate reduction and allow a period following reduction for some oedema to resolve before the ewe is aware of any discomfort and stimulated to strain again. Retention is generally attempted with sutures of umbilical tape, either across the vulva or, preferably, in a *purse-string suture* lateral to the vulval labiae. These must be removed before the ewe lambs, so it is necessary to confine the ewe for observation after treatment. With nervous Merino ewes, it is necessary to confine them in a paddock or yard in the company of five or six other sheep at least; otherwise, the stress of isolation and panic when approached will not favour a successful outcome. Antibiotic therapy, locally and parenterally, is advisable.

Ewes should be culled after they have raised their lamb because of the risk of recurrence at the next lambing. If a number of ewes are affected, steps should be taken to reduce over-fatness, to reduce the bulkiness of the diet, and to increase exercise of the ewes, where these actions are appropriate, in future pregnancies.

Husbandry at lambing

In order for the lamb to survive, both the ewe and the neonatal lamb must adopt specific, appropriate behaviours within minutes of the birth to ensure prompt initiation of teat-seeking and sucking by the lamb and the formation of a strong ewe-lamb bond. The husbandry of the ewe leading up to and during the birth has a strong influence on her nutritional state, her demeanour and consequently her behaviour during and after the birth.

To facilitate high lamb survival rates, ewes should lamb in good nutritional condition (a condition score around 3 is optimal) and in sheltered, comfortable surroundings with which the ewe is familiar. Lambing paddocks should have shelter from cold winds, should be large enough that stocking densities are moderate only and ewes can seek isolation during lambing, and should provide easy access to the farm operator for inspection of the flock and assistance of ewes and lambs if necessary. Ewes should be placed in the lambing paddock several weeks before lambing starts so that they avoid any distress or disturbance in their feed intake during late pregnancy and so that they are settled and familiar with their environment and flock mates before lambing.

In many commercial flocks there is no surveillance of lambing. If management is more intensive, some degree of surveillance is both desirable and feasible. Twice-daily inspections by an experienced shepherd (in the early morning and late afternoon) provide the opportunity to intervene in any dystocias with good outcomes for the ewes at least, and possibly to save lambs. If surveillance is planned, the producer should walk or drive through the flock on several occasions in the two to three weeks prior to the start of lambing to accustom the ewes to inspections.

Intensive management at lambing

In principle, provided that sufficient time and resources are invested, it is possible to reduce lamb losses almost to zero. In practice, the producer must consider how much cost and effort is warranted to reduce lamb losses. The manipulation of lamb birth weights by feeding according to foetal number after pregnancy diagnosis, the provision of shelter and the practice of frequent surveillance have been mentioned. Less commonly, other procedures include drift lambing where, with suitable subdivisional fencing, the lambing flock is gently moved each day to the next paddock, with a three- or four-day rotation. This enables small groups of separated lambing or just-lambed ewes to be left behind and alone, until the main mob catches up with them a few days later. Pen lambing of individual or small groups of ewes is more intensive and expensive. Ewes giving birth to multiples may be confined individually in pens to facilitate bonding to all lambs; two days of confinement may be required.

Shearing and lambing

The annual timing for shearing and lambing should be adjusted such that lambing commences when ewes have two to nine months of wool growth. Shearing closer to lambing has two negative consequences; first, the ewes have the stress of handling, yarding and removal from feed superimposed on the nutritional stresses of late pregnancy; and, second, if pre-lambing shearing occurs in the cold months of the year, the additional nutritional requirements to maintain body temperature off-shears are added to the nutritional requirements of pregnancy and lactation.

Because lambing often occurs during cold, wet months of the year, there have been recommendations to shear ewes very close to lambing with the intention of encouraging ewes to find shelter — to keep themselves warmer — during and soon after lambing, with benefits for the survival of the lamb or lambs. While it is clear that lambing in shelter does improve lamb survival, it is difficult to induce ewes to naturally choose shelter for lambing. Shearing must be within four weeks of lambing in order to change the shelter-seeking behaviour of the flock[173] and shelter must be available and used by the flock, particularly for night shelter so that ewes lambing in the early hours of the morning may remain in a sheltered spot. Given that there are substantial risks to shearing ewes in late pregnancy, the strategy is no longer recommended in Australia.[173]

Shearing nine months or more pre-lambing means that ewes will be shorn within three months or less after lambing commences, which may involve mustering of the ewe flock before lambs are weaned. The presence of unweaned lambs at shearing can significantly increase the complexity of the husbandry activities and it adds to the stresses of ewes, lambs and flock managers. In addition, late pregnant ewes in long wool are prone to become cast, especially when they are wet.

Ewes with more than four or five months of wool growth are usually crutched before lambing commences to reduce the accumulation of birth discharges on the crutch and to facilitate the lambs attempts to find the teat soon after birth.

SECTION D

PERINATAL MORTALITY

Every year in Australia, between 15% and 30% of lambs die in their first week of life.[174] This loss represents a significant biological inefficiency in sheep production systems when one considers the large nutritional and physiological investment by ewes in the development and delivery of a neonate. The production of a full-term lamb has a significant cost in terms of the additional nutrition provided to pregnant ewes — usually by adjusting the stocking density of pastures to ensure that pregnant ewes are adequately fed — and the losses in wool productivity and wool quality through competition for nutrients. The economic value to the Australian industry of a 10% improvement in lamb survival has been estimated at over $50m (not counting the cost of achieving an improvement), which suggests that perinatal mortality of lambs is one of the top six major controllable diseases of sheep in Australia.[175] Putting aside the economic considerations, the strong contribution to lamb survival made by good husbandry and the current level of loss imply that there is, on some farms at least, a substantial gap in the attention given to the welfare of lambing ewes and their lambs. Both economic and welfare considerations are now driving efforts in Australia to improve lamb survival. Some of this may come from genetic selection programmes, but probably most important are improvements in the understanding by sheep producers of the important role of good nutrition and good husbandry in the probability of survival of lambs in their first critical hours of life.

The perinatal period is defined as the time from the initiation of the birthing process until the lamb or lambs are 7 days of age. Once lambs are 7 days old the risk of mortality before weaning falls dramatically; perinatal deaths typically account for over 90% of pre-weaning lamb losses.

The incidence of perinatal mortality is highly variable between farms and between years. The risk of mortality is strongly affected by litter size. Under Australian extensive commercial sheep-raising conditions, it is uncommon for the perinatal mortality rate to be less than 10% for single-born lambs or less than 20% for twin-born lambs. In some surveys, the average mortality rate of lambs on commercial farms has been found to be substantially higher than these figures.

While there are a number of reasons for lambs to die in the perinatal period, many of these are relatively minor on a national basis, although they can cause major losses sporadically or in some particular circumstances. By far the most common reason for lambs to die in the perinatal period is because they fail to ingest sufficient nutrients to survive in the first few hours and days of life.

The reasons that so many lambs ingest insufficient nutrients can be related to difficulties during the birthing process, aberrant behaviour of the ewe and lamb during and after birth, or severe environmental conditions at the time of birth. The first two of these factors are strongly interrelated and, additionally, the nutritional status of the ewe has a very strong bearing on the likelihood of each occurring. Lamb birth weight is another factor that interplays both with the likelihood of difficult birth (in the case of large lambs) and with the susceptibility of lambs to severe chilling at birth (in the case of small lambs). Each of these factors will be discussed in turn, but it is important to remember the strong interactions that exist between all of them in their effect on lamb survival, and the strong role that nutrition of the pregnant ewe plays in determining their relative importance.

Lamb birth weight

The mean birth weight of Merino lambs of a medium-sized type is around 5 kg for single lambs and 4 kg for twin lambs (Table 7.10). Birth weight is strongly influenced by the age and

Table 7.10: Under similar nutritional conditions, the birth weight of lambs is affected by litter size and breed. The mean birth weights are shown for lambs sired by Merino (Mo) and Poll Dorset (PD) rams over Merino and Border Leicester × Merino (BL-Mo) ewes. Based on data from Holst et al. (2003).[175]

Sire x Dam	Mean birth weight (kg)		
	Mo x Mo	PD x Mo	PD x BL-Mo
Single	5.1	5.3	5.5
Twin	4.0	4.3	4.6
Triplet	3.4	3.5	3.9

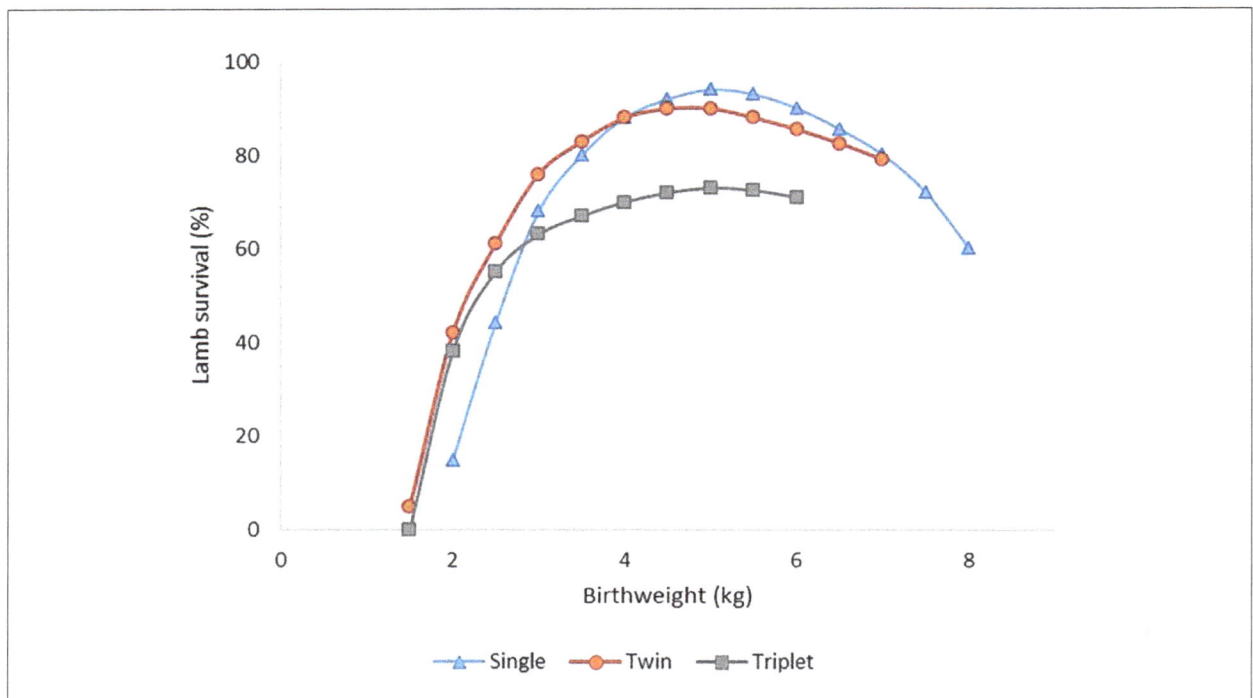

Figure 7.12: The relationship between lamb survival and birth weight is U-shaped, with the optimum birth weight around 4.5 to 5 kg. The relationship is illustrated for single-born, twin-born and for triplet (and higher order) lambs. Drawn by KA Abbott. Adapted from Holst et al. (2002)[175], with permission from CSIRO Publishing.

breed of the dam, her nutritional level before and during gestation, and litter size. Male lambs are usually about 0.2 kg heavier than female lambs. The breed of sire also influences birth weight.[176]

Birth weight has a profound effect on the survival rate of lambs. The relationship between the survival rate of lambs and their birth weight is often described as (inverse) U-shaped because the risk of mortality is highest for heavy lambs as well as light lambs (Figure 7.12). Lambs around 5 kg have the highest survival rate, although this varies with breed (mature size) and ewe age. In New Zealand, with Romney and Romney-type breeds, the optimum birth weight for survival is in the range from 5.5 to 6 kg, which is 0.5 to 1 kg higher than the mean lamb birth weight.[177]

Figure 7.13: The most common normal presentation of a lamb is anterior, with both forefeet and nose present at the vulva. The lamb is usually delivered within 30 to 60 minutes of the appearance of the feet at the vulva. Photograph courtesy of LA Abbott.

The birthing process

Ewes normally deliver their lamb within one hour of the commencement of straining in the second stage of labour. This stage of labour can be as short as a few minutes — as it is typically for the second-born lamb of twins — or, uncommonly, two to three hours.[178] About 50% of ewes deliver their lamb within 30 minutes of the first appearance of the lamb at the vulva, and 90% within one hour (Figure 7.13). Births are a little slower for primiparous ewes than for multiparous ewes, and slower for larger lambs than small lambs.[179]

The term 'dystocia' includes parturition that is difficult and prolonged but successful without assistance, or obstructed and, therefore, impossible without human assistance. Dystocia may lead to the delivery of a live lamb with or without human assistance, a dead lamb with or without assistance, or the lamb remaining in the birth canal of the ewe, usually leading to the death of the ewe. Ewes which deliver a live lamb after a prolonged labour are more likely to desert their lamb than ewes with a short labour, particularly if the ewe is primiparous or weak from poor nutrition, or both.[179] Lambs born after a prolonged labour are often slower to exhibit the normal behaviours of lambs and may suffer significant trauma during the birth, as discussed further below.

Compared to well-nourished ewes in a medium condition score, dystocia leading to prolonged labour is more common in ewes which are undernourished and become weak during labour and in ewes which are over-fat. Dystocia due to foeto-pelvic disproportion is more common with large singleton lambs than with twin or triplet lambs. The decline in survival rate of lambs over 5.5 kg (Figure 7.12) is largely a consequence of dystocia. Malpresentations leading to dystocia can occur with lambs of any size. The normal presentation for a lamb is anterior, with both forelimbs extended such that the lower jaw lies on the metacarpus as the feet and then nose are presented at the vulva. Ewes can deliver lambs unassisted with minor malpresentations, such as flexion of the shoulder and elbow, or one forelimb extended posteriorly, or posteriorly (hind feet first), but these presentations make delivery more difficult and generally slower.

Some degree of dystocia in ewes is reasonably common, perhaps occurring in 5% to 10% of births[180], but it can be much more common if the sire of the lambs is of a breed or type that is much larger than the breed or type of the ewe, or if the ewes are young and small, or if phyto-oestrogenic pastures have interfered with cervical relaxation. Dystocia is more common in some breeds than others. The incidence of dystocia in Poll Dorset ewes can be high and has been reported to be responsible for the death of over 15% of lambs born. The shape and dimensions of the ewe pelvis were considered to be a major contributing factor to the high incidence in the breed.[181] Rams of breeds selected for rapid growth — the terminal sire breeds in particular — may produce lambs which are very large or of a shape (broad-shouldered) which complicates delivery, but many progressive breeders include easy-birth characteristics in their genetic selection programmes.

Under extensive grazing conditions, daily inspections of lambing ewes can save the lives of many ewes with dystocia but the survival rate of lambs from assisted births is low because many births will have been prolonged before assistance is rendered. Frustratingly, when live lambs are delivered with human assistance in the field, ewes frequently desert the lamb immediately after being released, because of their fear of humans, the distress generated during capture and the difficult birth itself. The tendency of assisted ewes to desert their lambs can be overcome to some extent by penning the ewe and the lamb or lambs together, with portable hurdles, for example, and leaving them undisturbed for 24 hours. By the time they are released, a satisfactory ewe-lamb bond has often formed and the lamb or lambs have been fed by the ewe. Skilful and experienced shepherds are capable of catching and assisting ewes and confining them in a small pen with minimal disturbance of the remainder of the flock.

Nevertheless, some producers choose not to inspect lambing ewes closely because of the risk of causing mismothering of lambs by disturbing recently lambed or lambing ewes. The risk of disturbance can be reduced by habituating the ewes to the presence of a vehicle and shepherd before lambing starts.

Normal ewe and lamb behaviour at birth

Under extensive conditions, most ewes isolate themselves before lambing by either leaving the flock or allowing the flock to move away while grazing and remaining on the chosen birth site. As labour commences, amniotic fluid is spilled on the ground and ewes tend to remain in close proximity to that site. The odour of the birth fluids is a strong attractant to the ewe. Ewes will often alternate between recumbency and standing during the early stage of labour and usually deliver the lamb, at least to the hips, while recumbent. Most ewes stand as soon as the lamb is delivered or, if the hindquarters remain in the vagina, delivery is completed when the ewe stands.

Figure 7.14: Grooming of the lamb commences immediately after the lamb is delivered and, during this period, there are exchanges of olfactory and auditory cues between the ewe and lamb which are critical to the formation of a strong bond. Photograph courtesy of LA Abbott.

Grooming of the lamb by licking commences immediately and, in most cases, the ewe vocalises, uttering a low-pitched bleat or 'rumble' — a vocalisation almost exclusively reserved for the period immediately before and after parturition (Figure 7.14). As the newborn lamb shakes its head and starts to attempt to rise, it also commences bleating. Both olfactory and auditory cues exchanged between ewe and lamb in the few hours after birth are important in the formation of the ewe-lamb bond and in leading to the behaviours of the ewe, so that the ewe protects the lamb and facilitates its sucking in the immediate postpartum period and continues to recognise, protect and accept the lamb in the following weeks.[182]

Within minutes of birth, lambs start to right themselves onto their sternums, shake their heads and move onto their knees as a prelude to attempts at standing. These activities of the lamb occur while the ewe continues to groom the lamb and while both ewe and lamb may be vocalising. Lambs usually stand on all four feet within 30 minutes of birth and commence searching for the udder of the ewe by moving along the side of the ewe and nuzzling her flank and belly. The ewe may stand still, facilitating the lamb's search for the udder, or may retreat or circle, continuing to keep the lamb at her head and continuing to groom the lamb.[183] Circling may continue for some time, appearing to frustrate the lamb's attempt to find the teat, particularly in primiparous ewes, but usually ceases within a few hours.[179] Some circling behaviour on the part of the ewe may be normal and useful, as it continues to orient the lamb at her head, providing a position from which a search for the udder can begin.

When the ewe stands and the lamb's nuzzling approaches the udder, she will arch her back and position her hind leg in a way which raises and presents the teat towards the lamb.

Figure 7.15: Lambs usually stand and commence searching for the udder within 30 minutes of birth. In the case of multiple births, the ewe is likely to deliver subsequent lambs while attending to the first-born lamb or lambs. Photograph courtesy of LA Abbott.

The lamb is then able to locate the teat and suck. Most lambs suck successfully within two hours of birth (Figure 7.15). The speed at which a newborn lamb stands, seeks the udder and teat, locates the teat and sucks successfully is strongly related to the probability of its survival.

Bonding and lamb survival are improved by management practices which increase the time spent on the birth site by the ewe after parturition. If undisturbed, ewes typically remain on the birth site for three to five hours. Disturbance and movement of the ewe and lamb or lambs in the first six hours after parturition increase the risk of lamb desertion.[184] While the birth site is apparently made attractive to the ewe by the odour of amniotic fluid, the site itself may be less important than the fact that the ewe and her lambs are kept in close proximity to each other for several hours — the ewe remains attracted to the site by the odour of the birth fluids while she develops a bond to the lambs, and the lambs remain near the site due to their relative immobility and attraction to the shape, sound and smell of their dam.

Environmental conditions

The environmental conditions to which lambs are exposed immediately on birth can have a profound effect on their chances of survival. They have small reserves of energy to sustain themselves, maintain their body temperature and fund their activity in searching for the teat. If exposed to chilling effects (cold, wet, windy conditions) and in the absence of sufficient milk, their energy reserves may become exhausted and their continuing attempts to find the teat become slower and weaker.

The ability of unfed lambs to generate heat and maintain body temperature at or near 38 °C is due to their stores of brown fat (brown adipose tissue), principally in the perirenal and pericardiac regions. Metabolism of this tissue produces substantial amounts of heat, due to its rich content of mitochondria and the mitochondrial protein thermogenin. Newborn lambs are thus able to generate heat without shivering by catabolising brown fat until it is exhausted. If not metabolised, brown fat is gradually replaced by normal white adipose tissue over the first few days and weeks of life.

There are two questions to consider in regard to the ability of unfed or underfed lambs to survive. The first concerns the amount of energy that an unsuckled or inadequately suckled lamb can apply to activity and to thermoregulation — that is, how long it can survive without adequate milk intake in mild conditions; the second concerns the rate at which a newborn lamb can generate heat when the environmental conditions cause it to rapidly lose large amounts of heat from its body — that is, how long the lamb can survive in very cold conditions.

At birth, lambs have 2% to 3% of their body weight as fat — around 120 g in a 4 kg lamb — and lambs from well-nourished ewes have higher fat reserves than lambs from poorly nourished ewes. A starved lamb may be able to survive up to four days in mild, still weather but less than two days in cold, windy conditions. Survival time is related to body weight and the time to death has been predicted to be around 10 hours for every kg of body weight at 9 °C and 16 hours per kg at 23 °C. By the time lambs die from starvation under relatively mild environmental conditions their bodies contain about 1.25% fat (50 g in a 4 kg lamb) having used their meagre energy stores to produce 1.6 to 4 MJ of energy, depending on the extent of their reserves. Stores of glycogen and protein make some contribution to the substrates used for energy production but contribute less than 1 MJ. In a starved 4 kg lamb from a well-nourished ewe, brown fat provides two-thirds of the energy used before death.[185]

Brown fat has a particularly important role when lambs are born in very cold conditions and rapid heat loss and subsequent fall in body temperature are an immediate threat to survival. By catabolising brown fat, lambs can generate heat at up to five times the rate at which they would generate heat in thermoneutral conditions.[186] Under certain conditions, however, even this very high rate of heat generation is insufficient to maintain body temperature. A low air temperature alone (-5 °C, for example) is not usually the cause of lamb deaths from exposure, but the combination of low temperatures, wetness — from rain or from amniotic fluid on the birth coat — and wind can overwhelm the lamb's heat generation capacity, leading to death from hypothermia.

The mortality rate of lambs born on days and nights when the weather is harsh can be 50% or higher, but it can be dramatically improved by shelter from wind.[187] Larger lambs, lambs with hairy birth coats, lambs from well-nourished ewes and lambs which were delivered without any birth trauma (and which subsequently performed teat-seeking activities promptly after birth) have a higher chance of survival in cold, wet, windy weather than lambs without those characteristics. Larger lambs, provided their size was not an impediment to a quick and easy birth, have a greater chance of retaining heat because the ratio of their body mass to surface area is greater than that of small lambs.

Major causes of neonatal mortality

On a national basis, perinatal mortality rates of 10% to 30% are a common and seemingly inexorable source of reproductive wastage, and the major causes are dystocia, starvation and hypothermia, or a combination of these factors. Using the post-mortem examination of lambs and an estimation of the time of death, relative to birth, it is generally possible to classify mortalities of these types into one of two categories based on the time of death and other necropsy findings.[188] The categories, and major contributors, are as follows:

(1) deaths occurring during birth or within a few hours of birth — these are generally due to either *birth stress* as a consequence of dystocia or peracute hypothermia in lambs of low birth weight born in very cold conditions

(2) deaths in lambs which have survived for periods of a few hours up to seven days, but have died because they have received insufficient nutrients from the ewe — these deaths are considered to be caused by one or more of the following: starvation, mismothering and exposure (the *SME complex*); and they are characterised at necropsy by evidence of starvation, cold exposure or both.

Further emphasising the importance of dystocia is that difficult births, in addition to causing death of the lamb directly from birth stress, also predispose lambs to death from starvation, mismothering and exposure in the days after birth.

Birth stress

During the process of a difficult birth, lambs suffer from the effects of asphyxia, trauma to the central nervous system or internal organs, or a combination of these. This birth stress may lead to the lamb's death during delivery or may cause such interference with the lamb's behaviour after delivery that it is unable to satisfactorily feed and establish a bond with its dam.

Starvation, mismothering and exposure

Deaths of lambs from exposure may be due to primary or secondary hypothermia. With primary hypothermia, the chilling effect of the environment is so severe that lambs die before brown fat catabolism begins, or because brown fat catabolism is insufficient to prevent hypothermia or because hypothermia develops before sucking and colostral ingestion can produce sufficient substrate for thermogenesis. Secondary hypothermia occurs if the environmental effects are less severe but lambs do not ingest sufficient nutrients to maintain body temperature after brown fat and other substrates are exhausted.

One of the most common reasons for failing to ingest sufficient nutrients is a failure of the normal maternal and offspring behaviours which lead to the lamb quickly standing and finding the teat and establishing a maternal-offspring bond with an interested and co-operative dam. This failure may be due to CNS injuries to the lamb during a traumatic birth, or it may be due to failures of the ewe to adequately respond to the lamb. Ewes which have suffered a dystocia are more likely to show aberrant maternal behaviour, including lamb desertion, if they have endured a difficult and prolonged birth. Dystocia presents a high risk for lamb death because it adversely affects the behaviour of both ewe and lamb.

Other reasons for the starvation and death from exposure of lambs include agalactia, physical abnormalities of the udder and teat, and mismothering caused by disturbance of the ewe before she is adequately bonded with the lamb. The nutritional state of the ewe is important in several of the mechanisms leading to lamb survival. Poorly fed ewes are more likely to have light-weight lambs, lambs with low brown fat stores, little or no colostrum and prolonged births due to weakness. Additionally, poorly fed ewes are more likely to exhibit disinterest in their lambs and to desert their lambs.

Post-mortem examination of lambs dying in the perinatal period

A method for the systematic post-mortem examination of neonatal lambs, first described in detail by McFarlane (1965)[189], allows lamb deaths to be categorised by time of death (ante-parturient, parturient or post-parturient) and, for the post-parturient deaths, into subcategories of immediate post-parturient, delayed post-parturient and late post-parturient deaths.

Lambs which die during parturition (or within three hours of birth) show localised oedema of the head, one or more forelimbs and possibly of the tail and perineum. The oedema occurs in lambs which are alive for a period of time during parturition — therefore having an active heart and arterial blood pressure — but which have been subject to venous constriction while remaining in the vagina for some time, causing fluid to escape into the distally presented subcutaneous tissues — the head and forelimbs in the case of anterior presentations. The lungs are not aerated and there may be abdominal haemorrhage from liver rupture. The degree of renal cortical autolysis, which occurs much more rapidly when kept within the warm body of the ewe than when dropped on the ground, and the degree of subcutaneous oedema can be used together to infer at what point during parturition death occurred.

Examination of the feet, lungs, gastrointestinal and fat depots in the epicardium and pericardium around the coronary groove and near the left coronary artery, and in the perirenal areas, allows an estimate of the stage of post-parturient life achieved by the lamb before it died (Table 7.11). The fat depots before any fat metabolism occurs are firm and white. As metabolism of the fat occurs the fat becomes red, soft and gelatinous.

Table 7.11: Deaths of lambs in the post-parturient period can be categorised into time periods based on an examination of lungs, feet, intestinal contents and brown fat reserves. Adapted from McFarlane (1965).[188]

Time of post-parturient death	Observations from examination of lungs, feet, intestine and fat stores	
Immediate post-parturient	Not breathed	Body fat not metabolised
	Breathed not walked	
	Walked not fed	
	Food not beyond small intestine	
Delayed post-parturient	Breathed not walked	Body fat partially metabolised
	Walked not fed	
	Food not beyond small intestine	
Late post-parturient	Walked not fed	Body fat fully metabolised
	Food not beyond small intestine	
	Body fat not metabolised	Food has passed through whole of gastrointestinal tract
	Body fat partially metabolised	
	Body fat fully metabolised	

Further information about the factors leading to perinatal lamb death can be gathered from a post-mortem examination of the brain and spinal cord.[190] Lambs which die in the parturient and post-parturient period have a high incidence of subdural haemorrhage, particularly under and around the posterior brain stem, leptomeningeal haemorrhages, bloodstained cerebrospinal fluid in the sub-arachnoid spaces and in the spinal cord and, in the vertebral canal, epidural haemorrhages. The lesions can be observed by examining the brain and spinal cord by removing — with pointed footrot shears, for example — the bones and dura mater forming the roof of the cranium and cutting the arches of the vertebrae from the atlas to the sacrum. The vascular abnormalities are readily observed macroscopically and, in some cases, the extravasation and clot formation can be extensive.

In addition to the grossly visible lesions of the brain and spinal cord, there is evidence that hypoxia of lambs during dystocic deliveries causes damage to the central nervous system which is only detected by microscopic examination.[191] The hypoxia-ischaemia may result from a prolonged parturition; a parturition during which the lamb is subject to prolonged high pressures from the birth canal and myometrial and abdominal contractions; premature rupture or occlusion of the umbilical cord; placental insufficiency; or premature placental detachment.

Minor causes of perinatal mortalities

Minor causes of perinatal lamb mortality include primary predation, infections, nutritional deficiencies and lethal congenital conditions. While these causes can be very important in some flocks or some regions (particularly predation), they tend to be isolated or sporadic. Perinatal mortalities from causes other than birth stress and the SME complex usually amount to only a small proportion of deaths. Nevertheless, on isolated occasions, perinatal mortalities from congenital malformations, infectious causes or trace element deficiencies can be serious.

Predation

In southern Australia, foxes (*Vulpes vulpes*, the European fox) are common and often suspected of predation of lambs. Foxes are frequently seen in flocks of lambing ewes and consume placentae and dead lambs. Most foxes are very cautious and will rapidly retreat from a ewe or even an active lamb, but some individual foxes are less circumspect and may attack and kill an isolated but healthy and otherwise viable lamb. Killings of lambs by foxes generally appear to be opportunistic rather than the result of active hunting.[192]

Dingoes, wild dogs and feral pigs are active primary predators of lambs and can cause very severe losses in parts of Australia where they coexist with sheep. Uncontrolled domestic dogs are also capable of killing adult sheep and lambs and causing the death of lambs by mismothering while chasing or attacking a flock.

Lethal congenital malformations

Vaccination against bluetongue, Rift Valley fever and Wesselsbron disease and teratogenic plants can cause congenital malformations.

Congenital infections

These are effectively caused by the same organisms that cause abortion in ewes.

Infections acquired after birth

- *Clostridium septicum*, *C chauvoei* and *C novyi*, which cause gangrene around the umbilicus and peritonitis
- *Pasteurella haemolytica* and *P multocida*, which cause pneumonia and peritonitis
- *Staphylococcus aureus*, *Streptococcus* spp, *Corynebacterium* spp, *Fusobacterium necrophorum* and other bacteria which cause pyaemia with multiple purulent foci in the viscera
- *Escherichia coli*, which causes syndromes characterised by enteritis, septicaemia or leptomeningitis
- *Erysipelothrix rhusiopathiae* and *Chlamydia* spp, which cause polysynovitis.

Trace element deficiencies

Copper, iodine and selenium deficiencies can cause heavy mortalities of lambs under certain circumstances.

SECTION E

MANAGEMENT AND DISEASES OF LACTATING EWES

Uterine prolapse

Uterine prolapse occurs occasionally in ewes after a difficult lambing, particularly where straining has been prolonged or the vagina has been damaged during the birth process. Uterine prolapse also occurs as a consequence of phyto-oestrogenism in ewes.

The prolapse can be replaced and the success rate is high, at least in terms of ewe survival if not in terms of future breeding, provided that the damage done to the organ is minimal before and during treatment. Gross contamination should be removed and the organ replaced and completely reverted. The foetal membranes, if still present, should be left *in situ*. Antibiotic therapy, with intra-uterine pessaries, seems appropriate but pessaries are often lost soon after placement. Oxytocin should be administered intramuscularly to encourage uterine involution. A purse-string suture of umbilical tape may be placed around the vulva but probably does not add to the success rate in ewes which are ambulatory and in which involution commences within a few minutes of treatment.

Hypomagnesaemia

This appears to be a relatively uncommon condition of sheep. When it does occur, it involves ewes in the first month of lactation and, like the disease in cattle, affects those ewes with the heaviest lactations. Deaths can occur at pasture without the disturbance of handling or driving, so on some occasions affected ewes are simply found dead. If seen alive, affected ewes have a stiff, unco-ordinated gait and readily collapse and show repeated tetanic spasms. They are hyperaesthetic and show tremor, particularly of the facial muscles. The course of the disease is probably only a few hours.

Treatment

If ewes are found alive, treatment with both calcium borogluconate and magnesium salts is advised and often successful. Relapse, however, is common, so treatment may need to be repeated once or twice.

Prevention

The pathogenesis of the disease is probably similar to that of cattle, so preventive strategies are similar. Oral supplementation with magnesium oxide or magnesium carbonate should be effective.

Nutrition through lactation

Ewes in good condition at lambing which have been provided with adequate nutrition after lambing are capable of producing 1.5-3 kg of milk per day over the first few weeks of lactation. Milk production usually peaks two to three weeks after lambing and then declines by 10 weeks to a level between one-quarter and one-half of the peak level (Figure 7.16).[193] Some breeds of sheep (East Friesian, Awassi, for example) have been selected for milk production and tend to have higher peak levels of milk production and longer lactations than Merino sheep or Border Leicester x Merino crossbred sheep.

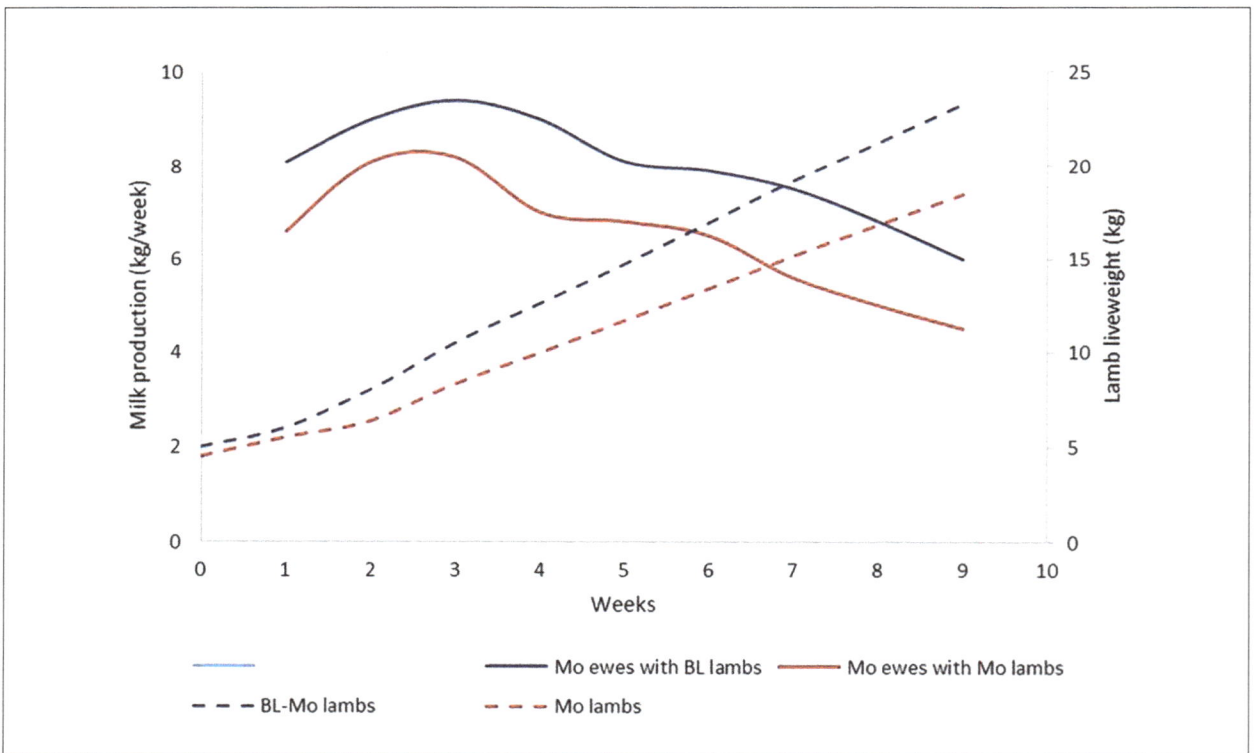

Figure 7.16: The milk production (solid lines) of Merino ewes raising either single Merino lambs or single Border-Leicester x Merino lambs, and the growth rates (dashed lines) of the lambs. Drawn by KA Abbott. Adapted from Lloyd Davies H (1963), The milk production of Merino ewes at pasture, Australian Journal of Agricultural Research 14, 824-38 (with permission of CSIRO Publishing).

When ewes produce 2-2.5 kg of milk per day in early lactation, lambs, particularly lambs of meat breeds, are capable of exhibiting very high growth rates (>350 g per day) and, in ewes which are in good condition and well fed, the growth rate of singleton lambs is not limited by the ewe's milking ability.[194] Ewes produce more milk when feeding multiple lambs than when feeding single lambs, reflecting the increased number of times that the ewe is sucked. Merino lambs are less vigorous feeders than Border Leicester x Merino lambs and ewes produce more milk when feeding crossbred lambs than when feeding Merino lambs (Figure 7.16). In well-fed ewes, milk production may not be limiting of lamb growth rate unless the ewe is nursing two or more lambs.

The peak, duration and total production of lactation is, however, profoundly affected by nutrition. In most commercial field conditions, the nutrition of the ewe limits her milk production and that in turn limits the growth rate of the lamb to 210-280 g per day. In approximate terms, lambs ingesting 1 kg of milk per day through the first four to six weeks of lactation will grow at 200-220 g per day and will, therefore, weigh up to 13 kg at 6 weeks of age.

When ewes are less well fed — which might occur when stocking density is high relative to the pasture production — they produce less milk (Figure 7.17), and the growth rates of their lambs are severely affected and the health and survival of the lambs are at risk. The effects of undernutrition of the ewe are greater for twin (or higher-order) lambs than for singletons.

The consequences of undernutrition of ewes, if it does not result in the death of the ewe through pregnancy toxaemia or the perinatal death of the lamb, are inadequate and short lactations and increased grazing activity of the lambs at an early age. Lambs normally commence

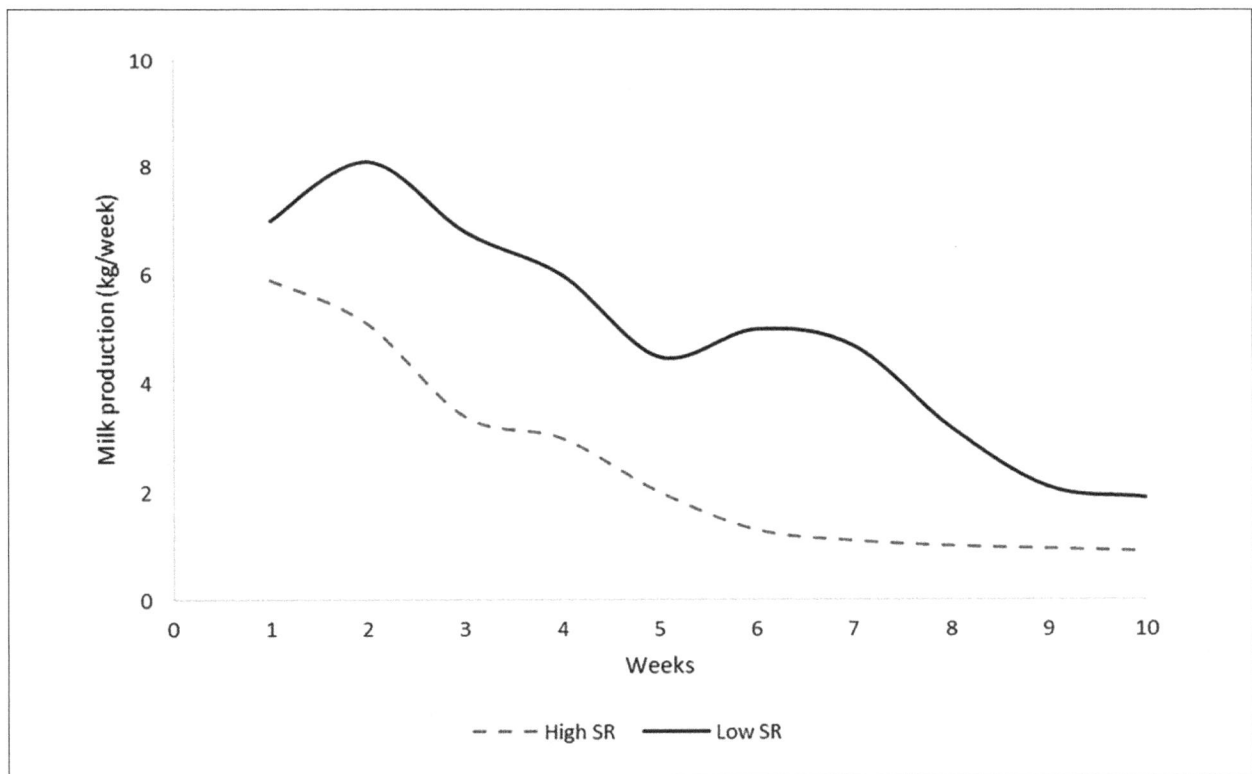

Figure 7.17: The milk production of Merino ewes raising single lambs at high stocking rate (dashed line) or low stocking rate (solid line). Drawn by KA Abbott. Based on data from Davies (1962).[192]

grazing at 3 weeks of age and, by the time they are aged 8 weeks and weigh 12 kg or more, they have sufficient rumen development to survive on good-quality pasture without milk. Below that age and weight, inadequate rumen development and high dietary protein requirements reduce the chance of survival of lambs without adequate milk intake. These lambs may die from undernutrition at any age from 7 days onwards (before that, the death is classed *perinatal*). If they survive to marking they will be very light (for example, 8 kg at 6 weeks of age) and will appear *poddy* or pot-bellied due to marked rumen development in a small body. They are weak and move slowly and lethargically compared to well-nourished lambs. They often develop intercurrent diseases and are likely to have a serious burden of intestinal parasites.

The same nutritional environment which leads to poor growth and poor health in lambs leads to poor health in the ewes. Despite the reduction in their lactation due to low feed intake, the ewes will produce some milk by using body reserves of energy and protein. Often, by the time the ewes cease lactating they are in low condition score (1½, for example) and are themselves at risk of other diseases, particularly internal parasitism.

There is a full range of conditions between the two extremes and it is likely that any one flock of ewes under normal commercial grazing conditions would have a few ewe-lamb combinations representing both ends of the spectrum. This is normal and does not require any action. It becomes abnormal and needs remedial action when a large proportion of the ewe flock is in poor condition and many lambs are poorly grown.

Mastitis

Mastitis is a cause of lamb mortality, low weight gain of lambs and losses of ewes from death and premature culling. It occurs sporadically in all flocks of sheep but few details of its epidemiology or economic impact are known. The most obvious mastitis of ewes in Australia is gangrenous mastitis, which is usually associated with *Staphylococcus aureus* infection, but acute mastitis caused by *Pasteurella haemolytica* is also common and often fatal. During outbreaks of pasteurella mastitis, cases of pasteurella pneumonia may also occur in lambs. A variety of other organisms cause mastitis which varies from acute, fatal infections to mild or even subclinical diseases.

Aetiology

Mastitis in ewes is predisposed by milk stasis (which occurs at lambing or weaning), teat damage, such as occurs from shearing or crutching wounds and CPD infection and, possibly, from udder damage caused by the bunting of lambs with or even without horns. It is possible that a site of transmission to predisposed to ewes may be sheep *camps*, where the flock congregates and lies at night, which are heavily contaminated with faeces.

The following organisms have been associated with mastitis in ewes: *Staphylococcus aureus*, *Pasteurella hemolytica*, *Escherichia coli*, *Corynebacterium pseudotuberculosis* (usually infecting the supramammary lymph nodes, but the subsequent abscessation may destroy udder tissue), *Actinobacillus pyogenes*, *Streptococcus uberis*, *Strep agalactiae*, *Actinobacillus seminis*[195,196] (which has caused subacute and gangrenous mastitis when introduced experimentally into the teat canal), *Histophilus ovis*[197] (the agent may have been *A seminis*[62]) and *Actinobacillus lignieriisi*.

Clinical signs

Acute mastitis — Ewes separate from the mob, walk with discomfort and are usually very lame on the affected side. They lose condition and their lambs appear thin and hungry. On examination, the udder will appear enlarged, painful and hot, with red skin. The milk, in the early stages, will show varying types of abnormality, including yellowness and clots. The ewes are febrile, with rectal temperatures exceeding 40.5 °C at some stages of the infection. If the infection leads to gangrenous mastitis, the udder will progress to become blue, cold and pitting when pressed. The milk becomes watery, may be bloodstained and scant. The ewe's temperature may become subnormal at this time. A large proportion of ewes with gangrenous mastitis die; the gangrenous half of the udder will slough in those that survive.

Treatment and control

Little is known about the effectiveness of treatment of ovine mastitis and, under Australian conditions, anything less than severe, acute mastitis will not be readily detected. One presumes that treatment of individual cases should proceed normally as for cases in dairy cattle. In cases of gangrenous mastitis, preventive treatment for flystrike should be used.

Intramammary antibiotic therapy at the time of weaning has been shown to be highly effective in lowering the incidence of post-weaning mastitis.[198] The ability of *Pasteurella* sp organisms to survive in the udder, possibly associated with palpable nodules in the udder, from one lactation to the next and to then cause an acute, fatal mastitis has been documented.[199] In extensively managed flocks, such treatment is unlikely to be accepted or economically advisable.

It has been suggested that mastitic ewes should be culled from flocks at weaning or before joining. There is evidence to indicate that ewes with bilaterally defective udders have a very low success rate in raising lambs, and that ewes with only one half affected (with teat injuries as well as mastitis) lose a much higher proportion of lambs than ewes with normal udders (18.5% vs 6.6%). Nevertheless, the presence of udder defects affecting only one half is not a powerful predictor of success at lamb rearing, and culling should be restricted to those ewes with bilateral defects.[200] The other reason suggested for culling of mastitic ewes is to remove a potential source of infection for other ewes, but the organisms causing mastitis are generally ubiquitous and there is no experimental evidence to support the recommendation. Nevertheless, culling of ewes with discharging mammary abscesses should reduce the contamination of camps.

Many udder defects and mastitides arise from injuries inflicted by the handpiece at shearing and crutching. The risk of injury is obviously proportional to the number of shearing/crutching events performed in the vicinity of the udder. Consequently, many producers request that persons performing the crutching operation *not* remove wool from the udder.

Contagious agalactia

This disease occurs only in Mediterranean, Adriatic and African countries, where it is of considerable economic importance. It is not present in Australia, New Zealand, Britain or the USA. Any one of three species of mycoplasma are involved — *M agalactiae*, *M mycoides* subsp *mycoides* and *M capricolum*. Ewes develop the disease soon after lambing and it involves mastitis, arthritis and keratoconjunctivitis. The mastitis is acute and severe and leads to total and permanent agalactia. The case fatality rate is between 10% and 30%.[201]

Lamb management at and after marking

By definition, marking rate, as an index of reproductive efficiency, is determined at marking time. Usually, lamb mortality rates between marking and weaning are low, and if deaths at that later age are occurring, the cause is usually readily distinguishable from the multifactorial *reproductive losses* discussed so far. Nevertheless, the job of reproducing the flock does not stop at marking, and some consideration should be given to management of lambs at and after marking. The task of keeping young sheep alive does not stop at weaning, either, and Chapter 12 discusses weaner management.

At marking, lambs are usually earmarked; tailed; castrated if male; vaccinated with all, some or none of CLA vaccine, a mixture of clostridial vaccines (with or without added selenium), and scabby mouth (CPD) vaccine; injected with vitamin B_{12}; and dosed with anthelmintic. In many Merino flocks, lambs are mulesed at lamb-marking time. The experience is a stressful one for lambs.

The variable need for, and manner of, administration of these treatments are discussed elsewhere. Lamb mortalities and serious disease, particularly arthritis, do occur after marking and mulesing, but a prevalence over 5% is uncommon. The ewe and lamb flock should be observed by the flock owner after marking, but major disturbances of the flock should be avoided to enable rapid healing of the mulesing wounds. The most common reasons for any mustering and treatment after marking is for treatment/prevention of flystrike or helminth parasites.

The nutritional requirements of the flock must be monitored after marking, as was necessary immediately after lambing. Despite the decline in energy requirements of the ewe for lactation, the requirements for dietary energy and protein of the lambs are increasing and, if more than 70% of the ewes rear lambs, the requirements of the flock will increase after lambing to a peak at weaning as the lambs grow. This increase should always be remembered when planning nutritional requirements for lambs and ewes — some advisers mistakenly calculate requirements for lactation as requirements for ewes alone.

SECTION F

INVESTIGATIONS OF POOR REPRODUCTIVE RATE IN COMMERCIAL SHEEP FLOCKS

Diagnostic techniques used in the field

Veterinary examination of the rams

Veterinary examination of the ram team prior to joining normally occurs for one of two reasons. First, on properties with good reproductive rates, annual or biennial inspection of the ram team is one of the procedures which assist in maintaining the flock's reproductive performance. The veterinary examination can identify any individual rams which may have developed a condition which reduces its fertility and can also detect any flock problems (a recent infection with *Brucella ovis*, for example) before they jeopardise the conception rates of the flock. The veterinary review of the ram team also provides an opportunity for the flock manager to discuss the general state of the ram team and plan the way that the rams are to be dispersed through the ewe flock for joining.

Second, in investigations of reproductive failures, examination of the rams is often the first and simplest step. Several specific diseases which could influence libido, semen production or semen quality can be detected during a physical examination of each ram.

Using ram harnesses to measure mating activity

Regular observation of the flock during joining will provide a useful insight into the level of mating activity. If considered necessary, however, better estimates of both flock-breeding activity and fertility can be obtained by the careful use of ram harnesses and crayons during joining. The procedure recommended is usually as follows:

Harnesses are applied to the rams and services are recorded during the first 18 days of joining. To do this, the ewes are yarded at least once, preferably twice, per week, and the number of marked ewes is recorded. The flock should be inspected daily for lost harnesses and crayons. The crayon colour is changed on day 19 for the rest of the joining period and observations continued. If joining is in spring, it is better to change the crayon colour only after the first 28 days of joining.

For autumn joinings, the proportion of ewes marked in the first 18 days measures mating activity. Flock fertility is assessed from this proportion and the proportion of marked ewes which are re-marked after the first 18 days. Note that a source of error here is that ewes marked with the first colour on day 18 may still be in oestrus on day 19 and may be marked with the second colour.

The interpretation of flock crayon data is most straightforward for autumn joinings. When ewes are joined out of season, the failure of some ewes to be re-marked may denote a return to anoestrus rather than a conception. The data obtained with crayons have other uses. For example, the flock can be subsequently divided into an *early lambing* and *late lambing or dry group*. However, bear in mind that the procedure outlined is laborious and hence costly, and may significantly interfere with joining performance. Apart from the mustering and yarding involved, harnesses can sometimes cause discomfort, wounds and lameness in the rams.

Pregnancy testing

Pregnancy testing in Australian sheep flocks has three major roles.

- It is used as a tool to investigate reproductive problems. Depending somewhat on the stage at which pregnancy diagnosis is carried out, the techniques enable a clear separation between problems associated with joining (prenatal wastage) and perinatal lamb loss. The techniques may also facilitate the identification of foetal loss in mid-late pregnancy in flocks where abortion is a significant problem.

- It is used as a management tool to allow producers to separate multiple-bearing ewes from single-bearing and non-pregnant ewes and to allow differential management, more efficient use of feed and, if successful, a higher survival rate of twin-born lambs.

- Veterinarians who carry out controlled breeding programmes use it as a tool (see Chapter 8).

Ram harnesses

Ram harnesses and crayons are sometimes employed to diagnose pregnancy in individual ewes. The technique cannot identify multiple-bearing ewes from single-bearing ewes but there are occasions when it is useful to identify non-pregnant ewes and remove them from the breeding flock. When crayons are used during joining, as previously discussed, the accuracy of detection of pregnant ewes should be high. The accuracy of detection of non-pregnant ewes, however, is highly variable and commonly low because marked ewes are not necessarily pregnant. The use of crayons after joining can provide the basis of a relatively inexpensive pregnancy diagnosis system. In this case harnessed vasectomised rams or androgen-treated wethers are run with the flock for 20-30 days immediately after the end of joining. If joining ends between early February and May the non-pregnant ewes should be cycling and will be marked. However, if joining ends between June and January this technique is not useful, since the non-pregnant ewes in the flock are probably in anoestrus.

Ultrasound pregnancy diagnosis

The diagnosis of both pregnancy and ewes bearing multiples is carried out by real-time ultrasound. This technique enables rapid and reliable diagnoses. Both linear and sector scanners can be suitable. Probes are usually applied externally to the abdominal wall, but rectal probes may be used. High accuracies and reasonable speed are achieved only with considerable experience. Ewes can be examined either standing or tipped, or by restraint in a sheep-handling device.

Some approximate accuracies of real-time ultrasound in relation to stage of pregnancy and number of foetuses, and as obtained by skilled operators, are shown in Table 7.12.

In the early stages fluid of the developing conceptus, or conceptuses, is the diagnostic principle, whereas after days 45-50, cotyledons and bony structures of the foetuses (especially rib cages and heads) are used.

As a non-pregnancy test, ultrasound is extremely accurate. As a pregnancy test, it should become 99% accurate by about day 55. The diagnosis of multiples takes considerably more skill, and accuracy peaks after day 55 at around 90%. Occasional operators may achieve higher accuracies than those above, especially around day 40, but anecdotal evidence suggests that the accuracy in diagnosing multiples amongst commercial operators is sometimes lower than expected. There is a tendency to underestimate the number of foetuses in ewes carrying multiples, especially in the case of triplets. In practice, commercial operators prefer to test ewe

Table 7.12: Approximate accuracies of real-time ultrasound.

Number of foetuses	Stage of pregnancy (days)		
	40-47	54-70	80-100
0	100	100	100
1	95	99	99
2	65	90	90
overall	93	98	98

Accuracies are defined as correct scans/total scans × 100

flocks about 12 weeks after the commencement of joining (that is, with an autumn joining, when most ewes are about 65-85 days pregnant). Diagnosis later than this remains accurate, but is less valuable, since less use can be made of the information obtained.

The benefits of pregnancy testing

It is useful to consider separately the potential benefits of pregnancy (yes/no) diagnosis and the diagnosis of multiples. Those for yes/no diagnosis include

(1) diagnosis as a normal management tool

- the more efficient use of resources allocated to lambing ewes, including feed, pre-lambing drenches, vaccination and crutching and saved or sheltered lambing paddocks
- increased management options, such as rejoining the non-pregnant ewes or using them for grazing management, like wethers

(2) and diagnosis as an investigatory tool

- monitoring joining performance while investigating the causes of low lamb marking percentages
- the later identification of ewes which have lambed and lost their lambs, especially when *wet and drying* cannot be performed soon after lambing.

The potential benefits of the diagnosis of multiples largely apply only to its application as a husbandry or management tool, and include

- better feeding of multiple-bearing ewes which leads to a reduction in lamb losses and in ewe losses due to pregnancy toxaemia. Several other production responses to the extra feed are also important. Ewes with multiples should have better fleece weights and less tender wool if fed at higher levels during late pregnancy. Higher weaner and hogget weights and increased weaner wool growth are achieved, mainly due to a marked reduction in the *tail* of weaners born as multiples. Ironically, of these several production responses, the reduction in lamb losses is sometimes the hardest to achieve. Indeed, the advent of ultrasound diagnosis has revealed that twinning rates in Merino flocks are often higher than was previously appreciated
- appropriate feeding for single-bearing ewes, which will comprise 60% to 80% of most Merino ewe flocks. Rather than feed the ewe flock to a level which represents a compromise between the needs of a single-bearing ewe and a twin-bearing ewe, the single-bearing ewes can be feed to a level appropriate to their needs. In many cases this will lead to an overall reduction in feed costs and, possibly, reduce the incidence of dystocia and mismothering in the single-bearing ewes.

Factors influencing the economics of real-time ultrasound

These are as follows:

- the true reproductive status of the flock: ultrasound scanning is more likely to be worthwhile if the rate of non-pregnant ewes exceeds 10% or the rate of twin-bearing ewes exceeds 20%
- the accuracy of detection of dry ewes, ewes bearing singles and ewes bearing multiples
- the extent of the various production responses achieved (number of extra lambs saved, increased income from wool, etc.)

- the speed and cost of testing: skilled operators can diagnose pregnancy (yes/no) and numbers of foetuses at rates of around 100-200 and 50 per hour, respectively
- the occurrence of an intangible benefit, often observed by a general lift in the quality of management, even if only to ensure that the benefits of ultrasound are realised.

Sometimes the numbers of lambs born in flocks do not seem to closely approximate the numbers of foetuses previously diagnosed. The possible explanations are

- poor accuracy of ultrasound diagnosis
- poor accuracy in counting lambs born (a major problem in larger flocks)
- loss of foetuses subsequent to ultrasound diagnosis but before term (specific causes of abortions have already been discussed earlier in this chapter).

Estimating lamb losses — the wet and dry technique

Estimates of losses based on observation of the lambing ewes are likely to be inaccurate, especially in large flocks typical of commercial sheep production in Australia. The most accurate measure available, especially where many ewes have multiples, is the difference between the number of foetuses counted with ultrasound and the number of lambs present at marking (corrected for ewe losses).

A simpler procedure, requiring no equipment but some experience, is *wet and drying*. In Merino flocks with relatively few twinning ewes, this method can provide reasonably accurate estimates. The lambed ewes are examined and scored as follows:

- Dry ewes: no significant udder development or tone and no lambing stain on the perineum. These ewes are usually in a higher-than-average condition score
- Lambed ewes: enlarged udders, often showing lambing stain on the crutch, with an average or lower condition score
    - ewes rearing at least one lamb: full, resilient udders containing milk, skin of teats and adjacent areas form clean *spectacles* and are soft and pliable as a result of the lamb's sucking activity
    - ewes which have lost all lambs born: variable udder development, with a tendency to cleavage between the two glands; teats stiff and dirty; udder secretion varies from normal milk to thin, watery or viscous, honey-coloured fluid.

There are some limitations to the procedure, as follows:

- It must be done within four or five weeks of birth.
- The estimate of lamb losses is minimal; the method cannot take account of situations where one or all of multiples die or of cross-mothering.
- It cannot distinguish between ewes which failed to establish pregnancy and those which became pregnant but subsequently aborted.
- Where the reproductive problem is sufficiently serious, *wet and drying* may be combined with prior ultrasound diagnosis of the pregnant ewes. This also reduces the need to *wet and dry* early after birth because it is no longer necessary to differentiate dry ewes from ewes which *lambed and lost*.

The consequences of increasing reproductive rates above normal levels

Veterinarians are sometimes asked to advise on ways to increase RR in flocks where it is already *normal* or *adequate* and the flock management is basically sound. Under these conditions, further increases in RR are usually associated with increases in the number of twin pregnancies and twin births in the flock. There are a few important consequences of increasing RR to high levels which are additional to those just discussed.

Ewe management at high reproductive rates

Increases in lambing rate come about from increases in the proportion of pregnant ewes *and* an increase in the incidence of multiple pregnancies. Ewes with multiple pregnancies are more difficult to manage than single-bearing ewes and more prone to pregnancy toxaemia. Multiple pregnancy and, more particularly, multiple suckling reduce fleece weight and wool quality. Lactation with a single lamb reduces annual wool production by 12% to 15%. Ewes rearing multiples lose more weight during lactation than single-rearing ewes and are, consequently, at increased risk of intercurrent disease. The cost of managing breeding ewe flocks rises when the proportion of multiple-bearing ewes in the flock increases. The extra difficulty in managing twin-bearing ewes is sometimes used to justify the use of ultrasound pregnancy testing so that multiple-bearing ewes can be managed separately from single bearing ewes. The cost of pregnancy testing must be included, then, in estimates of the costs of managing highly fecund flocks if pregnancy testing is considered necessary.

Lamb survival at high reproductive rates

A higher proportion of twin-born lambs die than single-born lambs. In one experiment in western Victoria[202], treatments which successfully increased lambing rate and the number of ewes bearing multiples had no effect on weaning rate because of the high death rate (29%) of multiple-born lambs. Even when treatment successfully increases weaning rate, the cost, in terms of lost production from multiple-bearing ewes which then do not raise two lambs, may be high.

Lamb growth rates at high reproductive rates

Lambs raised as twins have slower growth rates than single-raised lambs and this has implications for marketing of prime lambs and for the management of Merino and Merino type weaners. In prime lamb flocks, slow-growing lambs are more likely to miss early sales, which often have higher prices, and are likely to be older, with longer wool, and leaner, than faster grown flock mates. Additionally, slowly grown lambs are more likely to fail to achieve market weight before the growing season ends, necessitating more expensive forms of feeding to achieve market weights or, alternatively, an extended stay on the property. In self-replacing flocks, more slowly grown lambs will be lighter when the pasture-growing season ends, necessitating earlier and more extended supplementary feeding and involving a higher risk of mortality. It is likely that twin-raised lambs are overrepresented in drafts of cull hoggets and are therefore sold at discounted prices.

Economic considerations of high reproductive rates

There are very few field studies of the economic benefits of increasing RR of sheep. Obst and Thompson (1984)[203] found that a 40% increase in weaning rate had a negative effect on nett income per hectare in 1982 and, in 1983, a positive effect of $7 to $25 per hectare, $1.43 to $4.40 per ewe. Studies conducted with computer models are usually in general agreement with these estimates. White (1984)[204] estimated that a 10% increase in lamb numbers was worth $1.30 to $0.43 per ewe (decreasing with increasing stocking rate); Morrison and Young (1991)[205] estimated a benefit of $1.80 per ewe per 10% increase in lambing rate; for the same increase Morley and Peart (1988)[206] predicted an extra 5% to 7% of gross margin per ha.

The inescapable conclusion is that increasing RR is profitable provided little money is spent per ewe to achieve it. One also concludes that increases are more likely to be profitable when RR is low, say less than 70%, than when high, say over 90%, and more likely in flocks where a higher proportion of income is derived from the sale of sheep relative to wool income, and when prices for livestock are high.

It is instructive to consider whether any of the following methods which might be employed to increase weaning rates could be effected for less than $1 per ewe for a 10% gain in weaning rate:

- use of Ovastim®, Androvax® or Regulin® (see Chapter 8)
- additional supplementary feed to prevent weight loss or to increase liveweight
- specific supplementation with high-protein feed at joining
- lowering stocking rate to improve nutrition
- change of genotype to a more fecund strain/bloodline
- improvements in animal health and grazing management which simultaneously improve RR and other production parameters, such as wool production.

Under Australian conditions the last of these has the greatest potential for successful adoption because, when RR is low due to management problems, it is often very low, and because extra income will often flow from other sources, particularly improved wool income, when management problems are corrected. Consequently, veterinarians are usually asked to assist producers to raise RR in their flocks when husbandry strategies, predation, nutritional management or disease factors have caused RR to fall to low levels.

Strategies aimed directly at improving RR only, such as using hormonal treatments or nutritional supplementation at joining, should only be employed when RR is already as high as normal good management will allow. Such strategies should not be employed to compensate for failures in other areas — such as poor grazing management affecting ewe nutritional status, perinatal lamb death rates or poor weaner management. They are unlikely to work well and may even exacerbate the management problems. Such strategies, somewhat paradoxically, should only be employed when RR is already relatively high. Great care should be taken when considering changes in stocking rate to improve RR. Both RR and fleece weights are very sensitive to stocking rate, but the positive economic response to changes in RR is likely to be much smaller than the negative response in wool cut per hectare. One should also remember that increases in stocking rate will reduce production parameters recorded on a per head basis, like RR and fleece weight, but production parameters recorded on a per hectare basis, such as lambs per hectare and kg of wool per hectare, are likely to increase until stocking rates are very high. Economic analyses of these relationships are relatively complex and are often approached by using computer models.

Investigation of low reproductive rates

When asked to investigate a problem of low RR, the first actions which can be taken depend on the time of the year that the client reports the problem and requests assistance. In most cases of a novel failure, owners discover the problem at lambing or marking. Uncommonly, if an abortion outbreak occurs, aborted foetuses or ewes with retained membranes are observed. Even less commonly, in cases of fertilisation failure, ewes returning to service in abnormally large numbers are detected. Much more commonly, flock owners do not become aware of a problem until lambing time arrives or low marking rates are computed. In cases of chronically low weaning rate, the request for assistance could occur at any time in the reproductive cycle.

Two factors determine the course of action to take. First, the point in the reproductive cycle at which the producer requests an investigation limits the activities which can be undertaken at that time. Second, a good history, knowledge of the farm in question or, perhaps, just one farm visit, will allow a tentative diagnosis to be made. This will allow the preparation of a list of preferred steps for the investigation. The investigation should then be carried out in a logical progression pursuing first the avenues most likely to confirm the diagnosis, but opportunities to perform other diagnostic activities should not be totally ignored. Reproductive events, and the relevant diagnostic actions, can only be carried out at one point in every year and the opportunity to take some specific actions should not be missed, even if out of logical order of diagnostic possibilities.

Once a tentative diagnosis is made, preventive action should be put in place for the following year. This preventive action may be successful but may also prevent confirmation of the diagnosis. Although this is not entirely satisfying for the veterinarian, it is preferable to allowing another year of poor reproduction to occur just so that the diagnosis can be confirmed.

Deciding if there is a 'problem'

Producers often believe they have a reproductive rate problem in their flocks by comparing their weaning or marking percentage to the figures reported by their peers and neighbours. The consulting veterinarian must decide if the 'problem' is real or perceived. Reference has already been made (Chapter 3) to the sensitivity of reproductive rate to stocking rate. A producer who has a higher stocking rate than his neighbours can expect lower reproductive rates. It may still require an investigation to demonstrate that the reproductive rate is normal for the conditions under which the flock is managed, even though the subsequent recommendation is that no action be taken. A weaning rate which may be acceptable and 'normal' on one property may be correctly considered too low and a 'problem' on another. For example, a breeder of South Australian Merinos in southern Australia reported a 'disappointing' weaning rate of 75%. Such a result may be considered acceptable under certain conditions. Subsequent investigations revealed that the ewes were joined in April — the height of the breeding season — and were in high body condition score (3 to 3½) at joining. Considering these factors, the history did suggest that one or more specific factors were operating to depress reproductive rates. In fact, the ewes were joined while grazing dryland lucerne which had very high coumestan levels (see phyto-oestrogenism). A small change in grazing management resulted in 20% improvement in weaning rate.

Calculation of expected lambing performance

It is possible to calculate the expected lambing performance of a sheep flock. Where the season of mating and the duration of joining are known and the management of joining, pregnancy and lambing are adequate, reasonably precise estimates can be made. The point of this exercise is to provide a baseline from which to judge whether a particular client is experiencing a significant reproductive problem. As an example, consider the case of an autumn-joined Merino flock where the joining period is equal to one, two or three oestrous cycles. Reasonable assumptions for the particular district are as follows:

Percent ewes cycling	95%
Conception rate to one oestrus	70%
Mean number of lambs born per lambing ewe	1.1
Percent lambs lost	12%

The calculations are as shown in Table 7.13.

Note the assumption in these calculations that the conception rate remains at about 70% in ewes which did not conceive to the first cycle. At least in larger, commercial flocks infertility seems largely transient and the incidence of ewe sterility (permanent infertility) is quite low.

If the actual reproduction performance in a flock is, say, 15-20% below this estimated performance, you might conclude that a reproductive problem exists.

Planning an investigation of reproductive failure

Initial history collection and on-farm investigation should be directed at identifying in broad terms the stage of the reproductive cycle where the failure has occurred. The type of further diagnostic action is based on the outcome of initial categorisation. Figure 7.18 illustrates how a few initial conclusions can identify the area for more detailed investigations.

Some of the specific diagnostic procedures used for investigating reproductive failure were reviewed in detail at the beginning of Section F. At any time, the procedures which can

Table 7.13: The expected lambing performance of a sheep flock can be calculated with some assumptions about conception rates, fecundity and perinatal lamb mortality rates.

Duration of joining	% ewes lambing	% lambs born	% lambs marked
one cycle (17-18 days)	67*	74	65
two cycles (5 weeks)	89**	98	86
three cycles (8 weeks)	96	106	93

* 70% of 95%

** 67% plus 70% of 95% of 33%

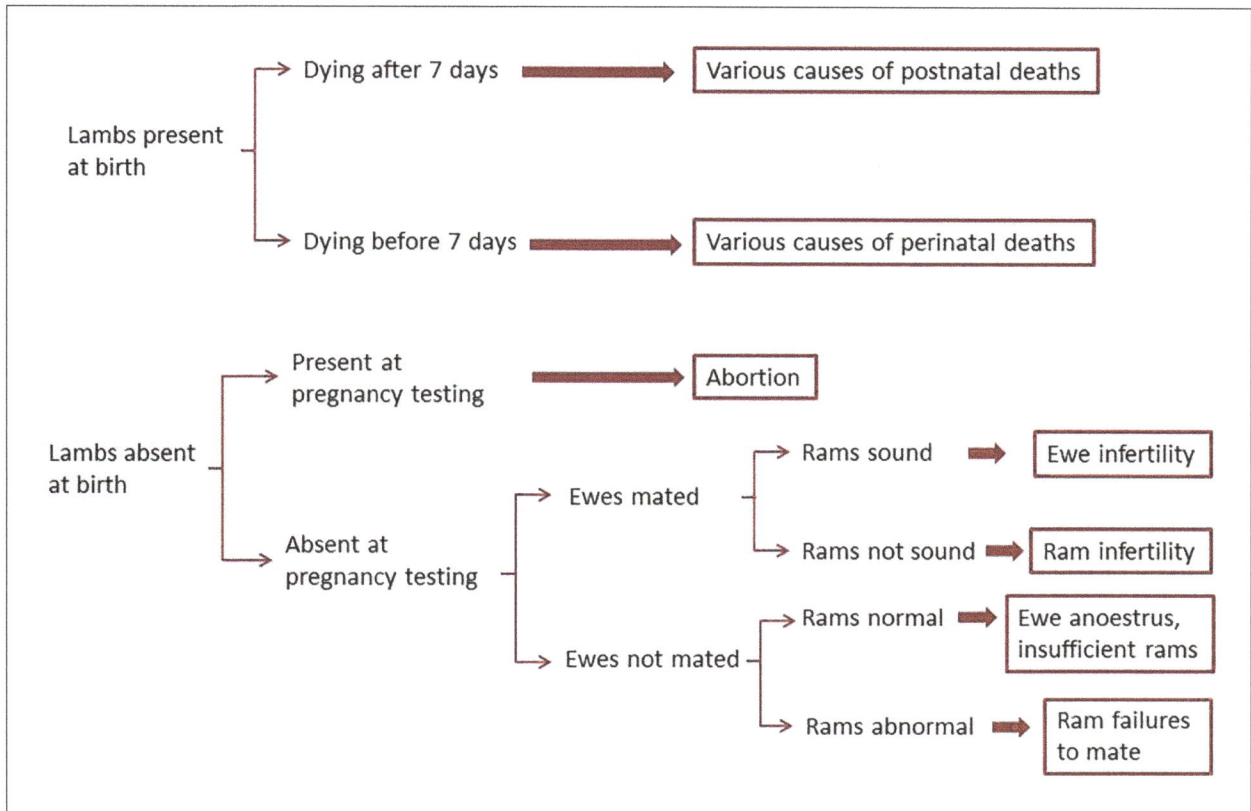

Figure 7.18: The categorisation of reproductive failure. Text in boxes indicates the likely general cause of the low reproductive rate. Investigations should aim first to identify the general area in which reproduction is failing. Subsequent investigation will then differentiate between the specific causes. Source: KA Abbott.

be carried out are determined by the stage of the reproductive cycle. As mentioned above, procedures cannot always be carried out in order of 'most likely cause'. Sometimes the approach has to incorporate some procedures which are unlikely to lead to diagnostic success but for which the opportunity is presented at the time. For example, it would be unwise to avoid an examination of the rams pre-joining when the opportunity presents, just because the history strongly suggests lamb predation. The client should be made aware of the likelihood that the veterinary procedure is useful and the cost of it. The client might, for example, approve a four-hour ram examination but decline a four-day session of pregnancy testing. Figure 7.19 tabulates the temporal relationship between reproductive events and veterinary diagnostic procedures.

History collection

This is one of the most important tools for investigating reproductive losses. Skill and experience are necessary to know what information the owner can provide which is reliable and meaningful. Ram-joining percentages and marking percentages by paddock are usually reliable. History of body weights, pregnancy-testing results, and so on are rarely available when problems are first reported, but useful information can be collected by asking questions about the season and the condition of the sheep and by using one's own knowledge about the

	Prejoining	Joining	Mid-pregnancy	Late pregnancy	Lambing	2-3 weeks post-lambing	Marking	2-3 weeks post-marking	Weaning
Investigations Rams	Breeding soundness exam; Harnesses								
Ewes	Condition score; Available pasture; Level of supplementary feed	Crayon marks	Pregnancy test; Surveillance (abortion)	Flock examination; Clinical examination	Difficult births		Wet and dry; Count		Count
Lambs					Count dead lambs; Post mortem exam dead lambs; Weigh dead and live lambs		Weigh. Count	Flock exam	Weigh. Count
Looking for:	Epididymitis etc; Age, condition of ewes; Season of joining (ewes and rams); Age, condition, percentage of rams; Heat stress in rams; Any husbandry procedure which might interfere with joining or conception; Relate nutrition and condition to reproductive indices calculated at all stages; (are indices close to the expected values considering the nutritional state, genotype, age of ewes and season of joining?)	Ewe anoestrus; Return to service; Ram libido; Ram-ewe contact	Estimate losses to date; Benchmark for estimate of subsequent losses; Count multiples	Evidence of pregnancy toxaemia	Lambs present or not; Lamb deaths pre- or post-parturient; Low/high birthweights	Early cessation of lactation	Udder development; Lamb growth rate; Did ewes lamb & lose?	Mulesing wounds healing?	Worm burdens; Other diseases of weaners
Expected values (mixed age Mo age flock):		90-95% of ewes marked in 1st 17 days (plus extra days if out of season)	85-95% of ewes pregnant; 10-50% with multiples (depending on age, condition, season, genotype)	Preg tox: <1%	Lambs dead: 10-20%; Birth weight: >3.5 kg; Dystocia: <1% of births	Lamb deaths after 7d: <3%	Ewes lambing: >90%; Ewes still lactating: >70%; Lactating ewes: full udders; Lamb gr rate: >1.5 kg/week		LW/El: >70%; Ewes present/ewes joined: >95%; Lamb 12 week wt: >16 kg
Diagnose at this point:	Brucellosis etc; Ram mismanagement	Seasonal infertility	Oestrogenic infertility; Embryonic death; Nutritionally low ovulation rate	Undernutrition in pregnancy; Abortion (foetuses found)	Predation; Perinatal mortality; Abortion (losses since pregnancy test)	Undernutrition in late preg/lactation; Internal parasite control failure			

Figure 7.19: The timing of reproductive investigations and points of the reproductive cycle at which specific investigations can occur. Source: KA Abbott.

farm and area. The collection of history effectively becomes an epidemiological investigation and should be planned carefully with the objective of forming a list of differential diagnoses. Based on the presumptive diagnosis, decisions for corrective action can be made. If the initial diagnosis and subsequent actions are correct, diagnostic procedures performed subsequently may not detect a problem.

Ram pre-joining examinations

See 'Veterinary inspection pre-joining' in Section B. The rams should be examined if the history suggests that reproductive failure could possibly be associated with ram infertility. The less time that has elapsed between joining and examination, the more likely it is that the problem can be identified as a ram problem if it *is* a ram problem. Unfortunately, rams lose much condition over the period of joining, restricting the conclusions which can be drawn if they are low in condition score. Lameness, infectious or traumatic lesions of the epididymis, or other disorders of the genitalia can be detected by clinical examination, supported by laboratory aids where necessary. Semen examination for live/dead sperm counts may be useful where one particular animal is suspect. Rams may still bear evidence of flystrike or other insults when examined several months after joining, and history will help decide whether any husbandry procedures before joining contributed to temporary ram infertility.

Ewe pre-joining examinations

When ewe flocks are inspected before joining, a sample should be condition scored and, if possible, weighed. A sample size of 50 is generally adequate. Estimates of subsequent ovulation rate and pregnancy rate can be based on condition score, the age and breed of the ewes and the time of joining.

Assessment of nutrition at joining

For many conditions, particularly nutritional ones, the circumstances operating when the ewes or rams are examined in the pre-joining period *after* the failure in RR has been detected are not the same as those operating *preceding* the failure. Consequently, attempts to retrospectively diagnose the nutritional state of the flock are very important. The pre-joining examination in the subsequent year is, however, still essential because

(a) the flock may again be in poor nutritional state

(b) the flock may be in adequate nutritional state and the reproductive failure may occur again.

The assessment should include pasture availability, pasture quality, estimated intake of pasture and supplementary feed and a comparison of that to energy and protein requirements for body weight maintenance. Ideally, body weights and rate of change of body weights (from two records two to three weeks apart) should be recorded.

Using harnesses on rams and pregnancy testing

These techniques are described earlier in this chapter.

Flock and farm inspections during pregnancy

The ewe flock and their pasture can be inspected during pregnancy to determine the need for supplementation.

Flock and farm inspections at lambing time

An inspection at this time affords an opportunity to assess environmental conditions, protection afforded by topography or vegetation, proximity to predator habitats and management practices which might disturb bonding. Dead lambs can be collected for necropsy. The prevalence of mismothered lambs which are still alive can be estimated and the maternal behaviour of lambing and just-lambed ewes can be observed.

Wet and dry ewes

The technique and its application were described under the heading 'Estimating lamb losses — "the wet and dry technique"' earlier in this section.

Assess lamb growth at marking and weaning

If ewes are satisfactorily nourished, single lambs should grow at the rate of 200 g/day to 6 weeks of age (Merino ewes and lambs) or more for crossbred ewes. The proportion of *poddy* lambs — those with marked rumen development indicating insufficient milk supply — should be <5%. The mean weight of a random draft of, say, 50 lambs at weaning provides an objective measure of ewe nutrition during late pregnancy and lactation. *Target* weaning weights are not appropriate in Australia because they ignore the large variation in pasture availability from year to year which is a feature of the Australian climate. The mean liveweight of the lambs should be considered in relation to the age and breed of the ewe, the prevalence of twins, the age of the lambs and the quality of the season's pasture growth. Corrective action should be taken if the mean weight of 11-week-old Merino lambs is significantly less than 16 kg with no obvious explanation in the history of the previous season.

Differential diagnosis of reproductive problems

Postnatal lamb deaths (lambs dying after seven days)

The major causes to differentiate are early cessation of lactation due to malnutrition or parasitism (or both) of ewes, other diseases of ewes, or primary diseases of lambs (such as white muscle disease), or mismanagement which has resulted in separation of ewes and lambs.

Perinatal lamb deaths

Perinatal lamb deaths should be broadly differentiated into those which are associated with undernutrition of ewes (slow births, low birth weights, failure of ewe-lamb bonding, mismothering, failure to initiate lactation) and may at times be associated with pregnancy toxaemia in ewes; those caused by primary predation, or severely adverse environmental conditions; those associated with management factors which impair maternal behaviour; or those resulting from other causes as discussed earlier in the chapter. Dystocia is a common cause of lamb death in some breeds (Dorset Horn and other British breeds) and in some forms of phyto-oestrogenism. It can also occur when lambs are very heavy (excessive nutrition of ewes) or when ewes are weak (undernutrition of ewes). Necropsy of lambs is a powerful diagnostic aid.

Abortion

Losses of lambs (foetuses) between implantation and term are nearly always infectious. Outbreaks of abortion in ewes are uncommon in Australia but do occur, particularly when ewes are managed at high stocking densities in rotational grazing systems. The differential diagnosis was discussed in detail earlier in the chapter.

Ewes mated but infertile or of low ovulation rate

The differential diagnosis of infertility in cases where ewes have been mated by apparently sound rams, but are not pregnant at pregnancy testing, includes phyto-oestrogenism, low body weights at joining, particularly in maiden ewes, and selenium deficiency (see Chapter 11).

Ewes mated: ram infertility

If ewes have been mated but have had lower than expected pregnancy rates, ram fertility should be suspected. Both deficiencies of general health (very low condition score) and specific genital diseases (such as ovine brucellosis) may reduce ram fertility. If rams are overworked, the frequency of mating of ewes declines with possible effects on fertility.

Failure to mate: ewe anoestrus, insufficient rams

If ewes were not mated but the rams were sound at joining, one should consider the season of joining in relation to the breed and age of the ewes, especially where British breed or crossbred ewes or British breed rams are employed. In the case of Merinos, activity is low in the earlier half of the non-breeding season. Body weight at joining, particularly of maidens, is important. Low body weight of mature ewes does not lead to anoestrus unless the ewes are extremely low in condition. Ewes will display postpartum anoestrus for 50 days or more.[207] If ram numbers are very low some ewes may not be mated at all.

Failure to mate: ram factors

Failure of rams to mate is a rare occurrence and, when it occurs, is usually related to physical disability such as lameness or flystrike. Clinical examination of the rams may detect ongoing problems or allow a retrospective diagnosis. Libido testing of rams (serving capacity test) in pens has been suggested, but the specificity of the test is low. Shearing of the ewes immediately before joining may have reduced the attractiveness of ewes for the rams. Maiden rams are often inexperienced and, if mated to maiden ewes, mating activity may be reduced.

RECOMMENDED READING

West DM, Bruere AN and Ridler AL (2009) Genital soundness in the ram and diseases of the genitalia. In: The sheep: Health, disease & production. 3rd ed. VetLearn Foundation: New Zealand, pp. 10-63.

Greig A (2007) Ram infertility. In: Diseases of sheep, ed ID Aitken. 4th ed. Blackwell Science: London, pp. 87-94.

GENERAL REFERENCES

Lindsay DR and Pearce DT, eds (1984) Reproduction in sheep. Australian Academy of Science: Canberra.

Haughey KG (1983) New insights into rearing failure and perinatal lamb mortality. In: Sheep production and preventive medicine. Proceedings No 67. University of Sydney Postgraduate Committee in Veterinary Science, Sydney, pp. 135-48.

Lynch JJ (1990) Physiology and behaviour of lambs in the perinatal period. In: Sheep Medicine. Proceedings No 141. University of Sydney Postgraduate Committee in Veterinary Science: Sydney, pp. 335-56.

Mellor DJ (1990) Constraints on lamb survival. In: Sheep Medicine. Proceedings No 141. University of Sydney Postgraduate Committee in Veterinary Science: Sydney, pp. 77-83.

Oldham CM, Martin GB and Purvis IW (eds) (1990) Reproductive physiology of Merino sheep: Concepts and consequences. The University of Western Australia: Australia.

Plant JW (1981) Field investigations of reproductive wastage in sheep. In: Sheep. Proceedings No 58. University of Sydney Postgraduate Committee in Veterinary Science: Sydney, pp. 411-38.

Plant JW (1981) Infertility in the ewe. In: Sheep. Proceedings No 58. University of Sydney Postgraduate Committee in Veterinary Science: Sydney, pp. 675-705.

REFERENCES

1 Afolayan RA, Fogarty NM, Gilmour AR et al. (2008) Reproductive performance and genetic parameters in first cross ewes from different maternal genotypes. J Anim Sci **86** 804-14. https://doi.org/10.2527/jas.2007-0544.

2 Restall BJ, Brown GH, Blockey et al. (1976) Assessment of reproductive wastage in sheep. 1. Fertilization failure and early embryonic survival. Aust J Exp Agric Anim Husb **16** 329-35. https://doi.org/10.1071/EA9760329.

3 Yeates NTM (1949) The breeding season of the sheep with particular reference to its modification by artificial means using light. J Agric Sci **39** 1-42. https://doi.org/10.1017/S0021859600004299.

4 Radford HM (1960) Photoperiodism and sexual activity in Merino ewes. Aust J Agric Res **11** 139-46.

5 Hart DS (1950) Photoperiodicity in Suffolk sheep. J Agric Sci **40** 143-9. https://doi.org/10.1017/S0021859600045597.

6 Underwood EJ, Shier FL and Davenport N (1944) Studies in sheep husbandry in WA V. The breeding season of Merino, crossbred and British breed ewes in the Agricultural Districts. J Agric WA **II** 135-143.

7 Smith ID (1967) The breeding season in British breeds of sheep in Australia. Aust Vet J 43 59-62.

8 Edgar DG and Bilkey DA (1963) The influence of rams on the onset of the breeding season in ewes. Proc NZ Soc Anim Prod **23** 79-87.

9 Schinckel PG (1954) The effect of the ram on the incidence and occurrence of oestrus in ewes. Aust Vet J **30** 189-95. https://doi.org/10.1111/j.1751-0813.1954.tb08198.x.

10 Pearce DT and Oldham CM (1984) The ram effect, its mechanism and application to the management of sheep. In: Reproduction in sheep, eds DR Lindsay and DT Pearce. Australian Academy of Science: Canberra, pp. 26-34.

11 Pearce GP and Oldham CM (1988) Importance of non-olfactory ram stimuli in mediating ram-induced ovulation in the ewe. J Reprod Fertil **84** 333-9. https://doi.org/10.1530/jrf.0.0840333.

12 Jorre de St Jorre T, Hawken PAR and Martin GB (2012) Role of male novelty and familiarity in male-induced LH secretion in female sheep. Reprod Fert Develop **24** 523-30. https://doi.org/10.1071/RD11085.

13 Dun RB, Ahmed W and Morrant AJ (1960) Annual reproductive rhythm in Merino sheep related to the choice of a mating time at Trangie central western New South Wales. Aust J Agric Res **11** 805-56.

14 Cumming IA (1977) Relationships in the sheep of ovulation rate with liveweight, breed, season and plane of nutrition. Aust J Exp Agric Anim Husb **17** 234-40. https://doi.org/10.1071/EA9770234.

15 Morley FHW, White DH, Kenney PA et al. (1978) Predicting ovulation rate from liveweight in ewes. Agric Syst **3** 27-45.

16 Kleeman DO and Walker SK (2005) Fertility in South Australian Merino flocks: Relationships between reproductive traits and environmental cues. Theriogenol **63** 2416-33.

17 Fletcher IC (1971) Effects of nutrition, liveweight and season on the incidence of twin ovulation in South Australian strong-wool Merino ewes. Aust J Agric Res **22** 321-30. https://doi.org/10.1071/AR9710321.

18 Behrendt R, van Burgel AJ, Bailey A et al. (2011) On-farm paddock scale comparisons across southern Australia confirm that increasing the nutrition of Merino ewes improves their production and the lifetime performance of their progeny. Anim Prod Sci **51** 805-12. https://doi.org/10.1071/AN10183.

19 Trompf JP, Gordon DJ, Behrendt R et al. (2011) Participation in Lifetime Ewe Management results in changes in stocking rate, ewe management and reproductive performance on commercial farms. Anim Prod Sci **51** 866-72. https://doi.org/10.1071/AN10164.

20 Coop IE (1966) Effect of flushing on reproductive performance of ewes. J Agric Sci Camb **67** 305-23.

21 Linday DR (1976) The usefulness to the animal producer of research findings in nutrition on reproduction. Proc Aust Soc Anim Prod **11** 217-24.

22 Knight TW, Oldham CM and Lindsay DR (1975) Studies in ovine infertility in agricultural regions in Western Australia: The influence of a supplement of lupins (*Lupinus angustifolius* cv. Uniwhite) at joining on the reproductive performance of ewes. Aust J Agric Res **26** 567-75. https://doi.org/10.1071/AR9750567.

23 Van Barneveld RJ (1999) Understanding the nutritional chemistry of lupins (*Lupinus* spp) seed to improve livestock production efficiency. Nutr Res Rev **12** 203-230.

24 Teleni E, King WR, Rowe JB et al. (1989) Lupins and energy-yielding nutrients in ewes. I Glucose and acetate biokinetics and metabolic hormones in sheep fed a supplement of lupin grain. Aust J Agric Res **40** 913-24. https://doi.org/10.1071/AR9890913.

25 Teleni E, Rowe JB, Croker KP et al. (1989) Lupins and energy-yielding nutrients in ewes. II Responses in ovulation rate in ewes to increased availability of glucose, acetate and amino acids. Reprod Fertil Dev **1** 117-25. https://doi.org/10.1071/RD9890117.

26 Somchit A, Campbell BK, Khalid M et al. (2007) The effect of short-term nutritional supplementation of ewes with lupin grain (*Lupinus luteus*), during the luteal phase of the estrous cycle on the number of ovarian follicles and the concentrations of hormones and glucose in plasma and follicular fluid. Theriogenol **68** 1037-46.

27 Brien FD, Cumming IA and Baxter RW (1977) Effect of feeding a lupin grain supplement on reproductive performance of maiden and mature ewes. J Agric Sci Camb **89** 437-43.

28 Alexander G, Bradley LR and Stevens D (1993) Effect of age and parity on maternal behaviour in single-bearing ewes Aust J Exp Agr **33** 721-8. https://doi.org/10.1071/EA9930721.

29 Quinlivan TD, Martin CA, Taylor WB et al. (1966) Estimates of pre- and perinatal mortality in the New Zealand Romney Marsh ewe I. Pre- and perinatal mortality in those ewes that conceived to one service. J Reprod Fertil **11** 379-90. https://doi.org/10.1530/jrf.0.0110379.

30 Allison AJ (1977) Flock mating in sheep. II Effect of number of ewes per ram on mating behaviour and fertility of two-tooth and mixed-age Romney ewes run together. New Zeal J Agr Res **20** 123-8. https://doi.org/10.1080/00288233.1977.10427315.

31 Tilbrook AJ and Cameron AWN (1989) Ram mating preferences for woolly rather than recently shorn ewes. Appl Anim Behav Sci **24** 301-12. https://doi.org/10.1016/0168-1591(89)90058-0.

32 Adams NR (1990) Permanent infertility in ewes exposed to plant oestrogens. Aust Vet J **67** 197-201. https://doi.org/10.1111/j.1751-0813.1990.tb07758.x.

33 Cox RI and Braden AW (1974) The metabolism and physiological effects of phyto-oestrogens in livestock. Proc Aust Soc Anim Prod **10** 122-9.

34 Adams NR (1986) Measurement of histological changes in the cervix of ewes after prolonged exposure to oestrogenic clover or oestradiol-17ß. Aust Vet J **63** p. 279-82. https://doi.org/10.1111/j.1751-0813.1986.tb08066.x.

35 Adams NR (1976) Cervical mucus changes in infertile ewes previously exposed to oestrogenic subterranean clover. Res Vet Sci **21** 59-63.

36 Maxwell JAL (1970) Field observations on four outbreaks of maternal dystocia in the Merino ewe. Aust Vet J **46** 533-6. https://doi.org/10.1111/j.1751-0813.1970.tb06640.x.

37 Braden AWH, Southcott WH and Moule GR (1964) Assessment of oestrogenic activity of pastures by means of increase of teat length in sheep. Aust J Agric Res **15** 142-52. https://doi.org/10.1071/AR9640142.

38 Little DL and Beale PE (1988) Renovation of Yarloop subterranean clover pastures with Trikkala. Aust J Exp Agric **28** 737-45.

39 Lightfoot RJ and Wroth RH (1974) The mechanism of temporary infertility in ewes grazing oestrogenic subterranean clover prior to and during joining. Proc Aust Soc Anim Prod **10** 130-4.

40 Morley FHW, Axelsen A and Bennet D (1966) Recovery of normal fertility after grazing on oestrogenic red clover. Aust Vet J **42** 204-6. https://doi.org/10.1111/j.1751-0813.1966.tb04690.x.

41 Coop IE (1977) Depression of lambing percentage from mating on Lucerne. Proc NZ Soc Anim Prod **37** 149-51.

42 Nancarrow CD (1994) Embryonic mortality in the ewe and doe. In: Embryonic mortality in domestic species, ed RD Geisert and MT Zavy. CRC Press: Boca Raton, Florida, pp. 79-97.

43 Edey TN (1979) Embryo mortality. In: Sheep breeding: Studies in the agricultural and food sciences, eds GL Tomes, DE Robertson and RJ Lightfoot. 2nd ed, revised by William Haresign. Butterworths: London, pp. 315-25.

44 O'Connell AR, Demmers KJ, Smaill B et al. (2016) Early embryo loss, morphology, and effect of previous immunization against androstenedione in the ewe. Theriogenol **86** 1285-93.

45 Bazer FW, Spencer TE and Ott TL (1997) Interferon Tau: A novel pregnancy recognition signal. Am J Reprod Immunol **37** 412-20. https://doi.org/10.1111/j.1600-0897.1997.tb00253.x.

46 Shorten PR, Peterson AJ, O'Connell AR et al. (2010) A mathematical model of pregnancy recognition in mammals. J Theor Biol **266** 62-9. https://doi.org/10.1016/j.jtbi.2010.06.005.

47 Edey TN (1967) Early embryonic death and subsequent cycle length in the ewe. J Reprod Fertil **13** 437-43. https://doi.org/10.1530/jrf.0.0130437.

48 Geisler PA, Newton JE and Mohan AE (1977) A mathematical model of fertilization failure and early embryonic mortality in sheep. J Agric Sci Camb **89** 309-17.

49 Wikins JF and Croker KP (1990) Embryonic wastage in ewes. In: Reproductive physiology of Merino sheep — concepts and consequences, eds CM Oldham, GB Martin and IW Purvis. University of Western Australia: Perth, pp. 169-77.

50 Lindsay DR, Knight TW, Smith JF et al. (1975) Studies in ovine fertility in agricultural regions of Western Australia: Ovulation rate, fertility and lambing performance. Aust J Agric Res **26** 189-98. https://doi.org/10.1071/AR9750189.

51 Lincoln GA, Lincoln CE and McNeilly AS (1990) Seasonal cycles in the blood plasma concentration of FSH, inhibin and testosterone, and testicular size in rams of wild, feral and domesticated breeds of sheep. J Reprod Fertil **88** 623-33. https://doi.org/10.1530/jrf.0.0880623.

52 D'Occhio MJ and Brooks DE (1983) Seasonal changes in plasma testosterone concentration and mating activity in Border Leicester, Poll Dorset, Romney and Suffolk rams. Aust J Exp Agric Anim Husb **23** 248-53. https://doi.org/10.1071/EA9830248.

53 Ryder ML (1983) Sheep and man. MPG Books Ltd: Cornwall, UK, p. 460.

54 Martin GB, Hötzel MJ, Blache D et al. (2002) Determinants of the annual pattern of reproduction in mature male Merino and Suffolk sheep: Modification of responses to photoperiod by an annual cycle in food supply. Reprod Fertil Dev **14** 165-75. https://doi.org/10.1071/RD02010.

55 Bremner WJ, Cumming IA and Winfield C (1984) A study of the reproductive performance of mature Romney and Merino rams throughout the year. In: Reproduction in sheep, eds DR Lindsay and DT Pearce. Australian Academy of Science: Canberra, pp. 16-19.

56 Lincoln GA and Davidson W (1977) The relationship between sexual and aggressive behaviour, and pituitary and testicular activity during the seasonal sexual cycle of rams, and the influence of photoperiod. J Reprod Fertil **49** 267-76. https://doi.org/10.1530/jrf.0.0490267.

57 Murray PJ, Rowe JB and Petrhick DW (1991) Effect of season and nutrition on scrotal circumference of Merino rams. Aust J Exp Agr **31** 753-6. https://doi.org/10.1071/EA9910753.

58 Masters DG and Fels HE (1984) Seasonal changes in the testicular size of grazing rams. Proc Aust Soc Anim Prod **15** 444-7.

59 Sutherland SRD and Martin GB (1980) The effect of a supplement of lupin seed on the testicular size and LH profiles of Merino and Booroola rams. Proc Aust Soc Anim Prod **13** 459.

60 Lino BF (1972) The output of spermatozoa in rams. II Relationship to scrotal circumference, tests weight, and the number of spermatozoa in different parts of the urogenital tract. Aust J Biol Sci **25** 359-66. https://doi.org/10.1071/BI9720359.

61 Foster RA, Ladds PW, Hoffmann D et al. (1989) The relationship of scrotal circumference to testicular weight in rams. Aust Vet J **66** 20-2. https://doi.org/10.1111/j.1751-0813.1989.tb09707.x.

62 Mattner PE and Braden AWH (1967) Studies in flock mating of sheep. 2 Fertilization and prenatal mortality. Aust J Exp Agric Anim Husb **11** 110-116. https://doi.org/10.1071/EA9670110.

63 Raadsma HW and Edey TN (1985) Mating performance of paddock-mated rams. I Changes in mating performance, ejaculate characteristics and testicular size during the joining period. Anim Reprod Sci **8** 79-99. https://doi.org/10.1016/0378-4320(85)90075-2.

64 Synott AL, Fulkerson WJ and Lindsay DR (1981) Sperm output by rams and distribution amongst ewes under conditions of continual mating. J Reprod Fertil **61** 355-61. https://doi.org/10.1530/jrf.0.0610355.

65 Salamon S (1962) Studies on the artificial insemination of Merino sheep: III The effect of frequent ejaculation on semen characteristics and fertilizing capacity. Aust J Ag Res **13** 1137-50.

66 Lindsay DR and Ellsmore J (1968) The effect of breed, season and competition on mating behaviour of rams. Aust J Exp Agric Anim Husb **8** 649-52.

67 Synnott AL and Fulkerson WJ (1984) Influence of social interaction between rams on their serving capacity. App Anim Ethol **11** 283-9.

68 Lightfoot RJ and Smith JAC (1968) Studies on the number of ewes joined per ram for flock matings under paddock conditions. Aust J Agric Res **19** 1029-42. https://doi.org/10.1071/AR9681029.

69 Allison AJ (1975) Flock mating in sheep. I Effect of number of ewes joined per ram on mating behaviour and fertility. New Zeal J Agr Res **18** 1-8. https://doi.org/10.1080/00288233.1975.10430379.

70 Dun RB (1955) Puberty in Merino rams. Aust Vet J **31** 104-6. https://doi.org/10.1111/j.1751-0813.1955.tb05515.x.

71 Skinner JD and Rowson LEA (1968) Puberty in Suffolk and cross-bred lambs. J Reprod Fertil **16** 479-88. https://doi.org/10.1530/jrf.0.0160479.

72 Belibasaki S and Kouimtzis S (2000) Sexual activity and body and testis growth in prepubertal ram lambs of Friesland, Chios, Karagouniki and Serres dairy sheep in Greece. Small Ruminant Res **37** 109-13.

73 Watson RH, Sapsford CS and McCance I (1956) The development of the testis, epididymis and penis in the young Merino ram. Aust J Agric Res **7** 575-90. https://doi.org/10.1071/AR9560574.

74 Swan AA, Brown DJ and Banks RG (2008) Genetic progress in the Australian sheep industry. Proc Assoc Advmt Anim Breed Genet **18** 326-9.

75 Murray PJ, Rowe JB, Pethick DW et al. (1990) The effect of nutrition on testicular growth in the Merino ram. Aust J Agric Res **41** 185-95. https://doi.org/10.1071/AR9900185.

76 De Blockey MA and Wilkins JF (1984) Field application of the ram serving capacity test. In: Reproduction in sheep, eds DR Lindsay and DT Pearce. Australian Academy of Science: Canberra. p. 53.

77 Purvis IW, Edey TN, Kilgour RJ et al. (1984) The value of testing young rams for serving capacity In Reproduction. In: Reproduction in sheep, eds DR Lindsay and DT Pearce. Australian Academy of Science: Canberra p. 59.

78 Galloway DB (1983) Reproduction in the ram. In: Sheep production and preventive medicine. Proceedings No 67. University of Sydney Postgraduate Committee in Veterinary Science: Sydney, pp. 163-95.

79 Kilgour RJ (1993) The relationship between ram breeding capacity and flock fertility. Theriogenol **40** 277-85.

80 Preston BT, Stenson IR, Pemberton JM et al. (2003) Overt and covert competition in a promiscuous mammal: The importance of weaponry and testes size to male reproductive success. Proc R Soc Lond B Biol Sci **270** 633-40. https://doi.org/10.1098/rspb.2002.2268.

81 Fowler DG and Jenkins LD (1976) The effects of dominance and infertility of rams on reproductive performance. Appl Anim Ethol **2** 327-37. https://doi.org/10.1016/0304-3762(76)90066-3.

82 Ladds PW (1993) The male genital system. In: Pathology of domestic animals, eds KVF Jubb, PC Kennedy and N Palmer. 4th ed. Vol 3. Academic Press: New York & London, p. 471.

83 Pulsford MF, Eastick BC, Clapp KH et al. (1967) Traumatic epididymitis of Dorset Horn and Poll Dorset rams. Aust Vet J **43** 99-101. https://doi.org/10.1111/j.1751-0813.1967.tb08893.x.

84 Foster RA, Ladds, PW, Hoffman D et al. (1989) Pathology of reproductive tracts of Merino rams in north western Queensland. Aust Vet J **66** 262-4.

85 Hughes KL (1972) Experimental Brucella ovis infection in ewes 1. Breeding performance of infected ewes. Aust Vet J **48** 12-17. https://doi.org/10.1111/j.1751-0813.1972.tb02200.x.

86 McGowan B and Schultz G (1956) Epididymitis of rams: Clinical description and field aspects. Cornell Vet **46** 277-81.

87 Plant JW (1982) Ovine brucellosis. Agfact A3.9.2 AGdex 432/653 NSW Agriculture.

88 Burgess GW, McDonald JW and Norris MJ (1982) Epidemiological studies on ovine brucellosis in selected ram flocks. Aust Vet J **59** 45-7.

89 Plant JW, Eamens GJ and Seaman JT (1986) Serological, bacteriological and pathological changes in rams following different routes of exposure to *Brucella ovis*. Aust Vet J **63** 409-12.

90 Hughes KL (1972) Experimental *Brucella ovis* infection in ewes 2. Correlation of infection and complement fixation titres. Aust Vet J **48** 18-22. https://doi.org/10.1111/j.1751-0813.1972.tb02201.x.

91 Haughey KG, Hughes KL and Hartley WJ (1968) *Brucella ovis* infection 2. The infection status in breeding flocks as measured by examination of rams and the perinatal lamb mortality. Aust Vet J **44** 531-5. https://doi.org/10.1111/j.1751-0813.1968.tb04920.x.

92 Searson JE (1986) Distribution of *Brucella ovis* in the tissues of rams reacting in a complement fixation test for ovine brucellosis. Aust Vet J **63** 30-1. https://doi.org/10.1111/j.1751-0813.1986.tb02872.x.

93 Searson JE (1987) The distribution of histopathological lesions in rams reacting in a complement fixation test for *Brucella ovis*. Aust Vet J **64** 108-9. https://doi.org/10.1111/j.1751-0813.1987.tb09640.x.

94 Burgess GW and Norris MJ (1982) Evaluation of the cold complement fixation test for diagnosis of ovine brucellosis. Aust Vet J **59** 23-5. https://doi.org/10.1111/j.1751-0813.1982.tb02706.x.

95 Laws L, Simmons GC and Ludford CG (1972) Experimental *Brucella ovis* infection in rams. Aust Vet J **48** 313-17. https://doi.org/10.1111/j.1751-0813.1972.tb02257.x.

96 Hughes KL and Claxton PD (1968) *Brucella ovis* infection 1. An evaluation of microbiological, serological and clinical methods of diagnosis in the ram. Aust Vet J **44** 41-7. https://doi.org/10.1111/j.1751-0813.1968.tb04951.x.

97 Webb RF, Quinn CA, Cockram FA et al. (1980) Evaluation of procedures for the diagnosis of *Brucella ovis* infection in rams. Aust Vet J **56** 172-5. https://doi.org/10.1111/j.1751-0813.1980.tb05673.x.

98 Lee K, Cargill C and Atkinson H (1985) Evaluation of an enzyme linked-immunosorbent assay for the diagnosis of *Brucella ovis* infection in rams. Aust Vet J **62** 91-3. https://doi.org/10.1111/j.1751-0813.1985.tb14147.x.

99 Rothwell JT, Searson JE, Links IJ et al. (1986) Examination of rams culled during an ovine brucellosis accredited free flock scheme. Aust Vet J **63** 209-11. https://doi.org/10.1111/j.1751-0813.1986.tb02996.x.

100 Gorrie CJR and Rushford BH (1969) Control of ovine brucellosis. Aust Vet J **45** 304. https://doi.org/10.1111/j.1751-0813.1969.tb01965.x.

101 Claxton PD (1968) *Brucella ovis* vaccination of rams. Aust Vet J **44** 48-54. https://doi.org/10.1111/j.1751-0813.1968.tb04953.x.

102 Ridler Al and West DM (2011) Control of *Brucella ovis* infection in sheep. Vet Clinics North America: Food Animal Practice **27** 61-6. https://doi.org/10.1016/j.cvfa.2010.10.013.

103 Rycroft AN and Garside LH (2000) Actinobacillus species and their role in animal disease. Vet J **159** 18-39. https://doi.org/10.1053/tvjl.1999.0403.

104 McGillivery DJ, Webber JJ and Dean HF (1986) Characterisation of *Histophilus ovis* and related organisms by restriction endonuclease analysis. Aust Vet J **63** 389-93. https://doi.org/10.1111/j.1751-0813.1986.tb15914.x.

105 Baynes ID and Simmons GC (1960) Ovine epididymitis caused by *Actinobacillus seminis*. Aust Vet J **36** 454-9. https://doi.org/10.1111/j.1751-0813.1960.tb03745.x.

106 Al-Katib WA and Dennis SM (2009) Ovine genital actinobacillosis: A review. NZ Vet J **57** 352-8. https://doi.org/10.1080/00480169.2009.64722.

107 Bruère AN, West DM, MacLachlan NJ et al. (1977) Genital infection of ram hoggets associated with a Gram-negative pleomorphic organism. NZ Vet J **25** 191-3. https://doi.org/10.1080/00480169.1977.34401.

108 Baynes ID and Simmons GC (1968) Clinical and pathological studies of Border Leicester rams naturally infected with *Actinobacillus seminis*. Aust Vet J **44** 339-43. https://doi.org/10.1111/j.1751-0813.1968.tb14399.x.

109 Watt DA, Bamford V and Nairn ME (1970) *Actinobacillus seminis* as a cause of polyarthritis and posthitis in sheep. Aust Vet J **46** 515. https://doi.org/10.1111/j.1751-0813.1970.tb09190.x.

110 Foster G, Collins MD, Lawson PA et al. (1999) *Actinobacillus seminis* as a cause of abortion in a UK sheep flock. Vet Rec **144** 479-80. https://doi.org/10.1136/vr.144.17.479.

111 Angen Ø, Ahrens P, Kuhnert P et al. (2003) Proposal of *Histophilus somni* gen. Nov., sp. Nov. for the three species *incertae sedis* 'Haemophilus somnus', 'Haemophilus agni' and 'Histophilus ovis'. Int J Syst Evol Microbiol **53** 1449-59. https://doi.org/10.1099/ijs.0.02637-0.

112 Philbey AW, Glastonbury JR, Rothwell JT et al. (1991) Meningoencephalitis and other conditions associated with *Histophilus ovis* infection in sheep. Aust Vet J **68** 387-90. https://doi.org/10.1111/j.1751-0813.1991.tb03104.x.

113 Walker RL and LeaMaster BR (1986) Prevalence of *Histophilus ovis* and *Actinobacillus seminis* in the genital tract of sheep. Am J Vet Res **47** 1928-30.

114 Saunders VF, Reddacliff LA, Berg T et al. (2007) Multiplex PCR for the detection of *Brucella ovis*, *Actinobacillus seminis* and *Histophilus somni* in ram semen. Aust Vet J **85** 72-7. https://doi.org/10.1111/j.1751-0813.2006.00098.x.

115 Constable PD and Webber JJ (1987) *Escherichia coli* epididymitis in rams. Aust Vet J **64** 123. https://doi.org/10.1111/j.1751-0813.1987.tb09651.x.

116 Smith KC, Brown PJ, Barr FJ et al. (2012) Cryptorchidism in sheep: A clinical and abattoir survey in the United Kingdom. Open Journal of Veterinary Medicine **2** 281-4.

117 Dolling CHS and Brooker MG (1964) Cryptorchism in Australian Merino sheep. Nature **203** 49-50. https://doi.org/10.1038/203049a0.

118 Miller SJ and Moule GR (1954) Clinical observations on the reproductive organs of Merino rams in pastoral Queensland. Aust Vet J **30** 353-63. https://doi.org/10.1111/j.1751-0813.1954.tb05398.x.

119 Gunn RMC, Sanders RN and Granger W (1942) Studies in fertility in sheep. Bull Coun Sci Industr Res Aust **148** 111-15.

120 Watt DA (1978) Testicular pathology of Merino rams. Aust Vet J **54** 473-8. https://doi.org/10.1111/j.1751-0813.1978.tb00291.x.

121 Dolling CHS (1970) Breeding Merinos. Rigby Australia: Adelaide.

122 Singh LB, Dolling CHS and Singh ON (1969) Inheritance of horns and occurrence of cryptorchism in indigenous, Rambouillet and crossbred sheep in India. Aust J Exp Agric Anim Husb **9** 262-6.

123 Bruere AN (1970) Some clinical aspects of hypo-orchidism (small testes) in the ram. NZ Vet J **18** 189-98. https://doi.org/10.1080/00480169.1970.33897.

124 Underwood EJ and Somers M (1969) Studies of zinc nutrition in sheep I. The relation of zinc nutrition to growth, testicular development and spermatogenesis in young rams. Aust J Agric Res **20** 889-97. https://doi.org/10.1071/AR9690889.

125 Webb RF and Chick BF (1976) Balanitis and vulvo-vaginitis in sheep. Aust Vet J **52** 241-2. https://doi.org/10.1111/j.1751-0813.1976.tb00091.x.

126 Watt B, Wait P and Slattery S (2016) Ulcerative balanitis in rams — 'An enigmatic disease of unknown aetiology'. Flock and Herd Case notes. Available from: http://www.flockandherd.net.au/sheep/reader/ulcerative-balanitis.html. Accessed 4 July 2018.

127 Rhodes AP (1975) Seminal degeneration associated with chorioptic mange of the scrotum of rams. Aust Vet J **51** 428. https://doi.org/10.1111/j.1751-0813.1975.tb15792.x.

128 Guo M, Dubey JP, Hill D et al. (2015) Prevalence and risk factors for *Toxoplasma gondii* infection in meat animals and meat products destined for human consumption. J Food Prot **78** 457-76. https://doi.org/10.4315/0362-028X.JFP-14-328.

129 Buxton D (1990) Ovine toxoplasmosis: A review. J R Soc Med **83** 509-11. https://doi.org/10.1177/014107689008300813.

130 Buxton D (1989) Toxoplasmosis and chlamydial abortion. In: Proceedings of the Second International Congress for sheep veterinarians, Sheep and Beef Cattle Society of the New Zealand Veterinary Association. 12-16 February, Massey University, New Zealand, p. 319.

131 Munday BL and Dubey JP (1986) Serology of experimental toxoplasmosis in pregnant ewes and their foetuses. Aust Vet J **63** 353-5. https://doi.org/10.1111/j.1751-0813.1986.tb02894.x.

132 Wilkins M, O'Connell E and Jonas W (1988) Toxoplasma vaccine trials. In: Proceedings of the 18th Seminar, Sheep and Beef Cattle Society of the New Zealand Veterinary Association. January 1988, p. 201.

133 Dennis SM (1967) The possible role of the raven in the transmission of ovine vibriosis. Aust Vet J **43** 45-8. https://doi.org/10.1111/j.1751-0813.1967.tb15061.x.

134 Dennis SM (1975) Perinatal lamb mortality in Western Australia 5. Vibrionic infection. Aust Vet J **51** 11-13. https://doi.org/10.1111/j.1751-0813.1975.tb14490.x.

135 Plant JW, Beh KJ and Acland HM (1972) Laboratory findings from ovine abortion and perinatal mortality. Aust Vet J **48** 558-61. https://doi.org/10.1111/j.1751-0813.1972.tb08011.x.

136 Pauling BA (1988) Campylobacter abortion update. In: Proceedings of the 18th Seminar, Sheep and Beef Cattle Society of the New Zealand Veterinary Association. January 1988, p. 205.

137 Murray CJ (2007) *Salmonella* serovars and phage types in humans and animals in Australia 1987-1992. Aust Vet J **71** 78-81. https://doi.org/10.1111/j.1751-0813.1994.tb03332.x.

138 Dennis SM and Armstrong JM (1965) Ovine abortion due to *Salmonella typhimurium* in Western Australia. Aust Vet J **41** 178-81.

139 Linklater KA (1991) Salmonellosis and *Salmonella* abortion. In: Diseases of sheep, eds WB Martin and ID Aitken. 2nd ed. Blackwell Scientific Publications: London, p. 65.

140 Wray C (1985) Is salmonellosis still a serious problem in veterinary practice? Vet Rec **116** 485-9. https://doi.org/10.1136/vr.116.18.485.

141 Buchrieser C, Rusniok C, Garrido P et al. (2011) Complete genome sequence of the animal pathogen *Listeria ivanovii*, which provides insights into host specificities and evolution of the genus *Listeria*. J Bacteriol **193** 6787-8. https://doi.org/10.1128/JB.06120-11.

142 Guillet C, Join-Lambert O, Le Monnier A et al. (2010) Human Listeriosis caused by *Listeria ivanovii*. Emerg Infect Dis **16** 136-8. https://doi.org/10.3201/eid1601.091155.

143 Dennis SM (1975) Perinatal lamb mortality in Western Australia 6. Listeric infection. Aust Vet J **51** 75-9. https://doi.org/10.1111/j.1751-0813.1975.tb09409.x.

144 Hughes KL (1975) *Listeria* as a cause of abortion and neonatal mortality in sheep. Aust Vet J **51** 97-9. https://doi.org/10.1111/j.1751-0813.1975.tb09415.x.

145 Seaman JT (1985) Chlamydia isolated from abortion in sheep. Aust Vet J **62** 436. https://doi.org/10.1111/j.1751-0813.1985.tb14140.x.

146 Rofe JC (1967) Suspected enzootic of virus abortion of ewes. Aust Vet J **43** 117-18.

147 Sachse K, Bavoil PM, Kaltenboeck B et al. (2015) Emendation of the family *Chlamydiaceae*: Proposal of a single genus, *Chlamydia*, to include all currently recognized species. Syst Appl Microbiol **38** 99-103. https://doi.org/10.1016/j.syapm.2014.12.004.

148 McCauley LME, Lancaster MJ, Butler et al. (2010) Serological analysis of *Chlamydophila abortus* in Australian sheep and implications for the rejection of breeder sheep for export. Aust Vet J **88** 32-8. https://doi.org/10.1111/j.1751-0813.2009.00536.x.

149 Yang, R, Jacobson C, Gardner G et al. (2014) Longitudinal prevalence and faecal shedding of *Chlamydia pecorum* in sheep. Vet J **201** 322-6. https://doi.org/10.1016/j.tvjl.2014.05.037.

150 Everett KDE (2000) *Chlamydia* and *Chlamydiales*: More than meets the eye. Vet Microbiol **75** 109-26. https://doi.org/10.1016/S0378-1135(00)00213-3.

151 Hughes KL (1972) Experimental *Brucella ovis* infection of ewes 1. Breeding performance of infected ewes. Aust Vet J **48** 12-17. https://doi.org/10.1111/j.1751-0813.1972.tb02200.x.

152 Hughes KL (1972) Experimental *Brucella ovis* infection in ewes 2. Correlation of infection and complement fixation titres. Aust Vet J **48** 18-22. https://doi.org/10.1111/j.1751-0813.1972.tb02201.x.

153 Haughey KG, Hughes KL and Hartley WJ (1968) *Brucella ovis* infection 2. The infection status in breeding flocks as measured by examination of rams and the perinatal lamb mortality. Aust Vet J **44** 531-5. https://doi.org/10.1111/j.1751-0813.1968.tb04920.x.

154 Beveridge WIB (1981) Mucosal disease. In: Viral diseases of farm livestock. Australian Government Publishing Service: Canberra, p. 60.

155 Beveridge WIB (1981) Akabane disease. In: Viral diseases of farm Livestock. Australian Government Publishing Service: Canberra, p. 3.

156 Parsonson IM, Della-Porta AJ, Snowdon WA et al. (1975) Congenital abnormalities in foetal lambs after inoculation of pregnant ewes with Akabane virus. Aust Vet J **51** 585-6. https://doi.org/10.1111/j.1751-0813.1975.tb09398.x.

157 Gorrie CJR (1962) Ovine abortion in Victoria. Aust Vet J **38** 138-42. https://doi.org/10.1111/j.1751-0813.1962.tb16029.x.

158 McDonald JW and Lancaster MJ (1991) Bracken fern, Onion grass and oestrogenic clover problems in north east Victoria. In: Proceedings of the Australian Sheep Veterinarians Conference, Australian Veterinary Association. 13-17 May, Darling Harbour, pp. 109-10.

159 Smith ID (1967) Breed differences in the duration of gestation in sheep. Aust Vet J **43** 63-4. https://doi.org/10.1111/j.1751-0813.1967.tb15065.x.

160 Reid RL (1968) The physiopathology of undernourishment in pregnant sheep with particular reference to pregnancy toxaemia. Adv Vet Sci **12** 162-238.

161 Caple IW and McLean JG (1981) Pregnancy toxaemia. In: Current veterinary therapy food animal practice, ed JL Howard. WB Saunders Company: Philadelphia, p. 348.

162 Reid RL (1968) The physiopathology of undernourishment in pregnant sheep, with particular reference to pregnancy toxaemia. Adv Vet Sci **12** 163-238.

163 Schlumbohm C and Harmeyer J (2008) Twin-pregnancy increases susceptibility of ewes to hypo-glycaemic stress and pregnancy toxaemia. Res Vet Sci **84** 286-99. https://doi.org/10.1016/j.rvsc.2007.05.001.

164 Wastney ME, Arcus AC, Bickerstaffe R et al. (1982) Glucose tolerance in ewes and susceptibility to pregnancy toxaemia. Aust J Biol Sci **35** 381-92.

165 Buswell JF, Haddy JP and Bywater RJ (1986) Treatment of pregnancy toxaemia in sheep using a concentrated oral rehydration solution. Vet Rec **118** 208-9. https://doi.org/10.1136/vr.118.8.208.

166 Weirda A, Verhoeff J, van Dijk S et al. (1985) Effects of Trenbolone acetate and propylene glycol on pregnancy toxaemia in ewes. Vet Rec **116** 284-7. https://doi.org/10.1136/vr.116.11.284.

167 Zamir S, Rozov A and Gootwine E (2009) Treatment of pregnancy toxaemia in sheep with flunixine meglumine. Vet Rec **165** 265-6. https://doi.org/10.1136/vr.165.9.265.

168 Rizos D, Kenny DA, Griffin W et al. (2008) The effect of feeding propylene glycol to dairy cows during the early postpartum period on follicular dynamics and on metabolic parameters related to fertility. Theriogenol **69** 688-99.

169 Nielsen NI and Ingvartsen KL (2004) Propylene glycol for dairy cows. A review of the metabolism of propylene glycol and its effects on physiological parameters, feed intake, milk production and risk of ketosis. Anim Feed Sci Technol **115** 191-213. https://doi.org/10.1016/j.anifeedsci.2004.03.008.

170 Foot JZ (1983) Nutrition of ewes pre- and post-lambing. In: Sheep production and preventive medicine. Proceedings No 67. University of Sydney Postgraduate Committee in Veterinary Science: Sydney, pp. 267-81.

171 Larsen JWA, Constable PD and Napthine DV (1986) Hypocalcaemia in ewes after a drought. Aust Vet J **63** 25-6. https://doi.org/10.1111/j.1751-0813.1986.tb02867.x.

172 Hay LA (1991) Hernia and prolapse. In: Diseases of sheep, eds WB Martin and ID Aitken. Blackwell Scientific Publications: London, p. 71.

173 Lynch JJ and Alexander G (1980) The effect of time of shearing on sheltering behaviour of Merino sheep. Proc Aust Soc Anim Prod **13** 325-8.

174 Hinch GN and Brien F (2014) Lamb survival in Australian flocks: A review. Anim Prod Sci **54** 656-66. https://doi.org/10.1071/AN13236.

175 Sackett D, Holmes P, Abbott K et al. (2006) Assessing the economic cost of endemic disease on the profitability of Australian sheep and cattle producers. Final Report to Meat and Livestock Australia. Project AHW.087. Meat and Livestock Australia: North Sydney.

176 Holst PJ, Fogarty NM and Stanley DF (2002) Birth weights, meningeal lesions, and survival of diverse genotypes of lambs from Merino and crossbred ewes. Aust J Agric Res **53** 175-81. https://doi.org/10.1071/AR01046.

177 Everett-Hincks JM and Dodds KG (2008) Management of maternal-offspring behaviour to improve lamb survival in easy care sheep systems. J Anim Sci **86** 259-70. https://doi.org/10.2527/jas.2007-0503.

178 Arnold GW and Morgan PD (1975) Behaviour of the ewe and lamb at lambing and its relationship to lamb mortality. Appl Anim Ethol **2** 25-46. https://doi.org/10.1016/0304-3762(75)90063-2.

179 Alexander G (1960) Maternal behaviour in the Merino ewe. Proc Aust Soc Anim Prod **3** 105-14.

180 George JM (1975) The incidence of dystocia in fine-wool Merino ewes. Aust Vet J **51** 262-5. https://doi.org/10.1111/j.1751-0813.1975.tb06931.x.

181 Hall DG, Gilmour AR and Fogarty NM (1994) Variation in reproduction and production of Poll Dorset ewes. Aust J Agric Res **45** 415-26. https://doi.org/10.1071/AR9940415.

182 Dwyer CM, McLean KA, Deans LA et al. (1998) Vocalisations between mother and young in sheep: Effects of breed and maternal experience. Appl Anim Behav **58** 105-19. https://doi.org/10.1016/S0168-1591(97)00113-5.

183 Vince MA (1993) Newborn lambs and their dams: The interaction that leads to sucking. Adv Stud Behav **22** 239-68. https://doi.org/10.1016/S0065-3454(08)60408-8.

184 Putu IG, Poindron P and Lindsay DR (1988) Early disturbance of Merino ewes from the birth site increases lamb separation and mortality. Proc Aust Soc Anim Prod **17** 298-301.

185 Alexander G (1961) Energy metabolism in the starved new-born lamb. Aust J Agric Res **13** 144-64. https://doi.org/10.1071/AR9620144.

186 Alexander G (1962) Temperature regulation in the new-born lamb. IV. The effect of wind and evaporation of water from the coat on metabolic rate and body temperature. Aust J Agric Res **13** 82-99. https://doi.org/10.1071/AR9620082.

187 Egan JK, McLaughlin JW, Thompson RL et al. (1972) The importance of shelter in reducing neonatal lamb deaths. Aust J Exp Agric Anim Husb **12** 470-2. https://doi.org/10.1071/EA9720470.

188 Haughey KG (1991) Perinatal lamb mortality — its investigation, causes and control. J Sth Afr Vet Assoc **62** 78-91.

189 McFarlane D (1965) Perinatal lamb losses I. An autopsy method for the investigation of perinatal losses. NZ Vet J **13** 116-35. https://doi.org/10.1080/00480169.1965.33615.

190 Haughey KG (1973) Vascular abnormalities in the central nervous system associated with perinatal lamb mortality. 1. Pathology. Aust Vet J **49** 1-8. https://doi.org/10.1111/j.1751-0813.1973.tb14663.x.

191 Dutra F, Quintans G and Banchero G (2007) Lesions in the central nervous system associated with perinatal lamb mortality. Aust Vet J **85** 405-13. https://doi.org/10.1111/j.1751-0813.2007.00205.x.

192 Alexander G, Mann T, Mulhearn CJ et al. (1967) Activities of foxes and crows in a lambing flock. Aust J Exp Agric Anim Husb **7** 329-36. https://doi.org/10.1071/EA9670329.

193 Davies HL (1962) The milk production of Merino ewes at pasture. Aust J Agric Res **14** 824-38. https://doi.org/10.1071/AR9630824.

194 Hunter TE, Suster D, DiGiacomo K et al. (2015) Milk production and body composition of single-bearing East Friesian x Romney and Border Leicester x Merino ewes. Small Ruminant Res **131** 123-9.

195 Watt DA, Bamford V and Nairn ME (1970) *Actinobacillus seminis* as a cause of polyarthritis and posthitis in sheep. Aust Vet J **46** 515. https://doi.org/10.1111/j.1751-0813.1970.tb09190.x.

196 Alenosy AM and Dennis SM (1985) Pathology of acute experimental *Actinobacillus seminis* mastitis in ewes. Aust Vet J **62** 234-7. https://doi.org/10.1111/j.1751-0813.1985.tb07320.x.

197 Roberts DS (1956) A new pathogen from a ewe with mastitis. Aust Vet J **32** 330-2.

198 Hendy PG and Pugh KE (1981) Prevention of post weaning mastitis in ewes. Vet Rec **109** 56-7.

199 Tunnicliff EA (1949) *Pasteurella* mastitis in ewes. Vet Med **44** 498-502.

200 Hayman RH, Turner HN and Turton E (1955) Observations on survival and growth to weaning of lambs from ewes with defective udders. Aust J Agric Res **6** 446-55. https://doi.org/10.1071/AR9550446.

201 Madel AJ (1983) Mastitis. In: Diseases of sheep, ed WB Martin. 1st ed. Blackwell Scientific Publications: Melbourne, p. 157.

202 Cummins LJ, Brockhus MA, Thompson RL et al. (1988) Lamb mortality limits effectiveness of hormonal treatment to improve reproductive performance of wool producing flocks. Australian Advances in Veterinary Science, Australian Veterinary Association: Artarmon, NSW, p. 149.

203 Obst JM and Thompson RL (1984) Economics of increased reproductive rate for production of wool. In: Reproduction in sheep, eds DR Lindsay and DT Pearce. Australian Academy of Science: Canberra p. 382.

204 White DH (1984) Economic values of changing reproductive rates. In: Reproduction in sheep, eds DR Lindsay and DT Pearce. Australian Academy of Science: Canberra, p. 371.

205 Morrison D and Young J (1991) Profitability of increasing lambing percentage in the Western Australian wheatbelt. Aust J Agric Res **42** 227-41. https://doi.org/10.1071/AR9910227.

206 Morley FHW and Peart G (1988) The profits from selection of Merino sheep. Proc Aust Assoc Anim Breed Genet **7** 86-94.

207 Bourke ME (1964) A comparison of joining systems for prime lamb production. Proc Aust Soc Anim Prod **5** 129-34.

8

CONTROLLED BREEDING

INTRODUCTION

This second chapter on reproduction is concerned with various direct interventions to manipulate and enhance reproductive function in sheep flocks. Collectively referred to as controlled breeding, these interventions may encompass the limitation of reproduction and control of its timing; the increase in the number of progeny per ewe; artificial reproductive technologies such as artificial insemination (AI) and multiple ovulation and embryo transfer (MOET); and the induction of abortion or parturition.

Synchronisation of oestrus, advancement of the breeding season, AI and MOET are of most interest to the Australian sheep industry, and as such will be the topics primarily discussed in this chapter.

CONTROL OF OESTRUS AND OVULATION

There are many reasons that a producer may seek to control oestrus and ovulation within a ewe flock. In Australia, the primary purpose of oestrus control is for either synchronisation prior to timed artificial insemination or the advancement of the breeding season. Producers may also wish to increase the number of ovulations per cycle so as to increase the fecundity of the flock.

Synchronisation of oestrus

Synchronisation of oestrus in sheep is achieved by either artificially extending the luteal phase of the oestrous cycle by administering progesterone or other progestagens, or by prematurely shortening it by inducing luteolysis with prostaglandins. The oestrous cycle lasts for 15-18 days in the ewe (17 days on average) and involves repetitive fluctuations in reproductive hormone levels in preparation for fertilisation and pregnancy (Figure 8.1).[1] For a limited period of approximately 20-35 hours the ewe will accept, and may actively seek out, the ram. This is the fertile window of the oestrous cycle when the animal is said to be in oestrus or 'in heat'. Unlike many other livestock species, the ewe shows no outward signs of oestrus in the absence of the ram. Ovulation occurs around the end of oestrus and the interval between the onset of oestrus and ovulation in Merinos is usually about one day, with considerable variation. Oestrus synchronisation enables the insemination of ewes at a fixed time, without the need to detect oestrus.

A rise in oestradiol acts via the central nervous system to stimulate oestrus and prepare the reproductive tract for pregnancy (Figure 8.1). The luteinising hormone (LH) peak occurs

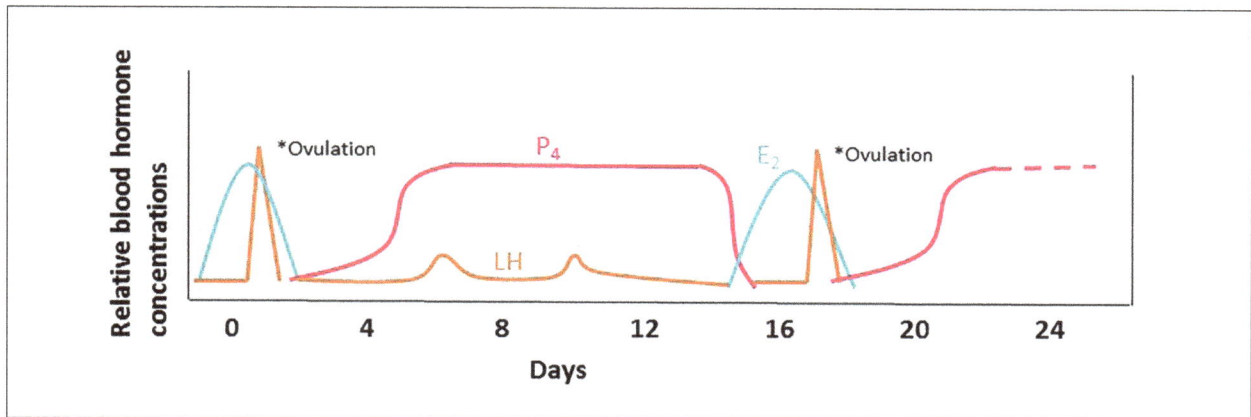

Figure 8.1: Relative changes in the level of luteinising hormone (LH), oestradiol (E2) and progesterone (P4) in the peripheral plasma of the ewe during the oestrous cycle. Source: SP de Graaf.

6-18 hours after the oestradiol peak and causes the Graafian follicle to rupture and ovulate the oocyte. The Graafian follicle is then transformed into a corpus luteum which begins to secrete progesterone. If mating and fertilisation occur, the luteal progesterone primes the uterine endometrium for normal development of the embryo until the luteoplacental shift, when the placenta takes over the responsibility for progesterone synthesis and release for the remainder of the pregnancy. If fertilisation does not occur then luteal progesterone ensures that the subsequent ovulation is accompanied by oestrus and the corpus luteum resulting from this ovulation has a normal life-span and function. The subsequent cycle is initiated by a rise in luteal phase oestradiol which primes the uterus to respond to ovarian oxytocin. This, in turn, results in uterine prostaglandin F2α (PGF2α) release, which causes the structural and functional degradation of the corpus luteum, a process termed luteolysis.[1]

Progestagens to synchronise oestrus

The use of progestagens is the preferred method of oestrus synchronisation in the ewe. This involves the insertion of an intravaginal pessary for at least 12 days, which artificially extends the luteal phase beyond the lifespan of the naturally occurring corpus luteum.[2,3] Following removal of the intravaginal pessary, oestrus commences approximately 24-36 hours (h) later with a peak in oestrus activity at 48 h and nearly all ewes having entered oestrus by 60 h.[4] Oestrus itself lasts 20-35 h.[4]

There are two types of intravaginal pessary — the sponge and the CIDR (Controlled/Constant Internal Drug Release) device. The polyurethane sponge was the original commercial method of oestrus synchronisation developed by Professor Terry Robinson of the University of Sydney during the 1960s.[3] The progestagen (usually 60 mg medroxyprogesterone acetate or 20-30 mg flugestone acetate) is dissolved in ethanol, injected into the sponge and dried to give a fine dispersion of crystals through the network of the sponge to maximise surface area and aid absorption. Until recently the only registered progestagen sponge available in Australia was the Ova-Gest (20 mg FGA; Vetoquinol, Brisbane Australia), but this product was discontinued in mid-2016. Similar sponges are still available direct from compounding pharmacies but it should be noted that these products are not currently registered. As such, the EAZI-BREED® CIDR Sheep and Goat Device (Zoetis, Australia) is the stand-alone pessary available for oestrus synchronisation of ewes in Australia. It is a silicone elastomer device (mounted on a nylon spine) containing 300 mg (9%) of progesterone. In 2016, 90% of all oestrus synchronisation of ewes within Australia was conducted with CIDRs despite their greater expense.[5] CIDRs are preferred to sponges because they are easier to insert; they are less frequently lost during synchronization; they result in reduced vaginal discharge at removal; and they provide a logistically simpler ovulation window post-removal. (AI starts approximately 48 h after device removal rather than 55 h for sponges; therefore, devices do not need to be removed in the middle of the night to generate AI windows during daylight hours.)

At sponge or CIDR removal, ewes are administered equine chorionic gonadotrophin (eCG), still commonly known by its original name of pregnant mare serum gonadotrophin (PMSG). The most effective dose of eCG that should be used is a topic of debate within the artificial breeding community but the standard recommendation is 400 IU.[4] This small dose of PMSG both advances and increases the precision in the time of onset of oestrus. It also slightly increases ovulation rate. For this reason, in some flocks which are particularly fecund or highly responsive to eCG, the amount of eCG may be reduced in order to minimise the frequency of triplets and the deleterious effect on lamb survival that these bring, but reduction below 300 IU is not recommended.

If control over the time of oestrus is sufficiently precise (as it usually is following pessary and eCG treatment), it is not necessary to use teasers or to observe oestrus. Instead, ewes are inseminated at a fixed time after pessary removal. Usually a minority of treated ewes fails to exhibit oestrus but may still become pregnant if inseminated. The precise timing for fixed-time inseminations varies considerably and is discussed later in this chapter.

Prostaglandins to synchronise oestrus

The second and less commonly used approach to controlling oestrus is to administer a single dose of prostaglandin (cloprostenol, 100 µg, Estrumate, Jurox Pty Ltd or dinoprost-PGF2α, 4-5 mg, Lutalyse, Upjohn Pty Ltd). This induces luteolysis, and the ewe returns to oestrus. Prostaglandins are only effective, however, when given more than four to five days after oestrus, so, in order to get all ewes into oestrus at the same time, a second prostaglandin treatment must be given, preferably about 10 days after the first.[4,6,7] Prostaglandins clearly can only be effective in ewes that are cycling regularly (as a corpus luteum must be present in order for synchronisation by luteolysis to occur), and they may cause abortions if given during the first 60 days of pregnancy (discussed later in this chapter). Oestrus occurs two to three days after treatment but not with sufficient precision to allow fixed-time inseminations. Further, these treatments will be quite unreliable if a significant portion of the ewes are not cycling regularly. As a consequence of these factors and in contrast to the case with cattle, prostaglandins are rarely used for controlled breeding in sheep.[8]

Advancing the breeding season

The seasonal nature of sheep reproduction restricts the sheep industry from accessing favourable seasonal markets, integrating lambing with other farm activities or implementing accelerated lambing programmes which aim to deliver three lambs every two years.[9] Because of this, producers may wish to join or use artificial reproductive technologies outside the breeding season. There are three basic approaches that can be used to induce oestrus and ovulation in the seasonally anoestrus ewe. These are

- using the *ram effect*
- mimicking seasonal changes in day length through manipulation of the light-dark cycle or the administration of melatonin implants
- administering progestagens and PMSG.

These techniques may be used in conjunction with each other to maximise the ovulatory response. The effectiveness of these techniques is quite variable and the response is influenced by breed, body weight/condition score, environment, sexual maturity and the month of joining. Of these, the seasonality of the sheep breed is the predominant concern and influences the time of year the techniques can be employed. For example, Merinos and Poll Dorsets are less seasonal than most other sheep breeds in Australia, making it possible to advance the breeding season of these breeds and their crosses in the middle of the non-breeding season. For most other British breeds, including the Border Leicester, Suffolk and Romney, these techniques are only effective closer to the breeding season.

The ram effect

Ram-induced ovulation in sheep (the *male* or *ram effect*) is a popular technique because it is a 'clean and green' method of controlling reproduction which is inexpensive and easy to apply. Socio-sexual signals from an intact male are used to advance the onset of breeding activity and provide a reasonable degree of female synchrony.[10] Olfactory signals from a pheromone secreted by the glands of the wool follicles of rams trigger LH secretion in the ewe which may lead to ovulation and oestrus. The secretion of this pheromone is androgen-dependent, so rams with greater testosterone levels and strong sexual libidos have a stronger ram effect than sexually inexperienced rams. Sexual experience of the ewe also increases the ovarian response to the ram effect, with young and sexually naïve ewes exhibiting lower ovulation rates than experienced or mature ewes.[11] These volatile chemicals act synergistically with other social-sexual stimuli such as the sight and sounds of rams. Anoestrus ewes have even been shown to respond to projected images of rams with an increase in LH secretion but the magnitude of the effect was small compared to that observed in ewes exposed to rams.[12] For the ram effect to work the ewes must be unaccustomed to the presence of the rams. It was originally thought that the ewes must be isolated from the sight and smell of rams for at least one month before joining, but more recent studies have shown that male novelty is more important than isolation, and no separation of ewes from rams is required if the rams are novel.[13]

Much out-of-season breeding of Merinos relies to a considerable extent on the ram effect. In Merinos, ram introduction in spring and summer (October-January) causes a high proportion of ewes to commence cycling. In the absence of conception, most will show one, two or three consecutive cycles, then revert to spasmodic breeding or full anoestrus. When the ram effect works well, 80% or more of Merino ewes should lamb as a result. In breeds with a more sharply defined breeding season and deeper anoestrus, the rams must be introduced within six weeks of the time at which breeding would normally commence. If introduced earlier, the ewes' responses are highly variable and, while the male stimulus may induce an LH response, ovulation may not occur.

The ram effect can also be induced by testosterone-treated wethers or with vasectomised rams. In either case, these sheep act as 'teasers' and can be used to stimulate ewes to cycle and, if desired, to identify oestrus ewes by marking them with crayon from chest-mounted harnesses.

Anoestrus ewes usually react to the introduction of rams in one of two ways. After the silent, ram-induced ovulation on days 1-2 (rams introduced day 0), approximately half the ewes experience a short cycle and have another silent ovulation on days 6-7 (type I response, Figure 8.2). This ovulation is silent because it is not preceded by an adequate period of progesterone priming. These ewes next experience a normal cycle, so that they ovulate a third time around days 24-25, and this ovulation is accompanied by oestrus. The balance of the ewes in the flock do not experience the short cycle, so that after the ram-induced ovulation they ovulate a second time around days 18-19 and this ovulation is accompanied by oestrus (type II response, Figure 8.2). The proportion of the flock exhibiting either response is variable but overall there should be a high incidence of oestrus around days 18-25.

Priming the ewes with progesterone overcomes the luteal insufficiency that causes ewes to have a short cycle following ram-induced ovulation. Ewes can be primed with a progestagen sponge or CIDR for 6-12 days prior to ram introduction. If this occurs the majority of ewes will follow the type II path shown in Figure 8.2 and come into oestrus around days 18-19 (or ovulate on days 19-20). This enables a relatively cheap approach to AI with potential for the whole flock to be inseminated over a four- or five-day period, using teasers for oestrus detection.

For natural mating, it is best, if feasible, to induce the ram effect with teaser wethers or vasectomised rams for the first 14-16 days, so that the entire rams do not take out their frustrations on each other waiting for the ewes to come into oestrus and damaging each other before most ewes are cycling. If wethers are to be used, they should be large, mature individuals. Testosterone propionate or testosterone enanthate (Ropel, Jurox Pty Ltd) is administered. A single dose of 400 mg testosterone enanthate given 14 days before use and 1% wethers are usually adequate. Alternatively, some veterinarians recommend using 2% wethers which have received 150 mg, 150 mg and 300 mg testosterone at 0, 7 and 14 days (inserting the teasers at 14 days and the fertile rams 14 days later with teasers removed at this time). Note that, in contrast to the preparation of teasers for AI if mounting of the ewes (to mark them) is required, the wethers used as teasers before natural mating are only required to exude pheromones and do not have to actively attempt mating with the ewes.

Gonadotrophic hormones to stimulate oestrus

When suitably administered, gonadotrophins with FSH and GnRH activity will induce ovulation in the anoestrous ewe, but the ovulation is not usually accompanied by oestrus. If treatment is preceded by progestagen application for a minimum of six days, the ovulation is accompanied by oestrus. In practice, the usual regime is to insert sponges or CIDRs for 12 days and give PMSG at the time of device removal. A slightly higher dose of PMSG

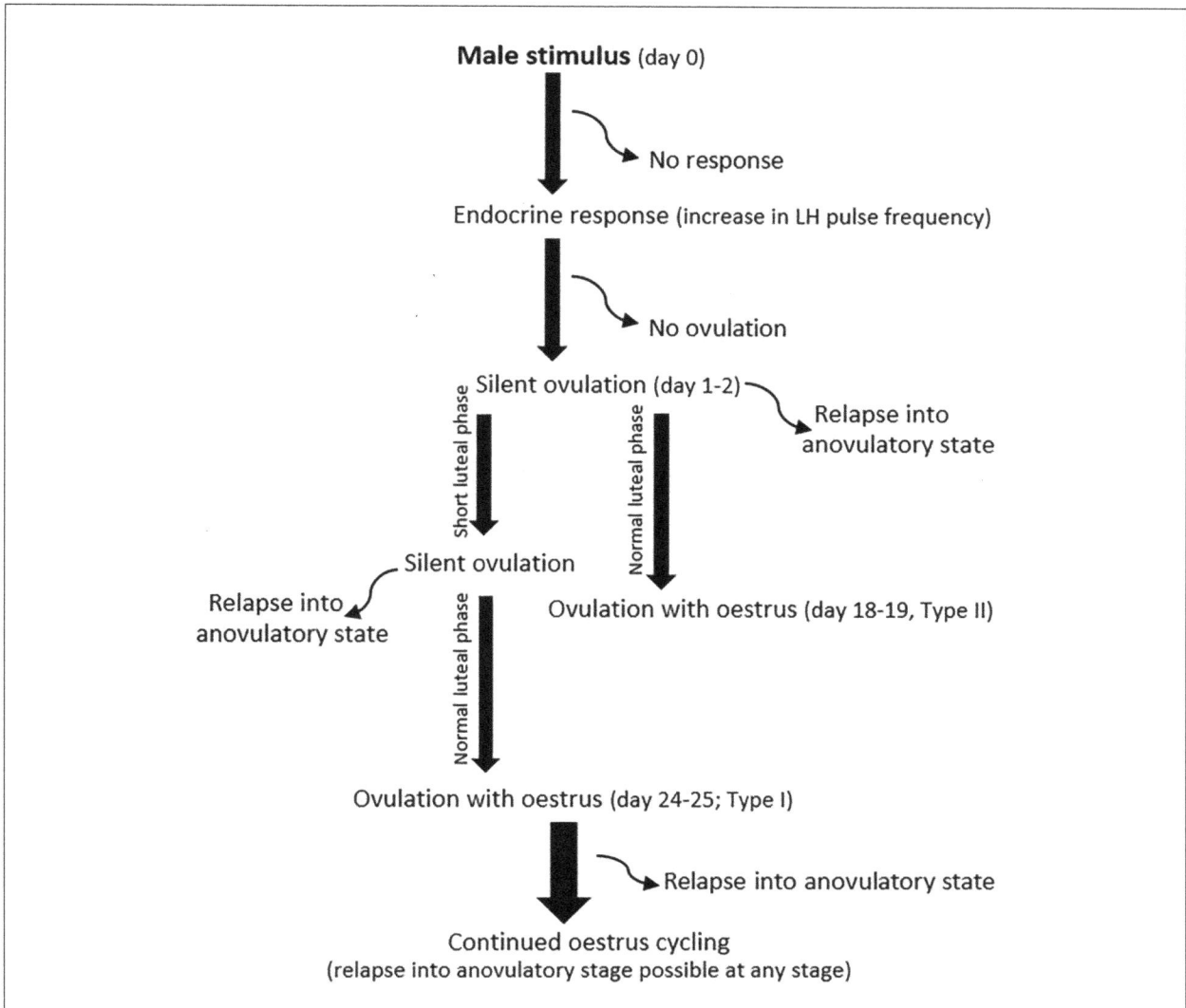

Figure 8.2: The ovarian response of anoestrus ewes to the introduction of rams. Approximately half the ewes that respond to the male stimulus will exhibit a type I response and half will exhibit a type II response. The ovarian response is dependent on continuation of the male stimulus and can be terminated at any time. Adapted from Walkden-Brown, Martin and Restall (1999).[14]

(500-700 IU) than that which is used in the cyclic ewe is required. The disadvantage of this hormone treatment is the expense of the drug regime. If natural mating is employed, the joining ratio of rams to ewes should be increased to ensure satisfactory conception rates, especially in the case of British breeds, and this also adds to the cost. For best results, the number of ewes in oestrus per day per ram should not exceed three if British breed rams are used. An alternative strategy to increasing the joining ratio is to divide the ewe flock into subgroups and stagger the hormone treatments by intervals of two or three days. The flock response is critically influenced by the stage of anoestrus of the ewes when joining begins (Table 8.1).

Melatonin implants to stimulate oestrus

Seasonal variation in breeding activity is controlled by seasonal changes in the photoperiod. Secretion of the hormone melatonin from the pineal gland is high at night and low during

Table 8.1: The response of Border Leicester X Merino ewes in the southern hemisphere treated with sponges or CIDRs for 12 days and injected with 500-700 IU of PMSG at different stages of the non-breeding season.

Stage of non-breeding season	Effectiveness of treatment		
	Ewes in oestrus	Ewes pregnant	Further breeding
Mid (Aug-Oct)	50% or less	50% or less	nil
Late (Nov-Dec)	50-80%	50-60%	possible
Very late (Jan-Feb)	80% or more	60% or more	probable

the day. Changes in day-length alter the pattern of melatonin secretion and serve as an endocrine signal that relays photoperiodic information to the hypothalamic-pituitary-gonadal axis. The transition from long days to short days can therefore be mimicked by artificial regulation of the light-dark cycle or through melatonin supplementation. Artificial regulation of the light-dark cycle can be achieved by yarding ewes into light-proof sheds for a portion of each day but this procedure is too costly and time-consuming to be of commercial use. A much more practicable alternative is the use of slow-release melatonin implants (Regulin, CEVA Animal Health Pty Ltd) which mimic changes in day length. The melatonin signal is only effective when a daily dose is administered over a lengthy period of time.

After 40 days of melatonin treatment anoestrus ewes show a significant increase in GnRH secretion with high GnRH and LH pulsatility occurring after 74 days.[15] With this in mind, it is recommended to implant the ewes with melatonin and isolate them from rams for six weeks before a joining period of six to eight weeks. This regime uses a combination of endocrine signals to manipulate the light-dark cycle, and environmental signals, through the use of the ram effect. In Merino ewes, Regulin treatment may result in a substantial stimulation of ovulation rate if used for October, November or December joinings, but considerable variation has been observed between flocks. The use of Regulin is not recommended for highly seasonal British breeds or in areas where ewes are already cycling in late spring or early summer.

Oestrus stimulation in ewes in postpartum or lactational anoestrus

Another form of anoestrus is postpartum anoestrus. In ewes which lamb in the latter half of the breeding season, or in spring-early summer, postpartum anoestrus will usually continue into seasonal anoestrus. If ewes lamb earlier in the breeding season, the first, silent postpartum ovulation occurs 20-30 days later. This first ovulation is usually followed by a short cycle and a second silent ovulation six to eight days later. This, in turn, is followed by a normal cycle and a third ovulation with oestrus at around 45-55 days postpartum. Note the similarity of this sequence of events in the postpartum ewe to that seen in dry, seasonally anoestrous ewes after the introduction of rams. Hence, during most of the year, treatment to induce oestrus and ovulation will be necessary in order to breed recently lambed ewes. The conception rate to a hormone-induced oestrus and ovulation in postpartum ewes is dependent upon the time after lambing, lactational status of the animal, the season of the year in which the ewe lambs and the degree of seasonality of the breed/genotype. The recovery of the uterus postpartum to provide an environment which allows fertilisation and embryo development (involving a regrowth of an epithelial cell layer back over the caruncles) takes longer in ewes which are lactating and/or lambing in spring. Table 8.2 gives a guide to the pattern of return to normal fertility in Border Leicester x Merino crossbred ewes.

If lambing occurs once annually, as is customary in Australia, then postpartum anoestrus is not a significant concern. It is only when lambing occurs at intervals of less than one year (for example, in accelerated lambing programmes) that it becomes a potential limitation to high fertility. Lambing more frequently than once per year is more common in some European countries but is occasionally used in Australia for prime lamb production. To be successful, such programmes require high standards of nutrition and management. It is important in such programmes that a high percentage of fertile rams are used (or, in the case of AI, ample doses of high quality semen). In practice, the minimal lambing interval achievable is about 200 days (that is, three lambs in two years or possibly five lambs in three years).

Increasing fecundity

An increase in the number of lambs born per ewe can be achieved by increasing the ovulation rate of the ewe. Useful increases in the number of lambs reared per ewe often result from treatments that cause only a modest increase in ovulation rate. For example, in Merinos, lifting the mean ovulation rate from 1.2 to 1.7 may result in 20-30% more lambs reared. Excessive ovulatory responses are to be avoided, since rates of embryo and perinatal mortality increase with increasing ovulation rate. Breed, hybrid vigour, season, body weight and nutritional flushing (especially with supplementation of lupin grains) all have a positive effect on ovulation rate, but are largely outside the scope of this chapter (see Chapter 7 and de Graaf (2010)[16] for a further discussion of these factors).

A simple method to increase ovulation rate is the administration of eCG. The FSH activity of this drug directly increases the number of follicles produced by the ovary in a given cycle and, in turn, the number of ova produced. The 400 IU dose rate — as used in oestrus synchronisation programmes — will generally increase ovulation rate by 0.5 but eCG is rarely used for this purpose since the cost and the variability in ovarian responses are unacceptably high. Other exogenous forms of FSH (Folltropin-V, Vetoquinol) will also boost ovulation rate but are only used for superovulation of ewes in MOET programmes (see below).

An alternative means of temporarily increasing ovulation rate is by active immunisation of ewes against polyandroalbumin (Ovastim, Virbac Australia, previously sold as Fecundin; Androvax Plus MSD Animal Health, New Zealand). The immunogen (polyandroalbumin) is a conjugate of the steroid hormone androstenedione and human serum albumin. Antibodies

Table 8.2: A guide to the days required for return to normal fertility in Border Leicester x Merino ewes post-lambing.

Season of lambing	Return to normal fertility post-lambing (days)	
	Non-lactating	Lactating
Autumn	50-60	60-70
Spring	60-70	>70

secreted in response to vaccination partially inhibit the normal negative feedback effect of ovarian androgens and oestrogens on the anterior pituitary gland. This results in higher than normal levels of gonadotrophin secretion and hence an increased ovulation rate and fecundity. Immunisation does not induce ovulation in anoestrus ewes, so responses only occur in ewes which are cycling. Rates of fertilisation failure and embryo mortality are often higher in treated ewes than untreated ewes — factors which tend to reduce the benefit of the higher ovulation rate.[17] There is a strong relationship between the average body weight of ewes at joining and the response to vaccination. In British breed ewes, an additional 20 kg of liveweight at joining may result in an increase in response from an extra 25% of lambs born to an extra 45% of lambs born per ewe as a result of immunisation.[18] British breeds and Merino crossbreds also respond better than pure Merinos.[19] In a large series of trials, vaccination increased lambing rates by 28% in crossbreds (range 11-48%) but only by 16% (0-33%) in Merinos.

The effect of immunisation is to increase the number of ewes having multiple births; thus fewer ewes have single lambs and more ewes have twins. In flocks in which the twinning rate is already high (in untreated ewes), there could be a significant increase in the number of triplet births. Ewes with triplet pregnancies are relatively difficult to manage and without careful nutritional management they are at high risk of pregnancy toxaemia, particularly in extensive Australian conditions. Furthermore, the survival rate of triplet lambs, without intensive husbandry, can be very low. Consequently, the use of Ovastim or Androvax in flocks with already high lambing rates (150% to 170%) is contraindicated.[18]

Ovastim or Androvax should not be considered as remedies for low lambing rates which are associated with low condition score in ewes at joining. Low condition score implies that ewes are on a restricted nutritional plane and, in such flocks, immunisation is likely to increase the number of twin pregnancies. Unless the level of nutrition of the flock is improved markedly, an increased twinning rate will increase the risk of metabolic disease in ewes and decrease lamb survival. The increase in the number of lambs raised per ewe may be disappointing and may not justify the cost of treatment.

The lower response to immunisation in Merinos compared to crossbreds and British breeds, combined with the poorer maternal abilities of Merinos which can adversely affect the survival of multiple-birth lambs, have led the manufacturer to recommend that the product not be used in Merinos.

In the first year of use the vaccine is given twice, first six to nine weeks and then three to four weeks before joining. In subsequent years, ewes require only a single booster dose three to four weeks before mating. In years when pasture availability is low and twinning is not desired, the booster vaccination can be omitted and ovulation rates are the same as in unvaccinated ewes of the same strain. Vaccination does not interfere with the practice of joining ewes out of season and does not alter the percentage of such ewes that exhibit premature luteal regression.

INDUCTION OF ABORTION

The maintenance of pregnancy in the ewe ceases to be dependent on the corpus luteum at around day 50 when placental progesterone dominates. Prior to this time, abortion can be attempted with a single luteolytic dose of prostaglandin (for doses, see the earlier section on oestrus synchronisation). Treatment before day 5 is ineffective, as prostaglandins have no luteolytic effect on a corpus haemorrhagicum or early-stage corpus luteum, but should be given as soon after day 5 as possible. In practice, induction of abortion by this method is erratic and often ineffective for reasons which are not clear. One possible explanation is that the luteolytic signal is not easily interpreted in animals that have already received maternal recognition of pregnancy. After day 50, abortion is even less easy to induce, even with large or repeated doses of prostaglandins. These treatments with higher doses are dangerous and the retention of foetal membranes following abortion is a serious side effect. Treatment with high levels of oestradiol benzoate (10-40 mg) has been shown to induce abortion in the final trimester of pregnancy, with one study reporting 40-70% of ewes induced to abort when injected between days 126-130 of gestation.[20] Glucocorticoids are unreliable before day 140 and should be considered as a treatment for induction of parturition rather than abortion.

INDUCTION OF PARTURITION

Parturition is usually induced with glucocorticoids. This should initiate most of the normal prepartum maturational events, thereby ensuring adequate foetal preparation for postnatal life. Treatment with dexamethasone (15-20 mg) or the more potent flumethasone (2 mg) should result in parturition some 24-48 hours later. The optimal treatment time depends on breed. Treatments on days 144 and 148 are recommended for British breeds and Merinos, respectively. Following a synchronised mating, the treatments are given when 5% of the flock has lambed.

Oestradiol benzoate may also be used to initiate parturition but is less commonly used. Recommended doses range from 2-20 mg administered in the last week of pregnancy, but some studies suggest that this induction regime may increase dystocia and perinatal mortality, particularly in comparison to glucocorticoid treatment. Other agents such as epostane (essentially an anti-progestin compound), RU486 (progesterone receptor blocker) and prostaglandin have been used, but are either unreliable (as is the case with PGF2α) or not registered for use in this species. Regardless of induction method, some increase in the rate of perinatal mortality should be expected.

ARTIFICIAL INSEMINATION (AI)

Around 250 000 of the 36 million ewes in the Australian flock are bred each year by artificial insemination.[5] Three methods of insemination are available for use in Australia: vaginal, cervical and uterine. The majority of these inseminations are carried out or supervised by veterinarians, although law permits primary producers to undertake some procedures normally considered acts of veterinary medicine in most states (such as uterine or laparoscopic AI) on their own stock. Almost all inseminations are intrauterine via laparoscopy (Figure 8.3) to a synchronised oestrus and are conducted anywhere from October to April depending on the state (Western Australia's season commencing the earliest) and region.[71.] There are currently 28 artificial breeding companies and/or veterinary practices in Australia that offer laparoscopic intrauterine insemination as a service to sheep producers. Cervical artificial insemination (sometimes referred to as *over the rail*, Figure 8.4) and vaginal (historically known as *shot-in-the-dark*) are still practised on some properties but are less common since the advent of laparoscopic AI, which facilitated the use of frozen semen.

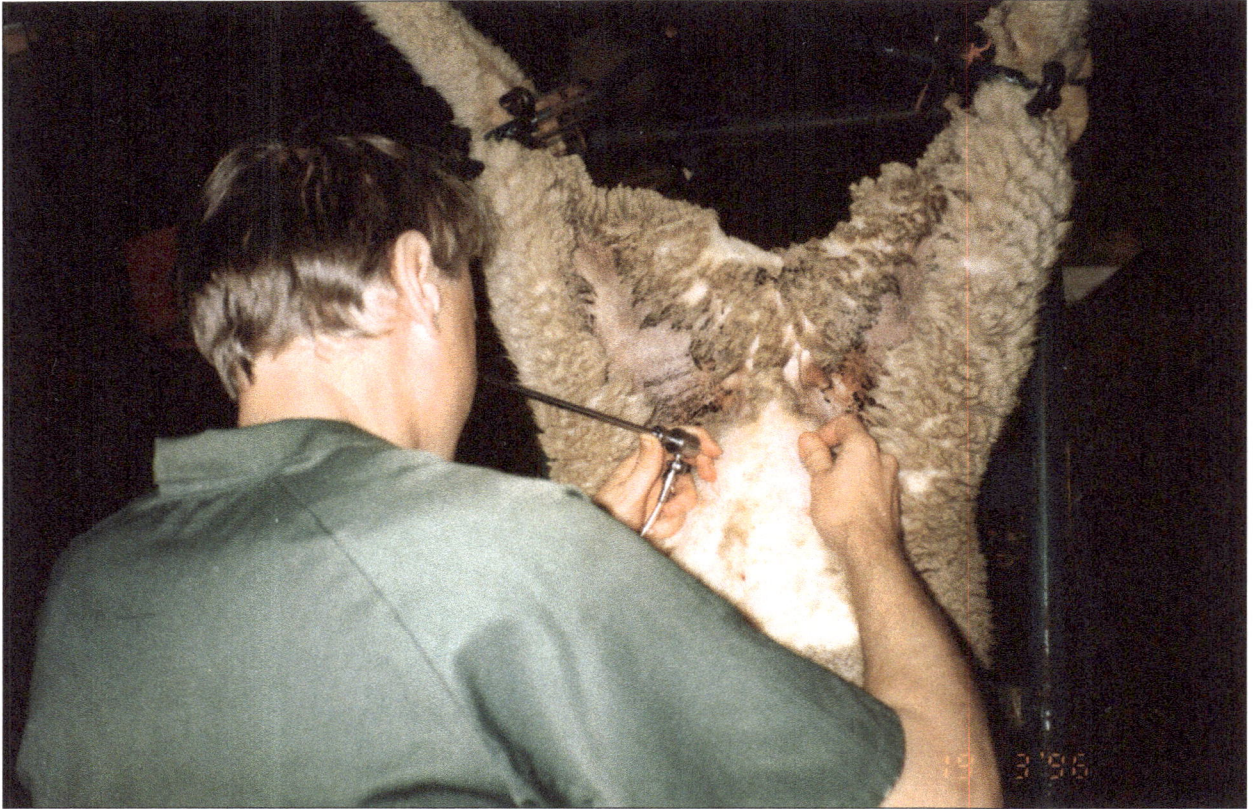

Figure 8.3: Laparoscopic intrauterine artificial insemination of a ewe. Photograph courtesy of LA Abbott.

A successful artificial insemination programme requires both considerable attention to detail and management of multiple facets of reproduction. Suitable preparation of rams, semen collection, processing and quality control, along with synchronisation of ewes and accurate delivery of the spermatozoa themselves, are required in order to maximise the chances of conception and pregnancy. Some of these factors are considered below.

Selection and preparation of rams

Prior to collection and processing of any semen for AI, rams brought into a breeding facility should have their testes checked for any obvious morphological defects and a blood specimen collected for brucellosis testing. Special housing and quarantine requirements are necessary for any animals whose semen is to be exported. If the rams have not previously been trained for collection by artificial vagina, sufficient time should be allowed to familiarise the ram to this process.

Collection of semen

Semen can be collected either by artificial vagina (AV) or electroejaculation. While electroejaculation is a suitable means for collection of semen from rams which have low libido or have not been trained to serve into an AV, there are few other points to recommend it. Ejaculates obtained can be of high quality but tend to be variable and are more likely to be contaminated by urine. The welfare implications of restraint and transrectal electrical

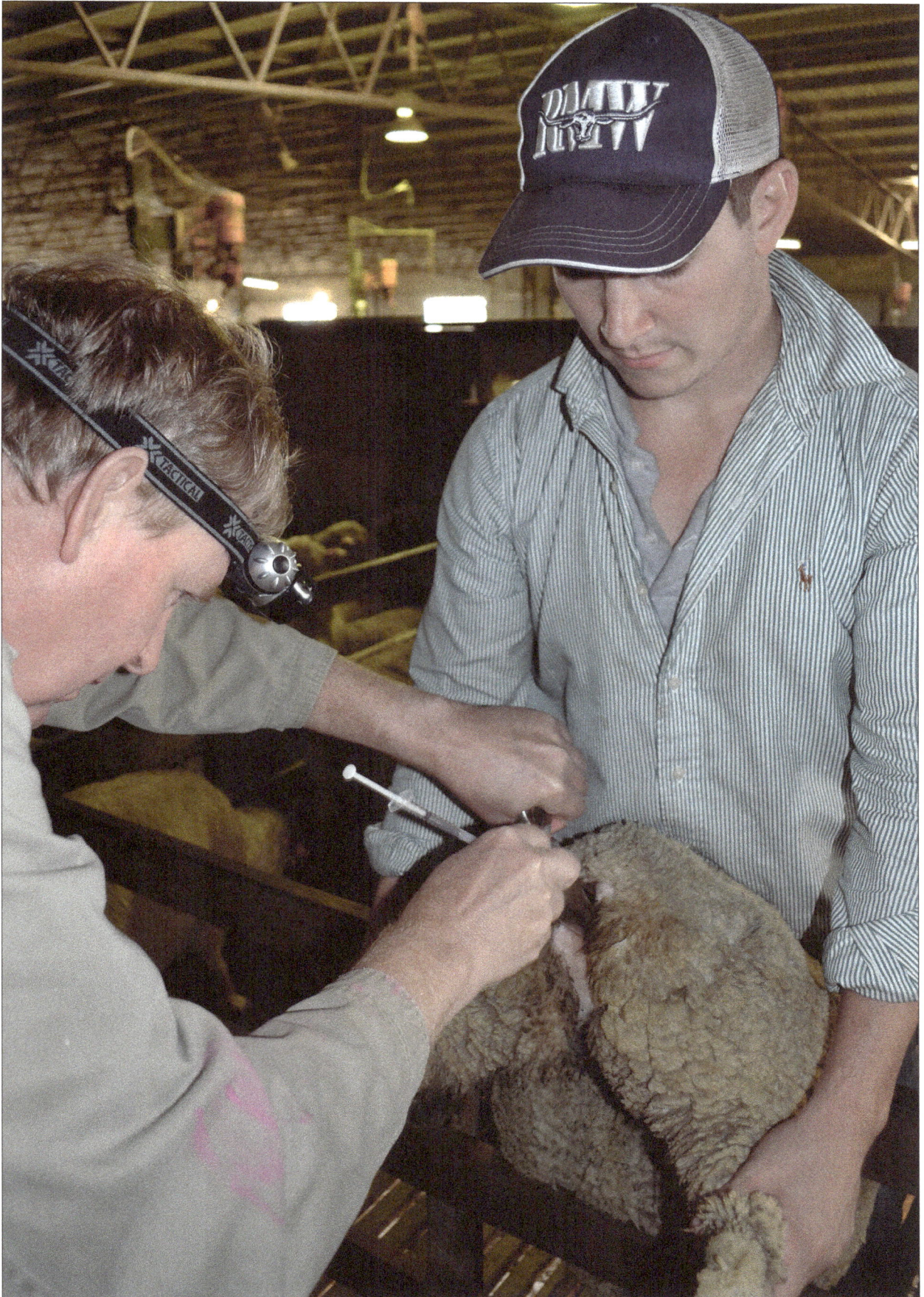

Figure 8.4: Cervical artificial insemination of a ewe. Source: SP de Graaf.

stimulation are also worth considering. For these reasons, semen collection is most commonly undertaken with the aid of an artificial vagina. This device uses temperature and pressure to simulate the conditions present inside the vagina of an oestrous ewe and stimulates the ram to ejaculate. A graduated glass placed in the artificial vagina serves to collect the semen at ejaculation as well as acting as an initial holding vessel prior to further processing. Care must be taken to ensure that the artificial vagina and collection glasses are clean, dry, warm and free of any other substance which may compromise the quality of the semen obtained.

Handling of semen

Following semen collection, spermatozoa remain susceptible to a variety of factors which may adversely affect their viability and subsequently harm their fertility should they be used in an AI programme.

Change of temperature

A sudden drop in temperature to less than 10 °C causes *cold shock* — an irreversible loss of viability. Exposure to temperatures greater than 42 °C causes rapid death of spermatozoa, whereas storage at greater than 30 °C reduces the fertilising life of the spermatozoa due to exhaustion of energy sources and a drop in pH (resulting from the accumulation of lactic acid). An ideal temperature for short-term storage (that is, for one to two hours) is 30 °C. Slow cooling over two to three hours to around 5 °C will maintain good viability and fertilising capacity for at least 12 hours, assuming egg yolk has been included in the diluent to protect sperm membranes during this cooling process. A similar cooling curve is used prior to the freezing of ram semen.

Sunlight

Exposure to direct sunlight damages spermatozoa, mainly due to the effects of ultraviolet radiation. Aluminium foil as a temporary cover for collection glasses prevents UV exposure to semen between collection and initial assessment.

Exposure to heavy metals

Copper, lead, mercury and cadmium are toxic to spermatozoa; glass or plastic containers should be used for semen processing and storage.

Contact with water

Water is extremely hypotonic and induces rapid change in osmotic pressure across the sperm membrane. Artificial vaginas and collection and storage vessels must be dry, otherwise spermatozoa will be killed.

Contaminants

Urine is hypotonic and like water, reduces sperm survival. In general, samples containing urine are discarded. The presence of micro-organisms within a semen sample is cause for investigation of the health status of the animal and/or the cleanliness of the collection equipment. Contaminants should be avoided and the reason for their presence investigated and addressed.

Disinfectants

Nearly all antiseptics, bactericides and detergents are spermicidal. For sterilisation, glass or plastic containers should be washed thoroughly, then rinsed repeatedly and thoroughly with distilled water to remove detergents. Sterilisation by dry heat, gas (ethylene oxide) or UV irradiation is effective. There is a danger with autoclaving of contamination with heavy metals or other materials which had previously been placed in the autoclave.

Evaluation of semen

The ultimate evaluation of semen quality is achieved by measuring pregnancy or lambing rates but earlier, indirect evaluation may avoid the cost of low conception rates caused by poor semen quality. The evaluation or assessment of semen in the field is a critical factor in the success of any AI programme. Sufficient training and standardisation is required in order to facilitate sufficient levels of quality control of semen produced by a breeding centre.

Estimates of probable fertilising capacity can be made by visual examination and are outlined in detail by Evans and Maxwell (1987).[4] Semen samples can be examined macroscopically for volume, colour, consistency (or concentration) and wave motion. Ejaculates obtained by collection with an artificial vagina have an average volume of around 1.0 mL (range 0.5-1.5 mL). The colour should be milky white; other colours indicate contamination. The consistency of the ejaculate can be scored subjectively to estimate concentration, according to the criteria in Table 8.3. Wave motion can be macroscopically assessed only roughly; swirling on the side of the collecting vessel indicates high motility.

Microscopic examination of semen may be used to determine wave motion, motility, concentration, morphology and viability. Motility can be scored subjectively using the criteria in Table 8.4.

Microscopic examination of wave motion and motility must be carried out on a warm slide (37 °C), preferably on a warm stage and at appropriate magnification (40-100x). Wave

Table 8.3: Scoring system based on macroscopic examination of the density of ram semen and its predicted sperm concentration.

Score	Consistency	Mean (range) in number of spermatozoa/mL ($\times 10^9$)
5	thick creamy	5.0 (4.5-6.0)
4	creamy	4.0 (3.5-4.5)
3	thin creamy	3.0 (2.5-3.5)
2	milky	2.0 (1.5-2.5)
1	cloudy	1.0 (0.3-1.5)
0	watery	<0.3

Table 8.4: Wave motion scoring system based on microscopic examination.

Score	Motility	Percent of sperm active	Description
5	excellent	>80	rapid, dense waves changing direction rapidly
4	good	65-80	good wave motion but slower changes in direction
3	fair	40-65	slow thin waves
2	poor	20-40	no waves; individual spermatozoa can be seen
1	very poor	<20	few spermatozoa show progressive movement
0	none	0	no movement

motion is assessed with an undiluted drop of fresh semen, while individual sperm motility is assessed following dilution to preferably 20-50 million spermatozoa/mL and requires phase or differential interference contrast optics. Concentration of ejaculates are determined by haemocytometer or colorimeter and morphology is assessed at much higher magnification (1000x) by subjective analysis. Live:dead ratios can be estimated using staining regimes such as Eosin/Nigrosin. Detailed discussion and explanation of these various techniques are beyond the scope of this chapter but can be pursued further in Evans and Maxwell (1987).[4]

Specialist semen assessment centres are able to offer objective assessment techniques by means of computer-assisted semen analysers (CASA). CASA systems are able to measure additional *in vitro* predictors of fertility such as progressive motility, velocity and straightness in addition to the aforementioned parameters of total motility, sperm concentration and morphology. Caution is still advised in assessing the practical significance of these various objectively measured semen characteristics, as significant correlations have yet to be demonstrated between objectively measured characteristics and fertility in sheep. No perfect *in vitro* predictor of fertility exists, no matter how many parameters of semen quality are measured for a given sample.

Dilution of ram semen

The dilution of ram semen is generally essential but may not be necessary if only a modest number of ewes are to be inseminated promptly after semen collection. Semen is diluted for both technical and biological reasons. The technical reason is to enable more ewes to be inseminated with a minimal volume of inseminate (in effect, to get the right number of motile spermatozoa into the right volume for insemination). The biological reason is to enhance the survival of spermatozoa, by providing energy substrates (for example, glucose, fructose), using buffers to stabilise pH, antibiotics to protect against micro-organisms present in semen or the female reproductive tract, and cryoprotectants to protect against cooling and freezing damage (usually glycerol and egg yolk).

Formulae for the preparation of various diluents suitable for ambient, liquid or frozen storage of ram semen can be found in Evans and Maxwell (1987).[4] Antibiotics should be added to all diluents (for example, penicillin, 200 IU/mL; streptomycin, 1.0 mg/mL). For the fresh use of ram semen for cervical or intrauterine AI, various synthetic diluents can be used, but the most common remains UHT cows' milk. When diluting the semen with UHT milk it is important to

- dilute promptly after collection and examination of the semen

- ensure that diluent and semen are held at the same temperature (25-30 °C), preferably in a water bath

- add the diluent slowly to the semen and mix gently

- dilute to a level appropriate to the initial sperm concentration and motility of the ejaculate, and the route of insemination. Highly motile and concentrated semen can be diluted further than samples with lower concentrations; and cervical AI doses require higher sperm concentration than inseminates to be used for intrauterine deposition.

Evans and Maxwell (1987)[4] provide recipes for diluents containing glycerol and egg yolk which are suitable for one-step freezing of ram semen in pellet form or in straws. The composition of these diluents may have to be adjusted for different dilution rates. All presently available freezing diluents seem to reduce the quality of frozen-thawed spermatozoa and there is ongoing research to devise superior diluents.

Storage of ram semen

Semen must be stored in a manner which decreases or arrests metabolism and hence prolongs the fertilising life of the spermatozoa. Semen is stored either

(a) chilled, at 2-5 °C (although the French method suggests 15 °C)

(b) frozen in pellets or straws and held at -196 °C.

For liquid or chilled storage the semen is cooled slowly over two to three hours. For use, it may be warmed promptly to 30 °C or inseminated without warming. Chilled semen maintains good motility (when rewarmed) for several days, but the fertilising ability of chilled semen declines at a faster rate. For acceptable fertility with cervical insemination (>45% conception rate), chilled semen should not be stored for more than 12-24 hours. If uterine insemination is used, the semen may be used after storage for up to three days.

For frozen storage, the semen is diluted with diluent containing cryoprotectant and chilled to 2-5 °C, as described above. Two methods of freezing and thawing are used. In the pellet method, the rate of temperature change is fast. Aliquots (0.2-0.25 mL) of chilled semen are dropped into holes made on the surface of a block of dry ice (solid CO_2) and the resulting pellets are transferred with forceps into liquid nitrogen. For use, the pellets are transferred promptly into a dry vessel at 37 °C to obtain rapid thawing and maximum recovery.

In the straw method, which is semi-fast, the straws are loaded with chilled semen and placed in liquid N_2 vapour, then stored in liquid N_2. For use, the straws or pellets are thawed at 37 °C. The proportion of spermatozoa which survive these procedures (as indicated by subsequent motility) varies widely — 40-70% should survive in good semen samples. There are differences between individual rams in this regard — the semen from some rams does not tolerate freezing and thawing. The inseminating dose must be adjusted according to the survival rate so that the *motile dose* is standardised. The conception rates obtained after cervical AI with frozen semen usually range from 10-30%. Spermatozoa which survive the freeze-thaw process have a decreased fertilising lifespan, and frozen semen should, by preference, only be used for intrauterine AI. Using this method, pregnancy rates of 60-70% should reliably be achieved with frozen semen.

Detection of ewes in oestrus

An important decision in cervical AI programmes is whether to inseminate after oestrus detection or at a fixed time after hormonal synchronisation. Even when the latter policy is adopted, it is probably worthwhile to run teasers wearing crayon-bearing harnesses with the ewes and inseminate first those ewes which came into oestrus early, since the inseminations may take many hours or all day. Use of teasers gives an indication of the success of the hormone treatments in the ewes and it is possible that the presence of teasers enhances the ewe response. If the timing of insemination is based on oestrus detection, the quality of detection is a critical determinant of the conception rate to AI. The selection, preparation and management of teasers were previously discussed. The following percentages of teasers are recommended:

- natural cycles: 2%

- synchronised at previous cycle: 4%

- synchronised at this cycle: 6-10%.

Rams are occasionally used as teasers. They may be entire rams fitted with aprons but this system is not reliable, as aprons may fail and unplanned pregnancies may occur. Alternatively, sexually active young rams are vasectomised at least six weeks before use and, immediately before use, semen samples are collected by electroejaculation to confirm the absence of spermatozoa. The teasers used are usually testosterone-treated wethers. If the wethers are given 150 mg testosterone cypionate (or enanthate) at fortnightly intervals, commencing four weeks before use and continuing until the end of the AI programme, they should exhibit good mounting activity.

Insemination of the ewe

During coitus, semen is deposited in the vagina. Ejaculation is rapid, with a small volume of semen containing a high concentration of spermatozoa deposited in the anterior vagina. As much as possible, AI should simulate natural insemination and preferably deposit semen in a more favourable position for fertilisation (cervix, uterus). In the ewe, however, it is very difficult and usually impossible to pass an inseminating pipette through the cervix.

The volume of inseminate can vary within limits. The upper limit will be determined by the capacity of the organ or site of insemination to retain the semen. Volumes below 50 µL are not practicable due to device delivery restrictions. Recommended volumes are:

- vaginal insemination: 0.3-0.5 mL

- cervical insemination: 0.05-0.20 mL

- uterine insemination (per uterine horn): 0.05-0.10 mL.

The minimum safe numbers of motile spermatozoa in the inseminate (required to obtain satisfactory conception rates) are determined by the route of insemination and the type of semen (fresh/liquid-stored/frozen-thawed) as well as by whether oestrus in the ewes is spontaneous or controlled. Some workers have obtained good fertility with considerably lower numbers of spermatozoa, but the numbers of motile spermatozoa shown in Table 8.5 are recommended for general use.

If ample semen is available, the number of spermatozoa can sensibly be increased somewhat above the relevant number shown above.

Table 8.5: Minimum safe numbers of motile spermatozoa per inseminate and appropriate method of AI according to semen and oestrous type. Adapted from Evans and Maxwell (1987)[4] and de Graaf (2010).[16]

Method of insemination	Type of oestrus	Type of semen		
		Fresh ($\times10^6$)	Liquid-stored ($\times10^6$)	Frozen-thawed ($\times10^6$)
Vaginal	Spontaneous	150	NR	NR
	Controlled	300	NR	NR
Cervical	Spontaneous	100	150	NR
	Controlled	200	300	NR
Uterine (total in two uterine horns)	Controlled	20	20	20
	Superovulated	20	20	30

NR = not recommended; conception rates generally <50%, may be very low.

The time of insemination is related to the time of ovulation. Ovulation occurs around the end of oestrus. Insemination should be performed at a time sufficiently before ovulation so that by the time of ovulation a large population of spermatozoa is established in the ampulla, the site of fertilisation. In general, this requires insemination 12-24 hours before ovulation. The time of insemination should be adjusted according to whether the semen is fresh or frozen-thawed, since the spermatozoa in fresh semen should have a fertilising life in the female reproductive tract well in excess of 24 hours, while for spermatozoa in frozen-thawed semen, the fertilising life may not be more than 12 hours. For vaginal or cervical AI of ewes in spontaneous oestrus, it is best to aim for 12-18 hours after the onset of oestrus, with twice-daily inspections for detection. With once-daily inspection in the morning, ewes in oestrus should be inseminated soon after inspection.

For cervical or uterine AI of ewes in controlled oestrus, insemination is usually performed at a fixed time in relation to the synchronisation treatment. The interval between progestagen device removal and time of insemination is influenced by season, type of intravaginal device (shorter for CIDR than for progestagen sponge), type of semen (fresh, frozen), the use of PMSG and the dose employed. If progestagen sponges are used in conjunction with a non-superovulating dose of PMSG and 5-10% teasers, the majority of ewes will be in oestrus within 36-48 hours and will ovulate about 60 hours after sponge withdrawal. Superovulated donor ewes for MOET programmes, which have been treated with large doses of FSH, will be in oestrus within 24-36 hours and will ovulate about 48 hours after sponge withdrawal. The recommended times of insemination, where a single insemination is employed, following progestagen sponge withdrawal for different hormonal treatments, type of insemination and semen types are outlined in Table 8.6.

Ewes are usually inseminated only once. Double inseminations are not recommended when uterine insemination is used. In the case of cervical insemination, the likely small (5-10%) increase in conception rate is probably not worth the extra costs associated with double insemination. If employed in ewes not receiving PMSG, inseminations should occur at 48-50 and 58-60 hours after sponge removal. The number of motile spermatozoa recommended for single insemination should always be used at each insemination.

Table 8.6: Time of insemination in relation to method of insemination, hormone treatment and semen type.

Method of insemination	Hormone treatment	Semen type	Insemination time[a]
Cervical	Progestagen + PMSG	Fresh	36-48 h
	Progestagen PMSG	Fresh	48-60 h
Intrauterine	Progestagen + PMSG (non-superovulated)	Fresh	36-48 h
		Frozen	60-66 h
		Sexed-Frozen	57-59 h
	Progestagen + PMSG (superovulated)	Fresh	24-48 h
		Frozen	44-54 h
		Sexed-Frozen	42-43 h[b]

[a] post-sponge removal

[b] optimum insemination time yet to be fully established

NOTE: These times can all be shortened a little if CIDRs are used instead of progestagen sponges.

Conception rates in the AI programmes described above should be about the same as those that could be obtained following natural mating with sufficient numbers of fertile rams, but they are commonly, in practice, somewhat less. Occasional claims from commercial operators of extraordinary conception rates (up to 90%) must be treated with caution. Rates of 60-70% using intrauterine AI are very acceptable and seem to be the nationally recognised average.

MULTIPLE OVULATION AND EMBRYO TRANSFER (MOET)

The ovaries of the ewe contain vast numbers of oocytes, only a minute proportion of which ever develop into lambs. Almost all of these oocytes become atretic and are lost. This represents a loss of potentially valuable progeny in genetically superior ewes. MOET provides a method whereby some of these otherwise wasted oocytes can be used and the genetics of superior dams can be more widely disseminated. MOET is the female gene dissemination equivalent of AI.

MOET programmes incorporate AI but are an order of magnitude more complex, with additional hormones required for superovulation, surgical recovery of embryos and their transfer into synchronised recipients. The costs involved are considerable and the technique is not widely used, with only around 7000 donors flushed in Australia each year. While the surgical procedures involved require extensive training, some discussion of the principles of superovulation and the management of MOET programmes is useful.

Superovulation of donor ewes

Various gonadotrophins rich in follicle-stimulating activity will induce superovulation when given during the latter stages of the luteal phase of the oestrous cycle. PMSG used to be the gonadotrophin most widely used because of its availability, relatively low cost and ease of use. When it is used alone at doses of 1200-1500 IU, however, a significant proportion of ewes show ovarian responses characterised by many persistent large follicles which fail to ovulate and relatively few ovulations.

Purified FSH of porcine origin has become the hormone most commonly utilised for superovulation of ewes. It leads to more ovulations and fewer persistent large follicles than PMSG but its use is more tedious and some ewes fail to exhibit any superovulatory response (regardless of the dose of FSH employed). FSH has a half-life in the sheep of around two hours, which is much less than the equivalent interval of about 20 hours for PMSG. PMSG is given by a single intramuscular or subcutaneous injection, whereas FSH, because of its much shorter half-life, has to be given as a series of injections, usually twice daily over three consecutive days.

In order to effectively programme the embryo transfer operations, it is essential to control the times of oestrus and ovulation in donor ewes, by using progestagen sponges or CIDRs for 12-13 days. A non-superovulatory dose of PMSG is administered 48 hours before removal of the intravaginal device in addition to the first of several injections of FSH which continue over the ensuing three days. The ewes should be in oestrus 24-36 hours after pessary removal and should ovulate around 42-54 hours after pessary removal. Several factors appear to influence the timing and spread of ovulations following a superovulatory treatment, and some experience is required in determining the optimal time for insemination. The uncertainty about the time of ovulation is a significant problem when frozen semen is employed, and GnRH (40 µg Fertagyl, Intervet; 24 h post-pessary removal) is sometimes administered in an effort both to reduce the interval from first to last ovulation and to shorten the time elapsing between pessary removal and the median time of ovulation. There is probably no need to use GnRH where fresh semen is employed. Suggested optimal times for insemination were listed previously.

Variation in the superovulatory response

As already noted, variability in ovarian responses (numbers of ovulations and persistent large follicles which fail to ovulate) remains substantial and it is normal for some treated donors to yield no transferable embryos. It is important to realise that, in the practice of MOET, it is relatively easy to master the surgical and manipulative skills required but considerably more difficult to gain precise control over the superovulatory responses. Factors which influence the ovulatory response in donors include

- the treatment regime, especially the type, batch and dose of gonadotrophin employed and timing of treatments. Variation in the relative amounts of FSH and LH activity in different gonadotrophin preparations is probably a factor here. Inhibitory factors produced by the ovaries (inhibin and other peptides) are presumably involved

- breed, genotype and age of donor. Booroola ewes and Merino ewes selected for multiple births are more responsive to PMSG than are unselected (non-Booroola) Merino ewes

- the stage of the oestrous cycle at which the synchronisation treatment commences. The ovulation rate is lower when treatment commences mid-cycle (days 8-10) and higher when treatment commences early (days 1-3) or late (days 13-15) in the cycle

- body weight and condition score of the donor as well as the environmental conditions during stimulation.

Mating

Neither natural mating nor cervical insemination are recommended for donors in MOET programmes because fertilisation is likely to fail due to faulty transport of spermatozoa through the cervix. The cervix is exposed to abnormally high levels of ovarian hormones, principally oestrogens (due to the higher number of follicles present), which increase secretory activity of this organ and dramatically heighten the volume of mucus produced. This problem is overcome by intrauterine insemination. Table 8.5 displays the number of spermatozoa recommended for delivery direct to the uterus in superovulated ewes.

Embryos

Embryos are usually collected six days after mating, by which time they have moved from the oviducts into the uterus. The original method of collection involved backward flushing of the uterine horns and oviducts with a catheter inserted into the infundibulum. Today, all commercial programmes flush from the tip of the uterine horn into a foley catheter inserted at the uterine bifurcation. This approach is less invasive and can be used repeatedly in the same ewe but it tends to yield slightly lower recovery rates than the original *retrograde* flushing method (which could also be used to collect embryos prior to day 4 and before they had entered the uterus). Commercially prepared complete media are used for flushing and storage, although, if necessary, balanced salt solutions (for example, Dulbecco's phosphate buffered saline, pH 7.3) supplemented with 10% sheep serum and antibiotics can be used and are equally as successful. Fresh medium is used each day and stored at 30 °C for use. Embryos should be located in the flushing medium and transferred to fresh medium soon after flushing. They can then be stored in the medium at 20-25 °C for three to four hours. A stereoscopic microscope (magnification 40-60x) is used to locate and assess the embryos in the flushings and fresh medium.

The assessment of embryo viability is done morphologically and is based on general appearance and stage of development. This method is quick, cheap and noninvasive, but it is subjective and the correlation with viability is certainly less than perfect. In the evaluation, size, fragmentation and granulation of blastomeres, symmetry of the cell mass and the appearance of the zona pellucida should be considered. For operators to achieve reliably high success rates, there is no substitute for a lot of practice and critical review of procedures in the light of subsequent lambing rates. Systems for scoring embryos for quality or viability are often used but are probably of limited value for inexperienced operators.

Variation in the numbers of normal embryos recovered from donors

The recoveries from individual donors can range from none to 20 or more. Assuming adequate surgical skill and timing of flushing, the most likely reason for variation in the number of normal embryos recovered from donors is the superovulatory response — 20 embryos cannot be recovered from a donor that has only ovulated three follicles. Recovery of a large number of unfertilised eggs would suggest either that semen quality is poor or that there is a problem with sperm transport within the female.

Transfers to recipients

Transfers of embryos to recipients should be carried out as soon as reasonably possible after collection. The recipients must be in oestrus within 12 hours of their respective donors. Note that this implies that synchronising intravaginal devices must be removed from recipients about

24 hours before they are removed from donors. Recipients need not be of the same breed as the donors. Because of their good maternal qualities, mature crossbred ewes are effective and popular recipients, especially if two embryos are transferred to each recipient. Transfers are best done using a cradle, sedation, local anaesthesia and laparoscopy. A uterine horn ipsilateral to an ovary containing a corpus luteum is identified and then about 5 cm of the ovarian end of this horn is exteriorised through the wound made by the trochar for the transfer. There is always debate about the relative merits of transferring one or two good-quality embryos per recipient. It seems reasonable to assume that pregnancy rates are higher after the transfer of two embryos but there may be a lower rate of conversion of embryos into live lambs. Thorough studies of this question are lacking. In commercial practice, decisions may be based on the relative numbers of embryos and recipients available and the viability scores allotted to embryos. For maximal speed and efficiency in commercial MOET programmes it is best to deposit two embryos per recipient. Some operators using this policy claim 60% survival of transferred embryos. It is advisable to programme five recipients per donor, thus allowing effective use of up to 10 embryos per donor. The general approach with retarded embryos is to transfer these last if recipients are available.

It should be self-evident that, even more than in the case of an AI programme, the subsequent feeding and management of the recipients during pregnancy, lambing and lactation should be excellent, in order to optimise lamb birth weights and to maximise lamb survival.

Expected results for a MOET programme

The results in Table 8.7 should be attainable in a well-managed programme, but in practice the average number of progeny per treated donor seems often to be less than four to six.

Schedule for the preparation of donor and recipient ewes

An example of the hormone treatments and timing used for both donors and recipients is given in Table 8.8.

Storage of embryos

Sheep embryos can be successfully stored frozen in liquid nitrogen for unlimited periods. At best, with present methods, about 70% of embryos survive frozen storage. Hence, in the case

Table 8.7: Expected results for a well-managed MOET programme. Adapted from de Graaf (2010).[16]

Number of ovulations (corpora lutea)	10-15
Recovery rate (ova recovered/corpus luteum)	70-90%
Percentage of recovered ova scored as normal	70-90%
Percentage of transferred embryos surviving to lambs*	60-80%
Average number of progeny per treated donor**	4-6

* Assumes transfer of one or two normal embryos freshly collected from a mature donor; the survival rate will be lower after transfer of (a) frozen-stored embryos, (b) fresh embryos scored as retarded and/or abnormal (c) embryos collected from lambs or where (d) the recipient was detected in oestrus more than 12 hours apart from the respective donor.

** This number can be increased to, say, 20 progeny per donor per annum, if the same donor is reprogrammed four times in the same year.

Table 8.8: A sample calendar for treating donors and recipients in a MOET programme. Adapted from de Graaf (2010).[16]

Day of programme	Donor ewes	Recipient ewes
0	Insert progestagen sponge or CIDR	Insert progestagen sponge or CIDR
10 (8 pm)	Treat with 400 IU PMSG Treat with 22 mg of FSH	—
11 (8 am) (8 pm)	Treat with 22 mg of FSH Treat with 22 mg of FSH	Remove sponges; treat with 400 IU PMSG; join harnessed teasers
12 (8 am) (8 pm)	Treat with 22 mg of FSH Treat with 22 mg of FSH Remove sponges, join harnessed teasers	—
13 (6 am) (6 pm)	Treat with 22 mg of FSH Treat with 0.5 µg of GnRH Isolate from feed and water	—
14 (2 pm)	Oestrus Laparoscopic intrauterine AI (frozen semen); remove teasers	Oestrus Remove teasers
19 (pm)	Isolate from feed and water	Isolate from feed and water
20	Collect and evaluate day 6 embryos	Transfer two normal embryos to a uterine horn ipsilateral to an ovary containing a corpus luteum
26-28	Treat with luteolytic dose of prostaglandin analogue; remove skin sutures	—
67-69	—	Score for pregnancy and twins using real-time ultrasound (equals days 53-55, if pregnant)
160	—	Pregnant ewes will all lamb over the next 1-7 days (if lambs are Merinos); review management of lambing

of frozen embryos, 40-50% are expected to survive to lambs after transfer to synchronised recipients. The techniques used for freezing and thawing are still evolving. Both slow and rapid procedures, as well as vitrification, are employed.

SPERM SEXING

The only effective means of accurately preselecting the sex of offspring is through the use of sex-sorted spermatozoa or *sexed semen*. Using a procedure known as *flow cytometry*[21] (Figure 8.5), it is possible to separate all of the spermatozoa within an ejaculate into X- and Y-chromosome-bearing populations (that is, sexed semen) and inseminate one of the populations to produce either a female or a male, respectively. The technical points of the sex-sorting procedure are complex and covered in considerable detail elsewhere[22,23], but, briefly, it is achieved by exploiting the difference in DNA content (4.2% for sheep) between spermatozoa carrying the X and Y chromosome.[24] Recent research has increased the fertility of sexed frozen-thawed ram semen to a comparable level with normal frozen-thawed semen in both standard AI programmes[25,26,27] and in superovulated ewes for MOET.[28] These improvements in fertility

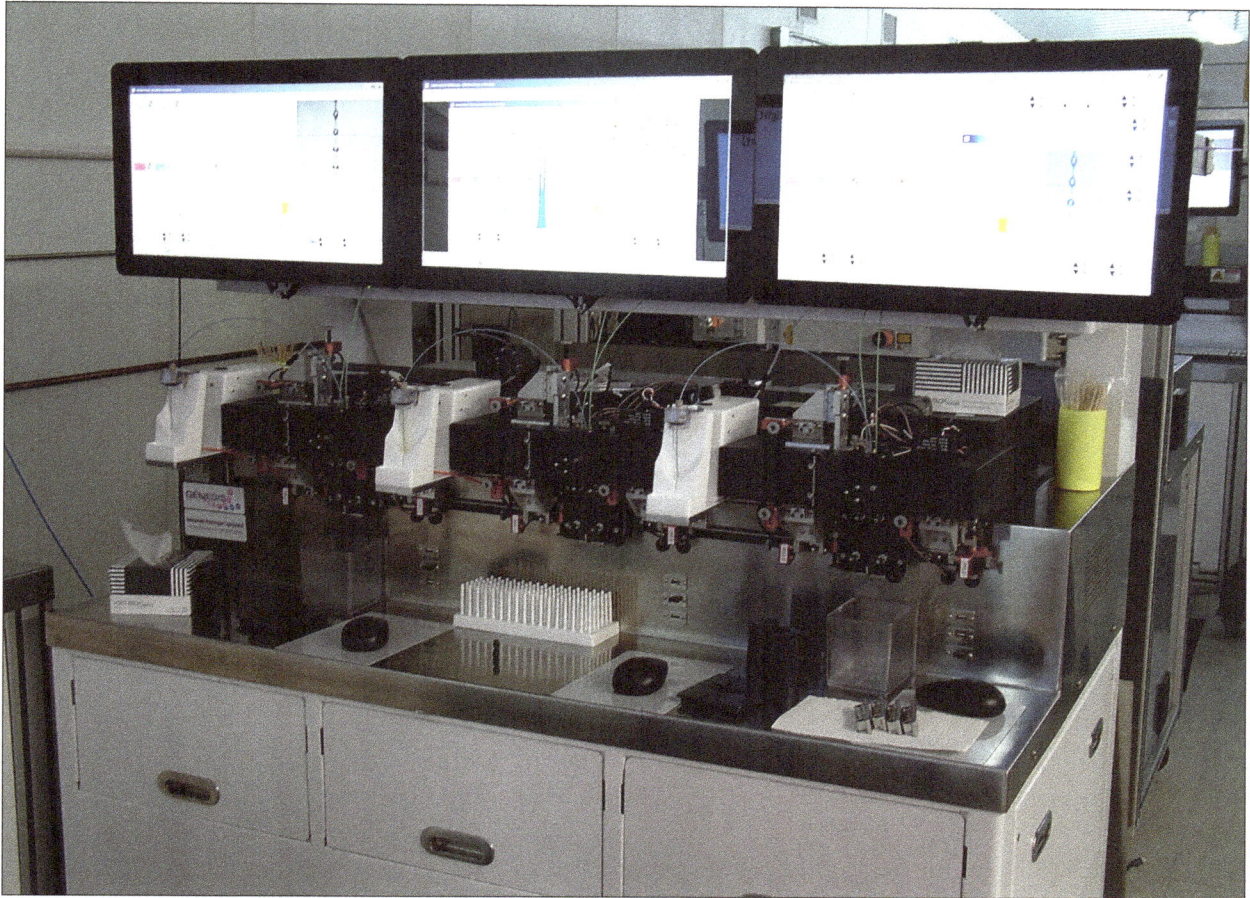

Figure 8.5: A state-of-the-art GENESIS III flow cytometer used for the sorting of sperm cells into X and Y populations. Photo courtesy KM Evans (Sexing Technologies), used with permission from ST Genetics.

have enabled the commercial release of *sexed ram semen* onto the worldwide market in 2016. Further detail on sperm sexing in sheep and future developments in this area are discussed in de Graaf et al. (2009, 2014).[29,30]

JUVENILE IN VITRO EMBRYO TECHNOLOGY (JIVET)

Transferrable embryos can also be produced *in vitro* (IVP) after oocyte pick up (OPU) from a superstimulated donor animal.[31] This is achieved by aspiration of pre-ovulatory follicles (often by laparoscopy) in donor animals treated with similar hormone regimens to those used for MOET and transfer of the resultant ova to a laboratory for *in vitro* maturation (IVM) and *in vitro* fertilisation (IVF). Embryos are grown to morula or blastocyst stage before transfer to recipients using the same technique described above for MOET. When the oocyte donors are mature ewes, this method of assisted reproductive technology is known as *mature* in vitro *embryo transfer* (MIVET). While this technique eliminates the variation in fertilisation rate and superovulatory response observed with MOET[32], embryos produced *in vitro* remain less viable than those created *in vivo*.[33] Oocyte donors need not be ewes which have reached puberty and sexual maturity. Pre-pubertal lambs as young as 3 weeks of age can be stimulated with gonadotrophins and their eggs harvested by OPU.[34] Known as *juvenile in vitro embryo transfer* (JIVET), this procedure offers the opportunity for rapid genetic improvement via a reduction of the generation interval to as little as six months.[35] Unfortunately, oocytes from juvenile animals do not respond as favourably to IVP as those from mature ewes[36] and embryo/ foetal survival post-transfer is also reduced.[37,38] Consequently, JIVET remains considerably less efficient than MIVET.[34,39] In reality, the expense of both JIVET and MIVET excludes most commercial breeders from utilising their benefits, and few artificial breeding companies offer such services in sheep.

REFERENCES

1   Bartlewski PM, Baby TE and Giffin JL (2011) Reproductive cycles in sheep. Anim Reprod Sci **124** 259-68. https://doi.org/10.1016/j.anireprosci.2011.02.024.

2   Evans G and Robinson TJ (1980) The control of fertility in sheep: Endocrine and ovarian responses to progestagen-PMSG treatment in the breeding season and in anoestrus. J Agric Sci **94** 69-88. https://doi.org/10.1017/S002185960002791X.

3   Robinson TJ (1965) Use of progestagen-impregnated sponges inserted intravaginally or subcutaneously for the control of the oestrous cycle in the sheep. Nature **206** 39-41. https://doi. org/10.1038/206039a0.

4   Evans G and Maxwell WMC (1987) Salamon's artificial insemination of sheep and goats. Butterworths: Sydney.

5   Rickard JP and de Graaf SP (2016) National Sheep Artificial Breeding industry 2016 survey report. The University of Sydney: Sydney

6   Fairnie IJ and Wales RG (1980) Fertility in Merino ewes in artificial insemination programs following synchronisation of ovulation using cloprostenol, a prostaglandin analogue. Proceedings of the Australian Society of Animal Production **13** 317-20.

7   Smith ID (1982) The synchronisation of oestrus and conception in Merino ewes by single or sequential administration of prostaglandin F2α. Proceedings of the Australian Society of Animal Production **14** 531-4.

8   Fierro S, Gil J, Viñoles C et al. (2013) The use of prostaglandins in controlling estrous cycle of the ewe: A review. Theriogenol **79** 399-408. https://doi.org/10.1016/j.theriogenology.2012.10.022.

9   Notter D (2002) Opportunities to reduce seasonality of breeding in sheep by selection. Sheep Goat Res J **17** 21-32.

10   Rosa H and Bryant M (2002) The 'ram effect' as a way of modifying the reproductive activity in the ewe. Small Ruminant Res **45** 1-16. https://doi.org/10.1016/S0921-4488(02)00107-4.

11   Fabre-Nys C, Kendrick KM and Scaramuzzi RJ (2015) The 'ram effect': New insights into neural modulation of the gonadotropic axis by male odors and socio-sexual interactions. Front Neurosci **9** 111. https://doi.org/10.3389/fnins.2015.00111.

12   Hawken PAR, Esmaili T, Scanlan V et al. (2009) Can audio-visual or visual stimuli from a prospective mate stimulate a reproductive neuroendocrine response in sheep? Animal **3** 690-6.

13   de St Jorre TJ, Hawken PAR and Martin GB (2012) Role of male novelty and familiarity in male-induced LH secretion in female sheep. Reprod Fertil Dev **24** 523-30. https://doi.org/10.1071/RD11085.

14   Walkden-Brown S, Martin G and Restall B (1999) Role of male-female interaction in regulating reproduction in sheep and goats. J Reprod Fertil-Supp **54** 243-57.

15   Viguié C, Caraty A, Locatelli A et al. (1995) Regulation of LHRH secretion by melatonin in the ewe. I. Simultaneous delayed increase in LHRH and LH pulsatile secretion. Biol Reprod **52** 1114-20. https://doi.org/10.1095/biolreprod52.5.1114.

16   de Graaf SP (2010) Reproduction. In: International sheep and wool handbook, ed DJ Cottle. Nottingham University Press: UK, pp. 189-222.

17 Scaramuzzi RJ and Martin GB (1984) Pharmacological agents for manipulating oestrus and ovulation in the ewe. In: Reproduction in sheep, eds DR Lindsay and DT Pearce,. Australian Academy of Science: Canberra, pp. 316-25.

18 Geldard H, Scaramuzzi R and Wilkins J (1984) Immunization against polyandroalbumin leads to increases in lambing and tailing percentages. NZ Vet J **32** 2-5. https://doi.org/10.1080/0048016 9.1984.35043.

19 Geldard H (1984) Field evaluation of Fecundin: An immunogen against androstenedione. Proceedings of the Australian Society of Animal Production **15** 185-91.

20 Gordon I (1993) Controlled reproduction in sheep and goats. Vol 2. CABI Publishing: Wallingford, UK.

21 Maxwell WMC, Evans G, Hollinshead FK et al. (2004) Integration of sperm sexing technology into the ART toolbox. Anim Reprod Sci **82-83** 79-95. https://doi.org/10.1016/j. anireprosci.2004.04.013.

22 Garner DL (2001) Sex-sorting mammalian sperm: Concept to application in animals. J Androl **22** 519-26.

23 Sharpe JC and Evans KM (2009) Advances in flow cytometry for sperm sexing. Theriogenol **71** 4-10. https://doi.org/10.1016/j.theriogenology.2008.09.021.

24 Pinkel D, Lake S, Gledhill BL et al. (1982) High resolution DNA content measurements of mammalian sperm. Cytometry **3** 1-9. https://doi.org/10.1002/cyto.990030103.

25 de Graaf SP, Evans G, Maxwell WMC et al. (2007) Birth of offspring of pre-determined sex after artificial insemination of frozen-thawed, sex-sorted and re-frozen-thawed ram spermatozoa. Theriogenol **67** 391-8. https://doi.org/10.1016/j.theriogenology.2006.08.005.

26 de Graaf SP, Evans G, Maxwell WMC et al. (2007) Successful low dose insemination of flow cytometrically sorted ram spermatozoa in sheep. Reprod Domest Anim **42** 648-53. https://doi. org/10.1111/j.1439-0531.2006.00837.x.

27 Beilby KH, Grupen CG, Thomson PC et al. (2009) The effect of insemination time and sperm dose on pregnancy rate using sex-sorted ram sperm. Theriogenol **71** 829-35. https://doi.org/10.1016/j. theriogenology.2008.10.005.

28 de Graaf SP, Beilby KH, O'Brien JK et al. (2007) Embryo production from superovulated sheep inseminated with sex-sorted ram spermatozoa. Theriogenol **67** 550-5. https://doi.org/10.1016/j. theriogenology.2006.09.002.

29 de Graaf SP, Beilby KH, Underwood SL et al. (2009) Sperm sexing in sheep and cattle: The exception and the rule. Theriogenol **71** 89-97. https://doi.org/10.1016/j.theriogenology.2008.09.014.

30 de Graaf SP, Leahy T and Vishwanath R (2014) Biological and practical lessons associated with the use of sexed semen. Reprod Domest Rum **VIII** 507-22.

31 Morton KM, de Graaf SP, Campbell A et al. (2005) Repeat ovum pick-up and in vitro embryo production from adult ewes with and without FSH treatment. Reprod Domest Anim **40** 422-8. https://doi.org/10.1111/j.1439-0531.2005.00603.x.

32 Maxwell WMC, Szell A, Hunton JR et al. (1990) Artificial breeding: Embryo transfer and cloning. In: Reproductive physiology of Merino sheep, concepts and consequences, eds CM Oldham, GB Martin and IW Purvis. School of Agriculture, Animal Science, The University of Western Australia, Nedlands: Perth, pp. 217-37.

33 Galli C and Lazzari G (2008) The manipulation of gametes and embryos in farm animals. Reprod Domest Anim **43** 1-7. https://doi.org/10.1111/j.1439-0531.2008.01136.x.

34 Morton KM (2008) Developmental capabilities of embryos produced in vitro from prepubertal lamb oocytes. Reprod Domest Anim **43** 137-43.

35 van der Werf J (2005) Applying new technologies in sheep breeding programs. Paper presented at the Merinotech Best Practice Sheep Breeding Forum, 3 August, Kojonup, Western Australia.

36 O'Brien JK, Dwarte D, Ryan JP et al. (1996) Developmental capacity, energy metabolism and ultrastructure of mature oocytes from prepubertal and adult sheep. Reprod Fertil Dev **8** 1029-37.

37 Kelly JM, Kleemann DO and Walker SK (2005) Enhanced efficiency in the production of offspring from 4- to 8-week-old lambs. Theriogenol **63** 1876-90. https://doi.org/10.1016/j.theriogenology.2004.09.010.

38 Ptak G, Matsukawa K, Palmieri C et al. (2006) Developmental and functional evidence of nuclear immaturity in prepubertal oocytes. Hum Reprod **21** 2228-37. https://doi.org/10.1093/humrep/del184.

39 Cognie Y (1999) State of the art in sheep-goat embryo transfer. Theriogenol **51** 105-16. https://doi.org/10.1016/S0093-691X(98)00235-0.

9

Helminth diseases of sheep

In Australia and New Zealand the helminth parasites of sheep which account for the greatest share of lost production are *Teladorsagia* (formerly *Ostertagia*) *circumcincta*, *Trichostrongylus* spp, *Haemonchus contortus* and *Fasciola hepatica*. The major nematodes, trematodes and cestodes of sheep are summarised in Table 9.1 and some less common nematodes in Table 9.2.

Gastrointestinal nematodes are one of the most serious causes of economic loss for sheep producers in the medium and high rainfall areas of Australia.[1,2] The ability to limit these losses improved markedly following the introduction of highly effective broad-spectrum anthelmintics; initially thiabendazole (1962), levamisole (1967) and ivermectin (1987), and then monepantel and derquantel (2010 and 2014, respectively). The efficacy of the new products is, however, potentially compromised by the development of resistance, which is invariably detected first in *Haemonchus contortus*, often within two to five years of release of the anthelmintic group, followed by the detection of resistance within the genera of *scour worms* (*Teladorsagia* and *Trichostrongylus*).

To slow the development of anthelmintic resistance, an integrated approach to nematode control, referred to as integrated parasite management (IPM), is recommended over control measures relying heavily, or solely, on anthelmintic treatments.[3] IPM programmes still use anthelmintics, with the timing of strategic treatments based on the epidemiology of the important parasites in each region, but they integrate these with other control strategies such as grazing management, effective combinations of anthelmintics, optimal nutritional management of young sheep and selection for enhanced resistance or resilience (tolerance) to nematodes.

Other strategies for the control of nematode infections, including the biological control of free-living larvae using nematode-trapping fungi such as *Duddingtonia flagrans*[4,5] or the strategic supplementation of susceptible sheep with high-quality protein[6,7], have yet to become commercially available and widely adopted. This is despite considerable research demonstrating that they have some benefits, although these are often not cost-effective, even when compared to anthelmintics with considerably reduced efficacy. Vaccination against the bloodsucking nematode *Haemonchus contortus* is feasible, and a commercial vaccine (Barbervax™) was released in Australia in 2014.[8] Because it is based upon a 'hidden' antigen from the gut of the worm, effective protection requires three priming doses before weaning, then booster vaccinations at six-weekly intervals. Consequently, a total of five vaccinations are needed during a typical risk period for haemonchosis in summer rainfall areas. Much work has been done in an attempt to develop vaccines against the other major production-limiting nematodes, but progress has been disappointing.[9]

Sheep and goats are definitive and intermediate hosts of a number of cestodes (Table 9.1), and the presence of masses of worm segments in the faeces of young animals can be a cause for concern amongst producers. Apart from the important role of *Echinoccocus* cysts in sheep spreading hydatid tapeworms to dogs, cestodes are generally of lesser economic or public health concern and will not be discussed further in this chapter. (See Chapter 4 for a discussion of cestodes affecting meat quality.)

This chapter is divided into four sections:

- Section A: Nematodiasis — diseases caused by nematodes of sheep
- Section B: Anthelmintics and anthelmintic resistance
- Section C: Epidemiology and control of nematode parasites
- Section D: Liver fluke.

SECTION A

NEMATODIASIS

Life cycles

The major nematodes of sheep, with the exception of *Trichuris* spp and *Strongyloides papillosus*, belong to the order *Strongylida* (Figure 9.1).[10] Consequently, the life cycles of the important sheep nematodes are broadly similar, most having a simple direct life cycle with no intermediate hosts. Exceptions are the lungworms *Muellerius capillaris* and *Protostrongylus rufescens*, which develop to the infective stage in slugs and snails, and snails, respectively.

The eggs of the gastrointestinal nematodes of sheep are of a similar shape and size, except those of *Nematodirus* spp, which have much larger eggs that can be easily differentiated from the others in a worm egg count. The eggs are passed in faeces and start to develop (embryonate) immediately at suitable temperatures and in the presence of sufficient moisture and oxygen. Development is strongly temperature dependent; at 26 °C, first-stage larvae (L_1) hatch within 24 hours but, at lower temperatures (generally below 9 °C, depending on the species), or with a lack of moisture or oxygen (for example, the anaerobic conditions at the centre of a faecal deposit), development is much slower.[11] For most of the important gastrointestinal nematodes in the family *Trichostrongylidae*, development from the egg to infective larvae (L_3) can occur within four to six days at 27 °C provided sufficient oxygen and moisture are available.

There is considerable variation in the capacity for egg production of the three main genera of gastrointestinal nematodes. In general, *Haemonchus* is the most fecund, *Trichostrongylus* is intermediate and *Teladorsagia* is the least fecund. Development within the egg occurs from the original egg cell to a morula, then a larva. The larva moults four times before adulthood, with the free-living stages (L_1 to L_3) on pasture and the L_4 and L_5 being parasitic. The cuticle is shed and replaced at each moult except for L_3 in which the sheath is generally retained and moulting (ex-sheathment) occurs in the sheep.

Larvae hatch from the egg at the L_1 stage except for *Nematodirus* and *Trichuris* spp. Free-living L_1 and L_2 have an active phase during which they feed, mainly on bacteria, and an inactive phase preceding the next moult. The larvae are infective to sheep only at the L_3 stage. If these are free-living, the cuticle of the L_2 is retained as a protective sheath in all except *Strongyloides papillosus*.

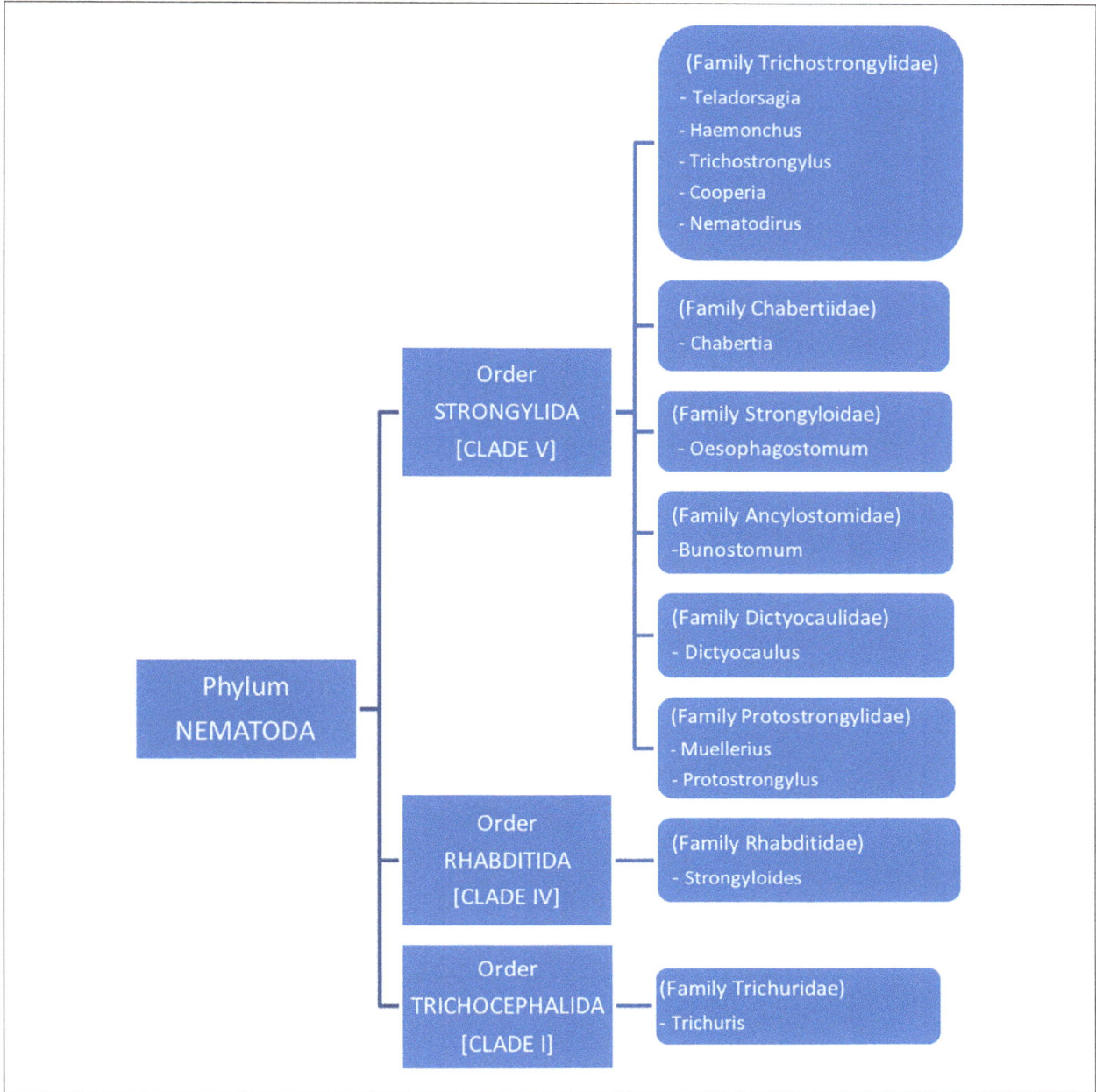

Figure 9.1: The major nematodes of sheep with taxonomy according to molecular phylogeny (clade/order) and the genus arranged according to their traditional morphological classification by family. Based on data from Blaxter et al. (1998).[10]

The ensheathed L_3 cannot feed and they die once they have exhausted their energy reserves. Infection occurs when grazing sheep ingest infective larvae with herbage. *Bunostomum trigonocephalum* (hookworm) can also infect sheep by penetrating skin, typically the interdigital skin of the feet. In this case, the pre-patent period (time before appearance of eggs in faeces) is eight to ten weeks, whereas for oral infections with the more important gastrointestinal nematodes the pre-patent period is much shorter — typically 17-21 days.

For *Nematodirus* spp, egg production is relatively low and the time taken to the pre-parasitic stage is about twice as long as for other trichostrongylids. Survival of the egg and larval stages of *Nematodirus* spp is relatively high and less dependent on environmental moisture. Because

Table 9.1: Common helminths of sheep in Australia.

Class	Location	Parasite Specific name	Common name (size of adult worms)
Nematodes	Abomasum	*Haemonchus contortus* *Teladorsagia circumcincta* *Trichostrongylus axei*	Barber's pole worm (12-17 mm) Small brown stomach worm (8-12 mm) Stomach hair worm (3.5-6 mm)
	Small intestine	*Trichostrongylus colubriformis* *T vitrinus* *T rugatus*	Black scour worm (3.5-6.5 mm)
		Nematodirus spathiger *N abnormalis* *N filicollis*	Thin-necked intestinal worm (8-20 mm)
		Strongyloides papillosus *Cooperia curticei* *Bunostomum trigonocephalum*	Threadworm (2-3 mm) Small intestinal worm (10-15 mm) Hookworm (12-26 mm)
	Large intestine	*Trichuris ovis* *T skrjabini* *Oesophagostomum columbianum* *Oes venulosum* *Chabertia ovina*	Whipworm (30-60 mm) Whipworm (30-60 mm) Nodule worm (12-20 mm) Large bowel worm (12-20 mm) Large-mouthed bowel worm (12-20 mm)
	Lungs	*Dictyocaulus filaria* *Muellerius capillaris* *Protostrongylus rufescens*	Large lungworm (white, 3-10 cm) Small lungworm (coiled, fine worms within 2-4 mm nodules on lung surface) Reddish (16-35 mm)
Trematodes	Liver	*Fasciola hepatica*	Liver fluke (30 × 13 mm, intermediate hosts are lymnaeid snails)
	Rumen and reticulum	*Calicophoron calicophoron* *Paramhistomum ichikawai* *Orthocoelium streptocoelium*	Stomach fluke (pear-shaped, 5-12 mm)
Cestodes	Small intestine	*Moniezia expansa* *M benedeni* *Thysaniezia giardi*	Tapeworm (up to 1 m, intermediate hosts are oribatid mites)
	Lungs, liver and other organs	*Echinococcus granulosus*[1]	Hydatid cysts
	Muscle	*Cysticercus ovis*[2]	Sheep measles (cysts 2-10 m)
	Abdominal cavity	*Cysticercus tenuicollis*[3]	Bladder worm (cysts up to 60 mm, attach via a long thin neck to liver or other organs)

[1] Intermediate stage of adult *Echinococcus granulosus* tapeworm in dogs; [2] Intermediate stage of adult *Taenia ovis* tapeworm in dogs; [3] Intermediate stage of adult *Taenia hydatigena* tapeworm in dogs.

development to the L$_3$ stage occurs within the egg, these are quite resistant to desiccation. Infective larvae of *Nematodirus* are resistant to cold and desiccation and heavy burdens can develop under conditions which do not favour other trichostrongylids, such as during droughts or in a paddock that has been used recurrently over a number of years for weaning young sheep.

Trichuris spp (whipworm) differ from the strongyles, as they are much slower to develop to L$_3$ and the larvae remain in the egg until ingested by sheep. The eggs remain viable for

Table 9.2: Less common species of nematodes of sheep in Australia.

Location	Parasite	Comment
Oesophagus	*Gongylonema pulchrum*	
Abomasum	*Haemonchus placei*	Presence associated with cattle
	Teladorsagia trifurcata	Presence associated with cattle
	Ostertagia ostertagi	Presence associated with cattle
	Teladorsagia davtiana	Presence associated with cattle
	Camelostrongylus mentulatus	Originally a parasite of camels
Small intestine	*Trichostrongylus probolorus*	Associated with cattle
	Nematodirus helvetianus	Infests sheep & cattle
	Cooperia oncophora	Previously *C mcmasteri*
	C pectinata (C surnabada*)	Associated with cattle
	C punctata	Associated with cattle

* morph types of a single species.

extended periods and the pre-patent period is relatively long — typically one to three months. The pre-patent periods of other large intestinal strongyles are 48 to 54 days for *Chabertia*, 41 days for *Oesophagostomum* and 30 to 56 days for *Bunostomum* spp.

The adult *Oesophagostomum* and *Chabertia* spp lie close to or are attached to the mucosa of the large intestine. Less is known about their free-living stages but *Oes columbianum* has a similar geographic distribution to *H contortus*, so its eggs and larvae are probably less tolerant of cold and desiccation. *Oes venulosum* is more widely distributed, which suggests that it is more tolerant of environmental conditions. Larvae of *Chabertia ovina* have a relatively high survival rate over winter, so the conditions suitable for its development are probably similar to those of *T circumcincta*.

Strongyloides spp have two adult forms, one parasitic and one free-living. The parasitic forms are parthenogenetic and eggs give rise to either infective larvae or males and females of a free-living generation which, in turn, gives rise to a parasitic generation. Infective larvae enter the host by ingestion or, more commonly, by skin penetration, after which they travel via the bloodstream to the lungs, then via the trachea and pharynx to the gut. The pre-patent period is five to seven days. Eggs passed in faeces contain fully developed embryos or may even have hatched.

The life cycles of the lungworms differ slightly from the gastrointestinal nematodes. Eggs of *Dictyocaulus filaria* hatch mainly in the gastrointestinal tract but occasionally in the lungs. L_1 are passed in the faeces and all free-living stages remain ensheathed in the cuticle of the previous stage. Consequently, none of the free-living stages of this parasite can feed, but sheep are still infected by ingesting L_3. Infection with the smaller lungworms (*Protostrongylus rufescens* and *Muellerius capillaris*) occurs when sheep ingest the L_3 within the intermediate host.

For those species infecting sheep directly from pasture, infective larvae require thin films of water in order to migrate (translate) onto herbage but excessive moisture can hinder their migration. The larvae are negatively geotropic and positively phototropic to mild light (although negatively phototropic to strong light). There is an element of randomness in the migration of larvae which limits the number that move up herbage on any one day to a low proportion of the total present.

Survival of free-living nematode stages

The survival of free-living stages of the parasitic nematodes is determined mainly by the available moisture and environmental temperature.[12,13] The ability of the eggs and larvae to survive varies with the species and the stage of development of the parasite when it is exposed to adverse conditions.[14] Desiccation is generally lethal to eggs which have not developed to the pre-hatch stage and to free-living L_1 and L_2. The sheath of free-living L_3 is, however, protective, so this stage of larva is more resistant to desiccation than earlier stages. This is particularly so for *T circumcincta* which, in southeastern Australia, survives better over summer than *Trichostrongylus* spp.[13,15] The L_3 of *Nematodirus* spp remain in the egg and are also quite resistant to prolonged hot and dry conditions.

The L_3 of *Haemonchus* spp are active but less able to survive adverse conditions than infective larvae of the other major strongyles. The L_3 of *Strongyloides papillosus* have no protective sheath and are quite susceptible to desiccation.

Factors that can influence the development and survival of larvae include

- the height and density of individual plants, which lead to differences in light and moisture within the sward

- the morphological characteristics of each plant species, which may alter the ability of L_3 to migrate up the leaves (for example, clovers or fodder crops, such as rape, compared to grasses)

- the relative proportion of each plant species present.

Many studies have examined the influence of the height and composition of the pasture sward on the survival and availability of the free-living stages of trichostrongylid parasites. Differences in larval recovery rates and infection acquired by grazing lambs on different pastures have been recorded but, generally, most larvae remain near the base of plant stems and tillers, with approximately 60% of those that climb located in the lowest 2.5 cm of herbage.[16] For example, on dry and irrigated pastures at Werribee, a winter rainfall area in Victoria, the length of the herbage (4-8 cm and 10-15 cm) had no bearing on the development or survival of the free-living larvae of *T circumcincta*.[11] In contrast, some studies in New Zealand found that a very high pasture sward did influence the density of the larvae, with more L_3 found 26 to 75 mm above ground level compared to the number below 26 mm.[17,18] Pastures grazed by sheep in Australia are rarely greater than 10 cm (2200 kg of dry matter (DM) per ha) and are more typically between 1-8 cm high (400-1900 kg DM/ha).

Each of the major climate zones in Australia where sheep are raised has different distributions of the major parasites. For example, *Haemonchus contortus*, *Trichostrongylus colubriformis* and *Oes columbianum* (now relatively uncommon) are predominant in coastal and warm, moist areas with summer-dominant rainfall[19,20,21,22], whereas *T circumcincta*, *Trichostrongylus vitrinus*, *Chabertia ovina* and *Dictyocaulus filaria* are predominant in moist but cooler, winter-dominant and uniform rainfall areas (Figure 9.2).[13,14] Knowledge about seasonal fluctuations in the availability of L_3 is critical to developing effective worm control programmes.[12] These patterns of infection and their influence on control programmes can vary considerably, both between and within regions, and are discussed further in Section C.

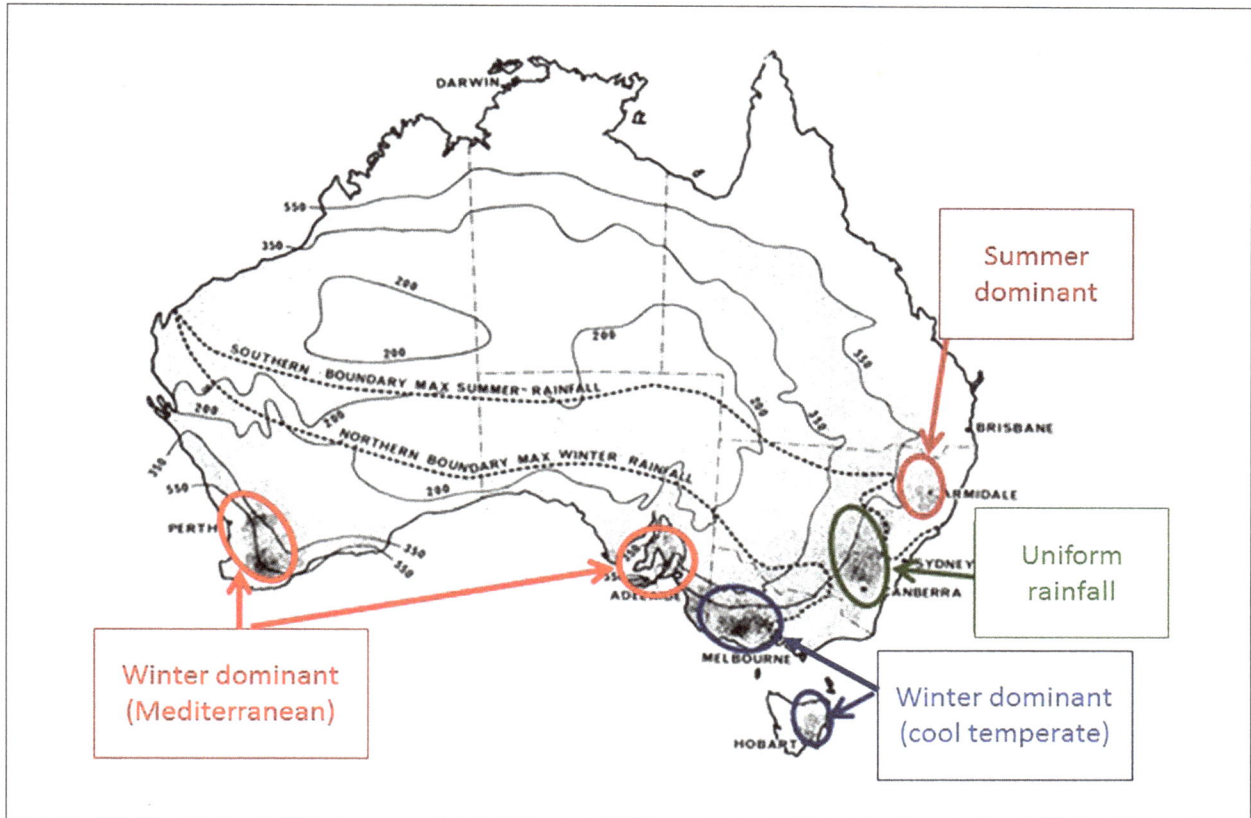

Figure 9.2: Distribution of sheep within the major climate zones that influence the epidemiology and control of gastrointestinal nematodes in Australia. Isohyets are in mm; each dot represents 30 000 sheep. Source: KA Abbott.

Host immunity and the pathophysiology of nematode infections

Sheep are not born with any significant innate resistance to gastrointestinal nematodes but exposure to these parasites, principally infective larvae but also immature and mature adults, leads to an acquired immunity against infection. Although this immunity is far from complete, it does enable a coexistence of the host and parasite. Successful worm control programmes include additional control measures that are needed for young animals while their acquired resistance develops and, occasionally, for ewes when their immune response is temporarily reduced during lactation. Adult, non-reproducing sheep generally have a solid immunity against gastrointestinal nematodes — a mechanism which is essential for sheep production in most environments in Australia. Without some level of host immunity, sheep farming, even with effective anthelmintics, would be impossible — especially in high rainfall areas where profitable sheep production systems are based upon grazing improved pastures at high stocking rates.

The development of immunity is complex and, while details of its nature vary according to breed, nematode species and intensity of infection, it generally increases with age and exposure to infection. It is slow to develop and may not be expressed to a practically useful degree for *Teladorsagia* and *Trichostrongylus* spp until sheep are 8-15 months of age and have been exposed to the parasite for at least four months. A functional immunity to *H contortus* may develop at a younger age.[23,24]

The manifestations of immune competence include[24,25]

- failure of infective larvae to establish and develop to adults in the gut
- inhibition of development at the early fourth stage and a relatively low proportion of the inhibited L4 subsequently developing to adults
- reduced fecundity and length of adult female worms
- expulsion of adult worms.

Regulation of worm burdens

Adult dry sheep have sufficiently strong immunity against most helminths to prevent clinical disease in the face of continued larval intake provided they do not suffer nutritional stress. They maintain generally stable and low worm burdens and produce fewer worm eggs than younger sheep or reproducing ewes. The immune response is generally strong for *Nematodirus* spp and intestinal *Trichostrongylus* spp but more labile for the abomasal parasites *T circumcincta* and *H contortus*.[24]

The effect of immunity is to prevent the accumulation of worm burdens despite the ongoing but seasonally variable ingestion of infective larvae. In the case of *T circumcincta*, the population is maintained by *turnover*, in which established adult worms are expelled and replaced by adults developed from recently ingested larvae. For *H contortus* the process of regulation is different. Once immunity develops, a relatively stable number of adult worms reside in the gut and any further ingested larvae are either rejected or have their development arrested. *Trichostrongylus* spp are regulated in a similar way to *H contortus* but the development of immunity often leads to the expulsion of adult worms and the establishment of a new but lower worm burden typical of the adult resistant animal.[12] The rate of development of immunity against *Trichostrongylus* spp appears to be dose-dependent.[12,26]

For immunity to develop in young sheep, exposure to parasites is necessary. The need for sensitising exposure must be considered when planning the administration of anthelmintics to control worm burdens in sheep grazing infective pastures. In non-immune animals, treatment will be followed by re-establishment, the rate depending on the infectivity of the pasture, and the development of immunity may be delayed. Without treatment, however, unacceptable losses may occur. For some parasites, particularly *Trichostrongylus* spp and *H contortus*, the timing of anthelmintic treatment to coincide with the development of immunity may be followed by the re-establishment of a significantly smaller worm burden.[23] By contrast, the treatment of lambs which have only partially developed immunity and are infected with relatively stable burdens of *H contortus* may dramatically increase their susceptibility and expose them to a higher risk of haemonchosis than would occur without treatment, unless they are moved to clean pasture or treated with long-acting anthelmintics.[23]

Pathophysiology

Gastrointestinal nematodes induce a severe inflammatory response within the gastrointestinal tract. In young animals, the severity of the inflammation is directly related to the number of parasites present. The immunity to nematode larval challenge developed by most adult sheep is characterised by a T-helper lymphocyte type 2 (Th2) response and increased populations of mast cells, globule leucocytes and eosinophils.[24,25] The immunity of some adult sheep, however, leads to an inflammatory response characterised by abnormally increased populations of eosinophils which, in winter rainfall and uniform rainfall areas, causes diarrhoea (*hypersensitivity scouring*) and faecal soiling (*dag*).[27,28]

The major pathophysiological effects of gastrointestinal nematodes are decreased food intake (inappetence), altered gut function (primarily the loss of protein into the gastrointestinal tract and increased secretion[29]) and, with *Haemonchus* infections, anaemia.

Ingested *H contortus* larvae usually develop to adult worms within two weeks. In temperate climates an increasing proportion of larvae ingested during the autumn and winter become arrested in the abomasum at the early fourth stage, with some of these resuming their development in early spring.[19] This *arrested development* allows the carry-over of infection at times which are less suited to the survival of the free-living stages of this parasite (temperatures <10 °C and frosts). Adult *H contortus* suck up to 0.05 mL of blood per day from the abomasal mucosa. Consequently, sheep with only moderate worm burdens (~2000 worms) can lose 2-3% of their total blood volume each day, causing anaemia, poor growth and mortalities. A self-cure phenomenon is described for *H contortus*, expressed as a sudden reduction of worm egg count (WEC) from high levels to zero or a low count. This was thought to follow a period of rapid reinfection (vaccination) or the ingestion of high-quality pasture (improved nutrition). The overall influence and benefits of self-cure, however, remain enigmatic because, although rapid reinfection is a regular feature of *H contortus* infections, self-cure does not always occur.

For *T circumcincta*, growth and development of L_3 occur predominantly within the gastric glands of the abomasum. This produces nodules 2-4 mm in diameter, which coalesce to give a *Morocco-leather* appearance in heavy infections. Undifferentiated mucous cells replace secretory cells, the pH increases to ≥5, and protein leaks into the gut lumen.[30] This is accompanied by inappetence, mild diarrhoea and decreased wool growth and body weight. Plasma pepsinogen levels rise due to the combined effects of elevated pH, accumulation of pepsinogen in the mucosa and increased mucosal permeability.[31,32] The accumulation of large numbers of early-fourth-stage larvae from late spring to early summer (hypobiosis or Type II ostertagiasis) is also a feature of *T circumcincta* infections.[33]

Three *Trichostrongylus* species inhabit the anterior third (3-4 m) of the small intestine and complete their life cycle within the crypts of the mucosa. The most pathogenic is *T vitrinus*, found predominantly in the cool high winter rainfall areas, followed by *T colubriformis*. These infections produce a severe inflammatory response, with immature and adult worms occupying mucosal tunnels that rupture during moulting. Severe villous atrophy is accompanied by leakage of plasma proteins into the gut lumen and decreased plasma concentrations of proteins and calcium.[34,35,36] *T rugatus* is restricted to warmer and drier areas, such as the sheep-cereal zones and East Gippsland in Victoria. It is much less pathogenic, causing only mild mucosal erosions, and it has little or no effect on body weight or skeletal development.[34,37]

Inappetence

Depression of food intake is a major factor in the pathogenesis of gastrointestinal infections, accounting for over 60% of the difference in weight gain between sheep infected with *T circumcincta* and uninfected sheep.[29,38,39,40] Despite this, the mechanisms which induce inappetence are poorly understood and cannot be manipulated to reduce the production losses from gastrointestinal parasites.[41]

Altered gut function

Although reduced feed intake accounts for a large proportion of production differences between infected and parasite-free animals, infected animals do have impaired uptake and utilisation of nutrients relative to pair-fed animals.[40] Gastrointestinal parasites can alter gut motility, change gut secretions and alter the digestion and absorption of nutrients.

Loss of protein

The loss of protein (creating a protein-losing enteropathy) is a feature of infection with gastrointestinal nematodes. This occurs through leakage of plasma, loss of erythrocytes, sloughing of epithelial cells and increased secretion of mucus into the gastrointestinal tract.[29,42] Infected sheep can lose up to 10% of their total plasma volume into the gut each day[43], of which 13% is irreversibly lost past the terminal ileum.[44] Although the losses from parasite infections are only a small proportion of the total protein turnover, they induce a deficiency because of the need to resynthesise the 30-50 g of protein lost each day.[44,45,46] Overall, this leads to decreased production and reduced dressing percentage, as the remaining protein is partitioned into essential homeostatic processes rather than growth and wool production.[47,48,49]

Experimental studies have often involved single species infections in pen trials, which rarely reflect the situation in grazing sheep exposed to fluctuating levels of reinfection with mixed populations of nematodes through the year. Decreased plasma protein (hypoalbuminaemia) and elevated pepsinogen are found in both lambs and adult sheep grazing pastures with mixed nematode infections.[32,50] The feeding of protected or increased amounts of protein (for example, cottonseed meal) to susceptible sheep, such as lambing ewes and weaned lambs, can increase resilience and compensate for lost protein, hence reduce production losses, although this is usually not a cost-effective strategy in commercial flocks.[51,52]

Energy metabolism

Reduced feed intake is the major factor limiting the energy available for growth in parasitised animals, although a small reduction in the efficiency of energy utilisation has been shown with experimental infections.[53,54] The cycling of large amounts of protein through the gut, increased urinary excretion of urea nitrogen and increased methane production, in conjunction with substantial increases in liver and gastrointestinal protein synthesis, account for most of this reduced efficiency.[40]

Mineral metabolism

Dramatic changes to macromineral metabolism of growing sheep, most notably osteodystrophy, are induced by gastrointestinal parasites. Lambs infected with *T colubriformis* and *T vitrinus* have impaired phosphorus absorption and increased losses of calcium and phosphorus[47,53,55,56], whereas the effects of *T circumcincta* on bone mineralisation are attributed to deficiencies in protein and energy rather than mineral absorption.[44,53,54,55,56]

Clinical signs and effects on production

Typical clinical signs of mixed *T circumcincta* and *Trichostrongylus* spp infections include diarrhoea, faecal staining of the breech (dag), reduced wool growth, a break in the wool and increased mortalities. Reduced wool growth can also occur in adult sheep with good immunity[57,58,59], while scouring and dag increase the risk of flystrike in sheep of all ages and mandate management interventions such as crutching and the application of insecticides.[60]

In contrast, *Haemonchus contortus* infections in summer-dominant and uniform rainfall areas can cause anaemia, sometimes rapidly fatal, with mortalities exceeding 30%.[12] Sheep with acute or subacute haemonchosis have pale mucosae and reduced capillary refill and may collapse when mustered. In chronic infections the anaemia may be associated with submandibular oedema (*bottle jaw*).

Heavy infections with *Trichostrongylus axei* are not common but can cause ulceration and erosions of the abomasal mucosa (*thumb-print* lesions), acute diarrhoea, reduced food intake and loss of body weight.[61] This parasite is unusual in that it also infects cattle and horses, and control programmes must account for cross-infection from these sources, especially from young cattle.

The initial development of larval stages of *Chabertia* and *Oesophagostomum* spp occurs in the small intestine. The immature stages of *Chabertia ovina* cause little damage but adult worms move from site to site in the colon, digesting a plug of mucosa in their large buccal capsule. Infected sheep can have soft faeces coated with excess mucus or flecks of blood. At necropsy, sheep with heavy infections have clearly visible worms attached to the gut mucosa, which is thickened with longitudinal ridges, blood spots and small ulcers.

With repeated infections, the immature stages of *Oes columbianum* (nodule worm of summer rainfall areas) can cause fibrous nodules with thick greenish pus in mucosal and muscular layers of the small intestine (*knotty gut* or *pimply gut*). Before the release of the benzimidazole anthelmintics, this parasite caused considerable economic loss because the affected gut tissue could not be used for sausage casings or catgut sutures. Larvae can also migrate through the mucosa of the caecum, causing severe calcified or caseous lesions and even peritonitis. Large numbers of *Oes venulosum* (large bowel worm) can cause ill-thrift in young sheep but this worm rarely causes disease and does not form nodules.

Heavy burdens of *Dictyocaulus filaria* can cause persistent coughing in young sheep (Figure 9.3). The smaller lungworms can cause surface nodules on the lung (*Muellerius capillaris*) and local inflammation in the bronchioles (*Protostrongylus rufescens*). Coughing and weight loss are not a feature of these infections, although they can cause economic losses through the rejection of offal at slaughter.

Young sheep

Young sheep (those <12 months old) are more susceptible to parasitism and so generally have higher worm burdens and excrete more eggs in their faeces than older sheep. Ingestion of a relatively low dose of infective larvae (around 2000 per week) can cause a 50% reduction in weight gain and decrease wool growth by up to 30% in 12-month-old Merinos and 18% in second-cross lambs.[62] In winter-dominant and uniform rainfall areas, spring-born Merino

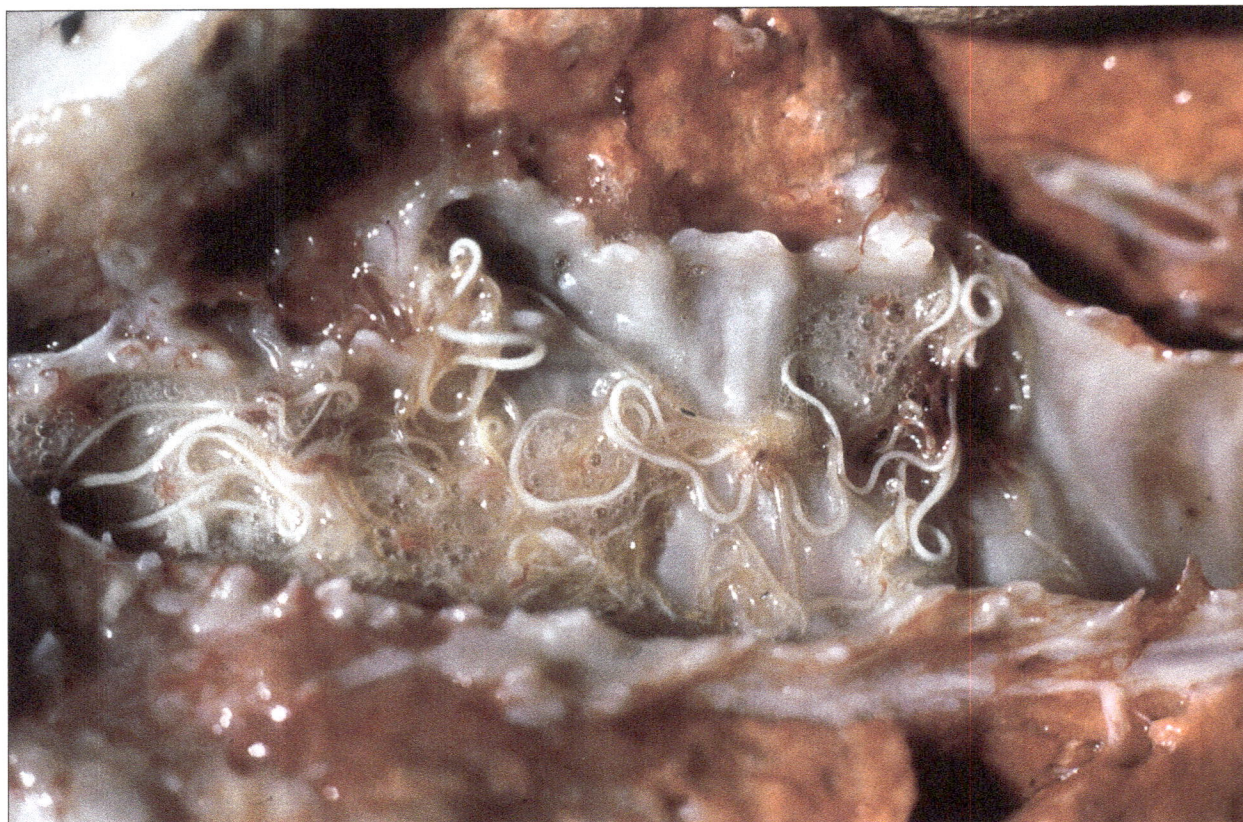

Figure 9.3: *Dictyocaulus filaria* in a bronchus. Photograph courtesy of KA Abbott.

lambs weaned in late spring or early summer tend to have low worm burdens during the summer but these increase after autumn rains and peak in winter. This is associated with scouring, reduced body weights and up to 30% mortalities in outbreaks of clinical parasitism.[63,64,65] From late spring onwards larval intake declines and a progressively increasing age- or body-weight-related immunity to gastrointestinal nematodes develops.[12,31,57,66]

In addition to decreased liveweight, decreased dressing percentages contribute to the cost of parasitism. For example, 10-month old Merinos with subclinical infections of *T circumcincta* and *T colubriformis* (no diarrhoea, no reduced feed intake or weight gain, and worm egg counts remaining below the treatment threshold) had a 1.3% lower dressing percentage than uninfected animals (39.5% vs 40.8%).[49]

Mature sheep

The worm burden of mature sheep tends to be lower than that of weaned lambs due to the development of immunity with age. Gastrointestinal nematodiasis still has a significant effect in these animals due to the effects of parasitism on the value of the ewe's fleece and carcase as well as milk supply and hence on the growth of lambs (a loss of up to $2.55 per ewe per annum, when compared to 'best practice' treatment).[59] Inappetence occurs with both *T circumcincta* and *Trichostrongylus* spp infections.[39] Merinos are often thought to be more susceptible than other breeds to the effects of gastrointesinal nematodes, but lactating Poll Dorset ewes infected with *T circumcincta* had a 16% reduction in food intake which led to increased weight loss and a 17% decrease in milk production in early lactation.[67]

The periparturient rise in worm egg count

An important feature of the epidemiology of gastrointestinal nematode infections in sheep is the peri-parturient rise (PPR) in worm egg count of late pregnant and lactating ewes, which typically peaks six to eight weeks after lambing.[68,69,70,71,72] The PPR is associated with increased worm burdens and ewe mortality, and increases the contamination of pasture, which may lead to high levels of infectivity for lambs that graze these pastures later in the season. Although the PPR is well documented, its pathophysiology is complex and incompletely understood. This temporary reduction in immunity is affected by the time of lambing within the year, the existing ewe worm burden, the amount and quality of pasture, and its level of contamination with infective larvae. It is also influenced by ewe genotype, parity and age, and is truncated when ewes lose their lamb.[71] The degree to which the PPR affects the ewe's immunity also varies with the infecting nematode species.

Although it would appear intuitive to drench ewes before lambing to reduce the PPR, the worm egg counts of ewes grazing contaminated pastures following such treatments become similar to untreated ewes within a few weeks unless they are shifted to pastures with low larval contamination (*low-risk* pastures).[73,74,75,76,77] It is difficult and impractical to prepare sufficient numbers of low-risk pastures for lambing ewes on most farms in the high rainfall areas, so anthelmintic treatment two to four weeks before lambing, or at lamb marking, will often confer no useful benefits. Nevertheless, anecdotal evidence from producers and veterinarians is that ewe deaths often occur in high rainfall areas if an anthelmintic is not given at this time, especially when ewes are below ideal condition score (2.7-3) or when pasture availability is less than optimum (<1200-1400 kg DM/ha).

Scouring and dag

The scouring and dag associated with gastrointestinal parasitism are significant problems because they increase the risk of flystrike of affected animals and increase animal health costs due to the need for additional crutching and applications of insecticides.[60,78,79] They also reduce the income from wool sales because crutchings are less valuable than fleece wool. For example, in 1995 the cost of scouring in adult sheep in high rainfall areas of Victoria was estimated to be about $10 million.[80]

Diarrhoea may be due to the presence in the gastrointestinal tract of a high burden of mature parasites but can also be caused by the ingestion of trichostrongylid larvae without the establishment of an adult worm population. This response is referred to as *hypersensitivity scouring*. The occurrence of diarrhoea in mature sheep associated with hypersensitivity to ingested larvae can lead to overuse of anthelmintics if sheep producers assume that scouring and dag indicate a high worm burden, requiring anthelmintic treatment. In a field study involving over 3000 untreated mature Merino ewes on three farms, the frequency of severe breech soiling was similar in ewes with a low WEC and those with a high WEC.[27] This was consistent with an earlier observation by Anderson (1972)[31] in which 30% of a mob of adult Merino wethers in western Victoria had diarrhoea despite their WECs being below 100 epg. That hypersensitivity diarrhoea follows the ingestion of trichostrongylid larvae was demonstrated when ewes treated with a controlled-release capsule containing albendazole were 12-16 times less likely to have severe dag during late winter and early spring than ewes not given a capsule.[27] A heightened inflammatory response occurs in the gut of affected sheep, with increased number of eosinophils in the abomasum and small intestine, and flattening of the villi in the jejunum.[27,28]

Hypersensitivity scouring is a highly repeatable and heritable trait distinct from resistance to gastrointestinal nematodes.[81,82,83,84] Only a relatively low dose of infective larvae (1000 per week) is required to induce diarrhoea in adult Merino sheep that are susceptible to scouring, which is typical of many hypersensitivity syndromes.[85] It follows that even highly effective worm control programmes will not provide adequate control of hypersensitivity scouring. The solution appears to depend on selective breeding — the use of rams and ewes with lower dag scores over several generations will ultimately reduce scouring and dag, and it is feasible to coselect for both parasite resistance (low WEC) and decreased scouring.[27,86]

Similar observations have been made with Romney sheep in New Zealand, with the estimates of heritability of dag score in Romneys (0.24 ± 0.8) consistent with those for 9-month-old Merinos (0.27 ± 0.1).[87,88] In summary, hypersensitivity scouring is regarded as an immune-pathological response to the ingestion of infective larvae in immunologically mature sheep, rather than the direct effects of the parasite, and probably affects all breeds of sheep in winter and uniform rainfall areas at least to some degree.[86]

Diagnosis of nematode infections

Grazing livestock are continually infected with nematodes and so the broad aim of diagnostic procedures is to determine what number and genera of nematodes are present.

Worm egg counts (WEC)

The WEC of a faecal sample is a reasonable indication of the number of worms present in young sheep (the *worm burden*) and, hence, the potential production loss. The value of WECs is less in older sheep (those >12 months old) because of the development of acquired immunity which affects both the number of worms present and the fecundity of female worms. Nevertheless, WECs are a useful way of monitoring nematode infections over time, plus they are used to assess the efficacy of anthelmintics. The results can be obtained rapidly; the test is easy to conduct and is relatively inexpensive.

Monitoring WECs is an important component of worm control programmes and is commonly used to determine if additional (tactical) treatments are required. Their use over a number of years can also detect seasonal variation in worm burdens and monitor the success of a nematode control programme in a sheep flock over time.

There are many variations of the basic flotation (McMaster) method.[89,90,91] Regardless of the laboratory method, it is important to sample enough animals to make an accurate and reliable estimate. A bulk WEC from 10 individual faecal samples was the original standard for monitoring but at least 20 randomly collected samples are now recommended to provide an accurate estimate of the mean WEC, especially for larger mobs (>400 sheep). Egg-counting methods in the winter rainfall areas employ a higher sensitivity (for example, 1 egg counted represents 15 to 40 eggs per gram (epg) of faeces), whereas those for the summer rainfall areas where *Haemonchus* is predominant are usually less sensitive because there are much higher counts (for example, 1 egg counted represents 60-100 epg).

Identifying genera of worms present

The common nematodes vary considerably in their fecundity and pathogenicity; the most fecund are *Haemonchus* (5000-10 000 eggs/female/day), *Chabertia* and *Oesophagostomum* spp (3000-5000/day), whereas *Trichostrongylus* and *T circumcincta* produce far fewer eggs (100-200/day). Of the intestinal *Trichostrongylus* spp, *T rugatus* is more fecund than *T vitrinus*. While the former is found in drier and uniform rainfall areas, the latter in cooler, wetter areas, their distribution can overlap and the latter is considerably more pathogenic. Consequently, knowing the proportion of genera or species represented by eggs in a faecal sample can refine the interpretation of WECs and drench resistance tests.

The eggs of the common nematodes are similar in shape and size and cannot be reliably differentiated. The genus, however, can be identified by culturing eggs and examining the infective larvae (larval culture), or by molecular assay on the eggs or larvae. Larval culture is relatively slow, requiring 7-10 days of incubation, and the results of larval differentiation can vary considerably depending on the technique and incubation temperature. Warmer temperatures (25-27 °C) favour the growth of *Haemonchus*[92] and there is considerable overlap in the morphological criteria used to categorise the larvae.[93] Quantitative polymerase chain reaction (PCR) assays on the DNA of trichostrongylid eggs are a more accurate, faster and cheaper way of estimating of the proportion of genera present.[94,95] A PCR test can be completed within one or two days of sample collection; the method does not favour any particular genus; and the interpretation of results is relatively straightforward.

Total worm counts

Counts of the immature and adult nematodes in the abomasum and the small and large intestine of sheep provide the most definitive measure of worm burden. A total worm count (TWC) provides information on the number and genera of worms present, and the proportion of immature and mature worms also provides an estimate of recent larval intake. Worm counts are labour-intensive and relatively expensive, requiring sheep to be killed and their gastrointestinal tract washed before samples of the contents are examined — either grossly in the field or microscopically in a parasitology laboratory.

In order to use the technique to estimate the worm burden in a flock, three to five sheep should be selected for necropsy. The selected sheep may be clinically affected but should not, in general, be the worst cases in the flock, nor moribund or with severe diarrhoea. The abomasum should be tied off at the omasal-abomasal junction and distal to the pyloric sphincter. The first 3-4 m of small intestine is dissected free from its mesentery and tied off at both ends. The organs can be either transported whole or their contents washed into sealed jars for processing later in the laboratory. Alternatively, a preliminary estimate of the abomasal and small intestinal nematode burden can be made in the field. The advantages of the field technique are that the result is available immediately and the flock owner is present and can directly observe both the skill of the veterinarian and the outcome. The field technique is described in Appendix 1 of this chapter.

A total count of worms in the large intestine is rarely undertaken because this requires large volumes of water and several sieves and large intestinal nematodes are less commonly a cause of gastrointestinal parasitic disease. Even without counting, however, the worms present can be identified if the caecum and colon are opened and the contents examined. In the caecum, *Trichuris ovis* and *Oesophagostomum* spp are easily differentiated by the characteristic *whipworm* appearance of the former. In the colon *Chabertia* is identified by its large buccal capsule, whereas *Oesophagostomum* spp taper at both ends. In heavy infections these two worms can also be found in the caecum.

SECTION B

ANTHELMINTICS AND ANTHELMINTIC RESISTANCE

Anthelmintics were first developed as therapeutic or tactical treatments, to reduce the clinical effects of gastrointestinal nematodes. They are now used for both tactical and strategic treatments, with the latter aiming to reduce the overall nematode population on the pasture and in animals (see Section C). Anthelmintics are integral to control programmes and, due to their relatively low price, will remain so even when the efficacy of individual compounds is reduced by anthelmintic resistance.

There are currently four groups of broad-spectrum anthelmintics registered in Australia, plus two compounds, naphthalophos and derquantel, which are used in conjunction with broad-spectrum drugs (Table 9.3). Resistance to most anthelmintic groups is common. Consequently, the use of combinations of anthelmintics, first suggested in 1988[91,96], is now standard practice, with 27% of anthelmintic treatments given to sheep in southeastern Australia in 2011 being a combination of three classes of anthelmintics.[97]

Long-acting anthelmintics

Most anthelmintics have a short duration of action with little or no residual effect. An exception is oral moxidectin, which has registered claims to be effective for at least seven days against newly ingested *T colubriformis* and at least 21 days against *T circumcincta*.

In Australia there are two forms of long-acting anthelmintics available: injectable moxidectin, which gives prolonged protection against new infections (for example, up to 91 days against *T circumcincta*), and controlled-release intraruminal capsules, which release a small dose of anthelmintic daily for 90-100 days. These capsules currently contain a macrocyclic lactone, either alone or in combination with a benzimidazole.

Both long-acting anthelmintic preparations can be used either strategically (for example, during the summer months in winter rainfall areas, such as western Victoria, when larval reinfection occurs sporadically but is generally low), or tactically (for example, to keep ewes or young sheep relatively free of gastrointestinal nematodes during the autumn and winter when larval infection rates are high).[15]

Anthelmintic resistance (AR)

A commonly accepted measure of resistance is the failure of an anthelmintic, or combination of anthelmintics, to reduce the worm egg count by at least 95% compared to an untreated control when measured 10-14 days after treatment, with an additional criterion being a lower

Table 9.3: Anthelmintic groups available for use in Australia, when they were released, the dose range of compounds within each group and their mode of action.

Group (year released)	Compounds (dose rate; year released)	Mode(s) of action on nematode
Benzimadazoles (from 1961)	Thiabendazole (44 mg/kg; 1961) Fenbendazole (5 mg/kg; 1974) Oxfendazole (5 mg/kg; 1975) Albendazole (7.5 mg/kg; 1976)	• Disruption of intracellular microtubular transport systems by binding to and damaging tubulin, preventing tubulin polymerisation and inhibiting microtubule formation • Inhibition and disruption of metabolic pathways
Levamisole (1968)	Levamisole (7.5 mg/kg)	• Stimulation of sympathetic and parasympathetic ganglia of worm (nicotinic receptor agonist) • Interference with carbohydrate metabolism
Macrocyclic lactones (1988)	Ivermectin (0.2 mg/kg; 1988) Abamectin (0.2 mg/kg)	• Paralysis and death through the increased release of GABA at pre-synaptic neurons, which prevents stimulation of the post-synaptic receptors
Milbemycin-related Macrocyclic lactones (1995)	Moxidectin (0.2 mg/kg)	• As above (increased GABA release) • Paralysis and death by binding receptors that increase membrane permeability to chloride ions, thus inhibiting nerve cell function
Organophosphates ('rereleased' in 1995)	Naphthalophos (12-47.5 mg/kg)	• Spastic paralysis through inhibition of cholinesterase in nerve synapses (an anticholinesterase)
Amino-acetonitrile derivative (2010)	Monepantel (2.5 mg/kg)	• Paralysis of parasite by inhibiting nicotinic acetylcholine receptors (a nicotinic receptor agonist)
Spiroindoles (2014)	Derquantel (2-deoxyparaherquamide) (2 mg/kg; combined with 0.2 mg/kg of abamectin in the sole commercial product)	• Flaccid paralysis through blocking acetylcholine, allowing expulsion of worm (a nicotinic receptor antagonist)

95% confidence limit around the estimate of reduction of <90%.[98] Although these may seem somewhat arbitrary figures, they were chosen because strategic treatments need to be at least this effective to capture their full benefits. Nevertheless, it is obvious that if a proportion of nematodes survive an anthelmintic treatment, then resistance is present, even with a 98% or 99% reduction in WEC. Consequently, a higher figure, as close as practicable to 100%, is now preferred for any combination of anthelmintics used for strategic treatments.

The frequent or incorrect use of anthelmintics, such as by underdosing or the continued use of an anthelmintic when resistance has developed, has hastened the development of AR.[99] In 2013, the results from over 100 Australian flocks found that resistance to the benzimidazole and levamisole anthelmintic groups was detectable in more than 95%; resistance to ivermectin was detectable in 85%; and resistance to abamectin was detectable in 75%. There was also resistance to moxidectin, the most potent macrocyclic lactone, in nearly 55% of flocks.[100]

Anthelmintics have a critical role in controlling and treating nematode infections, so resistance can reduce the efficacy of control programmes and increase costs — through lost production and the need for additional treatments, as well as the need to use more expensive compounds or combinations of anthelmintics. As an example, a study conducted in a winter-dominant rainfall area of Western Australia found that using a 65% effective anthelmintic compared with a completely effective one reduced wool and sale values in weaned Merino lambs by $2 and $4 per head, respectively.[101]

Because this is a significant potential loss, implementation of additional strategies to complement anthelmintics is a universal recommendation, usually referred to as *Integrated Parasite Management* (IPM).[3,102] The severity and consequences of anthelmintic resistance are more serious in summer rainfall areas, where *H contortus* is the dominant parasite. In other areas serious production losses are not an inevitable consequence of gastrointestinal parasitism. For example, in a study in western Victoria, six flocks with long-standing AR were able to maintain or increase their production intensity, measured by stocking rates and wool production per hectare, or increase their scale and profitability, despite the presence of AR.[102]

Testing for anthelmintic resistance

The most common test is a worm egg count reduction test (WECRT), also commonly called a drench resistance trial. The efficacy of an anthelmintic is assessed by comparing the WECs of treated sheep 10-14 days after anthelmintic treatment (*treatment groups*) to those of untreated sheep (*controls*). A number of single anthelmintics in each class, or combinations of anthelmintics, can be assessed at the same time.

The best sheep to use are weaners (3 to 6 months of age) that have never previously been treated with anthelmintic and have a relatively high WEC (at least 150 epg, preferably >300 epg).[98] At least 15 sheep are given each treatment and then faecal samples collected for individual WECs from 10 sheep in each group 10-14 days after treatment. The average WEC of each treatment group is calculated and compared to that of a group of untreated weaners, thus determining the average reduction in WEC for each treatment. Calculation of the upper and lower 95% confidence limits of each treatment describes the variability around the mean reduction, and helps refine the recommendations for anthelmintic use.[98,103] Ideally, a WECRT should be conducted every two to three years to determine which anthelmintics or combinations are effective and most appropriate to use as either strategic or tactical treatments.

The efficacy of a treatment is calculated according to the formula

$$\% \text{ Reduction} = 100 \text{ x } (1 - (X_t/X_c))$$

in which X_t and X_c are the mean egg counts for the treated and control groups, respectively.

A worked example of a WECRT is given in Appendix 2.

A simpler test is a pre- and post-treatment WEC on a mob of sheep (a *drench check*). This can indicate the efficacy of the current treatment but, as for the WECRT, is best performed on a mob of young sheep with a high WEC (>150 epg).

Total worm counts can also be undertaken 5 to 10 days after sheep have been treated with an anthelmintic drug or combination. This is a quick way to detect anthelmintic resistance, although it is relatively expensive, requiring the sacrifice of a number of sheep, and often has no untreated sheep as a comparison. For these reasons it is not usually undertaken unless there is an urgent need for a result and there are no suitable sheep available for a WECRT or drench check.

A number of *in vitro* tests of the efficacy of anthelmintics have been developed, including larval migration inhibition assays (for levamisole) and egg-hatch or larval development assays (for benzimidazoles and macrocyclic lactones). Apart from the DrenchRite™ larval development assay developed by CSIRO in the 1990s, which is no longer available, these tests have been used predominantly for research and are not offered commercially by diagnostic laboratories.

Refugia

Every sheep farm has a population of nematode parasites which include individuals represented by eggs, larvae or adult parasites. When the sheep flock or a portion of it is treated with anthelmintics, some of the parasite population is exposed to the drug while some of the population is not. The portion of the nematode population which is not exposed to anthelmintic treatment is considered to be *in refugia*.[104] This includes free-living stages of larvae on pasture, as well as immature and mature worms within sheep that have not been treated with anthelmintics.

The concept of refugia is important because the development of resistance within a nematode population is hastened if a low proportion of the total nematode population on a farm (both parasitic and free-living stages) is *in refugia*.[105,106,107] This was first demonstrated with a series of pen studies involving six generations of *Haemonchus* infections. Anthelmintic resistance developed rapidly when 0% or 10% of the population was *in refugia*, but more slowly when this proportion was 30% or 75%.[108] The rapid development of resistance to thiabendazole was associated with a zero or negligible refugia population on a farm at Seymour, Victoria, when ewes were treated with this anthelmintic and then grazed on pastures which had been grazed only by cattle for more than 10 years.[104]

In Mediterranean environments such as Western Australia, only 5-10% of the worm burden of lambs in winter is estimated to originate from contamination of pastures the previous spring, with the remaining 90-95% originating from nematode eggs deposited in the autumn.[15] This is because free-living larvae derived from eggs deposited during the late spring and early summer desiccate rapidly in this environment, which is characterised by short growing seasons (four to five months) and annual pastures with little or no ground cover during long, hot summers. Consequently, strategies such as delaying strategic drenching of adult sheep from December to March, or leaving a proportion of sheep untreated (targeted selective treatment), are recommended to preserve susceptible nematodes within some animals, hence increase refugia, in that strongly Mediterranean climatic region.[109]

In contrast, it has been shown that larval survival over the summer months in western Victoria, which has cooler summers, perennial pastures and a longer growing season (six to nine months), is much greater. An analysis of three separate studies found that from 30-58% of the larvae recovered from lambs during winter originated from larvae deposited during the previous spring and early summer period, before the most recent summer strategic treatments.[15] In this environment a proportion of infective larvae become trapped within the faecal mass but survive because they enter a state of reduced or minimal metabolism known as anhydrobiosis. When sufficient rainfall occurs the faecal pat is moistened and the L_3 rehydrate and resume their activity.[110] This proportion of over-summering larvae is consistently higher following dry summers, presumably because more infective larvae are released from faecal masses in a wet summer following rainfall events of 12-24 mm or more.[15,111]

At the other end of the scale, deliberately increasing refugia populations has also been demonstrated to delay the development of resistance in a cool, moist environment in New Zealand. In particular, strategies to maintain susceptible populations of nematodes within adult ewes at lambing, such as not drenching ewes during late pregnancy and lactation, are less selective for anthelmintic resistance than additional treatments given to lambs the following autumn.[106,112] Similar experimental investigations or modelling have yet to be undertaken in Australia but it would be surprising if comparable results were not found when prime lamb flocks are run under climatic conditions similar to those of New Zealand, such as those that occur in parts of southern Victoria and Tasmania.

SECTION C

EPIDEMIOLOGY AND CONTROL OF NEMATODE INFECTIONS

In Australia sheep are managed under diverse conditions, from semi-arid — where nematode infections are generally not a problem — to areas of higher rainfall (>450 mm) where control programmes are consistently needed. Within high rainfall zones, the distribution and amount of rainfall determines the patterns of infection and the predominant parasite species. *H contortus* and *T colubriformis* are of most concern in areas with summer-dominant rainfall, whereas *T circumcincta* infections assume more importance in winter and uniform rainfall zones and *T vitrinus* is consistently a problem in cold winter rainfall areas, such as western Victoria and Tasmania.

Because of this variation it is difficult to make universal recommendations about what constitutes a sustainable worm control programme. Some broad principles apply, and programmes have been developed from research undertaken from 1970-2000, principally by the CSIRO but also by universities and state Departments of Agriculture.[19,31,57,63,113] This was first summarised in farmer advisory programmes in the 1980s and 1990s for summer and winter rainfall areas in most mainland states (with programmes named Wormkill, Drenchplan, Wormplan, Wormcheck and CRACK). Subsequently, these have been updated and modified according to the experience of advisors and some additional research within each climate zone. The recommended programmes for the main areas are described at the advisory WormBoss website[114] and include a mix of

- grazing-management strategies, in order to create pastures of reduced worm risk for susceptible groups of sheep, such as recently weaned lambs or ewes around lambing, with maiden, aged or poor condition ewes the more susceptible ewe mobs

- strategic anthelmintic treatments (strategic drenches), the recommended timing of which varies considerably but is informed by previous research into the epidemiology of infections in the summer, uniform and winter-dominant rainfall areas. These areas can be further subdivided, with programmes described for two summer rainfall and seven winter rainfall areas at the WormBoss website

- additional *tactical* treatments given according to monitoring of worm egg counts (WECs). The suggested timing of WEC for monitoring in each area is given in detail at the WormBoss website but the important classes of sheep are peri-parturient ewes and young sheep after they are weaned

- strategies to monitor and manage anthelmintic resistance, including conducting a WECRT every two to three years and treating introduced sheep with an effective combination of anthelmintics (a quarantine drench)

- *Breed and feed* strategies, which include, if appropriate, the selection of rams for increased immunity (resistance) and decreased scouring using rams with negative Australian Sheep Breeding Values (ASBV) for WEC and dag score.[115] Sheep in better condition and grazing sufficient good-quality pasture have increased immunity and resilience to gastrointestinal nematodes. Consequently, meeting targets for body weight and growth of weaners, and condition score and feed availability for ewes, as defined by the Lifetime Ewe and weaner management programmes, will assist in worm control. These include a minimum weaning weight of 40% of adult body weight, maintaining post-weaning growth rate targets for Merino weaners and a condition score of 2.7-3.0 with adequate high-quality pasture (1200-1400 kg of green dry matter per ha) for late pregnant and lactating ewes.[116,117,118,119]

To be effective and profitable in the longer term, say 30 or more years, programmes should avoid practices that encourage rapid selection for anthelmintic resistance, which include

- drenching too frequently

- underdosing (check the calibration of drench guns and dose to the heaviest sheep in the mob)

- using less than highly effective anthelmintics for strategic treatments

- drenching sheep onto paddocks known to have low residual populations of worm larvae (low refugia).

The last point has been identified as a risk factor for increased anthelmintic resistance in South Africa, where *H contortus* is a predominant parasite, and for *T circumcincta* in Western Australia.[105,109,120] Care must be taken not to uncritically extrapolate the need for modified programmes to other areas. For example, strategies such as *targeted selective treatments* or leaving a proportion of sheep untreated at the strategic drenches as recommended for flocks in Western Australia, which has a pronounced Mediterranean climate, are not appropriate for areas of southeastern Australia which have longer growing seasons, shorter, cooler summers and much higher refugia populations.[15,121,122]

Finally, management decisions can profoundly influence gastrointestinal nematode infections within a flock. These decisions include[123]

- stocking rate

- timing of reproduction, especially joining and weaning

- pasture resting, including rotational grazing

- alternate grazing by cattle or immune sheep

- use of fodder crops, proportion of new pastures and aftermath from hay or silage

- flock structure, especially the proportion of young stock and reproducing ewes

- selection of pasture species.

Patterns of infection — winter rainfall areas

In winter rainfall areas, pasture larval availability is closely associated with rainfall. Populations of infective larvae increase in late autumn, peak in winter and then decline in late spring, with low but sporadic numbers available during the summer (Figure 9.4). At any time the infectivity of pasture depends upon the prevailing weather conditions, the timing and number of nematode eggs deposited previously and the proportion of nematode larvae present as infective larvae.[12] Consequently, the availability of infective larvae varies considerably, both within and between years.[13]

In western Victoria, the availability of *T circumcincta* and *Trichostrongylus* spp larvae peaks when the mean maximum temperature is below 15.5 °C and relative humidity above 60%, typically between June and October.[31,57] Larval development slows in cooler weather but large numbers of infective larvae derived from eggs deposited during late summer and autumn can persist through winter, spring and early summer. Most infective larvae found on pasture during the winter and early spring originate from eggs deposited in late summer and autumn, whereas those present after mid- to late spring may originate from eggs deposited during either the autumn, winter or early spring.[13]

The availability of larvae on pasture declines rapidly during late spring and early summer (October to December). This is due to the dilution effect from increased pasture growth, as well as the rise in maximum daily temperature and decrease in relative humidity.[31,57] Infective larvae of *T circumcincta* and *Trichostrongylus* spp require films of moisture to migrate from faeces to pasture, so their availability on pasture is related to the amount of moisture present within the pasture sward. Larval availability in summer is therefore heavily dependent on rainfall and dew.[12,111] Rainfall events that lead to significant numbers of *T circumcincta* and *Trichostrongylus* spp larvae migrating onto the sward in November do not have the same effect in January, when more rainfall is needed to make larvae available to grazing sheep, presumably because higher temperatures dry the sward more quickly.[111] For example, only 3.5% of

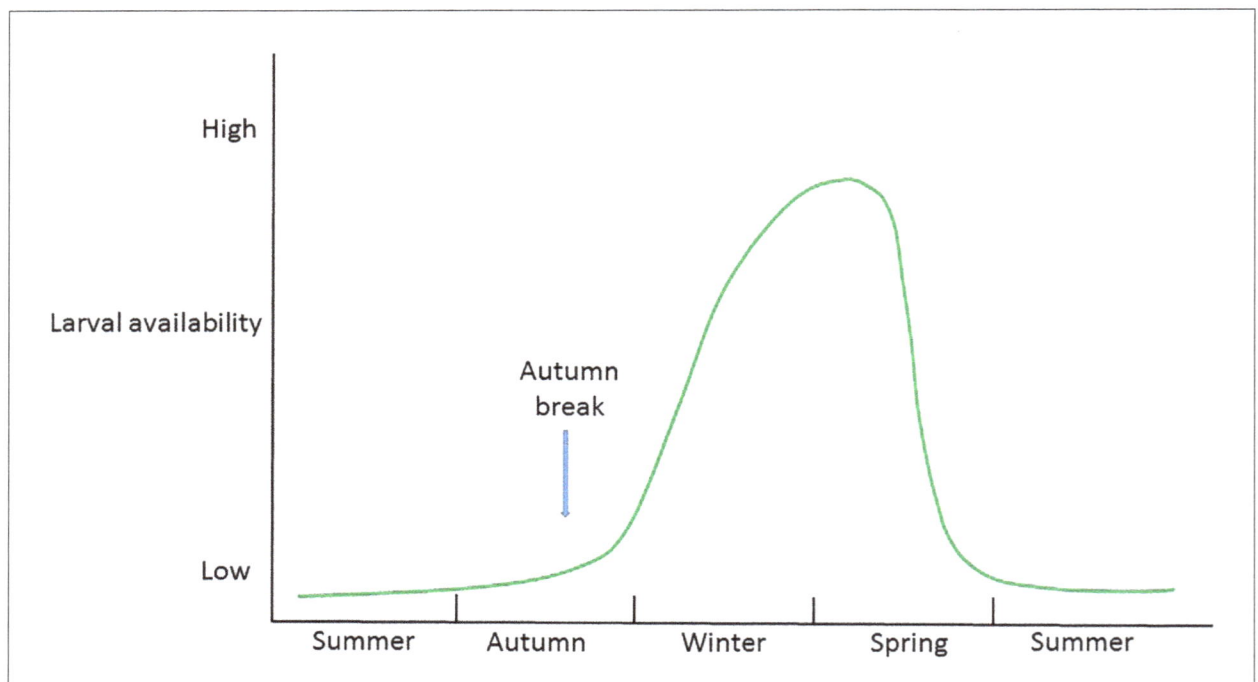

Figure 9.4: The pattern of availability of infective larvae (L$_3$) in a winter rainfall area.
Source: KA Abbott.

T circumcincta eggs deposited onto a dryland pasture in January were recovered on pasture and in soil, with most recovered 126 days after faeces were deposited.[11] Eggs deposited on pasture in January make very little contribution to worm burdens until the late autumn or winter.

Control programmes — winter rainfall areas

For the winter rainfall areas, the WormBoss site describes five components that can be integrated into a control programme:[114]

- strategic anthelmintic treatments appropriate to each region, including to lambs at weaning

- monitoring of WECs before additional tactical treatments. The core programme varies according to rainfall and region, but for the higher rainfall areas it includes ewes just before lambing and/or at lamb marking, all classes of sheep before the second summer treatment in February, and young sheep at four- to six-week intervals following the autumn break until early spring

- grazing management to create low worm-risk paddocks (discussed below)

- *breed and feed* programmes for enhanced immunity and resilience. Using rams with negative ASBV values for WEC and dag score is appropriate for many flocks in higher rainfall areas. Currently there are relatively few rams with breeding values for dag score and so an alternative is the phenotypic culling of rams or ewes with excessive dag.[28] Meeting target weaning weights, weaner growth rates and ewe condition scores specified by weaner and lifetime ewe management programmes is also appropriate[117,119]

- monitoring and management of anthelmintic resistance. Conduct a WECRT every two to three years, avoid unnecessary treatments and use effective combinations of short-acting treatments. If long-acting treatments are used, monitor WECs during the expected period of protection from reinfection (for example, at 3, 6 and 9 weeks) and give a *tail-cutter* treatment with a different short-acting anthelmintic group if significant WECs are detected.

The strategic double- or single-summer treatment programme aims to decrease the contamination of pastures with worm eggs in late summer and early autumn to reduce populations of infective larvae on pastures the following winter.

Two treatments were originally suggested for the western Victorian environment because reinfection can occur after the first treatment when sufficient rainfall (12-24 mm) enables the translation of L_3 from faeces onto pasture.[111] The first treatment decreases pasture contamination with worm eggs in the late spring and early summer and so is given at or just after pasture senescence (*haying off*), when mean ambient temperatures increase and average humidity decreases. This is typically in November or December.[31] The second summer treatment, if required, is given in mid-summer (early February) to reduce deposition of eggs in the late summer and autumn and reduce the peak availability of L_3 the following winter.[13,31] The optimum timing of these treatments can vary considerably with seasonal conditions, particularly in drier areas where a second treatment is not consistently required. Treatments are usually given at a convenient time, hopefully close to the time best suited for worm control, such as when sheep are mustered for weaning of spring-born lambs in November or December. The decision about a treatment in February should depend upon the results of WEC monitoring of a cross-section of mobs within a flock.

'Smart grazing' — integrating grazing management with summer treatments

In some high winter rainfall areas, unseasonal summer rainfall occurs often and sheep are consistently reinfected after the summer treatments. To extend the period when the deposition of eggs onto pastures is low or negligible, sheep can be either treated with long-acting anthelmintics[124] or grazing management can be integrated with the short-acting strategic treatments.[125]

The latter strategy, termed *smart grazing*, requires the grazing of selected paddocks at around two or three times the usual set-stocking rate for up to, but no longer than, four weeks after each strategic summer treatment using an effective anthelmintic. The four-week period is only slightly longer than the pre-patent period and so sheep are ingesting and removing L_3 but depositing few if any worm eggs onto the pasture because their WECs are negligible when they are removed (from 0-5 epg).[125] Consequently, these paddocks have 50-90% fewer L_3 the following winter than would be the case without treatment. This confers significant production advantages to weaner sheep in a self-replacing wool flock when they graze these paddocks, with significantly lower WECs, 7% (3.2 kg) increased body weight in October and 14% (255 g) more clean wool.[125]

Using this system, nematode eggs that would otherwise have been deposited on the smart-grazed paddocks during summer and autumn are effectively transferred from a small area of the farm (typically 15-20%) to other paddocks. This occurs over a relatively short period, and so *smart grazing* is unlikely to increase this contamination to the detriment of other sheep. Overall, the production advantages make it a simple and effective strategy for preparing low-risk paddocks for vulnerable classes of sheep. This could include joined hogget ewes in a prime lamb flock, although this system has not been tested experimentally.

Other forms of grazing management

Grazing cattle on pastures can considerably reduce pasture contamination for sheep because, apart from *T axei*, sheep and cattle are infected by different species of gastrointestinal nematodes. In the initial studies examining the benefit of cross-grazing, sheep and cattle were interchanged at six-monthly intervals in January and July because most L_3 derived from sheep nematodes would have died or been ingested by cattle during this time.[126] A significant reduction in the number of L_3 of *H contortus* and *T colubriformis*, but not *Nematodirus* spp, can be achieved by grazing cattle for as little as 6-12 weeks. Counts of *T circumcincta* were also significantly reduced when cattle grazed for 12 weeks.[127] *H contortus* can infect young cattle, so, in summer-dominant or uniform rainfall areas where this is a major parasite or on farms on which higher proportions of *T axei* infection are maintained in sheep[128], yearling or adult cattle should be used to provide safe pastures for sheep, rather than cattle under 1 year of age.[129]

In order to provide a sufficient area of low-contamination pastures during the winter for weaner sheep in a self-replacing wool flock, at least 20% of the grazing pressure on the farm must be provided by cattle. A frequent objection from producers is that cattle are disadvantaged when interchanged with sheep because the sheep pastures are much shorter than that required for acceptable levels of pasture intake by cattle. This can be partially overcome by moving the sheep about four weeks before the cattle to allow some regrowth.

Other forms of grazing management can also reduce pasture contamination with worm eggs, including grazing crop stubble paddocks or hay aftermath. Such pastures are of low infectivity because of the long period during which they have not been grazed by sheep and the relatively poor survival of free-living stages of nematodes after sowing of crops or tilling. Because such pastures may have extremely low populations of L_3, a *shift and treat* strategy, in which untreated sheep are allowed to deposit nematode eggs on stubbles or other areas with low refugia before the flock is treated, may reduce the risk of selection for anthelmintic resistance.[105,107] The risk, however, may not be high because the developmental success of eggs when stubbles are being grazed during summer and autumn in winter rainfall areas is extremely low.[11]

Modified strategies for areas with a Mediterranean climate

The summer treatment strategy, originally proposed for western Victoria, has been modified for the Mediterranean rainfall areas of Western Australia (WA) following concern over the high prevalence of resistance by *T circumcincta* to the macrocyclic lactone anthelmintics in these areas.[130] Strategies proposed include delaying the summer treatments for adult sheep until March[109] and *targeted selective treatment* by leaving a proportion of sheep untreated.[120] The aim is to increase the winter population of L_3 *in refugia* by depositing an increased proportion of eggs derived from worms not exposed to the most recent anthelmintic treatment close to the autumn break.

For targeted selective treatment (TST), sheep can either be randomly selected (for example, drench only 80% of each pen of sheep when yarded for treatment), or selected based on criteria that indicate they may have above-average worm burdens or may be more susceptible to the effects of infection. A well-known and proven example is the FAMACHA system for treating *H contortus* infections in South Africa, where conjunctivae are examined and only clinically anaemic sheep are treated.[131] This procedure is quite labour-intensive and is not practical for large flocks in Australia, nor applicable to areas where scour worms are the predominant parasites.

Within any mob of sheep, the distribution of WECs is over-dispersed, with most sheep having counts lower than the mean and only a relatively low proportion of sheep (~10-20%) having very high counts.[132] Using individual WECs to detect and treat sheep with high counts is neither practical nor cost-effective, so alternative criteria are needed for TST. For large flocks these include treating to maintain a target mean WEC, or treating sheep with low body weight or low condition score. The regular weighing and drafting of lighter sheep from large mobs require additional handling and, preferably, automated drafting equipment, so condition score is generally a more convenient and practical selection criterion for TST than body weight.[133]

In WA, a target WEC of 200 epg for ewes is used as a level below which worm burdens are not expected to cause significant production loss.[109] Experimentally, this guide was used in an examination of a TST programme on three farms.[120] Unfortunately the study was conducted for only 12 months, during which no clinical parasitism occurred in the TST mobs and only relatively small differences in body weight gain and greasy wool production occurred on two of three farms (2 kg and 300 g, respectively). The experimental sheep were in excellent condition score throughout the study (from 3⁻ to 4⁻) and so care is needed when extrapolating these results to sheep in lighter condition. This was demonstrated when TST ewes with a low condition score (CS) before lambing were at least three times more likely to fall to a critically low CS (<2.0) at lamb marking and weaning.[133]

Results from computer modelling studies using 20 years of climate data have generally supported the concept of leaving some sheep untreated and delaying summer treatments for adult sheep until March in WA.[121,122,134] While these strategies delay the development of anthelmintic resistance, the studies highlight that some reduction in the efficacy of worm control can occur. This was more pronounced in a higher rainfall area of WA[134] so these programmes should not be universally recommended for the high winter rainfall areas of southeastern Australia.[15]

Patterns of infection and control programmes — summer rainfall areas

The seasonal patterns of infection in summer rainfall areas differ between the three main genera of nematodes (Figure 9.5). For *H contortus*, numbers of infective larvae on pasture reach a peak in mid- to late summer, then decline rapidly in autumn and winter to be negligible in spring.[19] In autumn an increasing proportion of ingested larvae become arrested as early-fourth-stage larvae in the abomasal mucosa. In young sheep these resume development in spring but in older sheep a large proportion of larvae resuming development are rejected. There are low populations of over-wintering larvae; then, with increasing temperatures and moisture, the population of L_3 increases rapidly to peak again in late summer.[12]

Infection with *T circumcincta* has similar peaks in summer and spring, with most spring larvae arising from contamination the previous late summer and autumn.[19] Worm burdens of lambs closely follow larval availability. For *Trichostrongylus* spp, the larval population is highest in summer with a much smaller rise in spring but worm burdens of untreated lambs continue to increase until about 12 months age (Figure 9.5).

For the summer rainfall (tablelands and slopes) region, the WormBoss website describes five components to be integrated into a control programme.[114]

- Practise grazing management to create low worm-risk paddocks: for example, to prepare pastures of reduced risk for spring-lambing ewes, contamination with nematode eggs must be minimised in the six months before lambing. In March and April this can be achieved by spelling paddocks, grazing with cattle or grazing with sheep for up to 21 days after an effective anthelmintic treatment (an adaptation of the *smart grazing* concept).[22] There are then no grazing restrictions from May to August if maximum daytime temperatures are consistently <18 °C but, if it is warmer, the strategies described for March and April are used. Paddocks for weaning lambs in summer should have reduced contamination for three months prior to weaning.

- *Breed and feed* to enhance immunity and resilience: use rams with a negative WEC ASBVs and maintain good nutrition, such as a target condition score of 2.7-3.0 recommended by the lifetime ewe programme.[119]

- Monitor WECs before giving anthelmintic treatments: suggested times are before shearing for all classes of sheep and, for ewes, before lambing, at lamb marking and before weaning. After weaning, young sheep should be monitored at four- to six-week intervals in summer and six to eight weeks during winter.

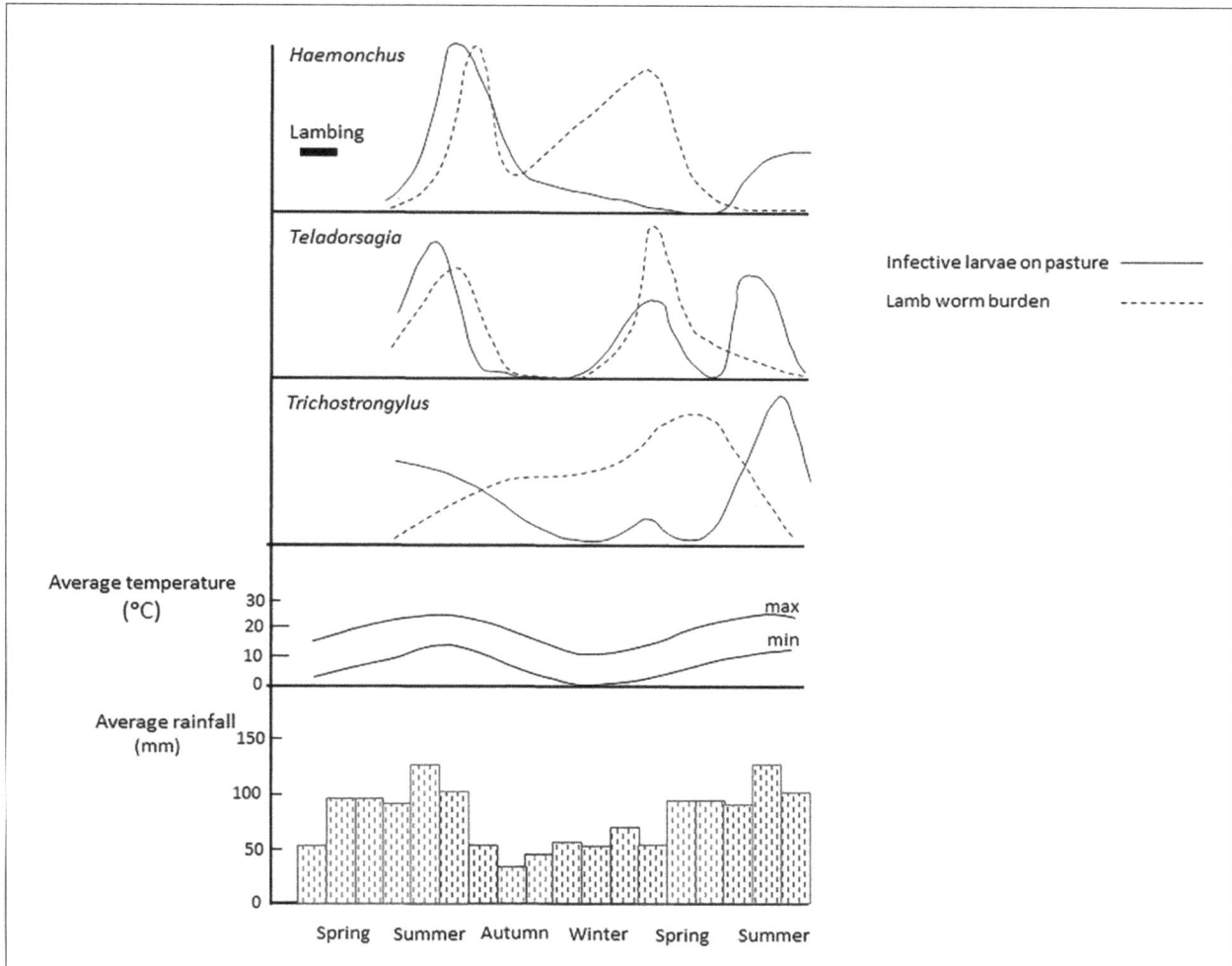

Figure 9.5: Seasonal patterns of the availability of infective larvae (L₃) and worm burdens in spring-born lambs for *Haemonchus, Teladorsagia (Ostertagia)* and *Trichostrongylus* spp in a summer rainfall area. Source: Redrawn by KA Abbott. Reproduced with permission from original images, CSIRO.

- Give anthelmintic treatments (or Barbervax® for weaners) at the recommended times: the treatment of ewes before lambing is routinely recommended, although this has been shown to increase selection for anthelmintic resistance in New Zealand.[106,112] Other standard treatments are to lambs at weaning and when sheep show signs of worm infections (anaemia, deaths or elevated WECs). The WormBoss site has a *Drench Decision Guide* to assist producers make decisions about when to treat.

- Monitor and manage anthelmintic resistance: conduct a drench resistance test every two to three years, avoid unnecessary treatments and use effective combinations of short-acting treatments. In general, avoid long-acting products or, if they are used at *high-risk* times, monitor treated sheep at three-weekly intervals and administer a *tail-cutter* treatment with a different short-acting anthelmintic group.[114]

The programme suggested for summer rainfall areas with lower or more variable rainfall (slopes and plains of NSW) differs slightly. The period to prepare lower-risk pastures is shorter (two to five months) and the strategic anthelmintic treatments given in October-November vary according to pasture conditions. If pastures are green and actively growing, all sheep should be treated or, if dry, only sheep <18 months of age should be treated. Monitoring of WECs is recommended four to six weeks after treatments or after any significant rainfall events (≥20 mm with follow-up rains of ≥10 mm) to assess whether further treatments, or possibly long-acting treatments, are needed.[114] In addition to *H contortus*, sporadic infections with scour worms may occur, with more breech soiling (dags) in central and southern NSW compared to the northern Tablelands.

Grazing management

In addition to cattle-sheep interchange and using the ~21-day pre-patent period following treatments with an effective anthelmintic to prepare lower risk pastures for lambing ewes[22], intensive rotational grazing is an effective strategy for controlling haemonchosis in wet tropical environments.[135] Here the warm, wet conditions favour rapid and continuous egg hatching and larval development, hence extremely high death rates of infective larvae which survive on heavily contaminated tropical pastures for no more than four to eight weeks following a peak one week after contamination. Consequently, grazing intervals longer than about 30 days achieve high mortality of *H contortus* L_3 and significantly reduce reinfection, such as that achieved by grazing each paddock for four days in a simple 10-paddock rotational system.[135]

In addition, intensive rotational grazing systems with grazing periods of five days and rest periods of around 100 days have been confirmed to assist in the control of *H contortus* in cooler summer-rainfall-dominant areas, such as the Northern Tablelands of NSW.[136,137] Overall, all sheep (lambs, ewes and hoggets) had a lower mean WEC (326 vs 536 epg), higher haematocrit, significantly lower proportion of *H contortus* in faecal cultures (60 vs 81%) and longer time between anthelmintic treatments (144 vs 78 days) compared to sheep managed under a typical grazing system.[136] These benefits are mediated by reduced larval challenge and were confirmed by a six-year farmlet study, with the WEC and treatment frequency of intensively grazed sheep significantly less than for a typical grazing system (444 and 3.1 vs 1122 epg and 4.3 treatments/year, respectively).[137] Improved host nutrition did not provide more effective worm control but, compared to a typical grazing system, did allow a 48% increase in stocking rate without any adverse effects on worm control by the end of the study.

Patterns of infection and control programmes — uniform rainfall areas

Most research studies relevant to the uniform rainfall zone of eastern Australia have been at the CSIRO Ginninderra Research Station near Canberra, which has an average rainfall of about 690 mm evenly distributed throughout the year, or Badgery's Creek, 50 km west of Sydney.[113,138,139] *T circumcincta* and *Trichostrongylus* are the most economically important parasites in these areas but outbreaks of haemonchosis occur in wet summers.[12] In the drier pastoral areas gastrointestinal nematodes are generally unimportant but outbreaks of clinical parasitism can occur in wet years.

The patterns of infection are generally a composite of summer and winter rainfall patterns.[113] In years when there is insufficient rainfall to maintain pasture growth, the patterns of infection for *T circumcincta* and *Trichostrongylus* spp are similar to winter rainfall areas. The period with dry pastures during summer is generally short (often four to six weeks), allowing *Haemonchus* populations to accumulate; and deaths from haemonchosis can occur between March and May in some years.[12] With cooler conditions the availability of *Haemonchus* L₃ decreases to very low numbers in late winter, whereas numbers of *T circumcincta* and *Trichostrongylus* spp larvae peak during winter.

Control programmes are based upon the principles described for the high winter rainfall areas; that is, double-summer treatments, grazing management to provide lower risk pastures for weaners and worm egg count monitoring to assess when additional treatments are needed.[12] Dry periods during summer are much shorter and so increased monitoring for *Haemonchus* infections and additional treatments may be needed in years when they become unacceptably high. Molecular tests offer a quicker and more accurate alternative to larval culture and differentiation and so are useful adjuncts for monitoring nematode infections in uniform rainfall areas.[95]

The benefits of integrating anthelmintic treatments with grazing management, particularly cross-grazing with sheep and cattle, are greater than for winter rainfall areas due to the wetter summers.

Near Canberra, young sheep given a drench at weaning and then moved three times in the following 12 months to pastures grazed by cattle grew as much wool as sheep kept virtually worm-free by drenching at two-weekly intervals.[129] The reduction in total worm count was greater for *Trichostrongylus* spp than for *T circumcincta*, which is consistent with other grazing management schemes used to control nematode infections, such as *smart grazing*, and estimates of increased refugia populations for the latter parasite.[15,125]

Selection of resistant sheep

The variation in parasite burdens and WEC between sheep within a mob has a genetic basis, with differences mainly due to acquired resistance rather than innate resistance to infection.[140] Sheep must be exposed to infection before acquired resistance is expressed, but genetically resistant sheep are able to mount an effective immune response more rapidly following this initial exposure. The effects of acquired resistance are subtly different between worm species — for *T circumcincta*, it reduces the size and fecundity of adult worms, arrests increased proportion at the L4 stage and prevents establishment of ingested L₃[141,142] whereas, for *T colubriformis*, it prevents establishment of ingested L₃ and causes the expulsion of adult worms.[143,144]

There is substantial cross-resistance between nematode species, with sheep selected specifically for resistance to *T colubriformis* having substantial resistance to *T rugatus*, *T axei* and *T circumcincta* and some resistance to *H contortus*, whereas sheep selected for resistance to *H contortus* also exhibit resistance to *T colubriformis*.[145] Ewes with enhanced resistance have a lower rise in WEC at lambing time and this can contribute to an epidemiological benefit through decreased contamination.[146,147,148] For example, lambs born to resistant ewes in the Rylington Merino selection demonstration flock in WA were up to 22% heavier at weaning[149] and significantly fewer L₃ were on pastures after being grazed by resistant Romney lambs after weaning in New Zealand.[150]

Overall estimates of the heritability of resistance to internal parasites from large databases are remarkably similar: 0.28 ± 0.072 for Merinos[88] and 0.28 ± 0.02 for NZ Romneys.[151] This moderate heritability, and a lack of serious adverse correlations with production traits for Merinos[152], make genetic progress for parasite resistance a realistic goal for ram breeders. Consequently, in high rainfall areas breeding sheep with enhanced resistance, using WEC as an indicator trait, is a useful and sustainable component of an integrated control programme[153], and Australian Sheep Breeding Values (ASBVs) for WEC are available for an increasing number of rams at the Sheep Genetics Australia website.[115] A slight increase in breech soiling (dag) was reported in initial analyses of the resistant Rylington sheep in WA, although this effect has since been shown to be small and often insignificant.[154] Selection that incorporates these genetically independent traits with selected production traits in an index will be the most efficient way to achieve productive, resistant sheep that also have reduced scouring.

Some negative correlations with production traits have been recorded in resistant lines of sheep in New Zealand, such as significantly decreased post-weaning growth rate and increased breech soiling of Romneys[151], and decreased growth rate and yearling fleece weights in resistant Perendales.[155] Consequently, selection for increased resilience (the ability to maintain production in the face of nematode challenge) has also been investigated. These traits appear independent of WEC but tend to have lower heritability (for example, 0.13 ± 0.02 for age at first drench), so progress should be relatively modest.[151] Nevertheless, experimental lines of Romneys established in NZ have similar WECs to unselected sheep but were 4.5 kg heavier and had less dag at 6 months of age.[151] Similar selection lines have not been established in Australia, although one study did identify that resilience to *Haemonchus* infections in Merinos, judged by greasy fleece weight, liveweight gain and haematocrit, involved a suite of moderately correlated traits.[156]

Finally, alternative methods of identifying sheep with enhanced immunity, such as using salivary swabs to assess the IgA response to nematode carbohydrate larval antigen after weaning (*CarLA*), are under experimental investigation.[157] These tests are not yet developed to the point of being a realistic alternative to WEC.

Prime lamb systems

The control programmes for prime lamb flocks are generally based upon recommendations from epidemiological studies in Merinos. There are, however, a number of important differences between prime lamb flocks and self-replacing wool flocks, and these differences can influence the choice of an integrated control programme for a prime lamb flock. They include

- different flock structures, with a greater proportion of ewes (often all-ewe flocks) compared to self-replacing Merino flocks
- breeds (often composite or cross-breeds) which are generally more resistant or resilient to internal parasites
- a different time of lambing, typically four to six weeks earlier than for Merinos[158]
- the use of different pastures, including forage crops such as rape, chicory and *Lotus* spp for finishing lambs
- reduced opportunity for providing low-risk pastures, and potentially reduced benefits from the anthelmintic treatment of periparturient prime lamb ewes compared to Merinos
- later weaning, hence uncertainty as to when to treat or wean suckling lambs older than 12 weeks old that are still grazing contaminated pastures.

Despite a dramatic increase in prime lamb production since the collapse of the reserve price scheme for wool in 1990 and dramatically increased exports of lamb, very little was known about the worm control programmes being used in specialist prime lamb systems until a study of 16 flocks in southeastern Australia, from 2004-08,[159] found that control measures generally included

- anthelmintic treatments during summer, although the timing of these was not always consistent with the most effective timing for that area

- regular use of WECs to make decisions on tactical anthelmintic treatments

- treatment of ewes two to four weeks before lambing

- treatment of lambs less than 8 weeks of age, which was generally shown to be ineffective

- treatment of lambs at weaning, although the timing of weaning was quite variable. The more profitable enterprises gave one or at most two anthelmintic treatments to lambs before sale.

Indicators of whether an anthelmintic treatment was successful in improving production were the efficacy of the anthelmintic on the farm, and whether or not animals could be moved onto lower-risk pastures after treatment, which was difficult to achieve on most farms. Suboptimal nutrition limited lamb growth more often than internal parasites, particularly as pasture quality decreased as temperatures increased in the spring and early summer.[159]

Subsequently, a major study was undertaken on 17 farms at four locations in eastern Australia.[160] The performance of sheep managed under either *integrated parasite management* (IPM) or *typical* worm control programmes specific to each region were compared with that of worm-suppressed twin-bearing meat-breed ewes and their lambs. Adoption of IPM resulted in lower WEC and fewer treatments, with the advantages more pronounced in the Northern Tablelands of NSW where *Haemonchus* infections are prevalent. For example, in this region the mean WEC and average number of treatments for ewes were 766 vs 931 epg and 4.5 vs 5.5 drenches/year, for IPM and typical programmes, respectively.[161]

Over all the regions, meat-breed ewes in good condition and grazing sufficient improved pastures were highly resilient to infection with gastrointestinal nematodes, with no differences between marking percentage or ewe mortality on IPM or typical farms.[160] There was a slightly greater loss of ewe liveweight on the IPM farms in the Northern Tablelands (-0.11 vs -0.01 kg) but this was of little practical significance.

The WEC of lambs at weaning was also lower for the IPM compared to typical farms on the Northern Tablelands (159 vs 322 epg).[161] Anthelmintic treatments were often given to lambs in all regions before weaning, but there was no benefit from these treatments if the lambs were growing at more than 200 g/day.[160,162] By contrast, in the two months after weaning, growth was significantly reduced by nematode infections (by 0.5 kg or 6.5%; 7.7 vs 7.2 kg for worm-suppressed and not-suppressed groups, respectively), even if lambs were grazed on low-risk pastures.[162]

In the Victorian flocks, ewes (+5.7 kg) and lambs (+5.9 kg) were significantly heavier on IPM compared to typical farms. It was proposed that factors other than worm control, such as the availability and quality of pasture and grazing management, contributed to this difference.[163] In this region, nutrition was more important for optimum lamb growth than any effect of internal parasites, with lambs having unsatisfactory weight gains when pastures started to senesce and their digestibility and quality decreased.

SECTION D

LIVER FLUKE (*FASCIOLA HEPATICA*)

Fasciola hepatica is endemic over large parts of eastern New South Wales, the Murray basin, central and eastern Victoria and northeastern Tasmania, but its presence is restricted by the distribution of the host snails. It occurs to a limited extent in South Australia but is not found in Western Australia. Sheep and goats are more susceptible than cattle but horses, pigs, rabbits and marsupials can also become infected.[164,165]

The production penalties from liver fluke are not as clearly defined as for nematodiasis but, in moderately or heavily infected flocks, they are probably similar in terms of reduced growth rate, body weight and wool growth.[165,166] A recent estimate of the overall cost to the Australian sheep industry was $25m.[2]

In contrast to the major gastrointestinal nematodes, sheep do not develop immunity against reinfection with *Fasciola hepatica*.[167]

Life cycle of *Fasciola hepatica*

The host snails for *Fasciola hepatica* in Australia are *Australopeplea* (formerly *Lymnaea*) *tomentosa* and *Pseudosuccinae* (formerly *Lynmnaea*) *columella*. The aquatic habitats of these snails are temporary or permanent springs which expand in wet years and provide refuge in long, dry spells. Larger creeks, rivers and lakes are not preferred habitats but adjacent backwaters and swamps provide suitable conditions.[164,165] When *P columella* was introduced into Australia there was concern that this could increase the range of *F hepatica* infection[168], as occurred in the North Island of New Zealand[169], but so far these concerns have not been realised.

Under ideal conditions of warmth (26 °C) and moisture, eggs of *F hepatica* hatch to form larvae (miracidia) within 10-12 days. Low temperatures (<10 °C) slow or prevent development, but eggs can remain viable for months. In Australia, hatching typically occurs in 21 days in summer but can take around 90 days in winter. The miracidia need sufficient moisture and temperatures above 5 °C to survive and must penetrate a host snail with 24-30 hours. Development and multiplication within the snail (miracidium→sporocyst→redia→cercaria) require temperatures over 10 °C, with a single miracidium capable of generating up to 4000 cercariae. Depending on temperature, cercariae emerge five to seven weeks after infecting the snail and encyst as metacercariae on herbage. After the metacercariae are ingested by grazing livestock, they migrate through the intestinal wall to the peritoneal cavity and then liver. The immature fluke migrate through the liver parenchyma for several weeks before reaching the bile ducts and becoming sexually mature fluke. Infections can become patent about eight weeks after infection with metacercariae, although some flukes may take several more weeks before arriving in the bile ducts and starting egg production. Individual flukes can live as long as 10 years and can produce up to 50 000 eggs per day.[170] This, plus the ability to multiply in the intermediate host snail, underlines the considerable biotic potential of liver fluke.

Clinical signs of liver fluke infection

Acute fascioliasis occurs five to six weeks after the ingestion of large numbers of metacercariae due to the liver damage caused by masses of young flukes. The lesions in the liver parenchyma are mainly traumatic but there is an element of coagulation necrosis associated with the tracts which may be related to toxins excreted by the flukes. If unusually large numbers of flukes invade the liver over a short period, the resulting tissue damage may cause acute hepatitis.[171] The migration of immature flukes also predisposes sheep to the development of infectious necrotic hepatitis (*Black disease* — see Chapter 14) by providing a suitable anaerobic environment for the sporulation and multiplication of *Clostridium novyi* spores. Thus, maintaining an effective clostridial vaccination programme is important on farms with high populations of the intermediate host snails and recurrent liver fluke infestations.

Subacute fascioliasis is associated with the ingestion of a lower dose of metacercariae. This causes similar but less severe hepatic damage and results in a syndrome reflecting the reduced activity of the liver, including reduced synthesis of albumin.

Chronic fluke disease occurs when small numbers of metacercariae are ingested over longer periods. Thus, acute and subacute disease may not occur but adult flukes in the bile ducts can cause cholangitis, biliary obstruction, fibrosis and anaemia. An adult fluke ingests up to 1 mL blood per day and burdens of 100 fluke in sheep are potentially lethal. Submandibular oedema (*bottle jaw*) and ascites occur in long-standing cases.

Epidemiology of liver fluke

Moisture and temperature are the key factors that influence the life cycle of *F hepatica*.[164,165] The development of larval stages in snails is slow and there is a higher mortality of infected snails over winter than non-infected ones. Consequently, eggs which survive on pasture over winter are a more important source of populations of metacercariae in spring than those which hatch in autumn and provide miracidia for over-wintering snails. The latter contribute to an early spring availability of metacercariae. Metacercariae can survive for long periods over winter and remain available and infective to grazing animals for 10 weeks or more, whereas they survive for only around a week in summer.[172]

As a consequence of the synchronous hatching of over-wintered eggs and improved environmental conditions for development within snails in early spring, availability increases from late spring and builds up until April in cold areas and until May in warmer areas. Thus, outbreaks of fascioliasis tend to occur during the summer-autumn period in southern winter rainfall regions and during winter in the northern summer rainfall regions. Entire paddocks may be suitable habitats for snails after irrigation so, in the warmer irrigation areas, outbreaks of fascioliasis may occur at any time from spring to late autumn. This seasonal pattern, and the occasional outbreaks of unexpectedly severe fluke disease which occur, are probably largely determined by sheep grazing swampy areas of paddocks which they would normally avoid when other areas of the paddock are green.[165]

Control programmes for liver fluke

The control of *Fasciola* is based on two strategies. First, strategic treatments are given before the main breeding season of the host snail to reduce the number of miracidia that can infect snails. The main breeding season is spring, and so a treatment in late winter or early spring will remove adult fluke and prevent pasture contamination for 4 to 12 weeks at this time, depending on the flukicide used. If pastures are highly contaminated and sheep routinely become reinfected from over-wintering metacercariae and sporocysts in snails, an additional treatment in late spring or early summer may be necessary to remove fluke infections acquired since the strategic treatment in winter.[165]

A secondary snail breeding season occurs in autumn, so treating sheep with a flukicide before then (that is, February-March) will remove any remaining adult flukes and reduce pasture contamination with eggs, hence reduce over-wintering populations of cercaria and reduce the infectivity of spring pastures. With continued treatments and more effective flukicides the infectivity of pastures will decline and the additional treatments can be gradually omitted.

Second, on farms where paddocks containing the lymnaeid snails are known and snail-free paddocks also exist, a system of rotational grazing can be used. Sheep graze the fluke-infected paddocks for a maximum of nine weeks, during which time they ingest metacercariae but deposit no eggs on pasture. They are then removed to a snail-free pasture where they may deposit fluke eggs which have no chance of completing their life cycle. After a period of 2 to 12 weeks, depending on the efficacy of the treatment against immature flukes, sheep can be treated and returned to the snail-infested pasture for nine weeks. This rotation continues, especially at the time of year when metacercariae are most available, and can gradually reduce the infectivity of the snail-infested pasture.[164,173]

Control or elimination of the intermediate host snails have been attempted with a range of molluscicides, such as copper sulphate, but are rarely successful.[174] The snails are capable of rapid multiplication so an incomplete removal of the snail population only temporarily reduces their numbers. A more practical solution is to drain or fence off snail-infested swampy areas but this will most likely be only affordable on a small scale. Measures which reduce the impact of snails without eliminating them are both practical and successful. For example, removing herbage from stream banks can remove a large portion of the available metacercariae, because snails shelter under this foliage beside the water.

Vaccination could contribute to an integrated control strategy and many target antigens based on the excretory-secretory products of liver fluke have been identified and assessed (mainly cathepsin-B and -L). It is proposed that vaccines which are only 50-80% effective may still provide useful benefits for control, especially if resistance to triclabendazole becomes more widespread and severe[175], but a commercial vaccine is unlikely in the near future.[176]

Flukicides and resistance

Flukicides in current or recent use in sheep fall into four main chemical groups:[177]

- benzimidazoles (triclabendazole, albendazole)
- salicylanilides (closantel, oxyclozanide, rafoxanide)
- sulphonamides (clorsulon)
- halogenated phenols (nitroxynil).

The efficacy of these drugs is summarised in Tables 9.3 and 9.4. Triclabendazole (Fasinex™ and several generic products) is a benzimidazole derivative with a unique structure and spectrum of activity that distinguishes it from other benzimidazoles. It lacks activity against nematodes and cestodes but has an extremely high efficacy against both adult and immature stages of *Fasciola*. This means it prevents pasture contamination for extended periods and can be used to treat acute fascioliasis.

When first released, triclabendazole (TCBZ) killed 95% of one-week-old fluke and 99% of all older stages when given at 10 mg/kg. Over-reliance on a single drug is likely to be unsustainable, especially for liver fluke where the life cycle features hermaphroditic self-fertilisation and parthenogenetic egg production in the definitive host (sheep and cattle) and clonal asexual reproduction in the intermediate host (lymnaeid snails). Many fluke populations also exhibit triploidy, which can increase the mutation rate by up to 50%.[178] Not surprisingly, resistance to TCBZ has been detected on many farms, especially those on which sheep or cattle are treated frequently, because they graze irrigated or inundated pastures.[179,180,181] As for anthelmintic resistance in nematodes, an alternative means of introducing resistant fluke is in large mobs of purchased sheep or cattle.

Another benzimidazole, albendazole, has some efficacy against mature fluke (around 80% against 12-week-old larvae)[182] but this is not sufficient to warrant its use solely as a flukicide. Oxfendazole, used in combination with TCBZ, can potentiate the efficacy of this compound against immature fluke.[177]

The salicylanilides induce spastic paralysis of *Fasciola*. Closantel (for example, Seponver™ and generic brands) and rafoxinide (for example, Ranizole™) have reasonable efficacy against 6-week-old fluke (50-90%) but not younger stages (Tables 9.3 and 9.4). Closantel was originally used for its persistent action and high efficacy against *Haemonchus contortus*, although resistance to closantel by *Haemonchus* is now widespread. Rafoxanide does have increased activity against four-week-old fluke at higher dose rates (10-15 vs 7.5 mg/kg), but this increased dose has a reduced safety margin. The long-term use of rafoxanide and closantel in sheep has selected resistant strains of *F hepatica*. Rafoxanide-resistant fluke have side-resistance to closantel (but not oxyclosanide) and cross-resistance to nitroxynil, a halogenated phenol.[177] Like closantel, rafoxanide is a persistent anthelmintic that is strongly bound to plasma protein for up to 90 days and is therefore more likely to select for resistance. In contrast oxyclozanide is rapidly absorbed and excreted.[177]

Clorsulon (a sulphonamide), oxyclosanide (for example, Nilzan™) and nitroxynil (for example, Trodax™) are available for use against *Fasciola* in cattle in Australia, but not sheep. To avoid bacterial degradation in the rumen, nitroxynil is given by subcutaneous injection but it stains wool and is less effective than either closantel or TCBZ. Consequently it has never been used extensively in sheep, although it does have reasonable (although erratic) efficacy against 6- to 8-week-old fluke (from 50-90%). Nevertheless, these drugs could become useful in combinations (with or without TCBZ) when resistance to TCBZ becomes more widespread and severe, and injectable drug combinations are already available for use against liver fluke in cattle (for example, TCBZ with oxfendazole, clorsulon with nitroxynil).

There is little recent published work on the use of combinations of flukicides in sheep but extensive studies were undertaken by Joseph Boray in Australia in the 1990s.[177] For example, TCBZ and clorsulon were highly effective against TCBZ-resistant *F hepatica* at six weeks,

Table 9.3: Comparative efficacy of drugs against *Fasciola hepatica* in sheep and cattle. Adapted from Fairweather and Boray (1999).[177]

Anthelmintic	Application	Dose rate (mg/kg)		Safety index at recommended dose rate for sheep	Minimum age of fluke (wks) for an efficiency ≥90%	
		Sheep	Cattle		Sheep	Cattle
Triclabendazole	Oral	10	12	20-40	1	1
Albendazole	Oral	4.75	10	8.0	>12	>12
Closantel	Oral	7.5-10	NR	4.0	8-6	NR
	SC	NR	3	NR	NR	>12
Rafoxanide	Oral	7.5	7.5	6.0	6	12
	SC				NR	12
Oxyclosanide	Oral	15	13-16	4.0	12	>14
Clorsulon	Oral	-	7	5.0		8
	SC	-	2			>12
Nitroxynil	SC	10	10	4.0	8	10

SC = subcutaneous injection; NR = not recommended

Table 9.4: Efficacy of flukicides against liver fluke of different ages in sheep. Adapted from Fairweather and Boray (1999).[177]

Flukicide	Age of fluke in weeks [stage of infection]													
	1	2	3	4	5	6	7	8	9	10	11	12	13	14
	[Pre-pathogenic]				[Acute/subacute]				[Subacute & chronic disease]					
Albendazole Oxyclosanide Clorsulon + ivermectin (SC)										50-70%		80-99%		
Clorsulon (oral)								90-99%						
Nitroxynil, Closantel							50-90%		91-99%					
Rafoxanide				50-90%		91-99%								
Triclabendazole		90-99%	99-100%											

When given orally except SC (= subcutaneous injection)

as was a combination of TCBZ and luxabendazole. Combinations of closantel (7.5 mg/kg) with TCBZ, clorsulon or luxabendazole were also highly effective against closantel- and luxabendazole-resistant fluke and a reduced dose of nitroxynil was effective against 6-week-old fluke resistant to both closantel and luxabendazole when used in combination with either closantel or clorsulon in injectable formulations.[177]

Diagnosis of liver fluke

Worm egg counts

Liver fluke worm egg counts (WEC) are useful in sheep, with counts generally being higher than for cattle. Recent infections do not become patent for at least eight weeks and peak around 17 weeks after infection, so considerable damage and production loss can occur before WECs become positive. Trematode eggs are heavier than nematode eggs so sedimentation techniques are more sensitive and the preferred diagnostic test rather than flotation in salt solutions. In one technique, a 5 g faecal sample is homogenised in 60 mL water, then shaken vigorously through a 100 µ mesh sieve fixed to the top of a jar. This is repeated three times, with the liquid filtrate (containing fluke eggs and fine particulate matter) collected in a 250 mL flask. The sediment from this flask is then collected into a 15 mL tube and counted, with the sensitivity of this method estimated to be 67% in cattle.[183] An alternative method is to shake the faecal suspension material through sieves of decreasing aperture (usually 150, 90 and then 45 µ). The material lodged on the 45 µ sieve is allowed to sediment in water in cylindrical flasks and then counted. These are obviously more laborious tests, so fluke WECs are generally more expensive than for nematodes. Bulking individual samples together reduces the cost but also lowers the sensitivity of test.

Antibody enzyme-linked immunosorbent assay (ELISA)

An ELISA to detect antibodies to excretory or secretory antigens of *F hepatica* in blood and milk has been validated and is commercially available. In sheep, titres appear four to six weeks after infection (six to eight weeks for cattle), and remain high for at least 12 weeks after successful treatment. The antibody ELISA can detect acute and subacute fascioliasis, which occurs in late spring, and from autumn to early winter, because antibodies remain high during these infections. The test is highly specific (0% false negatives in sheep, 2.5% in cattle) and more sensitive than WECs for the detection of chronic fascioliasis, in both sheep and cattle (about 20% and 30% more infected animals detected, respectively). To reduce costs, five blood or milk samples can be pooled in the laboratory.

Antigen ELISA

A capture ELISA, based upon a monoclonal antibody to excretory/secretory antigen, can detect trace amounts of fluke excretory/secretory antigen in faeces (0.3 ng/mL, equivalent to 1-2 fluke).[184] This faecal antigen (coproantigen) test can detect fluke five weeks after infection, hence three weeks before they start producing eggs, and therefore can detect infections with immature fluke.[185] The BIO K201 antigen used in a commercial kit (Bio-X Diagnostics, Belgium) is 100% specific and has a high sensitivity, which can be improved from 88% to 100% for sheep by using modified cut-off values.[186]

The antigen ELISA becomes negative two to three weeks after successful treatment, making a coproantigen reduction test a sensitive assessment of resistance to flukicides.

Pathology

In acute fascioliasis, there may be peritonitis, particularly on the visceral surface of the hepatic capsule. The migration of flukes in the liver leaves dark haemorrhagic streaks and foci. The liver is swollen, friable and has capsular perforations marked by haemorrhagic tags. Older tunnels appear as slight yellow streaks.[171,174]

Chronic fascioliasis is characterised by anaemia, oedema and emaciation, with mature flukes readily visible in larger bile ducts which are enlarged and thickened.[171] These lesions are often most obvious in the ventral lobe. In long-standing lesions there is also fibrosis of the hepatic parenchyma.

Around 100-200 adult flukes will produce signs of chronic disease, but in natural outbreaks of disease up to 1000 flukes of mixed ages can be found in adult sheep.

Testing for flukicide resistance

According to World Association for the Advancement of Parasitology guidelines, the most reliable test for detecting anthelmintic resistance in *F hepatica* is a dose and slaughter trial.[187] This is not a realistic option in most flocks, although necropsy of sheep treated with a flukicide within the past 10-14 days is an opportunity to monitor the efficacy of that treatment.

Either the modified fluke worm egg count reduction test (Fluke WECRT) or coproantigen reduction test (CRT) can be used in sheep.[185,188]

For a Fluke WECRT, faecal samples taken 14-17 days after treatment will assess the efficacy of a treatment against mature fluke, although the release of fluke eggs from the gall bladder can give false positive results even when all the adult fluke are killed.[189,190] These WECs can be compared with those on the day of treatment or an untreated control group on day 14-17.[188] To assess efficacy against the immature stages of liver fluke, additional faecal samples must be collected. The suggested timing of these varies between authors, but additional collections at days 35 and 56 after treatment will give a reasonable assessment of the efficacy of TCBZ against immature fluke providing there is no reinfection.

For the CRT, samples positive on the day of treatment with TCBZ (or any other flukicide) are retested 14 days later, with a negative result indicating that both the mature and immature stages have been removed.[185,188]

Bulking 5 or 10 individual faecal samples (that is, combining a similar quantity of each individual sample into a composite sample) is equally sensitive at detecting resistance, so this is a practical and cost-effective alternative to counting individual samples for both the fluke WECRT and CRT.[181,191]

FURTHER READING AND RESOURCES

Anderson N (1982) Internal parasites of sheep and goats. In: Sheep and Goat Production, ed IE Coop. Elsevier, pp. 175-191.

Anderson N and Waller PJ, eds (1985) Resistance in nematodes to anthelmintic drugs. CSIRO Div Animal Health: Canberra.

Bowman DD (2014) Georgis' parasitology for veterinarians. 10th ed. Elsevier: St Louis Missouri, USA.

Boray JC (1969) Experimental fascioliasis in Australia. In: Advances in Veterinary Parasitology, ed B Davies. Vol 7. Academic Press: New York & London, pp. 95-210.

Donald AD, Southcott WH and Dineen JK, eds (1978) The epidemiology and control of gastrointestinal parasites of sheep in Australia. CSIRO: Melbourne.

Sutherland I and Scott I (2010) Gastro-intestinal nematodes of sheep and cattle. Wiley-Blackwell: Chichester, West Sussex, UK.

Taylor MA, Coop RL and Wall RL (2016) Veterinary Parasitology. 4th ed. John Wiley & Sons Ltd: Chichester, West Sussex, UK.

WormBoss: www.wormboss.com.au/.

REFERENCES

1   Sackett D, Holmes P, Abbott K et al. (2006) Assessing the economic cost of endemic disease on the profitability of Australian beef cattle and sheep producers. Report AHW.087. Meat and Livestock Australia: North Sydney, Australia.

2   Lane J, Jubb T, Shephard R, Webb et al. (2015) Priority list of endemic diseases for the red meat industries. Report AHE.001. Meat and Livestock Australia: North Sydney, Australia.

3   Kelly GA, Kahn LP, Walkden-Brown SW (2010) Integrated parasite management for sheep reduces the effects of gastrointestinal nematodes on the Northern Tablelands of New South Wales. Anim Prod Sci **50** 1043-52. https://doi.org/10.1071/AN10115.

4   Kahn LP, Norman TM, Walkden-Brown SW et al. (2007) Trapping efficacy of *Duddingtonia flagrans* against *Haemonchus contortus* at temperatures existing at lambing in Australia. Vet Parasitol **146** 83-9. https://doi.org/10.1016/j.vetpar.2007.02.004.

5   Waller PJ (2003) Global perspectives on nematode parasite control in ruminant livestock: The need to adopt alternatives to chemotherapy, with the special emphasis on biological control. Anim Health Rev **4** 35-43. https://doi.org/10.1079/AHRR200350.

6   Steel JW (2003) Effects of protein supplementation of young sheep on resistance development and resilience to parasitic nematodes. Aust J Exp Agric **43** 1469-76. https://doi.org/10.1071/EA03004.

7   Kahn L (2003) Regulation of the resistance and resilience of periparturient ewes to infection with gastrointestinal nematode parasites by dietary supplementation. Aust J Exp Agric **43** 1477-85. https://doi.org/10.1071/EA02202.

8   Besier B, Kahn L, Dobson R et al. (2015) Barbervax — a new strategy for *Haemonchus* management. In: Proceedings of the Pan Pacific Veterinary Conference, Australian Veterinary Association and New Zealand Veterinary Association. 24-29 May, Brisbane, pp. 373-7.

9   Matthews JB, Geldhof P, Tzelos T et al. (2016) Progress in the development of subunit vaccines for gastrointestinal nematodes of ruminants. Parasite Immunol **38** 744-53. https://doi.org/10.1111/pim.12391.

10 Blaxter ML, De Ley P, Garey JR et al. (1998) A molecular evolutionary framework for the phylum Nematoda. Nature **392** 71-5. https://doi.org/10.1038/32160.

11 Young RR (1983) Populations of free-living stages of *Ostertagia ostertagi* and *O. circumcincta* in a winter rainfall region. Aust J Agric Res **34** 569-81. https://doi.org/10.1071/AR9830569.

12 Anderson N, Dash KM, Donald AD et al. (1978) In: The epidemiology and control of gastrointestinal parasites ofaSheep in Australia, eds AD Donald, WH Southcott and JK Dineen. CSIRO: Melbourne, pp. 9-22.

13 Anderson N (1983) The availability of trichostrongylid larvae to grazing sheep after seasonal contamination of pastures. Aust J Agric Res **34** 583-92. https://doi.org/10.1071/AR9830583.

14 O'Connor LJ, Walkden-Brown SW and Kahn LP (2006) Ecology of the free-living stages of major trichostrongylid parasites of sheep. Vet Parasitol **142** 1-15. https://doi.org/10.1016/j.vetpar.2006.08.035.

15 Larsen JWA (2014) Sustainable control of sheep worms in Australia. Small Ruminant Res **118** 41-7. https://doi.org/10.1016/j.smallrumres.2013.12.018.

16 Silangwa SM and Todd AC (1964) Vertical migration of Trichostrongylid larvae on grasses. J Parasitol **50** 278-85. https://doi.org/10.2307/3276286.

17 Moss RA and Vlassoff A (1993) Effects of herbage species on gastro-intestinal roundworm populations and their distribution. NZ J Agric Res **36** 371-5. https://doi.org/10.1080/0028823.1993.10417734.

18   Niezen JH, Charleston WAG, Hodgson J et al. (1998) Effects of plant species on the larvae of gastrointestinal nematodes which parasitise sheep. Int J Parasitol **28** 791-803. https://doi.org/10.1016/S0020-7519(98)00019-8.

19   Southcott WH, Major GW and Barger IA (1976) Seasonal pasture contamination and availability of nematodes for grazing sheep. Aust J Agric Res **27** 277-86. https://doi.org/10.1071/AR9760277.

20   Bailey JN, Kahn LP and Walkden-Brown SW (2009) The relative contributions of *T. colubriformis*, *T. vitrinus*, *T. axei* and *T. rugatus* to sheep infected with *Trichostrongylus* spp. on the northern tablelands of New South Wales. Vet Parasitol **165** 88-95. https://doi.org/10.1016/j.vetpar.2009.06.028.

21   Bailey JN, Walkden-Brown SW and Kahn LP (2009) Availability of gastro-intestinal larvae to sheep following winter contamination of pasture with six nematode species on the Northern tablelands of New South Wales. Vet Parasitol **160** 89-99. https://doi.org/10.1016/j.vetpar.2008.10.083.

22   Bailey JN, Walkden-Brown SW and Kahn LP (2009) Comparison of strategies to provide lambing paddocks of low gastro-intestinal nematode infectivity in a summer rainfall region of New South Wales. Vet Parasitol **161** 218-31. https://doi.org/10.1016/j.vetpar.2009.01.016.

23   Barger IA (1988) Resistance of young lambs to *Haemonchus contortus* infection, and its loss following anthelmintic treatment. Int J Parasitol **18** 1107-9. https://doi.org/10.1016/0020-7519(88)90082-3.

24   McRae KM, Stear MJ, Good B et al. (2015) The host immune response to gastrointestinal nematode infection in sheep. Parasite Immunol **37** 605-13. https://doi.org/10.1111/pim.12290.

25   Miller HRP (1996) Prospects for immunological immunological control of ruminant gastrointestinal nematodes: Natural immunity, can it be harnessed? Int J Parasitol **26** 801-11. https://doi.org/10.1016/S0020-7519(96)80044-0.

26   Barger IA (1987) Population regulation in trichostrongylids of ruminants. Int J Parasitol **17** 531-40. https://doi.org/10.1016/0020-7519(87)90129-9.

27   Larsen JWA, Anderson N, Vizard AL et al. (1994) Diarrhoea in Merino ewes during winter: Association with trichostrongylid larvae. Aust Vet J **71** 365-72. https://doi.org/10.1111/j.1751-0813.1994.tb00930.x.

28   Larsen JWA, Anderson N and Vizard AL (1999) The pathogenesis and control of diarrhoea and breech soiling in adult Merino sheep. Int J Parasitol **29** 893-902. https://doi.org/10.1016/S0020-7519(99)00050-8.

29   Holmes PH (1985) Pathogenesis of trichostrongylosis. Vet Parasitol **18** 89-101. https://doi.org/10.1016/0304-4017(85)90059-7.

30   Titchen DA and Anderson N (1977) Aspects of the patho-physiology of parasitic gastritis in the sheep. Aust Vet J **53** 369-73. https://doi.org/10.1111/j.1751-0813.1977.tb07953.x.

31   Anderson N (1972) Trichostrongylid infections of sheep in a winter rainfall region I. Epizootiological studies in the Western District of Victoria, 1966-67. Aust J Agric Res **23** 1113-29. https://doi.org/10.1071/AR9721113.

32   Yakoob AY, Holmes PH and Armour J (1983) Pathophysiology of gastrointestinal trichostrongyles in sheep: Plasma losses and changes in plasma pepsinogen levels associated with parasite challenge of immune animals. Res Vet Sci **34** 305-9.

33   Michel JF (1974) Arrested development of nematodes and some related phenomena. Adv Parasitol **12** 279-366. https://doi.org/10.1016/S0065-308X(08)60390-5.

34   Beveridge, I, Pullman AL, Phillips PH et al. (1989) Comparison of the effects of infection with *Trichostrongylus colubriformis*, *Trichostrongylus vitrinus* and *Trichostrongylus rugatus* in Merino lambs. Vet Parasitol **32** 229-45. https://doi.org/10.1016/0304-4017(89)90123-4.

35   Taylor SM and Pearson GR (1979) *Trichostrongylus vitrinus* in sheep. I. The location of nematodes during parasitic development and associated pathological changes in the small intestine. J Comp Pathol **89** 397-403. https://doi.org/10.1016/0021-9975(79)90030-6.

36   Taylor SM and Pearson GR (1979) *Trichostrongylus vitrinus* in sheep. II. The location of nematodes and associated pathological changes in the small intestine during clinical infection. J Comp Pathol **89** 405-12. https://doi.org/10.1016/0021-9975(79)90031-8.

37 Pullman AL, Beveridge I and Martin RR (1991) Effects of challenge with trichostrongylid larvae on immunologically resistant grazing yearling sheep in South Australia. Vet Parasitol **38** 155-62. https://doi.org/10.1016/0304-4017(91)90125-F.

38 Sykes AR and Coop RL (1977) Intake and utilisation of food by growing sheep with abomasal damage caused by daily dosing with *Ostertagia circumcincta* larvae. J Agric Sci **88** 671-7. https://doi.org/10.1017/S0021859600037369.

39 Symons LEA (1985) Anorexia; occurrence, pathophysiology and possible causes in parasitic infections. Adv Parasitol **24** 103-33. https://doi.org/10.1016/S0065-308X(08)60562-X.

40 Holmes PH (1987) Pathophysiology of parasitic infections. Parasitol **94** S29-S51. https://doi.org/10.1017/S0031182000085814.

41 Coop RL and Holmes PH (1996) Nutrition and parasite interaction. Int J Parasitol **26** 951-62. https://doi.org/10.1016/S0020-7519(96)80070-1.

42 Coop RL and Kyriazakis I (1999) Nutrition-parasite interaction. Vet Parasitol **84** 187-204. https://doi.org/10.1016/S0304-4017(99)00070-9.

43 Holmes PH and Coop RL (1994) Workshop summary: Pathophysiology of gastrointestinal parasites. Vet Parasitol **54** 299-303. https://doi.org/10.1016/0304-4017(94)90101-5.

44 Bown MD, Poppi DP and Sykes AR (1991) The effect of post-ruminal infusion of protein or energy on the pathophysiology of *Trichostrongylus colubriformis* infection and body composition in lambs. Aust J Agric Res **42** 253-67. https://doi.org/10.1071/AR9910253.

45 Steel JW, Jones WO and Symons LEA (1982) Effects of a concurrent infection of *Trichostrongylus colubriformis* on the productivity and physiological and metabolic responses of lambs infected with *Ostertagia circumcincta*. Aust J Agric Res **33** 131-40. https://doi.org/10.1071/AR9820131.

46 Abbott EM, Parkins JJ and Holmes PH (1988) Influence of dietary protein on the pathophysiology of haemonchosis in lambs given continuous infections. Res Vet Sci **45** 41-9.

47 Bown MD, Poppi DP and Sykes AR (1989) The effect of a concurrent infection of *Trichostrongylus colubriformis* and *Ostertagia circumcincta* on calcium, phosphorus and magnesium transactions along the digestive tract of lambs. J Comp Pathol **101** 11-20. https://doi.org/10.1016/0021-9975(89)90072-8.

48 Symons LEA (1989) Pathophysiology of endoparasitic infection compared with ectoparasitic infestation and microbial infection. Academic Press: Sydney.

49 Jacobson CL, Bell K and Besier RB (2009) Nematode parasites and faecal soiling of sheep in lairage: Evidence of widespread potential production losses for the sheep industry. Anim Prod Sci **49** 326-32. https://doi.org/10.1071/EA08251.

50 Yakoob AY, Holmes PH and Armour J (1983) Plasma protein loss associated with gastro-intestinal parasitism in grazing sheep. Res Vet Sci **34** 58-63.

51 Kahn LP (2003) Regulation of the resistance and resilience of periparturient ewes to infection with gastrointestinal nematode parasites by dietary supplementation. Aust J Exp Agric **43** 1477-85. https://doi.org/10.1071/EA02202.

52 Steel JW (2003) Effects of protein supplementation of young sheep on resistance development and resilience to parasitic nematodes. Aust J Exp Agric **43** 1469-76. https://doi.org/10.1071/EA03004.

53 Sykes AR and Coop RL (1976) Intake and utilisation of food by growing lambs with parasite damage of the small intestine caused by daily dosing with *Trichostrongylus colubriformis* larvae. J Agric Sci **86** 507-15. https://doi.org/10.1017/S0021859600061049.

54 Sykes AR and Coop RL (1977) Intake and utilisation of food by growing sheep with abomasal damage caused by daily dosing with *Ostertagia circumcincta* larvae. J Agric Sci **88** 671-7. https://doi.org/10.1017/S0021859600037369.

55 Coop RL and Field AC (1983) Effect of phosphorus intake on growth rate, food intake and quality of the skeleton of growing lambs affected with the intestinal nematode *Trichostrongylus vitrinus*. Res Vet Sci **35** 175-81.

56 Wilson WD and Field AC (1983) Absorption and secretion of calcium and phosphorus in the alimentary tract of lambs infected with *Trichostrongylus colubriformis* and *Ostertagia circumcincta* larvae. J Comp Pathol **93** 61-71. https://doi.org/10.1016/0021-9975(83)90043-9.

57 Anderson N (1973) Trichostrongylid infections of sheep in a winter rainfall region II. Epizootiological studies in the Western District of Victoria, 1967-68. Aust J Agric Res **24** 599-611. https://doi.org/10.1071/AR9730599.

58 Barger IA and Southcott WH (1975) Trichostrongylosis and wool growth 3. The wool growth response of resistant grazing sheep to larval challenge. Aust J Exp Agric Anim Husb **15** 167-72. https://doi.org/10.1071/EA9750167.

59 Morris RS, Anderson N and McTaggart IK (1977) An economic evaluation of two schemes for the control of helminthiasis in breeding ewes. Vet Parasitol **3** 349-63. https://doi.org/10.1016/0304-4017(77)90021-8.

60 Morley FHW, Donald AD, Donnelly JR et al. (1976) Blowfly strike in the breech region of sheep in relation to helminth infection. Aust Vet J **52** 325-9. https://doi.org/10.1111/j.1751-0813.1976.tb02398.x.

61 Abbott KA and McFarland IJ (1991) *Trichostrongylus axei* infection as a cause of deaths and loss of weight in sheep. Aust Vet J **68** 368-9. https://doi.org/10.1111/j.1751-0813.1991.tb00741.x.

62 Donald AD (1979) Effects of parasites and disease on wool growth. In: Physiological and environmental limitations to wool growth, eds JL Black and PJ Reis. University of New England: Armidale, pp 99-114.

63 Pullman AL, Beveridge I and Martin RR (1988) Epidemiology of nematode infections of weaner sheep in the cereal zone of South Australia. Aust J Agric Res **39** 691-702.

64 Brown TH, Ford GE, Miller DH et al. (1985) Effect of anthelmintic dosing and stocking rate on the productivity of weaner sheep in a Mediterranean climate environment. Aust J Agric Res **36** 845-55. https://doi.org/10.1071/AR9850845.

65 Beveridge I, Brown TH, Fitzsimmons SM et al. (1985) Mortality in weaner sheep in South Australia under different regimes of anthelmintic treatment. Aust J Agric Res **36** 857-65. https://doi.org/10.1071/AR9850857.

66 Greer AW and Hamie JC (2016) Relative maturity and the development of immunity to gastrointestinal nematodes in sheep: An overlooked paradigm? Parasite Immunol **38** 263-72. https://doi.org/10.1111/pim.12313.

67 Sykes AR, Henderson AE and Leyva V (1982) Effect of daily infection with *Ostertagia circumcincta* larvae on food intake, milk production and wool growth in sheep. J Agric Sci **99** 249-359. https://doi.org/10.1017/S0021859600030008.

68 Salisbury JR and Arundel JH (1970) Peri-parturient deposition of nematode eggs by ewes and residual pasture contamination as sources of infection for lambs. Aust Vet J **46** 523-9. https://doi.org/10.1111/j.1751-0813.1970.tb06637.x.

69 O'Sullivan BM and Donald AD (1970) A field study of nematode parasite populations in the lactating ewe. Parasitol **61** 301-15. https://doi.org/10.1017/S0031182000041135.

70 Connan RM (1976) Effect of lactation on the immune response to gastrointestinal nematodes. Vet Rec **99** 476-7. https://doi.org/10.1136/vr.99.24.476.

71 Barger IA (1993) Influence of sex and reproductive status on susceptibility of ruminants to nematode parasitism. Int J Parasitol **23** 463-9. https://doi.org/10.1016/0020-7519(93)90034-V.

72 Beasley AM, Kahn LP and Windon RG (2010) The periparturient relaxation of immunity in Merino ewes infected with *Trichostrongylus colubriformis*: Parasitological and immunological responses. Vet Parasitol **168** 60-70. https://doi.org/10.1016/j.vetpar.2009.08.028.

73 Arundel JH and Ford GE (1969) The use of a single anthelmintic treatment to control the post-parturient rise in faecal worm egg count of sheep. Aust Vet J **45** 89-93. https://doi.org/10.1111/j.1751-0813.1969.tb01881.x.

74 Donnelly JR, McKinney GT and Morley FHW (1972) Lamb growth and ewe production following anthelmintic drenching before and after lambing. Proc Aust Soc Anim Prod **9** 392-6.

75 Donald AD, Axelson A, Morley FHW et al. (1982) Effects of reproduction, genotype and anthelmintic treatment of ewes on *Ostertagia* spp. populations. Int J Parasitol **12** 403-11. https://doi.org/10.1016/0020-7519(82)90069-8.

76 Waller PJ, Axelson A, Donald AD et al. (1987) Effects of helminth infection on the pre-weaning production of ewes and lambs: Comparison between safe and contaminated pasture. Aust Vet J **64** 357-62. https://doi.org/10.1111/j.1751-0813.1987.tb09600.x.

77 Waller PJ, Donnelly JR, Dobson RJ et al. (1987) Effects of helminth infection on the pre-weaning production of ewes and lambs: Evaluation of a pre- and post-lambing drenching and provision of safe lambing pasture. Aust Vet J **64** 339-43. https://doi.org/10.1111/j.1751-0813.1987.tb06062.x.

78 Watts JE and Marchant RS (1977) The effects of diarrhoea, tail length and sex on the incidence of breech strike in modified mulesed Merino sheep. Aust Vet J **53** 118-23. https://doi.org/10.1111/j.1751-0813.1977.tb00132.x.

79 Watts JE, Dash KM and Lisle KA (1978) The effect of anthelmintic treatment and other management factors on the incidence of breech strike in Merino sheep. Aust Vet J **54** 352-5. https://doi.org/10.1111/j.1751-0813.1978.tb02491.x.

80 Larsen JWA, Vizard AL and N Anderson N (1995) Production losses in Merino ewes and financial penalties caused by trichostrongylid infections during winter and spring. Aust Vet J **72** 58-63. https://doi.org/10.1111/j.1751-0813.1995.tb15332.x.

81 Larsen JWA, Vizard AL, Webb-Ware JK et al. (1995) Diarrhoea due to trichostrongylid larvae in Merino sheep during winter: Repeatability and differences between bloodlines. Aust Vet J **72** 196-7. https://doi.org/10.1111/j.1751-0813.1995.tb03512.x.

82 Jacobson C, Bell K, Forshaw D and Besier B (2009) Association between nematode larvae and 'low worm egg count diarrhoea' in sheep in Western Australia. Vet Parasitol **165** 66-73. https://doi.org/10.1016/j.vetpar.2009.07.018.

83 Woolaston RR, Ward JL (1999) Including dag score in Merino breeding programs. Proc Assoc Advmt Anim Breed Genet **13** 512-15.

84 Karlsson LJE, Pollott GE, Eady SJ et al. (2004) Relationship between faecal worm egg counts and scouring in Australian Merino sheep. Anim Prod Aust **25** 100-3. https://doi.org/10.1071/SA0401026.

85 Larsen JWA and Anderson N (2000) The relationship between the rate of intake of trichostrongylid larvae and the occurrence of diarrhoea and breech soiling in adult Merino sheep. AustVet J **78** 112-16. https://doi.org/10.1111/j.1751-0813.2000.tb10537.x.

86 Williams AR and Palmer DG (2012) Interactions between gastrointestinal nematode parasites and diarrhoea in sheep: Pathogenesis and control. Vet J **192** 279-85. https://doi.org/10.1016/j.tvjl.2011.10.009.

87 Bisset SA, Vlassoff A, Morris CA et al. (1992) Heritability of and genetic correlations among faecal egg counts and productivity traits in Romney sheep. NZ J Agric Res **35** 51-8. https://doi.org/10.1080/00288233.1992.10417701.

88 Pollott GE, Karlsson LJE, Eady S et al. (2004) Genetic parameters for indicators of host resistance to parasites from weaning to hogget age in Merino sheep. J Anim Sci **82** 2852-64. https://doi.org/10.2527/2004.82102852x.

89 Wormboss. Worm egg counting. Available from: http://www.wormboss.com.au/tests-tools/tests/worm-egg-counting.php. Accessed 4 April 2017.

90 Gibbons LM, Jacobs, DE and Fox MT. McMaster egg counting technique: Principle. Available from: http://www.rvc.ac.uk/review/parasitology/eggcount/Principle.htm. Accessed 4 April 2017.

91 Anderson N, Martin PJ and Jarrett RG (1988) Mixtures of anthelmintics: A strategy against resistance. Aust Vet J **65** 62-4. https://doi.org/10.1111/j.1751-0813.1988.tb07355.x.

92 Dobson RJ, Barnes EH, Birclijin SD et al. (1992) The survival of *Ostertagia circumcincta* and *Trichostrongylus colubriformis* in faecal culture as a source of bias in apportioning egg counts to worm species. Int J Parasitol **22** 1005-8. https://doi.org/10.1016/0020-7519(92)90060-X.

93 van Wyk JA, Cabaret J and Michael LM (2004) Morphological identification of nematode larvae of small ruminants and cattle simplified. Vet Parasitol **119** 277-306. https://doi.org/10.1016/j.vetpar.2003.11.012.

94 Bott NJ, Campbell BE, Beveridge I et al. (2009) A combined microscopic-molecular method for the diagnosis of strongylid infections in sheep. Int J Parasitol **39** 1277-87. https://doi.org/10.1016/j.ijpara.2009.03.002.

95 Roeber F, Jex AR, Campbell AJD et al. (2011) Evaluation and application of a molecular method to assess the composition of strongylid populations in sheep with naturally acquired infections. Infect Genet Evol **11** 849-54. https://doi.org/10.1016/j.meegid.2011.01.013.

96 Anderson N, Martin P and Jarrett RG (1991) The efficacy of mixtures of albendazole sulphoxide and levamisole against sheep nematodes resistant to benzimidazole and levamisole. Aust Vet J **68** 127-33. https://doi.org/10.1111/j.1751-0813.1991.tb03154.x.

97 Reeve I and Walkden-Brown S (2014) Benchmarking Australian sheep parasite control: Cross sectional survey report. Final report to Australian Wool Innovation and Meat and Livestock Australia. University of New England and Institute for Rural Futures: Australia.

98 Australian Agricultural Council, Standing Committee on Agriculture (1989) Anthelmintic resistance: Report of the working party on anthelmintic resistance. CSIRO: Melbourne.

99 Martin PJ (1987) Development and control of resistance to anthelmintics. Int J Parasitol **17** 493-501. https://doi.org/10.1016/0020-7519(87)90125-1.

100 Playford MC, Smith AN, Love S et al. (2012) Prevalence and severity of anthelmintic resistance in ovine gastrointestinal nematodes in Australia (2009-2012). Aust Vet J **92** 464-71. https://doi.org/10.1111/avj.12271.

101 Besier RB, Lyon J and McQuade N (1996) Drench resistance — a large economic cost. West Aust J Agric **37** 2.

102 Larsen JWA, Anderson N, Webb Ware J et al. (2006) The productivity of Merino flocks in south-eastern Australia in the presence of anthelmintic resistance. Small Ruminant Res **62** 87-93. https://doi.org/10.1016/j.smallrumres.2005.08.008.

103 Vizard AL and Wallace RJ (1987) A simplified faecal egg count reduction test. Aust Vet J **64** 109-10. https://doi.org/10.1111/j.1751-0813.1987.tb09641.x.

104 Martin PJ, Anderson N and Jarrett RG (1985) Resistance to benzimidazole anthelmintics in field strains of Ostertagia and Nematodirus in sheep. Aust Vet J **66** 236-40. https://doi.org/10.1111/j.1751-0813.1989.tb13578.x.

105 van Wyk, JA (2001) Refugia — overlooked as perhaps the most potent factor concerning the development of anthelmintic resistance. Onderstepoort J Vet Res **68** 55-67.

106 Leathwick DM, Miller CM, Atkinson DS et al. (2006) Drenching adult ewes: Implications of anthelmintic treatments pre- and post-lambing on the development of anthelmintic resistance. NZ Vet J **54** 297-304. https://doi.org/10.1080/00480169.2006.36714.

107 Waghorn TS, Miller CM, Oliver A-MB et al. (2009) Drench-and-shift is a high risk practice in the absence of refugia. NZ Vet J **57** 359-63. https://doi.org/10.1080/00480169.2009.64723.

108 Martin PJ, Le Jambre LF and Claxton JH (1981) The impact of refugia on the development of thiabendazole resistance in *Haemonchus contortus*. Int J Parasitol **11** 35-41. https://doi.org/10.1016/0020-7519(81)90023-0.

109 Woodgate RG and Besier RB (2010) Sustainable use of anthelmintics in an Integrated Parasite Management Program for sheep nematodes. Anim Prod Sci **50** 440-3. https://doi.org/10.1071/AN10022.

110 Chylinski C, Lherminé E, Coquille M et al. (2014) Desiccation tolerance of gastrointestinal nematode third-stage larvae: Exploring the effects on survival and fitness. Parasitol Res **113** 2789-96. https://doi.org/10.1007/s00436-014-3938-1.

111 Niven PG, Anderson N and Vizard AL (2002) Trichostrongylid infections in sheep after rainfall during summer in southern Australia. Aust Vet J **80** 567-70. https://doi.org/10.1111/j.1751-0813.2002.tb11041.x.

112 Leathwick DM, Miller CM, Atkinson DS et al. (2008) Managing anthelmintic resistance: Untreated adult ewes as a source of unselected parasites, and their role in reducing parasite populations. NZ Vet J **56** 184-95. https://doi.org/10.1080/00480169.2008.36832.

113 Donald AD, Morley FHW, Waller PJ et al. (1978) Availability to grazing sheep of gastrointestinal nematode infection arising from summer contamination of pastures. Aust J Agric Res **29** 189-204. https://doi.org/10.1071/AR9780189.

114 WormBoss (2018) Worm control program for sheep. Available at: http://www.wormboss.com.au/programs/sheep.php. Accessed 4 April 2017.

115 Sheep Genetics. ASBVS and indexes explained. Available at: http://www.sheepgenetics.org.au/Getting-started/ASBVs-and-Indexes. Accessed 4 April 2017.

116 Hatcher S, Eppleston J, Graham RP et al. (2008) Higher weaning weight improves postweaning growth and survival in young Merino sheep. Aust J Exp Agric **48** 966-73. https://doi.org/10.1071/EA07407.

117 Campbell AJD, Vizard AL and Larsen JWA (2009) Risk factors for post-weaning mortality of Merino sheep in south-eastern Australia. Aust Vet J **87** 305-12. https://doi.org/10.1111/j.1751-0813.2009.00457.x.

118 Thompson AN, Ferguson MB, Campbell AJD et al. (2011) Improving the nutrition of Merino ewes during pregnancy and lactation increases weaning weight and survival of progeny but does not affect their mature size. Anim Prod Sci **51** 784-93. https://doi.org/10.1071/AN09139.

119 Trompf JP, Gordon DJ, Behrendt R et al. (2011) Participation in Lifetime Ewe Management results in changes in stocking rate, ewe management and reproductive performance on commercial farms. Anim Prod Sci **51** 866-72. https://doi.org/10.1071/AN10164.

120 Besier RB, Love RA, Lyon J et al. (2010) A targeted selective treatment approach for effective and sustainable sheep worm management: Investigations in Western Australia. Anim Prod Sci **50** 1034-2. https://doi.org/10.1071/AN10123.

121 Dobson RJ, Barnes EH, Tyrrell KL et al. (2011) A multi-species model to assess the impact of refugia on worm control and anthelmintic resistance in sheep grazing systems. Aust Vet J **89** 200-8. https://doi.org/10.1111/j.1751-0813.2011.00719.x.

122 Dobson RJ, Hosking BC, Besier RB et al. (2011) Minimising the development of anthelmintic resistance, and optimising the use of the novel anthelmintic monepantel, for the sustainable control of nematode parasites in Australian sheep grazing systems. Aust Vet J **89** 160-6. https://doi.org/10.1111/j.1751-0813.2011.00703.x.

123 Morley FHW and Donald AD (1980) Farm management and systems of helminth control. Vet Parasitol **6** 105-34. https://doi.org/10.1016/0304-4017(80)90040-0.

124 Larsen JWA, Anderson N and Preshaw AR (2009) The use of long-acting moxidectin for the control of trichostrongylid infections of sheep in south-eastern Australia. Aust Vet J **87** 130-7. https://doi.org/10.1111/j.1751-0813.2009.00395.x.

125 Niven PG, Anderson N and Vizard AL (2002) The integration of grazing management and summer treatments for the control of trichostrongylid infections in Merino weaners. Aust Vet J **80** 559-66. https://doi.org/10.1111/j.1751-0813.2002.tb11040.x.

126 Barger IA and Southcott WH (1978) Parasitism and production in weaner sheep grazing alternately with cattle. Aust J Exp Agric Anim Husb **18** 340-6. https://doi.org/10.1071/EA9780340.

127 Southcott WH and Barger IA (1975) Control of nematode parasites by grazing management — II. Decontamination of sheep and cattle pastures by varying periods of grazing with the alternate host. Int J Parasitol **5** 45-8. https://doi.org/10.1016/0020-7519(75)90096-X.

128 Larsen JWA and Anderson N (2009) Worm infections in high and low bodyweight Merino ewes during winter and spring. Aust Vet J **87** 102-9. https://doi.org/10.1111/j.1751-0813.2009.00396.x.

129 Donald AD, Morley FHW, Axelsen A et al. (1987) Integration of grazing management and anthelmintic treatment for the control of nematode infections in young sheep. In: Temperate pastures, their production, use and management, eds JL Wheeler, CJ Pearson and GE Robards. Australian Wool Corporation/CSIRO: Melbourne, pp. 567-9.

130 Besier RB and Love SCJ (2003) Anthelmintic resistance of sheep in Australia: The need for new approaches. Aust J Exp Agric **43** 1383-91. https://doi.org/10.1071/EA02229.

131 Van Wyk JA and Bath GF (2002) The FAMACHA© system for managing Haemonchosis in sheep and goats by clinically identifying individual animals for treatment. Vet Res **33** 509-29. https://doi.org/10.1051/vetres:2002036.

132 Barger IA (1985) The statistical distribution of trichostrongylid nematodes in grazing lambs. Int J Parasitol **15** 645-9. https://doi.org/10.1016/0020-7519(85)90010-4.

133 Cornelius MP, Jacobsen C and Besier RB (2014) Body condition score as a selection tool for targeted selective treatment-based nematode control strategies in Merino ewes. Vet Parasitol **206** 173-81. https://doi.org/10.1016/j.vetpar.2014.10.031.

134 Cornelius MP, Jacobsen C, Dobson R et al. (2016) Computer modelling of anthelmintic resistance and worm control outcomes for refugia-based nematode control strategies in Merino ewes in Western Australia. Vet Parasitol **220** 59-66. https://doi.org/10.1016/j.vetpar.2016.02.030.

135 Barger IA, Siale K, Banks DJD et al. (1994) Rotational grazing for control of gastrointestinal nematodes of goats in a wet tropical environment. Vet Parasitol **53** 109-16. https://doi.org/10.1016/0304-4017(94)90023-X.

136 Colvin AF, Walkden-Brown SW, Knox MR et al. (2008) Intensive rotational grazing assists control of gastrointestinal nematodosis of sheep in a cool temperate environment with summer-dominant rainfall. Vet Parasitol **153** 108-20. https://doi.org/10.1016/j.vetpar.2008.01.014.

137 Walkden-Brown SW, Colvin AF, Hall E et al. (2013) Grazing systems and worm control in sheep: A long-term case study involving three management systems with analysis of factors influencing faecal worm egg count. Anim Prod Sci **53** 765-79. https://doi.org/10.1071/AN13037.

138 Donald AD (1968) Ecology of the free-living stages of nematode parasites of sheep. Aust Vet J **44** 139-44. https://doi.org/10.1111/j.1751-0813.1968.tb09057.x.

139 Donald AD and Waller PJ (1973) Gastro-intestinal nematode parasite populations in ewes and lambs and the origin and time course of infective larval availability in pastures. Int J Parasitol **3** 219-33. https://doi.org/10.1016/0020-7519(73)90027-1.

140 Albers GAA, Gray GD, Piper LR et al. (1987) The genetics of resistance and resilience to *Haemonchus contortus* infection in young Merino sheep. Int J Parasitol **17** 1355-63. https://doi.org/10.1016/0020-7519(87)90103-2.

141 Stear MJ, Strain S and Bishop SC (1999) Mechanisms underlying resistance to nematode infection. Int J Parasitol **29** 51-6. https://doi.org/10.1016/S0020-7519(98)00179-9.

142 Smith WD (2007) Some observations on immunologically mediated inhibited *Teladorsagia circumcincta* and their subsequent resumption of development in sheep. Vet Parasitol **147** 103-9. https://doi.org/10.1016/j.vetpar.2007.03.026.

143 Dobson RJ, Waller PJ and Donald AD (1990) Population dynamics of *Trichostrongylus colubriformis* in sheep: The effect of infection rate on the establishment of infective larvae and parasite fecundity. Int J Parasitol **20** 347-52. https://doi.org/10.1016/0020-7519(90)90150-L.

144 Harrison GB, Pulford HD, Hein WR et al. (2003) Immune rejection of *Trichostrongylus colubriformis* in sheep; a possible role for intestinal mucus antibody against an L$_3$-specific surface antigen. Parasite Immunol **25** 45-53. https://doi.org/10.1046/j.1365-3024.2003.00602.x.

145 Woolaston RR, Barger IA and Piper LR (1990) Response to helminth infection of sheep selected for resistance to *Haemonchus contortus*. Int J Parasitol **20** 1015-18. https://doi.org/10.1016/0020-7519(90)90043-M.

146 Woolaston RR (1992) Selection of Merino sheep for increased and decreased resistance to *Haemonchus contortus*: Peri-parturient effects of worm egg counts. Int J Parasitol **22** 947-53. https://doi.org/10.1016/0020-7519(92)90052-M.

147 Williams AR, Greef JC, Vercoe PE et al. (2010) Merino ewes bred for parasite resistance reduce larval contamination onto pasture during the peri-parturient period. Animal **4** 122-7. https://doi.org/10.1017/S1751731109990802.

148 Morris CA, Bisset SA, Vlassoff A et al. M (1998) Faecal nematode egg counts in lactating ewes from Romney flocks selectively bed for divergence in lamb faecal egg count. Anim Sci **67** 283-8. https://doi.org/10.1017/S1357729800010043.

149 Greeff JC and Karlsson LJE (2006) Breeding for worm resistance — whole farm benefits. Int J Sheep Wool Sci **54** 7-14. https://doi.org/10.1016/S0304-4017(96)01148-X.

150 Bisset SA, Vlassoff A, West CJ et al. (1997) Epidemiology of nematodosis in Romney lambs selectively bred for resistance or susceptibility to nematode infection. Vet Parasitol **70** 255-69. https://doi.org/10.1016/S0304-4017(96)01148-X.

151 Morris CA, Vlassoff A, Bisset SA et al. (2000) Continued selection of Romney sheep for resistance or susceptibility to nematode infection: Estimates of direct and correlated responses. Anim Sci **70** 17-27. https://doi.org/10.1017/S1357729800051560.

152 Pollott GE and Greeff JC (2004) Genotype × environment interactions and genetic parameters for fecal egg count and production traits of Merino sheep. J Anim Sci **82** 2840-51. https://doi.org/10.2527/2004.82102840x.

153 Karlsson LJE and Greeff JC (2006) Selection response in faecal worm egg counts in the Rylington Merino parasite resistant flock. Aust J Exp Agric **46** 809-11. https://doi.org/10.1071/EA05367.

154 Karlsson LJE, Pollott GE, Eady SJ et al. (2004) Relationship between faecal worm egg counts and scouring in Australian Merino sheep. Anim Prod Aust **25** 100-3.

155 Morris CA, Wheeler M, Watson TG et al. (2005) Direct and correlated responses to selection for high or low nematode faecal egg count in Perendale sheep. NZ J Agric Res **48** 1-10. https://doi.org/10.1080/00288233.2005.9513625.

156 Kelly GA, Kahn LP and Walkden-Brown SW (2013) Measurement of phenotypic resilience to gastro-intestinal nematodes in Merino sheep and association with resistance and production variables. Vet Parasitol **193** 111-17. https://doi.org/10.1016/j.vetpar.2012.12.018.

157 Shaw RJ, Morris CA and Wheeler M (2013) Genetic and phenotypic relationships between carbohydrate larval antigen (CarLA) IgA, parasite resistance and productivity in serial samples taken from lambs after weaning. Int J Parasitol **43** 661-7. https://doi.org/10.1016/j.ijpara.2013.03.003.

158 Warn LK, Geenty KG and McEachern S (2006) What is the optimum wool-meat enterprise mix? Int J Sheep Wool Sci **54** 40-9.

159 Carmichael I (2009) Parasite control in southern prime lamb production systems. Report AHW.0045. Meat and Livestock Australia Limited: North Sydney, Australia.

160 Kahn L, Eppelston J, Larsen J et al. (2015) Lifting the limits imposed by worms on sheep meat production. Report AHE.0045. Meat and Livestock Australia Limited: North Sydney, Australia.

161 Dever ML, Kahn LP and Doyle EK (2017) Integrated parasite management improves control of gastrointestinal nematodes in lamb production systems in a high summer rainfall region, on the Northern Tablelands, New South Wales. Anim Prod Sci **57** 958-68. https://doi.org/10.1071/AN15805.

162 Dever ML, Kahn LP and Doyle EK (2017) Growth is impeded by gastrointestinal nematodes in weaned rather than suckling meat-breed lambs in a high summer rainfall region, on the Northern Tablelands, New South Wales. Anim Prod Sci **57** 969-74. https://doi.org/10.1071/AN15806.

163 Kirk B (2015) Internal parasitism and production in prime lamb flocks. MVSc thesis, University of Melbourne Faculty of Veterinary and Agricultural Sciences.

164 Boray JC (1981) Fasciolosis in sheep. In: Sheep. Proceedings No 58. University of Sydney Postgraduate Committee in Veterinary Science: Sydney, pp. 508-36.

165 Barger IA, Dash KM and Southcott WH (1978) Epidemiology and control of liver fluke in sheep. In: The epidemiology and control of gastrointestinal parasites of aheep in Australia, eds AD Donald, WH Southcott and JK Dineen. CSIRO: Melbourne, pp. 65-74.

166 Hawkins CD and Morris RS (1978) Depression of productivity in sheep infected with *Fasciola hepatica*. Vet Parasitol **4** 341-51. https://doi.org/10.1016/0304-4017(78)90020-1.

167 Meek AH and Morris RS (1979) The effect of prior infection with Fasciola hepatica on the resistance of sheep to the same parasite. Aust Vet J **55** 61-4. https://doi.org/10.1111/j.1751-0813.1979.tb15163.x.

168 Boray JC, Fraser GC, Williams JD et al. (1985) The occurrence of the snail *Lymnaea columella* on grazing areas in New South Wales and studies on its susceptibility to *Fasciola hepatica*. Aust Vet J **62** 4-6. https://doi.org/10.1111/j.1751-0813.1985.tb06030.x.

169 Harris RE and Charleston WAG (1980) Fascioliasis in New Zealand: A review. Vet Parasitol **7** 39-49. https://doi.org/10.1016/0304-4017(80)90008-4.

170 Happich FA and Boray JC (1969) Quantitative diagnosis of chronic fascioliasis 2. The estimation of daily total egg production of *Fasciola hepatica* and the number of adult flukes in sheep by faecal egg counts. Aust Vet J **45** 329-31. https://doi.org/10.1111/j.1751-0813.1969.tb05012.x.

171 Cullen JM and Stalker MJ (2016) Liver and biliary system. In: Jubb, Kennedy and Palmer's pathology of domestic animals, ed MG Maxie. 6th edition. Elsevier, pp. 320-22. https://doi.org/10.1016/B978-0-7020-5318-4.00008-5.

172 Meek AH and Morris RS (1979) The longevity of *Fasciola hepatica* metacercariae encysted on herbage. Aust Vet J **55** 58-60. https://doi.org/10.1111/j.1751-0813.1979.tb15161.x.

173 Osborne HG (1967) Control of fascioliasis in sheep in the New England District of New South Wales. Aust Vet J **43** 116-17. https://doi.org/10.1111/j.1751-0813.1967.tb08907.x.

174 Seddon HR (revised by Albiston CBE) (1967) Fascioliasis. In: Diseases of domestic animals in Australia, Part 1 Helminth Infestations. Commonwealth of Australia Department of Health: Canberra, pp. 7-27.

175 Toet H, Piedrafita DM and Spithill TW (2014) Liver fluke vaccines in ruminants: Strategies, progress and future opportunities. Int J Parasitol **44** 915-27. https://doi.org/10.1016/j.ijpara.2014.07.011.

176 Molina-Hernández V, Mulcahy G, Pérez J, Martínez-Moreno Á et al. (2015) *Fasciola hepatica* vaccine: We may not be there yet but we're on the right road. Vet Parasitol **208** 101-11. https://doi.org/10.1016/j.vetpar.2015.01.004.

177 Fairweather I and Boray JC (1999) Fasciolicides: Efficacy, actions, resistance and its management. Vet J **158** 81-112. https://doi.org/10.1053/tvjl.1999.0377.

178 Fletcher HL, Hoey EM, Orr N et a. (2004) The occurrence and significance of triploidy in the liver fluke, *Fasciola hepatica*. Parasitol **128** 69-72. https://doi.org/10.1017/S003118200300427X.

179 Overend DH and Bowen FL (1995) Resistance of *Fasciola hepatica* to triclabendazole. Aust Vet J **72** 275-6. https://doi.org/10.1111/j.1751-0813.1995.tb03546.x.

180 Brockwell, YM, Elliott TP, Anderson GR et al. (2014) Confirmation of *Fasciola hepatica* resistant to triclabendazole in naturally infected Australian beef and dairy cattle. Int J Parasitol: Drugs and Drug Resistance **4** 48-54. https://doi.org/10.1016/j.ijpddr.2013.11.005.

181 Elliott TP, Kelley JM, Rawlin G et al. (2015) High prevalence of fasciolosis and evaluation of drug efficacy against *Fasciola hepatica* in dairy cattle in the Maffra and Bairnsdale districts of Gippsland, Victoria, Australia. Vet Parasitol **209** 117-24. https://doi.org/10.1016/j.vetpar.2015.02.014.

182 Johns DR and Dickeson SJ (1979) Efficacy of albendazole against *Fasciola hepatica* in sheep. Aust Vet J **55** 431-2. https://doi.org/10.1111/j.1751-0813.1979.tb05599.x.

183 Anderson N, Luong TT, Vo NG et al. (1999) The sensitivity and specificity of two methods for detecting *Fasciola* infections in cattle. Vet Parasitol **83** 15-24. https://doi.org/10.1016/S0304-4017(99)00026-6.

184 Mezo M, Gonzalez-Warleta M and Ubeira, FM (2007) The use of MM3 monoclonal antibodies for the early immunodiagnosis of ovine fascioliasis. J Parasitol **93** 65-72. https://doi.org/10.1645/GE-925R.1.

185 Flanagan A, Edgar HWJ, Forster F et al. (2011) Standardisation of a coproantigen reduction test (CRT) protocol for the diagnosis of resistance to triclabendazole in *Fasciola hepatica*. Vet Parasitol **176** 34-42. https://doi.org/10.1016/j.vetpar.2010.10.037.

186 Palmer DG, Lyon J and Forshaw D (2014) Evaluation of a copro-antigen ELISA to detect *Fasciola hepatica* infection in sheep, cattle and horses. Aust Vet J **92** 357-61. https://doi.org/10.1111/avj.12224.

187 Coles GC, Bauer C, Borgsteede FH et al. (2006) World Association for the Advancement of Veterinary Parasitology (WAAVP) methods for the detection of anthelmintic resistance in nematodes of veterinary importance. Vet Parasitol **44** 35-44. https://doi.org/10.1016/0304-4017(92)90141-U.

188 Flanagan A, Edgar HWJ, Gordon A et al. (2011) Comparison of two assays, a faecal egg count reduction test (FECRT) and a coproantigen reduction test (CRT), for the diagnosis of resistance to triclabendazole in *Fasciola hepatica* in sheep. Vet Parasitol **176** 170-6. https://doi.org/10.1016/j.vetpar.2010.10.057.

189 Mitchell GBB, Maris L and Bonniwell MA (1998) Triclabendazole-resistant liver fluke in Scottish sheep. Vet Rec **143** 399.

190 Sargison ND and Scott PR (2011) Diagnosis and economic consequences of triclabendazole resistance in *Fasciola hepatica* in a sheep flock in south-east Scotland. Vet Rec **168** 159. https://doi.org/10.1136/vr.c5332.

191 Daniel R, Jenkins T, van Dijk J et al. (2012) A composite faecal egg count reduction test to detect resistance to triclabendazole in *Fasciola hepatica*. Vet Rec **171** 153-7. https://doi.org/10.1136/vr.100588.

Appendix I

FIELD TECHNIQUE FOR A WORM COUNT OF THE ABOMASUM AND SMALL INTESTINE

1. Remove the entire gastrointestinal tract and, to prevent leakage, tie the gut with string at both ends of the abomasum, the start of the small intestine and the ileo-caecal valve, then divide into the abomasum, small intestine and large intestine.

2. Cut the abomasum along the greater curvature and empty the contents into a bucket (marked at 1.0, 1.5 and 2.0 L).

3. Wash the abomasal mucosa with water (warm if possible) and vigorously rub with the back of the hand to remove worms from the abomasal folds. This process is repeated a few times until the mucosa is free from contents.

4. Add clean water such that the volume of material in the bucket is made up to the next mark (preferably 1 or 2 L, as this makes the counts derived from the dilution factors easier to calculate in the field).

5. Mix the contents using a 'figure of 8' action (rather than a simple circular stirring motion, as the centrifugal force results in the nematodes being distributed in a 'band').

6. Take a 100 mL subsample using a small jar or beaker and ladle it into a glass or plastic jar with a 60-mesh brass sieve (openings of 0.25 mm) securely glued to a hole cut into the lid. Secure the lid (Figure 9.6).

7. Shake the contents out through the mesh in the lid, then refill the jar with clean water and shake again. This process is repeated two or three times to remove fine particulate matter (Figure 9.7).

8. Run water back through the sieve to dislodge any worms inside the mesh and make the volume in the jar up to 200 mL.

9. Mix the contents of the jar, then take two 20 mL aliquots, placing them into separate white polystyrene trays (15 x 15 cm) (Figure 9.8).

10. Stain with parasitological iodine (30 g iodine, 40 g potassium iodide; add water to 100 mL) (Figure 9.9).

11. After a few minutes, add sodium thiosulphate to clear the iodine stain from any remaining plant matter (Figure 9.10).

12. Count the number of worms in each tray and calculate the number of worms in the whole organ by multiplying by the appropriate dilution factor (for example, if contents were made to a total of 1000 mL, then a 100 mL subsample taken, the sample includes $\frac{1}{10}$ of the total volume. Subsequently, each 20 mL subsample of the 200 mL in the jar is again $\frac{1}{10}$ of the total volume, so the dilution factor applied to the count in each tray is

100 (alternatively, the dilution factor would be 200 if the total volume of contents in the bucket was made up to 2000 mL).

For the small intestine, the first third (3-4 m) is stripped from the mesentery and the contents run into a similarly marked bucket by squeezing the outside of the intestine between thumb and forefinger; 50-100 mL of (warm) water is then run into the small intestine using a funnel, and washed through the small intestine, again squeezing the intestine firmly to dislodge worms embedded in the mucosa. Washing should be performed from the distal end because most of the small intestinal nematodes are in the proximal section of the duodenum. This process should be repeated two or three times until the washings are clear. The contents are then processed and counted as for the abomasum.

Immature worms lose their stain more quickly and may be difficult to accurately identify in the field, but a total count is usually sufficient to make a diagnosis and demonstrate very clearly to the producer that nematodes are indeed the problem. It is possible to visualise and count the worms without staining with iodine or clearing with sodium thiosulphate, but staining is generally preferable.

Alternatively, if more accurate identification is required or there is insufficient time, 5 mL of 37% formalin can be added to a 200 mL subsample of the separate abomasal and small intestinal washings to preserve the worms for counting later in a parasitology laboratory. Again, to increase the ease and accuracy of counts, the subsamples of gut contents are sieved before counting. Using a sieve with a smaller mesh opening (0.04 mm) will capture all immature worms, including recently ex-sheathed L_4. For increased accuracy, 5-10 $\times$ 5 mL aliquots are taken and stained for

Figure 9.6: A field technique to estimate the worm burden in a sheep. A known volume (such as 100 mL) is taken as a subsample of the diluted abomasal contents. Photograph courtesy of KA Abbott.

Figure 9.7: The contents are washed by shaking out through the sieve top. Photograph courtesy of KA Abbott.

Figure 9.8: Two 20 mL aliquots are taken, placed into separate white polystyrene trays ready for staining. Photograph courtesy of KA Abbott.

counting under a stereomicroscope at 10-20 times magnification. The total worm count is again calculated by multiplying the number of worms counted by the dilution factor. In this case, counting 10 × 5 mL aliquots (a total of 50 mL) from a subsample derived from either 1 L or 2 L of contents makes the dilution factors 20 and 40, respectively. A definitive identification of the species present requires examination of the bursa and spicule morphology of up to 100 males, but female and immature worms can be quickly identified to genera by an experienced operator.

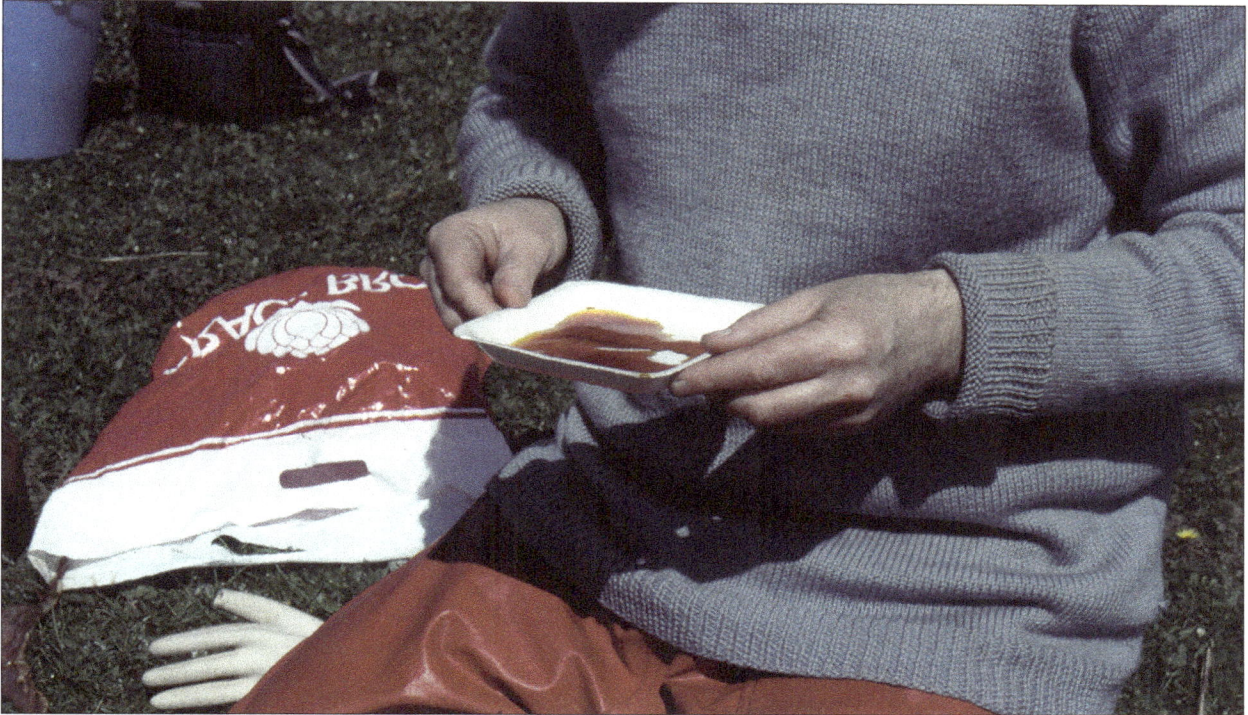

Figure 9.9: The contents are stained with parasitological iodine for a few minutes before partially clearing with sodium thiosulphate. Photograph courtesy of KA Abbott.

Figure 9.10: The stained nematodes can be counted in the tray. In this case, numerous *Teladorsagia* are present in the sample. Photograph courtesy of KA Abbott.

Appendix 2

ASSESSING ANTHELMINTIC RESISTANCE USING A WORM EGG COUNT REDUCTION TRIAL (WECRT)

1. The ideal sheep are 12-13 week-old Merino lambs that are about to be weaned (or slightly older prime lambs if weaning is delayed in these flocks). These should not have been treated previously with an anthelmintic.

2. If such a mob is not available, young sheep (those <6 months old) with a sufficient worm egg count (WEC >200 epg) are the next best alternative. Sheep older than this should not be used.

3. The WEC of the trial mob needs to be at least 200 epg. This should be confirmed using a bulk WEC of the trial mob at least seven days before the trial is set up. This avoids the wasted time and expense of a WECRT that cannot be interpreted or that, at best, provides only equivocal results because the WEC of the Control group is too low.

4. In areas which experience only sporadic infections with *Haemonchus* (Barbers Pole Worm), the trial mob can be treated with closantel seven days prior to the setup of the WECRT to eliminate *Haemonchus* infections. This makes the interpretation of the results more accurate for the scour worms (*Teladosdagia, Trichostrongylus* and large intestinal parasite species).

5. The farm's history will dictate which groups are included in the WECRT. Typically these will include

   a. an untreated control (always!)

   b. a macrocyclic lactone (ML); ivermectin is the least potent ML, but if ivermectin resistance is already known to exist in a flock then abamectin or moxidectin will be more appropriate

   c. a benzimidazole (BZ) and levamisole (LEV), either alone or in combination

   d. an organophosphate (OP), alone or in combination with BZ and/or LEV

   e. the combination or product the producer has been using most recently (for example, Abamectin + BZ + LEV).

6. Sheep should be clearly identified to their group by a coloured raddle or eartags (including the untreated controls). It is best to devise and use a standard scheme for the common groups, but this should be clearly recorded on day 0.

7. On day 0 (the setup of the WECRT), aim to include 13 or 14 sheep in each treatment group. This makes it easier to collect faecal samples from at least 10 individual sheep in each group 10-14 days later (that is, before the pre-patent period of 17-21 days).

8. Long-acting products are rarely included in a formal WECRT, but their efficacy can be assessed by monitoring the WECs of treated sheep at appropriate intervals after

treatment (for example, at 2, 4 and 6 weeks for long-acting moxidectin injections, which is claimed to prevent reinfection for up to 49 days against *T colubriformis*, and for 91 days against reinfection with *Teladorsagia circumcincta* and *Haemonchus contortus*). Suggested monitoring intervals after treatment with controlled-release capsules are 2, 6, 8, 10 and 12 weeks (these prevent reinfection for up to 90-100 days)

9. Results from a typical WECRT are given below.

Table 9.5: Worm egg counts (epg) in faeces collected at the end of a worm egg count reduction trial.

				WEC of sheep in each group:					
	Control	BZ	LEV	BZ + LEV	ML	OP	OP+BZ	OP+ LEV	OP+BZ +LEV
	540	0	90	0	0	0	0	0	0
	570	0	210	0	0	15	0	0	0
	720	60	210	0	0	15	0	0	0
	750	120	390	0	30	15	15	15	0
	1890	120	480	0	45	15	30	30	0
	2130	240	480	90	60	30	30	30	0
	2430	300	540	90	60	30	45	45	15
	2430	720	540	180	105	45	45	45	15
	3450	1260	630	300	150	60	60	45	15
	3660	No sample	870	480	165	270	450	45	15
Group Average WEC:	1857	313	444	114	62	50	68	26	6
% Reduction in WEC: [A]		83	76	94	97	97	96	99	99.7
Lower 95% CI [B]		53	59	83	93	92	85	97	99
Upper 95% CI		94	86	98	98	99	99	99	100

[A] The percent reduction is calculated as $(WEC_{Control} - WEC_{treat}) \div WEC_{Control}$. A figure of 95% or greater was originally regarded as indicating sufficient efficacy, but that figure is now preferred to be as close to 100% as possible, particularly for strategic treatments.

[B] The lower 95% confidence interval is an additional measure of the efficacy of an anthelmintic treatment. Generally a value <90% is taken as indicating that the treatment is not effective, and this increases the statistical probability of declaring resistance.[98] This calculation is best done using a spreadsheet, but can be undertaken manually (see p. 26 of *Anthelmintic Resistance: Report of the working party on anthelmintic resistance*.[98])

10

Diseases of the integument and eye

This chapter is divided into four sections:
- Section A: Diseases of the eye
- Section B: Bacterial and viral diseases of the integument (fleece rot, dermatophilosis, actinobacillosis, contagious pustular dermatitis or scabby mouth, capripox)
- Section C: Non-infectious diseases of the integument (photosensitisation, squamous cell carcinoma, burns, gangrene)
- Section D: External parasites (flies, lice, mites).

SECTION A

DISEASES OF THE EYE

Ovine infectious keratoconjunctivitis (pink eye)

In sheep, pink eye is associated with infection with a number of different species of bacteria. *Mycoplasma conjunctivae*[1] and *Chlamydia pecorum* are probably the most important. Other organisms — *Branhamella* (*Moraxella*) *ovis* (formerly *Neisseria ovis*), *Colesiota* (formerly *Rickettsia*) *conjunctiva* and *Listeria monocytogenes* — are less commonly involved in the condition.

Branhamella ovis is present in the conjunctival sacs of many normal, healthy sheep and, without other pathogens or trauma to the eye, is generally not responsible for significant ocular disease. There appear to be exceptions, however, possibly depending on the virulence of the strain — the organism was found to be responsible for an outbreak of keratoconjunctivitis in a herd of goats which, in some cases, involved severe corneal lesions. In that case, the disease was reproduced experimentally with isolates from affected goats introduced into the conjunctival sacs of otherwise healthy kids.[2] Generally, however, *B ovis* appears not to be a significant pathogen except to have a role in increasing the severity of disease in eyes infected with *M conjunctivae*.[3]

M conjunctivae is capable of initiating the development of keratoconjunctivitis either alone or in combination with *B ovis*.[4,5] Close contact between sheep appears to enhance the spread of mycoplasmal keratoconjunctivitis and dusty conditions probably contribute to the risk of developing disease.[6] Insect vectors are likely to be involved in transmission. Outbreaks in a flock often follow the introduction of sheep with mild or inapparent infection[7] and the organism can persist in the conjunctival sacs of sheep for at least three months after recovery from clinical disease. Both adult and young sheep are affected.

In Australia, keratoconjunctivitis has been reported in crossbred lambs and weaners as an occasional additional manifestation of acute chlamydial infection in which arthritis, lethargy and depression are the major presenting signs.[8] In the USA, the occurrence of chlamydial keratoconjunctivitis in lambs soon after entry into feedlots was reportedly similar but at a very much higher incidence — up to 90% — with polyarthritis occurring in 10% to 85% of lambs, depending on the feedlot.[9] Chlamydial keratoconjunctivitis has also been reported as the predominant sign of morbidity in adult sheep.[10] Attempts to reproduce keratoconjunctivitis in sheep by the introduction into the conjunctival sac of *C pecorum* have not been successful.[4] *C pecorum* can be isolated from the conjunctival sac of apparently normal animals as well as from animals with lesions of keratoconjunctivitis. Diagnosis of a chlamydial role in keratoconjunctivitis is difficult — isolation of the organism does not prove that it caused disease and serology for chlamydial infection in sheep is unreliable because infected animals frequently have negative titres.[11] Sheep with chlamydial keratoconjunctivitis usually respond promptly to treatment with intramuscular oxytetracycline, but relapse after days or a few weeks is common.

Trachoma — conjunctivitis of humans caused by *Chlamydia trachomatis* — is endemic in some developing countries and in parts of remote Australia. The way in which the disease spreads has been extensively researched in order to provide insights for control measures and the findings are probably relevant to the epidemiology of ocular infections in animals. *C trachomatis* is transmitted from infected people on fomites (clothing, face cloths), by hands, eye or nasal secretions and by eye-seeking muscid flies. Measures which have been proved to reduce the incidence of trachoma include fly control, access to clean water for face washing and maintaining the face cleanliness of children — factors which reduce the spread of the organism to hands and fomites and reduce the presence of flies on faces.[12,13]

Clinical signs

Ovine keratoconjunctivitis, regardless of the aetiological agent, is characterised by swelling and reddening of the conjunctivae, excessive lacrimation and staining of the face, corneal opacity, follicular hyperplasia and, in some cases, corneal ulceration and purulent ocular discharge. Development of corneal ulceration and subsequent pannus are usually accompanied by blindness which may persist for several weeks. In most cases, the condition resolves without treatment in about two weeks but, in severe cases, corneal scarring, hypopyon or corneal perforation may lead to permanent blindness. Blindness, whether temporary or permanent, can lead to isolation from the flock and trauma, even death, due to physical injury or becoming trapped due to lack of vision.

Outbreaks of keratoconjunctivitis can be sudden and marked by rapid spread within a flock.[14] The disease usually runs its course over a two- to three-month period, although it tends to be less serious, and to resolve faster, in lambs compared to adults. During the course of the condition reproductive performance and growth rate may be impaired.[15]

Keratoconjunctivitis caused by *C pecorum* is usually bilateral and usually accompanied, at a flock level, by signs of systemic illness, particularly arthritis. Pregnant ewes may abort.

Management of keratoconjunctivitis

Most cases resolve without treatment, although some severely affected animals may be blind for a period of days or a few weeks before recovery. Foreign bodies, particularly the awned spikelets of grass seed heads, can enter the eyes and cause outbreaks of keratoconjunctivitis in flocks of sheep with significant effect on the growth rate of weaners. When outbreaks of eye disease occur the sheep should be yarded and examined to determine whether foreign bodies are the cause and, if they are, the material should be removed and the animals placed on a safer pasture. Wigging the sheep may also reduce the chance of grass seed awns entering the eyes.

If it is established that the keratoconjunctivitis is primarily infectious and not caused by plant material or other foreign bodies, affected mobs of sheep should generally be left at pasture without treatment. Repeated yarding may exacerbate the condition and increase transmission. When treatment of individual sheep is desired, systemic and topical antibiotics, particularly tetracyclines[16], are appropriate.

SECTION B

BACTERIAL AND VIRAL DISEASES OF THE INTEGUMENT

Fleece rot

This is a superficial dermatitis caused by prolonged wetting of the skin and the multiplication of bacteria, particularly *Pseudomonas aeruginosa*, but also including other bacteria normally present in the fleece.[17] *P aeruginosa* produces a bright green pigment and appears to be responsible for the most severe forms of fleece rot, but other bacteria, some producing pigments of blue, orange, brown or pink, are also common.

Some of the colouration of the wool is *non-scourable*. Consequently, the wool is lowered in value. More important economically is the fact that fleece rot is the major predisposing condition for flystrike. In a study performed in the 1980s fleece rot was estimated to cost approximately $1 per hogget in moderately affected flocks. Approximately half of this cost was associated with flystrike control measures and half with discounts applied to coloured wool.

Fleece rot is endemic predominantly in the high and medium rainfall areas of Australia, particularly in spring and autumn and following rains in summer. Fleece rot is most common following periods of heavy rainfall (100 mm in one month) or eight days of rain.[18] In low rainfall areas fleece rot can be a serious problem following prolonged rain because it predisposes sheep to the devastating effects of flystrike, the management of which is often difficult in pastoral areas. Within flocks, young sheep are the most susceptible and generally suffer the highest incidence of fleece rot.[19]

Prolonged wetting — over several days — and warm temperatures are necessary for the disease to commence. *P aeruginosa* multiplies rapidly in moist, well-aerated fleeces, and within three days it is able to establish itself as the dominant if not sole species of bacteria present. It grows profusely on the fleece constituents, wax, suint and insoluble nitrogenous material and can hydrolyse wool wax which normally forms a hydrophobic layer on the wool fibres and skin surface.

Association with wool length

Sheep are more susceptible to fleece rot when they have four to six months' wool than at any other time off-shears[20] and this may have significant influence on some management decisions. For example, shearing lambs in spring can increase their susceptibility to fleece rot in the following autumn. In flocks where fleece rot is a problem or where time of shearing is under review, the association must be considered. Fleece rot occurrence, however, is just one of several factors which influence the determination of shearing time and it is not necessarily one of the most important (see Chapter 3).

Association with body conformation

Conformational characteristics which tend to trap and retain moisture on parts of the body are associated with an increased incidence of flystrike. The two most important ones are a wither conformation which tends to hold moisture in a dip between the shoulder blades (*devil's grip*) and excessive wrinkle in the skin, particularly around the breech.

Resistance to fleece rot

There is genetic variation in resistance to fleece rot both within and between flocks. Between strains of Merino, strong wools are more susceptible than fine wools. Within strains, some bloodlines are more susceptible than others. Within flocks there is genetic variation in susceptibility; the heritability of fleece rot incidence and severity score is of the order of 0.2 to 0.4, while the underlying *liability* to fleece rot has a heritability near 0.4.[21]

The variability in expression of fleece rot between seasons and between different environments has encouraged the search for indirect measures of resistance. Of these, *fibre diameter variability* and *greasy wool colour* are the best.[22] For greasy wool colour to be a useful indirect selection criterion, it is necessary for it to be measured objectively with a colorimeter rather than subjectively.[23] Although the use of fleece characteristics (like colour, *character* and *handle*) are potentially useful as indirect selection criteria for fleece rot resistance, the gains from performing any selection in commercial flocks which buy rams from an external source are very small and genetic improvement in the trait is more profitably left to the stud ram breeder.[24] The Australian national ram evaluation program (MERINOSELECT) produces estimated breeding values for the coefficient of variation of fibre diameter (CVFD) and for breech wrinkle — both traits which are associated with susceptibility to fleece rot and, therefore, flystrike.

Prevention of fleece rot

The following steps can be considered to reduce the incidence of fleece rot within a flock:
- selection of strains and bloodlines which perform best in the particular environment
- selection of shearing time
- minimisation of the yarding of wet sheep.

Vaccination has been investigated[25] and early studies showed some possibilities, but no commercial product exists.

Dermatophilosis (dermo, lumpy wool)

This is a dermatitis caused by *Dermatophilus congolensis* which affects sheep, particularly young sheep, after an episode of wetting and delayed drying. The predominant clinical sign is the formation of a hard scab in the wool. It causes economic loss to producers through deaths, lowered production, lowered wool values[26,27], treatment costs and restriction of management options. The prevalence of the disease is very high in some areas of Australia. A survey in Western Australia indicated a flock prevalence over 60% and a prevalence within flocks from zero to 75% — with hoggets being the most frequently affected. The disease is more common on strong-wool properties in medium and high rainfall areas but can occur in medium- and even fine-wool flocks.[28] It is, with fleece rot, one of the very important predisposing causes of flystrike.[29]

At this stage there are no useful indirect selection criteria to facilitate breeding for resistance in Merino sheep.

Close contact following episodes of wetting, such as dipping[30,31], jetting, transport or yarding (for example, for marking) in wet weather, is the most effective method of spreading the disease from infected to uninfected sheep. Zoospores are harboured in lesions on affected or *carrier* sheep — in the latter, the lesions may be confined to the ears and face. The addition of some bacteriostats (for example, zinc sulphate at 0.5%) to dips will reduce the *transmission* of dermatophilosis in dips but will have no curative effect on existing lesions.[30,31]

The disease can be treated by the administration of antibiotics. Penicillin-streptomycin combinations were commonly used but streptomycin has been withdrawn from veterinary use in Australia. Penicillin alone is ineffective. Tetracycline, oxytetracycline and erythromycin will cure some sheep.[32] Treatment of affected sheep is usually reserved for sheep which are so badly affected that they cannot be shorn. If antibiotic treatment is successful, then, approximately eight weeks after treatment, sufficient wool will have grown under the scabs (thus *lifting* the scabs) to allow a shearing comb between the skin and the scabs and the sheep to be shorn. Treatment should be restricted to the most severe cases. Most mild or moderately affected sheep will cure themselves and develop some resistance to the disease — a process which may be delayed by antibiotic treatment.

Actinobacillosis

Actinobacillus lignierisii in sheep affects the skin of the face, lips, nose, lower jaw and lower part of the neck, rather than the tongue, as it does commonly in cattle.[33] The disease, when it affects the lips and muzzle, is called *leather lips*[34] — a condition which has been associated with the compaction of barley grass seeds into the lower lips of sheep grazing annual pastures in summer.[35]

Contagious pustular dermatitis

Contagious pustular dermatitis (CPD or contagious ecthyma) is a common disease of sheep and goats in all countries where sheep are raised. It is colloquially known in Australia and New Zealand as *scabby mouth* and in Britain and many other countries as *orf*. The disease is caused by a parapoxvirus and is most commonly seen in lambs and weaners. Sheep of any age are susceptible if they have not developed immunity through vaccination or prior exposure. Typically, infection occurs on the lips and muzzle, but it also occurs on the lower leg, the oral mucosa or the udder of ewes. In all of these sites the lesions can interfere with the health and productivity of sheep. Infections also occur in other sites on the body where usually they are of no clinical consequence.

The lesions commence with one or more small focal areas of inflammation which develop into vesicles and pustules, and then rupture and form scabs. Scab formation does not occur with lesions in the oral mucosa. The disease normally runs a course of four to six weeks before resolving completely, although most animals become pain-free (apparently) and are able to resume normal function about three weeks post-infection.

Infection with the virus of CPD requires a break in the epithelium. Trauma to the skin can be caused by abrasive feed material such as the thorns of thistles or grass seed awns. On the legs, lesions occur above the coronary band, behind the pastern or on the interdigital skin. Infections are often preceded by persistent wetting that comes from walking through very wet pasture or in flooded paddocks. The resultant maceration of the skin provides the opportunity for the virus to enter the skin. Infection of the site of application of ear tags can occur, leading to an enlargement of the hole and a loss of the ear tag.[36]

Lesions on the lips of lambs can infect ewes' teats and udder skin; the ewes then refuse to allow the lambs to suck. Mastitis and necrosis of the quarter is an occasional consequence. CPD can also occur on the scrotum. Rarely (but not in Australia), the virus has caused severe systemic infection with significant mortality. Forms of CPD in which the mouth and lip lesions are extensive and severe have been reported in Australia[37] but the systemic forms which have been reported elsewhere have not.

CPD is seen most commonly in summer months. The virus may persist in sheep which carry the virus with or without clinical signs[38] and the virus can survive for at least one year in a dry environment. Dried scabs are a rich source of the virus. The virus does not persist in wet scabs or wet environments.[39]

The virus is widespread across Australian sheep flocks and probably occurs on most sheep farms, although significant outbreaks of infection may be restricted to farms where environmental conditions predispose sheep to the disease. Outbreaks occur in conditions which produce high levels of challenge (such as in feedlots or other heavily infected physical environments[37]), or following exposure to a source of skin trauma which simultaneously predisposes many animals in the flock. In one Australian survey, a quarter of flocks were found to have weaners with CPD lesions, with prevalence ranging up to 100% of the inspected animals.[40] In New Zealand, approximately 0.5% of lambs slaughtered at meat works were found to have lesions of CPD.[41] Some lines of lambs had no lesions; the average prevalence in affected consignments was 13%.

Economic consequences of CPD

The results of infection are generally trivial but at times can be serious. The presence of CPD lesions on sheep at particular times may lead to management complications such as delays in shearing or sale of rams. Limb lesions can confuse a diagnosis of benign footrot as well as cause some lameness.

Lip and mouth lesions on weaners lead to reduced feed intake and reduced weight gains or increased rates of weight loss. As a consequence of the reduced body weight, affected young sheep may suffer an increased rate of mortality or may need to have an increased level of supplementation in order to regain a satisfactory weight and condition score.

The infection has been transmitted rapidly between sheep in *Sharlea* sheep sheds[a] and between sheep on ships en route to the Middle East. The resultant inappetence can lead to increased mortalities but, more importantly, the presence of lesions on arrival has on occasion led to the refusal of importing countries to accept delivery.

Other hosts

Goats and some other species of ruminants are also susceptible to infection with CPD. Humans are readily infected and the disease is an occupational hazard amongst sheep farmers, shearers and others who handle sheep. Reinfection does occur but there is some evidence for a protective immunity in people. Lesions in humans are most commonly solitary lesions on a finger (*target-like* lesions) which heal spontaneously in six or seven weeks, but more severe reactions occur in a small proportion of cases, developing *erythema multiforme* and even widespread vesicular eruptions.[42]

Vaccination and immunity

A vaccine is available for use in sheep. It is applied with a small, double-pointed fork which holds a small volume of vaccine suspension and simultaneously scratches the skin and dispenses the fluid. The usual site for vaccination is the inside of the thigh, and the operation can be done conveniently at lamb marking. This site is chosen because the skin is not wool-bearing, facilitating the skin scratch, and because CPD lesions in this site cause little if any inconvenience to the animal. It is advisable to examine the site of inoculation on a number of sheep five to seven days after treatment to ensure a successful *take*. Pustules should be evident along the site.

Immunity following the deliberate infection of the skin of the upper leg protects animals from any subsequent natural exposure, although immunity after one vaccination or the first natural infection appears to be incomplete.[43] Sheep can become infected again, although repeat infections tend to be milder and of shorter duration than new cases.[44]

The protection afforded by vaccination develops within 16 to 21 days of vaccination[40] and lasts for at least six months.[45] Repeated exposure to the virus probably provides frequent boosting of immunity for most sheep grazing under natural conditions, leading effectively to lifelong protection for most vaccinated or previously infected animals.

The vaccine is prepared from live virus which has been passaged *in vivo* or in cell culture and so is capable of causing the disease.[46] Consequently, any animals which miss vaccination are likely to develop the disease naturally from the contamination of the environment caused by vaccinates with active vaccination site lesions.

Vaccination of a flock of young sheep will provide a useful level of protection against infection but will not protect 100% of the flock. This may be due to the incomplete immunity generated by vaccination or it may be due to inadequate application of the vaccine to every susceptible animal.

---

a Sharlea sheep production refers to the growing of ultra-fine Merino wool from sheep which are maintained in sheds to enable tight control of diet and to minimise fleece contamination. The high level of dietary control leads to the production of very fine wool of high tensile strength. The system is named for a farm in Victoria (*Sharlea*) where this form of wool production was first developed.

Diagnosis

The diagnosis is generally made on clinical grounds but can be confirmed by the demonstration of virus in lesion material by electron microscopy. Limb lesions must be distinguished from strawberry footrot. *Dermatophilus* infection causes scab formation and can occur simultaneously with CPD infection.[47] In the woolled areas of the body the persistent scabs are readily distinguished from CPD but, on the haired skins, the two infections can appear similar. Photosensitisation of the face and exposed skin leads to the production of dried, necrotic skin which may resemble CPD infection, but the diffuse nature of photosensitisation differentiates it from the focal, discrete lesions of CPD.

Several diseases exotic to Australia have closer similarities to CPD. Sheep with bluetongue are febrile. The mucosa of the mouth and nose become hyperaemic and there is profuse salivation and nasal discharge which contains pus and becomes bloodstained. The severe systemic illness and high mortality — which would be expected in an outbreak in the naïve Australian sheep flock — helps to distinguish it clinically from CPD.

Sheep pox is also expected to have a high mortality rate if introduced into a naïve population. The disease is acute or peracute, with high fever, profuse salivation, discharges from nose and eyes. Pox lesions develop on all parts of the body including the oral mucosa.

Foot-and-mouth disease (FMD) is often mild in sheep but can cause sudden and severe lameness, with blisters on the foot and interdigital skin and lesions on the tongue and dental pad. Cases of FMD in sheep involving severe systemic illness are relatively straightforward to distinguish from CPD but there is a risk during an FMD outbreak that typically mild cases of FMD in sheep may be easily missed or misdiagnosed as CPD.

Vesicular stomatitis can present a similar clinical presentation, with the formation of vesicles in the mouth and on limbs and teats, but it is rarely contracted by sheep or goats.

Treatment

There is no effective treatment. The disease resolves spontaneously after two to three weeks. In the case of valuable animals with limb lesions, secondary infection may warrant antibiotic treatment.

Capripox infection (sheep pox and goat pox, SGP)

There are numerous strains of capripox virus which can infect sheep and goats. Some strains are more pathogenic in one host species than the other, while some strains are of similar pathogenicity in both. Some strains of capripox virus infect cattle, but transmission of infection from cattle to sheep or goats appears to be uncommon.[48] The viruses do not occur in Australia but are endemic in central and north Africa, the Middle East, Afghanistan, Pakistan, Bangladesh, India, Nepal, parts of China and Vietnam. There have been several outbreaks in Cyprus and Greece which have been controlled. Merino and European breeds of sheep are very susceptible to infection and much more severely affected than native African and Middle Eastern breeds.[49]

Epidemiology

Sheep pox is highly contagious. It is spread usually during close contact with infected animals, through abrasions, inhalation and possibly by arthropod bites. Fomites and areas contaminated by virus can also provide a source of infection because the virus is very resistant in the environment. It can survive months in dry scab material and on hair and wool out of sunlight. Skin lesions are the main source of virus.

In endemic areas, the prevalence of infection is often low for extended periods with periodic epidemics, possibly because immunity in the population declines in the absence of clinical cases.

Pathogenesis

Following infection there is a viraemia after about seven days, peaking at days 10 to 14 and persisting for one to two weeks. During this time the virus is distributed widely throughout the body, including to the skin. The skin lesions are characteristic pox lesions.

Clinical signs

The severe, acute form of the disease occurs in lambs and fully susceptible animals. There is marked depression and prostration, high fever and ocular and nasal discharges. Affected animals may die at this time or develop skin lesions on the non-woolled (hairy) areas of skin, on the nares, in the mouth and on the vulval mucosa within one or two days. In susceptible animals the disease is very severe, with a mortality rate of 50% to 100%.

In adult animals of resistant breeds, or in partially immune animals, the disease is milder with no systemic reaction. Skin lesions occur and are often concentrated under the tail. Severe losses can occur in ewes from acute secondary mastitis if the virus invades the udder. Healing of skin lesions is slow.

Diagnosis

Bluetongue and CPD also have buccal, nasal or skin lesions, but clinical signs differ from sheep pox. CPD lesions are usually more proliferative. Bluetongue lesions also have a different distribution. Clinical pathology for both virus detection and serology will confirm a diagnosis.

Treatment and control

There is no effective treatment. Control in endemic areas is achieved by vaccination and biosecurity practices.

SECTION C

NON-INFECTIOUS DISEASES OF THE INTEGUMENT

Photosensitisation

Photosensitisation refers to the abnormal and exaggerated sensitivity of unpigmented skin to sunlight caused by the presence of a photodynamic agent in the skin cells. In the case of grazing animals, the phototoxic substance or their precursors enter the body by ingestion and accumulate in body tissues following distribution through the circulatory system. In the unpigmented, unprotected skin these agents respond to sunlight of a particular wavelength — the absorption spectrum depending on the photodynamic agent — and the subsequent chemical reactions interfere with cellular functions, causing severe damage to skin cells and tissues.

The photosensitising compounds are chemically diverse but are usually pigments with natural fluorescence. Their common characteristic is that they are photodynamic — that is, they consist of molecules which can absorb light energy and become activated. In living tissues energy is then transferred from the photo-activated molecule to other molecules in the immediate environment, leading to disruption of normal cellular processes. As a result, there may be interference with membrane permeability, cell division, active transport mechanisms, glycolysis and cellular respiration. Ultimately, cell death may be the outcome.

Photosensitisation differs from sunburn in several ways. For photosensitisation to occur, a photodynamic agent is required, the effect on the skin containing the phototoxin occurs immediately the sunlight hits the skin, and the wavelength of light causing the skin damage is usually in the visible spectrum (cf. ultraviolet light, which causes sunburn).

Photosensitisation in sheep occurs through one of three different mechanisms:

(i)   primary photosensitisation, in which the photodynamic agent is ingested directly

(ii) secondary or hepatogenous photosensitisation, in which the failure of the liver to excrete phylloerythrin leads to its accumulation

(iii) photosensitisation caused by a congenital metabolic defect leading to the accumulation of phylloerythrin.

Clinical signs

When exposed to sunlight, animals with photosensitised skin react with pain and discomfort which encourages them to seek shade. Only unpigmented skin is affected and, in sheep, the fleece can provide substantial protection depending on its density and staple length. In Merino sheep, 14 weeks or more of wool growth is protective of those parts of the skin normally covered by wool.[50] If unpigmented, the skin of the face, including lips, eyes and ears, becomes erythematous, pruritic and oedematous. There may be corneal ulceration. The swelling of the head leads to a range of common names for the condition — different for different regions and countries, but often referring to the *big head*. If exposed and susceptible, the skin of the perineum, udder and teats may be involved. Photosensitive sheep at pasture will attempt to remove themselves from direct sunlight by shading their heads and other affected parts under any available object. If disturbed they will attempt to quickly find shade elsewhere.

In severe cases the skin becomes exudative and necrotic and, after two to three weeks, it sloughs. If unable to escape the sunlight affected animals may die in the early stages from shock or, in the later stages, from secondary infection. Pregnant ewes may abort. In surviving sheep, the skin lesions will ultimately heal with scar formation.

Hepatogenous photosensitisation

When photosensitisation is hepatogenous in origin, the initial disease process is damage to the liver — specifically the biliary system — caused by ingestion of a toxin produced by plants, algae or plant-associated fungi, by administration of toxic chemicals, or by parasitic damage. Phylloerythrin (also called phytoporphyrin) is a product of the ruminal fermentation of chlorophyll and is normally absorbed from the rumen and excreted by the liver through the biliary system. Failure of the hepatic excretion system leads to the accumulation of phylloerythrin in bodily tissues, including the skin, where its action as a photodynamic agent leads to lesions of photosensitisation. Some specific examples of hepatogenous photosensitisation are discussed below, grouped by the toxic principle which causes the hepatic insult.

Sporidesmin

Facial eczema is the most common and important example of hepatogenous photosensitisation in Australia and, particularly, New Zealand, where the disease has an economic impact on sheep production second only to nematode parasitism.[51] The liver damage is caused by the ingestion of sporidesmin, a toxin present in the spores of the saprophytic fungus *Pithomyces chartarum*. Sporidesmin damages the cells of the biliary epithelium, leading to cholestasis and the retention of phylloerythrin. *P chartarum* proliferates on dead pasture material under conditions of warmth and high humidity. Suitable climatic conditions occur frequently in the late summer and autumn in the North Island of New Zealand and serious, widespread outbreaks are reported from that region every few years.[51] Facial eczema is also reported from Gippsland in eastern Victoria, Australia.[52]

There is no treatment for facial eczema, but a range of preventive approaches can be used. These include recognition and avoidance of pastures which pose a high risk for facial eczema, particularly when climatic conditions are suitable for an outbreak, the application of fungicide to pastures and, in New Zealand, breeding of sheep genetically resistant to the disease. Zinc salts provide protection against the liver injury caused by sporidesmin and frequent or continuous provision of zinc to animals grazing high risk pastures provides effective prevention.[53] In both New Zealand and Victoria, facial eczema warnings are published for producers during periods when weather conditions favour spore formation, and when spore counts on sentinel farms are high.

Saponins

Geeldikkop is the common name given to a disease which occurs as outbreaks of hepatogenous photosensitisation in the low rainfall Karoo region of South Africa in sheep grazing the summer-growing prostrate herb *Tribulus terrestris*. The plant occurs in other countries including Australia, where it known as *Caltrop* or *Puncture-vine*. Cases of photosensitisation similar to the South African disease have been reported from NSW in both sheep[54] and goats[55] grazing *T terrestris*. The photosensitisation of Geeldikkop is caused by the accumulation of phylloerythrin in the skin — as in facial eczema — but the bile duct occlusion in Geeldikkop appears to be caused by crystalloid substances which are metabolites of saponins[b] which occur in the *Tribulus* plants. It had been suspected that sporidesmin was involved in the pathogenesis of Geeldikkop, but the hepatic pathology of facial eczema and Geeldikkop is different, and Geeldikkop can be reproduced in the absence of sporidesmin.[56]

Other plants containing steroidal saponins have also caused hepatogenous photosensitisation of sheep, including several representatives of the millet family. Millets are a group of

---

b   Saponins are a group of naturally occurring chemicals which are widely distributed in plants of many families. The saponin molecule consists of a *sapogenin* component conjugated to a glycoside. Sapogenins are usually based on a steroid molecular framework. The biliary crystals which occur in Geeldikkop and related conditions are insoluble salts (usually calcium) of the β-D-glucuronides of sapogenin compounds.

grasses of several different genera, including *Panicum*, *Paspalum* and *Echinochloa*. In Australia, *Panicum effusum* (hairy panic)[57], *P coloratum* (coolah grass or bambatsi)[58], *P schinzii* (sweet grass, also recorded as *P laevifolium*) and *Echinochloa utilis* (Japanese millet) have been identified as causes of photosensitisation.[59] In New Zealand, grazing of *Panicum dichotomiflorum* (smooth witch grass) and *P miliaceum* (proso millet) has led to saponin-derived biliary obstruction and photosensitisation outbreaks in sheep.[60]

In Britain, a hepatogenous photosensitisation associated with ingestion of *Narthecium ossifragum* (a lily, commonly called *bog asphodel*), is known by a variety of local common names including *plochteach*.[61] The condition is called *alveld* in Norway. The plant contains sapogenins, which may be responsible for the hepatic damage, although the pathogenesis is still uncertain.[62,63] Up to 50% of lambs may be affected in outbreaks, but the disease is rare in adult sheep. In Britain the disease is particularly prevalent in the west and northwest highlands of Scotland. It is also reported from the Faroe Islands.

Lantadenes

Lantana camara is an invasive weed in Queensland and NSW that is toxic to grazing animals, causing hepatic damage, cholestasis and subsequent photosensitisation due to the biliary occlusion. The hepatotoxins are pentacyclic triterpenoids called lantadenes.[64] Intoxication is accompanied by ruminal stasis which results in retention of the toxic material in the rumen, extending the time for which toxins are absorbed from the digestive tract. Removal of the toxin from the rumen will allow sheep to recover. Laxatives, gut stimulants and oral fluids do not appear to aid recovery but physical emptying of the rumen — by rumenotomy — followed by rumen repopulation may be effective. Alternatively, oral administration of activated charcoal (500 g in 4 L of electrolyte solution) is recommended for intoxicated animals. Supportive treatments, including further rehydration as necessary, should be instituted.[65]

Pyrrolizidine alkaloids

Pyrrolizidine alkaloids (PA) occur in a number of plants but PA poisoning of sheep is most commonly associated with *Heliotropium europaeum* (common heliotrope), *Echium plantagineum* (Paterson's curse, Salvation Jane) and *Senecio* spp (fireweed, ragwort). These chemicals are hepatotoxic, causing damage to hepatocytes and loss of liver function (see Chapter 17). Chronic ingestion of PA predisposes sheep to hepatogenous chronic copper poisoning, or they may develop clinical signs of wasting and emaciation as a result of primary PA poisoning, with photosensitisation as an occasional accompanying sign. The PAs are not themselves phototoxic, nor is the effect on the liver one of cholestasis. Photosensitisation occurs as a result of the failure of the normal hepatic metabolism and excretion of phototoxic substances in the diet, including phylloerythrin.

Phomopsins

The narrow-leafed lupin (*Lupinus angustifolius*) is a grain legume crop planted extensively in southern Australia, particularly in WA. The lupin grain is a valuable animal and human feed and the stubbles, available in summer, are frequently grazed by sheep which benefit from the nutrition supplied by the dead plant tissue and, particularly, any spilled seed.

The fungus *Diaporthe toxica* (anamorph = *Phomopsis leptostromiformis*[c])[66] grows on the stems of lupin plants and causes the plant disease *phomopsis stem blight*. The fungus produces metabolites — phomopsins — which are toxic to sheep, causing the condition *lupinosis*. Sheep grazing affected lupin stubbles may ingest sufficient mycotoxin to suffer liver atrophy and fibrosis. Weight loss, associated with inappetence, and jaundice are the most prominent presenting signs. Photosensitisation is sometimes seen in sheep which have chronic lupinosis-induced liver damage and gain sudden access to green feed.[67,68]

Microcystins

Cyanobacteria (blue-green algae) have a worldwide distribution, occurring in both freshwater and marine environments. They produce a variety of toxins which are classified functionally as cytotoxins, neurotoxins or hepatotoxins. The hepatotoxins are further categorised as microcystins or nodularin. The genera of cyanobacteria of veterinary interest in Australia are *Microcystis*, *Anabaena* and *Nodularia*. Microcystins are produced by species of *Microcystis* and *Anabaena*.[69] Nodularin is found only in *Nodularia* spp.

In Australia, sheep deaths have been reported due to a bloom of *Nodularia spumigenia* in Lake Alexandrina, South Australia[70], *Microcystis aeruginosa* (= *Anacystis cyanaea*)[71] and *Anabaena circinalis*[72,73] on farm dams in NSW. Rapid multiplication of the bacteria — producing an algal bloom — frequently occurs in still waters when nitrogen and, particularly, phosphorus concentrations in the water are high and when pH and water temperatures increase. Blooms are most likely to occur in farm dams or lakes in summer as the water temperature rises and, if there is no flow or mixing of water, a warm upper layer of water can develop. The algal toxins are released as the algae die. Animals can be poisoned by drinking water contaminated by algal toxins or by ingesting algal scum blown to the edge of water bodies. The dead algal material can remain toxic to animals and humans for up to five months. Cyanobacterial hepatotoxins are capable of causing such sudden and severe liver damage that animals ingesting the material may die within 24 to 72 hours. Hepatogenous photosensitisation can occur in those animals which survive for some days but incur sufficient liver damage to incapacitate the excretion of phylloerythrin.[74]

Physical and parasitic damage to the biliary ducts

Diaphragmatic hernia is a rare condition in sheep but has been reported in Merino and Texel lambs. In one case (a 9-month-old Texel lamb), the entrapment caused obstruction of the common bile duct resulting in obstructive jaundice and photosensitisation.[75] Biliary obstruction caused by infection with the trematode parasite *Dicrocoelium dendriticum* has been reported to cause hepatogenous photosensitisation in 14-month-old ewes in Scotland.[76]

Primary photosensitisation

Primary photosensitisation occurs when the sensitising photodynamic agent is ingested directly. A number of plants contain compounds which can cause primary photosensitisation and

---

c   The asexual reproductive form of a fungus is the anamorph. The sexual reproductive form is the teleomorph.

St John's Wort (*Hypericum perforatum*) is one of the most studied of these. It is native to Europe, Asia and north Africa and has established itself in many other countries, including Australia and New Zealand. The plant contains a number of bioactive products of which hypericin, pseudohypericin and several related analogues (collectively called hypericin for convenience) are the photosensitising agents. The hypericins are at highest concentrations in the soft growth tops of the plant, rather than the stalks, and concentrations are highest when the plant is in flower. The narrow-leaf biotype of the plant contains higher concentrations of the toxins than the broad-leaf biotype and both biotypes have higher concentrations in the October to June period (summer-autumn) than in the July to September (winter) period.[77] Photosensitisation which follows ingestion of St John's Wort is not dependent on any damage to the liver or other organs — the ingested hypericin compounds are the agents which accumulate in the skin and cause the lesions when exposed to bright sunlight.[78] If sheep are not in sunlight or if they are adequately protected by shade, skin pigment or fleece, no clinical condition occurs as a result of hypericin ingestion. Depending on the amount of hypericin ingested, sheep may start to show signs of photosensitisation within three days of grazing the plant. Hypericin persists in the blood for several days after removal from the offending pasture.[50]

Primary photosensitisation also occurs in sheep grazing buckwheat (*Fagopyrum esculentem*).[57,79] The photodynamic agents are fagopyrins — fluorescent compounds with a chemical structure similar to hypericin — and the clinical manifestation of photosensitivity is known as fagopyrism.

Chemicals of the furocoumarin (or furanocoumarin) group occur in a number of plant species and cause photosensitisation when ingested.[d] *Ammi majus* (Bishop's weed or Queen Anne's lace), in which the phototoxic furocoumarins are psoralens[80], occurs in Australia but photosensitisation in sheep has not been reported from this country. Spring parsley (*Cymopterus watsonii*), a cause of photosensitisation of sheep in the desert rangelands of southwestern United States, contains two phototoxic furocoumarin compounds.[81,82]

Photosensitisation with the family of three-leafed plants including clover (*Trifolium* spp), lucerne and medics (*Medicago* spp) is known colloquially as trefoil dermatitis.[83] The outbreaks tend to occur in spring and there is often evidence of aphid attack of the plants[84,85], although the significance of that observation is unknown. A primary photosensitisation of lambs grazing birdsfoot trefoil (*Lotus corniculatus*) has also been reported in New Zealand (NZ).[86]

Primary photosensitisation is also reported following grazing of biserrula (*Biserrula pelecinus*).[87] This annual, self-regenerating legume was introduced to Australia as a pasture plant in 1997 and there have been several reports of photosensitisation in sheep grazing pastures dominated by biserrula during the spring (August to October) in WA and NSW. The toxic principle has not yet been identified. Currently, producers are advised to avoid grazing shorn sheep on pastures composed of more than 30% biserrula, and to monitor sheep on the biserrula pastures for signs of photosensitivity so that they can be removed promptly if signs appear.

Erodium spp are widespread weeds of southern Australian pastures. *Erodium moschatum* (musk storksbill) photosensitisation has been reported in sheep in South Africa[88] but poisoning

---

d These weeds are members of the family *Umbelliferae* (also called *Apiaceae*). Other plants in the family include celery, parsnip and parsley. Furocoumarins in these commonly consumed vegetables can also cause phototoxic dermatitis in humans and other animals following ingestion or direct contact with the skin.

has not been recorded in Australia. Musk storksbill is a similar plant to common storksbill (*E cicutarium*).

Grazing of *Brassica* spp crops (turnips, rape, kale, swedes and commercially developed hybrids) is commonly associated with photosensitisation in cattle. The disease in that species appears to be one of hepatogenous photosensitisation with clinical chemistry very similar to that of facial eczema.[89] Grazing of immature rape (*B napus*) (before it becomes purple-red in colour) can lead to photosensitisation of sheep described colloquially as *rape scald*. While the photodynamic agent is unknown, it appears that the disease is one of primary photosensitisation in sheep because of the rapid recovery after the animals are removed from the crop.[90] In one instance reported from NZ[91], photosensitisation may have predisposed lambs to dermatophilosis when the grazing of *Brassica* spp occurred during periods of heavy rain. Brassicas can also cause acute and fatal poisoning of sheep with respiratory, digestive, urinary or nervous signs, probably unrelated to any photosensitising activity.

Congenital photosensitisation

An inherited congenital photosensitivity of Corriedale and Southdown[92] sheep and their crosses has also been described in which phylloerythrin accumulation follows a defect in bile excretion. In lambs maintained under natural conditions signs of photosensitisation begin to appear at 3 to 4 weeks of age. The condition is inherited as a single recessive gene.

Squamous cell carcinoma

The increased exposure to sunlight of the vulva and bare perineal skin of ewes that have had a radical Mules operation with a short-docked tail leads to an increased incidence of squamous cell carcinoma of the vulva, tail and perineum.[93] The condition is commonly called *rear-end cancer*. The carcinomas develop in ewes as young as 2 years of age and the incidence increases with age. On farms where the condition occurs, around 3% of older ewes may be affected.[94] Necrosis of the lesions or haemorrhage from trauma increases the susceptibility to flystrike. Affected ewes are often in poor condition and are usually culled or killed when the lesion is detected.

Preventive treatment involves using the *modified Mules* technique rather than the radical Mules operation. The former leaves a V-shaped piece of wool-growing skin on the proximal third of the dorsum of the tail. The wool that grows on the restricted woolled-skin area on the tail then provides protection of the bare skin part of the tail but is not so extensive that it traps faeces and increases the risk of flystrike. The tail should be amputated at the third coccygeal joint (and no shorter) to provide shading of the vulva from sunlight.

Burns

Sheep are common victims of bushfires in Australia and often large numbers are killed when they are trapped in corners of paddocks and are unable to escape downwind. Veterinarians are often involved in the assessment and management of sheep which survive the immediate effects of fire. It is recommended that burnt sheep be individually examined as soon as possible after fires and allocated to groups according to their prognosis for recovery without extensive treatment.[95]

All sheep should be tipped up and their feet, legs, belly and udder inspected. Severe burns to the lower leg, including the knee, hock and hooves are the most significant lesions. Severe burns may not be obvious at first but after two days the skin appears dry, scorched and leathery. Sheep with respiratory distress usually have a poor prognosis because they are likely to develop lung abscessation.

Sheep classed as *likely to survive* should be placed in a paddock with soft soil (such as sand), good feed and easy access to water and shade. Many will be inappetent for up to five days and lose condition for two weeks before starting to recover. Burns to the prepuce, scrotum and teats recover well provided that they are not too severe and the passage of urine is unaffected.[96] Semen quality may be affected for up to six months following scrotal burns.

Gangrene

Distal dry gangrene caused by tall fescue (*Festuca arundinacae*) occurs in cattle and has been recorded in sheep in New Zealand.[97]

SECTION D

EXTERNAL PARASITES

Flystrike (cutaneous myiasis)

Flystrike is one of the most important disease conditions of sheep in Australia. The annual cost to industry has been estimated at $173m.[98] It is a significant cause of mortality and lost productivity in sheep — factors which account for 60% of the estimated cost of disease. Nearly $70m is expended annually on prophylactic and therapeutic treatment. The disease is caused by the larvae of flies which deposit their eggs in the fleece of sheep under certain conditions. The larvae abrade the skin of the sheep, causing a severe exudative dermatitis which can extend to involve large parts of the skin area of the sheep. Untreated, the resulting bacteraemia and septicaemia can kill sheep within days of the initial strike.

Primary flies are those which can initiate a strike; *secondary flies* are those which normally do not initiate but can extend a struck area. *Lucilia cuprina* is by far the most important primary fly, although strikes also occur by *Lucilia sericata* and *Calliphora* spp. Secondary flies are of the *Chrysomya* and *Sarcophaga* genera. These mainly breed in carrion but also in strikes initiated by primary flies.[99]

L cuprina is most active in spring and autumn, when daily temperatures exceed 17 °C. They lay eggs (*oviposit*) on sheep and the eggs hatch in 7-12 hours in warm, humid conditions; the hatching process takes longer or does not happen at all at lower temperatures or lower humidity. The larvae go through two moults and the third instar maggots — about 12 mm long and aggressive feeders — are active on the third day after hatching. Once fully fed, these maggots fall from the sheep and pupate in the ground. At soil temperatures of 15 °C to 30 °C, a fly will emerge from the pupa in 6-25 days — the faster development under warmer conditions.[99] Female flies survive for about two weeks, during which time they may oviposit twice.

The completion of the life cycle is dependent on the availability of sheep with a bacterial skin infection resulting from moisture, whether that be from rain, urine, diarrhoea or wound exudation. *L cuprina* is attracted to sheep by an inherent attractiveness of sheep to the fly and, additionally, the odour of putrefaction which accompanies the moist exudate from fleece rot or dermatophilosis lesions. The fly can breed in carrion but this is a relatively unimportant source of flies and almost all flies present on a property have arisen from larvae on struck sheep within a 2 km radius of their emergence.

When soil temperatures fall below about 15 °C (around April in southeastern Australia) *L cuprina* maggots that fall to the ground to pupate suspend their development until soil temperatures rise. They resume development as the soil becomes warmer (in September-October in southeastern Australia) but the mortality of the pupae in soil over winter is high. The dependence on warm temperatures and moisture-affected sheep for reproduction leads to significant seasonality in the incidence of flystrike. Each year, depending on weather conditions, there is normally a double wave of primary flies, their prevalence rising to a peak in spring, declining in summer, rising again in autumn and falling to zero over winter.[100]

If preventive treatment of sheep has been applied early in the spring or if seasonal conditions do not favour the fly's life cycle, fly numbers on a farm may be very low. If, however, there are susceptible sheep present in the spring, one or more cycles of reproduction of the flies can lead to a rapid increase in fly numbers. High fly numbers and weather conditions which predispose susceptible sheep to flystrike can lead to a fly wave (in which the incidence of struck sheep can be very high), sometimes overwhelming the ability of a sheep producer to take control of the outbreak before many sheep are dead or severely affected by strike. Fly waves are most likely to occur in young Merino sheep in warm spring weather with persistent or repeated rain events, if no prophylactic chemical strategies have been applied.[101]

There are five main forms of flystrike.

- Breech strike: The major predispositions to breech strike are *urine soiling* of the skin and wool below the vulva of ewes; *excess skin wrinkle* in ewes; and *diarrhoea*, which may occur in either sex but is more common in young sheep. Three common sheep management operations — tail docking, crutching and mulesing — reduce the prevalence of this form of strike from sometimes disastrous proportions to a more tolerable level.

- Body strike: This is strike anywhere other than poll, pizzle or breech, most common around the withers, shoulders and flanks. The sheep are usually rendered attractive and susceptible to flies by fleece rot or dermatophilosis. This form of strike is predisposed by rainfall in warm weather, when flies are active, in sheep which are predisposed by fleece type, age or skin wrinkle and the development of fleece rot or dermatophilosis. Young Merino sheep are most at risk.

- Poll strike: This is usually confined to rams and results from the accumulation of *sweat* at the base of the horns or from wounds suffered from fighting.

- Pizzle strike: Wethers or rams with belly wool stained by urine or by discharges from posthitis, particularly if the wool is matted with burrs, are susceptible to strike in the area around the preputial orifice.

- Wound strike: This can occur in infected wounds — such as discharging abscesses and, less often, in clean wounds such as shearing wounds. The mulesing wounds of lambs are sufficiently susceptible in early spring to warrant protective treatment of winter- or spring-born lambs when they are mulesed at marking.

Prevention of breech and body strike

Jetting (strictly, *hand* jetting) is the application of insecticidal chemical to sheep through a wand with approximately six nozzles which are combed through the wool of the dorsum of the head, neck, back and rump under sufficiently high pressure to wet the sheep to the skin.

Insecticides can also be applied through automatic races or plunge or shower dips. Automatic jetting races are quicker but less effective than hand jetting; plunge or shower dipping is effective but more expensive because more chemical is applied per sheep than is necessary to prevent strike.

Some insecticides are produced in formulations which can be sprayed onto sheep at low pressure (*spray-on* treatments) and are applied to the backs of the sheep — from poll to breech — in a band. The volume applied is low and the products are very convenient and easy to apply. The chemical cost is relatively high, compared to jetting, but the saving in labour costs may justify the expense.

Crutching is the removal of wool from the tail, perineum and breech of the sheep. By ensuring that the wool in that area is short, the accumulation of faeces and urine (in ewes) is reduced. Crutching also removes any faeces (dag) or urine-stained wool that may have accumulated. The timing of the procedure is related to the time of shearing and the risk of flystrike. Consequently, crutching is performed if sheep are entering a flystrike season and have not been shorn (or crutched) for three to four months or more. For rams and wethers, the removal of belly wool around the pizzle (ringing) is usually done at the same time to reduce the accumulation of urine.

Mulesing is the surgical removal of some woolled skin from the edge of the perineum; it extends the width of the bare area beside the anus and, in ewes, the vulva. The Mules operation has been very widely used in Merino sheep because some sheep of the breed have skin with many small folds (wrinkles) and, in the breech area in particular, the skin wrinkle can trap and hold moisture from faeces and urine in ewes. Mulesing confers lifelong protection to breech strike. It is normally carried out on lambs at marking time when they are aged 2 to 8 weeks. Selective breeding of Merino sheep for less breech and body wrinkle (*plain* sheep, as opposed to wrinkly sheep) is gradually reducing the need for mulesing.

Tail docking is also practised to reduce the risk of the accumulation of faeces in the wool in the perineal area. Tail docking is performed usually at marking time, when lambs are 2 to 8 weeks of age. The length at which the tail is amputated is very important in conferring susceptibility to breech strike. The tail should be amputated at the third palpable intercoccygeal joint — level with the tip of the vulva in ewes. Amputation of the tail at shorter lengths increases the risk of wound infection, rectal prolapse and perineal cancer. Furthermore, sheep with short tails are unable to elevate the tail stump adequately when defecating and urinating and are more likely to collect faeces and urine in the wool of the tail than lambs with three-joint tails.[102]

Despite these three procedures, sheep with diarrhoea may still accumulate faeces (dags) around the breech and be at risk of breech strike. Prevention of diarrhoea remains, therefore, a very important strategy in sheep husbandry — see also Chapters 9 and 16.

Treatment

The treatment of struck sheep requires the close-clipping of wool from the affected area and an additional area about 5 cm beyond the struck area. Insecticide should then be directly applied to the wound and the healthy skin and wool surrounding it. The removal of wool is the most important part of the procedure and, in some cases, will be effective without insecticide. This method of treatment is termed hand dressing. For individual sheep of higher value, assessment should be made of the degree of toxaemia of the animal, particularly when the area of skin involved in the strike is large. Some cases may benefit from antibiotic therapy.

Insecticides used for prevention and treatment

*   Organophosphates: These chemicals are used for lice control and were used extensively for flystrike prevention before resistance to them became widespread. Their use for blowflies is now restricted to hand dressing, for which purpose they are a moderately effective treatment for strike when combined with wool removal. Products include diazinon, propetamphos and chlorphenvinphos.

*   Synthetic pyrethroids: These chemicals are widely used for lice treatment and some are registered for use against flystrike. They act by inhibiting oviposition.

*   Insect growth regulators (IGRs): There are two unrelated classes of IGRs used as insecticides in sheep. For prevention of flystrike, the two products used most commonly are cyromazine and dicyclanil, which both belong to the IGR class of *triazine and pyrimidine derivatives*. Cyromazine exists in formulations for either spray-on or jetting application and dicyclanil is used as a spray-on treatment. The IGRs are not larvicidal — their action is to interrupt moulting — so they are less efficacious as a treatment for active strikes than ivermectin. The IGRs have extended withholding periods before shearing or sale for meat. Protection against flystrike is prolonged: up to 14 weeks for cyromazine and 24 weeks for dicyclanil if applied correctly to sheep with sufficient wool growth to retain the chemical.

*   Macrocyclic lactones: One product — ivermectin — is available in Australia as a jetting or dipping fluid. It is effective against both lice and flies, providing up to 12 weeks of protection against flystrike if applied correctly to sheep. Unlike the IGRs, ivermectin is very effective against maggots and is therefore a suitable dressing for struck sheep.

*   Spinosad: This product is available as a jetting fluid and can be used for jetting or hand dressing. It offers relatively short protection against flystrike (four to six weeks) but has no withholding period for wool or meat and can be used on organic farms.

Bovicola ovis

There are three species of lice which infest sheep in Australia; two are sucking lice (*Linognathus* spp) and are relatively unimportant. Only one — *Bovicola ovis*, the chewing louse or biting louse of sheep — is important and, in the discussion here (and in general discussion about sheep lice in the industry) when reference is made to lousy sheep or lice in sheep, it is *Bovicola ovis* that is relevant. This louse species, formerly known as *Damalinia ovis*, is distributed widely throughout Australia.

In sheep of Merino or Merino-derived breeds, a lice infestation causes significant economic cost and must be controlled. Insecticides are used to treat lousy sheep. The most profitable management strategy for lice in almost all flocks in Australia is to eliminate lice from the flock and to maintain lice-freedom without further treatment. This goal is achievable in most flocks but the continuing presence of some infested flocks means that there is an ever-present risk of introduction of lice from stray sheep or purchased sheep. When the risk of reintroduction is high or treatment strategies have been unsuccessful in eliminating lice from the flock, some producers choose to treat the flock every year at the time of shearing whether lice are detected or not. With such an approach, elimination of lice from the flock is not necessary, provided that lice numbers can be reduced to such low levels that infestations will not cause significant problems before the next shearing. There is, however, growing concern about chemical use in agriculture — in terms of resistance development, risks to human health and cost — and it is becoming increasingly desirable to avoid routine annual treatment and, instead, to treat only when lice are detected and to treat with 100% efficacy.

The flock prevalence in Australia is around 30%.[103,104] Approximately 70% of flocks are, therefore, completely free of lice at any one time but do not necessarily remain so for long periods, due to the unwitting introduction of sheep infested with lice. A survey in Western Australia (WA) estimated that 27% of flocks become infected with lice each year and 35% of properties with lice at any one time will fail to eradicate lice within the next 12 months.[105] The annual cost to industry of lice infestations in sheep is estimated to be $81m.[98]

Epidemiology

B ovis is effectively an obligate parasite of sheep. (It can reproduce on goats which can, therefore, remain as vectors for lice from sheep to sheep[106], but goats are not considered a significant source of *B ovis* for sheep.) *B ovis* cannot survive off sheep (or goats) for more than a few hours unless protected from temperature extremes and light. In shorn wool in sheds lice can survive for up to 16 days or, possibly, longer, if conditions are highly suitable for survival.[107] Female lice lay eggs singly at the rate of approximately one egg every two to three days. Eggs hatch after 10 days. There are three nymphal stages and adult females commence egg-laying 24 to 25 days after hatching. In suitable conditions lice can survive one to two months on sheep, and some may survive for up to five months.

Transmission of lice occurs principally by direct contact — when lousy sheep are introduced to an uninfested flock or when infested sheep mix with uninfested sheep in yards, for example. Transmission can occur if uninfested sheep follow an infested mob through yards at a very short time interval but, generally, direct contact is necessary. Lice numbers build up relatively slowly on newly infested sheep. As a general rule, infestation in a flock becomes apparent (by fleece derangement) about four months after lice are introduced, but the time lag is affected by seasonal conditions (temperature, rainfall), host factors (breed, wool length) and intervening shearing events, if any occur.

There are significant differences in susceptibility to lice infestations between individual sheep within a flock, between strains of sheep within breeds[108], and between breeds. Merino sheep are more likely to carry high numbers of lice than sheep of other breeds.

Seasonal pattern of lice numbers

Solar radiation, temperature and rainfall all have a profound effect on lice numbers. Shearing has the most dramatic effect because many lice are physically removed with the wool and the exposure of lice to climatic conditions without the protection of a fleece leads to the death of many more. Lice numbers may decline by up to 90% of lice following shearing.

Solar radiation strongly influences the temperature of the host's skin and wool. Wool tip temperatures can be as high as 75 °C in direct sun on hot days, while skin surface temperatures may be 40° to 45 °C. Temperatures over 42.5 °C reduce oviposition and at temperatures over 45 °C some adults and nymphs die — and do so at increasing rates with higher temperatures and more sustained high temperatures. Lice will distribute themselves along the wool fibre and around the body to find temperatures closer to their preferred range. In freshly shorn sheep, the ventral neck may provide shelter, particularly if there are tufts of longer wool as a result of uneven shearing.

Saturation of the fleece by heavy and prolonged rain also kills many lice. Eggs exposed to a relative humidity over 90% fail to hatch.

There is a general pattern of increase in lice numbers from late autumn to early winter and a decline over summer, but this pattern is influenced strongly by the timing of shearing and the removal of the protective effect of the fleece. Long wool can provide a relatively stable microclimate in either winter or summer. Thus sheep shorn in late spring/summer will remain with very low lice numbers until winter; sheep shorn in autumn may have lice numbers increasing relatively quickly in summer.

Cost of lice infestation

The presence of moderate and heavy lice infestations causes a reduction in shorn-wool production of 0.3 to 0.8 kg of clean wool weight.[109] In addition to lowered productivity, the infested wool has a lower sale value due to lowered yield and yellow discolouration. Lice control costs include the expense involved in maintaining freedom from infestation and treatment of infested sheep. Treatment costs include the cost of chemicals, the cost of owning and maintaining plunge and shower dips and the losses of productivity or deaths of sheep as a result of dipping.

Clinical signs

B ovis is a chewing louse and feeds on the stratum corneum as well as scurf, sebaceous secretions and skin bacteria. Irritation of the sheep is probably derived from the biting activities of the lice and possibly some immune-based sensitivity. Infested sheep rub against objects like fences and fence posts and bite and chew their fleeces. The fleeces therefore appear deranged and have a *pulled* and ragged appearance, particularly the areas on the sides behind the shoulder which they can reach with their mouths.

Diagnosis

The diagnosis of lice infestation is usually carried out when sheep show fleece derangement and the flock owner wishes to know the cause. The definitive diagnosis of lice infestation is based on their presence in the fleece. Sheep should be examined for lice in a good light and the fleece parted to the skin in a number of areas of the upper flank, mid-flank and under the neck. Standard counting techniques include a systematic parting of the wool along the flanks (10 to 20 partings per side with the fingers are used to create an exposed line of skin

10 cm long). The numbers of lice seen per parting are averaged and the level of infestation categorised according to the values in Table 10.1. Light infestations are hard to detect and it may be necessary to examine several sheep thoroughly to find any lice at all. The detection of one live louse is sufficient for a positive diagnosis. A group of sheep cannot be declared free of lice with any acceptable degree of confidence unless multiple sheep are examined.

The best opportunity (possibly the only opportunity in any one year) to eliminate lice from a flock occurs at shearing. Unfortunately, the discovery of lice in a flock often occurs months before the next shearing event, even though lice were probably present, albeit in low numbers, when the sheep were last shorn. The detection of lice at such a time often leads to the application of insecticidal products in order to reduce, but not eliminate, infestation and the continuing presence of lice in the flock until shearing and, hopefully elimination, can occur.

It would be very useful to have a practical and accurate test for the presence of lice which could be applied at shearing time or a little before. A flock owner could then decide if treatment should be given in the immediate off-shears period when the chance of successful elimination from the flock is high.

Visual inspection, using the approach described above, has only a low sensitivity for detection of lice when the infestation is low (few lice per sheep) or when only a small proportion of flock is infested. For example, if the infestation is light and inconsistent across the flock (mean of one louse per parting and only 10% of the flock infested), the probability of detecting any lice if 10 sheep are examined with 20 partings per sheep is around 60%. Greater improvements in sensitivity can be made by increasing the number of sheep inspected than by increasing the number of partings.[110] If any sheep in the flock show signs of fleece derangement, these should be examined preferentially — when infestations are at a low level, differences in the degree of fleece derangement are correlated with the numbers of lice detected on a sheep.

Other tests which can assist with detection of lice at the time of shearing have been developed but are not widely used. These include a lamp test, which relies on the negative phototaxis and negative thermotaxis of lice in wool samples, and a table locks test.[111] The sensitivity of the lamp test on 10 fleeces, read after 10 minutes, was 63% but the test accuracy was sensitive to the temperature obtained under the lamp. The table locks test had a sensitivity of 87% but involved a greater degree of experience than the other two tests.

Laboratory-based lice detection tests — based on the presence of lice exoskeletons in lines of wool sent to wool stores[112,113] or on lice proteins in shearing comb debris[114] — have been developed but their use has been abandoned because of limited uptake by producers. A significant drawback with off-farm tests conducted on specimens collected at shearing time is the time between shearing and test reporting. The lack of a sufficiently fast response, perhaps

Table 10.1: Relationship between average lice counts and degree of infestation.

Number of lice per parting (mean of 20 partings)	Severity of lice burden	Expected size of lice burden per sheep
<2	Light	<5000
2 to 5	Moderate	5000 to 250 000
>5	Heavy	>250 000

combined with a reluctance to add more procedures to an already busy period on the farm, are probably the main reasons that the tests were not taken up by many producers.

Ultimately, on commercial sheep farms, the diagnosis of lice infestation is based on visual detection of lice. It is important that lice are observed and that diagnosis is not based on fleece derangement alone. Other important causes of fleece derangement include grass seed irritation and, to a lesser extent, itchmite, dermatophilosis and flystrike.

Treatment of lice infestations

Principles of managing lice infestations

There are several principles which underpin the approaches to management of lice in sheep flocks.

- In the event of a lice infestation in a flock of sheep, the ultimate aim of treatment is elimination of the lice from the entire flock. The term *control* is often used to mean the reduction of the severity of a lice infestation, rather than its elimination, but this should be considered to be an unacceptable long-term goal within a flock.

- Elimination of lice from a flock is very difficult to achieve if the sheep are more than six weeks off-shears.

- Treatment of sheep for lice with more than six weeks of wool growth is generally a *holding* operation, intended to control the infestation and keep it at a low level until another treatment is able to achieve elimination after the next shearing.

- Lice treatments are all based on the application of insecticidal chemicals to the fleece.

- Resistance to one or more chemical groups is common in lice populations.

Application methods

There are broadly two methods of applying lousicidal treatments to sheep which can achieve elimination of lice from a flock. These are

- back-line or pour-on applications
- saturation dips.

Back-line treatments involve the application of a low-volume, high-concentration lousicide to the dorsum of the sheep at low pressure, usually from a handheld applicator designed for each product. In the days and weeks following application, the chemical spreads over the entire surface of the sheep and achieves a gradient of concentration over the skin with the lowest concentration furthest from the site of application. The position of the line of application is important in the spread of chemical around the body.

Saturation methods include shower dipping, plunge dipping and immersion cage dipping. The sheep is wetted completely by showering or immersion in a high-volume solution of the lousicide in a purpose-built structure — a shower dip, plunge dip or immersion cage dip.[115] Total saturation of the sheep has, theoretically, the best chance of applying lethal concentrations of the insecticide to all parts of the sheep. Nevertheless, failures to eradicate lice are still common. Compared to back-line treatments, dipping is time-consuming and labour-intensive, and it involves additional musters and predisposes sheep to a number of diseases.

Hand jetting is also used at times with the same chemical products that are used in dips. Hand jetting, when done thoroughly, can achieve high levels of lice reduction but elimination of all lice from all treated sheep is unlikely. There are generally too many areas of the skin surface of the sheep which are not wetted by hand jetting and some lice can survive in those regions of the body.

Automatic jetting races are less effective than hand jetting and should not be considered an adequate control method for lice.

Timing of treatment

Depending on the product, the application of the ectoparasiticide is performed at different times in relation to shearing, in the following ways:

- off-shears treatment: some chemical products applied as pour-on products are only effective if applied within the 24 hours following shearing. These products are nonsystemic (that is, they are not absorbed through the skin) but are spread through the integument around the body of the sheep from the area where the product is applied. The mechanism of spread of the chemicals around the body is not fully elucidated but probably involves diffusion through skin lipids, contact between sheep and physical movement influenced by gravity[116]
- short-wool treatment: application during the six weeks immediately following shearing is called *short-wool* treatment. Chemicals which are applied by saturation are only likely to be effective if applied within six weeks of shearing. The preferred time for saturation dipping is two weeks after shearing, when the fleece is still very short but shearing cuts have healed
- long-wool treatment: application of lousicide to sheep with more than six weeks of wool growth is termed *long-wool* treatment. Products which are intended for use in sheep with more than six weeks of wool growth will usually only achieve short-term control of a lice population, rather than eradication of the lice. Wool-withholding periods (time before shearing) must be respected to ensure that residues are not detectable in shorn wool.

Lousicidal products

The ectoparasiticides used to treat lice infestations in sheep fall into seven categories based on their chemical structures (Tables 10.2 and 10.3). Some chemical products exist in different formulations prepared for either back-line or saturation dipping application.

Resistance to lousicides

Lice treatments based on the synthetic pyrethroid class of chemicals have been available in Australia since 1981 and were widely used, particularly in pour-on formulations, throughout the following two decades. A significant number of field strains of lice are now resistant to synthetic pyrethroids.

Products based on the insect growth regulator chemicals triflumuron and diflubenzuron became available in 1993 and these products effectively replaced the synthetic pyrethroid (SP) products as SP resistance became more widespread and more severe. Resistance to the IGRs was suspected in some lice populations just 10 years after their release on the market and ultimately proven in a laboratory assay.[117]

Table 10.2: Chemical groups used for treating lice.

Chemical group	Characteristics
Neonicotinoid	Neuroactive, chemically related to nicotine. Bind to nicotinic acetyl-choline receptors. Acts by paralysing lice.
Spinosyns	Novel chemical compounds discovered as fermentation products of the soil bacterium *Saccharopolyspora spinosa*. Disrupt acetylcholine neurotransmission leading to hyperexcitation of the insect nervous system. Considered safe for humans and environment. Spinosad has a nil withholding period for wool.
Organo-phosphate (OP)	Acetylcholinesterase inhibitors. Toxic to humans (and other animals). Use restricted to two products only. Temephos is fast-acting (treated sheep will be lice-free within 24 hours of treatment), relatively stable against stripping and, therefore, easier to maintain at effective concentrations in dip solutions.
Macrocyclic lactone (ML)	Bind to glutamate-gated chloride channel receptors in nerve cells, causing paralysis of lice. Considered safe products for human operators.
Insect Growth regulator (IGR)	Inhibit chitin production, and therefore interfere with exoskeleton production and moulting, adversely affecting the maturation of the juvenile stages of the insect life cycle. Consequently, live lice may be detected in decreasing numbers for up to 18 weeks after treatment. Considered safe products for operators.
Synthetic pyrethroid (SP)	Cause paralysis by interfering with sodium channels in insect nerve membranes. Considered safe for humans, although skin irritation can occur. Environmental impacts, particularly in waterways, are of concern.
Mg fluorosilicate/ rotenone/sulphur	Kill lice by desiccation. Rotenone is a naturally occurring insecticide. Relatively safe for human operators.

Table 10.3: Chemical products and application methods available for treating lice.

	Chemical class						
Application method	Neonicotinoid	Spinosad	OP	ML	IGR	SP	Mg fluorosilicate
Off-shears back-line	Imadocloprid	Spinosad	Diazinon	Abamectin	Diflubenzuron Triflumuron	Deltamethrin Cypermethrin	
Short-wool saturation	Thiacloprid	Spinosad	Temephos		Diflubenzuron		Mg fluorosilicate
Long-wool back-line		Spinosad			Diflubenzuron		
Long-wool jetting		Spinosad		Ivermectin	Diflubenzuron		

Resistance of lice to diazinon is uncommon[118] and, for practical purposes, can be considered an effective treatment for lice resistant to SPs or IGR.

Elimination or control

In most circumstances, the objective of a lice treatment programme is the elimination of lice from the flock. Acknowledging the risk of chemical resistance with SP and IGR classes, the strategies most likely to achieve elimination are

- plunge or immersion cage dip with a product of the neonicotinoid, OP or spinosad class, with all sheep treated two weeks off-shears

- back-line treatment with a product of the neonicotinoid, OP or spinosad class applied within 24 hours of shearing.

The chances of elimination of lice from the entire flock are improved if all sheep are shorn together so that treatment can be administered to all sheep off-shears over a brief period of days. When this is not possible, strict isolation of shorn and unshorn mobs becomes essential.

There are occasions when control (with perhaps 95% lice kill) is a satisfactory goal — usually when lice are detected in sheep with long wool and elimination of all lice is planned for an off-shears treatment. Achieving control and a high kill rate of lice can be achieved with a variety of strategies, including

- back-line treatment with spinosad or diflubenzuron

- hand jetting with ivermectin, spinosad or diflubenzuron

- immersion cage dipping with a product of the neonicotinoid, OP or spinosad class.

With all products except spinosad, careful attention must be given to wool-withholding periods to prevent the presence of chemical residues in the wool when it is shorn.

Failures to eliminate lice

For IGR products, there is a risk that a lice infestation could be transferred from treated, infested sheep to untreated sheep (pregnant ewes to their lambs after birth, for example) for up to 14 weeks following application. There is uncertainty, however, about whether adult lice exposed to the IGR are capable of transferring or, if they do, whether they can establish a patent infestation. If it is found that the lice are incapacitated by treatment even if still alive, long quarantine procedures after treatment may not be warranted.

When applied correctly to sheep infested with susceptible strains of lice, off-shears back-line products will eliminate lice on sheep. Failures to do so are generally caused by

- incorrect dosage

- incorrect application (either not a central stripe or the stripe is too short)

- treatment given too late off-shears

- some sheep omitted or lambs born soon after treatment

- excessively long wool left after shearing, particularly under neck

- some sheep affected by dermatophilosis.

Similarly, plunge dipping or immersion cage dipping is highly effective unless

- dip concentration falls too low, usually through *stripping*

- the sheep spends too short a time in the dip.

Stripping is the selective removal of active insecticide from the dipping solution. Dipping fluid is recycled from draining pens (where the sheep stand on release from the dip) back into the sump. The solubility of dipping chemicals in wool grease leads to a depletion of concentration of the active ingredient as more sheep are dipped. Mixing dip chemicals in cloudy water or dam water containing clay particles does not reduce the effectiveness of dip solutions provided that the dip sump is thoroughly stirred at least once per hour and that the clay-insecticide complex is not selectively filtered out of suspension by the sheep's fleeces.[120] The addition of zinc sulphate to such dip suspensions may, however, reduce their effectiveness.

Psorobia ovis (itchmite)

Psorobia (formerly *Psorergates*) *ovis* is an obligate parasite of sheep and all stages of the life cycle are completed on the host. The mites infest sheep of all ages; the size of the population generally increases with the age of the host. Young sheep are usually relatively free, but can develop heavy infestations after weaning, particularly if under nutritional stress. Peak numbers of mites on infested sheep generally occur in late winter/spring and are lowest in late summer. Rainfall does not have a pronounced effect on mite numbers. Spring shearing has little effect on mite numbers because mites are still reproducing actively and continue to do so until the hot weather arrives. The plane of nutrition of the host may also influence mite numbers, sheep with poor nutrition having more mites, more scurf and more fleece damage.

Adult mites mainly occur beneath the surface of the outer stratum corneum. It appears that they feed on epidermal lipid, probably suctorially. It is unlikely that mechanical stimuli are the source of irritation; it is more likely that a mite product is a sensitising antigen for an immune-mediated hypersensitivity response.[121] Itchmites are associated with a thickening of the stratum corneum and increased scurf — which is apparently a host response to itchmite infestation.

Transmission

Close contact, particularly at shearing, facilitates spread between sheep. Generally, spread between sheep is slow. Vertical spread occurs irrespective of wool length but may be facilitated by shearing ewes just after or just before lambing.

Prevalence

Until the use of macrocyclic lactone anthelmintics became widespread, itchmite infestation was a common cause of fleece derangement of sheep. In one NSW survey of 41 flocks, selected on the basis of the presence of sheep with deranged fleeces, itchmite infestation was considered the sole cause in 26 flocks (63%). The number of sheep with derangement associated with itchmite in each flock was, however, low.[122] This is a characteristic of itchmite infestation — they are present in many flocks, but the incidence of fleece derangement and the proportion of sheep with moderate or high (and therefore detectable) infestations, are both low. Another NSW survey of randomly selected flocks estimated the prevalence of infested sheep to be 5%[123], but it is not clear whether that figure is low simply because some flocks are completely free of itchmite. It is clear from other studies[124] that the sensitivity of the diagnostic test for itchmite is significantly less than 100%. The other main causes of fleece derangement, *Bovicola ovis* and grass seeds, attain higher prevalences of affected sheep within flocks, as recorded by both deranged fleeces and detectable infestations, but in fewer flocks.

The cost of itchmite infestation

Itchmite-infested sheep may produce less wool than uninfested sheep. The main cause of economic loss, however, is the lower value of wool due to fleece damage caused by rubbing. On a flock basis, the cost of the disease depends on the prevalence of serious fleece derangement.

Clinical signs

These range from a small area of the fleece which has a bleached appearance on the tips, to a generally ragged and tangled fleece with strings of wool hanging from the sides. The wool may be cotted and may have yellow discolouration. Sheep rub their fleeces particularly, but they also bite and chew their wool. The degree of fleece derangement appears to bear little relationship to mite numbers but the degree of scurfiness varies directly with mite numbers. The presence of both fleece derangement and excess scurf has been associated with the detectable presence of itchmite in over 60% of sheep.

Diagnosis

When fleece derangement is the presenting sign, B ovis, grass seed infestation, dermatophilosis, fleece rot and a wool break should be considered as potential causative agents as well as itchmite. All of these can be detected, if present, by careful examination of several sheep with deranged fleeces. Scurf is the most consistent grossly visible evidence of itchmite infestation. Definitive diagnosis requires the positive identification of mites in skin scrapings, but the presence of mites is not necessarily proof that they are causing the derangement. The presence of fleece derangement, excess scurf and mites, and the absence of other causes are strongly suggestive that itchmite is at least partly responsible for the fleece derangement.

Skin scraping

Johnson et al (1990)[122] describe the technique below (p. 117):

> Prepare 2 sites on the side between a line from the point of the ileum to the top of the shoulder and a line joining the bare skin areas of fore and hind limbs. The two sites are clipped with electric clippers with a size 40 blade. 2 ml of Shell Ondina medicinal oil is applied to the skin which is then scraped with a blunt, fixed blade scalpel until there is an erythematous reaction but no bleeding. Oil and debris are collected into vials and examined or deep frozen within 4 hours. The scrapings are examined at 25x magnification.

The test is highly specific, but the sensitivity varies with the mite density. Mite densities in summer may be very low and they vary between sheep at any time of the year. At least 20 sheep should be examined before any estimate of the flock prevalence is made.

Treatment and control

The mites are well controlled by oral or injectable treatment with one of the macrocyclic lactone anthelmintics (ivermectin, moxidectin, abamectin) and use of these chemical for control of nematodes has reduced the incidence of itchmite infestations markedly. Before the widespread use of these drugs a number of chemicals were used topically for control of itchmite in sheep, including amitraz, cypermethrin, rotenone and diazinon. Mites are most susceptible to acaricides when their reproductive activity is at its peak — usually between winter and spring. Where treatments are used in shower or plunge dips[125], treatment should be given within four weeks of shearing. Itchmite control after this time is much less effective.

Chorioptes bovis

C bovis causes an allergic, exudative dermatitis and is seen most frequently on the scrotum of rams. Rams are more heavily infested than ewes, but ewes do become infested and transmit the mites to their lambs during suckling. In heavy infestations on rams there is a thick, yellow crust and skin thickening which may interfere with temperature regulation of the testes and cause some testicular degeneration. Treatments used to reduce the severity of infection include amitraz and ivermectin.

Other external parasites of sheep

Linognathus pedalis — the foot louse

Infestations with *L pedalis,* a sucking louse which lives predominantly in the hairy parts of the legs, are often unnoticed unless they are very heavy, in which case they appear as dark patches on the legs, belly, scrotum or crutch. Light infestations have little, if any, effect on sheep but heavy infestations cause stamping and biting of the affected areas. The lice are controlled by treatment in shower or plunge dips but retreatment is necessary after two weeks, in order to kill lice which have hatched since treatment. Organophosphates are effective against adult stages, but must be applied by direct contact, not through back-line treatments. *L pedalis* can survive off sheep for 13 to 18 days, so it is necessary to depopulate the paddock of origin from the time of the first treatment until several days after the second treatment if eradication is to be achieved.[126]

Linognathus ovillus — the face louse

These lice live in the hairy wool of the face — often near the wool-hair junction. Treatment is the same as for *L pedalis* but these lice can only survive four days off the sheep. Both the foot louse and the face louse have been seen more frequently since organophosphates have been used less frequently for fly and body lice control — cyromazine and SP back-line products are ineffective against sucking lice. Ivermectin and closantel[127] have some activity against sucking lice.

Melophagus ovinus — the sheep ked

M ovinus is a wingless hypoboscid fly, 4 to 7 mm long. Keds are blood suckers and cause irritation, resulting in rubbing and biting. The sheep ked has declined in flock prevalence to very low numbers following the introduction of modern insecticides. It is killed by OPs, SPs including back-line products and ivermectin, and is now rare.

Ticks

Ticks which infest sheep include *Ixodes holocyclus, Boophilus microplus* (the cattle tick) and *Haemaphysalis longicornis* (the bush tick).

Other mite infestations

Sarcoptes scabei can cause mange in sheep, although it is apparently rare in Australia. There are numerous reports of sarcoptic mange in sheep from other countries. Recommended treatments include two doses of moxidectin, by injection, ten days apart.[128]

Tarsonemid mites are a family of mites which include genera and species which feed on plants, fungi and algae and which occasionally infest sheep and cause mild irritation.

Sheep scab

Psoroptes ovis is the cause of psoroptic mange in sheep — a disease colloquially known as sheep scab. The disease occurs in many countries throughout the world, but in Australia and some other countries it no longer occurs. The disease was introduced into Australia with the First Fleet (1788) and became a very significant problem — leading to government-led control programmes in the 1830s and, finally, eradication of the disease by 1896. New Zealand, Norway and Canada have also been free of the disease since the late 19th century. Eradication was achieved in the UK in 1952 but the disease was reintroduced in 1973, and continues to be a serious problem for sheep producers there — probably the most important ectoparasitic disease of sheep in that country.[129]

The mites cause a pyodermatitis marked by intense pruritus and loss of wool. The condition is so debilitating that affected sheep lose weight[130] and may die from the infection. Following introduction of mites onto the skin the condition remains subclinical for some time as mite numbers build up, before a visible lesion (a scab) develops. The lesions progressively increase in size and can become very extensive.

Epidemiology

The mites complete all stages of their life cycle on the sheep, although adult mites can survive off sheep for two to three weeks. They have a short life cycle (11-12 days) and can increase in population very rapidly. They increase in autumn and winter and tend to regress in summer to subclinical infections in protected body areas such as the groin, scrotum and interdigital fossa. Spread occurs by close contact between sheep but fomites and infested premises can be responsible for new infestations. The disease is highly contagious and treated sheep can be reinfected within a few weeks of treatment.

Clinical signs

The skin-puncturing habits of the mites cause intense irritation of the skin. Initially the lesions are small papules, oozing serum, often noticed first on the sides of the sheep. The wool is pulled and chewed by the sheep in response to the irritation. The lesions increase in size and coalesce and become covered in a thin yellow scab, often bleeding from the sheep's rubbing and biting. The wool over infected areas may contain large amounts of this scab material and is severely matted. Parts of the fleece are shed completely and severely affected sheep become very thin.

Diagnosis

Severe cases resemble dermatophilosis but the pruritus is more marked. Scrapie (a transmissible spongiform encephalopathy, or TSE) and infestations with lice or itchmite cause pruritus but have no visible skin lesions. Definitive diagnosis is made by demonstrating the mites in scrapings from the edge of lesions and from scabs collected from the base of wool fibres. Diagnosis is made by identifying the mites in skin scrapings but this diagnostic technique is unreliable, particularly in nonexpert hands, and early infections may be missed. There is interest in the use of an ELISA (enzyme-linked immunosorbent assay) to improve diagnostic sensitivity.[131]

Treatment and control

Treatment is carried out by prolonged dipping (two minutes or more) in an acaricide solution, repeated after 10 to 12 days if necessary, depending on the chemical used. Regional control programmes require the planned treatment of all flocks in one area. Both organo-phosphate chemicals applied topically (in dips) and injectable macrocyclic lactones are effective treatments.

RECOMMENDED READING

There are two comprehensive websites which provide reliable, detailed and practical advice about the management of *Bovicola ovis* and flystrike and which are updated frequently. They are strongly recommended to readers seeking more information on those topics.

Liceboss: http://www.liceboss.com.au/
Flyboss: http://www.flyboss.com.au/.

REFERENCES

1   Surman PG (1973) Mycoplasma aetiology of keratoconjunctivitis ('Pink-eye') in domestic ruminants. Aust J Exp Biol Med Sci **51** 589. https://doi.org/10.1038/icb.1973.56.

2   Bankemper KW, Lindley DM, Nusbaum KE et al. (1990) Keratoconjunctivitis associated with *Neisseria ovis* infection in a herd of goats. J Vet Diagn Invest **2** 76-8. https://doi.org/10.1177/104063879000200116.

3   Åkerstedt J and Hofshagen M (2004) Bacteriological investigation of infectious keratoconjunctivitis in Norwegian sheep. Acta Vet Scand **45** 19-26. https://doi.org/10.1186/1751-0147-45-19.

4   Dagnall GJR (1994) The role of *Branhamella ovis*, *Mycoplasma conjunctivae* and *Chlamydia psittaci* in conjunctivitis of sheep. Br Vet J **150** 65-71. https://doi.org/10.1016/S0007-1935(05)80097-1.

5   Van Halderen A, Van Rensburg WJ, Geyer A et al. (1994) The identification of *Mycoplasma conjunctivae* as an aetiological agent of infectious keratoconjunctivitis of sheep in South Africa. Onderstepoort J Vet Res **61** 231-7.

6   Cooper BS (1967) Contagious conjunctivo-keratitis (CCK) of sheep in New Zealand. NZ Vet J **15** 79-84.

7   Hosie BD (1988) Keratoconjunctivitis in a hill sheep flock. Vet Rec **122** 40-3. https://doi.org/10.1136/vr.122.2.40.

8   Walker E, Moore C, Shearer P et al. (2016) Clinical, diagnostic and pathologic features of presumptive cases of *Chlamydia pecorum*-associated arthritis in Australian sheep flocks. BMC Vet Res **12** 193. https://doi.org/10.1186/s12917-016-0832-3.

9   Hopkins JB, Stephenson EH, Storz J et al. (1973) Conjunctivitis associated with chlamydial polyarthritis in lambs. J Am Vet Med Assoc **163** 1157-60.

10 Andrews AH, Goddard PC, Wilsmore AJ et al. (1987) A chlamydial keratoconjunctivitis in a British sheep flock. Vet Rec **120** 238-9. https://doi.org/10.1136/vr.120.10.238.

11 Walker E, Lee EJ, Timms P et al. (2015) *Chlamydia pecorum* infections in sheep and cattle: A common and under-recognised infectious disease with significant impact on animal health. Vet J **206** 252-60. https://doi.org/10.1016/j.tvjl.2015.09.022.

12 Emerson PM, Lindsay SW, Walraven GE et al. (1999) Effect of fly control on trachoma and diarrhoea. Lancet **353** 401-3. https://doi.org/10.1016/S0140-6736(98)09158-2.

13 Stocks ME, Ogden S, Haddad D et al. (2014) Effect of water, sanitation and hygiene on the prevention of trachoma: A systematic review and meta-analysis. PLoS Med **11(2)** e1001605. https://doi.org/10.1371/journal.pmed.1001605.

14 Motha MXJ, Frey J, Hansen MF et al.2003) Detection of *Mycoplasma conjunctivae* in sheep affected with conjunctivitis and infectious keratoconjunctivitis. NZ Vet J **51** 186-90. https://doi.org/10.1080/00480169.2003.36362.

15 Axelson A (1961) Effect of contagious ophthalmia on multiple lambing and sheep liveweight. Aust Vet J **37** 60-2. https://doi.org/10.1111/j.1751-0813.1961.tb03855.x.

16 Hosie BD and Greig A (1995) Role of oxytetracycline dihydrate in the treatment of mycoplasma-associated ovine keratoconjunctivitis in lambs. Br Vet J **151** 83-8. https://doi.org/10.1016/S0007-1935(05)80067-3.

17 Kingsford NM and Raadsma HW (1997) The occurrence of *Pseudomonas aeruginosa* in fleece washings from sheep affected and unaffected with fleece rot. Vet Microbiol **54** 275-85. https://doi.org/10.1016/S0378-1135(96)01287-4.

18 Hayman RH (1955) Studies in fleece-rot of sheep: Some ecological aspects. Aust J Agric Res **6** 466-75. https://doi.org/10.1071/AR9550466.

19 Norris BJ, Colditz IG and Dixon TJ (2007) Fleece rot and dermatophilosis in sheep. Vet Microbiol **128** 217-30. https://doi.org/10.1016/j.vetmic.2007.10.024.

20 Raadsma HW (1988) Flystrike. In: Sheep health and production. Proceedings No 110. University of Sydney Postgraduate Committee in Veterinary Science: Sydney, pp. 317-37.

21 McGuirk BJ and Atkins KD (1984) Fleece rot in Merino sheep. I. The heritability of fleece rot in unselected flocks of medium-wool Peppin Merinos. Aust J Agric Res **35** 423.

22 McGuirk BJ and Watts JE (1983) Associations between fleece, skin and body characteristics of sheep and susceptibility to fleece rot and body strike. In: 2nd National Symposium on the Sheep Blowfly and Flystrike in Sheep, University of New South Wales. 6-8 December, Kensington, Sydney, p. 367.

23 Raadsma HW and Wilkinson BR (1990) Fleece rot and body strike in Merino sheep. IV. Experimental evaluation of traits related to greasy wool colour for indirect selection against fleece rot. Aust J Agric Res **41** 139. https://doi.org/10.1071/AR9900139.

24 James PJ, Warren GH, Ponzoni RW et al. (1989) Effect of early life selection using indirect characters on the subsequent incidence of fleece rot in a flock of South Australian Merino ewes. Aust J Exp Agric **29** 9-15. https://doi.org/10.1071/EA9890009.

25 Burrell DH (1985) Immunisation of sheep against experimental *Pseudomonas aeruginosa* dermatitis and fleece-rot associated body strike. Aust Vet J **62** 55-7. https://doi.org/10.1111/j.1751-0813.1985.tb14235.x.

26 Edwards JR (1985) Sale and processing of wool affected with dermatophilosis. Aust Vet J **62** 173-4. https://doi.org/10.1111/j.1751-0813.1985.tb07284.x.

27 Bateup BO and Edwards JR (1990) Processing of wool contaminated with dermatophilosis scab. Aust Vet J **67** 154-5. https://doi.org/10.1111/j.1751-0813.1990.tb07742.x.

28 Edwards JR, Gardner JJ, Norris RT et al. (1985) A survey of ovine dermatophilosis in Western Australia. Aust Vet J **62** 361-5.

29 Gherardi SG, Monzu N, Sutherland SS et al. (1981) The association between body strike and dermatophilosis of sheep under controlled conditions. Aust Vet J **57** 268-71. https://doi.org/10.1111/j.1751-0813.1981.tb05809.x.

30 Le Riche PD (1967) The activity of dipping fluids in the treatment and prevention of mycotic dermatitis in sheep. Aust Vet J **43** 265-9. https://doi.org/10.1111/j.1751-0813.1967.tb04860.x.

31 Le Riche PD (1968) The transmission of dermatophilosis (mycotic dermatitis) in sheep. Aust Vet J **44** 64-7. https://doi.org/10.1111/j.1751-0813.1968.tb04958.x.

32 Roberts DS (1967) Chemotherapy of epidermal infection with *Dermatophilus congolensis*. J Comp Pathol **77** 129-37.

33 Beveridge WIB (1983) Infection with *Actinobacillus lignierisi*. In: Bacterial diseases of cattle, sheep and goats. Vol 4. Australian Bureau of Animal Health/Australian Government Publishing Service: Canberra, p. 4.

34 Brightling A (1988) Sheep diseases. Inkata Press Pty Ltd: Melbourne & Sydney, p. 75.

35 McGregor BA (2010) Influence of stocking rate and mixed grazing of Angora goats and Merino sheep on animal and pasture production in southern Australia. 2. Liveweight, body condition score, carcass yield and mortality. Anim Prod Sci **50** 149-57. https://doi.org/10.1071/AN09129.

36 Allworth MB, Hughes KL and Studdert MJ (1987) Contagious pustular dermatitis (orf) of sheep affecting the ear following ear tagging. Aust Vet J **64** 61-2. https://doi.org/10.1111/j.1751-0813.1987.tb16134.x.

37 Gardiner MR, Craig J and Nairn ME (1967) An unusual outbreak of contagious ecthyma (scabby mouth) in sheep. Aust Vet J **43** 163-5. https://doi.org/10.1111/j.1751-0813.1967.tb04827.x.

38 Nettleton PF, Gilray JA, Yirrel DL et al. (1996) Natural transmission of orf virus from clinically normal ewes to orf-naïve sheep. Vet Rec **139** 364-6. https://doi.org/10.1136/vr.139.15.364.

39 McKeever, DJ and Reid H W (1986) Survival of orf virus under British winter conditions. Vet Rec **118** 613-16. https://doi.org/10.1136/vr.118.22.613.

40 Higgs ARB, Norris RT, Baldock FC et al. (1996) Contagious ecthyma in the live sheep export industry. Aust Vet J **74** 215-20. https://doi.org/10.1111/j.1751-0813.1996.tb15407.x.

41 Robinson AJ (1983) Prevalence on contagious pustular dermatitis (orf) in six million lambs at slaughter: A three year study. NZ Vet J **31** 161-3. https://doi.org/10.1080/00480169.1983.35008.

42 Buchan J (1996) Characteristics of orf in a farming community in mid-Wales. Br Med J **313** 303. https://doi.org/10.1136/bmj.313.7051.203.

43 Haig DM and McInnes CJ (2002) Immunity and counter-immunity during infection with the parapoxvirus orf virus. Virus Res **88** 3-16. https://doi.org/10.1016/S0168-1702(02)00117-X.

44 McKeever DJ, Jenkinson DMcE, Hutchison G et al. (1988) Studies of the pathogenesis of orf-virus infection in sheep. J Comp Pathol **99** 317-28.

45 Nettleton PF, Brebner J, Pow I et al. (1996) Tissue culture-propagated orf virus vaccine protects lambs from orf virus challenge. Vet Rec **138** 184-6. https://doi.org/10.1136/vr.138.8.184.

46 Gilray JA, Nettleton PF, Pow I et al. (1998) Restriction endonuclease profiles of orf virus isolates from the British Isles. Vet Rec **143** 237-40.

47 Cooper BS, Lynch RE and Marshall PM (1970) An outbreak of contagious pustular dermatitis associated with *Dermatophilus congolensis* infection. NZ Vet J **18** 199-201.

48 Animal Health Australia (2011) Sheep pox and goat pox disease strategy. Available from: https://www.animalhealthaustralia.com.au/our-publications/ausvetplan-manuals-and-documents/. Accessed 1 August 2018.

49 OIE World Organisation for Animal Health (2008) Sheep pox and goat pox. In: Manual of diagnostic tests and vaccines for terrestrial animals (Terrestrial manual). OIE, pp. 1058-68.

50 Bourke CA (2003) The effect of shade, shearing and wool type in the protection of Merino sheep from *Hypericum perforatum* (St John's Wort) poisoning. Aust Vet J **81** 494-8. https://doi.org/10.1111/j.1751-0813.2003.tb13370.x.

51 Morris CA, Towers NR, Hohenboken WD et al. (2004) Inheritance of resistance to facial eczema: A review of research findings from sheep and cattle in New Zealand. NZ Vet J **52** 205-15. https://doi.org/10.1080/00480169.2004.36431.

52 Hore DE (1960) Facial eczema. Aust Vet J **36** 172-6. https://doi.org/10.1111/j.1751-0813.1960. tb15310.x.

53 Munday R, Thompson AM, Fowke EA et al. (1997) A zinc-containing intraruminal device for facial eczema control in lambs. NZ Vet J **45** 93-8. https://doi.org/10.1080/00480169.1997.36002.

54 Glastonbury JRW, Doughty FR, Whitaker SJ et al. (1984) A syndrome of hepatogenous photosensitisation, resembling geeldikkop, in sheep grazing *Tribulus terrestris*. Aust Vet J **61** 314-16. https://doi.org/10.1111/j.1751-0813.1984.tb07135.x.

55 Glastonbury JRW and Boal GK (1985) Geeldikkop in goats. Aust Vet J **62** 62-3. https://doi. org/10.1111/j.1751-0813.1985.tb14238.x.

56 Miles CO, Wilkins AL, Erasmus GL et al. (1994) Photosensitivity in South Africa. VII. Chemical composition of biliary crystals from a sheep with experimentally induced geeldikkop. Onderstepoort J Vet Res **61** 215-22.

57 Hurst E (1942) The poison plants of New South Wales. Smelling Printing Works: Sydney, p. 19.

58 Regnault TRH (1990) Secondary photosensitisation of sheep grazing bambatsi grass (*Panicum coloratum* var makarikariense). Aust Vet J **67** 419. https://doi.org/10.1111/j.1751-0813.1990. tb03040.x.

59 Button C, Paynter DI, Shiel MJ et al. (1987) Crystal-associated cholangiohepatopathy and photosensitisation in lambs. Aust Vet J **64** 176-80. https://doi.org/10.1111/j.1751-0813.1987. tb09677.x.

60 Miles CO, Munday SC, Holland PT et al. (1991) Identification of a sapogenin glucuronide in the bile of sheep affected by Panicum dichotomiflorum toxicoses. NZ Vet J **39** 150-2. https://doi.org/ 10.1080/00480169.1991.35684.

61 Ford EJH (1964) A preliminary investigation of photosensitization in Scottish sheep. J Comp Pathol **74** 37-46.

62 Flaoyen A (1996) Do steroidal saponins have a role in hepatogenous photosensitization diseases of sheep? Adv Exp Med Biol **405** 395-403. https://doi.org/10.1007/978-1-4613-0413-5_34.

63 Pollock ML, Wishart H, Holland JP et al. (2015) Photosensitisation of livestock grazing *Narthecium ossifragum*: Current knowledge and future directions. Vet J **206** 275-83. https://doi. org/10.1016/j.tvjl.2015.07.022.

64 Sharma OP, Sharma S and Pattabhi V (2007) A review of the hepatotoxic plant *Lantana camara*. Crit Rev Toxicol **37** 313-52. https://doi.org/10.1080/10408440601177863.

65 Pass MA (1986) Current ideas on the pathophysiology and treatment of lantana poisoning of ruminants. Aust Vet J **63** 169-71. https://doi.org/10.1111/j.1751-0813.1986.tb02965.x.

66 Williamson PM, Highet AS, Gams W et al. (1994) *Diaporthe toxica* sp nov, the cause of lupinosis in sheep. Mycol Res **98** 1364-8. https://doi.org/10.1016/S0953-7562(09)81064-2.

67 Gardiner MR (1965) The pathology of lupinosis of sheep. Vet Pathol 2 417-45.

68 Allen JG, Moir RJ and Mackintosh JB (1983) Ovine lupinosis resulting from the ingestion of lupin seed naturally infected with *Phomopsis leptostromiformis*. Aust Vet J **60** 206-8. https://doi. org/10.1111/j.1751-0813.1983.tb09584.x.

69 Dawson RM (1998) The toxicology of microcystins. Toxicon **36** 953-62. https://doi.org/10.1016/ S0041-0101(97)00102-5.

70 Francis G (1978) Poisonous Australian lake. Nature **18** 11-12. https://doi.org/10.1038/018011d0.

71 McBarron EJ and May V (1966) Poisoning of sheep in New South Wales by the blue-green alga *Anacystis cyanaea* (Keutz.) Dr. and Dail. Aust Vet J **42** 449-53. https://doi.org/10.1111/ j.1751-0813.1966.tb14471.x.

72 McBarron EJ, Walker RI, Gardner I et al. (1975) Toxicity to livestock of the blue-green alga *Anabaena circinalis*. Aust Vet J **51** 587-8. https://doi.org/10.1111/j.1751-0813.1975.tb09400.x.

73 Negri AP, Jones GJ and Hindmarsh M (1995) Sheep mortality associated with paralytic shellfish poisons from the cyanobacterium *Anabaena circinalis*. Toxicon **33** 1321-9. https://doi. org/10.1016/0041-0101(95)00068-W.

74 Van Halderen A, Harding WR, Wessels JC et al. (1995) Cyanobacterial (blue-green algae) poisoning of livestock in the Western Cape province of South Africa. Tydskr S Afr Vet Ver **66** 260-4.

75 Edwards GT and Schock A (2010) Obstructive jaundice and photosensitization in a lamb secondary to diaphragmatic hernia. J Comp Pathol **142** 205-7.

76 Sargison ND, Baird GJ, Sotiraki S et al. (2012) Hepatogenous photosensitisation in Scottish sheep caused by *Dicrocoelium dendriticum*. Vet Parasitol **189** 233-7.

77 Southwell IA and Bourke CA (2001) Seasonal variation in hypericin content of *Hypericum perforatum* L. (St John's Wort). Phytochemistry **56** 437-41. https://doi.org/10.1016/S0031-9422(00)00411-8.

78 Araya OS and Ford EJH (1981) An investigation of the type of photosensitization caused by the ingestion of St John's Wort (*Hypericum perforatum*) by calves. J Comp Pathol **91** 135-41.

79 Mulholland JG and Coombe JB (1979) A comparison of the forage value for sheep of buckwheat and sorghum stubbles grown on the Southern Tablelands of New South Wales. Aust J Exp Agric Anim Husb **19** 297-302.

80 Ivie GW (1978) Linear furocoumarins (psoralens) from the seed of Texas *Ammi majus* L. (Bishop's weed). J Agric Food Chem **26** 1394-1403. https://doi.org/10.1021/jf60220a023.

81 Binns W, James LF and Brooksby W (1964) *Cymopterus watsoni*: A photosensitizing plant for sheep. Vet Med/Small Anim Clin **59** 375-9.

82 Williams MC, Kreps LB and Cronin EH (1970) Control of spring parsley on rangeland. Weed Sci **18** 623-5.

83 Bull LB and McIndoe RHF (1926) Photosensitization in sheep: Trefoil dermatitis. Aust Vet J **2** 85-91. https://doi.org/10.1111/j.1751-0813.1926.tb05332.x.

84 Dodd S (1916) Trefoil dermatitis: or the sensitisation of unpigmented skin to the sun's rays by the ingestion of trefoil. J Comp Pathol Therap **29** 47-62.

85 Ferrer LM, Ortin A, Loste A et al. (2007) Photosensitisation in sheep grazing alfalfa infested with aphids and ladybirds. Vet Rec **161** 312-14. https://doi.org/10.1136/vr.161.9.312.

86 Stafford KJ, West DM, Alley MR et al. (1995) Suspected photosensitisation in lambs grazing birdsfoot trefoil (*Lotus corniculatus*). NZ Vet J **43** 114-17. https://doi.org/10.1080/00480169.1995.35866.

87 Kessel AE, Ladmore GE and Quinn JC (2015) An outbreak of primary photosensitisation in lambs secondary to consumption of *Biserrula pelecinus* (biserrula). Aust Vet J **93** 174-8. https://doi.org/10.1111/avj.12318.

88 Stroebel JC (2002) Induction of photosensitivity in sheep with *Erodium moschatum*. J Sth Afr Vet Assoc **73** 57-61.

89 Collett MG (2014) Bile duct lesions associated with turnip (*Brassica rapa*) photosensitization compared with those due to sporidesmin toxicosis in dairy cows. Vet Pathol **51** 986-91.

90 Vermunt JJ, West DM and Cooke MM (1993) Rape poisoning in sheep. NZ Vet J **41** 151-2. https://doi.org/10.1080/00480169.1993.35759.

91 Allworth MB, West DM and Bruere AN (1985) Ovine dermatophilosis in young sheep associated with the grazing of *Brassica* spp crops. NZ Vet J **33** 210-12. https://doi.org/10.1080/00480169.1985.35238.

92 McGavin MD, Cornelius CE and Gronwall RR (1972) Lesions in Southdown sheep with hereditary hyperbilirubinemia. Vet Pathol **9** 142-51.

93 Vandergraaff R (1976) Squamous-cell carcinoma of the vulva in Merino sheep. Aust Vet J **52** 21. https://doi.org/10.1111/j.1751-0813.1976.tb05364.x.

94 Hawkins CD, Swan RA and Chapman HM (1981) The epidemiology of squamous cell carcinoma of the perineal region of sheep. Aust Vet J **57** 455-7. https://doi.org/10.1111/j.1751-0813.1981.tb05764.x.

95 Coghill K (1979) Saving burnt livestock. Agdex 400/29. Department of Agriculture: Victoria.

96 Brightling A (1988) Burns. In: Sheep diseases. Inkata Press: Melbourne & Sydney, pp. 35-8

97 Simpson BH (1975) Fescue poisoning in sheep. NZ Vet J **23** 182. https://doi.org/10.1080/0048 0169.1975.34232.

98 Lane J, Jubb T, Shephard R et al. (2015) Priority list of endemic diseases for the red meat industries. Final Report to Meat and Livestock Australia. Report AHE.0010. Meat and Livestock Australia: North Sydney, Australia.

99 Vogt WG and Woodburn TL (1979) Ecology, distribution and importance of sheep myiasis flies in Australia. In: National symposium on the Sheep Blowfly and Flystrike in Sheep. Department of Agriculture: New South Wales, pp. 23-32.

100 De Cat S, Larsen JWA and Anderson N (2012) Survival over winter and spring emergence of *Lucilia cuprina* (Diptera: Calliphoridae) in south-eastern Australia. Aust J Entomol **51** 1-11. https://doi.org/10.1111/j.1440-6055.2011.00835.x.

101 Watts JE, Murray MD and Graham NPH (1979) The blowfly strike problem in New South Wales. Aust Vet J **55** 325-34.

102 Watts JE and Luff RL (1978) The importance of the radical mules operation and tail length for the control of breech strike in scouring Merino sheep. Aust Vet J **54** 356-7. https://doi.org/10.1111/j.1751-0813.1978.tb02493.x.

103 Popp S, Eppleston J, Watt BR et al. (2012) The prevalence of lice (*Bovicola ovis*) in sheep flocks on the central and southern Tablelands of New South Wales. Anim Prod Sci **52** 659-64.

104 James PJ and Riley MJ (2004) The prevalence of lice on sheep and control practices in South Australia. Aust Vet J **82** 563-7.

105 Morcombe PW and Young GE (1993) Persistence of the sheep body louse, *Bovicola ovis*, after treatment. Aust Vet J **70** 147. https://doi.org/10.1111/j.1751-0813.1993.tb06110.x.

106 Hallam GJ (1985) Transmission of *Damalinia ovis* and *Damalinia caprae* between sheep and goats. Aust Vet J **62** 344-5. https://doi.org/10.1111/j.1751-0813.1985.tb07659.x.

107 Crawford S, James PJ and Maddocks S (2001) Survival away from sheep and alternative methods of transmission of sheep lice (*Bovicola ovis*). Vet Parasitol **94** 205-16. https://doi.org/10.1016/S0304-4017(00)00374-5.

108 James PJ, Carmichael IHC, Pfeffer A et al. (2002) Variation among Merino sheep in susceptibility to lice (*Bovicola ovis*) and association with susceptibility to trichostrongylid gastrointestinal parasites. Vet Parasitol **103** 355-65.

109 Wilkinson FC, de Chaneet GC and Beetson BR (1982) Growth of populations of lice, *Damalinia ovis*, on sheep and their effects on production and processing performance of wool. Vet Parasitol **9** 249-52. https://doi.org/10.1016/0304-4017(82)90059-0.

110 James PJ, Moon RD and Karlsson LJE (2001) Optimising the sensitivity of sheep inspection for detecting lice. In: Proceedings of the fly and lice integrated control strategies conference, ed S Champion. June, Launceston, pp. 331-40.

111 Morcombe PW, Young GE, Ball MD et al. (1996) The detection of lice (*Bovicola ovis*) in mobs of sheep: A comparison of fleece parting, the lamp test and the table locks test. Aust Vet J **69** 170. https://doi.org/10.1111/j.1751-0813.1996.tb10020.x.

112 Wilkinson FC and Buckman PG (1989) State sheep lice control — the Lice Detection Test. In: Australian advances in veterinary science, ed P Outteridge. Australian Veterinary Association: Sydney, p. 172.

113 Morcombe PW (1992) The sheep lice detection test. West Aust J Agric **33** 100.

114 Michalski WP, Young P, Shiell B et al. (2001) Development of a lice detection test for 'on-farm' use. In: Proceedings of the fly and lice integrated control strategies conference, ed S Champion. June, Launceston, pp. 341-5.

115 Chandler D and Richards G (2001) Low residue Diprite® dipping of sheep with the Richards immersion-cage sheep dip. In: Proceedings of the fly and lice integrated control strategies conference, ed S Champion. June, Launceston, pp. 200-5.

116 Lowe LB, Hacket KC, Rothwell JT et al. (2006) Mode of action of back-line application of spinosad, a non-systemic parasiticide, on sheep. Aust J Exp Agric **46** 857-61. https://doi.org/10.1071/EA05333.

117 James PJ, Cramp PF and Hook SE (2008) Resistance to insect growth regulator insecticides in populations of sheep lice as assessed by a moulting disruption assay. Med Vet Entomol **22** 326-30. https://doi.org/10.1111/j.1365-2915.2008.00753.x.

118 Levot GW (1994) A survey of organophosphate susceptibility in populations of *Bovicola ovis* (Schrank) (Phthiraptera: Trichodectidae). J Aust Entomol Soc **33** 31-4. https://doi.org/10.1111/j.1440-6055.1994.tb00913.x.

119 Griffin L (1993) Insect growth regulators for the control of Damalinia ovis on sheep. In: Proceedings of the Australian Sheep Veterinary Society, ed DA Hucker. 16-21 May, Gold Coast. Australian Veterinary Association: St Leonards, Australia, pp. 117-18.

120 Morcombe PW, Hide DF, Young GE et al. (1995) Settling of insecticide from dip wash mixed with dam water and zinc sulphate and used to control sheep lice (*Bovicola ovis*). Aust Vet J **72** 411-14. https://doi.org/10.1111/j.1751-0813.1995.tb06190.x.

121 Sinclair AN (1990) The epidermal location and possible feeding site of *Psorergates ovis*, the sheep itch mite. Aust Vet J **67** 59-62. https://doi.org/10.1111/j.1751-0813.1990.tb07696.x.

122 Johnson PW, Plant JW, Boray JC et al. (1990) The prevalence of itchmite, *Psorergates ovis*, among sheep flocks with a history of fleece derangement. Aust Vet J **67** 117-20. https://doi.org/10.1111/j.1751-0813.1990.tb07725.x.

123 Johnson PW, Boray JC, Plant JW et al. (1993) Prevalence of the causes of fleece derangement among sheep flocks in New South Wales. Aust Vet J **70** 220-4.

124 Sinclair AN and Gibson AJF (1970) Distribution of the itch mite (*Psorergates ovis*) on some Merino sheep. Aust Vet J **46** 311-16. https://doi.org/10.1111/j.1751-0813.1970.tb07907.x.

125 Johnson PW and Boray JC (1989) Efficacy of potential acaricides against the sheep itchmite (*Psorergates ovis*) with different application techniques. In: Veterinary therapeutics, Proceedings of a Scientific Meeting. Chapter of Veterinary Pharmacology: Australian College of Veterinary Scientists, p. 129.

126 Arundel JH and Sutherland AK (1988) Ectoparasitic diseases of sheep, cattle, goats and horses. In: Animal health in Australia. Vol 10. Australian Government Publishing Service: Canberra, pp. 69-72.

127 Butler RA (1986) Observations on the control of ovine face lice (*Linognathus pedalis*) with closantel. Aust Vet J **63** 371. https://doi.org/10.1111/j.1751-0813.1986.tb02901.x.

128 Fthenakis GC, Papadopoulos E, Himonas C et al. (2000) Efficacy of moxidectin against sarcoptic mange and effects on milk yield of ewes and growth of lambs. Vet Parasitol **87** 207-16.

129 Van den Broek AH and Huntley JF (2003) Sheep scab: The disease, pathogenesis and control. J Comp Pathol **128** 79-91.

130 Sargison ND, Scott PR, Penny CD et al. (1995) Effect of an outbreak of sheep scab (*Psoroptes ovis* infestation) during mid-pregnancy on ewe body condition and lamb birthweight. Vet Rec **136** 287-9. https://doi.org/10.1136/vr.136.12.287.

131 Ochs H, Lonneux J-F, Losson BJ et al. (2001) Diagnosis of psoroptic sheep scab with an improved enzyme-linked immunosorbent assay. Vet Parasitol **96** 233-42.

DEFICIENCIES OF TRACE ELEMENTS AND VITAMINS

INTRODUCTION

Trace elements are those minerals required by animals in amounts measured in micrograms and milligrams per kilogram of dry matter ingested. They include iron, manganese, zinc, copper, iodine, cobalt, molybdenum and selenium. In Australia and New Zealand, deficiency syndromes of grazing sheep are well recognised for copper, iodine, cobalt and selenium.[1] Poisoning is chiefly restricted to copper and, less commonly, selenium, although molybdenum excess can induce a syndrome of copper deficiency.

Vitamin nutrition of sheep is occasionally deficient, inadequate intake of vitamins A, D and E being responsible for the most common deficiency conditions. Vitamin B_{12} deficiency occurs when cobalt intake is inadequate. Deficiency of thiamin (vitamin B_1) leads to the development of polioencephalomalacia and is associated most commonly with the ingestion of substances which interfere with thiamin metabolism; it is discussed further in Chapter 15 as a cause of disease of the central nervous system. Deficiency of vitamin D occurs in sheep which have insufficient exposure to sunlight, leading to osteomalacia and rickets. It is discussed in Chapter 13 as a cause of lameness.

The major deficiencies considered here are those of copper, selenium, cobalt, iodine and vitamin E. The grouping together of these common trace element and vitamin deficiencies is not intended to suggest that the deficiency syndromes have in common any pathogenesis or clinical features, or that they are likely to occur together. What they do have in common is a tendency to cause subclinical and insidious losses of productivity in sheep, particularly young sheep; strategies for monitoring and preventing deficiency in each of these are also similar. While trace element and vitamin deficiencies are relatively unimportant causes of economic loss to the sheep industry compared to internal parasitism and nutritional deficiencies of energy and protein, the correction of trace element deficiencies does at times result in spectacular improvement in the health and productivity of grazing animals. Veterinarians working with sheep flocks should make themselves aware of the risk of trace element deficiencies in their local region and establish reliable and cost-effective ways to monitor, predict, detect and prevent the occurrence of clinical or subclinical syndromes that might result from inadequacies in the natural diet.

COPPER (Cu)

The copper nutrition of ruminants, particularly sheep, is complicated by a number of factors, including a relatively narrow margin between insufficiency and toxicity, variations between breeds in susceptibility to high intakes, complex metabolic interactions of copper with dietary molybdenum (Mo), sulphur (S) and other elements, and plant intoxications, which alter the way that the ovine liver accumulates copper, sometimes with fatal consequences.

Sheep have a lower requirement for dietary copper than many other species of animals, including cattle, and are efficient at storing copper ingested above their immediate requirements. They are very susceptible to copper toxicity and can be poisoned by the levels of copper which are commonly added to prepared feeds for other species, especially pigs and poultry. Some breeds of sheep are more sensitive than others — an extreme example is the North Ronaldsay sheep on the Orkney Islands of UK which have adapted to a unique diet (seaweed) with very low copper availability — and these sheep are very susceptible to copper toxicity if exposed to dietary copper levels which would be tolerated by other breeds of sheep. Some other breeds, including the Texel and Suffolk, are also considered to be highly susceptible to toxicity.

Despite their efficiency at storing copper, sheep in some regions are exposed to very low levels of dietary copper or are exposed to diets high in other minerals which interfere with copper absorption. Under these conditions, clinical signs of deficiency may then occur.

Physiology

Copper is an essential element in a number of enzymes which catalyse oxidase-type reactions in both plants and animals. In animals these enzymes are required for body, bone and wool growth, for pigmentation of wool and hair, myelination of nerve fibres and leucocyte function. *Ceruloplasmin* and *hephaestin* are copper-protein complexes which are important in copper transport in the blood and the mobilisation of iron for haemoglobin synthesis from intestinal mucosal cells, liver parenchymal cells and reticulo-endothelial cells. Copper is essential for the *lysyl oxidase* enzymes, which contribute to the cross-linking of bone collagen which, in turn, affects collagen solubility and bone strength. The enzymes *cytochrome oxidase*, which is involved in cellular respiration and phospholipid synthesis, and *tyrosinase*, which catalyses conversion of tyrosin to melanin, are also copper-dependent.[2]

Dietary sources of copper

With two major provisos, pastures in temperate regions of Australia generally contain adequate levels of copper for grazing ruminants. Plants typically contain 5-20 mg Cu per kg DM.[3,4] Legumes, including clovers, have concentrations generally at the higher end of the range, while grasses are at the lower end. Levels above 7 mg/kg DM will provide sufficient copper to grazing sheep under most circumstances. The two major factors that may interfere with copper nutrition in grazing sheep are the availability of copper in the soil to plants, and the interference with copper absorption in the gastrointestinal tract caused by sulphur, molybdenum, iron and other elements.

The availability of copper to plants is influenced by a number of factors including the levels of copper in the soil. Following the recognition of copper deficiency in particular regions of Australia, application of copper to soils has generally ensured that an uncomplicated soil deficiency of copper is now uncommon. The presence of copper in the soil does not, however, ensure that there are adequate concentrations of copper in plants. Almost all copper in soils is in the form of chemical complexes in organic matter and the element is not readily mobile through the soil profile.[5] Most of the copper presented to plant roots is derived from root interception — the growth of roots making contact with soil colloids — rather than through diffusion in the soil solution.[6] It is likely that the increased incidence of copper deficiency syndromes in sheep in wet seasons[7] is at least partly a consequence of poor root development of the pasture during winter and limited contact between the plant root system and the copper-containing fractions of the soil. Another important contributing factor to the occurrence of copper deficiency in grazing livestock in wet seasons is the level of molybdenum in the diet.

Molybdenum is an essential element for plant growth and many Australian soils are deficient in the mineral. Molybdenum availability to plants decreases as pH declines, so evidence of molybdenum-responsiveness in plants is usually associated with acid soils. On such soils there can be very substantial responses in pasture production, particularly of legumes, to the application of the mineral.

There are several factors which can lead to high levels of molybdenum in plants. Some soils — particularly peat soils — are naturally high in molybdenum. In other cases where soils are naturally low in molybdenum, the availability of the element to plants may be markedly increased by its direct application with fertiliser or by the liming of pastures with the intention of raising the soil pH. Unlike copper, molybdenum becomes more readily available to plants in wet conditions, and plants will accumulate molybdenum to high levels if the concentration in the soil solution is high.[8,5]

Copper which is released from the diet in the rumen reacts with both sulphur and molybdenum, forming insoluble copper sulphide and thiomolybdate salts of copper. The absorption of copper from the diet is markedly reduced by molybdenum over the range 0.5-2 mg molybdenum per kg DM in pastures, an effect which is exacerbated at higher sulphur intakes. Depending on the sulphur intake, the absorbability of copper may fall from 4% to 1% when Mo intake increases from 0.5 to 2 mg per kg DM.[2]

Iron, in herbage and in soil inadvertently ingested with herbage, also can reduce the absorbability of copper from the diet, probably through an interaction with dietary sulphur. If copper absorption is already reduced by the presence of molybdenum and sulphur, iron does not aggravate the effect.[2] Dietary manganese, zinc (Zn), cadmium (Cd) and, possibly, calcium (Ca) may also reduce the absorption of copper.

Copper which is absorbed travels in the portal circulation to the liver where it is either stored, incorporated into ceruloplasmin or excreted through the biliary system. The sheep readily accumulates copper in the liver when dietary intake and absorption is high, although biliary excretion rates do increase as liver copper levels rise. Copper is delivered to tissues as circulating ceruloplasmin, where it is incorporated into enzymes which function at intracellular level.

Unlike other ruminant species, the blood and liver levels of copper in foetal and newborn lambs are lower than in the dam.[9] There is a significant risk, therefore, if ewes are deficient in copper during pregnancy, that lambs will be born with clinical signs of copper deficiency or will exhaust their liver copper stores within a few days of birth and develop clinical signs soon after. Milk is a poor source of dietary copper for lambs and lambs born with low copper reserves may not accumulate satisfactory levels until they are 2 months old or more.[9]

Clinical signs of copper deficiency

Changes in wool growth and character

The loss of wool crimp and increased lustre of wool, producing *steely wool*, is one of the first clinical signs of copper deficiency to develop in sheep. Affected wool has crimps at three to four times the normal width and greatly reduced tensile strength. In black-woolled sheep on low copper diets, failure of pigmentation (achromotrichia) during growth in the wool follicle leads to the formation of bands of grey or white wool in an otherwise black staple, reflecting periods during which copper nutrition was inadequate. There may be several narrow bands interspersed with normally pigmented wool indicating that, over a period of months, the animal dipped in and out of 'normality' for a few days or weeks at a time. The copper-responsive aberrations in wool growth are particularly likely when dietary molybdenum and sulphur are at high levels.

Enzootic ataxia

Enzootic ataxia occurs in lambs born of ewes which have been copper-deficient during pregnancy. Affected animals have posterior paresis which is particularly apparent when the flock is disturbed or gathered (Figure 11.1). They appear weak and unco-ordinated in the hind limbs, knuckling over on the fetlocks and swaying from side to side as they attempt to rise or stay upright. The syndrome was first described from restricted areas in Western Australia where it affected lambs aged 6 weeks to 4 months, occasionally younger.[10]

Figure 11.1: A Merino lamb with enzootic ataxia. Affected lambs were first noticed when the flock was mustered for lamb marking. Source: KA Abbott.

In the earliest reports of the condition from Western Australia, it was noted that the condition in lambs was accompanied by the appearance of 'stringy' wool, anaemia and scouring in the ewes and general malaise in the sheep during June, July and August. All of the signs were prevented by, or responded to, the oral administration of copper sulphate.[11]

The disease is similar to *swayback* of lambs, recorded in Great Britain and other countries, which is also prevented by the administration of copper to ewes during pregnancy.[12] In flocks in which swayback occurred, lambs were often affected at birth or born dead as a consequence of demyelination lesions in the cerebral cortex. It is now clear that both swayback and enzootic ataxia represent different manifestations of the same nutritional deficiency, with the different presentations reflecting the severity of deficiency, the degree to which it is exacerbated by molybdenum and sulphur ingestion, and the timing of the deficiency with respect to foetal development. Enzootic ataxia may be akin to the condition known as *delayed swayback* in Britain[2] and the later appearance of the ataxia may be due to the deficiency occurring in late pregnancy, rather than in mid- and late pregnancy, with consequent impact on spinal cord myelination rather than myelination in the brain.

Lamb mortality

Treatment with copper of ewes which were deficient during pregnancy has reduced the mortality rate of lambs up to 6 months of age.

Bone disorders

Osteoporosis in lambs due to copper deficiency results in bone fragility and an increased incidence of fractures of long bones and rib bones. The osteoporosis arises from depressed osteoblastic activity, leading to reduced or absent deposition of bone matrix on the cartilaginous spicules. Metaphyseal trabeculae are severely reduced and very delicate and the bone cortices very thin. Osteoporosis involves the long bones of the limbs and the costo-chondral junctions. The bone fragility becomes evident when lambs aged 3 to 6 months are handled for husbandry procedures, such as shearing, and fractures of the limb bones are observed to occur at a high frequency with handling that is not unduly rough. The condition appears to occur in lambs which were copper-deficient at birth, rather than lambs which become deficient after weaning. Possibly, the competition for copper for growth and development lessens between birth and a few months of age, enabling normal bone growth to occur in the older lamb despite the low copper status.[13]

Subclinical copper deficiency

Loss of wool crimp, enzootic ataxia and osteoporosis occur relatively early in copper deficiency syndromes while achromotrichia, congenital swayback, anaemia and retardation of growth only occur when deficiency is severe.[13] Subclinical depression of body weight and wool growth is likely to occur at marginal levels of copper deficiency before clinical signs are seen. Reduced growth rate and anaemia is particularly associated with copper deficiency secondary to high intakes of dietary molybdenum and sulphur.

Seasonal variations in copper availability

The level of copper nutrition depends on the herbage copper concentration, the amount of herbage on offer and the availability of copper from the herbage. The concentration of copper in herbage is likely to be lowest in winter and it varies between seasons in response to variations in rainfall and temperature. The amount of herbage available is usually reduced in winter, and the relative availability of copper from herbage is lower in winter than in summer. For these two reasons, copper deficiency, when it occurs, does so seasonally in southern Australia in the winter period and resolves itself by summer.

High dietary intakes of Mo, S, Zn, Fe, Cd, Ca and soil all decrease the availability of dietary copper. Molybdenum application to pastures can exacerbate a copper deficiency. Liming of pastures can decrease copper availability to plants and animals and increase the availability of molybdenum. Some of the clinical syndromes associated with low copper intake occur only when dietary molybdenum levels are high. Sulphur has a highly complex interaction with molybdenum and a direct effect on copper absorption.[14] Increased dietary sulphur (either organic sulphur or inorganic sulphate) depresses the absorption of copper and depresses it more when molybdenum intake is higher. Increasing molybdenum intake has a depressing effect on copper absorption, an effect which itself is sulphur-dependent.

Clinical pathology

If a copper deficiency is suspected, animals of any age can be tested but animals which have grazed dry pasture (over a summer) are less likely to be deficient than young animals born in late autumn, winter or spring. Young sheep therefore provide the most sensitive indicator of a flock's copper nutritional status.

Which animal specimens to test

The copper nutritional status of an animal can be inferred by measurements of the levels of copper in the liver (a measure of the amount that is in reserve), in the blood (the amount in transport complexes) or at the tissue level, where the copper-containing enzymes and proteins are functional.

In blood, either plasma copper or plasma ceruloplasmin levels can be measured. Of the copper in plasma, most is bound in ceruloplasmin.

Provided that there is an adequate liver store or that the dietary source of copper is equal to or greater than the sheep's requirements, sheep maintain blood copper levels within a narrow range.[15] When copper absorption exceeds tissue requirements, it accumulates in the liver. When intake is less than requirements, liver copper levels are mobilised in order to maintain serum copper concentrations.[16]

When mobilisation of copper from the liver can no longer meet the demands for copper, ceruloplasmin synthesis declines and the plasma ceruloplasmin concentration falls. Tissue levels of copper-containing enzymes may, however, remain at functional levels for some weeks after ceruloplasmin levels fall.

Measurement of liver copper levels will both confirm an existing deficiency and enable a prediction of impending deficiency syndromes, and so, whenever possible, the copper status of a flock of sheep should be based on assessment of liver copper concentrations from several sheep.

Plasma copper levels — an indicator of ceruloplasmin concentrations — do not fall until liver stores are exhausted and therefore offer only a limited perspective of copper nutrition. A finding of low plasma copper levels will confirm a diagnosis of copper insufficiency. A finding of normal plasma levels, however, does not preclude the existence of a period of copper inadequacy before testing or the likelihood of a copper deficiency syndrome in the near future.

During extended periods of inadequate copper nutrition, the last tissues to become depleted are those where copper-containing enzymes function. *Copper-superoxide dismutase* (CuSOD) is an example of such enzymes. CuSOD is present in erythrocytes and appears to be synthesised at erythrocyte initiation. Measurement of the erythrocyte CuSOD activity reflects copper nutrition two to three months earlier than the time of testing and directly assesses the impact that copper deficiency is having at tissue level. Marginal deficiencies of copper may deplete liver copper stores and lead to low plasma copper levels, but erythrocyte copper concentrations may remain within normal ranges.[17]

The question arises, therefore, which tissues should be tested when investigating the copper status of a flock of sheep. For routine monitoring of young sheep in winter, blood tests may not be useful in predicting a period of reduced productivity over the ensuing weeks. Animals with normal plasma copper or erythrocyte CuSOD levels may become pathologically deficient before copper nutrition improves in the late spring. For monitoring at that time of year, liver specimens from three healthy animals selected from the flock should be tested for liver copper levels. However, during an investigation of a clinical syndrome, such as osteoporosis and limb fractures, it would be expected that plasma copper and erythrocyte CuSOD levels would be low and would assist with the confirmation of a diagnosis. Wherever possible, a group of otherwise healthy animals, perhaps 10, should be sampled to increase the validity of any test result.

Care should be taken when collecting blood samples to prevent the formation of a clot — ceruloplasmin levels are variably lower in serum than in plasma due to sequestration in the clot.[18]

Testing of pasture or soil

Provided pasture samples are also assayed for molybdenum, sulphate and iron, they can provide additional useful information about the pathogenesis of a copper deficiency that has been diagnosed in a grazing flock.[19] The information will allow the determination of the cause of the deficiency and the best way to avoid a recurrence. If pasture copper levels are low, for example, application of copper in fertilisers would be appropriate. Copper in soil is not particularly subject to leaching losses and, if adequate amounts of copper have been applied to soils (2 kg Cu/ha), there is evidence that the soil will provide adequate copper for pasture and sheep production for at least 23 years.

Where copper concentrations in pasture approach 10 mg/kg and copper deficiency still occurs, application of additional copper to the soil is not the most appropriate way to increase copper availability to the grazing animals. In such cases the animals should be treated directly with copper.[20]

Measurement of the level of copper in soils provides little, if any, useful information about the copper nutrition of grazing animals. The amount of copper absorbed by plants is difficult to predict and the amount of soil ingestion by animals is variable between seasons and, depending on the level of other elements on the soil, may enhance or reduce copper absorption in the gastrointestinal tract.

Prevention and treatment

Oral dosing of copper as oxidised copper wire particles (COWP) is the most effective method for prophylaxis of copper deficiency in sheep. The particles, administered in a gelatin capsule (Permatrace®, Coopers Animal Health), are retained for extended periods in the reticulo-rumen and abomasum and release copper slowly as a result of acid solubilisation of the copper. Administration to a sheep of 2.5 g COWP will raise liver copper levels for 10 weeks and develop sufficient stores to maintain adequate copper status for a further 20 weeks.[21]

Lambs cannot be administered COWP capsules until they are 8 weeks of age. Treatment of lambs before that age is practically difficult and, if copper solutions are given orally, there is a risk of copper poisoning. To ensure lambs have sufficient copper reserves when they are born, ewes should be treated with COWP or some other, long-acting form of copper, in early pregnancy or before joining.

Treatment should be timed to ensure satisfactory copper nutrition during the expected period of low dietary copper (usually winter). COWP treatment is a safe method of providing supplementary copper to sheep, but poisoning has occurred in sheep which were already receiving high levels of dietary copper in prepared concentrate feeds before dosing with the capsules.[22] COWP treatment also provides some anthelmintic activity against abomasal parasites, particularly *Haemonchus contortus*[23], which should be considered a useful side-benefit of copper supplementation of lambs in *Haemonchus*-endemic regions, rather than a reason to administer the capsule. The effect of COWP treatment on parasite burdens appears to be unpredictable and unreliable.[24]

Copper is also administered as an oral drench, either alone or mixed with anthelmintics. Treatment in this way gives only short-term prophylaxis — perhaps as little as two weeks — and there is a danger of poisoning. Addition to water supplies is practised in some areas but has similar risks of toxicity.

COBALT AND VITAMIN B$_{12}$

Vitamin B$_{12}$, also known as cobalamin or, strictly, cyanocobalamin, is synthesised by rumen microbes from dietary cobalt (Co). Adult ruminants have no natural sources of dietary vitamin B$_{12}$ and rely on ruminal synthesis to meet their needs. A dietary deficiency of cobalt leads to a deficiency of vitamin B$_{12}$. For the purposes of discussion below, the deficiency syndromes (cobalt or vitamin B$_{12}$) are considered to be the same.

Physiology

Vitamin B$_{12}$ is essential for two co-enzymes: methyl-cobalamin, which acts with the enzyme *methionine synthase* to promote methionine synthesis, and adenosyl-cobalamin, which acts with *methylmalonyl coenzyme A mutase* — a mitochondrial enzyme required for propionate metabolism.

Propionate is one of the three major volatile fatty acids derived from the fermentation of plant cellulose in the rumen. It is the major source of energy in the adult ruminant and the only volatile fatty acid which contributes significantly to gluconeogenesis. The gluconeogenic pathway for propionate is through carboxylation to methylmalonyl-CoA and then conversion to succinyl-CoA, an intermediary in the TCA cycle and therefore a substrate for glucose synthesis. The conversion of methylmalonyl-CoA to succinyl-CoA is dependent on *methylmalonyl-CoA mutase* (Figure 11.2), which is, in turn, dependent on adenosyl-cobalamin.

Vitamin B_{12} is stored in the liver but not accumulated above concentrations of 500 to 1100 nmol/kg (0.68 to 1.5 mg/kg) of freshweight even when cobalt intake is high.[26,25] Plasma is an important store of the vitamin, and plasma levels can rise to high levels (>2000 pmol/L) when cobalt nutrition is plentiful.[26] The mechanism by which vitamin B_{12} is stored in the body and released as required differs from that of copper homeostasis. Liver stores of Vitamin B_{12} are only slowly mobilised when dietary sources are inadequate. Plasma provides the major and most labile reserve of the vitamin and plasma levels fall quickly when absorption from the gastrointestinal tract declines. A reduction in liver concentrations lags behind that of plasma.

Vitamin B_{12} is transferred to foetal lambs across the placenta but only to a limited extent. The level of B_{12} stores in lambs at birth is influenced by the reserves in the ewe during pregnancy[27] but the neonate liver concentration of the vitamin is usually less than half that of the dam.

Colostrum can be a rich source of the vitamin if the ewe has adequate B_{12} stores. Non-colostral milk contains only low levels[28] even if the ewe is supplemented with vitamin B_{12} or cobalt in late pregnancy.[29]

Suckling, preruminant lambs do not synthesise vitamin B_{12} but their requirements for the vitamin are relatively low while milk comprises the larger part of their diet — presumably because their source of glucose is dietary lactose, from milk, rather than from absorption of propionate from ruminal digestion of cellulose.

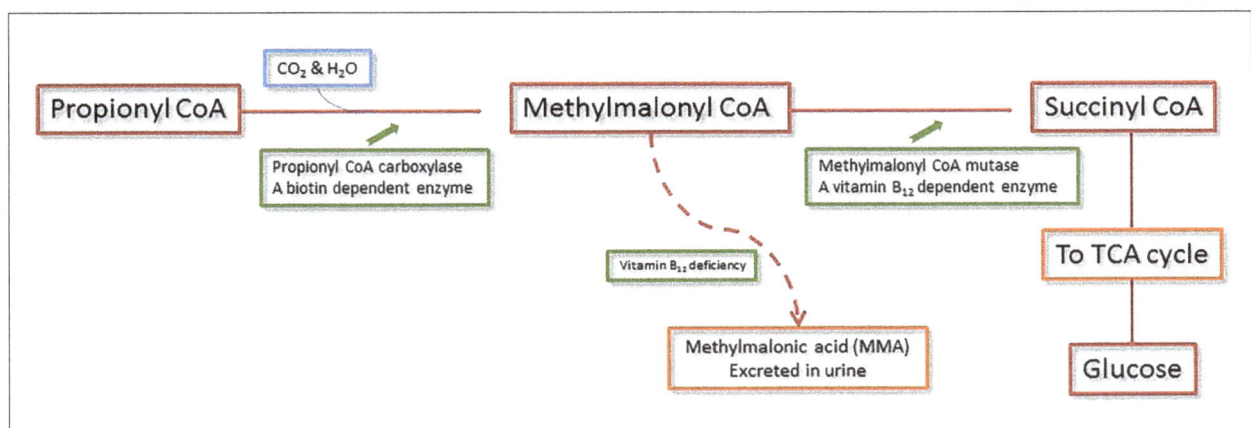

Figure 11.2: Propionate is converted to propionyl-CoA and then to succinyl-CoA, which can enter the TCA cycle and contribute to gluconeogenesis. One of the steps in the conversion is dependent on the enzyme methylmalonyl-CoA mutase and a cofactor that contains Vitamin B_{12}. Source: KA Abbott.

As lambs develop rumen function over the first few weeks of life they begin to absorb vitamin B_{12} from ruminal production. Despite this, their tissue stores tend to decline over the first three to four months of life and they may become critically low in vitamin B_{12} from 5 weeks of age until the rumen is fully functional at or around weaning age — assuming dietary cobalt is adequate.[30] A period of vitamin B_{12} inadequacy in lambs of 5 to 12 weeks of age may happen even if the ewes have adequate stores of vitamin B_{12} themselves.[29]

Dietary sources of cobalt

Sheep require diets containing at least 0.08 mg Co/kg DM and pastures typically contain between 0.05 and 0.12 mg/kg. Sheep normally ingest substantial amounts of soil while grazing and soil provides a more concentrated source of cobalt to the ruminant than does pasture.[17] Testing of pasture to ascertain cobalt nutrition to the grazing animal will therefore underestimate cobalt intake.

Dietary deficiencies of cobalt occur in regions where the soil is naturally low in cobalt. The soils of some areas, such as the southeast coast of South Australia, are profoundly deficient in cobalt, and deficiency syndromes can be anticipated nearly every year unless preventive action is taken. In other inland regions the deficiency is less marked and clinical or subclinical syndromes occur only in some years and with a severity which varies from year to year.

The seasonal variations in cobalt nutrition are significant. Seasons with lush pasture growth favour the development of cobalt deficiency — perhaps because the sheep ingest less soil. The concentration of cobalt in pastures and of B_{12} in plasma is lowest in spring.

Pathophysiology of cobalt deficiency

Deficiency of vitamin B_{12} leads to an inadequate synthesis of glucose from propionate, an accumulation of methylmalonic acid (MMA) in the blood and liver, and a high level of propionate in the blood. MMA is excreted in the urine at high levels. Urinary and serum levels of MMA can be measured to estimate the degree of interference with propionate metabolism — serum MMA levels are normally less than 7 μmol/L but can rise to levels of 20 to >70 μmol/L in sheep deficient in vitamin B_{12}.[31] The increased impairment of propionate metabolism during episodes of vitamin B_{12} deficiency is accompanied by a progressive loss of appetite, although the mechanism leading to the inappetence is not entirely clear.[31]

When the diet is deficient in cobalt, ruminal propionate production declines and ruminal succinate concentrations rise.[32] Propionate-producing rumen microbes are also dependent on vitamin B_{12} for the production of propionate from succinate — the reverse of the pathway that occurs in the host. Increased absorption of succinate from the increased rumen concentrations can effectively provide an alternative source of succinate for systemic glucose production. The degree to which cobalt deficiency leads to an impairment of energy production in the sheep is therefore influenced by diet, particularly the amount of roughage or concentrate in the diet, and the relative production of each of the volatile fatty acids from the diet.

Deficiency also interferes with the other important B_{12}-dependent biochemical pathway — that of homocysteine to methionine via the enzyme *methionine synthase* — which requires methyl-cobalamin as a cofactor. Failures in this pathway lead to a reduction of DNA replication in red blood cells and, ultimately, a megaloblastic anaemia. Because of the long half-life of erythrocytes, the anaemia associated with vitamin B_{12} deficiency takes several months to develop and may not be observed before other, more pressing, clinical signs are observed.

Signs of deficiency

Before cobalt deficiency was recognised as a cause of disease, there were areas of New Zealand, the south coast of Western Australia and South Australia and restricted regions in other countries, where sheep and cattle could not be grazed without developing *enzootic marasmus*. In South Australia, the combined deficiency of cobalt and copper led to a condition known as *bush sickness* or *coast disease*. Affected ruminants became weak and emaciated, progressively anaemic and died. Oral, but not parenteral, treatment with cobalt was ultimately found to prevent the disease, although copper was also necessary for complete treatment and prevention of coast disease. Now that the roles of cobalt and copper in sheep diets are better understood, these dramatic conditions are rarely seen. Nevertheless, cobalt deficiency syndromes still occur. The most common manifestation is a subclinical form in which the growth of young sheep is suppressed. In some cases, the impact is not observed and detection of the deficiency only occurs through careful and strategic monitoring of the cobalt status of the animals. In other cases the depression of growth and health is so obvious as to present as a clinical condition of *cobalt-responsive ill-thrift* or as *white liver disease*.

Ill-thrift

A severe and persistent dietary deficiency of cobalt, unless supplemented with parenteral vitamin B_{12}, leads to a syndrome of reduced appetite and weight loss. The wasting is accompanied by lethargy, ocular discharge and a mild anaemia. Affected lambs are in poor condition and are listless, despite the availability of feed which may even be abundant. The watery eye discharge leads to the formation of wet patches of matted wool below the eyes. There are disturbances of wool production leading to the development of open, 'dry' fleeces. The degree to which growth may be suppressed is striking; responses to supplementation of 7 kg (liveweight) and 0.5 kg (annual greasy wool production) have been recorded in Merino weaners in a cobalt-deficient region of South Australia.[33]

White liver disease

Ovine white liver disease is reported from New Zealand[34], Victoria[35], Western Australia[36] and Tasmania.[37] The disease appears to be a vitamin B_{12} deficiency complicated by liver damage. It principally affects lambs 2 to 6 months of age causing severe ill-thrift, hepatopathy, depression, serous ocular discharge, and crusty lesions on the ears, probably as a result of photosensitisation. There is a high morbidity, with mortality rates of 10% to 15% or, rarely, over 80%.[36] Affected sheep lose weight rapidly. Clinical signs are present for up to 14 days before death. Vitamin B_{12} or cobalt therapy leads to rapid improvement except in advanced cases. At necropsy, the liver is pale and, in some cases, almost white, and very swollen. Lipidosis is usually present.

Subclinical deficiency

Lambs from severely vitamin B_{12}-deficient ewes are of lower birth weight, are less vigorous at birth, are slower to suckle, are immunoglobulin-deprived and are subject to higher mortality, compared to lambs from ewes with adequate vitamin B_{12} nutrition.[38,39]

Large body weight and wool growth responses have followed supplementation of lambs of low vitamin B_{12} status that were not showing overt signs of deficiency.

Detection of cobalt deficiency

The subclinical deficiency of vitamin B_{12}, leading to reduced growth of lambs and reduced wool production, is the most important economic consequence of poor cobalt nutrition and may remain undetected in flocks over many seasons. The extent of the depression in growth rate varies with the degree of cobalt undernutrition and may, therefore, vary from year to year, from a syndrome causing lambs to be severely underweight in one year, to a negligible effect in another year. In areas where cobalt deficiency is likely to occur, a planned programme of monitoring of the blood levels of vitamin B_{12} in lambs is essential to establish the need for supplementation and the most cost-effective method of preventing deficiencies.

Clinical pathology in diagnosis

Plasma or serum B_{12} levels are both satisfactory measures of cobalt nutrition in sheep. Liver assays are used extensively in experimental studies but, for flock monitoring or investigations of ill-thrift, liver specimens are generally less available, or available from fewer sheep, than blood specimens. (Note that, for cattle, liver B_{12} is preferred because the relationship between cobalt nutrition and plasma levels is less precise than in sheep.) Pasture cobalt levels can be helpful in predicting the minimum dietary level of cobalt but, because sheep ingest substantial amounts of cobalt with soil, low pasture levels do not necessarily infer that the animals' diet is deficient. Soil levels of cobalt provide little or no information about the likelihood of deficiency in animals.

Preruminant lambs and lambs with recently developed rumens have low plasma B_{12} levels. In general, there is little benefit in measuring vitamin B_{12} levels in suckling lambs because levels are normally low and the significance of a low test is difficult to predict. By 4 months of age young sheep should be fully ruminant and producing vitamin B_{12} in their rumens — testing of plasma B_{12} levels will then indicate the dietary availability of cobalt.

For monitoring purposes, or as part of an investigation into a clinical presentation, blood samples should be collected from at least 10 animals in the flock. The mean and range of values for plasma B_{12} from the sampled animals can be used to predict the likelihood of responses in liveweight if the flock is supplemented with oral cobalt or parenteral vitamin B_{12}.

Reference curves to facilitate the prediction of response to treatment have been produced for sheep in New Zealand[40] (Figure 11.3). The curves have been produced from an analysis of numerous response trials which were conducted with sufficient statistical merit to produce reliable results.

The curves indicated that there was no statistically significant weight gain response in trials with mean serum B_{12} levels greater than 500 pmol/L. Using a weight gain response of 10 g/day

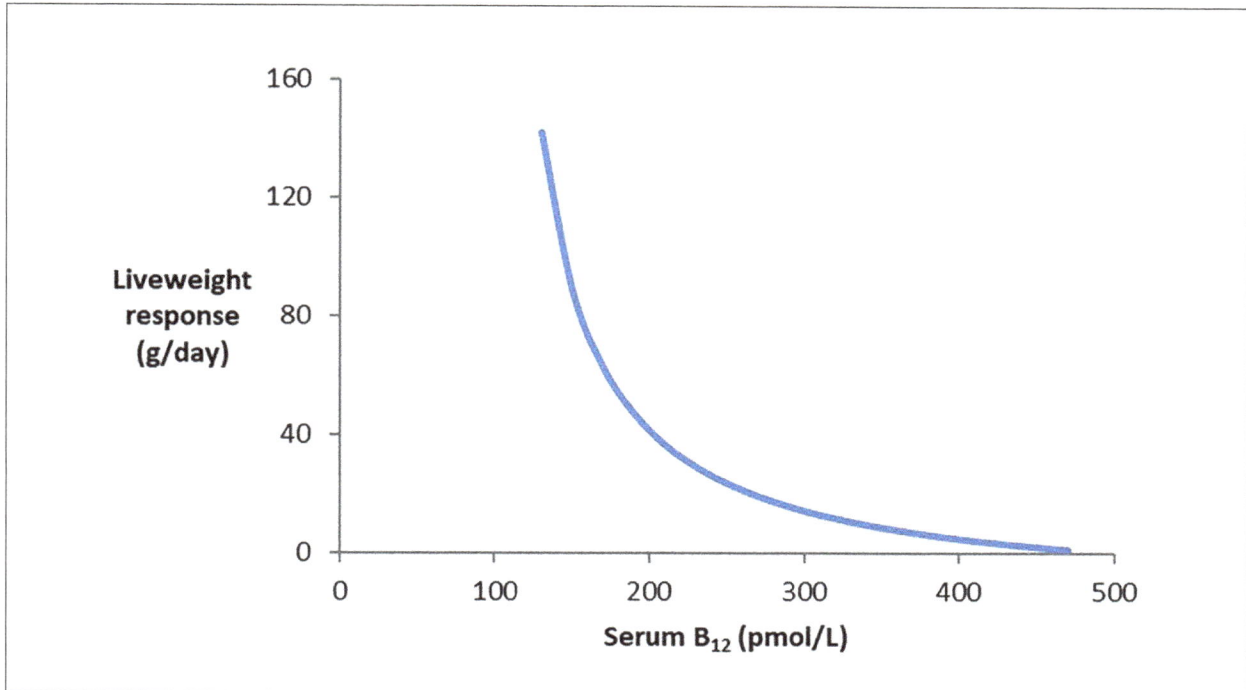

Figure 11.3: A reference curve for cobalt, predicting the increase in liveweight gain which can be expected by supplementation of animals at varying levels of cobalt nutritional status. Based on data from Clark et al. (1989).[40]

as an economic threshold, below which treatment is not economically justified, a cut-off level of 335 pmol/L was proposed, implying that treatment of animals with mean values in the *marginal* range given in Table 11.2 is unlikely to be justifiable, at least on immediate economic grounds. For individual animals, the serum B_{12} level above which responses will not occur may be as low as 220 pmol/L. When using reference curves or intervention values such as these it is important to remember the substantial variation in B_{12} levels between sheep within a flock and within sheep between sampling points.[30] While the mean value of a sample of animals may fall within the normal (unresponsive) range, animals within the flock with very low values may show a marked response to treatment — sufficient to justify the cost of treatment of the whole flock. In cases where 10 or more individual animal results are known, the proportion of individual results in the low-marginal or deficient range provides a useful indication of the possibility of a response to treatment in at least some members of the flock.[26]

Suckling lambs are more tolerant of low vitamin B_{12} levels than weaned lambs and adult sheep, probably because they obtain sufficient glucose from milk and are not dependent on a normal (adult) propionate metabolism pathway for which vitamin B_{12} is required.[31] In milk-fed lambs, responses to supplementation are only likely to occur at serum levels well below 220 pmol/L.[26]

As many cases of cobalt deficiency syndromes are recognised in spring and summer, early spring is often a good time to test a flock of young sheep, if there is a risk of inadequate nutrition. On farms where deficiency syndromes have been detected previously it may be considered prudent to provide preventive therapy every year, rather than incur the cost of repeated testing. Preventive therapy is not without cost, however, so the initial testing programme should be thorough and the diagnosis based on firm evidence before recommendations for routine treatment are made.

Treatment and prevention

The subcutaneous injection of 2 mg of soluble vitamin B_{12b} (hydroxocobalamin) is recommended for the immediate treatment of deficient animals and for short-term prevention of deficiency. Hydroxocobalamin is used over the cyano form of cobalamin because it is considered to produce a more prolonged benefit.[41] It is the preferred method of prophylaxis for preruminant lambs and is often administered at marking time, when lambs are 1 to 8 weeks of age. The effective period of protection is relatively short. In cobalt-deficient areas, injection with vitamin B_{12} can be expected to give no more than one to two months' protection.[42] Repeated treatments may therefore be necessary. A microencapsulated form of hydroxocobalamin is available in New Zealand and provides three to four months' protection against deficiency in lambs.[30]

Intraruminal cobalt pellets (*bullets*) are the most efficient method of long-term prevention in endemically deficient areas. They provide adequate cobalt nutrition for one to three years, depending on the formulation.[43] A small percentage of pellets may be regurgitated and lost. They are usually administered to young sheep at 6 months of age or more, because sheep with limited ruminal activity (under 10 weeks of age) do not benefit from dietary cobalt and because larger sheep are easier to dose with pellets. In flocks grazing calcareous soils the pellets become coated with a calcium phosphate deposit which reduces the rate of cobalt dissolution. The coadministration of a metallic *grinder* prevents coating. In some regions the administration of two grinders is necessary to prevent the deposition of salts on the pellets.[43]

Oral drenches of cobalt salts may be used but are of very limited value because they provide only short-term protection.[33] Salt licks and mineral blocks may be effective, provided that sheep ingest at least 0.05 mg of cobalt per day.

Addition of cobalt to pastures with fertilisers and foliar sprays can be useful, particularly when cobalt is being administered as an aid to preventing *phalaris staggers*.[a] Phalaris staggers is caused by ingestion of a toxin produced by the plant under certain growth conditions when sheep are grazing pastures on low-cobalt soils. High rumen concentrations of cobalt are protective against the toxin. The effect is apparently local — within the gastrointestinal tract — so parenteral vitamin B_{12} is not protective.

Ewes grazing cobalt-deficient pastures should receive cobalt bullets at least eight weeks before lambing to ensure adequate colostral vitamin B_{12} and to provide adequate foetal liver reserves of the vitamin. Lambs should then receive 1 to 2 mg of hydroxocobalamin parenterally at marking. For sheep which will be retained into adulthood this treatment should be followed by a cobalt pellet, plus grinder if necessary, at weaning. In the case of lambs to be sold for meat, a bimonthly injection of 2 mg vitamin B_{12} should provide sufficient protection against deficiency.[33,42]

---

a   Grazing of *Phalaris aquatica* can lead to outbreaks of *Phalaris staggers* (see Chapter 15) or *Phalaris sudden death* (see Chapter 14). Cobalt is not protective against *Phalaris sudden death*.

SELENIUM

Selenium is an essential trace element for all animals. Selenium deficiency occurs in all states across southern Australia and is most common in regions with medium and higher rainfall (>500 mm per year) and acid soils. Deficiency syndromes are associated with clover-dominant pasture for several reasons — these pastures sustain high growth rates of young animals; clovers tend to acidify soils by nitrogen fixation; and clovers are high in polyunsaturated fatty acids (PUFA), which predispose animals to clinical syndromes of selenium deficiency.

There are some similarities between the disease conditions caused by selenium deficiency and those caused by vitamin E deficiency. This chapter therefore discusses the physiology of the two nutrients together but the clinical syndromes of deficiency separately.

Physiology

The best-known biological activity of selenium is in the seleno-enzyme *glutathione peroxidase* (GSHPx). Vitamin E and GSHPx have similar and complementary physiological roles in protecting cells from damage caused by endogenous peroxides. These products, which are formed as part of the essential synthesis of prostenoids from polyunsaturated fatty acids[44], can damage lipid-rich cell membranes and intracellular organelles, leading to cellular degeneration and necrosis. GSHPx acts to destroy peroxides before they attack cell membranes, while vitamin E acts within the membrane to attract unsaturated fatty acid molecules and form loose chemical complexes until they are metabolised during cell respiration, thus preventing the formation of fatty acid hydroperoxides.[45] The peroxides can also denature cellular proteins and, while GSHPx is protective, Vitamin E is not. This may explain why selenium therapy of deficient animals is able to prevent some forms of muscular degeneration which cannot be prevented by administration of vitamin E.[46] Similarly, vitamin E may have a unique preventive role in clinical syndromes associated with peroxidation in tissues which are inherently low in GSHPx, regardless of the adequacy of selenium nutrition. The similarities in the myopathies associated with selenium and vitamin E deficiency originally led to the conclusion that the two nutrients had interchangeable roles. This is now known not to be true; selenium will not protect sheep from *weaner nutritional myopathy*, but vitamin E will; vitamin E will not protect sheep from *white muscle disease*, but selenium will.

Unsaturated fatty acids, particularly if polyunsaturated, form both a part of the energy supply and an essential part of tissue membranes. The unsaturated double bonds of membrane PUFAs are inherently unstable and are readily attacked by peroxides and other forms of active oxygen. It follows that the greater the level and input of PUFA to the cells, the greater the amount of antioxidants needed to prevent the reaction.[45] Animals with high-PUFA diets require satisfactory nutrition with selenium and vitamin E to avoid the risk of myopathy.

In species other than sheep, selenium appears to have a role in resistance to disease. In sheep there is evidence that selenium-deficient animals are no less able to produce antibodies against bacterial antigens[47] and no less able to prevent the establishment of internal parasites[48] than sheep with adequate selenium intake.

Signs of selenium deficiency

In young sheep, two selenium-responsive conditions are recognised — one a myopathy of lambs, *white muscle disease* (WMD), the other a syndrome of lowered productivity varying in severity from subclinical depression of wool production to a clinical condition of poor growth and increased mortality rates, *selenium-responsive unthriftiness* (SRU). In adult sheep, a selenium-responsive infertility of ewes has been reported. While the diseases are clearly associated with selenium-deficient areas, they have been difficult to reproduce experimentally or to predict on the basis of dietary selenium or animal tissue selenium levels. The possibility exists that other factors are involved in the expression of disease.

White muscle disease (WMD)

This syndrome is also called *nutritional* or *enzootic muscular dystrophy*. The myopathy of WMD can affect heart muscle or different skeletal muscle groups, clinical signs varying with the muscle groups involved. Muscles of the upper limbs are frequently involved, causing affected lambs to walk stiffly, if they can rise at all. Cardiac myopathy can cause sudden death. WMD has been classified as *congenital*[49] or *delayed*, and *acute* or *subacute*.[46] In the congenital form, affected lambs either are born dead or they die suddenly after a period of exertion, such as that required for suckling, within a few days of birth. Lesions are usually confined to the myocardium. The *delayed* and *subacute* forms occur predominantly in lambs from 3 to 6 weeks of age, sometimes up to 3 months of age. Affected lambs have a stiff, stilted gait (leading to an alternative name, *stiff lamb disease*) and an arched back. They prefer to remain in sternal recumbency, and then become prostrate and die within a few days. The myopathy involves several or many skeletal muscle groups.

Selenium-responsive unthriftiness (SRU)

A selenium-responsive syndrome of reduced weight gain and increased mortality rates, particularly in sheep 3 to 18 months of age, has been recorded in NZ and Australia. Until effective preventive measures were widely used, SRU was considered the most widespread and economically important of all the selenium-responsive diseases of New Zealand livestock. In sheep, it varies from a subclinical inability to maintain expected growth rate to a clinical unthriftiness which may, in some seasons, lead to a heavy mortality.[50] Affected animals lose condition, become weak, develop dry, open fleeces, and may also have diarrhoea.[51] There is usually a history of WMD and/or selenium-responsive infertility on the same property.

SRU is less commonly observed in Australia, but there are regions where the disease is endemic. In a response trial in the Strathbogie Ranges of central Victoria, treatment with selenium (0.1 mg/kg *per os*) of lambs at marking and weaning improved weaning weight by 2 kg (10%) and reduced mortality rate from 17.5% to nil. Despite the selenium responses, no WMD occurred during the trial, but it had occurred on the same property in the previous year. Fleece weight responses were also large. In other regions of Victoria, SRU is generally considered unlikely because pasture selenium levels are rarely sufficiently low (< 0.02mg/kg) for more than a few weeks or months in spring.[52]

Selenium-responsive infertility

In New Zealand, selenium treatment has resulted in marked improvements in fertility in ewes in regions where WMD and SRU occur.[53] The losses appeared to be associated with higher embryonic mortality in untreated ewes (24% to 26%) than in treated ewes (2% to 5%), as detected at slaughter four to eight weeks after mating.

In Australia, there are only a few reports of selenium responses in fertility of ewes. Godwin et al. (1970)[54] reported increased percentages of fertile ewes and increased numbers of lambs marked per ewe mated following selenium treatment of ewes on Kangaroo Island (SA) on a property where WMD had occurred. Wilkins and Kilgour (1982)[55] recorded improvements in fertility in 12 of 14 flocks on eight properties in the northern Tablelands of NSW following treatment with 5 mg selenium orally three weeks before joining. In that trial, 9% of treated ewes and 16% of untreated ewes failed to lamb. Of the ewes mated in the first six weeks, 46% of the untreated ewes and 5% of the treated ewes did not return to service yet failed to lamb. This result supported the observation from the NZ study reported above and other studies in the northern Tablelands of NSW[56] that the failure is one of embryonic loss.

Most attempts to obtain fertility responses to selenium in Australia have failed. For example, another study in the northern Tablelands of NSW[57] found that selenium supplementation had no effect on weaning rate at a low stocking rate. There was a positive effect at a high stocking rate but it was due to higher lamb survival rates, rather than improved fertility.

Subclinical deficiency of selenium

Selenium-responsive unthriftiness is effectively the severe and clinically apparent manifestation of a spectrum of disease which includes, as a mild form, subclinical losses of production. The most significant subclinical selenium response is an increase in fleece weight. In an experiment in the northern Tablelands over four years from 1984 to 1987, selenium-supplemented ewes, which had mean blood selenium concentrations over 0.05 µg/mL, had higher liveweights (1 to 2 kg higher) through most of the experiment, and higher fleece weights in three years (3.8% to 7.5% higher), with corresponding increases in fibre diameter (0.5 µm higher) than unsupplemented ewes (mean blood selenium concentration around 0.02 µg/mL).[58] Despite these responses, no clinical signs of deficiency, including depressed fertility, occurred in the ewes, and the incidence of WMD in lambs was very low. Lambs of untreated ewes had lower liveweights at birth and at weaning, more so at a high stocking rate (3.1 kg) than at a low SR (1.7 kg), and this was more pronounced in lambs of Merino ewes than in lambs of crossbred ewes.[59]

In general, responses to selenium are most likely to occur in young animals. Fleece weight responses of the order of 3% to 8% in WA[60], 4% to 12% in northern NSW[61], 9% and 17% in the southern Tablelands of NSW[62] and 14% in Victoria have been reported for sheep supplemented in their first year of life on commercial farms. Liveweight responses also occurred in many cases but these are generally of lesser economic significance, unless they were sufficiently severe to be associated with SRU. A feature of district trials which include a number of farms in each district is that responses are not usually detected on all farms, despite blood selenium concentrations which suggest responses might occur.

Seasonal variation in selenium nutrition

Selenium nutrition varies throughout the season with the lowest levels in spring[52] and rising levels in autumn. Blood selenium concentrations are also inversely related to the previous 12-month rainfall, so deficiency syndromes are more likely in wet years. Selenium levels in pasture are lower in high rainfall areas, possibly as a consequence of leaching, and in association with soils which tend to be low in minerals, such as granitic, basalt and sandy soils. Pasture improvement, particularly the introduction of clovers and subsequent soil acidification, is an additional risk factor for the depression of pasture selenium concentrations. Superphosphate application is usually a precursor or concurrent activity to pasture improvement but may of itself exacerbate marginal selenium availability to plants by raising soil sulphate levels, with consequent decline in the uptake by plants of selenite salts.[51] High stocking rates (leading to lower pasture availability) have also been associated with decreased blood selenium concentrations.

In improved pastures, clover growth is usually greatest in spring. Clover is similar or lower than grasses in selenium concentration, is high in PUFA, and promotes rapid growth of lambs. Tests on animal tissues performed in late winter will enable early warning of the risk of selenium-responsive diseases. Tests performed in spring require interpretation in light of the seasonal variations in selenium nutrition.

Clinical pathology

If selenium deficiency is suspected, the animals at risk should be tested, whatever their age. For routine monitoring of selenium nutrition on a farm, lambs 3 to 6 months of age are the most suitable age group to examine.

Both blood GSHPx and blood or plasma selenium concentrations give useful indications of selenium nutrition. Over 95% of blood enzyme activity is associated with erythrocytes, so blood samples must be collected with anticoagulant. Selenium is incorporated into erythrocyte GSHPx at the time of erythropoiesis, so measurements of erythrocyte selenium or GSHPx reflect selenium nutrition at a time up to two months in the past. Plasma selenium levels are preferable when one wishes to estimate current selenium nutrition.

Paynter et al. (1979)[63] described a relationship between blood GSHPx levels and liveweight responses to selenium. These authors suggested that GSHPx levels may have a better predictive value than blood selenium concentrations. As liveweight may be less responsive than fleece weight, values in the 'marginal' range between 'deficient' and 'normal' will also justify treatment in many situations.

Plasma concentrations of creatine kinase (CK) and aspartate amino transferase (AST, formerly SGOT) have frequently been used to assist with the diagnosis of WMD. The comments below in relation to the diagnosis of weaner nutritional myopathy (WNM) with muscle enzymes are probably also applicable to diagnosis of WMD.

Liver can also be used for testing selenium status. Soil testing is too insensitive to be useful. Pasture testing is reliable but will not necessarily give any additional information. SRU is likely to occur when pasture selenium concentrations fall below 0.02 mg/kg and is not likely if levels are above 0.03 mg/kg.[64] Responses in fertility cited previously[55] were obtained in northern NSW where, on one property, pasture selenium levels in March were 0.015 mg/kg.

Treatment

The reference curve for predicting growth responses in lambs to selenium, also developed in NZ[65], is reproduced in Figure 11.4. The authors considered that economically significant growth responses (>10 g/day) occurred only when blood selenium levels of a sample of lambs in the flock were less than 130 nmol/L. Again, reference to Table 11.2 suggests that treatment of animals in a flock with a mean in the 'marginal' range is unlikely to be justified unless it is anticipated that selenium levels will continue to fall.

Prevention and treatment

Selenium can be supplied to animals in a number of forms. Oral drenches, as sodium selenite concentrate for addition to anthelmintic drenches or for dilution with water and drenching, are effective therapeutic and preventive treatments in the short term (for up to three months). Injectable selenium, usually included with clostridial vaccine, is frequently used to protect lambs at marking. For adult sheep, intraruminal selenium pellets (5% elemental selenium, 95% iron) are effective in raising and maintaining tissue levels of selenium-containing enzymes, including GSHPx.[66] Pellets will provide adequate selenium nutrition for three to four years[67], although they may need to be accompanied by a steel grinder to prevent coating of the pellet.

Selenium can be applied to deficient pastures as encapsulated selenium (Selcote, Mintech NZ Ltd). When added to phosphatic fertilisers and applied at the rate of 10 g/hectare of selenium, it is a safe and effective method of supplementation which lasts for over 12 months.[68]

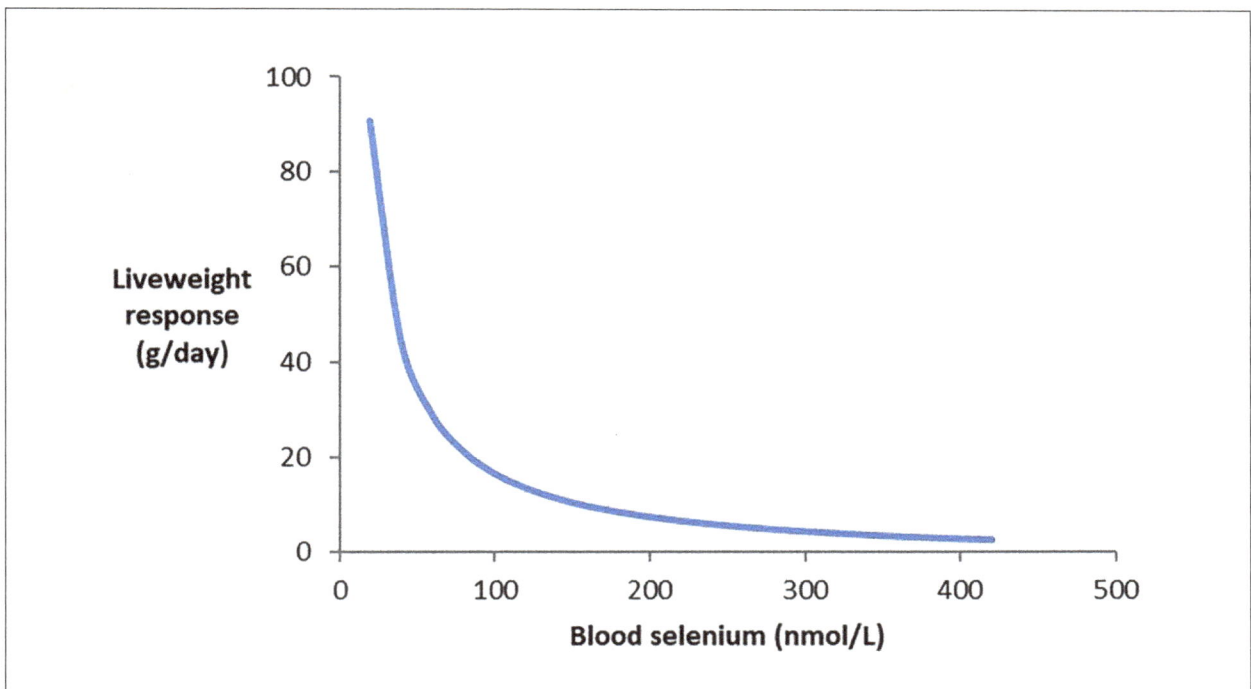

Figure 11.4: A reference curve for selenium, predicting the increase in liveweight gain which can be expected by supplementation of animals at varying levels of selenium nutritional status. Based on data from Grace and Knowles (2002).[65]

VITAMIN E

Clinical signs of vitamin E deficiency

Vitamin E deficiency can lead to WNM, which has clinical and necropsy signs similar to the myopathy associated with selenium deficiency but occurs despite an apparently adequate selenium status.[69] It has been mainly reported in weaner sheep grazing dry cereal stubble paddocks in southwestern Western Australia. Deaths occur after mustering and yarding, or they can occur spontaneously in the paddock. Clinical signs include weakness and a staggering gait, frothing from the mouth and nostrils, inability to stand, and paddling in lateral recumbency. The condition can occur in sheep which are in satisfactory condition. Lesions involve skeletal and cardiac musculature[70], particularly muscle groups containing Type II fibres.[71]

Subclinical deficiency of vitamin E

A subclinical myopathy in sheep on diets low in vitamin E has been identified.[71] Lesions in the muscles are detectable at autopsy but no production responses to treatment have been reported.

Seasonal variation in vitamin E nutrition

Deficiencies of vitamin E occur over summer and autumn in winter rainfall zones when sheep have no green feed in their diet. Most fresh green forages contain about 50 mg α-tocopherol (a form of vitamin E) per kg DM. Losses occur during sun-curing and hays generally contain 10 mg or less per kg. Cereal grains typically contain about 8 mg/kg DM.[45] There is little point in testing sheep which have a significant intake of green feed.

Clinical pathology

Both plasma and liver α-tocopherol concentrations provide good indication of vitamin E nutritional status. Plasma levels of muscle enzymes are frequently used to assist in the diagnosis of nutritional myopathies. Plasma creatine kinase (CK) in particular is used in all species because the enzyme is normally distributed only in muscle. Plasma levels of this enzyme fluctuate widely in sheep with subclinical WNM, and the enzyme has a half-life of only a few hours in sheep and cattle. Consequently, measurement of plasma levels of alanine amino transferase (ALT), alone or in conjunction with CK, is recommended in diagnosis of WNM. Aspartate aminotransferase (AST) and lactate dehydrogenase (LDH) are also sensitive indicators of muscle damage but are not specific; they are also raised when liver damage is present.[72,73]

Treatment and prevention

The oral route is probably preferable to parenteral routes of administration of vitamin E, particularly when treating clinically affected animals. Intramuscular injection of vitamin E in an oily base has been reported to give longer periods of elevated plasma α-tocopherol in most field situations than oral administration does, suggesting that IM injections are preferable to oral routes for prophylaxis.[74] Intramuscular injections, however, may cause a local myopathy and the vitamin may become sequestered in regional lymph nodes for extended periods.[75] 120 mg/kg liveweight of dl α-tocopherol acetate given orally will adequately raise vitamin E liver reserves for two months but subcutaneous treatment will not.

Diagnosis of myopathy

The myopathy associated with vitamin E can be differentiated from that of selenium deficiency on the history of access to green feed and with the aid of clinical pathology, particularly estimations of plasma selenium, GSHPx and vitamin E. Lupinosis, although primarily a hepatotoxicity, has also been associated with a myopathy in sheep in Western Australia (*lupinosis-associated myopathy*, LAM).[76] LAM is not preventable with selenium or vitamin E, although it has been proposed that the pathophysiology of the muscle damage is similar to that of WMD and WNM and is a consequence of the high PUFA and low vitamin E and selenium levels in lupin crop residues and lupin grain, in sheep which are predisposed by the hepatotoxic effects of phomopsin.[77] LAM has a history of recent exposure to lupin stubbles and jaundice is usually present in at least some of the flock with similar exposure to phomopsin. At autopsy, varying degrees of liver damage are detectable either grossly or histologically in sheep with LAM. Sudden death, while a feature of outbreaks of WMD/WNM, is not so common with LAM.[76]

The estimation of muscle and liver enzymes and blood biochemistry can also assist the differentiation between syndromes of myopathy with liver damage and those without. In lupinosis, liver damage leads to elevation of plasma gamma-glutamyl transpeptidase (GGT) and bilirubin, both conjugated and unconjugated. In myopathies without liver damage, GGT and bilirubin are normal, while CK and ALT are elevated. Plasma AST is raised with both muscle and liver damage.[78]

Sheep with *congenital progressive ovine muscular dystrophy* also develop a stiff hind limb gait. Affected sheep fail to thrive and their locomotor deficit worsens, so that they are culled or die of other causes before they are 2 years of age.[79] The condition is a true dystrophy rather than a degeneration as in WMD, WNM and LAM, and a distinctive feature at post-mortem is the involvement of the vastus intermedius muscle, which is not involved in the myopathy of vitamin E deficiency. The disease is rare.

Exertional rhabdomyolysis has also been reported in sheep, following prolonged energetic chasing or mustering with dogs.[80] Chapman (1990)[81] also includes monensin toxicity, poisoning with *Cassia* spp and *Ixidaena* spp, ischaemic myopathy, trauma, snake bite, clostridial myositis and *cereal grain associated myopathy* in the diagnostic considerations for myopathies.

IODINE

In Australia, iodine deficiency syndromes of sheep occur predominantly in inland mountainous areas on soils that are naturally low in iodine. There is a higher risk of deficiency on alluvial and sandy soils.[82] Deficiency is most commonly reported from Victoria[83] and Tasmania, particularly in the Derwent Valley, northern and southern Midlands[82], and parts of New South Wales. Iodine deficiency syndromes also occur in association with the grazing of some plant species which contain substances (goitrogens) which interfere with iodine metabolism.

Physiology

Iodine is essential for thyroid hormone production. Thyroid hormones (thyroxine (T4) and its more active derivative tri-iodothyronine (T3)) strongly influence thermoregulation and have wide-ranging effects on growth, immune competence and the circulatory system. They are important in the normal development of the foetal brain and foetal integument and are particularly important in the adaptation of the newborn animal to life outside the uterus.

Adult sheep receive sufficient iodine with feeds containing 0.2 to 0.5 mg iodine per kg DM, reflecting intakes of about 1 mg per day for an adult sheep. Requirements are higher in winter than in summer.

Selenium-containing enzymes (*5'-deiodinases*) are involved in the conversion of T4 to T3 and a deficiency of selenium in ewes and their foetal lambs can result in the birth of lambs which have low levels of circulating T3.[84,85] The clinical significance of this is not yet clear.

When animals are severely restricted in iodine intake or when thyroid hormone production is inhibited in other ways, there is a hyperplastic response of the thyroid gland tissue and a subsequent increase in the size of the thyroid gland. Thyroid gland enlargement is called goitre.

Goitrogens

Goitrogens are substances which inhibit the production of thyroid hormone. White clover (*Trifolium repens*) contains high levels of hydrogen cyanide which is converted to thiocyanate after ingestion. Thiocyanate is goitrogenic; it inhibits *thyroperoxidase* and thereby blocks the incorporation of iodine into thyroxine. Plants of the *Brassica* genus that are used as fodder crops (kale, choumoellier, turnips, swedes) also contain substances which release hydrogen cyanide after harvesting or ingestion. Ewes which graze these plants during pregnancy may give birth to lambs which are iodine deficient.

Plant goitrogens are considered to be an important cause of iodine deficiency in New Zealand. Goitrogens of the thiocyanate type can be counteracted by supplementation with iodine. Goitrogens can also be of the thiouracil type. Their disruption of iodine metabolism cannot be overcome by iodine supplementation.

Clinical signs of deficiency

A prolonged dietary deficiency of iodine leads to increased mortality of young animals, an increase in thyroid size and a depression of wool and milk production in adult animals. The presence of goitre in newborn animals, often in association with high neonatal mortalities, is the characteristic sign of severe iodine deficiency. Affected lambs lack vigour, and they have a reduced covering of wool and an increased susceptibility to death by hypothermia.

The thyroid enlargement is often hard to detect and a careful dissection and weighing of the glands is necessary. The presence of goitre which was detectable by palpation but not visually obvious was associated with lamb mortalities in Tasmania. Palpably goitrous lambs of an otherwise normal birth weight were more than twice as likely to die in the immediate perinatal period as non-goitrous lambs in the same flock.[86] Goitrous lambs which survive for the first few days of life usually recover and develop normally.

In ewes, iodine deficiency can lead to a reduction in fertility and fecundity, increased gestation length, reduced lamb birth weights[87] and reduced wool production.

Iodine supplementation of ewes grazing white clover pasture has led to an improvement in the numbers of lambs born and a small improvement in lamb survival in a Romney flock in New Zealand. Supplemented ewes produced 1.71 lambs per ewe, while unsupplemented ewes produced 1.57. The effect of supplementation on foetal numbers occurred in the first third of pregnancy, suggesting that iodine supplementation may have enhanced embryo survival.[88]

Supplementation of ewes in an iodine deficient area of northern Tasmania resulted in a 6% increase in fleece weight of the ewes.[89] The response occurred in the absence of any effect on perinatal lamb survival or subsequent lamb growth.

Diagnosis

The presence of enlarged thyroid glands in newborn lambs is a strong indicator of iodine deficiency but, in cases where the gross enlargement is unconvincing, the gland should be dissected and weighed. A ratio of thyroid weight to body weight >0.8 g/kg provides strong evidence that a lamb was born to an iodine-deficient ewe and ratios between 0.4 and 0.8 are considered marginally deficient.[90] Histology of affected thyroid glands provides additional information to assist an aetiological diagnosis.

It should be noted that goitre can occur in lambs without increases in mortality[87] and that iodine-responsive conditions in ewes and lambs can occur in the absence of grossly observable goitre.

Blood specimens can be assayed for iodine concentrations, total thyroxine concentrations, T4 concentrations and T3 concentrations, but interpretation is difficult. Low T3 levels may be the best indicator of a likely response to supplementation but responses may also occur in flocks where monitoring had indicated T3 levels in the normal range.

Seasonal variations in availability

There is a marked seasonal variation in intake of iodine. It increases during November (in Victoria), peaks in summer, and declines rapidly within days after autumn rain. Dietary intake continues to decline during winter. In Victoria, outbreaks of goitre and associated lamb mortalities are restricted to the high rainfall districts and to lambs born between August and October.[17] Seasonal conditions which favour iodine deficiency are those with early rains, good pasture growth in May and June, and rainfall in May, June and July of over 80 mm/month. The deficiency is probably associated with a reduction in soil ingestion (an important source of dietary iodine) in winters of good pasture growth.[82]

Treatment and prevention

Affected animals can be treated with oral solutions of potassium iodide (KI) (20 mg per lamb) or with parenteral iodised poppy seed oil containing 40% w/v iodine (Lipiodol®).

The need for preventive treatment is difficult to predict and a number of trials have failed to detect any response to iodine supplementation, even when there were strong grounds for suspecting that a deficiency syndrome may occur. Consequently, a decision to provide supplementation to ewes should be based on a thorough analysis of the risk of production losses or lamb mortality in the flock in question, and the cost of providing additional iodine. The challenge is made more difficult by the lack of a reliable measure of the iodine nutritional status of sheep. Thyroid hormone concentrations in blood are a poor indicator of the likelihood of responses to supplementation.[88]

In flocks where there is a significant risk of iodine deficiency affecting the viability of newborn lambs, pregnant ewes can be given preventive treatment in mid- and late pregnancy with two oral doses of KI (280 mg) or potassium iodate (KIO_3) (360 mg) approximately one month apart. If the ewes are grazing goitrogenic feeds, treatment at that level may not completely prevent the occurrence of goitre in lambs but is expected to prevent neonatal mortalities associated with iodine deficiency.[87] A premating injection of ewes with 1 mL of iodised oil will prevent congenital goitre in their lambs for two years.[89] Alternatively, salt licks containing KIO_3 (15 to 25 g per 100 kg of salt) can be provided.[17] KI in salt licks is unstable and iodine is lost from these preparations over time. KIO_3 is more stable and is the preferred source of iodine in licks.[87]

In regions where iodine deficiency has occurred previously, the provision of brassica forage crops to ewes as winter feed should be routinely accompanied by iodine supplementation.[90]

IRON, MOLYBDENUM, MANGANESE AND ZINC

Iron

Dietary iron deficiency does not occur in grazing ruminants. Grazing animals ingest sufficient iron in natural diets from both plants and soil. Milk is a poor source of iron for lambs but they obtain iron from accidental ingestion of soil from the dam's udder. An iron deficiency anaemia has been reported in lambs raised indoors, in which 2% of the lambs died with anaemia amongst other pathological changes.[91] Compared to lambs raised outdoors, the indoor-raised lambs on three farms had, on average, lower haemoglobin concentrations (8.4 to 9.4 cf. 11.7 g/dL) and lower haematocrit (28 to 32 cf. 38%).

In contrast to deficiencies of the trace element, excess dietary iron can interfere with copper absorption.

Molybdenum

Molybdenum (Mo) is required for several metalloenzymes in animals but a deficiency syndrome in animals has not been reported. The importance of Mo to the sheep industry in Australia is the widespread problem of low soil levels of Mo affecting pasture growth, and the interference with copper metabolism of sheep caused by high dietary levels of Mo. In many cases, molybdenum-induced copper deficiency occurs as a result of Mo being applied to pastures to improve the growth of plants, particularly leguminous plants like clovers (*Trifolium* spp).

Manganese

Manganese is an essential trace element for carbohydrate and lipid metabolism in animals but there are no reports of deficiency syndromes in naturally grazed sheep. There is one report of bone and joint disorders in lambs fed diets artificially low in manganese.[92] In plants, however, both deficiency syndromes and toxicity caused by excess manganese are reported. Manganese toxicity in grazing animals has been reported in Australia, associated with the grazing of lupins (*Lupinus albus*) which accumulate manganese to high levels.[17]

Zinc

Lambs fed artificial zinc-deficient diets have a reduced feed intake as a result of a loss of appetite. Sustained and severe deficiency leads to the development of parakeratotic skin lesions around the eyes, nose, feet and scrotum.[93] Wool growth is profoundly affected and, in lambs, wool may be shed.[94] The fleece becomes loose and, in one report, is heavily stained with a red-brown pigment.[93] In rams, zinc deficiency suppresses sperm production, and testicular weight declines. In pubescent rams, deficiency completely blocks testicular growth — partly through its effect of appetite suppression and partly due to a specific effect which disrupts testicular secretion.[95]

There is one report of an increase in fertility in Dorset Horn ewes following zinc and manganese supplementation.[96] In that instance, the sheep grazed pastures on light sandy loams with calcareous profiles and topsoil pH values of 6.8 to 8.3. There are several sets of environmental conditions which are known to be associated with zinc deficiency in plants, including calcareous soils of high pH.[97] There is no other evidence that zinc deficiency causes clinical or subclinical disease in sheep grazing natural pastures in Australia but there are reports from other countries.

CLINICAL PATHOLOGY IN THE INVESTIGATION OF DEFICIENCIES

Trace element nutrition, particularly dietary deficiencies, can be investigated by the collection of a varying range of tissue and soil samples. The choice of the most appropriate samples depends on the circumstances prompting the investigation and the time of the year it is performed.

Reasons for using clinical pathology

Clinical pathology can be used
- when specific clinical signs are evident, such as an outbreak of long bone fractures or myopathy
- when ill-thrift is suspected or when sheep (particularly young sheep) fail to grow and produce wool as well as expected
- when routine monitoring is being carried out to determine trace element nutritional status of a flock of sheep; this can detect deficiencies which are completely subclinical or which only occur in occasional years.

Plant and soil testing

Testing of plants to predict the availability of trace elements to grazing animals ignores the importance of soil ingestion as a source of nutrients. Ingestion of soil is an important source of cobalt and iodine and, to a lesser extent, selenium.[98] More soil is ingested when pastures are short (autumn and winter, and at high stocking densities) than when pastures are longer, such as in spring. Soil is not necessarily a useful source of copper and some soils may reduce the availability of copper ingested in plant material because of the levels of Fe, S or Mo.[99]

Seasonal variations in trace element availability

As pasture dries off in late spring, a number of nutritional changes occur, summarised in Table 11.1.

In general, winter to early spring is a good time for routine monitoring or investigations to determine the probability of occurrence of a significant deficiency of the four trace elements in the left column of Table 11.1. Additional factors, as discussed under the heading for each element, should be borne in mind when planning an investigation.

INTERPRETATION OF RESULTS

'Normal' levels for trace element concentration in animal and plant tissues are published in a number of sources and it is the usual practice of most laboratories to provide ranges which are considered to reflect *adequate* levels and those in which responses to supplementation may occur (the *marginal* range) or are very likely to occur (the *deficient* range). With all trace element nutrients, the lower the level in the marginal and deficient ranges, the greater the likelihood of a response to supplementation occurring.

When clinical syndromes of deficiency are apparent, generally the tissue levels in the animals are very low — well below the cut-off for the deficient band — and confirmation of a diagnosis is straightforward. Difficulties arise, however, when trying to predict whether there

Table 11.1: Changes in dietary nutrients from spring to summer.

Nutritional elements which increase in availability after spring	Nutritional elements which decline in availability after spring
Copper	Vitamin E
Selenium	Crude protein
Cobalt	Digestible energy
Iodine	Phosphorus
	Sulphur

Table 11.2: Normal and deficient ranges for blood clinical biochemistry in sheep. These ranges should be interpreted carefully and with reference to local experience and, in the case of selenium and B$_{12}$ status, response reference curves.

Nutritional element	Deficient	Marginal	Adequate
Copper			
Plasma copper (µmol/L)	<5	5-8	>8
Plasma ceruloplasmin (U/L)	<5	5-40	>40
Erythrocyte CUSOD (U/g Hb)	<200	200-450	>450
Selenium			
Blood selenium (nmol/L)	<250	250-500	>500
Serum selenium (nmol/L)	<200	200-400	>400
Blood GSHPx (units)	<20	20-50	>50
Vitamin B$_{12}$			
Serum B$_{12}$ (pmol/L)	<400 (0.54 µg/L)	400-700	>700

will be responses to supplementation in groups of animals in which the tissue level is reported to be in the marginal band. Trial work over several decades in Australia, New Zealand and the UK have led to a reconsideration, and general downward movement, of the levels below which responses can be reliable expected. The review of Suttle (2005)[26] is recommended reading in relation to this topic.

Animals which are marginally deficient in trace elements may show significant responses to supplementation by improvements in growth rate, wool growth or fertility. The signs of marginal deficiency are subclinical, so the potential for improvements in productivity is only detectable by either testing of tissues for trace element levels or conducting a response trial and measuring the improvements in supplemented animals compared to untreated controls.

Response trials, however, are time-consuming and laborious to conduct and it may take several months before results are detected. Fortunately, there is now a body of literature based on many response trials which can be used to predict the likelihood of economically significant responses to supplementation, based on the measured level of a trace element or trace-element-dependent tissue in a sample of animals in a flock. The meta-analysis of the response trials has allowed the creation of reference curves for selenium and cobalt nutrition. The probability, with confidence limits, and magnitude of liveweight responses can be predicted by consulting a reference curve.[100]

The original publications used to produce Figures 11.2 and 11.3 included curves for the probability of responses greater than 10 g/day, as an indicator of the uncertainty around predictions of responsiveness.

Veterinarians investigating cases of ill-thrift or lower-than-expected productivity in flocks of animals must exercise caution when assigning the cause of the poor performance to a trace element deficiency when tests indicate that the animals are in a marginal range. A diagnosis of trace element deficiency in a flock is often followed by the adoption of routine prophylactic treatment of animals for many years into the future, without any reassessment of the need. This may be a cost-effective strategy if deficiency syndromes would otherwise occur every few years but, if not, supplementation may be a waste of money. It is as important to avoid the expense of unnecessary treatment as it is to forego the improvements in productivity which may arise from permanent correction of the deficiency. It is also important that other causes of poor productivity are not overlooked because a trace element deficiency has been diagnosed on insubstantial grounds.

The following strategies are recommended when investigating the status of trace element nutrition in a flock in relation to clinical or subclinical deficiency:

1. Remember that clinical syndromes (enzootic ataxia (Cu), white muscle disease (Se), coast disease (Co), etc.) are associated with profoundly low tissue levels of the trace element or vitamin. These conditions are readily confirmed by testing of blood or liver specimens. Tissue levels of the nutrient are far below the cut-off value given for the bottom of the marginal range. The clinical syndromes will not be evident if the levels are in the marginal range or higher.

2. If the mean tissue levels of a random sample of animals from the flock fall within the deficient range, the animals are likely to respond to supplementation, and consideration should be given to introducing routine prophylaxis for future years, without repeated

testing. This decision may be influenced by reference to response curves, if available, particularly if the testing results are approaching the marginal band.

3. If the mean tissue levels of a random sample of animals from the flock fall within the marginal range, consideration should be given to supplementation of the animals under test, but a decision about future routine prophylaxis should be delayed until testing is repeated in the following year. Future testing should be timed in such a way that any deficiency can be predicted before it can significantly reduce the productivity of the flock.

Rather than remembering particular cut-off levels, one should try to appreciate that in general a large number of factors influence the response or lack of response to supplementation with particular trace elements. The results of tissue testing give useful assistance in determining the risk and likely cost of any clinical or subclinical deficiency.

Before making recommendations to reduce the risk of deficiency syndromes in the future, finding answers to the following questions is recommended:

Copper

- Are levels about to rise without treatment?
- Is deficiency due to a primary copper deficiency or secondary to excess molybdenum, etc.? If the deficiency is primary, is the problem one of low soil copper or failure by plants to pick it up?
- Is deficiency severe? Will it occur again? If yes, how often?

Cobalt

- Are levels low because lambs are too young, or because cobalt nutrition is low? B_{12} injection will not lead to extended maintenance of plasma levels.
- Are cobalt pellets already present in the sheep but not working?
- Are levels low enough to warrant treatment every year or should further testing be advised?

RESPONSE TRIALS

A response trial is, ultimately, the only 'proof' that a deficiency exists. There are a number of problems, however, associated with using response trials for diagnosis of deficiency. First, failure to demonstrate a response in one year does not mean that a response will not occur in other years. Second, trials must be set up and conducted carefully to give meaningful results, and this adds to the cost of diagnosis. Reference levels have been calculated from trial work, so one has to ask if a response trial is in fact an exercise in rediscovering the wheel.

PREVENTION AND TREATMENT

Mineral supplements are frequently provided to sheep in proprietary and homemade licks and blocks. Blocks are solid 'bricks', approximately 30 cm cubes, of salt, molasses or composite material in which trace and macronutrients are mixed. Blocks are placed in paddocks where sheep graze with the intention that they will lick, nibble or chew the block and ingest useful quantities of nutrients missing from the pasture feed. Licks are powder mixtures prepared with the same intention but are 'fed' in drums or troughs or tractor tyres cut in half to form circular troughs. These preparations are probably ineffective in the face of significant dietary deficiencies. Commercial blocks produced for widespread sale must contain very low levels of copper and selenium, for example, because increased consumption of these elements by sheep in some areas could lead to poisoning. Those products designed for specific areas and specific degrees of deficiency are, however, potentially useful. A limitation of their use is the variable intake of sheep within a flock. The subject is further discussed by McDonald (1983).[101] Other, more direct methods of prophylaxis are discussed above.

POISONING

Poisoning with copper is a significant problem in many areas of Australia and supplemental treatment should not be given without consideration of the risk of poisoning. The subject is discussed further in Chapter 17.

Selenium can also be toxic and poisoning can occur naturally (NIA), by treating animals which already have high selenium stores or by overdosing. Doses of 15 mg have been fatal in lambs of 10 kg liveweight. Young lambs may be particularly susceptible to selenium poisoning due to incomplete rumen development.[102] Signs of acute selenium toxicity include blindness, abdominal pain, excessive salivation, paralysis and death after one to seven days. In a number of field trials, high doses of selenium (>0.1 mg/kg liveweight) have resulted in negative responses while, in the same trials, doses of 0.1 mg/kg or less resulted in positive responses.[51,63]

Allen et al. (1986)[103] identified three therapeutic uses of zinc which could lead to intoxication of sheep with this element. These are the use of zinc against facial eczema, against lupinosis and in the treatment of footrot. They reported the death of 19 of 100 treated weaners, 14 within 24 hours of oral treatment with 3 g of zinc. At necropsy, there was marked necrosis and a lime-green discolouration of the mucosa of the abomasum and duodenum.

RECOMMENDED READING

Caple IW (1990) Trace element deficiencies. In: Sheep medicine. Proceedings No 141. University of Sydney Post-graduate Committee in Veterinary Science: Australia, p. 367.

Caple IW (1990) Myopathies: Selenium and vitamin E. In: Sheep medicine. Proceedings No 141. University of Sydney Post-graduate Committee in Veterinary Science: Australia, p. 359.

Grace ND and Knowles SO (2012) Trace element supplementation of livestock in New Zealand: Meeting the challenges of free-range grazing systems. Veterinary Medicine International 2012 Article ID 639472. http://dx.doi.org/10.1155/2012/639472.

Hosking, WJ, Caple IW, Halpin CG et al. (1986) Trace elements for pastures and animals in Victoria. Department of Agricultural and Rural Affairs: Melbourne, Victoria.

Suttle N (2005) Assessing the needs of sheep for trace elements. Practice 27 474-83.

Suttle N (2010) Mineral nutrition of livestock. 4th ed. CABI: Oxfordshire UK. https://doi.org/10.1079/9781845934729.0000.

REFERENCES

1 Grace ND and Knowles SO (2012) Trace element supplementation of livestock in New Zealand: Meeting the challenges of free-range grazing systems. Veterinary Medicine International 2012

Article ID 639472. http://dx.doi.org/10.1155/2012/639472.

2   Suttle N (2010) Mineral nutrition of livestock. 4th ed. CABI: Oxfordshire UK. https://doi. org/10.1079/9781845934729.0000.

3   Bowen HJM (1979) cited by Jarvis SC (1981) Copper concentration in plants and their relationship to soil properties. In: Copper and soils, eds JF Loneragan, AD Robson and RD Graham. Academic Press: Sydney, pp. 265-85.

4   Brown AJ (1982) cited by Hosking, WJ, Caple IW, Halpin CG et al. (1986) Trace elements for pastures and animals in Victoria. Department of Agricultural and Rural Affairs: Melbourne, Victoria.

5   Gartrell JW (1981) Distribution and correction of copper deficiency in crops and pastures. In: Copper and soils, eds JF Loneragan, AD Robson and RD Graham Academic Press, Sydney, pp. 313-49.

6   Jarvis SC (1981) Copper concentration in plants and their relationship to soil properties. In: Copper and soils, eds JF Loneragan, AD Robson and RD Graham. Academic Press: Sydney, pp. 265-85.

7   Hannam RJ and Reuter DJ (1977) The occurrence of steely wool in South Australia 1972-75. Agric Record **4** 26-9.

8   Hannam RJ, Judson GJ, Reuter DJ et al. (1982) Current requirements of copper for pasture and sheep on sandy soils in the upper south-east of South Australia. Aust J Exp Agric Anim Husb **22** 324.

9   Grace ND, Knowles SO, West DM et al. (2004) Copper oxide needles administered during early pregnancy improve the copper status of ewes and their lambs. NZ Vet J **52** 189-92. https://doi. org/10.1080/00480169.2004.36427.

10 Bennetts HW (1932) Enzootic ataxia of lambs in Western Australia. Aust Vet J **8** 137-83.

11 Bennetts HW and Chapman FE (1937) Copper deficiency in sheep in Western Australia: A preliminary account of the aetiology of enzootic ataxia of lambs and an anaemia of ewes. Aust Vet J **13** 138-49. https://doi.org/10.1111/j.1751-0813.1937.tb04108.x.

12 Dunlop G, Innes JRM, Shearer GD et al. (1939) 'Swayback' studies in north Derbyshire. J Comp Pathol **52** 259-65. https://doi.org/10.1016/S0368-1742(39)80023-6.

13 Suttle NF, Angus KW, Nisbet DI et al. (1972) Osteoporosis in copper-depleted lambs. J Comp Pathol **82** 93-7.

14 ARC (Agricultural Research Council) (1980) Trace elements. In: The nutrient requirements of ruminant livestock. Commonwealth Agricultural Bureaux: Farnham Royal, Slough, England, pp. 227-9.

15 Woolliams JA, Wiener G, Woolliams, C et al. (1985) Retention of copper in the liver of sheep genetically selected for high and low concentrations of copper in plasma. Anim Prod **41** 219-26. https://doi.org/10.1017/S0003356100027884.

16 Woolliams JA, Suttle NF, Wiener G et al. (1983) The long-term accumulation and depletion of copper in the liver of different breeds of sheep fed diets of differing copper content. J Agric Sci **100** 441-9. https://doi.org/10.1017/S0021859600033608.

17 Hosking, WJ, Caple IW, Halpin CG et al. (1986) Trace elements for pastures and animals in Victoria. Department of Agricultural and Rural Affairs: Melbourne, Victoria.

18 Laven RA and Smith SL (2008) Copper deficiency in sheep: An assessment of the relationship between concentrations of copper in serum and plasma. NZ Vet J **56** 334-8. https://doi.org/10. 1080/00480169.2008.36856.

19 Judson GJ and McFarlane JD (1998) Mineral disorders in grazing livestock and the usefulness of soil and plant analysis in the assessment of these disorders. Aust J Exp Agric **38** 707-23. https:// doi.org/10.1071/EA97145.

20 McFarlane JD, Judson GJ and Gouzos J (1990) Copper deficiency in ruminants in the south east of South Australia. Aust J Exp Agric **30** 187-93.

21 Judson GJ, Brown TH, Gray D et al. (1982) Oxidised copper wire particles for copper therapy in

sheep. Aust J Agric Res **33** 1073. https://doi.org/10.1071/AR9821073.

22 Porter WL and Watson PW (2011) Distribution category for oral copper supplements for sheep. Vet Rec **169** 54-5. https://doi.org/10.1136/vr.d4258.

23 Burke JM, Miller JE, Olcott DD et al. (2004) Effect of copper oxide wire particles dosage and feed supplement level on *Haemonchus contortus* infection in lambs Vet Parasitol **123** 235-43. https://doi.org/10.1016/j.vetpar.2004.06.009.

24 Waller PJ, Bernes G, Rudby-Martin L et al. (2004) Evaluation of copper supplementation to control *Haemonchus contortus* infections of sheep in Sweden. *Acta Vet Scand* **45** 149-60.

25 Gardiner MR (1977) cited by Hannam et al. (1980). Effect of vitamin B_{12b} injections on the growth of young Merino sheep Aust J Agric Res **31** 347-55. https://doi.org/10.1071/AR9800347.

26 Suttle N (2005) Assessing the needs of sheep for trace elements. Practice **27** 474-83. https://doi.org/10.1136/inpract.27.9.474.

27 Grace ND (1999) The effect of increasing the Vitamin B_{12} status of Romney ewes on foetal liver Vitamin B_{12} and liver Vitamin B_{12} concentrations in suckling lambs. NZ Vet J **47** 97-100. https://doi.org/10.1080/00480169.1999.36121.

28 Caple IW and McDonald JW (1983) Trace element nutrition. In: Sheep production and preventive medicine. Proceedings No 67. University of Sydney Post-graduate Committee in Veterinary Science: Australia, p. 236.

29 Grace ND, Knowles SO and West DM (2006) Dose-response effects of long-acting injectable vitamin B_{12} plus selenium (Se) on the vitamin B_{12} and Se status of ewes and their lambs. NZ Vet J **54** 67-72. https://doi.org/10.1080/00480169.2006.36614.

30 Grace ND, Knowles SO, Sinclair GR et al. (2003) Growth response to increasing doses of microencapsulated vitamin B_{12} and related changes in tissue vitamin B_{12} concentrations in cobalt-deficient lambs. NZ Vet J **51** 82-92. https://doi.org/10.1080/00480169.2003.36345.

31 Gruner TM, Sedcole JR, Furlong JM et al. (2004) A critical evaluation of serum methylmalonic acid and vitamin B_{12} for the assessment of cobalt deficiency of growing lambs in New Zealand. NZ Vet J **52** 137-44.

32 Kennedy DG, Young PB, McCaughey WJ et al. (1991) Rumen succinate production may ameliorate the effects of cobalt-vitamin B_{12} deficiency on methylmalonyl Co-A mutase in sheep. Br J Nutr **121** 1236-42.

33 Hannam RJ, Judson GJ, Reuter DJ et al. (1980) Effect of vitamin B_{12b} injections on the growth of young Merino sheep. Aust J Agric Res **31** 347-55. https://doi.org/10.1071/AR9800347.

34 Sutherland RJ, Cordes DO and Carthew GC (1979) Ovine white liver disease — an hepatic dysfunction associated with vitamin B_{12} deficiency. NZ Vet J **27** 227. https://doi.org/10.1080/00480169.1979.34658.

35 Mitchel PJ, McOrist S, Thomas KW et al. (1982) White liver disease of sheep. Aust Vet J **58** 181-4. https://doi.org/10.1111/j.1751-0813.1982.tb00648.x.

36 Richards RB and Harrison MR (1981) White liver disease in lambs. Aust Vet J **57** 565-8. https://doi.org/10.1111/j.1751-0813.1981.tb00437.x.

37 Mason RW and McKay R (1983) Ovine white liver disease. Aust Vet J **60** 219. https://doi.org/10.1111/j.1751-0813.1983.tb09590.x.

38 Duncan WRH, Morrison ER and Garton GA (1981) Effects of cobalt deficiency in pregnant and post-parturient ewes and their lambs. Br J Nutr **46** 337-43.

39 Fisher GEJ and MacPherson A (1991) Effect of cobalt deficiency in the pregnant ewe on reproductive performance and lamb viability. Res Vet Sci **50** 319-27. https://doi.org/10.1016/0034-5288(91)90132-8.

40 Clark RG, Wright DF, Millar KR et al. (1989) Reference curves to diagnose cobalt deficiency in sheep using liver and serum vitamin B_{12} levels. NZ Vet J **37** 7-11. https://doi.org/10.1080/00480169.1989.35537.

41 Hogan KG, Lorentz PB and Gibb FM (1973) The diagnosis and treatment of vitamin B_{12} deficiency

in young lambs. NZ Vet J **21** 234-7. https://doi.org/10.1080/00480169.1973.34115.

42 Judson GJ and Babidge PJ (2002) Vitamin B$_{12}$ injection for preventing cobalt deficiency in lambs. Aust Vet J **80** 777-8. https://doi.org/10.1111/j.1751-0813.2002.tb11352.x.

43 Judson GJ, Woonton TR, McFarlane JD et al. (1995) Evaluation of cobalt pellets for sheep. Aust J Exp Agric **35** 41-9. https://doi.org/10.1071/EA9950041.

44 Rice D and Kennedy S (1988) Vitamin E: Function and effects of deficiency. Br Vet J **144** 482. https://doi.org/10.1016/0007-1935(88)90089-9.

45 Putnam ME and Comben N (1987) Vitamin E. Vet Rec **121** 541.

46 Radostits OM, Blood DC and Gay CG (1994) Veterinary medicine. 8th ed. Bailliere Tindall: New York, p. 1415.

47 Ellis TM, Masters HG, Hustas L, Sutherland SS et al. (1990) The effect of selenium supplementation on antibody response to bacterial antigens in Merino sheep with a low selenium status. Aust Vet J **67** 226-8. https://doi.org/10.1111/j.1751-0813.1990.tb07767.x.

48 Jelinek PD, Ellis T, Wroth RH et al. (1988) The effect of selenium supplementation on immunity, and the establishment of an experimental *Haemonchus contortus* infection, in weaner Merino sheep fed a low selenium diet. Aust Vet J **65** 214-7. https://doi.org/10.1111/j.1751-0813.1988.tb14461.x.

49 Grant AB, Drake C and Hartley WJ (1960) Further observations on white muscle disease in lambs. NZ Vet J **8** 1-3. https://doi.org/10.1080/00480169.1960.33362.

50 Andrews ED, Hartley WJ and Grant AB (1968) Selenium-responsive diseases of animals in New Zealand. NZ Vet J **16** 3-17. https://doi.org/10.1080/00480169.1968.33738.

51 McDonald JW (1975) Selenium responsive unthriftiness of young Merino sheep in central Victoria. Aust Vet J **51** 433.

52 Caple IW, Andrewartha KA, Edwards SJA et al. (1980) An examination of the selenium nutrition of sheep in Victoria. Aust Vet J **56** 160-7.

53 Hartley WJ (1963) Selenium and ewe fertility. Proc NZ Soc Anim Prod **23** 20.

54 Godwin KO, Kuchel RE and Buckley RA (1970) The effect of selenium on infertility in ewes grazing improved pastures. Aust J Exp Agric Anim Husb **10** 672-8.

55 Wilkins JF and Kilgour RJ (1982) Production responses in selenium supplemented sheep in northern New South Wales 1. Infertility in ewes and associated production. Aust J Exp Agric Anim Husb **22** 18-23. https://doi.org/10.1071/EA9820018.

56 Piper LR, Bindon BM, Wilkins JF et al. (1980) The effect of selenium treatment on the fertility of Merino sheep. Proc Aust Soc Anim Prod **13** 241.

57 Langlands JP, Donald GE, Bowles JE et al. (1991) Subclinical selenium insufficiency 2. The response in reproductive performance of grazing ewes supplemented with selenium. Aust J Exp Agric **31** 33-5.

58 Langlands JP, Donald GE, Bowles JE et al. (1991) Subclinical selenium insufficiency 1. Selenium status and the response in liveweight and wool production of grazing ewes supplemented with selenium. Aust J Exp Agric **31** 25-31.

59 Langlands JP, Donald GE, Bowles JE et al. (1991) Subclinical selenium insufficiency 3. The selenium status and productivity of lambs born to ewes supplemented with selenium. Aust J Exp Agric **31** 37-43.

60 Gabbedy BJ (1971) Effect of selenium on wool production, body weight and mortality of young sheep in Western Australia. Aust Vet J **47** 318-22.

61 Wilkins JF and Kilgour RJ (1982) Production responses in selenium supplemented sheep in northern New South Wales 2. Liveweight gain, wool production and reproductive performance in young Merino ewes given selenium and copper supplements. Aust J Exp Agric Anim Husb **22** 24-8.

62 Dove H, Axelsen A and Watt R (1986) Selenium responses in grazing ewes and their lambs. Proc Aust Soc Anim Prod **16** 187.

63 Paynter DI, Anderson JW and McDonald JW (1979) Glutathione peroxidase and selenium in sheep. II. The relationship between glutathione peroxidase and selenium-responsive unthriftiness in Merino lambs. Aust J Agric Res **30** 703-9.

64 Andrews ED, Hartley WJ and Grant AB (1968) Selenium responsive diseases of animals in New Zealand. NZ Vet J **16** 3.

65 Grace ND and Knowles SO (2002) A reference curve using blood selenium concentration to diagnose selenium deficiency and predict growth responses in lambs. NZ Vet J **50** 163-5. https://doi.org/10.1080/00480169.2002.36303.

66 Paynter DI (1979) Glutathione peroxidase and selenium in sheep. I Effect of intraruminal selenium pellets on tissue glutathione peroxidase activities. Aust J Agric Res **30** 695. https://doi.org/10.1071/AR9790695.

67 Judson GJ, Ellis NJS, Kempe BR et al. (1991) Long-acting selenium treatments for sheep. Aust Vet J **68** 263-5.

68 Halpin CG, Hanrahan P and McDonald JW (1987) Selenium fertilizer for the control of selenium deficiency in grazing sheep. In: Temperate pastures: Their production, use and management, eds JL Wheeler, CJ Pearson and GE Robards. Australian Wool Corporation-CSIRO: Melbourne, p. 386.

69 Steele P, Peet RL, Skirrow S et al. (1980) Low alpha-tocopherol levels in livers of weaner sheep with nutritional myopathy. Aust Vet J **56** 529-32.

70 Allen JG and Steele P (1980) Distribution of lesions in ovine weaner nutritional myopathy in Western Australia. Aust Vet J **56** 560.

71 Allen JG, Steele P, Masters HG et al. (1986) A study of nutritional myopathy in weaner sheep. Aust Vet J **63** 8-13.

72 Fry JM, Allen JG, Speijers EJ et al. (1994) Muscle enzymes in the diagnosis of ovine weaner nutritional myopathy. Aust Vet J **71** 146-50.

73 Smith GM, Fry JM, Allen JG et al. (1994) Plasma indicators of muscle damage in a model of nutritional myopathy in weaner sheep. Aust Vet J **71** 12-17.

74 Doncon GH and Steele P (1988) Plasma and liver concentrations of α-tocopherol in weaner sheep after vitamin E supplementation. Aust Vet J **65** 210-13. https://doi.org/10.1111/j.1751-0813.1988.tb14460.x.

75 Dickson J, Hopkins DL and Doncon GH (1986) Muscle damage associated with injections of vitamin E in sheep. Aust Vet J **63** 231-3. https://doi.org/10.1111/j.1751-0813.1986.tb03005.x.

76 Allen JG (1978) The emergence of a lupinosis-associated myopathy in sheep in Western Australia. Aust Vet J **54** 548-9.

77 Allen JG, Steele P, Masters HG et al. (1992) A lupinosis-associated myopathy in sheep and the effectiveness of treatments to prevent it. Aust Vet J **69** 75-81. https://doi.org/10.1111/j.1751-0813.1992.tb15554.x.

78 Allen JG and Randell AG (1993) The clinical biochemistry of experimentally produced lupinosis in the sheep. Aust Vet J **70** 283-8. https://doi.org/10.1111/j.1751-0813.1993.tb07975.x.

79 Dent AC and Richards RB (1979) Congenital progressive ovine muscular dystrophy in Western Australia. Aust Vet J **55** 297.

80 Peet RL, Dickson J, Masters H et al. (1980) Exertional rhabdomyolosis in sheep. Aust Vet J **56** 155-6. https://doi.org/10.1111/j.1751-0813.1980.tb05666.x.

81 Chapman HM (1990) Myopathies that do not respond to selenium. In: Sheep medicine. Proceedings No. 141. University of Sydney Post-graduate Committee in Veterinary Science: Australia, p. 483.

82 Statham M and Bray AC (1975) Congenital goitre in sheep in southern Tasmania. Aust J Agric Res **26** 751-68. https://doi.org/10.1071/AR9750751.

83 Caple IW, Andrewartha KA and Nugent GF (1980) Iodine deficiency in livestock in Victoria. Vic Vet Proc **38** 43-4.

84 Donald GE, Langlands JP, Bowles JE et al. (1994) Subclinical selenium insufficiency 6. Thermoregulatory ability of perinatal lambs born to ewes supplemented with selenium and iodine. Aust J Exp Agric **34** 19-24. https://doi.org/10.1071/EA9940019.

85 Rock MJ, Kincaid RL and Carstens C (2001) Effects of prenatal source and level of dietary selenium on passive immunity and thermometabolism of newborn lambs. Small Ruminant Res **40** 129-38.

86 King CF (1976) Ovine congenital goitre associated with minimal thyroid enlargement. Aust J Exp Agric Anim Husb **16** 651-5. https://doi.org/10.1071/EA9760651.

87 Sinclair EP and Andrews ED (1958) Prevention of goitre in new-born lambs from kale-fed ewes. NZ Vet J **6** 87-95. https://doi.org/10.1080/00480169.1958.33296.

88 Sargison ND, West DM and Clark RG (1998) The effects of iodine deficiency on ewe fertility and perinatal lamb mortality. NZ Vet J **46** 72-5. https://doi.org/10.1080/00480169.1998.36060.

89 Statham M and Koen TB (1981) Control of goitre in lambs by injection of ewes with iodized poppy seed oil. Aust J Exp Agric Anim Husb **22** 29-34. https://doi.org/10.1071/EA9820029.

90 Knowles SO and Grace ND (2007) A practical approach to managing the risks of iodine deficiency in flocks using thyroid-weight: Birthweight ratios of lambs. NZ Vet J **55** 314-8. https://doi.org/1.1080/00480169.2007.36787.

91 Green LE, Berriatua E and Morgan KL (1993) Anaemia in housed lambs. Res Vet Sci **54** 306-11. https://doi.org/10.1016/0034-5288(93)90127-2.

92 Lassiter JW and Morton JD (1968) Effects of a low manganese diet on certain ovine characteristics. J Anim Sci **27** 776-9. https://doi.org/10.2527/jas1968.273776x.

93 Underwood EJ and Somers M (1969) Studies of zinc nutrition in sheep. 1. The relation of zinc to growth, testicular development, and spermatogenesis in young rams. Aust J Agric Res **20** 889-97. https://doi.org/10.1071/AR9690889.

94 Masters DG, Chapman RE and Vaughan JD (1985) Effects of zinc deficiency on the wool growth, skin and wool follicles of pre-ruminant lambs. Aust J Biol Sci **38** 355-64. https://doi.org/10.1071/BI9850355.

95 Martin GB, White CL, Markey CM et al. (1994) Effects of dietary zinc deficiency on the reproductive system of young male sheep: Testicular growth and the secretion of inhibin and testosterone. J Reprod Fert **101** 87-96. https://doi.org/10.1530/jrf.0.1010087.

96 Egan AR (1975) Reproductive responses to supplemental zinc and manganese in grazing Dorset Horn ewes. Aust J Exp Agric Anim Husb **12** 131-5. HTTPS://DOI.ORG/10.1071/EA9720131.

97 Nielsen FH (2012) History of zinc in agriculture. Adv Nutr **3** 783-9. https://doi.org/10.3945/an.112.002881.

98 Grace ND (2006) Effect of ingestion of soil on the iodine, copper, cobalt (vitamin B_{12}) and selenium status of grazing sheep. NZ Vet J **54** 44-6. https://doi.org/10.1080/00480169.2006.36603.

99 Suttle NF, Abrahams P and Thornton I (1984) The role of a soil x dietary sulphur interaction in the impairment of copper absorption by ingested soil in sheep. J Agric Sci **103** 81-6. https://doi.org/10.1017/S0021859600043343.

100 Clark RG, Wright DF and Millar KR (1985) A proposed new approach and protocol to defining mineral deficiencies using reference curves. Cobalt deficiency in young sheep is used as a model. NZ Vet J **33** 1-5. https://doi.org/10.1080/00480169.1985.35132.

101 McDonald JW (1983) Mineral licks. In: Sheep production and preventive medicine. Proceedings No 67. University of Sydney Post-graduate Committee in Veterinary Science: Australia, p. 283.

102 Lambourne DA and Mason RW (1969) Mortality in lambs following overdosing with sodium selenite. Aust Vet J **45** 208.

103 Allen JG, Morcombe PW, Masters HG et al. (1986) Acute zinc toxicity in sheep. Aust Vet J **63** 93-5.

12

MANAGEMENT AND DISEASES OF WEANER SHEEP

WEANER ILL-THRIFT

It is generally true that young animals are at greater risk than adults of disease brought on by infectious agents or nutritional deficiencies, and sheep are no exception. Some of the conditions which can afflict lambs up to weaning age — usually around 3 to 4 months of age — are described in Chapters 7 and 16. In the months following weaning, young sheep remain at high risk of ill health unless close attention is paid to their nutrition and their protection from disease. The term *weaner ill-thrift* is often used to describe the unexpected poor health of weaner sheep when all other age groups of sheep on the farm appear to be well.[1] The term is not usually applied to individual cases but is used to describe the occurrences of a significant morbidity, perhaps 10% or more, of affected animals within a large flock of weaners.

Weaner ill-thrift is indicated by poor growth rates, weight loss, weakness and susceptibility to intercurrent disease. It primarily affects weaners of liveweights less than 20 kg and has a declining risk as liveweight approaches 30 kg. Those animals which recover often remain at relatively low body weight for months, with adverse effects on future productivity.[2] Mortality rates can be very high. The death rate of Merino weaners across Australia varies considerably between districts and farms but frequently exceeds 10% for the 12-month period following weaning.[3]

Weaner ill-thrift is seen in sheep between 3 months and 15 months of age and usually has undernutrition at its root. Most commonly, the condition occurs on summer-autumn pastures of low digestibility. While adult sheep have sufficient bodily reserves of energy, stored as fat, to remain in good health on similar pastures even while losing liveweight, weaner sheep quickly exhaust their energy stores.

The critical preventive strategy in most cases of weaner ill-thrift is the provision of adequate supplementary feed (grain) to the weaners or the provision of highly nutritious fodder crops.

Weaner ill-thrift can also be caused by deficiencies other than dietary energy and protein. Deficiencies of cobalt, selenium and copper may be the sole cause of the condition in some flocks, or they may be a contributing factor with energy and protein undernutrition. Weaners are more prone to mineral and vitamin deficiency syndromes because of their limited stores, particularly in the case of trace minerals which accumulate in the body over summer rather than during winter or spring.

Infectious agents — particularly those causing chronic malaise — may further complicate weaner ill-thrift, or they may precipitate the syndrome. Some infectious diseases are more likely to occur in young sheep than they are in adults — helminthiasis, yersiniosis and eperythrozoonosis are good examples — while some are equally likely to affect adult sheep as weaners. Footrot is an example of these.

Veterinary involvement in cases of weaner ill-thrift usually follow the occurrence of significant mortalities in a flock of weaner sheep, accompanied by the existence of a *tail* in the flock of animals which are small, weak and obviously not doing well. In many such cases the flock owner suspects that the condition is caused by agents or syndromes other than a lack of dietary energy.[4] The veterinary investigation should proceed methodically to rule in or out any of the other possible contributing factors while simultaneously examining the quality and quantity of the diet offered to the young sheep.

While weaner ill-thrift is discussed in this brief chapter as if it were a distinct condition, it is in fact the result of several conditions or management strategies which are described in other chapters of this book. In particular, the reader should be familiar with the issues which lead producers to select a particular production objective and time of lambing (Chapter 3), and with the nutritional value of pastures and their variation through seasons as described in Chapter 6. The factors affecting the health of ewes in late pregnancy and lactation — which dramatically affect lamb weaning weight — are discussed in Chapter 7. The objectives in this chapter are to describe why weaner sheep, particularly in Merino-type production systems, are predisposed to ill-thrift, and to outline management strategies to treat and prevent the problem.

Predisposition to weaner ill-thrift

Merino sheep are more often affected than crossbred or British breed sheep. The reasons for this are a reflection partly of the Merino sheep's relatively poor adaptation to high rainfall and partly of the management systems in which purebred Merino sheep are run. Crossbred lambs are usually prepared for sale as meat animals before 8 months of age and so it is in the interests of producers to ensure that the lambs continue to grow after weaning or that they are sold to other producers who will maintain their growth rates until they reach marketable liveweights — around 45 to 60 kg.

Merino weaners, on the other hand, are often retained to 18 months of age or older before sale, or retained as breeding ewes to be joined for the first time at 19 months of age. Consequently, there is often no imperative to achieve marketable liveweights in their first year of life. In addition, young Merino sheep have lower growth rates than crossbred lambs and weaners, so it is often unrealistic to expect Merino weaners to achieve weights of 45 kg early in the first year of life, even if they are destined for the abattoirs. Many producers therefore will aim to achieve low growth rates for Merino weaners through the summer, autumn and winter following weaning, with the expectation of rapid weight gain in the weaners' second spring, when they are 12 to 16 months of age — their age depending on their season of birth (late autumn, winter or spring).

And therein lies the risk. If events transpire to produce weaners which are of low body weight when weaned, and which then fail to achieve even moderate growth rates in the subsequent summer and autumn, the scene is set for at least some of the weaner flock to be unthrifty and for some of them to die unless effective and prompt remedial action is taken. In many circumstances the events which precipitate occurrences of serious weaner ill-thrift are related to the season and failures of pasture growth. In other cases there may be outbreaks of disease — in the ewes which produced the lambs or in the lambs and weaners themselves — which contribute to the problem either alone, or in conjunction with undernutrition.

It should be noted that the comments about breed predisposition are generalisations only and do not always hold true. The condition can also be seen in crossbred lambs or British-breed lambs[1] and, conversely, many Merino flocks are capable of achieving high growth rates from Merino lambs and have no difficulty in achieving lamb liveweights of 40 to 50 kg by 8 or 9 months of age. This is particularly true of the *dual-purpose* types of Merino sheep, including the Dohne, SAMM and stronger-woolled type of the South Australian bloodlines.

Factors contributing to the liveweight of weaners

In Australia, lambs are most commonly born in the months between and including April to September. April-May lambing is called *autumn lambing*, June-July *winter lambing* and August-September *spring lambing*. In general, autumn and winter lambing are practised in districts with short growing seasons and lower annual rainfalls. Spring lambing is more commonly practised in districts with longer growing seasons and more extended spring growth, such that green pasture remains available in most years until late December.

Whatever the region, when spring-grown pastures senesce and the nutritional value of the pasture declines, sheep will stop growing and, after a period of liveweight maintenance, start to lose liveweight as pasture nutrition continues to decline. For healthy adult sheep, this presents little problem. They have built up bodily reserves of energy — as fat — during the spring and can sustain good health even while losing liveweight at rates of, for example, 1 kg per week, for several months.

For weaner sheep, however, this change in pasture nutritional value is much more threatening to their health. Depending on their liveweight, weaner sheep have very low body fat reserves and cannot sustain extended periods of weight loss without risking their health and, ultimately, survival.

At 20 kg liveweight, weaner Merino sheep have only approximately 1 kg of fat, but fat reserves increase rapidly with increasing liveweight over 20 kg.[5] Dietary energy deficiencies in lambs under 20 kg will result in the mobilisation of body protein stores for energy production, with consequent deleterious effects on the strength and resilience of the animal.

The reason for the weaner sheep's difficulty in maintaining good health when feed quality declines is not related to any differences between young sheep and adult sheep in their ability to capture energy and protein from roughage diets. Lambs and weaners over 8-10 weeks of age are capable of consuming at least as much dietary organic matter from pastures as mature sheep when calculated per unit of metabolic liveweight ($W^{0.75}$), with similar changes in liveweight.[6] Lambs can be weaned successfully at 11 kg liveweight provided that

(1) they have been ingesting roughages for a sufficiently long period for rumen development to have occurred

(2) the diet that they are offered is relatively high in protein, such as occurs in legume-rich, dense, green pastures.

The question to be addressed, therefore, is: What liveweight will the weaners achieve by the time that pastures will no longer provide sufficient nutrients to at least maintain liveweight? The answer is determined by a number of factors, the first of which is the weaners' age when pastures senesce or, to be answered another way, the relationship between the time of lambing and pasture senescence.

Consider a lamb born early in a six-week lambing period running from 1 June to mid-July. If weaning occurs 13 weeks after the start of lambing (1 September) and pasture senescence (to a degree no longer providing for liveweight maintenance) occurs on 1 December, there is a nine-week, post-weaning period during which the lamb may grow at rates of 50 to 120 g/day (possibly faster in mid-spring, but slower in late spring and early summer), thus adding 4 to 9 kg to its weaning weight. If the lamb was weaned at 20 kg, it may achieve a healthy and robust 24 to 29 kg before its further growth on pasture alone ceases for the season.

Some lambs, however, are not born early in the lambing season and are only 7 weeks of age when weaned. An illustration of the type of spread in weaning weight which can be expected from a six-week lambing period is shown in Figure 12.1.

In the example shown, the particular shape of the distribution is a result of the long period of growth of single-born lambs born early in the lambing period, plus a second group of lambs of moderate to low weaning weight which were later-born single lambs and twin-born and twin-raised lambs, with a *tail* to the right composed of late-born, twin-raised lambs. Despite

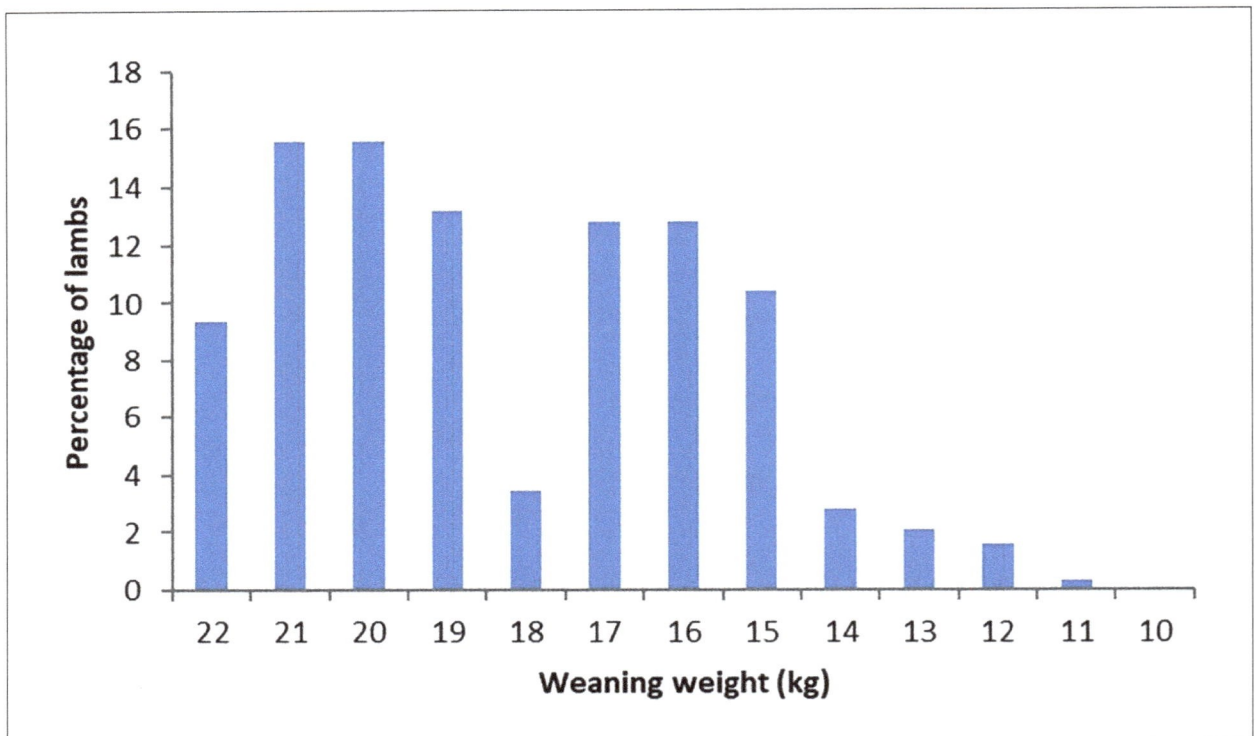

Figure 12.1: An example of a distribution of weaning weight of Merino lambs. Assumptions:
- single lambs born at 4.5 kg and growing at 200 g/day
- twin lambs born at 3.5 kg and growing at 160 g/day
- 20% of lambs are twins
- lambing over a 6-week period with 76% of lambs born in first 17 days;
- weaning when the oldest lamb is 13 weeks.

The average weaning weight is 19 kg. This stylised illustration makes no allowance for ill health in either ewes or lambs affecting their growth rate, and still there are 17% of the lambs under 16 kg at weaning. The light-weight lambs are twins or single lambs born in the second half of the lambing period. Source: KA Abbott.

an average weaning weight of 19 kg, there are 17% of the lambs weaned at 15 kg or less, and 7% which are 14 kg or less.

This example is a simplification. It makes no allowance for some factors which may increase the liveweight of younger lambs (later-born lambs usually have higher average growth rates to weaning); nor, importantly, does it allow for a number of factors which may adversely affect the mean weaning weight and increase the number of low weaning weight lambs. For example:

- Lambing may not have been so quick to begin. Particularly with autumn- and winter-lambings (spring and summer joining), there may be a one- to two-week delay before most ewes commence cycling. The peak of lambing activity will consequently move to the right and more lambs will be born in the later part of the lambing period.

- Some lambs will have suffered unfortunate luck during the pre-weaning period, such as their own ill health or their dam's ill health, or mismothering at marking time. This will affect the growth rate and weaning weight of these lambs.

- There may have been flock-wide effects on the growth of lambs, either generally or affecting a significant proportion of lambs. Such events include eperythrozoonosis in the lambs after marking, undernutrition of the ewes in late pregnancy (a poor autumn break, for example — see the discussion of pregnancy toxaemia in Chapter 7) — or an outbreak of haemonchosis in the ewes after lambing started.

The lambs of low liveweight at weaning make up a group of weaners which is often called a *tail*. These weaners, composed of late-born lambs, twin-raised lambs and lambs which have suffered misadventure, are likely to be at low liveweight when pastures senesce. In the example described above, with lambs gaining 4 to 9 kg between weaning and pasture senescence, a lamb which was 11 kg at weaning is still less than 20 kg when weight gains from pastures cease for the year.

The association between liveweight of weaners and mortality

Observational studies have shown that the peak in the weaner mortality rate occurs in the two to three months post-weaning[3,7], although this is strongly influenced by the level of feed offered to the weaners through the summer-autumn period. The weaners that are at greatest risk of death are those which have the lowest weaning weight. Lambs which comprise the lightest 25% of the weaner flock are typically two to three times more likely to die than lambs in the heavier 50% of the flock.[3,7] The high risk of mortality associated with the tail of the flock is reduced when the flock's average weaning weight is high and the distribution of weaning weights is tight. Whatever the relationship between a weaner's liveweight and that of its flock mates, those under 20 kg are at particularly high risk of mortality from undernutrition.

The risk of death for low liveweight weaners is reduced if the post-weaning growth rate is satisfactory. Weaners gaining 1 kg per month or more have a reduced risk of mortality compared to those growing more slowly.[3] The two factors — weaning weight and post-weaning growth rate — combine to influence the weaner's weight when feed levels decline.

Weaners in the liveweight range of 20 to 30 kg remain at significant risk of death if their post-weaning growth rates are too low or negative. It is clear that high survival rates of weaners are a function of both weaning weight and post-weaning growth rate.[2]

The association between time of lambing and weaner ill-thrift

Time of lambing has an important relationship with the risk of weaner ill-thrift. When lambing is timed for later in the season, the time period between weaning and pasture senescence is reduced and the opportunity for weaners to gain liveweight in the post-weaning period, before meeting nutritional challenges, is reduced. Consequently, it is expected that spring-born lambs are at higher risk of weaner ill-thrift than winter-born lambs. Spring lambing is often chosen by producers because it better matches the nutritional requirements of pregnant and lactating ewes to pasture quality and pasture availability, but it is also likely to demand high levels of management of the weaners through the summer-autumn period because of their reduced opportunity for growth in the immediate post-weaning period.

The association between the management of the ewe flock and weaner ill-thrift

The duration of the lambing period is the critical factor determining the number of lambs in the tail of the weaner flock. Joining for five weeks is recommended, particularly with autumn joining. For joining in summer, a longer joining period may be necessary, but the use of teasers and the ram effect will allow an effective joining period of five weeks.

Ewe nutrition before and during lactation and control of parasitism of ewes and lambs before weaning all have consequences for the size and health of the lambs at weaning. Provision of adequate pastures to the ewes during lactation has both an indirect effect (through higher ewe milk production) and a direct effect (through the pasture intake of the lambs) on lamb growth rates and weaning weights, with benefits for the survival rates of the lambs post-weaning.[8]

Introduction of grain feeding

As pasture quality declines in the post-weaning period, weaners should be supplemented with high-energy, high-protein feeds. In most cases this is provided by grain (oats, barley, lupins), although in some cases it can be provided by fodder crops or perennial pastures such as lucerne. If it is expected that weaners will be offered grain, it is advisable to train them to eat grain before weaning, so that they will learn to eat grain from the ground or from a feeding device with their mothers.

Paradoxically, the provision of grain to weaners can occasionally be the cause of an ill-thrift syndrome if there is a significant number of lambs developing grain poisoning. Such occurrences often occur if cereal grains like barley, triticale or wheat are offered and there is an insufficient period of adaptation or training of the weaners. It can also occur if very large flocks of weaners are trail-fed infrequently in large paddocks and not all weaners come to the trail after the feed is provided.

Age at weaning

The benefits of weaning 13 weeks after the start of lambing are, for the lambs, removal from parasite-contaminated pastures and, for the ewes, the cessation of lactation and increased production of wool. If lambing runs over a six-week period, most lambs will be 10 to 13 weeks of age when weaned and, provided they are placed on suitable high-quality pastures, they will be capable of growing at least as fast after weaning as before weaning. The few lambs aged 7 to

9 weeks may be slightly disadvantaged by weaning but the effect is small and short-lived. The benefit of weaning for the lambs, in addition to that of parasite control, is their removal from the competition with the ewe flock for the best pasture.

Post-weaning management of weaners

At the time of weaning, lambs are young, possibly recently mulesed, distressed about their maternal separation, still learning to graze efficiently and, in late winter or early spring, often wet and cold and grazing short pastures with very high water content. By virtue of these stressors, their immunological inexperience and other predisposing factors, weaners may have a high prevalence of helminthosis, enteric infections, fleece rot, dermatophilosis and flystrike or contagious pustular dermatitis unless effective preventive measures are taken. Some of the conditions which are likely to occur in weaners are listed in Table 12.1. The husbandry of weaners can also contribute to the risk of ill-thrift. Some of these husbandry strategies are listed in Table 12.2.

Preventive medicine programme for Merino weaners

In summary, the requirements of a preventive medicine, or *health maintenance*, programme for Merino weaners in the medium and high rainfall districts of Australia can be divided into five interrelated steps.

1. *Plan for an uninterrupted weight gain from birth to a reasonably robust size: 22 kg for fine-wool Merinos, 26 kg for medium-wools, 30 kg for strong-wools:* Under most Australian conditions in which Merinos are raised, there is at least one period in their first year of life when weaners are likely to lose weight for a period of weeks or months, unless special steps are taken to prevent weight loss. It is during this period of weight loss that the young sheep are at greatest risk. Their ability to withstand the effects of intercurrent diseases and nutritional deprivation is enhanced if they have achieved these liveweights. Planning involves consideration of the time of lambing, the nutritional management of the ewe flock and the expected seasonal variation in feed availability on the farm in question.

2. *Establish reliable practices to prevent or control diseases for which the cost-benefit clearly favours preventive action every year:* Helminthiasis and trace element deficiencies are two notable examples of this group.

3. *Introduce monitoring programmes and* action triggers *for preventive programmes for other diseases which occur from time to time:* Flystrike or grass seed infestations may fall into this category. Monitoring programmes may be as simple as a *reminder* system; *action triggers* for flystrike prevention may include 'rainfall after November' or the presence of fly activity detected in twice weekly inspections.

4. *Attend to the* tail *of the mob:* Around 20% of the flock of weaners will be significantly lighter or weaker than the rest and will probably experience the highest mortality from any disease processes which affect the mob. Identification of this subgroup for special attention may reduce the expense of preventive action and improve the economic efficiency of preventive programmes. If this group includes weaners under 20 kg they must be fed well enough to gain at least 1 kg of liveweight per month.

5. *Plan the timing and level of supplementary feeding in advance:* Have access to sufficient suitable supplements and introduce them before health or production problems emerge.

Table 12.1: Disease conditions which can cause or exacerbate the weaner ill-thrift syndrome.

Eperythrozoonosis (caused by *Mycoplasma ovis*, formerly *Eperythrozoon ovis*)	This condition causes anaemia in lambs and weaners, often around the time of weaning, and reduces the weight gain of affected animals. (See Chapter 19.)
Fleece rot, dermatophilosis and flystrike	Young sheep are more susceptible than adults to fleece rot and dermatophilosis, the two major predisposing factors for flystrike. Flystrike can lead to massive loss of weight in weaners. Dermatophilosis can be sufficiently severe to reduce weight gain or cause weight loss. (See Chapter 10.)
Helminthosis	Young sheep are particularly susceptible to gastrointestinal parasitism. (See Chapter 9.)
Yersiniosis	Yersiniosis can contribute to the *tail* of a weaner flock. Affected animals lose weight or fail to gain weight. (See Chapter 16.)
Coccidiosis	Outbreaks of coccidiosis can occur when weaners are closely confined, particularly under wet conditions. Affected animals recover but often lose body weight through the course of the disease. (See Chapter 16.)
Vitamin and trace element deficiencies	Cobalt and selenium deficiencies are the most likely trace element deficiencies, while vitamin E deficiency can occur in extended dry conditions. (See Chapter 11.)
Grain poisoning	This may occur in weaners fed high-starch, low-fibre grains like barley, triticale or wheat. (See Chapter 16.)
Contagious pustular dermatitis (scabby mouth, or CPD)	The virus of CPD can cause teat lesions in ewes (which prevent feeding of lambs) and several conditions in young sheep in both wet conditions (infections of the limbs) or dry conditions (infections of the lips). (See Chapter 10.)
Lupinosis	This can occur in weaners on lupin stubbles or, rarely, weaners being fed lupin grain. (See Chapter 17.)
Grass seed lodgement in the fleece and skin.	The fleeces of unshorn weaners grazing annual pastures in early summer can become heavily contaminated with grass seeds, potentially leading to irritation and penetration of the skin. (See Chapter 4.)
Footrot	Footrot can affect sheep of all ages. In lambs and weaners it can have a profound effect on liveweight. (See Chapter 13.)
Pink eye	In summer, kerato-conjunctivitis may be a minor contributor to failure of some weaners to thrive. (See Chapter 10.)
Arthritis	Lambs and weaners affected by chronic arthritis are usually part of the *tail* of the flock. (See Chapter 13.)
Pneumonia	Viral and bacterial pneumonias are discussed in Chapter 20. *Dictyocaulus filaria* (Chapter 9) may lead to pneumonia, but this is rare.

Table 12.2: Husbandry strategies which can cause or exacerbate the weaner ill-thrift syndrome.

Access to water	Immediately after weaning lambs may be confined to unfamiliar paddocks and unable to find water. Producers sometimes add adult sheep to the weaner flock to provide some wisdom. In long grass pastures, it is sometimes advisable to mow a pathway to the water trough.
Provision of grain	The ability to find and eat grain from a trail on the ground or lick feeder may not be shared across all weaners in the flock. Again, inclusion of a few adults may help.
Flock size	In very large flocks of weaners (over 500 sheep, for example) in large paddocks, it can be difficult to ensure that the whole group are accessing water and grain and are available for inspection.
Variation in size within the flock	If the weaner flock is dependent on supplementary grain, it may be necessary to separate the small weaners from the larger weaners to ensure equal access to the supplements.
Shearing	The time off-feed associated with shearing leads to a prolonged recovery period. Weaners can lose significant liveweight in the weeks after shearing, crutching or other events which take them off-feed.

RECOMMENDED READING

Wilkinson FC (1981) Lamb survival from marking and weaner illthrift. In: Refresher course on sheep. Proceedings No 58. University of Sydney Post-graduate Committee in Veterinary Science: Australia, pp. 193-209.

REFERENCES

1   Mulhearn CJ (1958) Unthriftiness of weaner sheep in South Australia. Aust Vet J **34** 383-90.

2   Hatcher S, Eppleston J, Thornberry KJ et al. (2010) High Merino weaner survival rates are a function of weaning weight and positive post-weaning growth rates. Anim Prod Sci **50** 465-72. https://doi.org/10.1071/AN09187.

3   Campbell AJD, Vizard AL and Larsen JWA (2009) Risk factors for post-weaning mortality of Merino sheep in south-eastern Australia. Aust Vet J **87** 305-12. https://doi.org/10.1111/j.1751-0813.2009.00457.x.

4   Allworth MB (1990) Weaner illthrift — treatment and prevention. In: Sheep medicine. Proceedings No 141. University of Sydney Post-graduate Committee in Veterinary Science: Australia, 435-9.

5   Allden WG (1970) The body composition and herbage utilization of grazing Merino and crossbred lambs during periods of growth and summer undernutrition. Aust J Agric Res **21** 261. https://doi.org/10.1071/AR9700261.

6   Egan JK and Doyle PT (1982) The effect of stage of maturity in sheep upon intake and digestion of roughage diets. Aust J Agric Res **33** 1099. https://doi.org/10.1071/AR9821099.

7   Hatcher S, Eppleston J, Graham RP et al. (2008) Higher weaning weight improves postweaning growth and survival in young Merino sheep. Aust J Exp Agric **48** 966-73. https://doi.org/10.1071/EA07407.

8   Thompson AN, Ferguson MB, Campbell AJD et al. (2011) Improving the nutrition of Merino ewes during pregnancy and lactation increases weaning weight and survival of progeny but does not affect their mature size. Anim Prod Sci **51** 784-93. https://doi.org/10.1071/AN09139.

13

LAMENESS

OSTEODYSTROPHIES

Osteodystrophies include all diseases in which there are failures of normal bone development or normal bone maintenance. Clinically, there may be distortion of the skeleton, a predisposition to fracture or disturbance of gait or posture. The pathology can arise because of changes in bone quantity, bone quality or both. In general, the term *osteoporosis* describes conditions where bones have lost volume and mass but remain normally mineralised, while *osteomalacia* describes conditions in which bones are soft and lacking in mineralisation. The more general term *osteopenia* is also used to describe a loss of bone mass without indicating whether mineralisation is normal or not. To appreciate the pathways that lead to these different pathological presentations, it is necessary to briefly revise the physiological processes leading to bone formation and maintenance.

In the growth plates of long bones and vertebrae, new growth occurs through the deposition of a cartilage matrix in which osteoblasts deposit osteoid, the collagenous material which, when mineralised, will constitute new bone. On the surfaces of bones, osteoid is deposited by cells in the periosteum and then mineralised with calcium and phosphorus from the extracellular fluid. Bones are undergoing constant remodelling in the growing animal and, to a lesser extent, in mature animals. Remodelling involves the resorption of mineralised old bone by osteoclasts and the deposition of new osteoid by osteoblasts.

Chemical analysis of bone can be used as a diagnostic and research tool. Bone with a low mineral content has a low ash:volume (A:V) ratio. This can be due to a deficiency of osteoid and its associated minerals or to poor mineralisation of the osteoid. Bone with insufficient osteoid has a low organic matter:volume (R:V) ratio and a low A:V ratio. Bone in which the osteoid is poorly mineralised has a low ash:organic matter (A:R) ratio and a low A:V ratio.

Osteoporosis

Bones become osteoporotic if resorption exceeds the rate at which new bone is laid down or if there are insufficient nutrients available to create bone during growth in young animals. The R:V ratio is low.[1] Osteoporotic bones are light and brittle and, in advanced cases, the thinness of the cortices is readily apparent radiographically. By the time that radiographic changes are evident, bones have lost 30% or more of their mass. Osteoporosis, when it occurs, is most evident in the metaphyses of long bones and in the vertebral bodies, scapula and flat bones of the skull.

Some degree of osteoporosis in farm animals, particularly young, growing animals, is common and does not necessarily lead to problems.

The causes of osteoporosis are usually nutritional and, in sheep in Australia, arise most commonly from deficiencies of calcium complicated by disturbances in vitamin D or phosphorus intake, or a general undernutrition in which inadequacy in dietary protein and energy reduces the availability of protein for the formation of bone matrix. Dietary deficiency of phosphorus alone can cause osteoporosis but is uncommon in Australia — the northern areas of the country where phosphorus deficiency occurs are not usually grazed by sheep.

Osteoporosis can occur in young sheep as a consequence of intestinal parasitism. Chronic infections with *Trichostrongylus colubriformis*[2] or *T vitrinus*[3] markedly reduce absorption of phosphorus and, to a lesser extent, calcium. The reduced availability of phosphorus, complicated by reduced calcium absorption and reduced protein anabolic activity as a consequence of the parasitism, is responsible for the reduction in bone growth and mineralisation of bone. Some evidence of osteomalacia, indicated by a low A:R ratio, may accompany the osteoporosis associated with intestinal trichostrongylosis.

Subclinical infections with *Teladorsagia circumcincta* cause osteoporosis through reduced availability of energy and protein in addition to reduced deposition of minerals. At least part of the effect is due to reduced feed intake of infected animals.[4] Mixed infections of lambs with *T circumcincta* and intestinal parasites can cause reductions in bone mineral density of 30%, and 9% reduction in bone size.[5]

Deficiency of copper is a common cause of osteoporosis and susceptibility to bone fractures of lambs and weaners in some regions of Australia and other countries. Copper is an essential component of an enzyme required in collagen formation, and copper-deficient lambs fail to produce sufficient normal osteoid. Consequently, the cortices of the long bones become very thin and fragile. Copper deficiency is discussed in Chapter 11.

Osteomalacia and rickets

In cases of osteomalacia the bones are soft because there has been a failure of mineralisation of osteoid. The A:R ratio is low. In young animals osteomalacia presents clinically as the condition known as rickets. The principal lesions result from a failure of mineralisation in the cartilage and osteoid laid down as a part of the growth of new bone.[6] In cases of rickets, the gross lesions are evident first in the growth plates of the proximal humerus, distal radius, ribs and femur, both ends of the tibia and distal metacarpal and metatarsal bones. Bones become shorter and broader and, because of the softness of the cortex, become curved as a result of weight bearing and are prone to fracture. Joints become enlarged as a result of widening of the metaphysis and epiphysis. The most common causes of osteomalacia in sheep are vitamin D deficiency and phosphorus deficiency.

Vitamin D deficiency

Cholecalciferol (vitamin D_3) is produced in the skin of ruminants following exposure to sunlight, as it is in many other mammalian species. Vitamin D_2 (ergocalciferol) occurs in plants. In the liver and kidney these vitamins undergo further chemical change before becoming biologically active products. The general term *vitamin D* can be used to include both vitamin D_2 and vitamin D_3, although vitamin D3 is more important and more effective metabolically than vitamin D_2.

Vitamin D deficiency occurs almost always because of the reduced exposure of grazing animals to ultraviolet (UV) light. Pastures provide very little vitamin D and nearly all of the vitamin D nutrition of ruminants is derived from synthesis of cholecalciferol in the skin by the action of the UV component of sunlight on 7-dehydrocholesterol.

Sunlight contains a relatively small component of UV light of the lower wavelength which is effective in vitamin D synthesis. The effective UV component is reduced by passage through the atmosphere and, the greater the angle of the sun, the greater the distance through the atmosphere which the rays must penetrate before hitting the earth. Consequently, the potency of sunlight for vitamin D synthesis is highest when the sun is most directly overhead (in summer rather than winter; at high altitudes rather than at sea level; at low latitudes rather than high; around midday ±2 hours rather than early or late in the day) and when there are no barriers to light transmission (on clear days rather than overcast; through clean air rather than polluted). Sheep with full fleeces and pigmented heads and legs synthesise less vitamin D than shorn sheep or white-faced sheep.[7]

Grains, oilseeds and short, actively growing pastures contain very little vitamin D. As plants mature and the leaves become paler in colour, UV light is able to penetrate the leaves and cause conversion of ergosterol to ergocalciferol. Consequently, leafy hay that has been dried with plentiful exposure to sun can be high in vitamin D_2.[8] Lucerne hay that has been well sun-cured and contains mature and dead leaves, is a good source of the vitamin. For supplementation at times when animals may lack sun exposure, vitamin D_2 is effective, but vitamin D_3 is more efficiently used by ruminants than vitamin D_2 and more active hormone will be produced from a dose of vitamin D_3 than from the equivalent amount of vitamin D_2.[7]

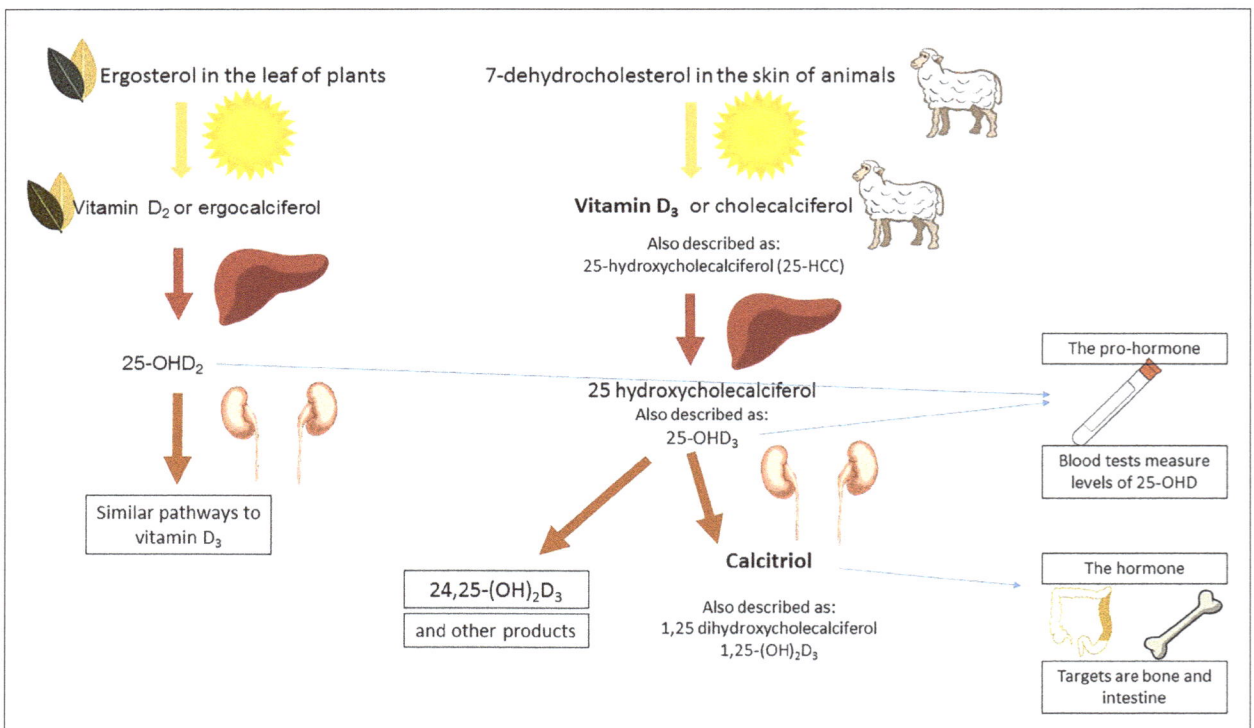

Figure 13.1: The principal metabolic pathways of vitamin D. Source: KA Abbott.

Vitamin D_2 and vitamin D_3 are converted in the liver to 25-OH vitamin D (25-OHD$_2$ and 25-OHD$_3$) which are the principal forms of the vitamin circulating in the blood, the levels of which can be measured to determine an animal's vitamin D nutritional status (Figure 13.1). In the kidney 25-OHD$_3$ is converted to 1,25-(OH)$_2$D$_3$ (calcitriol) and other related products, some of which are also likely to be important in calcium and phosphorus metabolism. These are the biologically active forms of the vitamin, and calcitriol is considered a hormone. Low levels of calcium, phosphorus, parathyroid hormone or calcitriol itself will all stimulate increased renal production of calcitriol from its precursor. Parathyroid hormone has a critical regulatory function in the production of calcitriol.

In addition to the kidney, calcitriol is produced in the placenta of the pregnant sheep.

Vitamin D is stored in the body in the hydroxylated and the unhydroxylated form — in the liver, adipose tissues and, particularly, the blood. Unlike vitamin A, liver stores of the vitamin are relatively small in mammals. (In fish, liver stores of the vitamin are high. Cod liver oil is a noted source of the vitamin for human and animal nutrition.) Vitamin D turnover in the body is relatively slow and adult sheep are able to store sufficient vitamin D to meet their needs for a six-week period without further dietary intake or biosynthesis.[9]

Placental transfer of cholecalciferol, 25-OHD$_3$ and calcitriol occurs in the sheep. Lamb foetuses maintain higher blood calcium levels and higher calcitriol levels than the dam and are able to maintain their blood calcium levels independently of the ewe, in the short term at least, by production of calcitriol in the foetal kidney.[10] Newborn lambs from adequately nourished ewes have sufficient stores of the vitamin to meet their vitamin D requirements for six weeks, after which time their vitamin D needs must be met by photobiosynthesis. Milk also contains some calcitriol and its precursors in the fat fraction, provided the ewe has satisfactory blood levels, but milk contains insufficient vitamin D to meet all of the lamb's requirements.

Calcitriol and calcium (Ca) and phosphorous (P) homeostasis

Calcitriol works in conjunction with calcitonin and parathyroid hormone to maintain blood levels of calcium and phosphorus in a narrow range. Calcitonin suppresses blood calcium levels by reducing intestinal absorption, bone demineralisation and calcium reabsorption in the kidney. The general effect of calcitriol is to raise blood levels of calcium. It achieves this in two ways: by increasing the efficiency of absorption of calcium from the intestinal tract and by stimulating osteoclastic activity to increase mobilisation of minerals from bone.

Absorption of calcium in the gastrointestinal tract occurs through both active and passive mechanisms. Passive absorption occurs throughout the intestine and depends primarily on the level of calcium in the diet. Active (transcellular) transport occurs through the wall of the rumen and through the intestinal wall. The mechanism of active absorption in the rumen is not regulated by blood calcium or calcitriol levels but, in the small intestine, absorption is calcitriol-responsive.[11]

The efficiency with which calcium is absorbed from the diet can increase markedly when an animal's requirements increase. The amount absorbed ultimately reaches a maximum value, above which further increases in absorption do not occur. In one study, at intakes ranging from

<100 to 400 mg of calcium per day per kg of liveweight[a], young sheep absorbed 22% to 30% of their dietary calcium. Levels above 400 did not result in further increases in absorption. Over the same range of dietary intake mature sheep had less efficient absorption and reached a maximum level of absorption when the diet included 200 mg of calcium per day per kg liveweight.[12] The higher efficiency of absorption in young sheep matches the high demand for calcium for bone growth. A similar increase in absorptive efficiency occurs in pregnant and lactating ewes and aids the recovery of skeletal calcium stores in the latter parts of lactation. The difference in efficiency of absorption between sheep with different levels of calcium requirements is controlled by calcitriol[13] and its action on receptors in the small intestinal wall. Provided that dietary levels of calcium are adequate, calcitriol has a major role in controlling blood calcium levels by regulating the absorption from the intestinal tract.

Under conditions of low dietary calcium intake or during periods of high demand, such as late pregnancy and lactation, calcitriol levels rise (provided vitamin D nutrition is adequate). This may increase the absorption efficiency of dietary calcium but, if this response is insufficient to return blood calcium levels to normal, it will also increase the rate of mobilisation of calcium from bone. There are differences between animal species in the extent to which each mechanism is employed — increases in efficiency of absorption of calcium from the intestine are much less marked in sheep than in goats, but sheep mobilise calcium from bones to a much greater extent than goats.[14,15] In sheep on low-calcium diets, the primary mechanism by which calcium homeostasis is maintained is by an increased rate of mobilisation of minerals from bones, rather than an increased efficiency of absorption of dietary calcium.[16]

When the supply of calcium to the osteoid tissue is adequate, calcitriol no longer stimulates osteoclastic activity. Bone resorption slows and the balance between bone resorption and bone formation becomes a net anabolic effect. The reduction of osteoclastic stimulation by calcitriol is probably mediated by changes in calcitonin and parathyroid hormone levels.[13]

Calcitriol has a role in bone growth

In sheep with adequate dietary calcium, vitamin D contributes to the mechanisms maintaining skeletal homeostasis. In addition to its role in calcium homeostasis by intestinal absorption and bone resorption, vitamin D is essential for the normal mineralisation of osteoid. The mechanism involved is unclear but appears to be more complex than simply ensuring adequate levels of calcium in the blood and hence in the extra-cellular fluid. The role of vitamin D in new bone growth is least important when the dietary Ca:P ratio is within the optimum range of 1.2:1 to 1.5:1. Outside that range, vitamin D has an increasingly important role in normal bone growth and deficiencies of vitamin D are increasingly likely to lead to bone pathology.

Clinical presentations of deficiency of vitamin D

Vitamin D deficiency leads to rickets in young sheep and osteomalacia in mature sheep. Sheep with rickets are lame, ill-thrifty and prone to recumbency. Growth rates are reduced or growth may cease altogether. Early signs may be detectable only by careful observation of the animals while they graze, where a tendency to stumble and a slight stiffness of movement may be

---

a   For a 30 kg young sheep, this range, therefore, is from <3-12 g of dietary calcium per day.

evident. With more advanced cases the epiphyses of the limb bones are enlarged and the long bones may be bowed — outward (bow-legged) or inward (knock-kneed). The joints most commonly affected by the enlargement of the epiphyses and metaphyses are those of the distal radius, distal tibia and distal metacarpal and metatarsal bones.[17,18] The ribs may show marked swelling at the costo-chondral junctions. The enlargements, which are visible grossly, are a result of the laying down, in excessive amounts, of non-mineralised osteoid and cartilage in the regions of the growth plates of the bones.

Blood biochemistry reveals a marked hypophosphataemia. Hypocalcaemia may be present or may occur only late in the disease. Serum alkaline phosphatase levels are raised.[19]

There are two major contributory factors to the occurrence of hypovitaminosis D. One is a lack of UV exposure; the other is the grazing of particular feeds. In a study comparing several forage crops and feeds used for winter grazing in New Zealand, rickets occurred severely and most frequently on green oats, least frequently on choumoellier, turnips and pasture. Italian ryegrass was intermediate in its effect on the development of rickets (its *rachitogenic effect*).[17] Rickets occurred despite adequate levels of calcium in the cereal crop and in the absence of high phytate levels or phosphorus. The association between rickets and the grazing of cereal oats is considered to be due to the rachitogenic effect of carotene in the oat plant.[20] Grazing of green barley provides a similar risk to that of green oats, but barley is used much less frequently than oats as a forage crop.

The effects of latitude and season and the subsequent altitude[b] of the sun on vitamin D production are pronounced. In Australia, rickets occurs in July and August and is more commonly reported from southern Victoria (38° South) and Tasmania than NSW. In Britain, it occurs in sheep more commonly in the late-winter, early-spring period, and is reported more commonly in northern England and Scotland (Edinburgh is 56° North) than in southern Britain. Regions of the world where the noon-day altitude of the sun is below 35° for some months of the year receive insufficient effective UV light to prevent the risk of severe rickets in humans without a dietary source of vitamin D.[21] In Australia, at places north of 31° S (Gilgandra, NSW, for example), the altitude of the midday sun is never less than 35°. At a latitude of 36° S (Shepparton, Victoria or Bordertown, SA, for example) the altitude of the sun is below 35° for about 66 days from mid-May to the end of July. Further south, the period of low sun altitude is even longer. No mainland part of WA is at latitudes of 36° S or greater.

The association between the incidence of rickets and the grazing of rachitogenic crops was made first in relation to sheep in the south island of New Zealand.[18] Since that time there have been numerous reports from southern Victoria (Australia)[22] involving spring-born hoggets or autumn-born lambs grazing oats in winter, and in 1-year-old sheep in northern England grazing lush ryegrass.[9] Cases associated with the grazing of green oats over winter have also been reported from the western slopes of central New South Wales[23,24] (33° South). The association with green oats, young lambs and winter grazing suggests that a factor in the feed (probably carotene) creates a requirement for higher levels of vitamin D than can be met from sunlight even at those relatively low latitudes. Rickets has been reported in Canada in very young lambs (3 to 4 weeks old) born to ewes which had inadequate vitamin D stores during pregnancy.[25] Cases of rickets and osteomalacia, both alone and in conjunction with osteoporosis, have been

---

b   The altitude of the sun is the angle between the horizon and the centre of the sun; it is also called the solar elevation angle.

described in a series of studies on sheep, both young sheep and adult ewes, in Scotland.[19,26] Contributing factors to the osteodystrophies included lack of sunshine, particularly in sheep with long fleeces, low phosphorus ingestion in sheep on extensive, unimproved pasture (hill-grazing), lactation in ewes and general undernutrition.

Rickets may be subclinical or mild as a clinical condition, but the effects of vitamin D deficiency on the liveweight and wool production of young sheep may be significant even if the signs of rickets are unremarkable. Franklin (1953)[27] readily demonstrated increases in liveweight gain and wool production in Corriedale, Merino and crossbred lambs grazing green oats over winter when they were supplemented with vitamin D. Responses occurred in experiments performed in Tasmania and near Canberra (latitude 35° South). Noting the low cost of vitamin D supplementation, he made a set of guidelines to help decide when the prophylactic use of vitamin D supplementation would be rational. The guidelines included the following:

- If weaner sheep are to be grazed on oats for several weeks over winter, vitamin D supplementation should be considered.

- A greater response to supplementation can be expected if grazing conditions are good and will allow rapid growth of the sheep.

- A greater response is expected if the sheep are in poor condition, rather than strong condition, at the start of winter.

- If winter begins with, or is likely to continue to have, wet, overcast or foggy weather conditions, greater responses can be expected.

- More valuable sheep, such as young seedstock, could be treated preferentially.

- Autumn-born lambs may benefit even in latitudes as far north as southern NSW.

If sheep graze pastures without rachitogenic factors (that is, if they are not grazing oats), liveweight responses to supplementation may not occur, even if monitoring of blood levels of vitamin D and calcium indicate an apparent deficiency.[28]

Diagnosis

Osteodystrophies of nutritional origin should be suspected where groups of young sheep show signs of gait abnormalities, recumbency, lameness or a high incidence of limb fractures. A useful diagnostic test of osteoporosis in young sheep is to press with the thumbs on the frontal bones. In affected animals the bones will fracture with little pressure but, provided the procedure is done carefully, no permanent harm is done to the animal.

Investigations into the cause of osteomalacia and osteoporosis should seek to determine the nutritional histories of the animals, with regard, in particular, to copper, calcium, vitamin D, energy and protein. Both blood specimens (from at least 10 sheep) and liver specimens (from at least three sheep) can be tested for copper levels. Blood can be tested for levels of vitamin D, calcium, phosphorus and alkaline phosphatase, but results within the normal range for any parameter do not exclude a diagnosis of an osteodystrophy. In some cases, tissue levels will be rising before the clinical condition is investigated, so a detailed exploration of the age, supplementation and grazing history of the affected animals will be an important aid to a diagnosis. The history of pasture treatments with copper or molybdenum should be explored (see molybdenum in relation to copper deficiency, Chapter 11). Histopathology on affected bones and analysis of bone ash of a foot bone[29] will confirm the osteopenia and aid the differentiation between conditions causing osteoporosis and those of osteomalacia.

Treatment and prevention

Young sheep should receive diets containing 0.4 to 1.2% calcium and sufficient phosphorus to achieve a Ca:P ratio of about 1.25:1. Sheep can tolerate a higher ratio (*Ca excess*) better than a low ratio (*P excess*). A 30 kg young sheep, growing at 200 g/day, requires about 3 g of calcium (absorbed) per day. Grass and grass-clover pastures and hays contain about 7.5 g Ca per kg DM, 3 g/kg P; and so a young sheep ingesting 1.5 kg to 2.0 kg of pasture dry matter per day should meet its calcium requirements for growth if it can absorb 25% to 30% of its dietary calcium intake. Dietary absorption rates in that range are readily achievable on pasture-based diets.

Cereal grains are low in calcium and high in phosphorus. Oats, for example, typically contain less than 1.2 g Ca per kg DM and 3-4 g P. Growing lambs offered a diet of pasture and oats in order to achieve a weight gain of 200 g/day will receive insufficient calcium unless the diet is supplemented. Limestone (calcium carbonate, 28% calcium by weight) should be added to the cereal grain at the rate of 2% (20 kg per tonne of grain) to provide sufficient calcium. The limestone should be finely ground (feed grade, 70 μm particle size) to ensure reasonable levels of absorption. While low rates of supplementation with cereal grain may not require calcium supplementation, higher rates of supplementation intended to promote growth require additional calcium in order to achieve the expected liveweight responses[30] and to prevent osteodystrophies.

Supplementation with vitamin D is only considered necessary in sheep which are at risk of prolonged low sun exposure and likely to be grazing high-risk feeds (green oats in particular). Treatment of lambs with 200 000-400 000 IU[c] of vitamin D (1500 to 2000 IU/kg) by injection is recommended. Sheep can also be treated orally (5 IU/kg/day), but frequent retreatment is necessary in high-risk periods. Oral treatment with a massive prophylactic dose of vitamin D_2 — 1×10^6 IU — is also effective in preventing rickets in young sheep over winter.[17]

If pregnant ewes are likely to be deficient in vitamin D, parenteral administration of 300 000 IU two months before lambing will ensure that the ewe and the newborn lamb have adequate levels of 25-OHD$_3$.[31]

Cases of rickets in lambs should be treated with vitamin D orally or parenterally, although several reports suggest that dietary calcium supplementation is also effective.

Other causes of osteodystrophies

An inherited form of osteomalacia has been reported in Corriedale sheep in New Zealand. Affected lambs are normal at birth but gradually develop signs of rickets as they grow. Biochemistry reveals hypocalcaemia and hypophosphataemia but elevated vitamin D levels, suggesting that the vitamin D receptor sites in intestine and bone are refractory to calcitriol.[32]

Bentleg (or Bowie), with extreme outward bowing of the forelimbs, occurs as a result of ingestion of *Trachymene* spp (wild parsnip). The condition is reported to occur sporadically in

c   1 mg of vitamin D contains 40 000 IU.

sheep in southwestern Queensland and northwestern NSW following exposure to the plant, which grows after late summer or early autumn rains in areas that have been bared of other plants. Ewes which graze wild parsnip may produce lambs which develop the limb deformities any time from a few days to several weeks after birth.[33]

VIRAL DISEASES ASSOCIATED WITH LAMENESS

Contagious pustular dermatitis (CPD, contagious ecthyma, scabby mouth)

CPD usually causes lesions on the lips, adjoining skin, muzzle and oral mucosa. Occasionally, similar lesions occur on the legs, around the coronary band and palmar surface of pastern and the interdigital skin (IDS), particularly between the bulbs of the heel. Lesions can also extend to the tarsal or carpal areas and be accompanied by a painful cellulitis and secondary infection. CPD is discussed more fully in Chapter 10.

Foot-and-mouth disease (FMD)

FMD does not occur in Australia. The disease is characterised by vesicles in the mouth and on the feet and teats, although oral lesions are not prominent in sheep. Feet lesions commonly occur on the coronet, the interdigital skin and the bulbs of the heel. FMD foot lesions can resemble footrot, particularly if there is secondary bacterial infection. Lameness is severe and the morbidity is high.

Bluetongue

Bluetongue infection has not been reported in sheep in Australia. During the initial stages of infection with the virus of bluetongue, there is hyperaemia of the mucous membranes of the mouth and of the skin of the feet around the coronet. Coronitis is severe and prominent and linear haemorrhages may be visible in the hooves. Lameness is severe.

BACTERIAL ARTHRITIS

The presentations of infectious arthritis in sheep can be divided into two categories — those infections characterised by the accumulation of pus in the affected joint (suppurative arthritis) and those in which the inflammatory process in the joint is marked by accumulation of fibrin rather than pus. The two forms of arthritis have different clinical and necropsy presentations and are associated with different species of bacteria, so it is helpful to discuss them separately, even though there are some similarities in the epidemiology and predisposing factors between the two forms. Whichever form it takes, infectious arthritis in sheep generally arises from haematogenous spread, is polyarticular and occurs more commonly in young animals than in adolescent or adult sheep. An exception to this is the condition known as foot abscess, which is a suppurative arthritis of the distal interphalangeal joint, discussed below in diseases of the foot.

Fibrinous arthritis

Fibrinous arthritis is most commonly associated with either *Erysipelothrix rhusiopathiae* or *Chlamydia pecorum* and is a cause of significant loss to the sheep industry. Infection with these organisms principally affects lambs aged 1 to 4 months and cases often start to appear one to two weeks after marking.[34] Many infected lambs recover spontaneously from infection, possibly after an illness which lasts only a few days and is rarely observed. A proportion of lambs, perhaps as many as 20% of those infected, develop chronic arthritis in at least one of the infected joints and are culled or killed because of their poor growth, poor condition, deformity and chronic lameness. Milder cases are detected at the abattoirs, where their carcases may be subject to partial or total condemnation.

The prevalence of arthritis in lambs at abattoir can be estimated with reasonable precision, but the number of cases which are euthanased on the farm of origin can only be inferred from questionnaire data. In WA, it was estimated in 1997 that, of 1 million lambs produced in a year, 1% had evidence of arthritis at the abattoir and a further 1.4% were culled on the farm.[35] The cost to the WA industry was estimated to be over $1m in 1997. Farquharson (2007) examined data from export abattoirs in all states of Australia and estimated the rate of total carcase condemnations to be around 0.02%, with an estimated partial condemnation rate of around 0.08%. He also identified a similar percentage of total condemnations due to arthritis in adult sheep. From a survey of producers, he estimated that 0.6% of lambs were killed or died on-farm as a result of arthritis.[36]

Clinical signs

The first sign of arthritis in an infected lamb is lameness in one or more limbs. The joints most often affected are the higher limb joints — hip, stifle and hock, shoulder, elbow and carpus. Initially, swelling is not marked but the affected joint is warm and painful when manipulated. There is generally no indication of systemic illness. In cases which do not resolve spontaneously or in response to treatment, the lameness becomes chronic. There is muscle wasting in the affected limb or limbs, and the animal is less active than its flock mates and grazes less efficiently, spending more time recumbent. These limitations affect its growth rate and wool production. Chronically lame animals are most obvious when the flock is gathered or moved — the lame animals dropping to the back of the group, forming a 'tail' in the mob (Figure 13.2).

Epidemiology

Several studies have demonstrated the ability of *E rhusiopathiae* to lodge in joints and cause arthritis following introduction through the umbilicus of newborn lambs, castration and tail-docking wounds, shearing cuts and other routes (reviewed by Farquharson (2007)[36]). Spread to joints occurs haematogenously. Mulesing wounds and shearing wounds have been identified as highly significant risk factors for the development of erysipelas arthritis. Flocks in which mulesing is practised are seven times more likely to have a high prevalence of arthritis (>4%) in lambs than unmulesed flocks. Shearing of lambs increases the odds of a high prevalence of arthritis by 4.3 times.[35] Shearing has also been associated with cases of erysipelas arthritis in young adult sheep.[37]

Chlamydia pecorum also causes a nonsuppurative arthritis in lambs and weaners. The route of infection is not known, although it is possibly oral in at least some cases. *C pecorum* bacteria are shed in the faeces of 30% of young sheep in southern Australia.[38] Affected lambs are lame, depressed and anorexic, and they often also show conjunctivitis.[39]

Figure 13.2: A crossbred lamb with arthritis. The lamb is lame and poorly grown compared to his peers of the same age. Source: KA Abbott.

Pathology

At necropsy of cases of erysipelas arthritis, the infected joints are swollen, with gross thickening of the periarticular tissues, including the joint capsule. The synovial fluid is increased in volume, turbid but not suppurative (that is, pus is not a feature). The synovial membrane appears roughened, and pits in the cartilage give the appearance of ulceration. As lesions progress, granulation tissue spreads from the perichondrium across the articular surface, producing pannus. The joints may become free of infection, but inflammatory changes persist, leading to organisation of the granulation tissue on the articular surfaces, fibrosis and possibly adhesions between articular surfaces.[39] In chronic cases, the joints have greatly reduced mobility and are enlarged by the periarticular fibrosis and development of osteophytes around the margins of the joint surfaces. The enlargement of the joints is made more obvious by the wasting of the musculature.

Treatment

Treatment is only likely to be successful if commenced early — before chronic joint changes occur. *C pecorum* is susceptible to tetracyclines, *E rhusiopathiae* to penicillin and tetracyclines.

Prevention

Given the evidence associating mulesing with erysipelas arthritis, it could be assumed that the incidence of arthritis in lambs could be reduced by ceasing mulesing. Mulesing provides lifetime benefits to Merino and Merino-cross sheep in protection against breech strike, but its use in lambs destined for slaughter before 12 months of age is questionable. Protection against breech strike can be provided effectively by other means for lambs that are sold in their first year of life.

In sheep where mulesing is considered necessary, there may be benefit in administering topical antiseptic and analgesia to the mulesing wound immediately after the procedure. Currently Tri-solfen® (Bayer Australia Ltd) is widely recommended for all sheep after mulesing, principally because of the benefits it provides in pain control. There may be additional benefits of reduced risk of wound infection as a result of the antiseptic in the dressing or the behaviour of the lambs after mulesing — treated lambs are less likely to lie down after the surgery and therefore may keep their wounds cleaner. This aspect of its use has not been examined.

Reduction in other procedures which cause wounding of lambs is more difficult. Castration of male lambs and tail docking of lambs of both sexes is considered routine and necessary under Australian conditions, although individual producers may be able to review that proposition, depending on their market for lambs and the age at which lambs are sold. Tail docking should be done at the third palpable intercoccygeal joint of the tail (a point level with, or slightly longer than, the tip of the vulva in ewe lambs) to provide most benefit against dagginess and flystrike in lambs. Docking at shorter lengths increases the risk of rectal prolapse, dagginess and, hence, flystrike, and also increases the risk of vulval cancer in retained animals. Docking of tails should also be done in lambs as young as possible, within reason. The increased wound size associated with shorter tail lengths and older, larger lambs is associated with a higher risk of wound infection and longer healing times.[40] This in turn may increase the risk of bacteraemia and, therefore, arthritis.[41]

In the past, it has been recommended that lambs be marked in temporary yards rather than permanent yards in order to reduce the risk of wound infection, and that instruments used for tail docking and castration be disinfected between sheep. The only study to examine this to date failed to detect any benefit from taking these steps to improve hygiene at marking.[35] While good hygiene should still be encouraged, the evidence so far suggests that efforts should be concentrated on reduction in wounding of lambs, rather than trying to protect the wounds from infection after they have occurred.

There is a vaccine against *E rhusiopathiae* which is available and registered for use in sheep (Eryvac®, Zoetis Australia). The manufacturer recommends two vaccinations to ewes, the second being given before lambing starts, in order to provide passive protection to lambs. Ewes vaccinated in one year require only one vaccination each year, pre-lambing. The vaccine should be used only in flocks where a clear and definitive diagnosis of erysipelas arthritis has been made because the vaccine will not protect lambs against other forms of arthritis. Producers should be aware that there have been no clinical trials to demonstrate the efficacy of the vaccine in sheep.

Prophylactic antibiotics have also been recommended as a control measure. Lambs can be given one injection of penicillin or oxytetracycline at marking. Again, the efficacy of this approach is unknown.

Suppurative arthritis

Arthritis conditions marked by purulent infections of one or more joints can be caused by a number of organisms, including *Fusobacterium necrophorum*, *Trueperella pyogenes*, *E coli*, *Histophilus somni* (formerly *H ovis*), *Actinobacillus seminis*[42], *Staphylococcus* spp and *Streptococcus* spp. In very wet years, post-mulesing infections with *Fusobacterium necrophorum* may result in a high prevalence of arthritis in some flocks.[43]

Staphylococcus aureus arthritis in lambs also occurs following the bacteraemia of tick pyaemia, arising from a bite by *Ixodes ricinus*. The disease occurs in the UK as occasional outbreaks, with a morbidity of around 5% in lambs between 2 and 10 weeks of age. There are a number of species of ticks which belong to the *Ixodes ricinus* complex, but none occurs in Australia.

In the UK, the colloquial term *joint ill* is used to describe a suppurative arthritis in lambs up to 1 month of age, usually caused by a streptococcus species, although other bacteria including coliforms, *T pyogenes* or *F necrophorum* may be involved.[44] This is the most common form of joint infection in lambs in the UK, often affecting multiple lambs in the flock with morbidity up to 15%. Infection localises in the joint following a bacteraemia, which may also result in spread to the cerebrospinal meninges, muscles and valves of the heart and the kidneys.

Initially lambs are lame in one or more limbs, with joint swelling becoming more noticeable after one to two weeks. The relatively high incidence in the UK may be associated with the practice of indoor lambing, although the disease also occurs in lambs born outside. The epidemiology of the disease is not well understood, but lambs are believed to be infected either through the navel or from the environment of the pens in which they are born — possibly contaminated by ewes which carry the bacteria. Lambs may also be infected when passing through the birth canal or from the skin or saliva of the ewe. Disinfection with iodine of the navel of the newborn lamb is a recommended control measure, as well as cleanliness and low stocking densities in the environment where lambs are born. A prolonged course of penicillin, rather than tetracyclines, is the recommended treatment for affected lambs, but recovery is rarely complete.[45]

BACTERIAL INFECTIONS OF THE LIMBS

Post-dipping lameness

This condition is a cellulitis of the lower leg caused by an infection with *Erysipelothrix rhusiopathiae*. The disease occurs when sheep with skin wounds are subject to plunge or shower dipping in a dip fluid contaminated with sheep faeces and other organic material. Multiple skin punctures caused by grass seeds in the fleece are one of the worst predisposing factors. If the wounds are not given sufficient time to heal after shearing, infection enters through the skin abrasions and develops into a cellulitis which extends to the coronary band and foot, causing a laminitis. Some sheep may develop a bacteraemia and subsequent polyarthritis, but this is uncommon.

Clinical signs

There is usually a sudden onset two to seven days post-dipping with a high morbidity. Affected animals are depressed and usually persistently febrile for some days. The sheep are observed to be grazing less than expected and they are slow to rise and move off when disturbed. Weight loss, or failure to recover weight following shearing, is marked. Lameness occurs in one to four legs, causing the sheep to have a short, 'proppy' gait when walking. The affected limbs are hot and slightly swollen. A history of recent dipping is important in establishing a diagnosis.

Prevention

Prevention involves the correct use of a bacteriostat in the dip, a reduction in faecal contamination of dip solutions and, most importantly, delaying dipping until shearing wounds heal (at least seven days post-shearing, but preferably 14 days). Dip bacteriostats cannot be relied upon to prevent post-dipping lameness in sheep dipped soon after shearing. Bacteriostats are most effective when added to dip solutions when a day's dipping is finished, in order to exert a bacteriostatic or bactericidal effect in the dip solution overnight.

Treatment

Penicillin is effective. Most untreated cases recover but may lose considerable body weight before recovering fully if not treated.

Strawberry footrot

This disease is a proliferative dermatitis of the lower limbs in the region from the coronet to the hock or knee, associated with the production of scabs and open bleeding areas, with a strawberry-like appearance when the scabs are removed.

The causative agent is believed to be *Dermatophilus congolensis* (which also causes lumpy wool). The lesions resemble those caused by CPD virus and many cases of strawberry footrot diagnosed in Australia may in fact be CPD.[d] Strawberry footrot differs from CPD infection in that lesions of strawberry footrot may persist longer (up to five weeks) and sheep can suffer repeated bouts of infection, with no apparent immunity.[46]

Infection is established in the skin following a small wound, such as that caused by prickly plants. The condition is often associated with pasture conditions which keep the skin of the lower limbs wet, so it is possible that maceration by water could cause sufficient skin damage to allow infection to commence.

Affected flocks of sheep should be moved to drier pastures if wetness is considered to be a contributory factor in an outbreak of strawberry footrot. *D congolensis* is sensitive to a number of antibiotics, including penicillin, so antibiotics can be used to reduce the recovery period.

BACTERIAL INFECTIONS OF THE FOOT

Four further important bacterial infections are confined to the hoof. These are *toe abscess*, *ovine interdigital dermatitis* (OID), *foot abscess* and *footrot* (see Table 13.1). Footrot exists in a range of forms; the least severe is called *benign footrot* (BFR) and the most severe is called *virulent footrot* (VFR). Another bacterial infection of the foot (contagious ovine digital dermatitis) is considered to be an emerging disease in Britain but has not been reported from Australia or New Zealand.[47]

Virulent footrot, foot abscess and toe abscess are economically important and have adverse effects on the welfare of affected animals.

---

d CPD is discussed in more detail in Chapter 10.

Table 13.1: Major bacterial infections of the hoof of sheep.

Disease	Tissue	Agents	Typical flock morbidity
OID	Interdigital skin (IDS)	F necrophorum	<30%
BFR	IDS	D nodosus F necrophorum	<80%
VFR	IDS + sensitive laminae	D nodosus F necrophorum	<80%
Foot abscess	Joint and joint capsule	F necrophorum T pyogenes	1% to 10%
Toe abscess	Sensitive laminae	Various	< 5%

Lamellar suppuration (toe abscess or white line abscess)

Toe abscess is an acute, purulent infection usually involving only one digit of the hoof. Unlike foot abscess, the infection is confined to the sensitive laminae of the hoof, usually in the toe region. The infection probably enters through cracks in the hoof, often at the white line — the junction between the wall of the hoof and the sole. Alternatively, infection may enter through a crack in the sole. Lameness is acute and severe but responds quickly to drainage.

The site of infection can be established by tapping or squeezing the digit, eliciting a pain response, until the site which is most painful is located. Hoof parers can be used to squeeze the hoof (without cutting) to localise the pain. Excess horn can then be pared away, if necessary, and the sole or hoof wall pared with a hoof knife until the abscess is exposed and drained. Paring should cease when effective drainage is achieved. Excessive paring may damage the vascular tissues of the hoof (the corium and sensitive laminae) and lead to delayed healing and, possibly, granuloma formation.

In cases which are not treated, the abscess will usually track upwards under the hoof wall and burst at the coronet. Most cases then resolve naturally.

The bacteria causing the abscess are those commonly found in the environment, mud and faeces of sheep. Antibiotics should not be used in lieu of paring and drainage, because such treatment will usually just delay the development of an abscess. Antibiotics can be used once drainage has been established but are usually not necessary.

Shelly hoof

Toe abscessation can also follow from *shelly hoof* (also called *white line disease* or *white line degeneration*), which is a common condition of sheep. The shelly hoof lesion is a separation of the hoof wall from the sole and the underlying laminae of the hoof, producing a pocket which packs with dirt and, under the pressure of weight bearing, progressively extends upwards.[48] Because the impacted material is in close contact with the corium of the hoof, pus formation and abscessation can occur, leading to a lesion which behaves like toe abscess.

Sheep with shelly hoof are generally not lame unless the sensitive laminae are infected, at which point pus formation occurs. Shelly hoof can sometime cause confusion with footrot because of its similarity to a chronic footrot lesion if the latter is not clearly associated with under-running of the sole (see footrot, below). During footrot eradication programmes, when sheep are being closely inspected for signs of footrot, a proportion of the sheep which are classed as *suspicious* and culled from the flock will be those suffering only from shelly hoof. Unfortunately, some unnecessary culling may be unavoidable so as not to compromise the footrot eradication programme.

Shelly hoof should be treated by paring the outer wall over the impacted material so that the impacted material can be removed. Paring should be continued sufficiently high up the wall of the hoof so that further impaction cannot occur. Shelly hoof is reported to be very common in some breeds (and possibly in most breeds) and to have a significant genetic component, leading to the possibility of selective breeding to reduce the incidence in flocks.[49] At this time, however, the condition is not considered sufficiently serious to justify inclusion in breeding objectives.

Ovine interdigital dermatitis (OID)

Ovine interdigital dermatitis is an infection of the interdigital skin (IDS) of the hoof associated with *Fusobacterium necrophorum* and a range of other environmental organisms. OID was precisely described in 1967 by Parsonson et al. (p. 309)[50] as 'an acute necrotizing infection restricted mainly to the posterior interdigital skin and associated with an intense epidermal invasion by *Fusiformis necrophorus'*. (*Fusiformis necrophorus* is now classified as *Fusobacterium necrophorum*.)

OID develops in the IDS after the skin has become hydrated and macerated, usually as a consequence of sheep being exposed to wet pasture for a period of weeks or months. Under those conditions the epidermis of the IDS, which is normally a thick and resilient tissue, loses its natural defences and succumbs to infection by *F necrophorum* — a bacterial species which is common in the faeces of sheep.

With OID, the interdigital skin becomes hairless and appears red and moist. The lesions vary from mild erythema to heavily inflamed and thickened skin, with a moderate amount of moist grey necrotic product from the infected epidermal tissue. There may be slight separation of the horn at the skin-horn junction in the posterior interdigital space but no under-running. In severe or more chronic cases, the lesions are painful and the sheep may resist the separation of the claws during examination.

In the field, lameness is moderate only and will often disappear when the sheep are disturbed, to reappear when the sheep are grazing quietly.

Lesions of OID are indistinguishable from those of benign footrot or the early stages of virulent footrot. Benign strains of *D nodosus*, the bacterium associated with footrot, is so commonly associated with OID that it is probably pointless to attempt to distinguish OID from benign footrot on the basis of the clinical appearance of the lesions or the presence or absence of *D nodosus* in swab specimens. Nevertheless, OID can occur in the absence of *D nodosus*.[50]

In southern Australia OID occurs typically in winter and early spring when flocks of sheep are grazing wet pastures. The lesions respond well to topical treatment, such as footbathing in 10% zinc sulphate solutions, but the disease will resolve spontaneously when the pastures become drier. OID is most important as a predisposing factor for foot abscess and footrot, and as a differential diagnosis during a footrot investigation. Occasionally, grass seeds may become trapped and held in the interdigital space because of the moisture associated with OID, and abrasion from the grass seeds may exacerbate the OID lesion, even as pastures become dry. Under those conditions, intervention may be necessary to remove the seeds, footbathe the sheep and return them to better pastures until the OID is resolved.

The tem *scald* or *footscald* is often used to describe lesions of OID, benign footrot or the early stages of more severe footrot. The term is poorly defined, however, and its use should therefore be avoided by veterinarians.

Foot abscess

Foot abscess is an acute infection involving the distal interphalangeal (coffin) joint. The condition is also known as *deep sepsis of the pedal joint* or *pedal septic arthritis*, which are accurate descriptive titles. The disease should not be confused with toe abscess. The term *foot abscess* should be used to describe infections which involve, primarily, the P2-P3 joint. The terms *heel abscess* and *infectious bulbar necrosis* are sometimes used but, because they imply an infection of the soft tissue of the heel alone, they are somewhat misleading.

The disease is characterised by severe lameness. Affected sheep may carry the limb entirely, especially when walking undisturbed by humans or dogs. The disease occurs in many flocks under suitable environmental conditions, although morbidity is generally low and usually below 2%. Foot abscess tends to affect heavier sheep (pregnant ewes, rams) and is associated with wet and muddy pasture conditions.

Pathogenesis

The development of the disease has been described in detail by West (1983)[51] from a study of 53 cases. The infection is commonly caused by *Fusobacterium necrophorum* and *Trueperella pyogenes,* together or separately, and often follows on from ovine interdigital dermatitis, presumably by extension through the compromised interdigital skin into the subcutaneous tissues, joint capsule and, ultimately, the P2-P3 joint itself. The distal interphalangeal joint is very close to the interdigital skin at the level of the axial coronary band and consequently is vulnerable to infection if the interdigital skin is no longer protective.

A suppurative infection of the joint develops and the accumulated pus eventually discharges through a sinus at the coronary band. The discharge is initially bloody and necrotic, then becomes more purulent. Granulation tissue forms at the point of discharge, often interfering with drainage. The axial collateral and interphalangeal ligaments are often involved and, if they rupture, the joint becomes unstable and the third phalanx is displaced laterally.

The infection will, in most cases, slowly resolve over a period of eight weeks, but the joint is permanently damaged and most recovered animals will be left with chronic lameness to some degree.

While OID appears to be the predisposing infection which allows entry of bacteria to the periarticular tissues, other routes probably also occur, including lamellar infections (from toe abscess or excessive paring) or even haematogenous spread.[52]

Economic importance

The prevalence within a flock is generally low. Much of the importance of the disease arises because of the severity of lameness, the occurrence in ewes close to lambing and in rams close to joining, and the likelihood that the arthritis and damage to the ligaments of the joint will leave the animal with permanent disability. Ewes with foot abscess in late pregnancy are predisposed to pregnancy toxaemia. Rams with foot abscess during joining will have reduced serving capacity.

Predisposing factors

Some prior damage to the IDS is necessary to allow the bacterial invasion of the subcutaneous tissues. Experimentally, the disease has been reproduced by housing the sheep with moist dirty conditions underfoot following a treatment to cause damage to the IDS. Naturally occurring predisposing factors include stones and prickly or abrasive vegetable matter. The disease is most common in heavy ewes in wet conditions, particularly if the ewes are driven on roads or in stony areas.

Clinical signs

The predominant sign is acute lameness, often to the extent of complete disuse of the affected foot and extended periods of recumbency. Initially there is lameness, heat and swelling of the pastern over the coffin joint, marked oedema and inflammation of the interdigital area with necrosis in some areas. As the infection progresses sinus tracts develop, draining from the joint to the coronary border in the interdigital space or abaxially.

Usually, the digit heals within eight weeks to a stage where no infection remains, but some permanent deformity and disability often remain.

Treatment

In closely monitored flocks, foot abscess can be detected early and diagnosed on the basis of acute lameness and the heat and swelling over the pastern. At this stage the infection may respond to an intensive course of parenteral antibiotic therapy. Procaine penicillin should be effective. Once the infection has commenced to discharge externally it is unlikely that antibiotic treatment will return the foot to normal. The infection may resolve, with or without antibiotics, but some permanent damage to joint and associated ligaments will remain. If the joint is considered to be permanently damaged and the lameness to be incapacitating then amputation of the digit can be considered.

Amputation can be performed using embryotomy wire and cutting through skin, soft tissues and the second phalanx. Surgery is performed under intravenous regional anaesthesia. The technique is described in full by Scott (1995).[53] The animal's useful life after amputation may be limited if the remaining digit cannot sustain the increased load.[54]

Prevention

As the disease occurs usually under prolonged wet pasture conditions and following the development of ovine interdigital dermatitis, steps may be taken to reduce the incidence of foot abscess by ensuring that heavily pregnant ewes are not exposed to such conditions. Unfortunately, the pastures which are well suited nutritionally to ewes in late pregnancy, particularly in winter leading up to a spring lambing, are often in low-lying parts of the farm. Additional steps to reduce the risk of trauma to the interdigital skin include attention to areas where the sheep walk or are driven or yarded, so that the interdigital skin is not further damaged by stones.

The efficacy of *F necrophorum* vaccines for preventing foot abscess in sheep is unknown.

Footrot

Footrot is a contagious dermatitis of the interdigital skin and sensitive laminae of the hoof of sheep and goats caused by an infection with the bacterium *Dichelobacter nodosus* (formerly *Bacteroides nodosus*) in association with other bacteria and under suitable environmental conditions. Three forms of the disease are recognised — *benign footrot* (BFR), *intermediate footrot* (IFR) and *virulent footrot* (VFR) — although the differences between them are not distinct and the clinical manifestations of the disease cover a spectrum from mild to severe disease. Virulent footrot, the most severe form of the disease, can cause large economic losses when it occurs in Merino sheep flocks in the medium and high rainfall zones of Australia, losses which provide a strong incentive for flock owners to eliminate the disease from their properties.

In Australia, virulent footrot was previously very common in the high rainfall zones of New South Wales and Victoria but statewide eradication programmes have dramatically reduced the proportion of infected flocks to fewer than 5%, even in the higher rainfall areas. When outbreaks do occur in flocks it is now usual practice to attempt to eliminate the disease from the affected flock as quickly as possible.

The disease occurs in most sheep-rearing countries of the world. In New Zealand, the disease is common and eradication has been advocated and achieved in some flocks. In the UK, virulent footrot is present on the majority of sheep farms[55] but eradication is considered to be impractical for most farms because of the climate and the high risk of reinfection following any attempts to eliminate the disease from flocks.[56,57] Consequently, the disease is usually controlled on a case-by-case basis by treatment of individual animals.

Footrot lesion scoring system

The advance of a footrot infection in a sheep's foot is a consequence of a number of factors, which are discussed further below. The extent to which the infection is capable of advancing in the foot is a measure of the disease *severity* in an individual sheep and is used in diagnostic procedures. A simple, repeatable scoring system has been devised to give a semi-objective measure of the severity of footrot lesions and is shown in Table 13.2. Other more complex systems subdivide score 3 lesions and define the most severe, chronic lesions as score 5.

Aetiology

Footrot is an infectious disease, always associated with *D nodosus*, but that organism alone cannot cause the disease. Ovine interdigital dermatitis (see above) must occur first. OID is associated with *F necrophorum* infection and footrot will develop only as a result of a combined infection of *F necophorum* and *D nodosus*.

When *D nodosus* is present or is introduced into an OID lesion, a footrot lesion develops. New infections of footrot *always* start as infections of the IDS. If the strain of *D nodosus* is

Table 13.2: Classification of footrot lesion severity.

Score	Lesion
0	normal foot
1	non-specific mild inflammation and/or necrosis of the IDS
2	severe inflammation of the IDS
3	any lesion in any claw which results in under-running of the soft horn or the heels or sole
4	under-running of any hard horn of the claw

sufficiently virulent, and the host sufficiently susceptible, the interdigital lesion will extend to become the severe and chronic infection of the sensitive laminae underlying the horn of the hoof which is characteristic of virulent footrot. The cells which generate horn are destroyed and the hoof separates from its underlying structures (Figure 13.3). This separation is termed *under-running*.[58] The development of the footrot lesion, influenced by host, environmental and bacterial factors, is termed the *expression* of the disease.

Thus, the development of new cases of virulent footrot requires

- exposure to wet pastures and faeces — the source of *F necrophorum*
- environmental temperatures which favour bacterial growth
- introduction of a virulent strain of *D nodosus*.

Figure 13.3: A hoof with a score 4 lesion of footrot. One claw has been trimmed, showing the under-running of the sole anteriorly to the toe. The other claw is not trimmed but shows the overgrowth and deformity which frequently accompanies a footrot lesion. Source: KA Abbott.

Characteristics of *D nodosus*, including antigenicity

D nodosus is effectively a parasite, in that its only long-term habitat is the foot of an infected animal. It does not survive in the environment for more than seven days; survival for a few hours is probably more usual.[59,60] The organism is a gram-negative anaerobe. Under a light microscope it appears as a rod with polar caps, areas at each end which stain more intensely than the remainder of the cell, a characteristic which is more pronounced in organisms taken from footrot lesions rather than from cultures.[61] Electron microscopy has revealed the presence of filamentous appendages or pili (also called fimbriae) which are known to carry the important antigens on which seroclassification is based and which stimulate effective immunity after vaccination.[62,63]

In culture, *D nodosus* is strictly anaerobic and is usually grown in an atmosphere containing 10% hydrogen and 10% carbon dioxide. The organism is fastidious in growth requirements on media, requiring either 10% horse serum[61] or ground hoof material[64] in agar (hoof agar), or trypticase, arginine and serine (TAS agar) as additives to agar or broth.[65]

Concentrations of agar above 3% restrict the size and spreading of bacterial colonies other than *D nodosus*.[66] Consequently, 4% agar is currently recommended for primary isolation of the organism from lesion material, and 2% agar is used for subcultures.[67] The organism grows best at 37° C and, at that temperature, colonies appear on agar plates in four to six days.

Two serogrouping systems exist. One describes 10 serogroups (Table 13.3), some of which are divided into subtypes, making a total of 19 serotypes.[68,69,70] A system described in UK[71,72] classifies the organism into 17 serogroups (A to R with no I) with no subtypes. At least some of the serotypes in the Australian system, particularly B_2, B_3 and B_4, are classified as distinct serogroups in the alternative system.

Serogrouping is particularly important in vaccine production because cross-protection across serogroups is limited. In addition, the existence of a relatively simple classification system based on serological reactions has led to serogrouping being used as an epidemiological marker to identify infecting strains of *D nodosus* across time within flocks or between flocks. Its use, for this purpose, however, is limited because there are only relatively few (10) serogroups and because there is evidence that the serogroup of a strain of a particular virulence can change, through genetic recombination, in the course of an outbreak if a mixed-serogroup infection is present.[73,74]

The presence of mixed-serogroup infections in sheep flocks is usual.[69,71,75,76] Up to six serogroups[77] and nine serotypes have been reported in any one flock. To fully appreciate the diversity of serogroups in one flock it is often necessary to collect specimens from many sheep because some serogroups may dominate the infection and others may be detected at only a low frequency.

Table 13.3: Serogroups and serotypes of *D nodosus* currently recognised in Australia.

	Serogroup									
	A	B	C	D	E	F	G	H	I	M
Number of serotypes	2	4	2	1	2	2	2	2	1	1

Variation in virulence within strains of *D nodosus*

There are many genetic variants of *D nodosus* and the innate virulence of strains of the bacterium varies over a very wide range. The most virulent cause a disease of sufficient economic impact to warrant intervention aimed at eliminating or minimising the infection in the flock. At the other end of the virulence spectrum are the benign strains, which cause a relatively minor and self-limiting disease. Within sheep flocks there is often a mixture of *D nodosus* strains present and these strains are usually of different serotype and of different innate virulence.

The term *virulent* is used to describe both the innate disease-producing potential of the most pathogenic strains of *D nodosus* and the clinical expression of the most severe forms of the disease. Hence, *virulent footrot* is associated with *virulent strains* of *D nodosus*. While it would be more precise to say that severe footrot is associated with virulent strains of the organism, common usage is to use the term *virulent* to describe characteristics of both the disease expression and the infecting strain.

Virulent footrot is defined as that disease which involves under-running of the hard horn (score 4) of the hoof of more than 10% of sheep that are exposed to infection.[78] This level of severity will cause enough lameness and reduction in feed intake to measurably affect productivity.

Benign footrot is that form of the disease in which infections are restricted to the IDS in all or nearly all affected sheep. Some sheep in a flock, however, may be particularly susceptible to infection. As a general rule, it is accepted that fewer than 1% of the flock will develop score 4 lesions with benign strains of *D nodosus*.

Intermediate footrot typically causes 1% to 10% of an exposed flock to develop score 4 lesions.[79,80,81] With the exception of a small proportion of sheep, under-run lesions are qualitatively less severe than those of virulent footrot, with less necrosis evident. Overall, intermediate footrot is a milder disease than virulent footrot on a flock basis and all but the few severely affected sheep recover spontaneously when climatic conditions become dry. Score 3 lesions occur at a low prevalence in benign footrot, but may occur at a high prevalence in intermediate footrot. Fewer than 5%[74] or 10%[82] of sheep are expected to have score 3 or score 4 lesions in benign footrot, while around 25% of sheep may have score 3 lesions with intermediate footrot (Figure 13.7).

Laboratory tests are available to distinguish between isolates of different virulence (see below) but these tests are not generally necessary for diagnosis. The diagnosis of footrot, both clinically and bacteriologically, should be based on the examination of a representative sample of the flock.

There is no relationship between virulence and serogroup or serotype.

Environmental factors affecting footrot spread

Outbreaks of footrot occur when footrot spreads from one or more infected sheep to other, susceptible sheep. Provided *D nodosus* is present in a flock, outbreaks occur after a period of sustained rainfall and when environmental temperatures are sufficiently warm. For an outbreak to occur in spring, a sustained and reasonably consistent amount of rainfall, averaging about 50 mm or more per month, must fall over the preceding four months.[58] Then, as mean daily temperatures start to consistently exceed 10 °C, spread from infected sheep to uninfected sheep will commence. Typically, in districts of Australia suitable for footrot, these conditions occur in late August to early October, but not usually in every year.[83] If suitably moist pasture conditions continue, an outbreak of footrot involving a high proportion of the flock will occur. If there is not continuing rain, conditions suitable for sustained spread will only be maintained for three to four weeks and a lower proportion of the flock will develop footrot.

When environmental temperatures are consistently below a mean daily temperature of 10 °C — for example, during winter — the temperature of the feet of the sheep also falls and the growth of the bacteria is consequently inhibited. Transmission of infection between sheep is slowed or ceases completely.

High environmental temperatures do not directly inhibit transmission but, if accompanied by the drying of pastures and, therefore, the feet of the sheep, cases of OID will resolve and recovered sheep will no longer be susceptible to infection. While footrot lesions, particularly under-running ones, may persist in some sheep as conditions become drier, spread from sheep to sheep will cease.

For transmission of footrot to occur, pasture is a key environmental factor. In the absence of rain, provided that the sheep are predisposed by long periods of prior exposure to wet pastures and that conditions are cool but not cold, heavy dew can keep pastures wet all day and provide conditions suitable for spread. This is more likely to occur if pastures are long and dense.[84] Clover has a reputation for encouraging footrot spread, probably by providing a wet environment at the base of the sward even when rain is relatively infrequent.

Compared to spring, outbreaks occur less commonly in late summer and autumn and only in years when the average monthly rainfall for the four or five months preceding the outbreak exceeds 70 mm per month and exceeds 75 mm in the month of the outbreak. Footrot does not spread following isolated periods of heavy rain or irrigation of pastures during a hot, dry period, presumably because the conditions do not lead to the development of OID.

Warm moist conditions, high levels of soil moisture accumulated over months, rainfall and low evaporation rates all favour both pasture growth and footrot spread, if *D nodosus* is present in the flock. Thus footrot outbreaks are typically seen in 'good' springs, usually in August, September and October in winter rainfall areas and one to two months later in summer rainfall districts. Outbreaks at other times of the year are uncommon but will occur if seasonal conditions are suitable. In flocks where footrot has not been previously recognised, diagnosis may not be made until late in an outbreak or even until transmission has ceased and most of the sheep affected have relatively advanced and chronic infections.

Host effects

Within a flock, some sheep are more susceptible to footrot than others. Variation in susceptibility to footrot is expressed as variation in both the severity and the duration or persistence of lesions. Sheep with higher levels of natural resistance to footrot are less likely to develop more advanced lesions than more susceptible sheep and are likely to recover more quickly from footrot when environmental conditions favour resolution of lesions, or when the infection is treated topically, parenterally or immunologically, by vaccination.[85]

There is a genetic component to this variation in susceptibility within flocks of sheep and it is possible, therefore, to select for resistance to footrot. The genetic component of resistance has been demonstrated in Merino sheep[86], Scottish Blackface and Mules (crossbred) in the UK[87], Romney and Corriedale sheep in New Zealand[88,89] and Targhee sheep in the US.[90]

Merino sheep are widely considered to be more susceptible to footrot than British breeds and their crosses, although the evidence to support this from controlled trials is very limited. One study in which sheep of five breeds were exposed to identical challenge showed that Merino sheep developed more severe lesions and more persistent lesions than sheep of the Border Leicester, Dorset Horn and Romney breeds.[91] The number of sires represented in each breed was not stated and may have been so low that individual sire effects, unrelated to breed, may have contributed to the differences observed.

Nevertheless, given that the differences in susceptibility to footrot between individuals within breeds and within flocks is now proven, it is not unreasonable to assume that the frequent observations about differences in breed susceptibilities are true, even if definitive evidence is lacking. If so, we can expect that flocks of more resistant sheep or of sheep of more resistant breeds will, for the same level of challenge from strains of *D nodosus* of identical virulence, have fewer cases of footrot and fewer cases of severe lesions; will heal faster when environmental conditions are less favourable; will respond more quickly and more completely to topical treatments or antibiotic therapy; and will respond to vaccination with a greater and more persistent immunity.

The variation in host response to infection, along with the variability associated with environmental influences, mean that diagnosis of footrot, particularly its classification into BFR, IFR or VFR, requires examination of a large sample of an infected flock — so that animals with a range of susceptibility are examined. This also means that the categorisation of a footrot outbreak within a flock must account for the assumed differences in susceptibility between breeds of sheep. For example, footrot associated with a virulent strain of *D nodosus* may cause typical virulent footrot in a Merino flock but may cause a much less severe disease — on a flock basis — in a flock of British breed sheep.

In New Zealand it has been observed that outbreaks of footrot in flocks of British breed sheep in the drier parts of New Zealand (NZ) occur occasionally and consist of relatively minor, mostly interdigital, infections which are best managed by topical treatment of affected sheep and culling of the few, worst-affected sheep. By contrast, in flocks in wetter areas of NZ, or in flocks of Merino sheep, footrot is often managed by taking steps to eliminate it from the farm.[92] This observation highlights the nature of breed differences and individual animal differences in their resistance to footrot, of climatic effects on the expression of footrot, and how those factors then determine the optimum strategy for flock owners to adopt in response to the disease.

Alternate hosts

Cattle, goats and deer are suitable alternate hosts for *D nodosus*. Goats suffer severely from footrot. Some *D nodosus* isolates which appear to be virulent for goats are virulent for sheep; others are benign. Until there is any information to the contrary, deer should be considered potential reservoirs of infection for sheep. *D nodosus* occurs naturally in the feet of cattle and is associated with interdigital lesions and probably contributes to the severity of those lesions.[93,94,95,96] The interdigital lesions vary from mild, superficial erosions to chronic hyperkeratotic lesions with thickened folds of skin in the interdigital space and deep, cracked fissures. Rarely, the horn is under-run.[97,98,99] Lameness is uncommon. The bacterial flora of the disease in cattle appears very similar to that of ovine footrot, with spirochaetes and *F necrophorum*.

The strains of *D nodosus* which infect cattle feet appear to be the same as those which infect sheep, although, generally, strains isolated from cattle will only cause benign footrot in sheep.[100,101,102]

There are a few reports of virulent sheep strains infecting cattle and causing difficulties for eradication of the disease from sheep.[103,104,105] Nevertheless, VFR has been eradicated from many properties despite the continuing presence of cattle, so it seems reasonable to conclude that cattle are not often a risk to eradication programmes for ovine footrot. Flock owners should be aware, however, that there is a risk of VFR being introduced with cattle or persisting in cattle during an eradication programme and cattle should be considered a small but significant threat to management or exclusion of VFR in sheep flocks.

Differential diagnosis of *D nodosus* infections in flocks

Traditionally in veterinary medicine, diagnosis of disease has been based on recognition of a characteristic clinical case of the disease in question. In managing footrot this approach is less applicable because of the occurrence naturally of a range of different expressions of the disease. An outbreak of footrot can be of the benign, intermediate or virulent form and these outbreaks are due to genetically stable variants of *D nodosus*. Benign footrot in a flock will not become virulent footrot unless there is the introduction of a virulent strain of *D nodosus*. Flock owners and veterinarians less familiar with the disease often find it difficult to comprehend the concept of a spectrum of disease severity and are then inclined to classify all outbreaks as virulent if any sheep have score 4 lesions, or as benign if no sheep have severe lesions. In reality there is a range in severity of footrot associated with the virulence of the infecting strain and the natural resistance of the individual sheep. Efforts must be made to carefully assess the disease by inspecting a large, randomly selected sample of the infected flock to determine the relative proportion of severely affected sheep.

A careful assessment of the severity of disease on a flock basis is important because the form of the disease present will determine the extent of intervention which is justified. Benign infections may warrant only minor intervention, or none, whereas the more severe forms cause significant discomfort and production loss and justify programmes designed either for long-term minimisation of the disease (control) or elimination from the flock.

Responsible flock diagnosis entails making a quantitative assessment of the prevalence of the disease in the flock under investigation and the prevalence of severe infections. This quantitative assessment requires a count of the number of sheep affected in an adequate sample of the flock and a record of the severity of the disease in those sheep.

The first sign of footrot in a flock is usually lameness, but there are many other causes of lameness in sheep. In addition, footrot may occur concurrently with those other diseases. This possibility reinforces the need to examine representative samples of affected flocks.

The virulence classification of the disease should be based on the prevalence of sheep with score 4 lesions relative to the number of sheep exposed to footrot. An example of a recording sheet used in an assessment of a footrot outbreak is shown in Figure 13.4.

The epidemiology of footrot in endemically infected flocks

In an infected flock, footrot prevalence shows a strongly seasonal variation (Figures 13.5 and 13.6). When environmental conditions favour the transmission of the disease from sheep to sheep the prevalence rises until most of the susceptible sheep are affected. As environmental conditions become drier the prevalence declines as some sheep recover from infection and become free of the disease. Generally, within any flock of sheep, there are some individuals who remain infected throughout the following non-transmission period, and these sheep are a source of reinfection for the flock in the next transmission period.

There is no evidence for immunity against reinfection following natural infection, and sheep which have recovered from infection are as susceptible as naïve animals when exposed to infection again.[106]

In footrot-endemic areas, seasonal conditions do not necessarily allow transmission in every year, but there are generally some sheep which remain chronically infected and will be a source of reinfection in the next favourable season. If conditions remain unfavourable for several years, the disease may disappear from the flock completely. As a consequence, flocks in low rainfall areas of the country are less likely to have footrot, and the disease, if introduced, can be readily eliminated. In contrast, in high rainfall districts which have relatively short hot, dry periods, footrot can become endemic and difficult to control or eliminate.

For VFR, the chronically infected sheep which maintain the flock infection over non-transmission periods have advanced and chronic under-running lesions of the hard horn of the toe or wall of the hoof. In some cases, these lesions are obvious and cause deformity of the foot, severe lameness and flystrike. In other cases, the lesions may be restricted to small pockets of infection in the horn of the hoof with only small amounts of deformity. The sheep in which infections persist are the most susceptible sheep in the flock. The most resistant sheep in the flock are generally the last to become infected, develop the mildest lesions and heal spontaneously soon after conditions suitable for transmission and lesion expression have passed.

For BFR, infection is not maintained in under-run horn because few, if any, sheep develop under-running with BFR. Consequently, it is believed that infection persists in subclinical infections of the IDS.[107]

For IFR, sheep with chronically under-run infections do occur, although such cases occur at a lower frequency than in flocks with VFR. IFR can persist in a flock through a non-transmission period in chronic lesions in the horn of the hoof and, possibly, in subclinical lesions of the IDS, as benign strains are believed to do.

Laboratory aids to diagnosis

A number of laboratory-assessed characteristics of *D nodosus* isolates have been associated with virulence. These include colony morphology, twitching motility[109,110], agar corrosion[111] and the presence and nature of extracellular proteases. It is the latter characteristic, first described in

Footrot scoring record

Veterinarian	James Smith
Date of inspection	12 Dec 2017
Owner	Andrew Jones
Description of mob	2.5 year old ewes

Sheep (sequential number or ID)	LF	RF	LH	RH	Worst score
1	2	2	3	2	3
2	4	3	2	2	4
3	1	1	2	2	2
4	2	2	4	3	4
5	4	4	4	4	4
6	4	4	3	4	4
7	3	3	3	4	4
8	3	3	3	2	3
9	0	1	1	1	1
10	3	4	4	3	4
11	2	2	2	2	2
Sheep numbers 12 to 30 not shown					
31	4	3	2	2	4
32	3	3	2	2	3
33	1	1	0	1	1
34	3	4	3	3	4
35	2	2	2	3	3
36	2	2	4	3	4
37	3	2	2	2	3
38	4	3	2	2	4
39	3	4	3	3	4
40	3	4	4	4	4

Number of sheep inspected	40	
Number with footrot (scores of 2 or greater)	36	90%
Number with maximim scores of 3	10	25%
Number with maximum scores of 4	22	55%

Figure 13.4: Example of a footrot record sheet. In this case, 40 sheep from one mob were selected at random, tipped over and their feet examined. The lesion score of each foot was recorded. The high prevalence of footrot (90% of the sample were affected) and high prevalence of sheep with a worst-foot score of 4 suggest that this is virulent footrot, not intermediate or benign footrot. Source: KA Abbott.

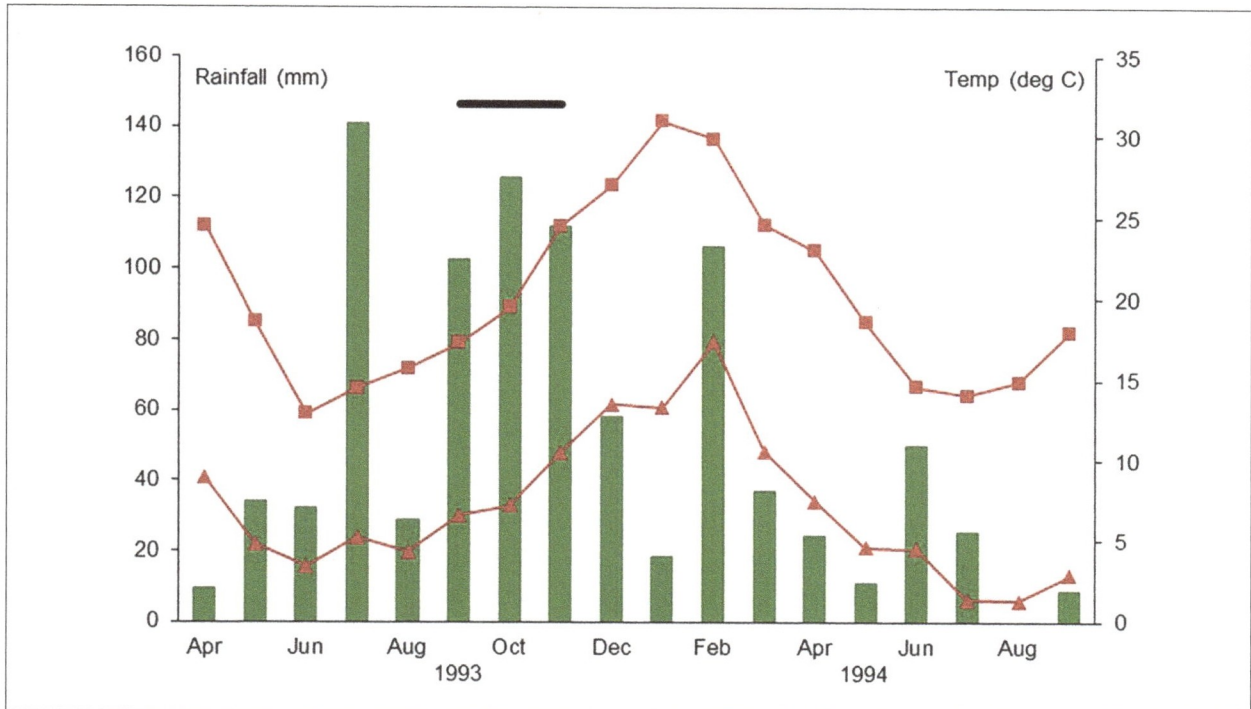

Figure 13.5: Rainfall and temperature data for Tarcutta, on the southwest slopes of NSW, in 1993 and 1994. The anticipated transmission period, predicted from monthly rainfall (green columns) and daily mean temperature data, is indicated with a black bar. Mean daily temperature (the average of daily maximum (red squares) and daily minimum (red triangles)) consistently exceeded 10°C from 24 August 1993.[108]

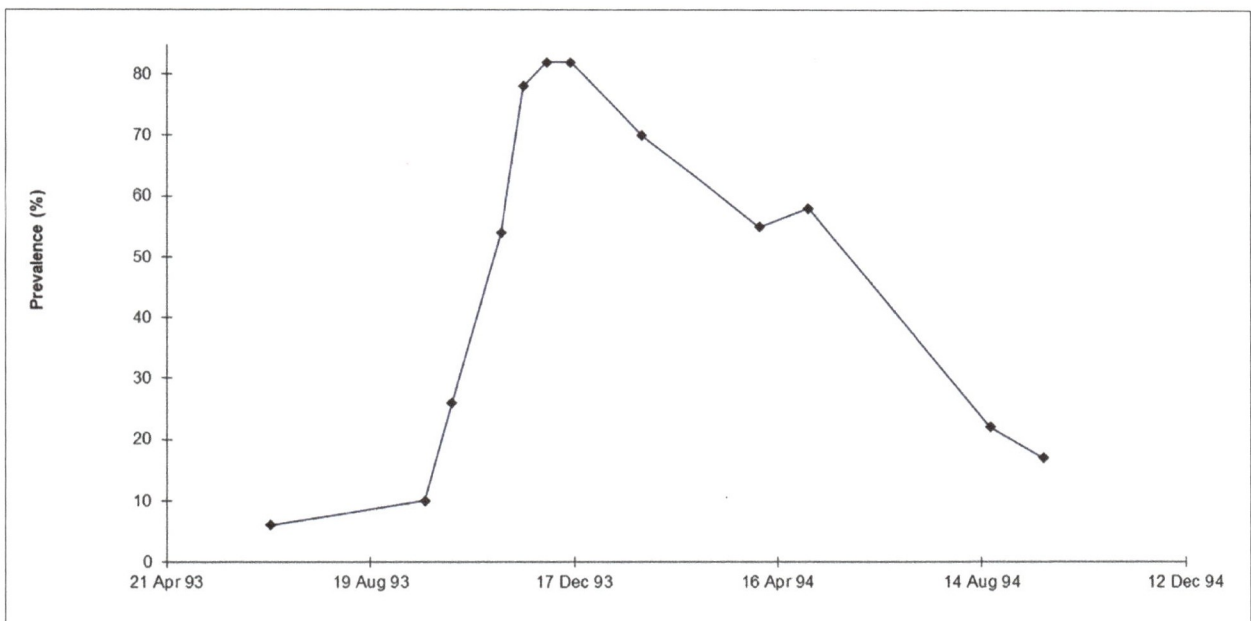

Figure 13.6: Prevalence of footrot in a 400 sheep flock at Tarcutta, NSW, subject to the climate shown in Figure 13.5. There was a low prevalence of footrot in the flock during winter (less than 5%) following an outbreak in the previous year. Transmission commenced in late August or early September and rapidly spread to most of the flock such that 80% had footrot by mid-December when hot, dry climatic conditions led to a decline in prevalence. The following season presented a drought and footrot did not spread.[108]

1962[112], which has received the most attention for its perceived ability to predict, *in vitro*, the *in vivo* virulence characteristics of strains of *D nodosus*.[113] The tests that have been developed to measure protease characteristics include the proteolytic index[114], the degrading proteinase test[115] and its derivatives, the elastase test[116], the zymogram test[117,118], and the protease ELISA.[119]

The degrading proteinase test was based on the tendency of the proteinases from virulent strains of *D nodosus* to remain stable during incubation in culture at 37 °C, while the enzymes produced by benign isolates were less likely to do so. It later became clear[120] that the differences in stability of the proteases from virulent and benign strains were detectable more quickly at higher temperatures and that the stability also varied with the concentration of calcium ions. Subsequently, the test incorporating these developments became known as the protease thermostability test.[121]

The protease thermostability test was further enhanced[122] by substituting gelatin for hide powder azure as the substrate used to detect protease activity. Gelatin, compared to hide powder azure, was cheaper, soluble and more easily standardised between batches. The gelatin-gel protease thermostability test is now frequently referred to as *the gelatin-gel test*, and the results are reported as stable (S) proteases, inferring virulence, or unstable (U) proteases, inferring benign characteristics.

The elastase test uses a solid culture medium containing elastin to compare the elastase activity of strains of *D nodosus* isolated from cases of virulent or benign footrot. Elastase-positive isolates produce a clearing of elastin particles in six to seven days of culture or slightly longer in some cases. Elastase-negative isolates produce no clearing within 21 or, in some cases, 28 days. There is generally close agreement between the elastase test result and the degrading proteinase test of the same isolates, and a strong agreement (but less than 100%) between the elastase test result and the reported clinical virulence of the outbreak from which each strain is isolated. At the time when these tests were developed, intermediate strains were not recognised.[123] Subsequently, it was shown that the elastase test could also distinguish at least some intermediate strains which showed rates of elastin clearing between those of virulent and benign strains.[111]

The zymogram test distinguishes between strains of *D nodosus* on the basis of the patterns produced by electrophoresis of their extracellular protease enzymes. Virulent strains produce some bands which benign strains do not produce, and vice versa.[124] The range of zymogram patterns was expanded to three for thermostable proteases (S1, S2 and S3) and six for unstable proteases (U1 to U6).[125]

The protease ELISA uses monoclonal antibodies against a virulent protease and a benign protease in an ELISA system on microtitre plates to demonstrate the presence or absence of the type of protease characteristic of virulent or benign strains.

A number of reports have compared the results from a range of laboratory tests of virulence. In general, there is good agreement between the tests, particularly in classification of benign or virulent strains. Agreement is less clear for isolates which are reportedly intermediate in virulence.[126] There are several problems associated with reliance on protease-based laboratory tests.[113] These include the following:

- There is no gold-standard determination of virulence. Definitions of virulence differ between states and clinical expression of virulence can be modified by environmental conditions.
- The isolates tested in a laboratory may not be from the dominant strain causing footrot in the field.

- There is an absence of controlled studies of *in vivo* virulence in sufficiently large groups of sheep in most reported evaluations of protease-based tests.

- There is a small but significant level of disagreement between the results of the various tests used.

- There is a small but significant level of disagreement between test results and reported field virulence.[127]

Genetic tests for virulence

A region of the chromosome of *D nodosus* which occurs at a high frequency in virulent strains but at a low frequency in benign strains has been identified and named the *vap* region (virulence-associated protein).[128,129] The *vap* regions are repeated in the genome of many of the strains that have been examined and, within each region, there are a number of *vap* genes.[130,131] The function of any of the products of the *vap* genes has not been determined.

The virulence-related locus or *vrl*[132] is a DNA sequence also present in some strains of *D nodosus*. The presence of the *vap* region and/or the *vrl* is a reasonable predictor of virulence, where virulence is determined by clinical evidence, elastase activity, protease thermostability, zymogram pattern or colony morphology.[133] Based on the presence or absence of these genomic regions, isolates of *D nodosus* can be placed into three major categories. Category 1 isolates contain both *vap* region and *vrl*; category 2 isolates contain only *vap* region; and category 3 isolates contain neither locus. In one study, 88% of isolates in category 1 were classified as virulent or high intermediate, 18% of category 2 isolates were classified as virulent but 70% were classified as intermediate. Of category 3 isolates, 83% were classified as benign or low intermediate. No isolates were detected with the *vrl* region only. Possibly, the presence of the *vap* region is essential either for the insertion or the maintenance of the *vrl* locus.

A technique using *D nodosus* genomic clones and dot-blot hybridisation[134,135] has been applied to panels of *D nodosus* isolates to differentiate strains of virulent, intermediate and benign virulence. The specificity of the virulent probe is less than 100% because 15% (3⁄20) of strains, determined to be benign by the elastase test, reacted with the virulent probe as well as the benign probe. The benign-specific probe reacted only with strains which had been characterised as benign or intermediate by elastase tests, demonstrating high specificity, but some intermediate strains did not hybridise with the benign-specific probe. The virulent probe is known to be derived from the *vap* region or *vrl* region.[133]

The identification of a series of chromosomal genetic elements including one named *intA* has led to the development of a further test to assist in predicting virulence.[136] Strains which behave as virulent in the field almost always include the *intA* element and, perhaps more importantly, strains which behave as benign in the field do not contain it. This latter group (field benign, *intA*-negative) can include isolates that are gelatin-gel stable.[127]

The *intA* test is not ratified as an approved diagnostic test by SCAHLS[e] and regulatory bodies in Australian states continue to use the gelatin-gel test, zymogram test or elastase test as the approved laboratory aids to diagnosis of virulence.[137]

---

e   SCAHLS: Subcommittee on Animal Health Laboratory Standards; see www.scahls.org.au.

Economic effects of footrot

Production losses associated with virulent footrot are due to

- reduced wool production
- reduced wool quality
- increased incidence of body strike
- reduced weaning percentages
- reduced value of sale sheep
- increased culling rates.

It is likely that footrot interferes with the productivity of affected sheep principally by reducing food intake. This happens in two ways. Severely affected animals spend increased amounts of time recumbent or grazing on their knees; thus their grazing efficiency is reduced. Even in pen trials in which feed was provided *ad lib*, footrot-affected sheep had lowered wool production and body weight than those without footrot, indicating a more direct effect of footrot on voluntary feed intake.[138]

There is no doubt that the presence of uncontrolled infections with virulent footrot reduces the wool production of sheep.[139] The estimate of 8% loss of annual wool weight by Marshall et al. (1991)[140] is the best available estimate of the effect of virulent footrot on average annual wool production in an infected flock, but there are two reasons why this is probably an underestimate of the losses which occur in field outbreaks. First, the comparison in their experiment was made between treated and untreated sheep. Footrot did occur in the treated sheep and so one can presume that these sheep would have produced more wool had they been completely free of footrot. Second, welfare concerns led to treatment of the infected sheep on two occasions during the experiment, and the withdrawal of one animal which was severely affected. It can therefore be assumed that the 'untreated' sheep would have produced even less had no treatment whatsoever been administered.

Marshall et al. (1991)[140] also found a significant relationship between the duration of severe lesions (score 3 or 4) and number of infected feet and fibre diameter, staple length and tensile strength. While the effect of footrot on fibre diameter may have a positive effect on wool price, lower fleece weights and lower staple strength may have a strong negative effect on fleece value.

Footrot-affected sheep are more likely to become flystruck. Their infected feet are very attractive to blowflies (such as *Lucilia cuprina*). The necrotic material from the feet, often accompanied by maggots, can be deposited in the fleece when the animals are recumbent and their feet press against their chest and belly. The material deposited on the body can then initiate a body strike.

Virulent footrot also reduces the rate of body weight gain, increases the loss of body weight or leads to the maintenance of lower body weights in affected sheep compared to uninfected sheep.[141,142] Estimates include a loss of 6.7% of body weight during an eight-week period of infection, maintenance of body weight at least 6 kg lower than that of uninfected adult Merino wethers during a 16-week period of infection, and a difference of 3.5% to 7% between infected and uninfected Downs breed-sired weaners. Marshall et al. (1991)[140] related the duration and severity of footrot lesions to the change in body weight and found that the more infected the sheep's feet were and the longer the period of active infection, the greater was the effect on body weight. They concluded that, for each foot continuously affected with footrot (score ≥3) for two years, the body weight of Merino wethers which weighed 54 kg before infection would fall by 12.3 kg. In fact, most affected sheep have periods of remission from infection during the year when conditions are dry or unsuitable for footrot.

It is clear from these experimental studies that the greatest changes in body weight occur during periods when footrot is spreading within the flock and when lesions are actively developing. At times when lesions are regressing with or without treatment, previously affected sheep may regain some lost weight. It is possible for sheep to regain much of the lost body weight rapidly after curative treatment, displaying some compensatory gain. Lowered wool production over the period of active infection will not be fully recovered following treatment, but, presumably, the rate of wool growth will return to normal about the same time that lesions heal. Thus the effect of virulent footrot on annual wool production will depend on the duration of the infection as well as the severity of lesions.

The effect of footrot on reproductive rates is a result of the effect of footrot on body weight and food intake. Lower body weights in ewes will lead to lower ovulation rates during joining and lowered lambing percentages, increased risk of pregnancy toxaemia, neonatal mortality and reduced weaning rates and weaning weights of lambs (discussed in Chapter 7).

The effect of footrot on the sale value of cull and cast-for-age sheep results from the restricted markets available to footrot-declared producers and the low body weights of affected sale sheep. Both lower reproductive rates and higher culling rates due to chronic footrot lesions reduce the amount of culling for productive traits that producers with affected flocks can practise.

Estimated losses associated with the presence of virulent footrot on a property where no control measures are practised are between $7 and $14 per sheep per annum (1990 dollars), depending on the suitability of the environment for footrot transmission and development[143] and the impact of footrot on the market options for cull and cast-for-age sheep. Where control measures are implemented, the costs are reduced to between $5.50 and $9.50.[144]

The severe and chronic lameness that can occur in uncontrolled outbreaks of virulent footrot also leads to concerns on animal welfare grounds. There are times when such concerns justify treatment or slaughter regardless of any economic justification. Intangible social effects also arise from damage to the reputation of the farm business and the effect that an outbreak of footrot can have on a producer's relationship with his or her neighbours.

Infection with less virulent strains of *D nodosus* causes less severe effects on wool weight and body weight than virulent strains. In a field study, Glynn (1993)[107] found that infection with a strain classified as intermediate in virulence led to a difference of 5% in greasy fleece weight between untreated sheep and sheep treated by footbathing to reduce the severity and prevalence of footrot lesions. There are few published estimates of the effect of benign footrot on body weight. Glynn (1993)[107] found that uncontrolled BFR decreased body weight of sheep in some parts of the year compared to sheep treated to reduce the effect of footrot, but that there were no significant differences at the end of the footrot spread season. Uncontrolled IFR did lead to significantly lower body weight than treated sheep, although the effect of the footrot was presumably exacerbated by grass seeds' penetration of the interdigital skin. One report estimated the cost of uncontrolled intermediate footrot in a high-risk environment to be $4 per sheep per annum and $0.20 for benign footrot, compared with $14 for virulent footrot.

Treatment

Topical treatments

Topical treatments are used to achieve both cure of affected individual sheep and the restriction of transmission within a flock, or 'control'. Curative treatment is generally attempted during a non-spread period, usually when pastures are dry, while control measures are implemented during times when transmission is expected. The solutions most commonly used are formalin (5%) and zinc sulphate (10%). Other products made specifically for footrot in sheep include a 20% zinc sulphate/sodium lauryl sulphate solution[145,146] and a 10% solution of CHF-1020 (copper nitrate trihydrate/copper chloride dihydrate in water) (Radicate® — Colbert Holdings).[147]

Traditionally, curative treatment has involved paring all affected feet and applying the antibacterial chemical in a footbath. The paring necessary to achieve good cure rates is labour-intensive and arduous and excessive paring causes additional pain and discomfort to affected sheep.

To control spread with or without prior paring, sheep are walked through a footbath six metres long containing either 5% formalin or 10% zinc sulphate at weekly intervals, or held in footbaths of zinc sulphate/sodium lauryl sulphate solution for one hour at three-weekly intervals, during the times of the year when transmission is expected to occur. These treatments will reduce the incidence of new infections. With the possible exception of zinc sulphate/sodium lauryl sulphate solutions, under-run lesions do not respond well to topical treatments without prior paring to expose infections in under-run horn, so bathing must be sufficiently frequent to intercept lesion development.

The following factors affect the success of topical treatments:

• Good cure rates of under-run lesions require paring before application of topical treatment.

• Formalin is not very effective unless sheep are placed in a dry environment after treatment.

• Zinc sulphate/sodium lauryl sulphate solution requires footbathing for one hour to achieve better results than zinc sulphate alone.

Injected antibiotics

One parenteral treatment with penicillin is effective in curing a high proportion of sheep affected with footrot, but the dose required is higher than that recommended by manufacturers for use in other disease conditions. Procaine penicillin at the rate of 70 000 units/kg has been used successfully[148,149] while rates of 50 000 units/kg or 300 000 units per sheep[150] have failed.

Streptomycin is ineffective alone but, in combination with procaine penicillin, has been shown in a number of studies to be highly effective and more effective than penicillin alone. It has been widely used in the field but streptomycin is no longer available for this purpose in Australia.

Long-acting oxytetracycline is effective at 20 to 24 mg/kg.[151] Intramuscular erythromycin has been found to be highly effective at 12[152] and 20 mg/kg. Combinations of lincomycin-spectinomycin (at 5 and 10 mg/kg of each antibiotic respectively, given intramuscularly) are effective[153] but the drugs are not registered for sheep in all states of Australia. No significant improvement in efficacy is achieved by repeating treatment with lincomycin-spectinomycin on the two following days, nor by using a three times higher dose.

Parenteral antibiotic therapy has advantages over topical applications for curative treatment. Extensive foot paring is not required, so treatment is faster. None of the antibiotics, however, is highly effective unless sheep are in a dry environment for at least 24 hours post-injection, although improvements in cure rates have been reported in UK in sheep returned immediately to pasture when parenteral long-acting oxytetracycline at 20 mg/kg was administered in addition to topical oxytetracycline spray, compared to the spray alone (91% *cf* 58%).[154]

Dry environments post-treatment may be provided by pasture in hot dry seasons or by placing sheep on battens (in a woolshed, for example).

In large flocks, antibiotic treatment is rarely appropriate for control of the disease during a spread period because it is expensive and gives no significant protection against reinfection. It is much more widely used in Australia as a treatment during a non-transmission period. Cure rates exceeding 90% can be expected and will frequently be as high as 95%. Care must be exercised when treating animals which could potentially be soon culled for slaughter: withholding periods following antibiotic treatment must be observed.

The antibiotic gamithromycin has been used in Denmark to eliminate footrot from sheep flocks, with a reportedly very high success rate. Ewes were treated with 600 mg (4 mL) administered subcutaneously. Of 48 flocks treated (the average size of the flocks being 200 adult sheep), 44 remained free of footrot one year later. Five flocks monitored for a longer period remained free after two years.[155] The antibiotic is not registered for sheep in Australia.

Vaccination

Natural infection with footrot produces a small antibody response in sheep which is not, however, protective and sheep can be repeatedly reinfected with homologous strains of *D nodosus*.[106.]By contrast, subcutaneous injection of adjuvanted piliated whole cells of *D nodosus* will stimulate a strong protective response against homologous infection.[156,157] The protective, agglutinating immunogen in vaccines is associated with pili, and vaccines of pure pili derived from *D nodosus* or from recombinant *Pseudomonas aeruginosa* will produce protection equal to or better than whole cell vaccines.[158,159,160,161]

There is very limited cross-protection between serogroups of *D nodosus*. In fact, within serogroups, particularly serogroup B, cross-protection between serotypes is also limited. Vaccination is, therefore, only effective against strains of *D nodosus* of the same serogroup (homologous challenge) that are included in the vaccine.

As noted earlier, mixed-serogroup infections in natural footrot outbreaks are normal and it is rarely practical to attempt to identify which serogroups are present in a sheep flock experiencing an outbreak of footrot. Nor is it practical for manufacturers to market vaccines which offer protection only against specific serogroups — it would make vaccines too expensive and, should the wrong vaccine be used, perceived 'failure' to protect sheep from footrot would bring the vaccine into disrepute.

Consequently, vaccine manufacturers sell multivalent vaccine — for example, Footvax®
(Coopers Animal Health, available in some states only), which contains ten serotypes of
D nodosus. Unfortunately, multivalent vaccines are much less effective than single-strain
vaccines, due to the phenomenon of *antigenic competition*.

Antigenic competition refers to the apparent competition for immune responses which occurs
when multicomponent vaccines are compared to any one component of the vaccine administered
alone. Antigenic competition significantly reduces the response by sheep to vaccination with
multistrain *D nodosus* vaccines, compared to the response to each component.[162,163,164]

Following administration of one dose of the commercial multistrain vaccine, the titre of
agglutinating antibody achieved is low and not sustained, so two doses of vaccine are necessary
to achieve useful protection. For intervals up to one year, the longer the time period between
the primary and secondary vaccination, the higher the titre achieved, but the faster the titre
declines after the second vaccination.[165] Whether the interval between vaccinations is six weeks
or longer (up to one year), the titre 12 weeks after the second vaccination is similar.

Vaccines containing antigens to protect against eight to ten serogroups provide only a
short duration of protection. In one report, two doses, nine weeks apart, of a multivalent
(eight-serogroup) vaccine protected Merino ewes for at least ten weeks during an outbreak of
footrot.[166] A number of other reports suggest that multivalent vaccination can be expected to
protect Merino sheep for up to 12 weeks after the second dose, although for some animals
protection may be inadequate by eight weeks.[167,168]

The use of multivalent vaccine in affected flocks has two important effects:

- Within a few weeks of the second vaccination, *most* unaffected sheep will be protected
  against footrot.

- Sheep already affected will heal more quickly and the severity of disease in those remaining
  affected will, mostly, be reduced.

There are some disadvantages of vaccination:

- At least six weeks must elapse between the first and second dose and, therefore, before
  sheep are effectively protected.

- Effective immunity after multivalent vaccination lasts is short-lived — at most 16 weeks
  and probably no more than 10 weeks in Merinos, or 12 weeks in crossbred sheep.

- Vaccine is relatively expensive.

- There is often a reaction at the injection site.

Despite these disadvantages, there are many conditions under which use of the commercial
multivalent vaccine offers the simplest and cheapest effective control measure. It will not
eradicate the disease, but it will reduce prevalence by up to 80%.

In flocks where the infecting strains of *D nodosus* are restricted in serogroup, the use of a
specific autogenous vaccine may be dramatically more effective. An occurrence of footrot in
Bhutan, following the introduction of infected sheep from Australia, was successfully eliminated
by the use of a monovalent vaccine prepared from the serogroup B isolates collected from the
Bhutan flock.[169] Subsequent trials in Australia with bivalent vaccines or monovalent vaccines have
demonstrated that this approach can be very successful. In flocks where multiple serogroups of
D nodosus exist, sequential use of different monovalent vaccines, at intervals of three to six months,

produced effective and persistent protection leading to elimination of infection from most flocks (four out of five) in which there were only three or fewer serogroups of *D nodosus* present.[170]

This strategy, however, currently remains a research tool rather than a readily available option for commercial producers.

Control of footrot in infected flocks

Control refers to strategies aimed at preventing or restricting the rate of transmission of footrot between sheep in an infected flock. Control measures, therefore, are normally applied immediately before or during a transmission period. In many Australian environments transmission occurs principally in spring and early summer, although in some districts and in some seasons, conditions are suitable for transmission at other times of the year. For the sake of this discussion, however, it will be assumed that transmission occurs in spring.

Broadly, two strategies are used for control, and often they are used together. The first is footbathing, normally in 10% zinc sulphate solution or 5% formalin solution. Formalin is generally slightly cheaper; zinc sulphate is more pleasant to use and safer for sheep, dogs (which might drink from the footbath) and operators. Alternative footbathing solutions have been mentioned above. The second strategy is vaccination.

When used to achieve control, footbathing does not require paring of the feet. Paring is not justified because control measures are aimed at limiting the establishment of infection — a process which occurs in the IDS and which therefore does not require reduction in the amount of horn tissue in order to allow contact between the footbath solution and the site of infection. Further, because of the extra time required to pare feet, paring will delay the footbathing by days or weeks.

Footbathing does not provide sustained protection against reinfection. If done effectively, footbathing will cure or ameliorate most early cases of footrot involving the IDS and superficial under-running. Consequently, the bathing must be repeated frequently. It is not feasible, during a transmission period, to attempt to separate infected sheep from uninfected sheep because many cases will be inapparent or mild. There is perhaps a small advantage in attempting to provide a 'clean' pasture to receive sheep after bathing, but successful control is not dependent on this. Whether clean pastures are used or not, some of the infected sheep will not respond fully to treatment and will be a source of infection allowing transmission to resume within a few days.

Given these facts, the recommended procedure for footbathing to reduce transmission rates is to walk the sheep through a footbath every five days while transmission is occurring — perhaps for three to six weeks, depending on the season. Bathing should commence as soon as transmission is likely to occur — perhaps in late August, September or early October depending on climatic factors.

The footbath should be at least 6 m long and the solution should be at least 50 mm deep. It is often preferable to have portable baths and to use them in portable yards, taking the facility to the sheep rather than droving sheep long distances to central permanent baths. The best approach will depend on each farm's physical infrastructure.

Zinc sulphate/sodium lauryl sulphate solution, if used as recommended, may offer some advantages in cases where frequent footbathing is difficult. If sheep are held in a zinc sulphate/

sodium lauryl sulphate solution for one hour, superior penetration of the horn of the hoof will occur, providing better cure rates and longer protection periods. Treatment frequencies may be reduced to once every two to three weeks.[146]

The most obvious limitation for footbathing is for treatment of lambing and lactating ewes and their lambs because disturbance could cause mismothering. Under these circumstances, vaccination may be the preferred option for control because it can be timed to occur before lambing or when ewes are yarded for lamb marking.

In sheep which have not previously been vaccinated against footrot, the first vaccination must be given at least seven weeks before the anticipated commencement of transmission. The second vaccination can then be given one week before transmission is expected to commence, and satisfactory protection can be expected for at least ten weeks from a time about one week after the second vaccination.

Should the first vaccination be given too late, footbathing can be used to control transmission in the period preceding protection from the second vaccination.

Sheep which have been vaccinated in the previous year will need only a booster, given once, about a week before transmission is anticipated.

Should transmission continue for longer than the protective period (say, 10 weeks), a further vaccination may be necessary to prevent an increase in footrot prevalence late in the season.

Vaccination has several obvious management advantages over footbathing and, in addition, its ability to cure chronically affected sheep without paring may be a distinct advantage over footbathing in cases where there are still significant numbers of sheep affected in this way — presumably following the previous season's uncontrolled outbreak. The most obvious disadvantage of vaccination is its price, but, for some farms, the cost of vaccination may be less than the cost of repeated footbathing.

In the UK, some workers recommend that antibiotic treatment of individual cases (long-acting oxytetracycline) can be used to reduce both the prevalence of disease (by direct treatment) and the incidence of new cases (by reducing the challenge) and, therefore, has a role in control. The recommendations are based on

(1)   surveys of farmers which indicated that those who promptly treated affected sheep with antibiotics and/or isolated them from the flock also reported generally lower levels of infection in their flocks[55]

(2)   a trial in one flock which related prompt treatment of affected sheep to short-term reductions in incidence of new cases.[171]

This approach may be feasible in small flocks where affected animals can be readily identified and caught, and when the incidence in a particular outbreak is not high. In outbreaks with a high incidence, like that described in Figure 13.3, it would be very difficult to treat and/or remove new cases as they occurred because of the rapidity of spread in the spring. Such an approach would be particularly difficult under the extensive sheep management systems common in Australia.

Eradication of footrot

The process of *control* can be an end in itself but is often the first step leading to the process of eradication of the disease from the flock and the farm. *Eradication* is based on the principle that elimination of cases will eliminate the virulent *D nodosus*, because the organism persists only in animal hosts and not in the environment.

Methods of eradication

Option 1

Non-selective disposal and replacement of the flock means that every sheep on the farm or every sheep at risk of footrot is sold, and 'clean' sheep purchased. The probability of success depends on a complete muster, the availability of footrot-free replacement sheep and the ability, through secure fencing, to prevent reinfection of the replacement sheep. A period of at least one week must elapse between the departure of the last infected sheep and the arrival of the replacement flock.

This method of eradication is often the cheapest and most reliable, particularly when the owner is confident that infection is restricted to a small part of the flock, such as a recently purchased and isolated 'mob'. The *changeover* price is a critical cost in deciding the value of this approach. Sheep at risk of footrot must often be sold relatively cheaply while 'clean' replacements may come with a price premium.

Option 2

By contrast, selective disposal requires the identification and sale or slaughter of only the infected animals from within a flock and the retention of sheep which, despite exposure to footrot, are clinically free of the disease. Clearly, this method requires the inspection of every foot of every sheep in the flock. Eradication based on selective disposal will be the preferred option in cases where the prevalence of footrot within the flock is low but when all or most of the flock is considered *at risk*.

The final steps in an eradication programme based on inspection and selective disposal — the removal of the last infected sheep — are most likely to be successful during a non-transmission period. Both the financial outcome and the probability of eradication are enhanced if control measures applied during the previous transmission period have been effective. In that case, fewer sheep will be sold, reducing the changeover cost, and the odds of misdiagnosing an infected sheep as uninfected are lower.

Option 3

A third method of eradication of footrot involves chemotherapy, which can be selective or can involve the whole flock. Selective treatment, like selective disposal, suffers from the risk of errors in identification of infected sheep (the *sensitivity* of the inspection).

Treatment can be highly effective in dry, summer conditions (>90%). There is no necessity for more than light paring or for separation of 'clean' and 'treated' sheep under those conditions. All sheep must be reinspected three weeks after treatment. Non-responders should be culled.

General procedure for eradication

Experience has shown that eradication based on inspection and selective disposal, with or without chemotherapy, is more likely to succeed if

- at least two, and preferably three, inspections of all sheep in the exposed flock are made during the non-transmission period, at intervals of three to four weeks
- the climate during the non-transmission period is hot and dry (usually summer), and the feet of the sheep are dry
- every infected sheep, or suspect sheep, is culled from the flock
- facilities for inspection are good and do not make the task difficult or excessively tiring for the operators — hence machines which invert the sheep, good lighting, a good environment for operators (usually requiring provision of shade and air movement) and pneumatic parers are all encouraged.

If either option 2 or 3 is adopted, infected animals are either culled or treated at the first inspection. At the second inspection, any affected animals are culled. This will include those sheep which fail to respond to treatment if option 3 is taken. At the third inspection, affected animals are culled, although, hopefully, at this inspection, there will be no clinical evidence of footrot in most or all of the individual mobs.

Surveillance through the months following the inspection and disposal procedures is highly recommended. Surveillance generally involves careful inspection of each mob at rest at pasture and then the individual examination of any lame sheep. Surveillance potentially allows the removal of any sheep which *breaks down* with clinical footrot from an undetected focus of chronic infection in the foot before transmission occurs. Alternatively, surveillance allows the isolation of any mob which breaks down and the prevention of transmission from that mob to other, uninfected mobs on the farm. If such breakdown occurs, control measures should be applied to the infected mob promptly, pending a decision about the best eradication option in the next non-transmission period.

For many sheep farmers, eradication is not successful in the first year and it is important that the reasons for failure are determined before proceeding with another attempt to eradicate. In many cases failure occurs because the summer inspection and selective disposal activities are undertaken too soon after footrot is diagnosed. Consequently, there may have been insufficient development of farm infrastructure to cope with the new activities (laneways, sheepyards, handlers, fence integrity) or inadequate control during the transmission period with, subsequently, a high prevalence of infection at the first inspection. Inevitably, these factors are better managed in the second year, so the 'failure' in the first year should be viewed as a preparatory step to final eradication rather than as a wasted effort.

Cost of eradication

In any eradication programme it is appropriate to present a realistic or even a maximum-possible list of all costs likely to be encountered. As a minimum list of costs, consider vaccine, antibiotic, mustering and handling time, veterinary consultative involvement, depreciation of equipment and low sale cost of cull sheep. It is unlikely that this cost will be less than $20 per head for a programme running in just one year.

Prevention of footrot

Managers of footrot-free flocks must take steps to prevent the introduction of footrot. Sound boundary fences are essential. Purchased stock can be inspected, preferably before delivery, and should be isolated from resident sheep until they have been through a transmission period without developing footrot.

It is important to consider the roles of goats, particularly, but also cattle, in introducing footrot into sheep flocks.

In flocks believed to be free of virulent or intermediate footrot, footbathing or vaccination should not be used as a way of providing 'extra biosecurity' against the risk of disease introduction. These strategies are not 100% effective at eliminating footrot and will not prevent the introduction of footrot from infected stray sheep, but they may mask signs in the first few cases and thus delay recognition of the disease, by which time it may have spread to many sheep on the farm. The cost of these activities is better directed at other steps to improve biosecurity, such as maintaining or improving secure boundary fencing.

For most commercial producers, it is necessary to buy some sheep into the flock, even if these are only rams, and it is often not feasible to hold the animals in isolation until a footrot transmission period occurs. To reduce the risk of introducing footrot with purchased sheep, it is important to ascertain the footrot status of the vendor's flock. Strategies such as footbathing purchased sheep will *not* prevent the introduction of the disease. Introduced rams should be held in isolated paddocks for as long as possible before sharing pastures with the new flock and they should be tipped up and examined individually before being allowed to mix with the resident flock. If such a strategy can be employed, then it is advisable to avoid footbathing on arrival (*off the truck*) and to isolate the new sheep in a paddock which is favourable for the development of footrot infections (a paddock with moist pasture is ideal) to provide the greatest likelihood that footrot will become detectable, if present.

Regional plans to reduce footrot prevalence

In 1988, NSW Agriculture, The University of Sydney, private veterinarians and representative industry groups developed a *Footrot Strategic Plan*[f] with the objective of reducing the prevalence of footrot-infected flocks to less than 1% in all 42 Rural Land Protection Board (RLPB[g]) areas of the state by December 2000. Once achieved, all of NSW could be deemed a Protected Area for footrot.

For the purposes of eradication in NSW, footrot was defined as existing in two forms only: virulent and non-virulent. By legislation (the *Stock Diseases Act 1923*) owners were required to notify the presence of virulent footrot in their sheep in protected and control areas. Private veterinarians, therefore, advised their clients accordingly.

At the time, in regions of the state with average annual rainfall exceeding 600 mm (the Albury, Holbrook and Gundagai regions), about half of all sheep flocks had virulent footrot and the prevalence in other, medium rainfall areas of the state was over 25%.[172] By 2008, over 6000 flocks had become free of VFR and fewer than 30 flocks were known to be still infected in the state.[173] The target set for the statewide prevalence was ultimately achieved.

Methods used to eradicate the disease from flocks included all three options described in the previous pages. The option of destocking and flock replacement was the quickest way

f   The NSW footrot strategic plan is published by NSW Agriculture, Orange, NSW.

g   These are now reformed into the 11 *Local Land Services* regions of New South Wales.

Table 13.4: Definition of disease control zoning based on prevalence of affected flock.

Status	Proportion of flocks affected with footrot
Residual area	More than 10%
Control area	1% to 10%
Protected area	Less than 1%

to eliminate the disease from the farm, but it was not the preferred method in larger flocks, probably because of the high cost and because, on farms with large flocks, income from sheep often formed a larger proportion of farm income than on farms with small flocks. There is some evidence that eradication took longer, on average, in flocks in which sheep were treated with footbaths rather than antibiotics when detected at the first inspection, but it took less time when infected sheep were culled (not treated) at the first inspection.[173]

Several additional aids to regional footrot control have been introduced. These include *vendor declaration forms* — standard forms which enable owners of flocks free of VFR to make a formal declaration to that effect when selling sheep. Virulent footrot is a notifiable disease in all states of Australia. It is illegal to travel flocks with VFR on public roads or to offer them for sale in public yards.

Emerging difficulties in categorisation of footrot outbreaks

The division of footrot into two forms only (benign and virulent) has been the approach taken in statewide programmes to control the disease. This categorisation ignores the continuity of the virulence scale which exists in the field. Increasingly, as the prevalence of VFR declines, IFR has become a more common finding and it seems likely that authorities will face greater resistance from producers to eliminate this milder, but non-benign, disease from their flocks.

Laboratory tests generally do not assist the separation of intermediate forms of footrot from virulent footrot. For example, the classification of isolates of *D nodosus* using the gelatin-gel test into stable (S) and unstable (U) tends to include many intermediate strains as VFR (caused by S strains), despite field evidence that the footrot is mild and causing serious lesions only in a small proportion of the flock.

It is true that sheep with score 3 and score 4 lesions warrant treatment, but it is doubtful if eradication of IFR from the entire flock can always be justified on economic grounds. Further, there is doubt whether some forms of IFR can be eradicated using programmes of inspection and selective disposal[174], although eradication of some forms has been achieved.[74,108] It may be necessary, therefore, for Australian regulatory authorities to adopt a more flexible approach to the categorisation of footrot so that only the more severe forms of IFR are included as targets for compulsory elimination. Figure 13.7 highlights the dilemma facing flock owners with IFR; the severity of the disease falls in a zone where laboratory tests may overestimate the virulence of the infecting strain, where there is uncertainty about the possibility of eradication of the disease without total destocking, and where there is uncertainty about the net economic benefit of eradicating the disease even if it is possible.

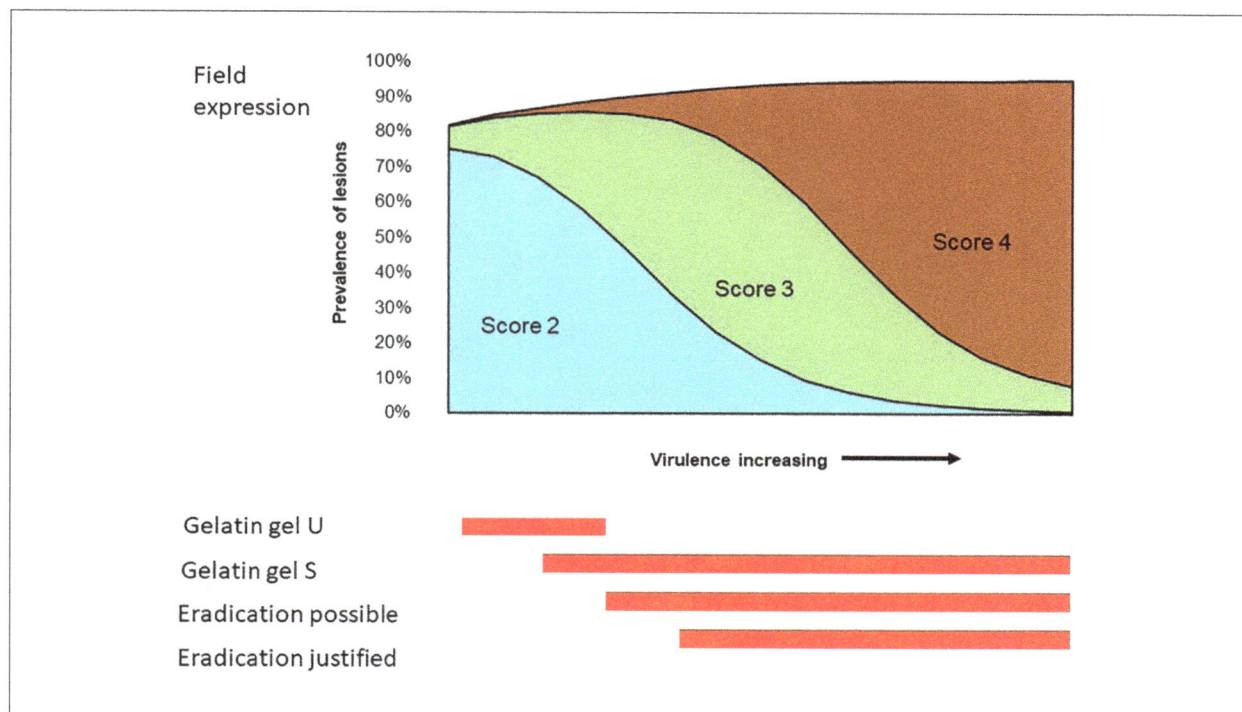

Figure 13.7: A schematic representation of the general relationship between the virulence of a form of footrot (top image) and four other characteristics of a footrot outbreak (red bars). Those characteristics are the laboratory assessment of the strains of *D nodosus* isolated from the outbreak, the likelihood that eradication can be achieved with inspection and selective disposal methods, and the economic justification of eradication of the disease. The figure can be interpreted by moving a vertical line from left to right, intersecting the severity of an outbreak with the four characteristics illustrated with red bars. In general, with increasing virulence of the infecting strain, it becomes increasingly likely that it will be gelatin-gel-stable and that it is both possible and economically wise to eliminate the infection from a flock. Source: KA Abbott.

Contagious ovine digital dermatitis (CODD)

CODD is a severe infectious disease of the hoof of sheep which was first reported in the UK in 1997.[175] It has not been reported from any country outside Britain. The condition was initially considered to be a severe form of ovine footrot and was, for a short time, given the name *supervirulent ovine footrot*. The disease differs from footrot in a number of respects. Unlike footrot, the condition commences at the coronary band, and the subsequent separation of the hoof wall, when it occurs, appears to progress distally from that area. In some cases, the entire shell of the hoof is lost (Figures 13.8 and 13.9). Also unlike footrot, the lesions are notably bloody and do not include the malodorous necrotic tissue which is characteristic of footrot.

The essential infectious agent is believed to be one or more species of *Treponema*, which underlines the similarity between the ovine disease and bovine digital dermatitis (BDD). It is not clear whether the spirochaetes cause the condition alone or in association with other agents, including *D nodosus*, or whether the species causing disease in cattle are the same as those associated with CODD, although there is now evidence showing very close relationships between isolates from sheep and cattle.[176]

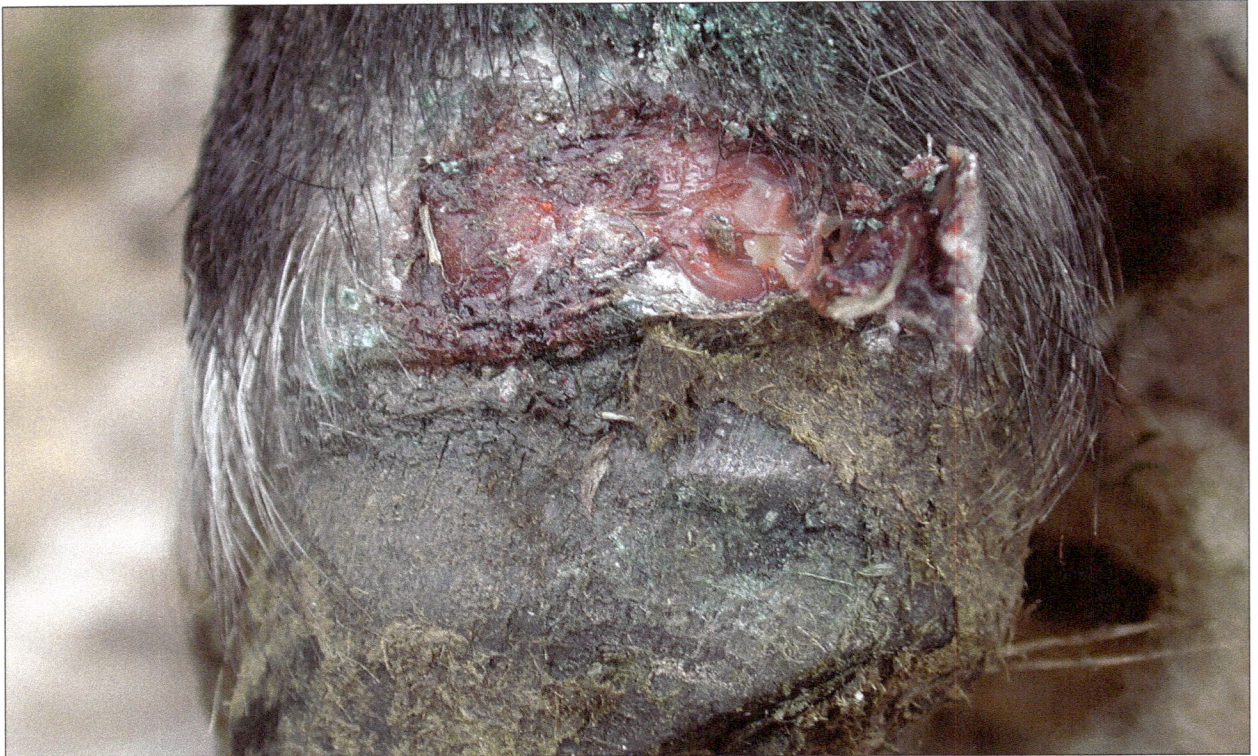

Figure 13.8: Contagious ovine digital dermatitis (CODD). The lesions generally commence at the coronary band and run distally under the hoof wall. This condition has not been reported from sheep in Australia. Source: KA Abbott.

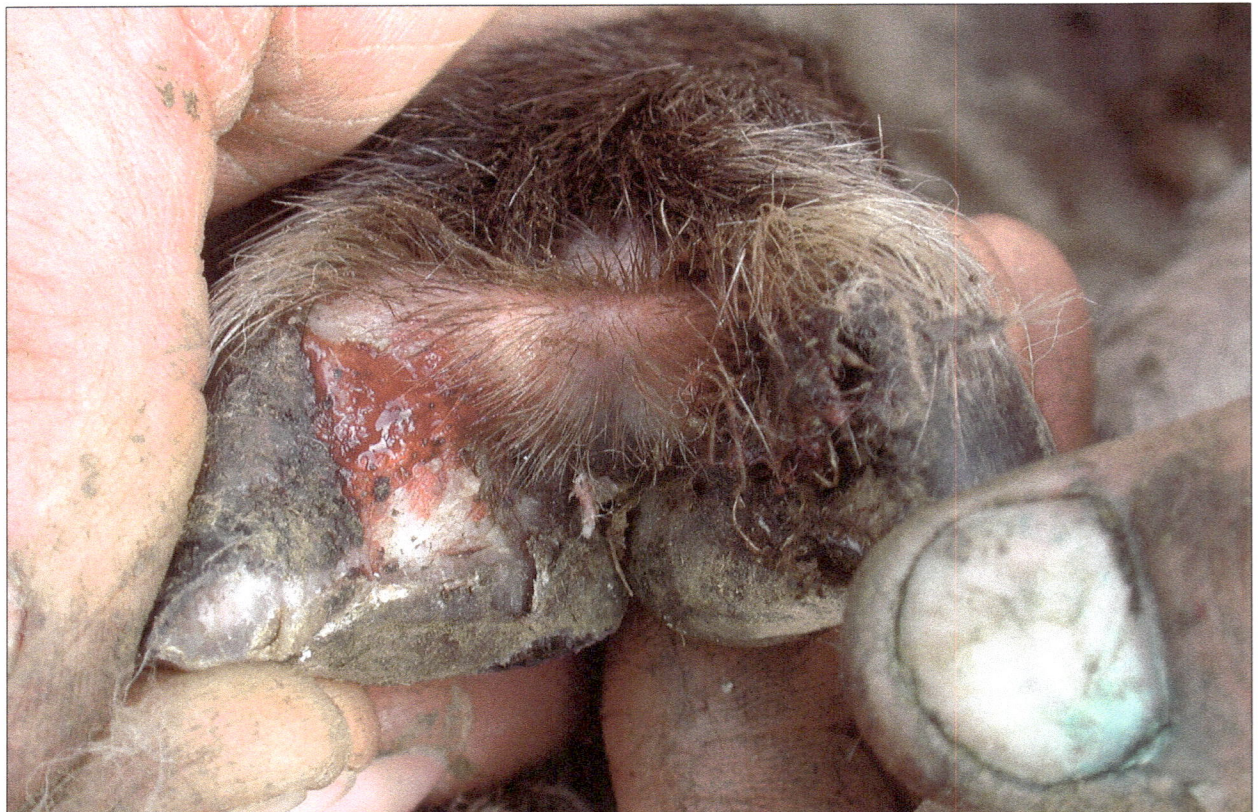

Figure 13.9: A CODD lesion showing partial loss of axial hoof wall. Source: KA Abbott.

RECOMMENDED READING

Toe abscess

Winter A and Arsenos G (2009) Diagnosis of white line lesions in sheep In Practice **31** 17-21.

Foot abscess

West DM, Bruere AN and Ridler AL (2009) Foot Diseases and Lameness. In: The Sheep: Health, Disease and Production 3rd ed. VetLearn Foundation, Wellington, pp. 262-82.

Footrot

Abbott KA and Lewis CJ (2005) Current approaches to the management of ovine footrot. The Veterinary Journal **169** 28-41.

Dhungyel O, Hunter J and Whittington R (2014) Footrot vaccines and vaccination. Vaccine **32** 3139-46.

Raadsma HW and Egerton JR (2013) A review of footrot in sheep: Aetiology, risk factors and control methods. Livestock Science **156** 106-14.

Lameness generally

Winter A (2004) Lameness in Sheep. The Crowood Press Ltd, Wiltshire.

REFERENCES

1   Sykes AR (2007) Deficiency of mineral macro-elements. In: Diseases of sheep, ed ID Aitken. 4th ed. Blackwell Publishing: London, pp. 363-70. https://doi.org/10.1002/9780470753316.ch53.

2   Sykes AR, Coop RL and Angus KW (1975) Experimental production of osteoporosis in growing lambs by continuous dosing with *Trichostrongylus colubriformis* larvae. J Comp Pathol **85** 549-59.

3   Sykes AR, Coop RL and Angus KW (1979) Chronic infections with *Trichostrongylus vitrinus* in sheep. Some effects on food utilization, skeletal growth and certain serum constituents. Res Vet Sci **26** 372-7.

4   Sykes AR, Coop RL and Angus KW (1977) The influence of chronic *Ostertagia circumcincta* infection on the skeleton of growing sheep. J Comp Pathol **87** 521-9.

5   Thamsborg SM and Hauge EM (2001) Osteopenia and reduced serum alkaline phosphatase activity in grazing lambs naturally infected with gastrointestinal nematodes. J Comp Pathol **125** 192-203.

6   Jubb KVF, Kennedy PC and Palmer N, eds (1993) Pathology of domestic animals. 4th ed. Vol 1. Academic Press: New York & London, pp. 65-7.

7   McDowell LR (2000) Vitamin D. In: Vitamins in animal and human nutrition. 2nd ed. pp. Iowa State University Press: US, pp. 96-113. https://doi.org/10.1002/9780470376911.

8   Smith BSW and Wright H (1984) Relative contributions of diet and sunshine to the overall vitamin D status of the grazing ewe. Vet Rec **115** 537-8. https://doi.org/10.1136/vr.115.21.537.

9   Mearns R, Scholes SFE, Wessels M et al. (2008) Rickets in sheep flocks in northern England. Vet Rec **162** 98-9. https://doi.org/10.1136/vr.162.3.98.

10 Ross R, Care AD, Robinson JS et al. (1980) Perinatal 1,25 dihydroxy-cholecalciferol in the sheep and its role in the maintenance of the transplacental calcium gradient. J Endocrinol **87** 17P-18P.

11 Hoenderop JG, Nilius B and Bindels RJ (2005) Calcium absorption across epithelia. Physiol Rev **85** 372-422. https://doi.org/10.1152/physrev.00003.2004.

12 Braithwaite GD (1975) Studies on the absorption and retention of calcium and phosphorus by young and mature Ca-deficient sheep. Br J Nutr **34** 311-24. https://doi.org/10.1017/S0007114575000359.

13 Braithwaite GD (1978) The effect of 1-α-hydoxycholecalciferol on calcium and phosphorus metabolism in the lactating ewe. Br J Nutr **40** 387-92. https://doi.org/10.1079/BJN19780135.

14 Wilkens MR, Mrochen N, Breves G et al. (2011) Gastrointestinal calcium absorption in sheep is mostly insensitive to an alimentary induced challenge of calcium homeostasis. Comp Biochem and Physiol Part B: Biochem and Mol Biol **158** 199-207. https://doi.org/10.1016/j.cbpb.2010.11.008.

15 Wilkens MR, Richter J, Fraser DR et al. (2012) In contrast to sheep, goats adapt to dietary calcium restriction by increasing intestinal absorption of calcium. Comp Biochem and Physiol Part A: Molec and Integrative Physiol **163** 396-406. https://doi.org/10.1016/j.cbpa.2012.06.011.

16 Wilkens MR, Mrochen N, Breves G et al. (2010) Effects of 1,25-dihydroxyvitamin D_3 on calcium and phosphorus homeostasis in sheep fed diets either adequate or restricted in calcium content. Domest Anim Endocrinol **38** 190-9.

17 Ewer TK and Bartrum P (1948) Rickets in sheep. Aust Vet J **24** 73-85. https://doi.org/10.1111/j.1751-0813.1948.tb04626.x.

18 Fitch LWN (1943) Osteodystrophic diseases of sheep in New Zealand I. Rickets in hoggets: With a note on the aetiology and definition of the disease. Aust Vet J **19** 2-20. https://doi.org/10.1111/j.1751-0813.1943.tb01492.x.

19 Nisbet DI, Butler EJ, Smith BSW et al. (1966) Osteodystrophic diseases of sheep: II. Rickets in young sheep. J Comp Pathol **76** 159-68. https://doi.org/10.1016/0021-9975(66)90018-1.

20 Grant AB (1953) Carotene: A rachitogenic factor in green feeds. Nature **172** 627. https://doi.org/10.1038/172627a0.

21 Tisdall FF and Brown A (1929) Relation of the altitude of the sun to its antirachitic effect. JAMA **92** 860-4. https://doi.org/10.1001/jama.1929.02700370008002.

22 Caple IW (1990) Vitamin D deficiency. In: Sheep Medicine. Proceedings No 141. University of Sydney Postgraduate Committee in Veterinary Science: Sydney, pp. 381-6.

23 Watt B (2006) Bone problems in lambs on grazing cereals. Flock and herd case notes. Available from: http://www.flockandherd.net.au/sheep/reader/bone-problems-lambs.html. Accessed 24 August 2018.

24 Edmonstone B (2012) Rickets in lambs on grazing cereal crops. Flock and herd case notes. Available from: http://www.flockandherd.net.au/sheep/reader/rickets.html. Accessed 19 July 2018.

25 Van Suan RJ (2004) Vitamin D-responsive rickets in neonatal lambs. Can Vet J **45** 841-4.

26 Nisbet DI, Butler EJ, Robertson JM et al. (1970) Osteodystrophic diseases of sheep: IV. Osteomalacia and osteoporosis in lactating ewes on West Scotland hill farms. J Comp Pathol **80** 535-42. https://doi.org/10.1016/0021-9975(70)90050-2.

27 Franklin MC (1953) Vitamin D requirements of sheep with special reference to Australian conditions. Aust Vet J **29** 302-9. https://doi.org/10.1111/j.1751-0813.1953.tb08131.x.

28 Caple IW, Babacan E, Pham TT et al. (1988) Seasonal vitamin D deficiency in sheep in south eastern Australia. Proc Aust Soc Anim Prod **17** 379.

29 Mason KW and Koen T (1985) Phalangeal bone ash as an aid in diagnosing some nutritional causes of lameness in sheep. Aust Vet J **62** 338-40. https://doi.org/10.1111/j.1751-0813.1985.tb07654.x.

30 Morcombe PW, Peet RL, Jacob RH et al. (1990) The effect of limestone added to oat grain on the growth of young sheep grazing wheat stubble. Proc Aust Soc Anim Prod **18** 308-11.

31 Smith BS, Wright H and Brown KG (1987) Effect of Vitamin D supplementation during pregnancy on the vitamin D status of ewes and their lambs. Vet Rec **120** 199-201. https://doi.org/10.1136/vr.120.9.199.

32 Thompson KG, Dittmer KE, Blair HT et al. (2006) Suspected inherited rickets in Corriedale sheep. NZ Vet J **54** 51. https://doi.org/10.1080/00480169.2006.36611.

33 Clark L, Carlisle CH and Beasley PS (1975) Observations on the pathology of bent leg of lambs in south-western Queensland. Aust Vet J **51** 4-5. https://doi.org/10.1111/j.1751-0813.1975.tb14489.x.

34 Murnane D (1938) Arthritis in lambs. Aust Vet J **14** 23-5. https://doi.org/10.1111/j.1751-0813.1938.tb14838.x.

35 Paton MW, Rose IR, Sunderman FM et al. (2003) Effect of mulesing and shearing on the prevalence of *Erysipelothrix rhusiopathiae* arthritis in lambs. Aust Vet J **81** 694-7. https://doi.org/10.1111/j.1751-0813.2003.tb12543.x.

36 Farquharson B (2007) Arthritis in prime lamb sheep, a review. Final Report to Meat and Livestock Australia. AHW.123. Meat and Livestock Australia: North Sydney, Australia.

37 Watt B (2007) Erysipelas arthritis in two year old crossbred ewes. Flock and herd case notes. Available from: http://www.flockandherd.net.au/sheep/reader/arthritis-erysipelas.html. Accessed 19 July 2018.

38 Yang, R, Jacobson C, Gardner G et al. (2014) Longitudinal prevalence and faecal shedding of *Chlamydia pecorum* in sheep. Vet J **201** 322-6. https://doi.org/10.1016/j.tvjl.2014.05.037.

39 Jubb KVF, Kennedy PC and Palmer N, eds (1993) Pathology of domestic animals. 4th ed. Vol 1. Academic Press: New York & London, pp. 164-70.

40 Watts JE, Murray MD and Graham NPH (1979) The blowfly strike problem of sheep in New South Wales. Aust Vet J **55** 325-34.

41 Lloyd J (2016) An investigation of the potential link between arthritis and tail length in sheep. Final Report to Meat and Livestock Australia. Report AHE.0238. Meat and Livestock Australia: North Sydney, Australia.

42 Watt DA, Bamford V and Nairn ME (1970) *Actinobacillus seminis* as a cause of polyarthritis and posthitis in sheep. Aust Vet J **46** 515. https://doi.org/10.1111/j.1751-0813.1970.tb09190.x.

43 Curran G (2012) Post mulesing arthritis. Flock and herd case notes. Available from: www.flockandherd.net.au/sheep/reader/arthritis-post-mulesing.html. Accessed 19 July 2018.

44 Angus K (1991) Arthritis in lambs and sheep. Practice **13** 204-7. https://doi.org/10.1136/inpract.13.5.204.

45 Watkins GH (2007) Arthritis. In: Diseases of sheep, ed ID Aitken. 4th ed. Blackwell Publishing: London, pp. 288-90.

46 Harriss ST (1948) Proliferative dermatitis of the legs ('Strawberry footrot') in sheep. J Comp Pathol **58** 314-28. https://doi.org/10.1002/9780470753316.ch41.

47 Duncan JS, Angell JW, Carter SD et al. (2014) Contagious ovine digital dermatitis: An emerging disease. Vet J **201** 265-8. https://doi.org/10.1016/j.tvjl.2014.06.007.

48 Winter A and Arsenos G (2009) Diagnosis of white line lesions in sheep. Practice **31** 17-21. https://doi.org/10.1136/inpract.31.1.17.

49 Conington, J, Nicoll L, Mitchell S et al. (2010) Characterisation of white line degeneration in sheep and evidence for genetic influences on its occurrence. Vet Res Commun **34** 481-9. https://doi.org/10.1007/s11259-010-9416-z.

50 Parsonson IM, Egerton JR and Roberts DS (1967) Ovine interdigital dermatitis. J Comp Pathol **77** 309-13.

51 West DM (1983) A study of naturally occurring ovine foot abscess in New Zealand. NZ Vet J **31** 152-6.

52 Winter AC (2004) Lameness in sheep 1. Diagnosis. Practice **26** 58-63. https://doi.org/10.1136/inpract.26.2.58.

53 Scott P (1995) Amputation of the ovine digit. Practice **17** 80-2. https://doi.org/10.1136/inpract.17.2.80.

54 Winter AC (2004) Lameness in sheep 2. Treatment and control. Practice **26** 130-9. https://doi.org/10.1136/inpract.26.3.130.

55 Wassink GJ, Grogono-Thomas R, Moore LJ et al. (2003) Risk factors associated with foot rot in sheep from 1999 to 2000. Vet Rec **152** 351-7. https://doi.org/10.1136/vr.152.12.351.

56 Green LE and George TRN (2008) Assessment of current knowledge of footrot in sheep with particular reference to *Dichelobacter nodosus* and implications for elimination or control strategies for sheep in Great Britain. Vet J **175** 173-80.

57   Winter AC (2009) Footrot control and eradication (elimination) strategies. Small Ruminant Res **86** 90-3. https://doi.org/10.1016/j.smallrumres.2009.09.026.

58   Graham NPH and Egerton JR (1968) Pathogenesis of ovine footrot: The role of some environmental factors. Aust Vet J **44** 235. https://doi.org/10.1111/j.1751-0813.1968.tb09092.x.

59   Beveridge WIB (1938) Investigations on the viability of the contagium of footrot in sheep. Journal of the Council for Scientific and Industrial Research **11** 4-13.

60   Laing EA and Egerton JR (1981) Aspects of *Bacteroides nodosus* infection of the feet of cattle. In: Ovine footrot, a report of a Workshop at University of Sydney, May, Sydney, NSW, pp. 195-9.

61   Beveridge WIB (1941) Footrot in sheep: A transmissible disease due to infection with *Fusiformis nodosus* (n.sp.). Journal of the Council for Scientific and Industrial Research Bull no 140.

62   Stewart DJ (1973) An electron microscopic study of *Fusiformis nodosus*. Res Vet Sci **13** 132.

63   Stewart DJ and Egerton JR (1979) Studies on the ultrastructural morphology of *Bacteroides nodosus*. Res Vet Sci **26** 227.

64   Thomas JH (1958) A simple medium for the isolation and cultivation of *Fusiformis nodosus*. Aust Vet J **34** 411. https://doi.org/10.1111/j.1751-0813.1958.tb05811.x.

65   Skerman TM (1975) Determination of some in vitro growth requirements of *Bacteroides nodosus*. J Gen Microbiol **87** 107-19. https://doi.org/10.1099/00221287-87-1-107.

66   Thorley CM (1976) A simplified method for the isolation of *Bacteroides nodosus* from ovine footrot and studies on its colonial morphology and serology. J Appl Bacteriol **40** 301-9. https://doi.org/10.1111/j.1365-2672.1976.tb04178.x.

67   Stewart DJ and Claxton PD (1993) Ovine footrot clinical diagnosis and bacteriology. In: Australian standard diagnostic techniques for animal diseases, eds LA Corner and TJ Bagust. Standing committee on agriculture and resource management, Subcommittee on animal health laboratory standards, CSIRO: Parkville, Australia, pp. 1-27.

68   Claxton PD, Ribeiro LA and Egerton JR (1983) Classification of *Bacteroides nodosus* by agglutination tests. Aust Vet J **60** 331. https://doi.org/10.1111/j.1751-0813.1983.tb02834.x.

69   Claxton PD (1986) Serogrouping of *Bacteroides nodosus* isolates In: Footrot in ruminants. Proceedings of a workshop, Melbourne 1985, eds DJ Stewart, JE Peterson, NM McKern et al. CSIRO Division of Animal Health/Australian Wool Corporation: Glebe, NSW, pp. 131-4.

70   Ghimire SC, Egerton JR, Dhungyel OP et al. (1998) Identification and characterisation of serogroup M among Nepalese isolates of *Dichelobacter nodosus*, the transmitting agent of footrot in small ruminants. Vet Microbiol **62** 217-33. https://doi.org/10.1017/S0950268899002290.

71   Thorley CM and Day SEJ (1986) Serotyping survey of 1296 strains of *Bacteroides nodosus* isolated from sheep and cattle in Great Britain and western Europe. In: Footrot in ruminants. Proceedings of a workshop, Melbourne 1985, eds DJ Stewart, JE Peterson, NM McKern et al. CSIRO Division of Animal Health/Australian Wool Corporation: Glebe, NSW, pp. 135-42.

72   Day SEJ, Thorley CM and Beesley JE (1986) Serotyping of *Bacteroides nodosus*: A proposal for 9 further serotypes (J-R) and a study of the antigenic complexity of *B nodosus* pili. In: Footrot in ruminants. Proceedings of a workshop, Melbourne 1985, eds DJ Stewart, JE Peterson, NM McKern et al. CSIRO Division of Animal Health/Australian Wool Corporation: Glebe, NSW, pp. 147-59.

73   Ghimire SC and Egerton JR (1999) PCR-RFLP of outer membrane protein gene of *Dichelobacter nodosus*: A new tool in the epidemiology of footrot. Epidemiol Infect **122** 521-8.

74   Allworth MB (1995) Investigations of the eradication of footrot. PhD thesis, University of Sydney.

75   Hindmarsh F and Fraser J (1985) Serogroups of *Bacteroides nodosus* isolated from ovine footrot in Britain. Vet Rec **116** 187-8.

76   Kingsley DF, Hindmarsh FH, Liardet DM et al. (1986) Distribution of serogroups of *Bacteroides nodosus* with particular reference to New Zealand and the United Kingdom In: Footrot in ruminants. Proceedings of a workshop, Melbourne 1985, eds DJ Stewart, JE Peterson, NM McKern et al. CSIRO Division of Animal Health/Australian Wool Corporation: Glebe, NSW, pp. 143-59.

77   Egerton JR (1983) Footrot control in drought. Aust Vet J **60** 315. https://doi.org/10.1111/j.1751-0813.1983.tb02824.x.

78 Egerton JR (1989) Control and eradication of footrot at the farm level — the role of veterinarians. In: Proceedings of the Second International Congress for Sheep Veterinarians, Sheep and Beef Cattle Society of the New Zealand Veterinary Association. 12-16 February, Massey University, New Zealand, pp. 215-18.

79 Roycroft CR (1986) Managing benign strains and intermediate strains of footrot. In: Sheep health and welfare, Department of Agriculture and Rural Affairs. Victoria Conference Proceedings series Number 8 153-6.

80 Dobson KJ (1986) The impact of *Bacteroides nodosus* strains of intermediate virulence on footrot control. In: Footrot in ruminants. Proceedings of a workshop, Melbourne 1985, eds DJ Stewart, JE Peterson, NM McKern et al. CSIRO Division of Animal Health/Australian Wool Corporation: Glebe, NSW, pp. 23-5.

81 Stewart DJ (1989) Footrot in sheep. In: Footrot and foot abscess of ruminants, eds JR Egerton, WK Yong and GG Riffkin. CRC Press: Boca Raton, Florida, USA, p. 11.

82 Allworth MB and Egerton JR (1999) An objective clinical assessment method for footrot diagnosis. In: Proceedings of the Australian Sheep Veterinary Society 1999 AVA Conference, Hobart, ed B Besier. Australian Veterinary Association: Indooroopilly, Qld, Australia, pp. 74-6.

83 Egerton JR, Ribeiro LA, Kieran PJ et al. (1983) Onset and remission of ovine footrot. Aust Vet J **60** 334-6. https://doi.org/10.1111/j.1751-0813.1983.tb02835.x.

84 Cummins LJ, Thompson RL and Roycroft CR (1991) Production losses due to intermediate virulence footrot. In: Proceedings of the Australian Sheep Veterinarians Conference. 12-18 May, Darling Harbour. Australian Veterinary Association: St Leonards, Australia, pp. 106-8.

85 Raadsma HW and Dhungyel OP (2013) A review of footrot in sheep: New approaches for the control of virulent footrot. Livest Sci **156** 115-25. https://doi.org/10.1016/j.livsci.2013.06.011.

86 Raadsma HW, Egerton JR, Wood D et al. (1994) Disease resistance in Merino sheep. III. Genetic variation in resistance following challenge and subsequent vaccination with an homologous rDNA pilus vaccine under both induced and natural conditions. J Anim Breed Genet **111** 367-90. https://doi.org/10.1111/j.1439-0388.1994.tb00475.x.

87 Nieuwhof GJ, Conington J, Bünger L et al. (2008) Genetic and phenotypic aspects of foot lesion scores in sheep of different breeds and ages. Animal **2(9)** 1289-96.

88 Skerman TM and Moorhouse SR (1987) Broomfield Corriedales: A strain of sheep selectively bred for resistance to footrot. N Z Vet J **35** 101-6. https://doi.org/10.1080/00480169.1987.35399.

89 Skerman TM, Johnson DL, Kane DW et al. (1988) Clinical footscald and footrot in a New Zealand Romney flock: Phenotypic and genetic parameters. Aust J Ag Res **39** 907-16.

90 Bulgin MS, Lincoln SD, Parker CF et al. (1988) Genetic-associated resistance to foot rot in selected Targhee sheep. J Am Vet Med Assoc **192** 512-20.

91 Emery DL, Stewart DJ and Clark BL (1984) The comparative susceptibility of five breeds of sheep to footrot. Aust Vet J **61** 85-8. https://doi.org/10.1111/j.1751-0813.1984.tb15524.x.

92 West DM, Bruere AN and Ridler AL (2009) Foot diseases and lameness. In: The sheep: Health, disease and production. 3rd ed. VetLearn Foundation: Wellington, pp. 262-82.

93 Shenman G (1962) A case of possible *F nodosus* infection in a cow. Aust Vet J **38** 306.

94 Gupta RB, Fincher MF and Bruner DW (1964) A study of the etiology of footrot in cattle. Cornell Vet **54** 66.

95 Thorley CM, Calder HA McC and Harrison WJ (1977) Recognition in Great Britain of *Bacteroides nodosus* in foot lesions in cattle. Vet Rec **100** 387. https://doi.org/10.1136/vr.100.18.387.

96 Richards RB, Depiazzi LJ, Edwards JR et al. (1980) Isolation and characterisation of *Bacteroides nodosus* from foot lesions of cattle in Western Australia. Aust Vet J **56** 517-21.

97 Morgan IR (1969) A survey of cattle feet in Victoria for *Fusiformis nodosus*. Aust Vet J **45** 264.

98 Laing EA and Egerton JR (1978) The occurrence, prevalence and transmission of *Bacteroides nodosus* infection in cattle. Res Vet Sci **24** 300-4.

99 Toussaint-Raven E and Cornelisse JL (1971) The specific contagious inflammation of the interdigital skin in cattle. Vet Med Rev **2-3/71** 223-47.

100 Egerton JR and Laing EA (1978) Characteristics of *Bacteroides nodosus* isolated from cattle. Vet Microbiol **3** 269-79.

101 Alexander TM (1962) The differential diagnosis of footrot in sheep Aust Vet J **38** 366-67. https://doi.org/10.1111/j.1751-0813.1962.tb04084.x.

102 Wilkinson FC, Egerton JR and Dickson J (1970) Transmission of *Fusiformis nodosus* from cattle to sheep. Aust Vet J **46** 382-4. https://doi.org/10.1111/j.1751-0813.1970.tb15578.x.

103 Egerton JR and Parsonson IM (1966) Isolation of *Fusiformis nodosus* from cattle. Aust Vet J **42** 425-9. https://doi.org/10.1111/j.1751-0813.1966.tb04646.x.

104 Mitchell RK, Moir DC, Bowden MS et al. (1992) Cattle transmit virulent footrot to sheep — new implications for eradication policy. In: Proceedings of Australian Sheep Veterinary Society, ed MB Allworth. 10-15 May, Adelaide. Australian Veterinary Association: St Leonards, Australia, pp. 49-53.

105 Trengove CL, Riley MJ and Saunders PE (1993) The role of cattle in the spread of ovine footrot. In: Proceedings of Australian Sheep Veterinary Society Conference, ed DA Hucker. 16-21 May, Gold Coast. Australian Veterinary Association: St Leonards, Australia, pp. 74-8.

106 Egerton JR and Roberts DS (1971) Vaccination against ovine footrot. J Comp Pathol **8** 1179-85.

107 Glynn T (1993) Benign footrot — an epidemiological investigation into the occurrence, effects on production, response to treatment and influence of environmental factors. Aust Vet J **70** 7.

108 Abbott KA (2000) The epidemiology of intermediate footrot. PhD thesis, University of Sydney.

109 Depiazzi LJ and Richards RB (1985) Motility in relation to virulence of *Bacteroides nodosus*. Vet Microbiol **10** 107-16. https://doi.org/10.1016/0378-1135(85)90012-4.

110 Depiazzi LJ, Henderson J and Penhale WJ (1990) Measurement of protease thermostability, twitching motility and colony size of *Bacteroides nodosus*. Vet Microbiol **22** 353-63. https://doi.org/10.1016/0378-1135(90)90022-N.

111 Stewart DJ, Peterson JE, Vaughan JA et al. (1986) The pathogenicity and cultural characteristics of virulent, intermediate and benign strains of *Bacteroides nodosus* causing ovine footrot. Aust Vet J **63** 317-26. https://doi.org/10.1111/j.1751-0813.1986.tb02875.x.

112 Thomas JH (1962) The differential diagnosis of footrot in sheep. Aust Vet J **38** 159-63. https://doi.org/10.1111/j.1751-0813.1962.tb16034.x.

113 Whittington RJ (1994) Protease-based diagnostic tests for footrot — a review. In: Proceedings of the Australian Sheep Veterinary Society Conference, ed D Jordan. 6-11 March, Canberra. Australian Veterinary Association: St Leonards, Australia, pp. 135-8.

114 Egerton JR and Parsonson IM (1969) Benign footrot — a specific interdigital dermatitis of sheep associated with infection by less proteolytic strains of *Fusiformis nodosus*. Aust Vet J **45** 345-9. https://doi.org/10.1111/j.1751-0813.1969.tb06606.x.

115 Depiazzi LJ and Richards RB (1979) A degrading proteinase test to distinguish benign and virulent isolates of *Bacteroides nodosus*. Aust Vet J **55** 25-8. https://doi.org/10.1111/j.1751-0813.1979.tb09541.x.

116 Stewart DJ (1979) The role of elastase in the differentiation of *Bacteroides nodosus* infections in sheep and cattle. Res Vet Sci **27** 99-105.

117 Every D (1982) Proteinase isoenzyme patterns of *Bacteroides nodosus*: Distinction between ovine virulent isolates, ovine benign isolates and bovine isolates. J Gen Microbiol **128** 809-12. https://doi.org/10.1099/00221287-128-4-809.

118 Kortt AA, O'Donnell IJ, Stewart DJ et al. (1982) Activities and partial purification of extracellular proteases of *Bacteroides nodosus* from virulent and benign footrot. Aust J Biol Sci **35** 481-9. https://doi.org/10.1071/BI9820481.

119 Links IJ, Stewart DJ, Edwards RD et al. (1995) Protease tests on *Dichelobacter nodosus* from ovine footrot — comparison of protease ELISA and Gelatin Gel tests. In: Proceedings of the Australian Sheep Veterinary Society AVA Conference, ed J Cox. May 1995, Melbourne. Australian Veterinary Association: St Leonards, Australia, pp. 45-8.

120 Depiazzi LJ and Rood JI (1984) The thermostability of proteases from virulent and benign strains of *Bacteroides nodosus*. Vet Microbiol **9** 227-36. https://doi.org/10.1016/0378-1135(84)90040-3.

121 Green RS (1985) A method to differentiate between virulent and benign isolates of *Bacteroides nodosus* based on the thermal stability of their extracellular proteinases. N Z Vet J **33** 11-13. https://doi.org/10.1080/00480169.1985.35135.

122 Palmer MA (1993) A gelatin test to detect activity and stability of proteases produced by *Dichelobacter (bacteroides) nodosus*. Vet Microbiol **36** 113-22. https://doi.org/10.1016/0378-1135(93)90133-R.

123 Stewart DJ, Clark BL and Jarrett RG (1982) Observations on strains of *Bacteroides nodosus* of intermediate virulence to sheep. In: Australian Advances in Veterinary Science. Australian Veterinary Association: Australia, pp. 74-6.

124 Gordon LM, Yong WK and Woodward CAM (1985) Temporal relationships and characterisation of extracellular proteases from benign and virulent strains of *Bacteroides nodosus* as detected in zymogram gels. Res Vet Sci **39** 165-72.

125 Depiazzi LJ, Richards RB, Henderson J et al. (1991) Characterisation of virulent and benign strains of *Bacteroides nodosus*. Vet Microbiol **26** 151-60. https://doi.org/10.1016/0378-1135(91)90051-G.

126 Liu D and Yong WK (1993) Use of elastase test, gelatin gel test and electrophoretic zymogram to determine virulence of *Dichelobacter nodosus* isolated from ovine footrot. Res Vet Sci **55** 124-9. https://doi.org/10.1016/0034-5288(93)90046-I.

127 Gaden CA, Cheetham BF, Hall E et al. (2013) Producer-initiated field research leads to a new diagnostic test for footrot. Anim Prod Sci **53** 610-7. https://doi.org/10.1071/AN11175.

128 Katz ME, Howarth PM, Yong WK et al. (1991) Identification of three gene regions associated with virulence in *Dichelobacter nodosus*, the causative agent of ovine footrot. J Gen Microbiol **137** 2117-24. https://doi.org/10.1099/00221287-137-9-2117.

129 Katz ME, Strugnell RA and Rood JI (1992) Molecular characterization of a genomic region associated with virulence in *Dichelobacter nodosus*. Infect Immun **60** 4586-92.

130 Cheetham BF, Tattersall DB, Bloomfield GA et al. (1995) Identification of a gene encoding a bacteriophage-related integrase in a *vap* region of the *Dichelobacter nodosus* genome. Gene **5** 53-8. https://doi.org/10.1016/0378-1119(95)00315-W.

131 Billington SJ, Johnstone JL and Rood JI (1996) Virulence regions and virulence factors of the ovine footrot pathogen, *Dichelobacter nodosus*. FEMS Microbiol Lett **145** 147-56. https://doi.org/10.1111/j.1574-6968.1996.tb08570.x.

132 Haring V, Billington SJ, Wright CL et al. (1995) Delineation of the virulence-related locus (*vrl*) of *Dichelobacter nodosus*. Microbiology **141** 2081-9. https://doi.org/10.1099/13500872-141-9-2081.

133 Rood JI, Howart PA, Haring V et al. (1996) Comparison of gene probe and conventional methods for the differentiation of ovine footrot isolates of *Dichelobacter nodosu.s* Vet Microbiol **52** 127-41. https://doi.org/10.1016/0378-1135(96)00054-5.

134 Liu D and Yong WK (1993) *Dichelobacter nodosus*: Differentiation of virulent and benign strains by gene probe based on dot blot hybridisation. Vet Microbiol **38** 71-9. https://doi.org/10.1016/0378-1135(93)90076-J.

135 Liu D (1994) Development of gene probes of *Dichelobacter nodosus* for differentiating strains causing virulent, intermediate or benign ovine footrot. Vet J **150** 451-62. https://doi.org/10.1016/S0007-1935(05)80196-4.

136 Cheetham BF, Tanjung LR, Sutherland M et al. (2006) Improved diagnosis of virulent ovine footrot using the *intA* gene. Vet Microbiol **116** 166-74. https://doi.org/10.1016/j.vetmic.2006.04.018.

137 Buller N and Eamens G (2014) Australian and New Zealand standard diagnostic procedure. Available from: www.scahls.org.au. Accessed 19 July 2018.

138 Symons LEA (1978) Experimental foot-rot, wool growth and body mass. Aust Vet J **54** 362-3. https://doi.org/10.1111/j.1751-0813.1978.tb02498.x.

139 Stewart DJ, Clark BL and Jarrett RG (1984) Differences between strains of *Bacteroides nodosus* in their effects on the severity of foot-rot, bodyweight and wool growth in Merino sheep. Aust Vet J **61** 348-52. https://doi.org/10.1111/j.1751-0813.1984.tb07153.x.

140 Marshall DJ, Walker RI, Cullis BR et al. (1991) The effect of footrot on body weight and wool growth of sheep. Aust Vet J **68** 45-9. https://doi.org/10.1111/j.1751-0813.1991.tb03126.x.

141 Littlejohn AI (1964) Foot-rot in feeding sheep. Vet Rec **76** 741.

142 Stewart DJ, Peterson JE, Vaughan JA et al. (1986) Clinical and laboratory diagnosis of benign, intermediate and virulent strains of *Bacteroides nodosus* In: Footrot in ruminants. Proceedings of a workshop, Melbourne 1985, eds DJ Stewart, JE Peterson, NM McKern et al. CSIRO Division of Animal Health/Australian Wool Corporation: Glebe, NSW, p. 81.

143 Egerton JR and Raadsma HW (1991) Breeding sheep for resistance to footrot. In: Breeding for disease resistance in farm animals, eds JB Owen and RFA Axford. CAB International: Wallingford, p. 347.

144 Allworth MB (1994) Flock health. In: Merinos, money and management, ed FHW Morley. Post-Graduate Committee in Veterinary Science: University of Sydney, p. 263.

145 Malecki JC, McCausland I and Lambell R (1983) A new topical treatment for footrot. In: Sheep production and preventive medicine. Proceedings 67. University of Sydney Postgraduate Committee in Veterinary Science: Sydney, pp. 63-70.

146 Malecki JC and Coffey L (1987) Treatment of ovine virulent footrot with zinc sulphate/ sodium lauryl sulphate footbathing. Aust Vet J **64** 301-4. https://doi.org/10.1111/j.1751-0813.1987.tb07331.x.

147 Reed GA and Alley DU (1996) Efficacy of a novel copper-based footbath preparation for the treatment of ovine footrot during the spread period. Aust Vet J **74** 375-82. https://doi.org/10.1111/j.1751-0813.1996.tb15449.x.

148 Egerton JR and Parsonson IM (1966) Parenteral antibiotic treatment of ovine footrot. Aust Vet J **42** 97-8. https://doi.org/10.1111/j.1751-0813.1966.tb04682.x.

149 Egerton JR, Parsonson IM and Graham NPH (1968) Parenteral chemotherapy of ovine footrot. Aust Vet J **44** 275-83. https://doi.org/10.1111/j.1751-0813.1968.tb04982.x.

150 Forsyth BA (1953) The experimental treatment of contagious footrot in sheep. Aust Vet J **27** 73-4. https://doi.org/10.1111/j.1751-0813.1953.tb05221.x.

151 Jordan D, Plant JW, Nicol HI et al. (1996) Factors associated with the effectiveness of antibiotic treatment for ovine virulent footrot. Aust Vet J **73** 211-15. https://doi.org/10.1111/j.1751-0813.1996.tb10037.x.

152 Webb Ware JK, Scrivener CJ and Vizard AL (1994) Efficacy of erythromycin compared with penicillin/streptomycin for the treatment of virulent footrot in sheep. Aust Vet J **71** 88-9. https://doi.org/10.1111/j.1751-0813.1994.tb03336.x.

153 Venning CM, Curtis MA and Egerton JR (1990) Treatment of virulent footrot with lincomycin and spectinomycin. Aust Vet J **67** 258-60. https://doi.org/10.1111/j.1751-0813.1990.tb07781.x.

154 Kaler J, Daniels SLS, Wright JL et al. (2010) Randomized clinical trial of long acting oxytetracycline, foot trimming and flunixine meglumine on time to recovery in sheep with footrot. J Vet Intern Med **24** 420-5. https://doi.org/10.1111/j.1939-1676.2009.0450.x.

155 Forbes AB, Strobel H and Stamphoj I (2014) Field studies on the elimination of footrot in sheep through whole flock treatments with gamithromycin. Vet Rec **174** 146-7. https://doi.org/10.1136/vr.102028.

156 Egerton JR and Burrell DH (1970) Prophylactic and therapeutic vaccination against ovine footrot. Aust Vet J **46** 517-22. https://doi.org/10.1111/j.1751-0813.1970.tb06636.x.

157 Egerton JR (1970) Successful vaccination of sheep against footrot. Aust Vet J **46** 114-15. https://doi.org/10.1111/j.1751-0813.1970.tb15936.x.

158 Stewart DJ, Clark BL, Peterson JE et al. (1982) Importance of pilus associated antigen in *Bacteroides nodosus* vaccines. Res Vet Sci **32** 140-7.

159 Elleman TC, Hoyne PA, Emery DL et al. (1986) Expression of pili from *Bacteroides nodosus* in *Pseudomonas aeruginosa*. J Bacteriol **168** 574-80. https://doi.org/10.1128/jb.168.2.574-580.1986.

160 Mattick JS, Bills MM, Anderson BJ et al. (1987) Morphogenetic expression of *Bacteroides nodosus* fimbriae in *Pseudomonas aeruginosa*. J Bacteriol **169** 33-41. https://doi.org/10.1128/jb.169.1.33-41.1987.

161 Egerton JR, Cox PT, Anderson BJ et al. (1987) Protection of sheep against footrot with a recombinant DNA-based fimbrial vaccine. Vet Microbiol **14** 393-409. https://doi.org/10.1016/0378-1135 (87)90030-7.

162 Schwartzkoff CL, Egerton JR, Stewart DJ et al. (1993) The effects of antigenic competition on the efficacy of multivalent footrot vaccines. Aust Vet J **70** 123-6. https://doi.org/10.1111/j.1751-0813.1993.tb06101.x.

163 Egerton JR, Raadsma HW, O'Meara TJ et al. (1994) Antigenic competition in host immune response to defined vaccines. In: Proceedings of the Australian Sheep Veterinary Society Conference, ed D Jordan. 6-11 March, Canberra. Australian Veterinary Association: St Leonards, Australia, pp. 37-40.

164 Raadsma HW, Attard GA, Nicholas FW et al. (1995) Disease resistance in Merino sheep IV. Genetic variation in immunogical responsiveness to fimbrial *Dichelobacter nodosus* antigens, and its relationship with resistance to footrot. J Anim Breed Genet **112** 349-72.

165 Schwartzkoff CL, Lehrbach PR, Ng ML et al. (1993) The effect of time between doses on serological response to a recombinant multivalent pilus vaccine against footrot in sheep. Aust Vet J **70** 127-9. https://doi.org/10.1111/j.1751-0813.1993.tb06102.x.

166 Lambell RG (1986) A field trial with a commercial vaccine against footrot in sheep. Aust Vet J **63** 415-18. https://doi.org/10.1111/j.1751-0813.1986.tb15921.x.

167 Skerman TM, Erasmuson SK and Morrison LM (1982) Duration of resistance to experimental footrot infection in Romney and Merino sheep vaccinated with *Bacteroides nodosus* oil adjuvant vaccine. NZ Vet J **30** 27-31. https://doi.org/10.1080/00480169.1982.34867.

168 O'Meara TJ, Egerton JR and Raadsma HW (1993) Recombinant vaccines against ovine footrot. Immun Cell Biol **71** 473-88.

169 Gurung RB, Dhungyel OP, Tshering P et al. (2006) The use of an autogenous vaccine to eliminate clinical signs of virulent footrot in a sheep flock in Bhutan. Vet J **172** 356-63.

170 Dhungyel O, Hunter J and Whittington R (2014) Footrot vaccines and vaccination. Vaccine **32** 3139-46. https://doi.org/10.1016/j.vaccine.2014.04.006.

171 Green LE, Wassink GJ, Grogono-Thomas R et al. (2007) Looking after the individual to reduce disease in the flock: A binomial mixed effects model investigating the impact of individual sheep management of footrot and interdigital dermatitis in a prospective longitudinal study on one farm. Prev Vet Med **78** 172-8. https://doi.org/10.1016/j.prevetmed.2006.09.005.

172 Locke RH and Coombes NE (1994) Prevalence of virulent footrot in sheep flocks in southern New South Wales. Aust Vet J **71** 348-9. https://doi.org/10.1111/j.1751-0813.1994.tb00919.x.

173 Mills K, McClenaughan P, Morton A et al. (2012) Effect on time in quarantine of the choice of program for eradication of footrot from 196 sheep flocks in southern New South Wales. Aust Vet J **90** 14-19. https://doi.org/10.1111/j.1751-0813.2011.00872.x.

174 Allworth MB (2014) Challenges in ovine footrot control. Small Ruminant Res **118** 110-13. https://doi.org/10.1016/j.smallrumres.2013.12.007.

175 Harwood DG, Cattell JH, Lewis CJ et al. (1997) Virulent footrot in sheep. Vet Rec **140** 687.

176 Clegg SR, Carter SD, Birtles RJ et al. (2016) Multilocus sequence typing of pathogenic treponemes isolated from cloven-hoofed animals and comparison to treponemes isolated from humans. App Env Microbiol **82** 4523-36.

14

SUDDEN DEATH

This chapter covers
- infectious diseases causing sudden death, particularly clostridial diseases and anthrax
- intoxications causing sudden death, inorganic chemicals and plant poisonings
- environmental conditions causing sudden death, including exposure and lightning strike.

CLOSTRIDIAL DISEASE OF RUMINANTS

Clostridium is a genus of bacteria which are obligate anaerobes and are able to produce endospores. In the reproducing, vegetative form the bacterial cells are rod-shaped and gram-positive or gram-variable. There are around 90 *Clostridium* species, many of which are saprophytes and inhabit soils and decomposing organic material. Some species inhabit the intestinal tract of animals, including humans. The species of the *Clostridium* genus which cause disease in ruminants produce lethal toxins which, in most cases, are responsible for their pathogenicity. (They also produce *toxins* of low or zero toxicity which are antigenically distinct and which assist in identifying isolates of the organism.) Clostridial toxins are proteins which are strongly antigenic and the activity of the lethal toxins can be neutralised by specific antisera.

C tetani and *C botulinum* produce powerful neurotoxins which cause disease (tetanus, botulism) through their effect on nerves and nerve impulse transmission. They do not cause sudden death and are not discussed further in the chapter.

Five clostridial species produce exotoxins which cause severe local tissue necrosis at the site of infection, leading to systemic toxaemia and rapid death in many cases. These species — *C chauvoei, C septicum, C novyi, C haemolyticum* and *C sordellii* — are the histotoxic clostridia. *C perfringens* produces enterotoxins which are responsible for disease.

C perfringens and *C novyi* exist in different types, identified by the range of toxins they produce. The five types of *C perfringens* produce different combinations of the four major lethal toxins — alpha, beta, epsilon and iota — in addition to minor toxins which may or may not contribute to the pathogenicity of the organisms. *C novyi* exists in three types, only two of which (A and B) are pathogenic. *C haemolyticum* is sometimes classified as *C novyi* type D. The toxins bearing the same names but from different clostridial species are not the same.[1]

ENTEROTOXAEMIA (PULPY KIDNEY)

Under particular dietary conditions, *Clostridium perfringens* type D proliferates in the small intestine and produces toxins, the principal one being the necrotising, highly lethal, *epsilon*

Table 14.1: The major lethal toxins produced by *C perfringens* and *C novyi* determine the type classification of each bacterial species. Adapted from Hatheway (1990).[1]

Toxin	C perfringens type						Toxin	C novyi type			
	A	B	C	D	E			A	B	C	D
Alpha	+	+	+	+	+		Alpha	+	+		
Beta		+	+				Beta		+		+
Epsilon		+		+							
Iota					+						

toxin. Absorption of this toxin leads rapidly to diarrhoea, convulsions and death. Death may occur so quickly that the diarrhoea is never exhibited.

The conditions which lead to the proliferation of the organism in the intestine are complex. A common factor in many outbreaks is that the sheep are offered a diet to which they are not accustomed. Before ruminal flora adapt to the changed diet, partially digested food may spill into the intestine and, if that food includes starch, *C perfringens* may proliferate using the starch as a substrate. The fact that starch is a preferred substrate for this organism leads to the association between the disease, lush feed conditions and individual animals with high growth rates. The disease also occurs under pasture conditions which are anything but lush, for reasons which are not always clear.

The epsilon toxin produces a profuse diarrhoea, stimulation and then depression of the central nervous system (CNS). Damage to the vascular endothelial cells in a variety of organs leads to some characteristic necropsy findings.

Clinical signs of animals acutely affected include clonic convulsions, frothing and sudden death. If they survive more than a few hours there will be a pasty diarrhoea, staggering, opisthotonus and convulsions or, more commonly, struggling. Older sheep may survive for longer (24 hours) and some may even recover.

Necropsy findings

The disease is most frequently diagnosed by necropsy. Putrefaction occurs quickly. In animals examined soon after death, necropsy findings provide reliable diagnostic evidence of the disease.

There is an excess of straw-coloured pericardial fluid, lung oedema with froth and haemorrhages in the endocardium, epicardium and parietal peritoneum. The intestinal contents are creamy, particularly if death was rapid. *Pulpy kidney* is not a particularly useful sign because the 'pulpiness' is in fact a more-rapid-than-usual autolysis.

Glycosuria virtually always occurs and is a consequence of hepatic glycogenolysis and extreme hyperglycaemia. The finding of glycosuria is a useful contribution to diagnosis but not completely pathognomic; glucose is fermented in urine within hours of death. (Note that glycosuria is not a sign of enterotoxaemia in cattle.)

Brain histopathology is particularly useful to confirm the diagnosis except in the most acute cases. The most common findings are symmetrical areas of haemorrhage, oedema and softening, particularly in the basal ganglia but also in other areas.

Control by husbandry procedures

Outbreaks of enterotoxaemia are often controllable by changes to the management of the affected mobs which alter the dietary predisposition to the disease. Exercise, a change of pasture or a change of diet, particularly to one with more roughage, will often stop deaths dramatically. In lambs, tailing and marking and the check in growth rate which accompanies the procedures will usually stop an outbreak. In the case of prime lambs over 8 weeks of age, weaning will serve the same purpose.

The results of attempts to control enterotoxaemia in this way are not always successful; nor are the procedures necessarily practicable. Antiserum is commercially available and is effective in preventing the disease, but only for a few weeks. Antiserum is of no use in treatment of clinical cases of enterotoxaemia.

Enterotoxaemia caused by *C perfringens* types A, B, C and E

Type A has been recorded as causing a highly fatal haemolytic disease in sheep. Type A produces more *alpha* toxin than other types; this toxin is haemolytic. It is also associated with wound infections, causing gas gangrene. Type B causes an enterotoxaemia of lambs under 3 weeks of age known as *lamb dysentery*. The disease is uncommon in Australia. Toxins involved are *alpha*, *beta* and *epsilon*.

Type C causes a condition called *Struck* — an enterotoxaemia with haemorrhagic enteritis and peritonitis affecting mainly adult sheep, usually when feed is abundant.

Type E is reported as causing enterotoxaemia in lambs. It produces the *iota* toxin.

INFECTIOUS NECROTIC HEPATITIS (BLACK DISEASE)

Black disease is a fatal, peracute intoxication caused by *C novyi* type B which proliferates in the liver following tissue damage which produces an anaerobic environment. This anaerobic environment is most often produced by migrating immature *Fasciola hepatica*.

Spores of *C novyi* are continuously ingested by grazing animals where black disease occurs and some spores cross the intestinal wall and populate reticuloendothelial cells — including those in the liver. Immature flukes leave tunnels about 5 mm in diameter which are surrounded by necrotic zones. Any latent spores in these necrotic zones become vegetative, produce toxins which further extend the necrotic areas, and multiply.

The disease occurs mainly in adult sheep and less frequently in young sheep and goats. Morbidity rates are usually low (5% to 10%) but the disease is always fatal. It occurs in *most* areas where liver fluke occurs and is most likely to be seen when sheep are ingesting metacercariae — that is, in summer and autumn.

Clinical signs are rarely observed. The course of the illness is only a few hours and sheep are usually found dead with no signs of struggling. At necropsy, rapid putrefaction is obvious as well as, variably, the following signs:

- There is marked subcutaneous venous congestion (hence the name), straw-coloured fluid in serous cavities, often blood-tinged in the abdomen.
- The liver is generally engorged and will have at least one area of necrosis, 1 to 2 cm in

diameter, often under the capsule of the diaphragmatic surface, *but* the lesions may be within the deeper liver parenchyma and apparent only after serial slicing of the liver.

- Impression smears of sections taken from the periphery of liver lesions examined by fluorescent antibody will confirm the presence of large numbers of *C novyi* — care should be taken in interpreting such results from animals dead more than a few hours.

Vaccination with *C novyi* type B toxoid is protective.

MALIGNANT OEDEMA AND SWELLED HEAD

The condition known as *malignant oedema* embraces severe wound infections of the subcutaneous tissues and intoxications caused by a variety of clostridial organisms — including *C septicum* most commonly, but also *C novyi* type A, *C chauvoei*, *C perfringens* type A and *C sordellii*. (*C novyi* type A was previously *C oedematiens*.) *Swelled head* is the special case of malignant oedema which occurs in rams and is usually caused by *C novyi* type A.

The disease follows introduction of the organism and the production of at least a small area of anaerobic necrosis. Wounds caused by shearing, tailing, castration, mulesing, vaccination or *crow pick* can lead to clostridial infection. Lambed ewes, particularly if there has been dystocia, can develop malignant oedema of the perineal area. In rams, fighting causes wounds on the head which predispose the ram to the development of swelled head.

Clinical signs develop within 12 to 48 hours of wounding and include local swelling, crepitus, heat and redness, the signs varying with the infecting organism. In swelled head, *C ovyi* produces oedema rather than gaseous swelling, and it leads to severe oedema of the face, throat and neck.

At necropsy, the oedema fluid is typically clear and thin, or it may be slightly gelatinous. Gas is present, except with *C novyi* infections. There is no odour associated with the lesion except infections caused by *C perfringens* and *C sordellii*, which produce a putrid odour.

The longer course of this disease allows the possibility of treatment — and penicillin can be used. Success is only likely in very early cases or when used prophylactically. Ewes which have assisted births should be given penicillin at the time of the intervention. Vaccination against *C chauvoei* and *C septicum* is available. Because *C novyi* types A and B both produce the *C novyi* alpha toxin, the *C novyi* type B toxoid is protective against malignant oedema or swelled head caused by *C novyi* type A.

BLACKLEG

Gangrenous myositis or blackleg is caused most commonly by *C chauvoei*. It is mainly a disease of young cattle but also occurs in sheep. In an outbreak occurring in rams aged 8 to 12 months[2], the clinical findings were swelling of the head, lameness marked by a stiff gait and disinclination to move, subcutaneous oedema and red to purple discolouration of the skin in the region of the body overlying a deeper muscle lesion. Crepitus was not evident before death. Animals died within a few hours of the development of clinical illness.

The subcutaneous or intramuscular oedema is bloodstained but not gaseous. Affected muscles have areas of necrosis which are dark red or black. The muscles involved include those of the head, neck and limbs. Some lesions may be deep within large muscles of the limbs or trunk and may be overlaid with normal, healthy muscle tissue.

The condition differs from swelled head in that, with blackleg, the swelling of the head is sanguineous but clear or straw-coloured in the case of swelled head.

Little is known about the pathogenesis of the disease specifically in sheep, but it is likely to be similar to that of cattle. The bacteria which cause the disease are endogenous — the spores are present in the muscle for some time before they become vegetative and cause disease.

Vaccination with formalin-inactivated whole cells of *C chauvoei* is protective.

BACILLARY HAEMOGLOBINURIA

Bacillary haemoglobinuria, caused by *C haemolyticum* (also known as C novyi type D), is reported more commonly in cattle than in sheep.

BRAXY (BRADSOT)

This condition is caused by *C septicum* and occurs when sheep, particularly lambs, have access to heavily frosted pastures. The abomasal wall is damaged, developing an abomasitis, and colonised by the organism. The toxins cause local damage which quickly develops into a fatal toxaemia. The condition is best known in countries with cold winters but has been reported from Tasmania, where mortalities of up to 20% of Merino hoggets occurred in winter.[3] Necropsy examination of animals that are freshly dead reveal acute congestion of the abomasal wall, with some haemorrhage, and very oedematous thickening of the mucosal folds. Smears from the cut surface reveal many gram-positive bacilli. Carcases of animals which die from braxy undergo very rapid putrefaction. The disease is readily prevented by vaccination and is now rare.

CLOSTRIDIAL VACCINES

In relation to the previous discussion of important clostridial diseases of sheep, the following vaccines are available and commonly used. For sheep, it is customary to include caseous lymphadenitis (CLA) vaccine with the clostridial combination vaccines, and this adds little to the cost.

If blackleg, malignant oedema, swelled head or black disease are considered to be a risk to sheep on a farm, vaccination with a 6-in-1 vaccine should be recommended. Rams are at risk of swelled head and are valuable, so it is sensible to ensure that they are well protected by vaccination. Where those diseases do not occur with a significant incidence, vaccination with only 3-in-1 could be considered as a lower cost alternative. Vaccine reactions cause persistent swellings on some sheep and, although not generally serious, these reactions are smaller and less frequent with the 3-in-1 vaccine compared to the 6-in-1 product.

Clostridial vaccines are relatively low-priced ($0.30 for 3-in-1; $0.40 for 6-in-1). For a producer who is considering the economic wisdom of vaccinating lambs, the following considerations should be taken into account:

- There is a significant risk of tetanus in lambs following marking (tailing, castration) which can be virtually eliminated by vaccination of ewes and lambs.
- The risk of enterotoxaemia deaths in lambs or hoggets is variable across farms in Australia but probably higher in the medium and high rainfall zones.

Table 14.2: The constituent antigens in commonly available cheesy gland-clostridial vaccines for sheep.

Disease condition	*3-in-1* vaccine	*6-in-1* vaccine
Caseous lymphadenitis (CLA)	*Corynebacterium pseudotuberculosis* toxoid	*Corynebacterium pseudotuberculosis* toxoid
Pulpy kidney	*C perfringens* type D toxoid	*C perfringens* type D toxoid
Tetanus	*C tetani* toxoid	*C tetani* toxoid
Black disease Swelled head caused by *C novyi* type A		*C novyi* type B toxoid
Blackleg Malignant oedema caused by *C septicum* or *C chauvoei*		*C septicum* toxoid *C chauvoei* bacterin

- Use of clostridial vaccine combined with CLA vaccine will achieve and maintain a low prevalence of CLA in the flock, but only if all sheep in the flock receive primary and secondary vaccinations and annual boosters.

- For meat lambs, sold before 12 months of age, the risk of death from tetanus and pulpy kidney in the first year of life can be reduced to near zero in those receiving primary and secondary vaccinations. The risk of CLA in unshorn lambs is low but increases following shearing. The risk of exposure to CLA infection at shearing is low if the adult flock is effectively vaccinated.

- The annual cost of vaccinating lambs twice, ewe hoggets once and adult ewes once each year is around $1 per adult ewe present, assuming that 3-in-1 vaccine at $0.30 per dose is used.

- Additional protection against blackleg, malignant oedema and black disease can be achieved at a cost of around an extra 30%.

The protection afforded by one vaccination is short-lived. Strong protection will, therefore, only come from carefully timed, repeat doses of vaccine. The timing varies with the quality of protection desired.

Table 14.3: The conventions which are normally observed for describing the first and subsequent administrations of sheep vaccines.

Dose of vaccination	Name used to describe
1	Primary
2	Secondary
3	First booster
4	Second booster

Recommended vaccination programmes for enterotoxaemia

Protection of lambs to 14 weeks of age

Vaccination of ewes such that they receive a booster vaccination in the last two weeks of pregnancy will provide passive protection to most lambs up to 8 weeks of age, but fewer than 90% of lambs will still be protected by maternal antitoxin at 14 weeks of age. Estimates of the proportion of lambs protected vary from 30% to 90%[4,5,6,7] depending on the timing of the ewes' vaccination history.[8,9,10]

To achieve a high level of protection, vaccination of lambs should, therefore, commence before they are 8 weeks of age if the ewes received a booster vaccination in late pregnancy, earlier if the ewes did not. Vaccination of lambs can be done within a few days of birth and is effective. Colostral antibodies do not appear to prevent the development of the lamb's immune response to active immunisation, but very high levels of passively acquired antitoxin may reduce the response.

Thus, if lamb marking is performed eight weeks after the start of lambing when most lambs are 5 to 7 weeks of age, booster vaccination of the ewes two weeks before lambing starts and vaccination of the lambs at marking will ensure a high level of protection of lambs to 14 weeks of age. Most lambs will remain protected for six to eight weeks after their primary vaccination; 80% will remain protected after a further eight to ten weeks.[11] The timing of the primary vaccinations for lambs usually fits well with the handling of the lambs for marking.

Timing of the vaccination of the ewes is, however, not always straightforward. Recommendations for the administration of CLA vaccine are for an annual booster to be given a few weeks before shearing — considered to be the major event contributing to disease transmission. This rarely coincides with lambing. Furthermore, producers may be unwilling to handle heavily pregnant ewes in order to administer a vaccine. Consequently, a compromise for the timing of the ewes' annual booster is necessary and the final decision will be influenced by the perceived risk of enterotoxaemia or tetanus in lambs or of CLA in the flock as a whole.

Protection of lambs beyond 14 weeks of age

To maintain continuous and strong protection, a secondary vaccination should be given within six to eight weeks of the primary dose. It is effective in raising titres to a high level if given two weeks or more after the primary vaccination, but not if given after only one week.[12] The period between primary and secondary doses can be extended to 15 to 20 weeks[2] and the longer interval increases the response to the secondary vaccination.[8] As the period between the two initial doses lengthens, however, an increasing proportion of the flock becomes susceptible to the disease. The usual manufacturer's recommendation that at least four weeks should intervene between primary and secondary doses appears to be based on avoiding any possible interference with maternally derived antitoxin by the time of the secondary vaccination.

The duration of immunity following secondary vaccination is probably 15 weeks for 90% of the flock, but only 50% are still protected at 24 weeks. Some reports suggest shorter periods of protection. In Figure 14.1, note the wide range in immune responses and the increase with time in the number of susceptible animals. The minimum protective level here is considered to be 0.15 units/ml of serum.[11]

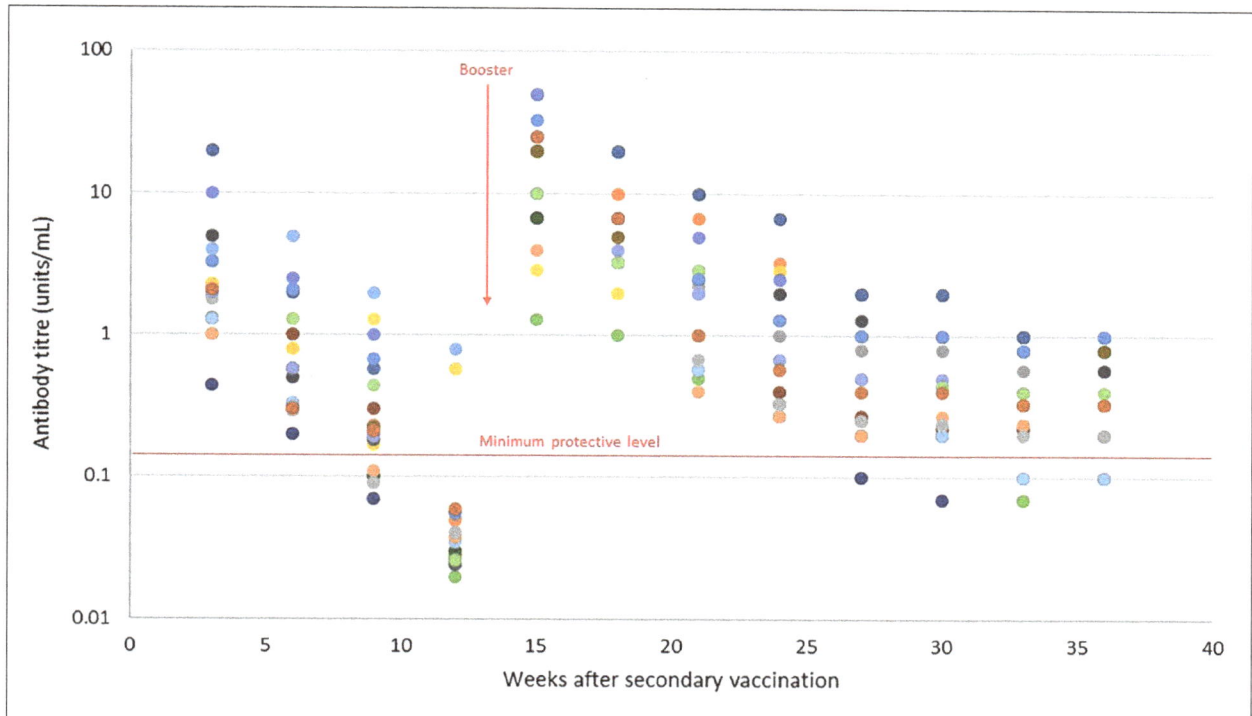

Figure 14.1: *C perfringens* epsilon antitoxin titres in a flock of sheep after receiving secondary and first booster doses of enterotoxaemia vaccine. Titres in some animals fall below protective levels five to ten weeks after the secondary vaccination and 25 to 30 weeks after the first booster vaccination. Drawn by KA Abbott, with data from Jansen (1967a).[11]

Protection to 1 year of age

Vaccination of ewes in late pregnancy and lambs at 8 and 14 weeks of age will ensure effective protection of most lambs to 7 months of age. Each booster vaccination given to animals under 1 year of age will afford a further four months' protection to nearly all of the flock and five to six months' protection to 80% of the flock.

Protection beyond 1 year of age

Booster vaccinations given at least one year after the primary vaccination lead to prolonged immunity of one year's duration at least. Further boosters given to animals in their third year of life will produce an immunity of at least three years' duration.[7]

Minimum effective protection

An alternative to the recommendations for the highest level of protection is to aim to protect sheep only at those times of the year and at those ages when the risk of disease is highest. Most deaths from enterotoxaemia occur in animals between 2 weeks and 18 months of age. It is more common during seasonal conditions which favour rapid growth and in British breed and British breed Merino cross sheep, rather than straight-bred Merinos.[13]

On any property, therefore, it may be possible to devise a vaccination programme which requires less intense treatment than that affording more continuous protection. For example, if enterotoxaemia causes the deaths of lambs before weaning, booster vaccination of ewes in late pregnancy and vaccination of lambs at marking only may completely resolve the problem. If deaths at hogget age occur, one vaccination early in life (coinciding with marking or weaning) and a secondary vaccination before the risk period some months later may be appropriate. Vaccination at this lower frequency, however, may prove to be ineffective in controlling CLA.

ANTHRAX

Most cases of anthrax in sheep are *peracute*, with sheep dying within one to two hours of first showing signs. Characteristically, there are discharges of tarry blood from all orifices.

The disease is caused by *Bacillus anthracis*. It is apparent in smears of blood from animals dead from anthrax as gram-positive rods which stain well with methylene blue — a distinguishing feature from clostridia. Spores form when the organisms are exposed to oxygen and the spores can survive in soil for many years. Animals are infected when spores enter the body by ingestion or through wounds. Soil, water or infected feeds (meatmeal, bonemeal) or bones from anthrax carcases provide sources of infection. In outbreaks, spread occurs from animals dead of the disease. The carrier state does not occur.

Within Australia, the disease occurs frequently only in NSW and Victoria and in NSW occurs most frequently between October and April. Within NSW, the higher rainfall country of the Tablelands and slopes are practically free of the disease.

The disease is readily transmissible to humans, causing respiratory, alimentary or cutaneous forms of anthrax. Fortunately, the cutaneous form is most common and responds to treatment with antibiotic therapy.

Diagnosis

When investigating sudden deaths of sheep or cattle, particularly in anthrax endemic areas, much care should be taken to avoid opening the carcases of animals dead from anthrax. Effective protective clothing should be worn for any necropsy if anthrax is considered a reasonable possibility.

Characteristic signs of death from anthrax include tarry blood discharges from any orifices, striking absence of *rigor mortis* and rapid gaseous decomposition. In such cases, a blood sample should be collected from a superficial vein. A smear is made, air-dried, fixed and stained with aqueous polychrome methylene blue.

If a necropsy is undertaken, there are a number of changes characteristic of anthrax. The blood does not clot, there are ecchymotic haemorrhages in many organs, blood-tinged serous fluid in body cavities, severe enteritis and enlargement of the spleen with liquefaction of its contents.

Anthrax must be differentiated from other causes of sudden death of sheep, particularly blackleg and from some of the following causes of sudden death.

INTOXICATIONS CAUSING SUDDEN DEATH

Poisoning with inorganic chemicals

Lead poisoning

Acute ingestion of toxic quantities of lead, leading to acute lead poisoning and, at times, sudden death, is much more common in cattle than in sheep. Due to their inquisitive habits and tendency to chew novel objects, cattle are more likely than sheep to ingest materials containing high levels of lead such as lead-acid car batteries, flaking old lead-based paint or used engine oils. While sheep can suffer from lead poisoning it is much more commonly associated with chronic, low-level ingestion of lead-containing soils.[14] Chronic lead ingestion in sheep may be asymptomatic or associated with anorexia, diarrhoea and abdominal pain.

Arsenic poisoning

Arsenical compounds were widely used as insecticides for control of ectoparasites of sheep and cattle until they were banned in Australia in 1987. Poisoning by ingestion or percutaneous absorption of arsenic was much more common when these products, and arsenical pasture insecticides, were widely used. Poisoning of grazing animals with arsenic is now uncommon.

Acute arsenic poisoning following the ingestion of toxic quantities of inorganic arsenic causes a severe gastroenteritis accompanied by severe abdominal pain, salivation and teeth grinding.[15] Animals are dull and recumbent and may die within a few hours of the appearance of signs or, in peracute cases, before premonitory signs occur. In animals which survive for more than a few hours, diarrhoea with straining and the passage of blood and mucus are evident. In subacute cases animals may survive for a few days.

Copper poisoning

Poisoning of sheep with copper takes a variety of forms but the acute disease, which can cause sudden death, is in fact the result of chronic ingestion and accumulation of copper in the liver followed by a sudden, massive release into the bloodstream and death from an acute haemolytic crisis marked by profound jaundice. The condition follows the chronic ingestion of the hepato-toxic pyrrolizidine alkaloids present in some plants and is now known as hepatogenous chronic copper poisoning but, from a time when the aetiology was unclear, is still referred to by some as *toxaemic jaundice*. The condition is discussed further in Chapter 17.

Phalaris sudden death

There are a number of plant toxicoses causing sudden death, most of which occur only sporadically when sheep have access to weeds or native plants in undeveloped grazing lands. The most common exception to this in southern Australia is the syndrome of sudden death associated with *Phalaris aquatica* — a productive pasture plant which is widely sown as a component of *improved pasture* in all southern states of Australia. Grazing of *Phalaris aquatica* (formerly *Phalaris tuberosa*) can lead to a syndrome of sudden death or to a staggers syndrome. The staggers syndrome is an acute, or more commonly, chronic nervous system disorder, associated with the presence of tryptaminic alkaloids in the phalaris plant. This form of phalaris intoxication is discussed further in Chapter 15.

In the late 1980s it was determined that phalaris sudden death is not caused by tryptamine alkaloids[16], despite earlier evidence that it was. Furthermore, outbreaks of phalaris sudden death seem to be of at least two types. The two most apparent classifications are the *cardiac* form and the so-called *polioencephalomalacic* form. Some forms of sudden death appear to be difficult to classify as either of these, and may be associated with cyanide or nitrate intoxication.[17]

The cardiac form is a cardiorespiratory disorder precipitated by a disturbance of the flock, particularly mustering, following a short period of grazing phalaris. Outbreaks are clustered in late summer, autumn and early winter. Usually fewer than 1% of the flock at risk are affected. Affected sheep collapse; some subsequently recover but most die, usually within minutes of the appearance of clinical signs. In animals examined alive the heart is in ventricular fibrillation and, although the sheep continue to breathe, cyanosis is marked.[18] There are no signs of nervous system derangement clinically or at necropsy.[17,19]

The polioencephalomalacic form of phalaris sudden death (*PE-like sudden death*) is more common than the cardiac form and losses are generally much higher.[20] This form need not be precipitated by disturbance.

Clinical signs can first appear within 4-12 hours of introduction to a *Phalaris* pasture, and deaths occur within 6-48 hours.[20] Any sheep seen alive show blindness, depression, aimless wandering, head pressing, opisthotonus, recumbency and convulsions or coma. Convulsing animals display head and body tremors, paddling and jaw champing. Frequently, dead animals are found entangled in, or pressed against fences or other objects, with evidence of terminal convulsions.[20] Morbidity can be high in some outbreaks, up to 14%, but more commonly around 5%. Mortality rates amongst affected animals are usually less than 15% but can exceed 90%.

The clinical signs are similar to the polioencephalomalacia of thiamine deficiency, albeit a very rapid form, but the name 'PE-like' is misleading. The necropsy findings are more consistent with ammonia toxicity than polioencephalomalacia, and it has been proposed[21] that this syndrome results from the sudden exposure of the animal to a high nitrogen dietary intake and a simultaneous incapacity in the urea cycle conversion of ammonia to urea.

Outbreaks of this syndrome typically occur in autumn in southern Australia when animals are introduced to rapidly growing *Phalaris* pasture, which is likely to be high in nitrogenous compounds. There is, possibly, an agent in the *Phalaris* pasture which inhibits the metabolism of ammonia until the animal is able to adapt to the diet, hence the occurrence of sudden deaths only in the first two to three days after introduction to the pasture.

The peak incidence of the nervous form of phalaris sudden death is autumn and early winter, following rainfall, particularly after a dry period, and the introduction to the pasture of animals which have not been exposed to *Phalaris* since the previous growing season, if ever. Outbreaks are less common in animals which are continuously stocked on the *Phalaris* pasture. The accumulation of nitrogen in the soil during the dry season before autumn rains is possibly a risk factor and the application of nitrogenous fertiliser may further increase the risk of an outbreak.

The earlier descriptions of field outbreaks of phalaris sudden death do not distinguish between cardiac and the 'PE-like' forms, so it is difficult to know if predisposing conditions are similar for both. It has been observed that highly fertile soils, particularly those improved with phosphate fertilisers and leguminous plants, seem to be more toxic than less fertile ones.[22]

Prevention of mortalities by grazing management is worthwhile but no strategy can be guaranteed. It would appear unwise to graze hungry sheep on *Phalaris* pastures and the risk seems to be higher if the pasture is freshly shooting and is on a soil likely to be high in nitrogen. Rotational grazing or cell grazing, in which hungry sheep may be exposed, suddenly, to rapidly grown fresh shoots of *Phalaris*, increases the risk of outbreaks compared to set-stocking. Wherever possible, *Phalaris* should be continuously grazed. Intraruminal cobalt pellets, which protect against phalaris staggers, do not protect against phalaris sudden death.[23] Phalaris sudden death occurs on a range of cultivars of *P aquatica* including low-alkaloid cultivars such as Sirolan.[24]

Poisoning with nitrate/nitrite

Some plants accumulate high levels of nitrate salts under certain conditions. In ruminants, ingested nitrate is converted to ammonia, but nitrite is intermediary in the reduction of nitrate. If sufficient nitrate is ingested in a short period the capacity of the rumen to produce ammonia is overwhelmed and nitrite accumulates and is absorbed. In the bloodstream nitrite competes with oxygen for haemoglobin, producing methaemoglobin. Administration of 25 g of sodium nitrate (18 g nitrate) to a 60 kg sheep can cause 60% of circulating haemoglobin to be converted to methaemoglobin within seven hours of administration.[25] Sheep can survive this level of methaemoglobin production but, if 80% of haemoglobin is lost to methaemoglobin, sheep become fatally anoxic. Adult sheep ingesting 1 kg of pasture containing 2% nitrate (on a dry matter basis) are effectively ingesting 20 g of nitrate.

Plants absorb nitrate from the soil and translocate it to growing parts of the plant where it is essential for the synthesis of protein. Most of the nitrogen absorbed by plants is in the form of nitrate and, if all other nutrients required for growth are not lacking, nitrate is rapidly assimilated into new tissue and does not accumulate to levels which might be toxic for grazing animals. Accumulation occurs when the rate of nitrate uptake from the soil exceeds the rate at which nitrate is incorporated into plant protein. New plant growth is generally lower in nitrates than older leaves and stems, where nitrate assimilation is reduced. Generally, within a plant, nitrate levels are highest in the bottom of stems and lowest in the youngest leaves.

The plants which are particularly prone to accumulation of nitrate include oats and other cereals, some forage *Sorghum* species and cultivars, Italian rye grass (*Lolium multiflorum*), perennial rye grass (*L perenne*)[26,27], capeweed (*Arctotheca calendula*), variegated thistle (*Silybum marianum*), button grass (*Dactyloctenium radulans*)[28] and others. Young green oats and oaten hay are notoriously high in nitrate, containing up to 7% nitrate — and a syndrome of nitrate/nitrite poisoning of cattle fed oaten hay has in the past been known colloquially as *oat hay poisoning*. Grazing of a pasture dominated by young, actively growing capeweed has led to deaths in sheep from nitrate/nitrite poisoning in Victoria; in that case, the accumulation of nitrate in the plant may have been associated with a period of drought which preceded the development of the toxic stand.[29]

Several factors increase the accumulation of nitrate in plants. When soil levels are high in nitrogen as a consequence of the application of nitrogenous fertilisers or the accumulation of heavy deposits of faeces and urine, such as occurs in areas where sheep gather in high densities, plants can absorb large amounts of nitrate. When soil conditions are dry, nitrate tends to accumulate in plants. Persistent overcast weather — for a period of days — and the consequent reduction in solar irradiation of plants also lead to nitrate accumulation, an effect which is exacerbated in dry soil conditions. Other factors which have at times been associated with nitrate/nitrite poisoning, such as the application of phenoxy herbicides, appear to be inconsistent and possibly only important when other soil or environmental conditions contribute to nitrate accumulation in plants.

Clinical signs

Animals suffering from the methaemoglobin-induced anaemic anoxia are dyspnoeic, weak, staggering and prone to collapse. The mucosae are pale or cyanosed and, in severe cases, show the chocolate-brown colour caused by the methaemoglobin. Severely affected sheep usually die within a few hours of developing signs and, in most cases, the owner will simply report the sheep as found dead.

Clinical pathology

In blood taken from affected live animals or recently dead animals, the diphenylamine blue test for nitrate/nitrite can be performed.[30] Stock solution is made up by dissolving 0.5 g of diphenylamine in 20 mL of water, then making up to 100 mL with sulphuric acid. The reagent is then made by adding 5 mL of stock solution to 40 mL of sulphuric acid. A few drops are added to a blood sample on a white dish — an intense blue colouration indicates the presence of nitrite or other substances which can oxidise diphenylamine in the presence of sulphuric acid. It is unlikely that blood would contain any other anion which would give this result, so the test used in this way is considered specific for nitrite or nitrate. The test is, however, highly sensitive, and very low concentrations of nitrite/nitrate can be detected. Positive reactions should therefore be interpreted conservatively.

Nitrite is stable in plasma for up to 48 hours and so the test can be applied to specimens some hours after collection, providing that the blood is collected from a live or recently dead animal and that plasma is promptly separated from red cells.[31] The diphenylamine blue test can also be applied to detect nitrate in urine from affected animals or in crushed leaf samples from plants which the affected sheep have been grazing.[29]

Nitrate test strips — developed for testing soil and water samples — are probably the most convenient method of testing for nitrate/nitrite, particularly in the field. They can be applied to aqueous humour, serum or urine. Nitrate persists for some hours in carcases after death, which makes this a practical test when investigating field cases. If the animal has been dead for some hours, aqueous humour is likely to be the least decomposed tissue to test.

If blood can be collected from affected live animals well before death and submitted promptly to a diagnostic laboratory, measurement of the methaemoglobin concentration will provide definitive and quantified evidence of the degree of the methaemoglobin-induced anoxia. Caution must be exercised in interpreting results if collection and testing is compromised. Methaemoglobin concentrations do not remain static after death but are affected by two opposing chemical processes which proceed at unpredictable rates. The methaemoglobin produced in life by the absorbed nitrite is reduced after death to haemoglobin, while post-mortem putrefactive changes cause a conversion of haemoglobin to methaemoglobin. If blood specimens cannot be assayed soon after collection, then stabilisation of methaemoglobin concentrations can be achieved for 48 hours by collecting blood with an anticoagulant and diluting the specimen 1:20 with 0.07M phosphate buffer at pH 6.6.[31]

Necropsy findings

In sheep which are examined soon after death caused by nitrite intoxication, the discolouration of the blood and tissues is obvious. The blood is dark red to chocolate-brown in colour. The circulatory collapse may lead to the presence of petechial haemorrhages in heart muscle and trachea and congestion of lungs and other tissues.

Treatment

Methylene blue is an effective treatment. It should be administered intravenously in a 1% solution (10 mg/mL) at the rate of 1-2 mg/kg (6-12 mL per 60 kg liveweight). In Australia, methylene blue can only be administered by registered veterinarians. The response to treatment is rapid but treatment may need to be repeated in 6-8 hours if nitrite absorption from ingesta continues after the first treatment.

(Methylene blue is also the antidote to para-aminopropiophenone (PAPP) — a poison used to kill foxes, wild dogs and dingoes in Australia and other pests in New Zealand, available commercially as FOXECUTE® and DOGABAIT® (Animal Control Technologies Australia Pty Ltd). After ingestion and absorption, PAPP is converted in the liver to the active metabolite which acts in erythrocytes to oxidise haemoglobin to methaemoglobin. Domestic dogs which ingest a PAPP bait will die unless methylene blue is administered with 45 to 90 minutes of ingestion.[32] The recommended dose of 1% methylene blue for dogs is 5-10 mL intravenously — administered slowly over 2-5 minutes. Veterinarians should have stocks of methylene blue immediately available for urgent use in the event of nitrate/nitrite poisoning of livestock or PAPP ingestion by a domestic dog.)

Prevention

Sheep exposed to high-nitrate diets are better able to avoid nitrite intoxication if their feed intake is high in soluble carbohydrates. Presumably the ready supply of dietary energy substrates facilitates the ruminal reduction of nitrate to ammonia with less accumulation and absorption of nitrite. Pastures containing weeds which have been sprayed with phenoxy herbicides (such as 2-methyl-4-chlorophenoxyacetic acid (MCPA)) should not be grazed for two to three weeks in order for nitrate levels in plants to decline. If pastures are considered to be a high risk for nitrate/nitrite poisoning, sheep should be introduced to the pasture after full feeding on other, low-nitrate feeds and removing them at night for several days. This allows close observation of the stock for signs of mild intoxication and also limits the nitrate ingestion for a period while the ruminal flora adapt to the high nitrate intake.

Fluoroacetate poisoning

Grazing animals are exposed to fluoroacetate as either a synthetic chemical — the sodium salt is known commonly as 1080 and produced as a poison for rabbits and other mammals — and as a naturally occurring substance in the leaves, flowers and seeds of some plants. Fluoroacetate exerts its toxic effect by interference with the citric acid cycle — specifically by preventing the activity of the enzyme aconitase which is essential for citrate metabolism. As a consequence, citrate accumulates and energy production from carbohydrate at the cellular level ceases.

Mammals intoxicated by fluoroacetate can die suddenly, with few premonitory signs, or show clinical signs for a few hours before death.

A number of leguminous plant species native to Australia produce fluoroacetate. All but one of these is in the *Gastrolobium* genus. (This genus now includes species which were formerly in the *Oxylobium* genus.) Most *Gastrolobium* spp are confined to southwest Western Australia, where they are known generally as *poison peas*, but also by a range of specific, local names. Australian wildlife native to those regions of Western Australia have evolved a degree of tolerance to fluoroacetate, enabling them to eat the plant with impunity. (This tolerance also confers protection from the effects of 1080.) Introduced ruminants, however, have no tolerance and there have been numerous occurrences of mortality in sheep grazing pastures or scrub land including *Gastrolobium* spp. In the first century of agriculture in Western Australia, these species formed the most important group of poisonous plants in the state.

The poisonous *Gastrolobium* species of Western Australia have been described in detail by TEH Aplin in a series of bulletins produced by the Western Australian Department of Agriculture in the period 1967-71.[33] Briefly, most species are shrubs or small trees to 1.5 m, some species up to 3 m. While most species are confined to the southwest region of WA, some occur in the north of the state, and some in the Coolgardie district. By 1967, there were 34 recognised poisonous species. The levels of fluoroacetate in the plants vary between species, between regions and with the stage of growth, being highest at flowering, during pod formation and when growing rapidly. Plants contain between 30 and 2500 ppm of fluoroacetate. At 1000 ppm, 25 g of plant material will kill a 50 kg sheep. Mortalities have also occurred in dogs and cats which have fed on carcases of animals which died by eating the plants.

Elimination of the plant from grazing land is the only control measure available. While this has been possible in the higher rainfall and more intensively managed regions of the state, it has not been a practicable control measure in the north and pastoral land, where the plants remain a risk to the grazing of sheep, cattle and goats.

Two plant species which occur in Northern Territory and Queensland also produce fluoroacetate. One of these is another *Gastrolobium* species — *G grandiflorum* — and the other is *Acacia georginae*. *G grandiflorum* (known variously as heartleaf poison bush, desert poison bush and wallflower poison bush in different regions) has a sporadic occurrence from the Hamersley Ranges of WA, through Northern Territory to central Queensland and Cape York Peninsula. The plant is highly toxic to grazing animals and cases of mass mortality have occurred as a result of grazing on the plant. In areas where it occurs, stockmen are assiduous in restricting the access of sheep or cattle to the plants until they can be physically removed. *A georginae* (Georgina gidgee) is a shrub or small tree which grows in the Georgina River watershed of northwestern Queensland and in eastern Northern Territory. The plant caused numerous outbreaks of mortalities of sheep and cattle in that district in the late 19th and early 20th centuries (described in detail by Bell et al., 1955[34]).

Both heartleaf and Georgina gidgee continue to cause occasional mortalities of cattle, particularly those animals which are recently introduced to the regions where the plants grow.

Sheep poisoned by fluoroacetate may simply be found dead or may be observed to stagger and fall or to suddenly run aimlessly before collapsing and dying — often within minutes of the appearance of altered behaviour. Sometimes, if left undisturbed, animals recover uneventfully.

Stockmen have noted that, with cattle, the ingestion of small amounts of the poisonous plants causes the animals distress sufficient to ensure that they learn to avoid the plant in future.

Cardiac glycoside poisoning

There are numerous plants in Australia which contain cardiac glycosides and which are potentially toxic to sheep, but poisoning with these plants is reported much more commonly in cattle than sheep. Many of the plants which contain cardiac glycosides are ornamentals that have been introduced into Australia and occur in gardens or have escaped from gardens and become naturalised. Some of the most familiar plants which are potentially toxic to sheep include *Convallaria majalis* (lily of the valley), *Digitalis* spp (foxgloves), *Nerium oleander* (common oleander)[35] and *Cascabela thevetia* (yellow oleander). Several *Bryophyllum* species which occur in Queensland contain cardiac glycosides and occasionally cause poisoning of cattle.[36]

The action of the cardiac glycosides is primarily on the heart muscle, causing severe disturbances of the heart rhythm and function. Affected sheep are depressed, frequently lying and rising, before becoming recumbent and convulsing. The course of the illness is short and, if sufficient material is ingested, animals die within one to four hours of first showing signs.

If animals are seen alive, then treatment can be considered. In valuable animals, activated charcoal can be given orally and is helpful in reducing the absorption of further glycosides from the alimentary tract. Antiarrhythmic drugs such as atropine and propranolol may be used to control or lessen cardiac disturbances.

Cyanogenic glycosides

Many plants contain cyanogenic glycosides and can produce hydrocyanic acid which has been considered to be a defence mechanism to reduce ingestion by herbivores. Indeed, herbivores tend to avoid eating plants containing high levels of hydrocyanic acid or cyanogenic glycosides. Nevertheless, these plants are consumed at times and the glycosides, when broken down by rumen micro-organisms, release hydrocyanic acid. Cyanide is absorbed into the bloodstream and transported to cells, where it combines with the mitochondrial enzyme cytochrome c oxidase, preventing further cellular respiration. Animals become hypoxic and, if the intoxication is sufficiently high, die very rapidly — sometimes within an hour or less of ingesting toxic feeds.

Cyanogenic plants which occur commonly in Australia and may poison grazing ruminants include the forage sorghums and species of *Eucalyptus*. *Eucalyptus cladocalyx* (sugar gum) is the best-known cyanogenic eucalypt and trees of this species are often planted on farms as windbreaks or along farm entrances. The leaves from felled trees, cut or fallen branches, or the new growth of young trees can poison grazing ruminants.[37] Over 20 species of *Eucalyptus* have been found to contain cyanogenic glycosides[38] but the levels are unpredictable and inconsistent. Poisoning of ruminants has not been reported from species other than sugar gums.

Treatment

The most commonly used treatment consists of the intravenous administration of sodium thiosulphate (660 mg/kg) and sodium nitrite (up to but not exceeding 22 mg/kg). For a 50 kg sheep, this amounts to 1 g of sodium nitrite and 33 g of sodium thiosulphate in 300 mL of sterile water or isotonic rehydration fluid. Lower doses of sodium nitrite are advised if there is a risk of concurrent nitrate toxicity, which may occur, for example, with sorghum forages.

The administration of sodium nitrite leads to the formation of methaemoglobin for which cyanide has a strong affinity, binding to form cyanmethaemoglobin. At the maximum recommended dose for sodium nitrite, approximately 40% of haemoglobin may be converted to methaemoglobin. Higher doses therefore may be dangerous if they lead to the formation of so much methaemoglobin that oxygen transport is impaired — further contributing to the tissue hypoxia which has resulted from the cyanide intoxication.

Sodium thiosulphate provides sulphur atoms in a form which can be readily used in the formation of thiocyanate in a reaction catalysed by the enzyme rhodanese. Thiocyanate is relatively harmless and is readily excreted. In the past, the recommended dose of sodium thiosulphate was 90% lower than that cited above but studies have demonstrated that the higher dose of thiosulphate is significantly more effective.[39]

Cobaltous chloride administered intravenously at the rate of 10.6 mg/kg may enhance the effect of sodium thiosulphate, with or without the use of sodium nitrite. The recommended dose rate of cobaltous chloride should not be exceeded.

The activity of sodium thiosulphate occurs relatively slowly, probably because rhodanese is an intracellular enzyme and the cyanide is acting within cells, while most thiosulphate circulates in extracellular fluids. Thus, in cases of acute, severe cyanide poisoning, thiosulphate alone may act too slowly to reverse the intoxication and sodium nitrite will be necessary to provide a quick reversal of the effect of cyanide at mitochondrial level. The thiosulphate acts more slowly to collect cyanide — including that which is gradually released from cyanmethaemoglobin and is responsible for the ultimate excretion of the cyanide from the body. Treatment may have to be repeated if absorption of cyanide continues, but care must be taken to avoid overdosing with nitrite and the formation of excessive amounts of methaemoglobin — effectively inducing a nitrite intoxication. The risk of this is enhanced if the cyanogenic plant material was also high in nitrates.

Prevention

In the case of the forage sorghums, there is variation between species in the cyanogenic potential, with Johnson grass (*S halapense*), Columbus grass (*S almum*), Sudan grass (*S sudanese*) and Sorghum-Sudan grass hybrids (*Sudax*™) presenting a higher risk than some other species. All sorghums can present the risk of poisoning, and regrowth after a check in growth — caused by drought, frost, grasshopper attack, heavy grazing, for example — is often the critical factor in leading to an outbreak of cyanide poisoning.

Given the choice, livestock do tend to avoid plants that are high in cyanide but, if there is no other forage available or if the animals are hungry, the low palatability of the plants may be ignored. Consequently, particular care should be taken if hungry livestock are likely to have access to sorghum forages, and even greater care if the forages include regrowth after a period of stress.

Green cestrum poisoning

Green cestrum (*Cestrum parqui*) is an introduced perennial shrub which has become established in areas of eastern Australia from southeast Queensland to South Australia. It grows in higher rainfall zones and along watercourses, particularly in neglected areas of urban environments. The leaves, branches and berries contain a toxin (carboxyparquin) which, if ingested, can cause rapid death with liver necrosis in sheep, goats, cattle and other animals, including humans. The plant is not readily eaten by livestock but, if other feed is in short supply, animals will consume the plant. A summary of six cases in cattle from four farms in Queensland described cattle which were either found dead or severely depressed, recumbent or staggering when ambulatory, with ruminal stasis, petechial haemorrhages of the conjunctivae and faeces of normal consistency but streaked with blood. The mortality rate of affected animals was over 90%.[40]

Blue-green algal poisoning

The dead algal material which accumulates in still waters following a bloom of blue-green algae contains potent hepatotoxins. If sufficient material is ingested to cause severe liver damage, death from liver failure can occur within 24 to 72 hours of exposure. Less acute intoxication leads to secondary photosensitisation — which is discussed further in Chapter 10.

ENVIRONMENTAL CONDITIONS CAUSING SUDDEN DEATH

Lightning strike

A lightning strike delivers an electric potential of many tens of kilovolts to the earth and can lead to the exposure of grazing animals to massive and often instantly fatal electric shocks. In some cases just one or a few sheep may be found dead after an electrical storm, but there are reports of cases where hundreds of sheep have been killed, usually within an area of about a 20 m radius from a tall object which is the focal point of a lightning strike. The electrical current can be transmitted to the animals in one of several ways[41]:

- The animal may receive a *direct strike*, particularly if it is exposed on high ground. In such cases, the entire current passes through the animal from the point of contact to the ground. This can also occur if the animal is in contact with the object which is struck, or in contact with other animals which are electrocuted.

- The animal may receive a *side flash*, when a tall object (usually a tree in the case of sheep) is hit by the lightning and an electric current may jump from the struck object to the animal standing near it.

- The animal receives a shock transmitted through the ground. When lightning hits an object in a field, such as a tree, the electricity is transmitted to the ground and spreads radially away from the base of the object through the ground. Due to the resistance offered by the earth, the potential declines markedly with increasing distance from the base of the object, leading to a potential drop between any two points on the ground which are oriented in a line away from the point of lightning strike. The potential gradient is highest if the electrical resistance of the earth is high. For quadrupeds this means that there can be a potential difference between the point of ground contact of the fore feet and the hind feet and a current will flow through the animal from the feet closest to the point of strike to the feet furthest from the point of strike. This form of lightning strike is called *step potential*.

This last form of electrocution from lightning strike is the most common form of death by lightning for sheep. Their natural instinct to gather as a flock means that many animals may be exposed to electrocution from step potential in one small area near a tree or other object which is struck by lightning.

At necropsy, there may be very few signs to indicate death from lightning strike, particularly if the deaths are the result of step potential. There may be burn marks on the skin, feet or muzzle depending on the point of contact with the source of current flow.

Diagnosis

When multiple deaths from lightning strike occur in a flock of sheep, the distribution pattern of the carcases, the history of an electrical storm and possibly the presence of a burnt or lightning-damaged tree are generally sufficient evidence of the cause of death. The other principal causes of simultaneous mass mortalities include plant poisonings, particularly from cyanogenic plants. On some occasions, when animals are found dead from lightning strike 6-24 hours after death, the post-mortem ruminal distension may initially suggest bloat as a cause of death. Electrocution from other sources, such as downed electrical cables, is possible but the cause is generally obvious.

Deaths in one or two animals only may be more difficult to ascribe to lightning unless there is evidence of lightning damage in the environment or burns to the animals affected. In such cases, other causes of sudden death should be considered and ruled out if possible. When deaths are discovered in sheep following storms, the possibility of exposure and hypothermia should also be considered — particularly in sheep which are in short fleece.

Exposure/hypothermia

Losses of sheep following shearing when exposed to low temperatures, rain and high wind chill factors can be very high, particularly in summer and particularly if the sheep have been losing weight prior to shearing. Losses as high as 90% of the recently shorn sheep have been recorded and mortalities can occur in sheep as long as 28 days off-shears.[42]

OTHER CAUSES OF SUDDEN DEATH

A number of other disease conditions which have been reviewed in other chapters can, under certain conditions, cause sudden death. These include salmonellosis, *red gut*, haemonchosis, acute fascioliasis, hypomagnesaemia and bloat.

RECOMMENDED READING

Radostits OM, Gay CC, Hinchcliff KW and Constable PD (2007) *Diseases associated with Clostridium spp*, In *Veterinary Medicine* 10th edition, Saunders Ltd, pp. 821-46.

REFERENCES

1   Hatheway CL (1990) Toxigenic clostridia. Clin Microbiol Rev **3** 66-98. https://doi.org/10.1128/CMR.3.1.66.

2   Seddon HR, Belschner HG and Edgar G (1931) Blackleg in sheep in New South Wales. Aust Vet J **7** 2-18.

3   Dumaresq JA (1939) Braxy in Tasmania. Aust Vet J **15** 252-5. https://doi.org/10.1111/j.1751-0813.1939.tb01241.x.

4   Hepple JR, Chodnik KS and Price EK (1959) Immunisation of Lambs against *Clostridium welchii* type D Enterotoxaemia (Pulpy Kidney Disease) with a purified toxoid aluminium treated. Vet Rec **71** 201-7.

5   Oxer DT, Minty DW and Leifman CE (1971) Vaccination trials in sheep with clostridial vaccines with special reference to passively acquired *Cl. welchii* type D antitoxin in lambs. Aust Vet J **47** 134-40. https://doi.org/10.1111/j.1751-0813.1971.tb02120.x.

6   Wallace GV (1963) Homologous passive protection of lambs against various clostridial diseases. NZ Vet J **11** 39-40. https://doi.org/10.1080/00480169.1963.33486.

7   Wallace GV (1964) Homologous passive protection of lambs against various clostridial diseases — Part 2. NZ Vet J **12** 61-2. https://doi.org/10.1080/00480169.1964.33550.

8   Cooper BS (1967) The transfer from ewe to lamb of clostridial antibodies. NZ Vet J **15** 1-7. https://doi.org/10.1080/00480169.1976.33676.

9   Thomson A and Batty I (1953) The antigenic efficiency of pulpy kidney disease vaccines. Vet Rec **65** 659.

10 Jansen BC (1960) The experimental reproduction of pulpy kidney disease. J S Afr Vet Med Assoc **31** 205-8.

11 Jansen BC (1967a) The duration of immunity to pulpy kidney disease of sheep. Onderstepoort J Vet Res **34** 333-43.

12 Jansen BC (1967b) The production of a basic immunity against pulpy kidney disease. Onderstepoort J Vet Res **34** 65-79.

13 Seddon HR (1965) Enterotoxaemia of sheep and goats. In: Diseases of domestic animals in Australia, Part 5: Bacterial diseases, ed HE Albiston. 2nd ed. Department of Health: Canberra, p. 71.

14 Koh TS and Judson GJ (1986) Trace-elements in sheep grazing near a lead-zinc smelting complex at Port Pirie, South Australia. Bull Environ Contam Toxicol **37** 87-95. https://doi.org/10.1007/BF01607734.

15 White IG, Blood DC and Whittem JH (1948) Arsenic poisoning in sheep. Aust Vet J **24** 331-4. https://doi.org/10.1111/j.1751-0813.1948.tb04614.x.

16 Bourke CA, Carrigan MJ and Dixon RJ (1988) Experimental evidence that tryptamine alkaloids do not cause *Phalaris aquatica* sudden death syndrome in sheep. Aust Vet J **65** 218-20. https://doi.org/10.1111/j.1751-0813.1988.tb14462.x.

17 Bourke CA and Carrigan MJ (1992) Mechanisms underlying *Phalaris aquatica* 'sudden death' syndrome in sheep. Aust Vet J **69** 165-7. https://doi.org/10.1111/j.1751-0813.1992.tb07503.x.

18 Gallagher CH, Koch JH and Hoffman H (1966) Diseases of sheep due to ingestion of *Phalaris tuberosa*. Aust Vet J **42** 279-84. https://doi.org/10.1111/j.1751-0813.1966.tb08836.x.

19 Gallagher CH, Koch JH, Moore RM et al. (1964) Toxicity of *Phalaris tuberosa* for sheep. Nature **204** 542-5. https://doi.org/10.1038/204542a0.

20 Bourke CA, Rendell D and Colegate SM (2003) Clinical observations and differentiation of the peracute *Phalaris aquatica* poisoning syndrome in sheep known as 'Polioencephalomalacia-like sudden death'. Aust Vet J **81** 698-700. https://doi.org/10.1111/j.1751-0813.2003.tb12545.x.

21 Bourke CA, Colegate SM, Rendell D et al. (2005) Peracute ammonia toxicity: A consideration in the pathogenesis of *Phalaris aquatica* 'Polioencephalomalacia-like sudden death' poisoning of sheep and cattle.. Aust Vet J **83** 168-71. https://doi.org/10.1111/j.1751-0813.2005.tb11631.x.

22   Moore RM and Hutchings RJ (1967) Mortalities among sheep grazing *Phalaris tuberosa* Aust J Exp Agric Anim Husb **7** 17-21. https://doi.org/10.1071/EA9670017.

23   Moore RM, Arnold GW and Hutchings RJ (1961) Poisoning of Merino sheep on *Phalaris tuberosa L.* pastures. Aust J Sci **24** 88-9.

24   Kennedy DJ, Cregan PD, Glastonbury JRW et al. (1986) Poisoning of cattle grazing a low-alkaloid cultivar of *Phalaris aquatica*, Sirolan. Aust Vet J **63** 88-9. https://doi.org/10.1111/j.1751-0813.1986.tb02938.x.

25   Lewis D (1951) The metabolism of nitrate in the rumen of the sheep. Biochem **48** 175-80.

26   Nicholls TJ and Miles EJ (1980) Nitrate/nitrite poisoning on cattle on ryegrass pastures. Aust Vet J **56** 95-6. https://doi.org/10.1111/j.1751-0813.1980.tb05640.x.

27   O'Hara PJ and Fraser AJ (1975) Nitrate poisoning in cattle grazing crops. NZ Vet J **23** 45-53. https://doi.org/10.1080/00480169.1975.34192.

28   McKenzie RA, Rayner AC, Thompson GK et al. (2004) Nitrate-nitrite toxicity in cattle and sheep grazing *Dactyloctenium radulans* (button grass) in stockyards. Aust Vet J **82** 630-4. https://doi.org/10.1111/j.1751-0813.2004.tb12612.x.

29   Fairnie IJ (1969) Nitrite poisoning in sheep due to capeweed (*Arctotheca calendula*). Aust Vet J **45** 78-9. https://doi.org/10.1111/j.1751-0813.1969.tb13697.x.

30   Householder GT, Dollatite JW and Hulse R (1966) Diphenylamine for the diagnosis of nitrate intoxication. J Am Vet Med Assoc **148** 662-5.

31   Watts H, Webster M, Chappel A et al. (1969) Laboratory diagnosis of nitrite poisoning in sheep and cattle. Aust Vet J **45** 492. https://doi.org/10.1111/j.1751-0813.1969.tb06605.x.

32   Southwell D, Boero O, Mewett O et al. (2013) Understanding the drivers and barriers to participation in wild canid management in Australia: Implications for the adoption of a new toxin, para-aminopropiophenone. Int J Pest Man **59** 35-46.

33   Aplin TEH (1967-1971) Series of Bulletins from Western Australian Department of Agriculture; Bulletins numbered 3483, 3554, 3672, 3706, 3778 and 3811.

34   Bell AT, Newton LG, Everist SL et al. (1955) *Acacia georginae* poisoning of cattle and sheep. Aust Vet J **31** 249-57. https://doi.org/10.1111/j.1751-0813.1955.tb05445.x.

35   Aslani MR, Movassaghi AR, Mohri M et al. (2004) Clinical and pathological aspects of experimental oleander (*Nerium oleander*) toxicosis in sheep. Vet Res Commun **28** 609-16. https://doi.org/10.1023/B:VERC.0000042870.30142.56.

36   McKenzie RA, Franke FP and Dunster PJ (1987) The toxicity to cattle and bufadienolide content of six *Bryophyllum* species. Aust Vet J **64** 298-301. https://doi.org/10.1111/j.1751-0813.1987.tb07330.x.

37   Webber JJ, Roycroft CR and Callinan JD (1985) Cyanide poisoning of goats from sugar gums (*Eucalyptus cladocalyx*). Aust Vet J **62** 28. https://doi.org/10.1111/j.1751-0813.1985.tb06041.x.

38   Gleadow RM, Haburjak J, Dunn JE et al. (2008) Frequency and distribution of cyanogenic glycosides in *Eucalyptus*. L'Hérit Phytochemistry **69** 1870-4.

39   Burrows GE and Way JL (1979) Cyanide intoxication in sheep: Enhancement of efficacy of sodium nitrite, sodium thiosulphate and cobaltous chloride. J Am Vet Res **40** 613-17.

40   McLennan MW and Kelly WR (1984) *Cestrum parqui* (green cestrum) poisoning of cattle. Aust Vet J **61** 289-91. https://doi.org/10.1111/j.1751-0813.1984.tb06013.x.

41   Gomes C (2012) Lightning safety of animals. Int J Biometeorology **56** 1011-23. https://doi.org/10.1007/s00484-011-0515-5.

42   Holm Glass M and Jacob H (1992) Losses of sheep following adverse weather after shearing. Aust Vet J **69** 142-3. https://doi.org/10.1111/j.1751-0813.1992.tb07486.x.

15

Diseases of the central nervous system (CNS)

POLIOENCEPHALOMALACIA (PEM)

A characteristic of sheep affected by this condition, also known as PEM, cerebrocortical necrosis (CCN) or polio, is the adoption of a stationary position with a head-high attitude, as though the animals were *star-gazing*. The disease occurs sporadically in sheep at pasture and in epidemic form in feedlot lambs. The disease process is an acute cerebral oedema with cerebral necrosis. Cases treated early with thiamin may respond rapidly. The disease occurs in animals of either sex and of any age. Occasionally, a change of diet to one of lower roughage content will lead to outbreaks in lambs or weaners which are confined and fed concentrates. The disease may also be precipitated by stressful events which prevent grazing for a day or two.

Role of thiamin

Thiamin (vitamin B_1) is normally produced by rumen microbes. PEM has been reproduced by feeding the thiamin antagonist amprolium. It is likely that natural cases of the disease are produced by thiaminase activity in the rumen rather than a deficiency of thiamin or its precursors. Thiaminase I appears to be the relevant enzyme, not thiaminase II. In addition to destroying thiamin, there is evidence that thiaminase I causes thiamin analogues to be formed which act as thiamin antimetabolites and accentuate the deficiency syndrome.[1] Thiaminase I is present in significant amounts in a number of ferns (see plant-associated toxicoses later in this chapter). Thiaminase activity can also occur through natural means in the alimentary tract of sheep, probably associated with the establishment of *Bacillus thiaminolytica* or other bacteria, but their role in the pathogenesis of field cases of PEM is unclear.[2] Sheep and goats with PEM may have elevated ruminal and faecal thiaminase activity and depressed erythrocyte transketolase activity.[3]

Thiamin is an essential cofactor in oxidative decarboxylation of some intermediate compounds in carbohydrate metabolism. Possibly, the high and specific requirement of the cerebral cortex for oxidative metabolism of glucose could explain the cerebral oedema and necrosis in PEM. Vitamin B_{12} also has a role in carbohydrate metabolism, and cobalt deficiency has been suggested as a contributing cause to PEM.

Diets high in sulphur have also been associated with cases of PEM in cattle and it seems likely that high levels of dietary sulphur will lead to reduced blood levels of thiamin.[4] In sheep, experimentally, high doses of sodium hydrosulphide have caused clinical and necropsy signs of PEM within 1 to 96 hours of treatment.[5] In sheep exposed to high-sulphur diets, thiamin may be protective of the effects of the high sulphur intake and clinical signs of PEM may be more likely to occur when animals with low thiamin status are exposed to diets high in sulphur.[6]

Clinical signs

Clinical signs include the sudden onset of blindness. Animals separate from the mob, wander aimlessly or stand in one place, often adopting a head-high or head-low posture or head-pressing against solid objects (Figure 15.1). There is a muscle tremor, particularly of the head, and jaw champing. Signs may be intermittent initially. After some hours the signs become continuous, and the animals lie down and demonstrate opisthotonus, nystagmus and convulsions, particularly if disturbed. Some animals recover without treatment, particularly those that develop signs following stressful events. Once recumbency intervenes, however, recovery is rare. Animals become comatose and die within two to three days, sometimes less. Most outbreaks involve fewer than 5% of a mob. Cases are often spread over several weeks.

Diagnosis

PEM must be differentiated from listeriosis, FSE, enterotoxaemia, tetanus, hypomagnesaemia, hypocalcaemia (including that caused by oxalate poisoning) and pregnancy toxaemia. When faced with a number of affected animals or a history of a number of deaths, sporadic diseases like brain abscess can be discounted. Response to treatment with thiamin is diagnostic.

At necropsy, brain lesions are characteristic and can be grossly visible. There are extensive, bilateral areas of yellow discolouration of the dorsal and lateral cerebrocortical grey matter.

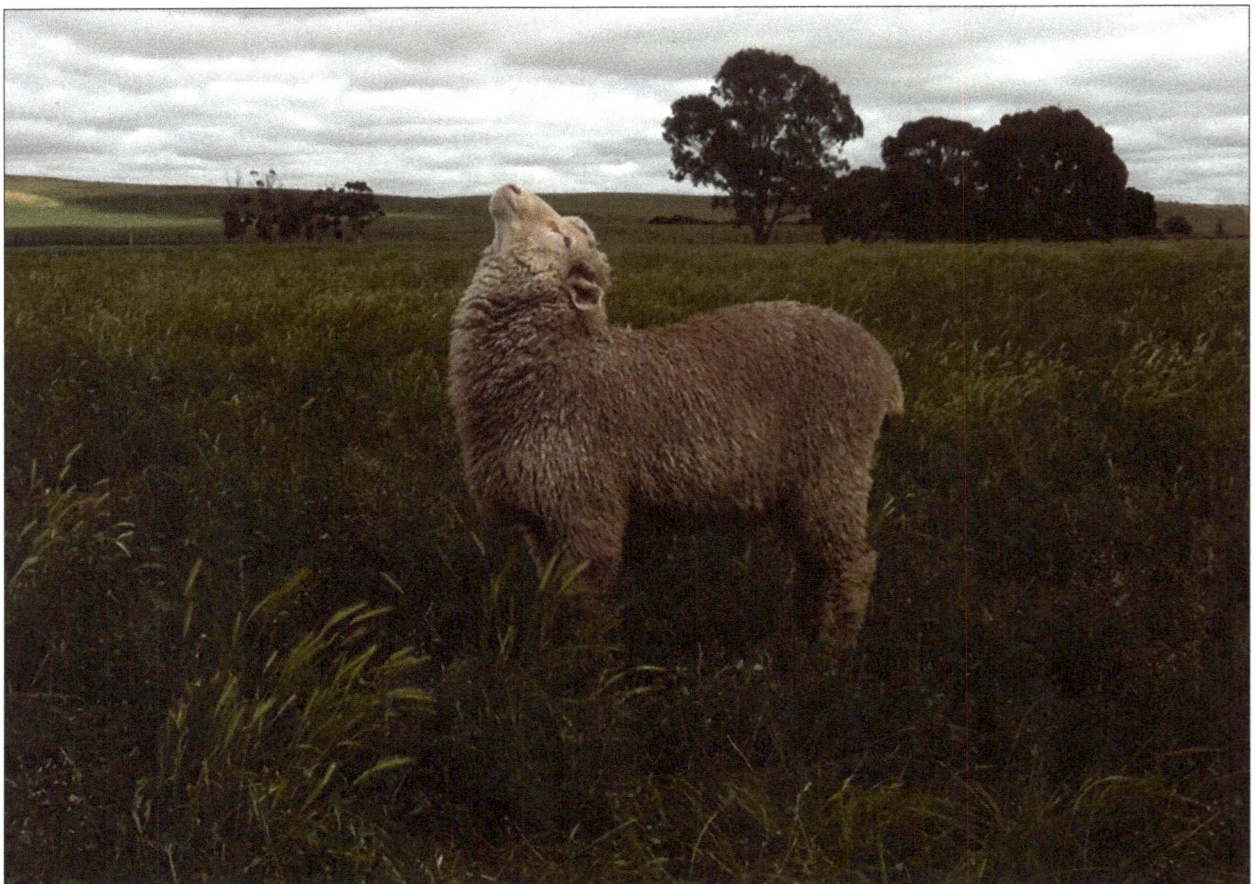

Figure 15.1: Sheep with polioencephalomalacia often adopt a 'star-gazing' posture. Photograph courtesy of Tom Trengove.

The lesions fluoresce under UV light. These latter signs are most obvious in animals that have been clinically affected for two days or more before death and necropsy. In some cases the brain swelling is sufficient for flattening of the gyri to be evident.

Laboratory confirmation can be based upon histopathology of fixed brain tissues, particularly cerebral hemispheres. Additionally, blood biochemistry may be useful — in cases of PEM, thiamin levels are depressed; pyruvate and lactate levels are raised; transketolase activity is lowered. Rumen and faecal samples may be tested for thiaminase activity.

Treatment

If given early, parenteral thiamin is highly effective in many cases, at 10 mg/kg IV initially, then repeated IM every 12 hours for a total of four treatments, if necessary. Response should be evident within two to six hours.

Prevention

Attempts to prevent the development of further cases include the introduction of more roughage, a change of pastures and the subcutaneous injection of thiamin.

FOCAL SYMMETRICAL ENCEPHALOMALACIA (FSE)

FSE is a form of enterotoxaemia caused by *Cl perfringens* type D in animals which have partial immunity or which suffer only partial intoxication. The signs are referable to chronic neurological damage. The disease is usually sporadic and typically affects animals under 12 months of age[7], but adults are also affected.[8]

Clinical signs

In acute cases, animals are in lateral recumbency and unable to stand even if assisted. Convulsions are uncommon — they mostly lie quietly but there may be intermittent paddling and head dorsiflexion. There is a pupillary light reflex but no eye preservation reflex, and they appear to be blind. Death occurs within 24 hours.

In more chronic cases affected animals are blind and nonresponsive to external stimuli and they wander aimlessly, occasionally circling and head-pressing before eventually becoming recumbent. The course of the disease varies between three and ten days, before death intervenes.

Diagnosis

At necropsy, brain lesions are grossly visible. These consist of bilaterally symmetrical haemorrhage and softening in the internal capsule, basal ganglia, dorsolateral thalamus, substanta nigra, mid-brain stem and cerebellar peduncles. These lesions are effectively an advance of the brain lesions histopathologically evident in enterotoxaemia.

Treatment and prevention

There is no effective treatment. Prevention is the same as for enterotoxaemia.

LISTERIOSIS

Listeria monocytogenes causes meningo-encephalitis in sheep of any age or sex. This latter form is characterised by the circling behaviour of affected sheep. The organism is a common gut inhabitant of many animals and can survive for extended periods in soil and vegetation. Outbreaks of the disease are often associated with silage feeding and the organism is capable of surviving, even multiplying, in poor-quality (pH above 5.5) silage, or in wet, muddy conditions. An outbreak involving sheep on 27 farms in one region of eastern Victoria occurred in the winter of 1978, associated with well-above-average early winter rainfall (three times higher than average) and sustained flooding of pastures.[9]

Clinical signs

Affected animals are dull, off-feed and febrile in the early stages of the disease, but the temperature may fall to normal before clinical cases are examined. They separate from the mob, walk in circles or stand with their heads to one side. There is unilateral facial nerve paralysis and sometimes ocular and/or nasal discharge. The animals become recumbent and die; the course of the disease is from a few days up to one week.

Diagnosis

The disease must be differentiated from pregnancy toxaemia in ewes in late pregnancy, PEM and FSE. A fever, if present, facial paralysis and circling, indicating a unilateral lesion of the brain, are presumptive evidence of listeriosis.

Confirmation generally comes from necropsy examination. Gross lesions of the brain are minimal, although the cerebrospinal fluid may be cloudy. Histologically there is microabscess formation, particularly in the mid-brain. Bacteriological isolation is difficult in encephalitic cases. Definitive diagnosis is not always possible but histological evidence of encephalitis, combined with the history, allow a presumptive diagnosis in most cases. From suspected field cases, brains should be removed either intact or in two halves, fixed in formol saline and dispatched for histological examination. Effort should be made to preserve the origin of the facial nerves at the level of the medulla oblongata, as this area is important in the histopathological diagnosis of listeriosis.

Treatment

Penicillin and tetracyclines are effective but must be given very early in the course of the disease to prevent irreparable brain damage. The incidence of further cases may be reduced or prevented by changing pastures or cessation of silage feeding.

BRAIN AND SPINAL CORD ABSCESSATION

Spinal cord abscesses are most frequent in lambs following ascending infection from tail docking. Affected lambs are often febrile and show signs characteristic of a space-occupying lesion of the thoracolumbar spinal canal, including posterior paresis.

Brain abscesses occur only sporadically. Rams are probably most often affected as a result of fighting and damage to the poll area. Septic emboli lodge in the brain and develop into abscesses, particularly in the pituitary area.

TETANUS

The disease is produced by the toxin of *Clostridium tetani* which produces hyperaesthesia, tetany and convulsions. The spores of *Cl tetani* can survive many years in the environment and are resistant to most disinfectants. They occur widely in soil, especially soil containing mammalian faeces. Boiling for 15 minutes kills most, but not all, spores. Disease results when spores enter the body, usually following a wound and the provision of an anaerobic environment in the tissues. The spores vegetate and produce their toxin, which has an affinity for nervous tissue and is transported within nerve axons to the CNS. The bacteria do not move from the site of original infection.

Tetanus usually occurs sporadically, but occasionally outbreaks in one group of animals may occur. Most frequently, outbreaks commence about one week after marking and may affect 1% to 5% of lambs, rarely more. The organisms enter wounds caused by castration and tail docking particularly, but also by shearing, dehorning, mulesing and areas of tissue necrosis following lambing. Lambs under 6 months old are most commonly affected. Older animals often have natural protective antibodies and are less often exposed to the type of wounds which are most likely to precipitate the disease.

Clinical signs

The clinical signs include muscular stiffness, spasms and tremors, and prolapse of the nictitating membrane. Affected animals are hyperaesthetic, and noise or touch precipitates tetany. While still able to walk the animals appear stiff-legged and they turn without flexing their spines. Ultimately, they cannot walk and lie in lateral recumbency with legs out stiffly and head extended back. Rectal temperature is raised by muscular activity.

Diagnosis

The clinical signs are usually diagnostic. PEM may be similar, but animals with PEM are blind; they also head-press and usually show periods of relaxation between convulsions. The history and presence of a recent wound, particularly marking or tail docking, are also suggestive.

Treatment

Treatment is expensive and seldom successful. Prophylactic treatment of unaffected animals in the same mob with antitoxin (100 IU for lambs) is more rational but only if given within 10 days of the inciting operation. After that, few new cases are likely to develop. Toxoid should be given simultaneously. Valuable affected animals can be treated with antitoxin (10 000 IU) and penicillin. Antitoxin is less effective in treatment than prevention because it cannot neutralise toxin bound to nervous tissue. If the site of infection is considered to be a tail docked with rubber rings, the tails of the unaffected lambs can be redocked above the position of the ring.

Prevention

Hygiene at marking should be attempted. Attempts should be made to avoid the contamination of wounds with soil and faeces. This can be best achieved by using temporary yards, but many producers use permanent yards without suffering significant losses from tetanus. Lambs should be released into the paddock or a large yard and allowed to return quickly to their dams.

Vaccination is commonly used to protect lambs against tetanus. Ewes should be given the first and secondary vaccinations as young sheep, or at the earliest opportunity after purchase, then given a booster before lambing. Vaccination of the lamb at marking will then protect the lamb as the maternal protection declines — usually about eight weeks after lambing. (Vaccination against clostridial diseases generally is discussed in more detail in Chapter 14.) Vaccination at marking confers no protection for 10 to 14 days but appears to be effective in preventing outbreaks of tetanus from marking wounds. Anaphylactic shock is considered a risk of vaccinating goats.

Antitoxin can be given at marking in cases of valuable animals when the risk is considered high. 100 IU is protective for lambs, 500 IU for adult sheep.

BOTULISM

In sheep, botulism is a moderately acute, often fatal ascending motor paralysis caused by the ingestion of the *preformed* toxin of *Clostridium botulinum*. The main sources of toxin are bones and decomposing carcases of any animal or bird. Rotting vegetation can also become toxic as can silage (particularly if oxygen has contacted any part of it). Steps should be taken to ensure that animal carcases do not contaminate silage or silage pits. Chicken manure has also produced botulism in cattle.

Clinical signs

In the early stages, affected sheep wag their tails as though flystruck. As the disease progresses they walk stiffly, separate from the mob, and stand with head down, dribbling saliva and urine, often with tongues protruding. They cannot eat. Paralysis progresses over three to five days and the animals become recumbent and die quietly. The flaccid paralysis which characterises the disease in other species appears late in the disease in sheep. Some sheep recover after showing signs for up to two weeks.

Treatment and prevention

Treatment is not feasible. Vaccination against the toxin is available and one dose will protect sheep for one year. Prevention involves correcting deficiencies which cause pica and preventing access to rotting plant and animal tissues.

PURPORTEDLY INHERITED CONDITIONS OF THE NERVOUS SYSTEM

Five chronically progressive diseases of the CNS have been described in Australian Merino sheep, each of which has been restricted to specific geographic areas and, more particularly, to genetically related flocks. The association with particular strains or bloodlines and the low but persistent annual incidence in affected flocks suggest an inherited basis for each disease which has not been confirmed by test matings. The principal features of these conditions are described in Table 15.1.

Table 15.1: Purportedly inherited conditions of the CNS of Merino sheep.

Name of condition	Epidemiology	Clinical signs	Pathological characteristics	Comments
Segmental axonopathy[10,11] (Murrurundi disease	Merino sheep, 1 to 5 years of age, all fine-wool sheep of one bloodline, incidence on affected properties between 0.1% and 1%	Posterior paresis, swaying or dropped stance, *dog-sitting* posture, toe-knuckling when walking. No head tremor or forelimb hypermetria.	Vacuolar appearance due to spheroid formation involving mainly large, myelinated, white matter tracts of thalamus, cerebellum, brain stem and spinal cord.	Pathology has resemblance to the transmissible spongiform encephalopathies, but the latter involve grey matter.
Cerebellar abiotrophy	Merino sheep, 3½-6 years, all fine-wool sheep from one stud or its daughter studs; 0.1% to 1% annual incidence.	Dysmetria of all limbs, pronounced forelimb hypermetria, fine head tremor, awkward running gait, frequent falls.	Diffuse cerebellar degeneration with severe loss of Purkinje neurons.	Cerebellar atrophy also occurs in some forms of *Daft lamb disease* — another probably inherited condition of sheep.
Neuroaxonal dystrophy[12]	Merino sheep, weaner age, medium-wool from one flock, 1% annually.	Posterior paresis and posterior hypometria worsening with exercise, progressing to sternal recumbency	Axonal swellings in grey matter of brainstem and mid-brain, particularly cuneate and gracile nuclei of medulla.	Similar to condition reported in the Romney, Perendale and Coopworth breeds in NZ.[13]
Degenerative thoracic myelopathy[14]	Merino sheep, weaner age, some surviving up to 3 years, medium-wool from one stud, 1% annually.	Posterior paresis and posterior hypometria, worsening with exercise. Hindlimb proprioceptive deficits.	Degeneration in tracts of thoracic spinal cord.	Spinal lesions similar to those of early tribulosis.
Thalamic cerebellar neuropathy[15]	Merino sheep, aged over 2 years, medium-wool from one stud or its daughter studs; up to 1% annual incidence.	Posterior paresis, forelimb hypermetria, fine head tremor, incoordinated gait, frequent falls.	Axonal swelling in thalamus, cerebellum, pons, medulla oblongata and spinal cord.	Neuropathology similar to lysosomal storage disorders.

Several other, possibly inherited, diseases of the CNS of sheep have been described overseas. These include ceroid lipofuscinosis in South Hampshire sheep in NZ, a glycogen storage disease in Corriedale sheep (with similarities to bovine generalised glycogenosis in Shorthorn cattle and Pompe's disease of man), globoid cell leukodystrophy in Poll Dorsets, and Daft Lamb disease (inherited cerebellar cortical atrophy) of Welsh Mountain, Border Leicester and Blackface breed sheep in UK and Corriedale sheep in Canada. Cerebellar abiotrophy is a disorder of cerebellar function acquired after a period of apparently normal development. Many of the degenerative diseases of the cerebellum are genetically determined and the term *abiotrophy* is commonly used for these disorders in the veterinary literature.

PLANT-ASSOCIATED TOXICOSES CAUSING PARESIS, PARALYSIS AND GAIT DISTURBANCES

Ryegrass staggers is a (usually) temporary incoordination of sheep and cattle grazing perennial ryegrass (PRG) (*Lolium perenne*), particularly in late summer and autumn and often following sufficient rain to stimulate limited plant growth. Affected sheep are difficult to drive and are predisposed to death by misadventure. The disease is caused by ingestion of lolitrem toxins (mostly lolitrem B) produced by the endophytic fungus *Acremonium lolii*. This endophyte is present in virtually all older stands of perennial ryegrass. As well as the lolitrem toxins, the endophyte produces a compound called peramine which improves the establishment and performance of the ryegrass plant. Unfortunately, the presence of high levels of endophyte in pastures has been associated experimentally with an increased lamb mortality rate in addition to the periodic occurrence of staggers. Consequently, current research is directed at reducing the lolitrem activity in the plant-endophyte association without losing the agronomic benefits of the endophyte.[16]

Animals affected with ryegrass staggers move stiffly and bound when attempting to move quickly. They fall readily when driven, collapsing with tetanic spasms of the limbs. If left undisturbed they relax and can stand again after a short time. Signs usually disappear following removal from the pasture or after the onset of autumn rains or cooler weather. The characteristic clinical signs in sheep, the low mortality and the presence in the pasture of perennial ryegrass plants with short green shoots from the base of an otherwise dry plant are generally sufficient for diagnosis.

Annual ryegrass toxicity (ARGT) is a fatal neurological disease caused by the ingestion of a preformed toxin of the tunicamycin type in the seed heads of annual ryegrass (*Lolium rigidum*) parasitised by the nematode *Anguina funesta* associated with the bacterium *Clavibacter toxicus*.[17] The disease is a serious problem in some sheep/cropping areas of South Australia and Western Australia and is spreading. The disease has not yet spread to annual ryegrass stands in Victoria or NSW.[18] Identical intoxications occur in *flood plain staggers* in sheep and, more particularly cattle[19], grazing the native blown-grass (*Agrostis avenacea*, also called blowaway grass) along the major river plains in western NSW[20] and in Stewart's Range syndrome in sheep and cattle grazing the introduced plant annual beard grass (*Polypogon monspeliensis*) in southeast SA.[21,22] All three diseases are caused by toxins elaborated by the same bacterial species carried into galls by an *Anguina* sp nematode which parasitises the seed head or, in the case of *P monspeliensis*, other parts of the plant, under particular conditions.

Clinical signs of ARGT include ataxia, hind limb incoordination and hypermetria when sheep are driven. Some sheep may collapse, demonstrate fine muscle tremors or suffer violent convulsions before rising again. Eventually, the sheep become persistently recumbent; they convulse, show nystagmus, opisthotonus and limb paddling; and they die after one to two days. Clinical signs may continue to develop for up to several weeks after sheep are removed from toxic pastures.[23] Diagnosis is best made on clinical signs, history of access to annual ryegrass in pasture or hay in the past few weeks, demonstration of bacterial galls in that ryegrass and histological evidence of hepatotoxicity and vascular lesions in the brain.[24]

Grazing of *Phalaris aquatica* by sheep can produce either a sudden death syndrome or a staggers syndrome. *Phalaris staggers* is caused by the ingestion of several structurally related indole alkaloids which accumulate in the CNS and have a tryptaminic effect on serotonin receptor sites in nuclei in the mid-brain and medulla and their synaptic contacts in the lower brain stem and spinal cord. At necropsy, characteristic greenish or golden-brown pigment can often be demonstrated histologically in the same sites in affected sheep. These pigments are probably post-effect metabolites of the ingested alkaloids concentrated in neuronal storage lysosomes[25], rather than accumulations of active alkaloids.

Clinical signs of intoxication develop after days, weeks or even months of grazing the toxic stand and signs may develop months after removal from pasture and persist for weeks or months.[26] Affected sheep show some or all of the following signs: hyperexcitability; tremors of the head and body; twitching of the lips, tail and ears; shaking and nodding of the head; limb paresis; limb stiffness; ataxia; bounding, hopping or jumping movements; kneeling; walking on the knees; knuckling over at the fetlock joints; falling down into sternal or lateral recumbency; vigorous struggling to stand up again; rapid fatigue; laboured respiration; and an intensified heart beat after exercise. Convulsions are not a characteristic of phalaris staggers, except terminally. Death or recovery can occur over the ensuing weeks or months.[27]

Cobalt, administered as a drench of cobalt sulphate, a cobalt bullet, or by pasture treatment, is useful as a preventive of the staggers syndrome (but not phalaris sudden death). The short-term treatments (drenching, pasture misting) are effective if given at the high-risk time of the year, usually autumn and early winter. This effect will occur in the absence of any other productive benefit from the administered cobalt and appears to be a consequence of better rumen destruction of the toxic principles.[28] In general, the best method to limit the losses from both syndromes of phalaris toxicity probably involves continual grazing of the pasture and avoiding the introduction of animals which are not accustomed to grazing phalaris pastures. Sirolan is a low-alkaloid cultivar of phalaris but sudden death and staggers still occur on Sirolan pastures.[23]

Chronic ingestion of *Tribulus terrestris* (*cathead* or *caltrop*) has been associated with outbreaks of a nervous system disease (*Coonabarabran staggers*) in the months following periods of drought in the Coonabarabran district of New South Wales. Signs include an irreversible, assymetrical hindquarter paresis.[29] It is suggested that the alkaloids harmane and norharmane accumulate in upper motor neurones of affected sheep and exert a tryptaminergic effect in those neurones.[30,31] This disease is quite distinct from the hepatogenous photosensitisation associated with ingestion of *T terrestris* and the sporidesmin-producing saprophytic fungus *Pithomyces chartarum*. Ingestion of *Tribulus micrococcus* (yellow vine) has been associated with a transient ataxia of sheep.[32]

Humpy back disease is characterised clinically by difficulties in moving the hind limbs, causing a short, stiff-legged gait with inadequate hock flexion, followed by lowering of the head, arching of the back and inability to continue walking. Signs are not apparent until the sheep are driven at least one to two kilometres. The disease mainly affects full-woolled wethers in western Queensland mustered for shearing in hot months.[33] Development of signs is accompanied by hyperthermia, and sheep which do not die at that stage appear to recover after shearing, although they may have relatively high mortality rates in subsequent months.[34] Histopathologically, there is Wallerian degeneration in the CNS, particularly of the spinal cord. The occurrence of the syndrome has been associated with the plant *Solanum esuriale* (Quena), which grows after summer rains, and with *Malvastrum americanum* (Spiked mallow) in northern NSW.[35] Attempts to reproduce the disease by feeding fruit of *S esuriale* have not been particularly successful.[36] Bourke[23] suggests that, if caused by *S esuriale*, the disease is probably a neurotoxicty; if caused by *M americanum,* it is probably a myopathy.

Swainsona spp (Darling peas) are widely distributed over NSW. They usually grow in autumn and winter and sheep grazing some species for several weeks develop signs referable to a neurological disorder. They develop a stiff, incoordinated gait and walk with a high head carriage. Sheep removed promptly from affected pastures will recover. Histopathologically, there is vacuolation in CNS cells associated with the accumulation of mannose rich oligosaccharides — a syndrome resembling the lysosomal storage disease mannosidosis of cattle.

Ergotism occurs uncommonly in Australia; the most common manifestation is the neurological syndrome of ataxia and tremor produced by the ergot of the *Paspalum* spp grasses *Claviceps paspali*.[37] (Gangrenous ergotism does not occur on *Paspalum* spp pastures and, in fact, gangrenous ergotism caused by *C purpurea* is very rare in sheep in Australia.) Clinical signs include hypersensitivity to noise and movement, and muscle tremor which can become sufficiently severe to cause shaking and to interfere with grazing. Ruminants affected by the ergot nervous syndrome develop tremorgenic nervous signs, not convulsions, and clinically, the syndrome closely resembles ryegrass staggers. Convulsions of sheep grazing grass-dominant pastures in which seed set is occurring or has finished are more likely to be caused by tunicamycin type toxicosis than ergot poisoning.[20]

Clinical signs referable to lesions of the CNS produced by plant intoxications have also been recorded in Australia as follows: neurological disease and lipofuscinosis in sheep grazing *Trachyandra divaricata* (branched onion weed) in WA[38]; poisoning by the tropane (atropine-like) alkaloids of *Datura stramonium* (thornapple) and *D ferox* (long-spined thornapple); poisoning by the nicotine alkaloids of *Nicotiana suavolens* (native tobacco) and *N glauca* (tree tobacco) despite low palatability[35]; ataxia, depression, blindness and death associated with *Stypandra* spp; nervous signs and wasting after five to six weeks grazing on *Ipomoea* spp in northern Australia and NSW north coast; irreversible, chronic, progressive paresis from ingestion of palms of the cycad group (*Cycas* spp, *Bowenia* spp, *Lepidozamia* spp and *Macrozamia* spp).[23]

In addition to toxins which have direct pathogenic effects, some plants can induce specific diseases through metabolic or dietary mechanisms. Thus polioencephalomalacia can be induced by grazing plants containing thiaminase: *Marsilea drummondii* (Nardoo), *Cheilanthes sieberi* (rock fern) and *Pteridium esculentum* (common bracken fern).[39] Similarly, hypocalcaemia may be induced by unaccustomed access to plants high in oxalate. *Oxalis pes-caprae* (soursob) is a common oxalate-rich plant in the medium rainfall districts of Australia and there are a number of other oxalate-containing plant species distributed in inland Australia. Although hypocalcaemia can be precipitated by soursob, nephrosis is a more common sequel to chronic exposure. Sheep affected by one of a number of hepatotoxic plants can demonstrate nervous signs which are in fact an hepatic encephalopathy rather than the result of a neurotoxin.

COMMON CHEMICALS RESPONSIBLE FOR CLINICAL SIGNS OF NEUROLOGICAL DISEASE

Sheep on commercial farms often have access to organophosphates, carbamates, urea and selenium, and poisoning with these, or overdosing in the case of selenium, can produce nervous signs. Although the risk was higher in the past than now, carbon tetrachloride, chlorinated hydrocarbons, cyanide, lead and nicotine can cause poisoning and these episodes are usually marked by neurological signs in affected animals.

DIFFERENTIAL DIAGNOSIS OF ATAXIA AND PARESIS IN SHEEP

Ataxia and paresis can be caused by trauma, particularly involving vertebral fractures, spinal cord inflammatory diseases, a number of plant intoxications (notably phalaris staggers and (perennial) ryegrass staggers) and, rarely, congenital or genetically determined disorders. Copper deficiency of ewes during pregnancy can lead to enzootic ataxia of lambs. This paretic syndrome usually occurs in lambs 1 to 2 months of age. Sporadic cases require differentiation from spinal cord infections and the rare conditions *neuroaxonal dystrophy* and *degenerative thoracic myelopathy*. Animals with recently acquired spinal cord infections are usually febrile.

Differentiation of plant intoxications may be based partly on the history of access to particular plants, some of which are restricted to reasonably limited geographic areas, and the clinical signs described above. Note should also be taken of the season in which signs occur. Ryegrass staggers usually occurs in late summer and early autumn; phalaris staggers usually occurs in autumn or winter, but it can occur at any time because of the often long lag time between exposure and the development of a clinical syndrome. ARGT and related tunicaminyluracil toxicoses and ergotism occur when plant seeds have set or soon after. The epidemiology of outbreaks should also be noted. Plant intoxications frequently involve 1% to 10% of the exposed flock; traumatic lesions usually involve only one or two animals.

Some diseases with signs of ataxia do not develop beyond ataxia in individual cases, or do so only slowly. Metabolic diseases of ewes tend to be rapidly progressive but hypocalcaemia can cause collapse and paresis in weaners, with most regaining their feet after a day or two. ARGT has a significant mortality; most other plant intoxications have low mortality rates.

DIFFERENTIAL DIAGNOSIS OF DUMMY SYNDROMES AND RECUMBENCY IN SHEEP

PEM and FSE are difficult to distinguish clinically. History, the usually sporadic nature of FSE and the response to thiamin treatment with PEM may assist, but necropsy is confirmatory. Hypocalcaemia and pregnancy toxaemia cause recumbency in ewes but the paralysis is flaccid and the animals are weak. Listeriosis is usually characterised by facial nerve paralysis and fever in the early stages. The clinical signs usually reflect the unilateral nature of the brain lesions. Hypomagnesemia also occurs with nervous signs in ewes but, as the clinical course is short, it is usually considered a cause of sudden death. Amongst chemical intoxications, overdosing with levamisole should be considered as a cause of muscle weakness, tremor, collapse and death or recovery. Signs usually appear soon after drenching, so history and clinical signs are usually diagnostic.

EXOTIC DISEASES WITH NERVOUS SIGNS

Rabies

In the USA, rabies occurs with a low sporadic incidence in sheep. It is transmitted only by the bite of saliva-infected rabid animals. A number of cases often occur together as a result of an attack by a rabid predator. There are three phases in the clinical course of rabies in sheep following an incubation period of 15 days to several months. The prodromal phase is short and usually undetected in sheep. During the second, excitative phase, affected animals refuse feed and water, move restlessly, salivate, act aggressively towards people, bleat frequently, bite and pull wool, rhythmically oscillate the tongue and faint, and rams show sexual excitement. In the final phase animals become depressed, weak and recumbent through paralysis. Death usually occurs within 12 hours of prostration; the total course of clinical illness is less than three days.[40,41]

Scrapie

Scrapie is endemic in Britain, parts of continental Europe, Scandinavia, parts of North and South America, Africa and Asia. It has not occurred in Australia since 1952, when it occurred in four of ten Suffolk sheep imported from the UK and was eradicated by a policy of slaughter and property quarantine.

Clinical signs

Affected sheep show either pruritis, incoordination or various combinations of both. Pruritis leads to the loss of wool from the flanks and hindquarters from rubbing or wool biting. Incoordination leads to an exaggerated gait, sometimes with a high-stepping or trotting fore limb action, but more often with ataxia of the hind limbs and frequent stumbling. Additionally, there are non-specific and highly variable behavioural changes. The clinical signs progress over a period of a week or up to several months and invariably end in death.

Epidemiology

Scrapie cases are extremely rare under 1 year of age and uncommon under 2 years. The modal age of development is 3 to 4 years. Most animals are infected while young, so the age of incidence reflects the incubation period of the disease.

Spread occurs vertically and laterally. Vertical spread, from dam to offspring, probably occurs predominantly via the alimentary tract, either before birth from infected amniotic fluid or after birth from a contaminated environment. Transplacental infection may occur and the possibility of infection of the ovum has not been excluded, with repercussions for embryo-importing strategies.[42] Lateral spread of scrapie between unrelated sheep is proven and has even occurred from sheep to goats. Placentae from infected ewes are rich sources of infection; entry to uninfected animals *per os* and through damaged skin appears to be the main route of infection. The infectious particle of scrapie is extremely resistant to attempts to destroy it and can persist in contaminated environments for at least one year.

The development of clinical scrapie is controlled, or at least strongly modified, by the genotype of the host. A gene, called *Sip* for *s*crapie *i*ncubation *p*eriod, has been identified with two alleles, sA and pA. Animals homozygous for sA are susceptible to natural scrapie, and the course of the disease is that described above. Heterozygous sheep, or those homozygous for pA, are resistant to the development of clinical signs, either for very long incubation periods (five to seven years) or forever. The identification and planned use of homozygous pApA rams is currently under study as a method for controlling scrapie in countries where the disease occurs. Unfortunately, it appears possible that *Sip* sApA or even *Sip* pApA animals could be infected carriers of the disease.

It is known that ewes can infect their offspring while they are still incubating the disease and that they are no more likely to do so whether they are early or late in the incubation phase. This, the presence of infected carriers, the inability to test sheep for the presence of scrapie infection and the environmental persistence of the agent make eradication extremely difficult. Slaughter programmes have so far failed to eradicate scrapie in any country where the disease has become endemic.

Diagnosis

The scrapie particle cannot be visualised by any microscopic techniques and evokes no immune or other host response which could be used to test for the presence of disease in live animals. Consequently, diagnosis is limited to necropsy examination of the brains of affected animals. Scrapie infection causes a normal brain protein to become modified so that it accumulates and acquires the ability to form abnormal, scrapie associated fibrils (SAFs). These fibrils can be seen in brain extracts by electron-microscopy, and methods of purification are now available.[43] A mouse-adapted scrapie agent has been imported into AAHL and local experience in procedures for detecting SAFs has been developed.[44]

Ovine encephalomyelitis (Louping ill)

Louping ill has long been recognised in UK and Ireland but recent evidence confirms its occurrence in parts of continental Europe. The disease is caused by a *Flavivirus* which is spread by the sheep tick *Ixodes ricinus*. No known tick vectors occur in Australia. In endemic areas the disease occurs principally in sheep under 2 years of age and in sheep brought in from nonendemic areas. Not all infected animals develop clinical signs but those that do develop incoordination, paralysis and coma, and they die within 24 to 48 hours. Vaccination is used, amongst other control strategies, to limit the losses from the disease in endemic areas. The disease also affects man and other domestic and wild animals.[45]

Coenurosis (Gid)

Coenurus cerebralis is the cystic larval stage of the tapeworm of carnivores *Taenia multiceps*. Disease results from the ingestion of tapeworm eggs and the migration and localisation of the larva or larvae in the brain. There is an acute form of the disease, principally affecting young lambs, and a chronic form typically affecting sheep 6 to 18 months of age. Clinical signs relate to the focal nature of the brain lesion, progressing as the cysts grow in size. Coenurosis is one of the most common causes of disease of the CNS of sheep in Britain[46] but it is apparently absent from Australia and New Zealand.

Aujeszky's disease

Aujeszky's disease (pseudorabies, mad-itch) does not occur in Australia but does occur in a number of other countries including NZ. Although it is principally a disease of pigs in those countries, sporadic cases occur in other mammals including sheep. The disease is almost invariably fatal. Clinically, the most striking feature is intense pruritus of a localised area of skin. The course of the disease is usually less than two days.

Visna

Visna is the CNS manifestation of infection with the maedi-visna virus. The disease occurs in many sheep-raising countries but not in Australia or New Zealand. After a long incubation period, sheep develop signs either of maedi (a chronic, slowly progressive respiratory syndrome marked by dyspnoea) or visna, with slowly progressive incoordination, ataxia, facial muscle trembling, paresis and paraplegia with rapid weight loss. The clinical course may be up to one year.[41]

Borna disease

This viral disease is principally limited to Germany and Switzerland. It affects horses and sheep and occasionally other species. Clinical signs relate to the characteristic meningo-encephalitis. Mortality rates are high.

RECOMMENDED READING

Beveridge WIB (1983) Bacterial diseases of cattle, sheep and goats. Australian Agricultural Health and Quarantine Service/Australian Government Publishing Service: Canberra.

Radostits OM, Blood DC and Gay CC (1994) Diseases caused by *Clostridium* spp. In: Veterinary medicine. 7th ed. Bailliere Tindall: London, p. 677.

West DM (1990) Nervous diseases of sheep. In: Sheep medicine. Proceedings No 141. University of Sydney Post-graduate Committee in Veterinary Science: Australia, p. 53.

REFERENCES

1   Chick BF (1990) Thiamin deficiency — Thiaminase. In: Feeding standards for Australian livestock Ruminants. Standing Committee on Agriculture, CSIRO: Melbourne, p. 190.

2   Thomas KW and Griffiths FR (1987) Natural establishment of thiaminase activity in the alimentary tract of newborn lambs and effects on thiamine status and growth rates. Aust Vet J **64** 207-10. https://doi.org/10.1111/j.1751-0813.1987.tb15183.x.

3   Thomas KW, Turner DL and Spicer EM (1987) Thiamine, thiaminase and transketolase levels in goats with and without polioencephalomalacia. Aust Vet J **64** 126-7. https://doi.org/10.1111/j.1751-0813.1987.tb09654.x.

4   Amat S, McKinnon JJ, Olkowski AA et al. (2013) Understanding the role of sulfur-thiamine interaction in the pathogenesis of sulfur-induced poliencephalomalacia in beef cattle. Res Vet Sci **95** 1081-7. https://doi.org/10.1016/j.rvsc.2013.07.024.

5   McAllister MM, Gould DH and Hamar DW (1992) Sulphide-induced Polioencephalomalacia in lambs. J Comp Path **106** 267-78. https://doi.org/10.1016/0021-9975(92)90055-Y.

6   Olkowski AA, Gooneratne SR, Rousseaux CG et al. (1992) Role of thiamine status in sulphur induced polioencephalomalacia in sheep. Res Vet Sci **52** 78-85. https://doi.org/10.1016/0034-5288(92)90062-7.

7   Hartley WJ and Kater JC (1962) Observations on diseases of the central nervous system of sheep in New Zealand. NZ Vet J **10** 128-42.

8   Gay CC, Blood DC and Wilkinson JS (1975) Clinical observations of sheep with focal symmetrical encephalomalacia. Aust Vet J **51** 266-9. https://doi.org/10.1111/j.1751-0813.1975.tb06932.x.

9   Vandergraff R, Borland NA and Browning JW (1981) An outbreak of listerial meningo-encephalitis in sheep. Aust Vet J **57** 94-6. https://doi.org/10.1111/j.1751-0813.1981.tb00457.x.

10 Harper PAW, Duncan DW, Plant JW et al. (1986) Cerebellar abiotrophy and segmental axonopathy: Two syndromes of progressive ataxia of Merino sheep. Aust Vet J **63** 18-21. https://doi.org/10.1111/j.1751-0813.1986.tb02865.x.

11 Hartley WJ and Loomis LN (1981) Murrurundi disease: An encephalopathy of sheep. Aust Vet J **57** 399-400. https://doi.org/10.1111/j.1751-0813.1981.tb00541.x.

12 Harper PAW and Morton AG (1991) Neuroaxonal dystrophy in Merino sheep. Aust Vet J **68** 152-3. https://doi.org/10.1111/j.1751-0813.1991.tb03162.x.

13 Nuttall WO (1986) Neuroaxonal dystrophy in sheep. Surveillance **13(1)** 20.

14 Harper PAW, Plant JW, Walker KH et al. (1991) Progressive ataxia associated with degenerative thoracic myelopathy in Merino sheep. Aust Vet J **68** 357-8.

15 Bourke CA, Carrigan MJ and Dent CHR (1993) Chronic locomotor dysfunction, associated with a thalamic-cerebellar neuropathy, in Australian Merino sheep. Aust Vet J **70** 232-3.

16 Cunningham P (1990) The significance of the ryegrass endophyte. Proceedings No 31. Conference of Grassland Society of Victoria: Parkville, p. 19.

17 McKay AC and Riley IT (1993) Sampling ryegrass to assess the risk of annual ryegrass toxicity. Aust Vet J **70** 241-3. https://doi.org/10.1111/j.1751-0813.1993.tb08038.x.

18 Jago MV and Culvenor CCJ (1987) Tunicamycin and corynetoxin poisoning in sheep. Aust Vet J **64** 232-5. https://doi.org/10.1111/j.1751-0813.1987.tb09689.x.

19 Bourke CA and Carrigan MJ (1993) Experimental tunicamycin toxicity in cattle, sheep and pigs. Aust Vet J **70** 188-9. https://doi.org/10.1111/j.1751-0813.1993.tb06131.x.

20 Davis EO and Curran GC (1991) cited by S Davidson (1992) Livestock poisonings: Exposing a familiar culprit. Rural Research **153** 9.

21 Trengove CL (1991) A neurological syndrome affecting livestock in the south-east of South Australia. In: Proceedings of the Australian Sheep Veterinary Society Conference 1991, Sydney, p. 118.

22 Finnie JW (1991) Corynetoxin poisoning in sheep in the south-east of South Australia associated with annual beard grass (*Polypogon monspeliensis*). Aust Vet J **68** 370. https://doi.org/10.1111/j.1751-0813.1991.tb00743.x.

23 Bourke CA (1990) Plant toxins affecting the nervous system of sheep in Australia. Proceedings No 141. University of Sydney Postgraduate Committee in Veterinary Science: Australia, p. 87.

24 Giesecke R (1990) cited in Sheep health. Newsletter No 27, ed JW Plant. NSW Agriculture: Australia.

25 Bourke CA, Carrigan MJ and Dixon RJ (1990) The pathogenesis of the nervous syndrome of *Phalaris aquatica* toxicity in sheep. Aust Vet J **67** 356-8. https://doi.org/10.1111/j.1751-0813.1990.tb07400.x.

26 Bourke CA, Carrigan MJ, Seaman JT et al. (1987) Delayed development of clinical signs in sheep affected by *Phalaris aquatica* staggers. Aust Vet J **64** 31-2. https://doi.org/10.1111/j.1751-0813.1987.tb06057.x.

27 Bourke CA, Carrigan MJ and Dixon RJ (1988) Experimental evidence that tryptamine alkaloids do not cause *Phalaris aquatica* sudden death syndrome in sheep. Aust Vet J **65** 218-20. https://doi.org/10.1111/j.1751-0813.1988.tb14462.x.

28 Lee HJ, Kuchel RE, Good BF et al. (1957) The aetiology of phalaris staggers in sheep. IV. The site of preventive action and its specificity to cobalt. Aust J Agric Res **8** 502-11. https://doi.org/10.1071/AR9570502.

29 Bourke CA (1984) Staggers in sheep associated with the ingestion of *Tribulus terrestris*. Aust Vet J **61** 360-3. https://doi.org/10.1111/j.1751-0813.1984.tb07156.x.

30 Bourke CA, Carrigan MJ and Dixon RJ (1990) Upper motor neurone effects in sheep of some beta-carboline alkaloids identified in zygophyllaceous plants. Aust Vet J **67** 248-51. https://doi.org/10.1111/j.1751-0813.1990.tb07778.x.

31 Bourke CA, Stevens GR and Carrigan MJ (1992) Locomotor effects in sheep of alkaloids identified in Australian *Tribulus terrestris*. Aust Vet J **69** 163-5. https://doi.org/10.1111/j.1751-0813.1992.tb07502.x.

32 Bourke CA and MacFarlane JA (1985) A transient ataxia of sheep associated with the ingestion of *Tribulus micrococcus* (yellow vine). Aust Vet J **62** 282. https://doi.org/10.1111/j.1751-0813.1985.tb14253.x.

33 O'Sullivan BM (1976) Humpy back of sheep, clinical and pathological observations. Aust Vet J **52** 414-18. https://doi.org/10.1111/j.1751-0813.1976.tb09514.x.

34 Pearse BHG, Peucker SKJ and Hoey WA (1992) Hyperthermia in Merino wethers affected with humpyback disease. Aust Vet J **69** 94-5. https://doi.org/10.1111/j.1751-0813.1992.tb15563.x

35 McBarron EJ (1983) Poisonous plants. Inkata Press: Sydney.

36 Dunster PJ and McKenzie RA (1987) Does *Solanum esuriale* cause humpyback in sheep? Aust Vet J **64** 119-20. https://doi.org/10.1111/j.1751-0813.1987.tb09648.x.

37 Seawright AA (1982) Chemical and plant poisons. Australian Agricultural Health and Quarantine Service/Australian Government Publishing Service: Canberra, p. 74.

38 Huxtable CR, Chapman HM, Main DC et al. (1987) Neurological disease and lipofuscinosis in horses and sheep grazing *Trachyandra divaricata* (branched onion weed) in south Western Australia. Aust Vet J **64** 105-8. https://doi.org/10.1111/j.1751-0813.1987.tb09639.x.

39 Chick BF, Carroll SN, Kennedy C et al. (1981) Some biochemical features of an outbreak of polioencephalomalacia in sheep. Aust Vet J **57** 251-2.

40 Jensen R and Swift BL (1982) Diseases of sheep. 2nd ed. Lea and Febiger: Philadelphia.

41 Geering WA and Forman AJ (1987) Exotic diseases. Australian Agricultural Health and Quarantine Service/Australian Government Publishing Service: Canberra.

42 Foster JD, Goldmann W and Hunter N (2013) Evidence in sheep for pre-natal transmission of scrapie to lambs from infected mothers. PLoS ONE **8(11)** e79433. https://doi.org/10.1371/journal.pone.0079433.

43 Kimberlin RH (1991) Scrapie. In: Diseases of sheep, eds WB Martin and ID Aitken. 2nd ed. Blackwell Scientific Publications: London.

44 Forman AJ (1993) Recent developments in diagnosis of exotic disease at the Australian Animal Health Laboratory. Aust Vet J **70** 161-3. https://doi.org/10.1111/j.1751-0813.1993.tb06118.x

45 Reid HW (1991) Louping-ill. In: Diseases of sheep, eds WB Martin and ID Aitken. 2nd ed. Blackwell Scientific Publications: London.

46 Skerritt GC (1991) Coenurosis. In: Diseases of sheep, eds WB Martin and ID Aitken. 2nd ed. Blackwell Scientific Publications: London.

16

Diseases of the alimentary tract

Gastrointestinal helminth parasites are the most important cause of alimentary tract diseases of sheep in Australia. They are endemic in all areas of the country and control of the parasites can present significant challenges to producers and advisors, particularly in the medium and high rainfall areas of Australia. The diseases caused by helminth parasites and their control are discussed in detail in Chapter 9. In this chapter, helminth parasites are mentioned only in conjunction with the differential diagnosis of gastroenteric disease, particularly when it is accompanied by diarrhoea.

This chapter presents information about the non-helminth diseases of sheep and is divided into four sections

- Section A: Gastrointestinal diseases of adult sheep (diseases restricted to adult sheep, of which Johne's disease is the most important)
- Section B: Gastrointestinal diseases of sheep of all ages
- Section C: Gastrointestinal diseases of young sheep (diseases generally restricted to lambs and weaners)
- Section D: Gastrointestinal diseases exotic to Australia (some important diseases of the alimentary tract of sheep, particularly bluetongue, foot-and-mouth disease and pestes des petits ruminants).

SECTION A

GASTROINTESTINAL DISEASES OF ADULT SHEEP

Ovine Johne's disease (OJD)

In sheep, Johne's disease or paratuberculosis is a chronic, progressive disease characterised by wasting and death of adult animals. The causative bacterium is *Mycobacterium avium* subsp *paratuberculosis* (MAP), also referred to as *Mycobacterium paratuberculosis* (*M ptb*). The disease is transmitted primarily by the faecal-oral route. The bacteria cause a granulomatous enteritis, primarily of the distal small intestine, which compromises the absorption of nutrients. Once clinical signs of disease appear, the condition is invariably fatal.

Mortality rates in non-vaccinating affected flocks are generally less than 6% of adult sheep annually. In infected flocks, there are usually several times as many infected sheep as there are sheep which are clinically affected. Not all infected sheep die of OJD — it is possible for sheep to contain the infection and remain in a preclinical stage for long periods.

While paratuberculosis has been recognised in cattle in Australia since 1925, it was not recorded in sheep until 1980.[1] Experience gained after that time has led to the widely accepted conclusion that the outbreak in sheep in Australia following that first report was the result of the introduction of a new strain of *M ptb* — the S strain — at some time in the preceding years or decades. This strain established readily in sheep flocks in the high rainfall zones of the central and southern Tablelands of New South Wales (NSW) and spread from infected flocks to uninfected flocks with the movement of sheep, facilitated by the movement of water, soil or sheep faeces between adjacent properties.

Following the first report in Australian sheep flocks, attempts were made to stop the spread of the infection to uninfected flocks. These efforts were frustrated by the insidious, chronic nature of the disease. Subclinically infected sheep are likely to excrete *M ptb* in faeces and therefore be capable of transmitting infection for months before the disease can be reliably detected by clinical pathology or the appearance of clinical signs.

Between 1980 and March 1993, the disease was detected in 37 flocks in the same geographic area of NSW — the Rural Lands Protection Board districts of Bathurst and Carcoar — and in three flocks outside that area. By late 1996 OJD was known to be present on 158 properties in NSW, 28 properties in Victoria and 6 properties on Flinders Island (Tasmania), but in no other states.[2] In 1997 it was first detected in South Australia — on a number of properties on Kangaroo Island. During that same year Victoria embarked on an eradication programme, compulsorily destocking all known infected properties, and the number of known infected flocks in NSW rose to 200. A control programme (the National Ovine Johne's Disease Control and Evaluation Program, or NOJDP) began and the possibilities for eradicating the disease from flocks, regions or the country were explored. Increasingly it became clear that it was very difficult to eliminate the infection from even just one farm[3] and that eradication on a larger scale would not be achievable.

By 2004, there were 1343 Australian sheep flocks that were known to be infected and OJD was officially recorded in all states except Queensland.[4] In that year, the NOJDP came to an end, the eradication programme in Victoria ceased and a less regulated approach to the management of OJD began — this based on the market assurance programme (MAP) which had begun in 1998 and, later, an assurance-based credits (ABC) scheme. Further deregulation and adoption of a more nationally consistent approach to OJD control led to the adoption of the Ovine Johne's Disease Management Plan 2013-2018 by most, but not all, states of Australia. This approach passed responsibility for control of the spread of OJD to individual sheep producers and was based largely around compulsory Sheep Health Statements accompanying every sale and purchase of sheep. The Ovine Johne's Disease Management Plan 2013-2018 concluded without extension in June 2018. National and state-based management of OJD continues as part of the Sheep Health Project, overseen by Animal Health Australia.[a]

The number of known infected sheep flocks in Australia has continued to rise, with over 2000 flocks known to be infected by 2010. The record of the number of known infected flocks is no longer maintained in Western Australia, Victoria, Tasmania or New South Wales.

a   The Sheep Health Project is described on the Animal Health Australia website: https://www.animalhealth australia.com.au/what-we-do/biosecurity-services/sheep-health-project/.

OJD has a worldwide distribution, occurring in New Zealand (NZ), the United States (USA), South Africa, Iceland, Norway and most European countries. In most countries of the world other than Australia the disease is considered to be endemic and of limited impact, and in individual flocks it is managed, if necessary, by vaccination.

Aetiology

M ptb belongs to a taxonomic group which includes two other subspecies of *M avium*: *M avium* subsp *avium* and *M avium* subsp *silvaticum* (wood pigeon mycobacteria). *M ptb* organisms are gram-positive, acid-fast rods which are extremely slow-growing on culture media. Isolates from cattle typically require up to 16 weeks for primary isolation. Isolates from sheep (of the S strain) are much more difficult to grow in culture than those from cattle. In common with a few other species of mycobacteria in the *Mycobacterium avium* complex, isolation of the organism usually depends on the presence of mycobactin in the media.[5] Laboratory confirmation of *M ptb* is now based on molecular genetic techniques to identify the presence of particular DNA elements — known as insertion sequences — which are unique, or nearly so, to *M ptb*.

Insertion sequences are sequences of nucleotides which are repeated a number of times within the bacterial chromosome and which can vary in number and location between different strains or isolates of the one bacterial species. A number of insertion sequence (IS) families have been described and members of the one IS family can be found across a number of bacterial species.

The genome of *M ptb* has approximately 15 to 18 copies of an insertion sequence called IS900 — a 1451 base-pair sequence which appears to be almost unique to *M ptb* and which therefore differentiates it from other mycobacteria of the *M avium* group. It was first described in 1989 by two research groups[6,7] who recognised its potential application as a specific diagnostic probe. IS900 is now frequently used to confirm the presence of *M ptb* and has also been used to detect genetic differences between isolates of *M ptb* obtained from different host species[8], as discussed further below.

Pathogenesis

Sheep can be infected with *M ptb* at any age but young animals, particularly neonates, are more susceptible to infection than older animals. Infection occurs usually by ingestion. The M cells overlying ileal Peyer's patches appear to be the major portal of entry for *M ptb* in calves and lambs. Both intact and degraded bacteria are endocytosed and transported to patch macrophages.[9,10] The decreased susceptibility to paratuberculosis of older animals is likely to be related, at least in part, to the age-related involution of ileal lymphoid tissue.[11]

After phagocytosis of the bacteria, macrophages form part of a cell-mediated immune response directed at elimination of the organisms. There are several possible outcomes of this immune response. Intracellular mycobacterial multiplication may eventually induce a cell-mediated response that successfully clears the infection. If, however, the bacteria resist killing, an excessive immune response, including a delayed-type hypersensitivity, may lead to a chronic granulomatous (tuberculoid) lesion with few bacteria (paucibacillary) or to a proliferation of bacteria (multibacillary) and lepromatous lesions along the intestine, characterised histologically by the presence of many acid-fast organisms in the lesions. Development of the multibacillary form usually coincides with an increase in specific antibody levels and a progressive increase in faecal shedding of organisms. Anergy, however, may occur in the late stages of the disease.

The variation in lesion type results from variability in the delayed-type hypersensitivity response, the type of cells and cytokines in the cell-mediated response, and the numbers of mycobacteria present. These factors, in turn, are influenced by the genetic characteristics of the host and the level of existing immunity.

Necropsy

One Australian study has reported the clinical and post-mortem examination of 50 sheep from 10 NSW properties.[12] Characteristically, findings include thickening of the intestinal wall, particularly of the ileum and caecum and, less frequently, the jejunum (Figure 16.1). In some sheep the gross changes are very mild. In approximately half the cases, the mucosa of the terminal ileum is formed into transverse ridges. Mesenteric lymph vessels are sometimes corded, up to 3 mm in diameter, and the mesenteric nodes are often enlarged and oedematous.

Histologically, there is granulomatous enteritis, typhlitis and colitis. The most severe lesions are found in the terminal ileum. Villus atrophy and invasion by epithelioid cells — macrophages which have taking on the appearance of epithelial cells — are characteristic features of the mucosal lesions. In most cases, large numbers of acid-fast organisms within the macrophages are readily visible (Figures 16.2 and 16.3). In some studies, cases have been grouped into two[13] or

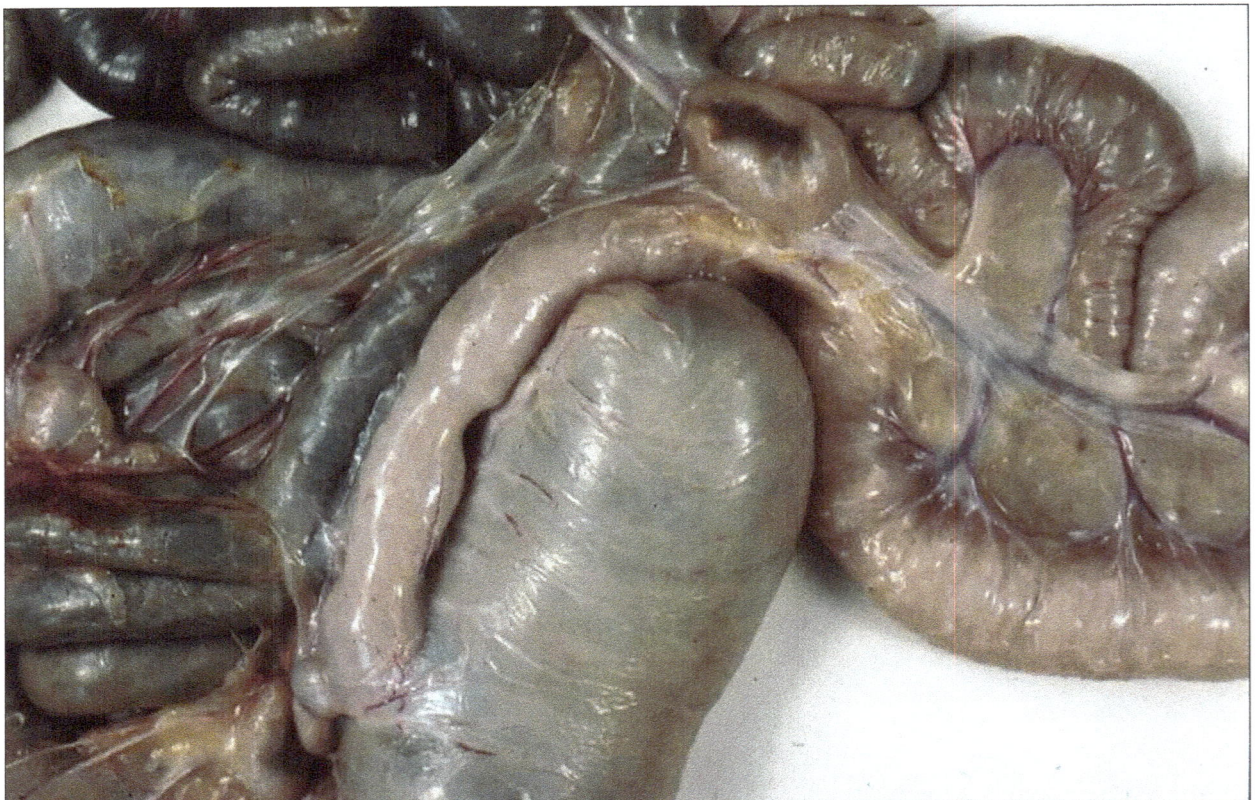

Figure 16.1: Thickening of the intestinal wall, particularly of the terminal ileum and caecum, is a characteristic necropsy finding in sheep clinically affected with paratuberculosis. Source: KA Abbott.

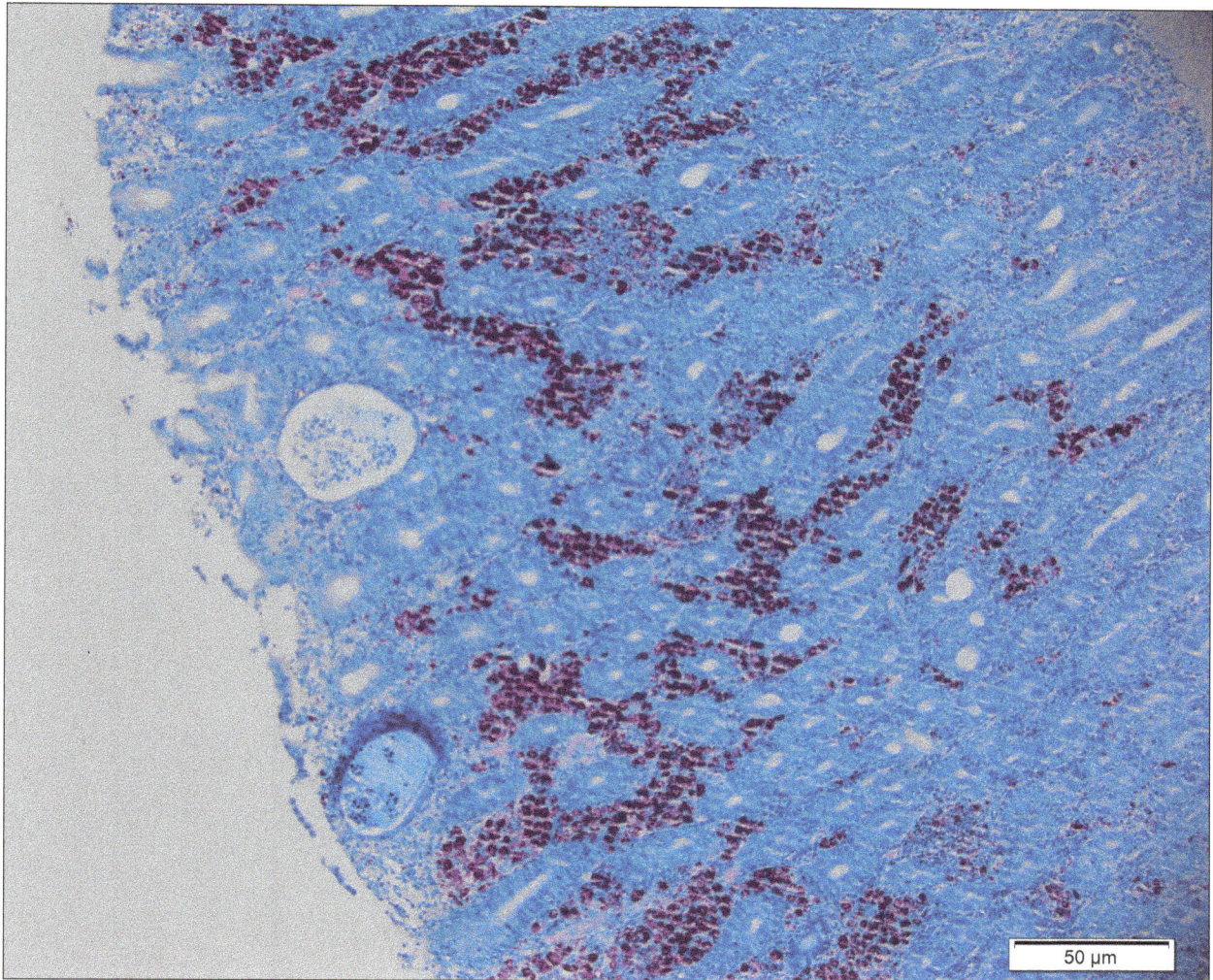

Figure 16.2: Intestine of a sheep with paratuberculosis. Macrophages containing large numbers of acid-fast organisms infiltrate the mucosa and submucosa of the intestine. Source: KA Abbott.

five[14] categories based on their histopathological appearance. Generally, when cases are selected on the basis of the appearance of clinical signs, most are found to have an extensive, diffuse, epithelioid cell infiltration in the mucosa and submucosa. These cells are packed with masses of acid-fast bacteria, particularly those in the mucosa (the multibacillary form). Cases in the alternate category have a marked cellular infiltration in the intestinal lamina propria and submucosa, and the epithelioid cells contain few, if any, acid-fast organisms (the paucibacillary form). The cellular infiltrate may be dominated by lymphocytes rather than macrophages. Serological tests are more likely to be positive in multibacilliary cases than in paucibacilliary ones.

Epidemiology

Strain typing based on IS900 and IS1311

A restriction digest of *M ptb* DNA will typically produce many DNA fragments but only 12 to 15 containing one or more copies of IS900, which can then be detected and visualised when a labelled hybridisation probe complimentary to portions of IS900 is applied in a Southern blot. Differences in the patterns indicate some of the differences in the positions of IS900

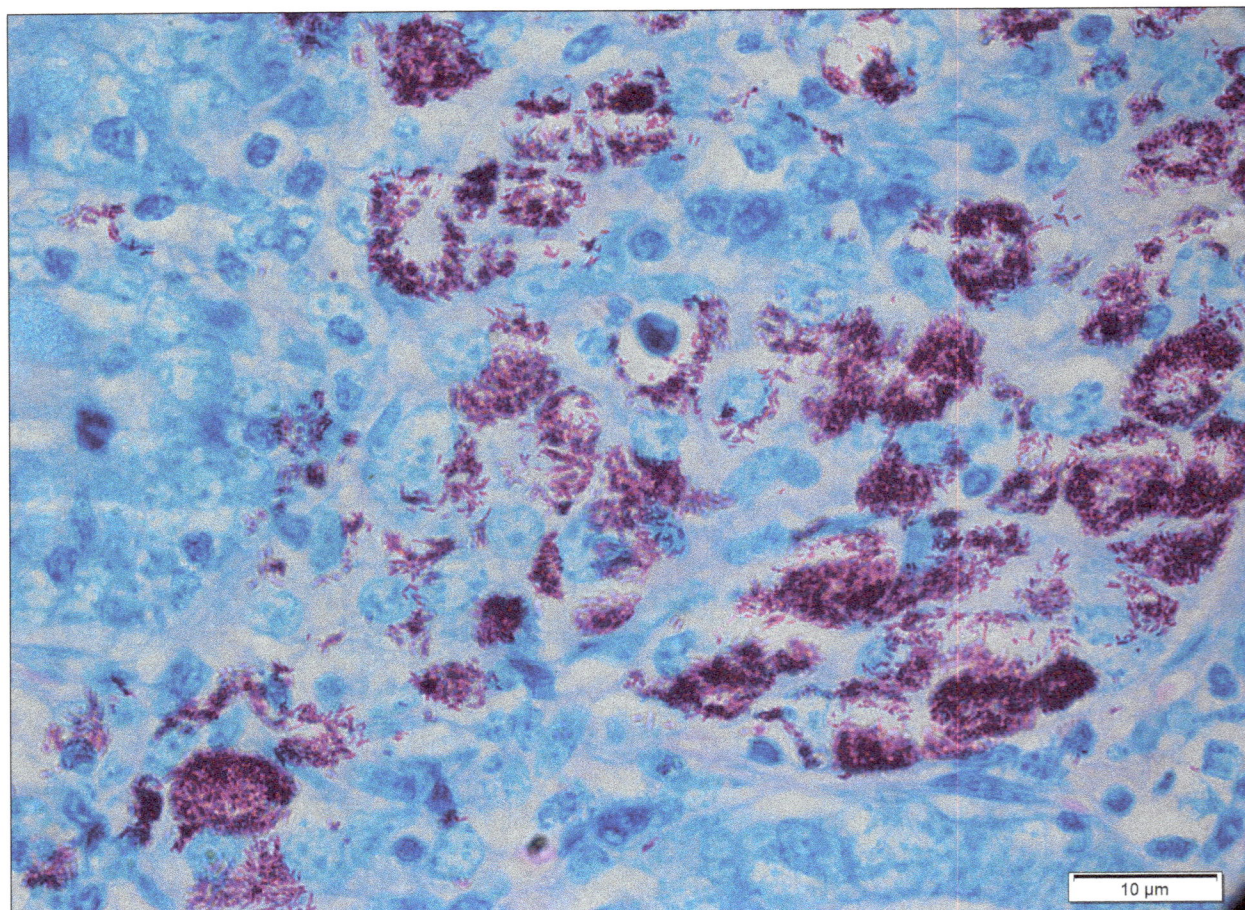

Figure 16.3: The same tissue as shown in Figure 16.2 at a higher magnification. With Ziehl-Neelsen staining, the acid-fast mycobacteria are readily apparent as masses of red organisms within the cytoplasm of the epithelioid macrophages. Source: KA Abbott.

within the genome of different strains of *M ptb*. This technique — restriction fragment length polymorphism or RFLP — simultaneously determines that the mycobacterial strain is almost certainly *M ptb* (because it contains IS900) and what strain of *M ptb* it is.

Using IS900 RFLP typing[b], Collins et al. (1990)[8] divided 48 strains of *M ptb* from cattle, sheep and goats and human and vaccine strains into three major groups — labelled by those authors as C, S and I, for cattle, sheep and intermediate. Within groups there were differences in band patterns but isolates in the same major group had 75% or more of their bands in common.

Since the late 1990s, IS900 RFLP strain typing of *M ptb* in Australia has been replaced by a simpler test based on another insertion sequence — IS1311. This insertion sequence is present in both *M avium avium* and *M ptb*, but there are differences in the structure of the insertion sequence between the two mycobacterial subspecies. There are also point mutations in the insertion sequence which differentiate C strains from S strains. A single PCR-REA[c] test based on IS1311 and developed in Australia[15] is able to confirm the presence of *M ptb* in

---

b RFLP: restriction fragment length polymorphism.

c PCR-REA: polymerase chain reaction and restriction endonuclease analysis.

cultures and to differentiate between the two major strain groups. Subsequent investigations have also identified another strain (B) isolated from bison in USA[16] and a range of host species in India[17] which can also be differentiated by IS1311 PCR-REA.

Strain typing based on other genotyping techniques

Pulsed-field gel electrophoresis (PFGE) has been used to categorise strains of *M ptb* into three major types — described as type I, II and III.[18] The alternate nomenclature was chosen to avoid any confusion between the genetic characteristics of the isolates and the species of host from which isolates were sourced but it is apparently consistent with the S, C and I nomenclature determined with IS900 RFLP typing[19] (Table 16.1).

Another technique — SNP[d] analysis of the *gyr* gene — also supports the categorisation of *M ptb* strains into the same three categories.[20] It appears that the type I/III and type II clusters of isolates represent the two major evolutionary lineages of *M ptb*, and the two differ in a number of phenotypic characteristics in addition to their genetic differences. Some of the differences are seen in culture — the type I/III strains being slower-growing and more fastidious, for example — and some are related to the immune response expressed by ruminant hosts.[21] Phenotypic differences between strains of type I and type III are not, however, apparent and these two types can be considered subtypes of the S strain group. The pigmented strains of *M ptb* — which occur occasionally in sheep in the UK and Faroe Islands — can be of type I or type III.[20]

Host preference for each strain

Studies of Australian isolates have demonstrated that Australian cattle, dairy-breed goats and alpacas are most commonly infected with C strains, sheep almost always with S strains.[22,23]

Exceptions to this general rule occur and are important if grazing with alternate species is used as a strategy to eliminate OJD infection on a farm. Cattle appear to be at least partly refractory to infection with S strain[24] but infection of beef cattle with S strain has now been reported in several herds in regions of southeastern Australia where sheep and beef cattle commonly share pastures.[25] S strain infection of cattle is also reported from Iceland[26] and New Zealand. Consequently, steps should be taken to reduce the risk of transmitting infection from sheep to cattle by avoiding the cograzing of infected sheep with young cattle and avoiding the handfeeding of cattle on the ground on pastures which have been contaminated by infected sheep.

Table 16.1: Nomenclature of *M ptb* isolate strains or types based on three alternate genetic techniques.

IS900 RFLP	PFGE	IS1311 PCR-REA
C	Type II	C, B
S	Type I	S
I	Type III	S

---

d   SNP: single nucleotide polymorphism.

Infections of sheep with C strains do occur in Australia but are reported much more commonly in other countries. It is not yet clear what forms the basis for the relative host specificity seen so far in Australia. It could be due to limited opportunities for cross-species transmission, but that seems unlikely, given that bovine paratuberculosis was recognised in Australia over 90 years ago[27] with only one report of an infected sheep[28] before the beginning of the OJD epidemic was reported in 1980.

Alternatively, the association of strains with specific hosts — or the lack of association in other countries — could be an artefact of the difficulty of culturing isolates of *M ptb* from sheep. Sheep isolates have long been renowned for very slow growth, or no growth at all, compared to cattle-derived isolates. Techniques for successfully culturing sheep strains, using radiometric culture[29] were developed and used widely in Australia during the first 20 years after *M ptb* was recognised in sheep and greatly facilitated epidemiological studies in this country.

Transmission of paratuberculosis from sheep to cattle under natural conditions has been reported in Iceland. The transmission of paratuberculosis from cattle to sheep has been reported in NZ, although it was not shown that the sheep represented a source of infection to other cattle or sheep.[30] The herd and flock prevalence of paratuberculosis is high in that country: 76% of sheep flocks and 42% of beef herds are infected with *M ptb*.[31] Cograzing of sheep and beef cattle is common in New Zealand and there is strong evidence for transmission between the two host species. A host preference of type I strains (S strains) for sheep is reported. Sheep and beef cattle are more commonly affected with type I strains than with type II strains (over 80% of isolates from sheep and beef cattle are type I strains) while deer and dairy cattle are most commonly infected with type II strains (around 90% of isolates from these two species are type II strains).[32] The incidence of clinical paratuberculosis in beef cattle is uncommon in New Zealand, suggesting that type I strains are less virulent for cattle than for sheep. For that reason and because of the clustering of type I strains in sheep and type II strains in dairy cattle and deer and the difference in prevalence between sheep flocks and beef cattle herds, it seems probable that sheep are the principal source of infection with type I strains for beef cattle in New Zealand.

Infection of goats with S strain is reported commonly in other countries, including Cyprus, Czech Republic, Spain and Greece.[33]

In goats in Australia, fibre-producing breeds are more likely to be infected with S strain *M ptb*, while dairy breeds of goats are usually infected with C strains. In NZ, feral goats have become infected with *M ptb* of bovine origin under conditions of natural grazing, and they develop clinical paratuberculosis and excrete *M ptb* in faeces.[34]

Paratuberculosis in other hosts

Paratuberculosis has been described in all of the domesticated ruminant species and a number of other ruminants in the wild and in zoos.[35] Monogastric animals can also be infected, although often only by experimental challenge. Natural infection of rabbits with *M ptb* has been reported from Scotland[36], where there was a strong association between the presence of infection in cattle and the detection of infection in rabbits on the same farm. The strains present in the two host species appeared to be the same and, although the rabbits were excreting fewer than 1% of the number of organisms excreted by cattle, they were still considered a possible source of infection to susceptible cattle.[37,38] In the Scottish studies, no evidence was found of an association between infection in sheep and infection in rabbits although the usual difficulties in culturing *M ptb* from sheep in the UK were encountered.

Infection in rabbits, but at a very low level, has also been reported in other countries[39] and associated with infection in cattle.

In Australia, two studies have examined the prevalence of paratuberculosis in rabbits on farms carrying sheep infected with OJD. In one study of 300 rabbits in NSW, none was found to be infected.[40] In Victoria 100 rabbits from an OJD-affected sheep farm were examined — again with all negative results. In the same study, 210 rabbits from two properties carrying paratuberculosis-infected cattle were also examined with negative results.[41]

Three studies have sought to estimate the prevalence of paratuberculosis in macropods exposed to pastures contaminated by OJD-affected sheep flocks. In Victoria[41] and NSW[40], a total of 400 Eastern Grey kangaroos were examined. No evidence of active infection was detected but one kangaroo in NSW had a positive faecal culture, despite the absence of any detectable histopathological lesion in the intestinal tissue.

On Kangaroo Island, 785 Tammar Wallabies and 55 Western Grey kangaroos were examined.[42] Two animals had histopathological lesions of paratuberculosis but all faecal cultures were negative. The authors of the study concluded that excretion of significant numbers of *M ptb* organisms from macropods is rare and that the animals do not represent a reservoir of infection for sheep.

Survival of the organism

M ptb is considered to be an obligate parasite which cannot multiply outside an infected host. Unlike other members of the *M avium* species, *M ptb* cannot produce the iron-chelating compound *mycobactin* that enables it to acquire iron from the environment. It is believed that this mycobactin dependence renders the organism dependent on the cells of an infected host, or mycobactin-enriched culture media, in order to multiply. (There is, however, some evidence that mycobactin is not always necessary for growth on media and that some media will support growth in the absence of mycobactin.[43])

M ptb can survive, presumably without multiplication, for extended periods in faecal pellets and on soil, pasture and in water. In common with other *M avium* subspecies, *M ptb* is well adapted to aquatic environments. Reports of its survival in spiked water samples, summarised by Collins (2003)[44], indicate maximum survival times up to 517 days (17 months). The longest survival times were associated with particularly favourable conditions — darkness, constant warm temperature and neutral pH.

Studies in Australia with the S strain of *M ptb* indicate that the maximum survival time in exposed and dry sites is typically less than 32 weeks[45] but up to 55 weeks in a dry, fully shaded site.[46,47] The organism was found to survive in the sediment of trough water, sourced from a dam, for 48 weeks.

The likelihood that an environmental site has sufficient living *M ptb* bacteria to cause an infection in a susceptible animal is influenced both by the environmental conditions and the number of bacteria which contaminate the site. The mortality of the bacteria in the environment follows a logarithmic curve, rather than a linear decline or a sudden contemporaneous disappearance after a critical time period. Estimates of the rate of decline vary around a rate of 1 log per month[46] over extended periods, but the possibility of a biphasic rate of decline — 5 logs in the first month then a much slower rate thereafter — has also been proposed.[45] As a consequence of the pattern of logarithmic decline, it can be confidently proposed that the greater the number of bacteria present in a site when it is first contaminated, the greater the chance that an infective dose of bacteria could be ingested from the site at any time in the subsequent months.

Factors which increase the environmental survival time are the presence of shade, water and aquatic sediments. Factors which decrease the survival time are direct exposure to the sun, high environmental temperatures and a lack of vegetation or shade.

In summary, studies in Australia and overseas, with S strain and C strain examples of the bacteria, support the view that survival of the organism outside a host and free in the environment is finite but prolonged. While the number of bacteria which survive declines very rapidly over the first few weeks, some organisms can survive for a year in favourable sites and represent potentially a source of infection for a susceptible host. Given that dams and water courses on farms are frequently heavily contaminated with sheep faeces and that the survival of the organism is prolonged in such environments, low-lying areas of farms should be considered as areas of high risk for survival of the bacteria. Nevertheless, based on the studies reported to date, survival of the organism at levels which could lead to new infections is unlikely for periods greater than 18 months.

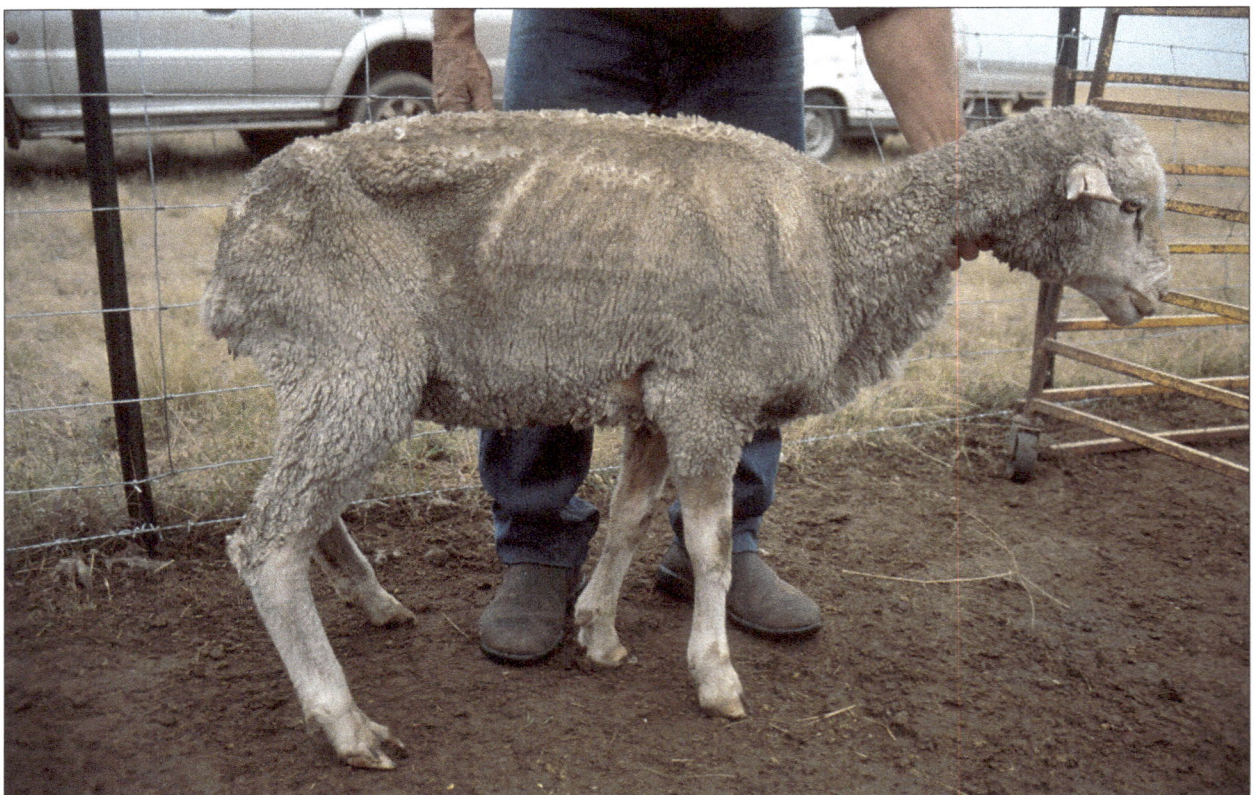

Figure 16.4: A sheep in the late clinical stages of paratuberculosis. Emaciation is the predominant sign of the disease in sheep. Source: KA Abbott.

Clinical signs

Following an incubation period of months or even years, infected sheep show ill-thrift and progressive emaciation (Figure 16.4). Within an age cohort, deaths from OJD typically begin in animals aged 2 years or more but occasional animals will die at younger ages. Deaths of sheep from OJD at 17 months of age have been recorded in at least two separate Australian studies.[48] The age at which the clinical condition appears is related to the age at which the host becomes infected. On farms where the disease is endemic and, therefore, the potential for infection of neonates exists, animals as young as 13 months of age can show clinical signs. Weight loss, relative to flock mates, begins about eight months before death. At the time of death sheep with OJD are, on average, over 30% lighter than expected if in good health.

The clinical phase is also accompanied by a reduction in serum albumin levels, probably as a consequence of interference with protein digestion and absorption.[49] Dependent oedema occurs occasionally. Some sheep may develop diarrhoea, but more commonly the faeces are normal or soft, but not fluid.

Sheep of all ages appear to be susceptible to infection with *M ptb*, but lambs are much more likely to develop patent infections following exposure than adults, and infections developed by lambs are more likely to lead to severe infections and clinical signs at shorter intervals post-infection than infections developed by adult sheep.

Merino sheep challenged as lambs are many times more likely to shed *M ptb* in faeces than sheep challenged for the first time as adults.[50] Earlier first-challenge also increases the risk of mortality from OJD — sheep exposed to *M ptb* before 6 months of age are two to five times more likely to die from paratuberculosis before 36 months of age than sheep exposed for the first time after 3 months of age or as adults.[48,50] Lambs challenged with a high dose of *M ptb* at 6 weeks of age are more likely to develop multifocal lesions (Pérez type 2 and type 3a) than adult ewes, which develop predominantly focal lesions (type 1) under the same challenge conditions.[51] The different age-related responses are associated with a more rapid and more efficient cell-mediated immune response in adult sheep than lambs.

The level of challenge also has an effect on the outcome in individual sheep. Sheep challenged with highly contaminated pastures are three times more likely to have paratuberculous lesions at 36 months of age and 10 to 20 times more likely to die than those exposed at low levels.[48,50]

Subclinical cases

A sheep is said to be infected with *M ptb* if the organism can be cultured from faeces, intestinal tissue or lymph node, or if organisms are detectable in characteristic histopathological lesions in the intestine or lymph nodes.[52]

Sheep which are infected but have not begun to lose weight (or show other clinical signs) are in the subclinical phase of the condition. There are several possible outcomes of subclinical infection. Based on a study[52] to 3 years of age of 77 young Merino sheep from a heavily infected flock in NSW, the following outcomes and the percentage of cases may be proposed:

- No infection develops before three years of age (40%).
- Subclinical infection develops (60%), which in turn leads to
  - rapid progression to clinical disease and death, within 12 months of the development of an infection in the intestine (16%)

   – progression slowly and at variable rates towards clinical disease, some animals remaining with subclinical infection for extended periods (36%)

   – full recovery, with complete freedom from infection (6%)

   – apparent recovery, then relapse to subclinical infection (1%).

Sheep in the early subclinical phases of infection generally have paucibacilliary infections, shedding *M ptb* in faeces but at low levels and, possibly, intermittently. In the later phases of a subclinical infection the lesions may remain paucibacilliary, but more commonly the infection becomes multibacilliary. Clinical disease rarely occurs unless the infection is multibacilliary. Multibacilliary disease is always accompanied by a clinical manifestation and, ultimately, death.

The progression to clinical disease accompanied by multibacilliary infection is linked to a failure of the cell-mediated immune response to contain the infection. Consequently, in a typical case, CMI responses diminish and humoral responses (antibodies) develop as the disease becomes multibacilliary and clinical signs develop.

Asymptomatic infection

A small proportion of sheep which are exposed to *M ptb* on infected farms may develop minor infections, detectable by culture of intestinal tissues but with no evidence of intestinal pathology. The infection is clearly subclinical but this form of OJD has also been described as asymptomatic. In one study, asymptomatic infection was distinguished from both clinical and other forms of subclinical infection by the type of immune response that developed. The outcome of asymptomatic infection is not clear — the infection may progress to a clinical form, may persist as a permanent low level of infection, or it may resolve.[53]

In a large Australian study[48], 12% of 3-year-old sheep, challenged since birth and slaughtered at the end of a field study, had no detectable histopathological lesion but had positive intestinal-tissue culture. Because the field site was *M ptb*-contaminated, it is possible that at least some positive results could have been from contaminated ingesta in the intestine, but the figure puts an upper limit on the possible frequency of asymptomatic infection. The distinction of asymptomatic infection from other forms of subclinical infection is somewhat technical. From a clinical and field management viewpoint, all subclinical cases appear to be asymptomatic.

Levels of infection in OJD-infected flocks

When *M ptb* is introduced at a low level into a non-infected flock (such as by the introduction then removal of a few infected sheep, or by pasture contamination through a boundary fence), the disease takes many years before it increases in prevalence to a level where it is detectable by clinical observations or even by routine testing. Initially, the disease is most likely to be transmitted to lambs born on the property, and adult sheep are least likely to develop infections. Furthermore, because of the low levels of contamination to which they are exposed, only a few lambs will develop infections which become patent and the infections will tend to become patent at ages over 3 years, rather than at younger ages. The contamination caused by this first generation of infected sheep will also remain at low levels for a year or more, leading to a second generation of infected sheep which, likewise, do not shed significantly until they are also 3 years of age or more. Patent infection is likely to remain clustered in one or two age groups and at a low frequency for up to seven years after the organism is introduced into the flock, and clinical cases of OJD are unlikely to appear within that time.[54]

Once infection becomes established in a flock, however, lambs are routinely exposed to *M ptb* organisms from birth and, unless steps are taken to reduce their exposure, the level of challenge is a consequence of the level of infection and shedding in the adult flock. Ultimately it is likely that, in flocks in higher rainfall zones in which OJD has become established, over 40% of the flock will be infected with *M ptb*.

The sensitivity of serological tests is known to be low and the true prevalence of infected sheep is always significantly higher than the seroprevalence. When OJD is established in a non-vaccinating flock, the prevalence of seropositive adult sheep is typically between 6% and 18%.[48,55,56,57]

Levels of clinical disease and mortalities in infected, non-vaccinating flocks

In Australia, for 10 to 15 years after the disease was first described in sheep in New South Wales in 1981, it was commonly expected that the mortality rate from the disease would be low (0.4 to 4% of adult sheep) and likely to remain so because of the climatic conditions common to sheep-raising areas of Australia.[58]

Subsequent investigations, however, indicated that this was not always the case and, before OJD vaccination became available, a number of Merino flocks reported high levels of clinical disease and mortalities. In one of these flocks[55], a structured survey including necropsy of a sample of dead sheep on the farm over a two-year period concluded that paratuberculosis caused or contributed to the deaths of 14.5% of the adult sheep in the first year and 13.2% in the second year.

Other OJD-infected Australian flocks reported OJD mortality rates below 10%, with a mean between 5% and 6% annually.[59]

A number of management strategies (other than, or in addition to, vaccination) can be employed to reduce the severity of OJD in a flock. These strategies were proposed following a three-year-long field study which showed that steps that are feasible for flock managers under commercial conditions can make a substantial difference to the number of sheep which develop clinical disease by 3 years of age. These steps include the immediate removal from the flock of any sheep showing the early signs of OJD, short joining periods, early weaning, and the preparation of pastures for lambing and, especially, for weaned lambs which are likely to be low in contamination.[60]

The need to manage OJD does, however, further limit and constrain the operation of a sheep flock, so the level of control described above is not without cost. Control of the disease through management is rarely used in Australia now that vaccine is available, but it could remain a consideration for flock managers to use in conjunction with vaccination when OJD is first diagnosed in a flock, to hasten the progression to a low-prevalence status. It could also be used by producers in environments which are not favourable to OJD spread who wish to exercise control without vaccination.

Published data on paratuberculosis-attributable mortality rates in other countries are scarce. In the most comprehensive study, Sigurdsson[61] presented data from over 6000 sheep in 141 flocks in Iceland indicating annual death rates of 8% to 12% attributable to paratuberculosis in 1950-51. Most other published figures are unsubstantiated estimates. In Cyprus, paratuberculosis death rates were estimated to be 4% to 5% in sheep flocks.[62] In the UK, OJD mortality rates are generally considered to be low, with occasional exceptions as high as 6%.[63] In New Zealand, reported mortality rates from paratuberculosis in sheep are typically around 1% per annum[64], although some flocks have reported losses of 4% of adult ewes.[65]

Clinical pathology

Identification of the organism

A faecal smear for acid-fast organisms is a moderately sensitive test for paratuberculosis but lacks specificity. The sensitivity is only moderate because the organism is not always detectable in faecal smears of infected, clinically affected animals. The specificity is low because other species of *Mycobacteria* are also acid-fast and commonly occur in faeces.

Smears from ileal and caecal mucosa of suspect cases at necropsy can also be used to contribute to a diagnosis because the organism is usually present in large numbers on the surface of those tissues, particularly in multibacilliary cases (Figure 16.5).

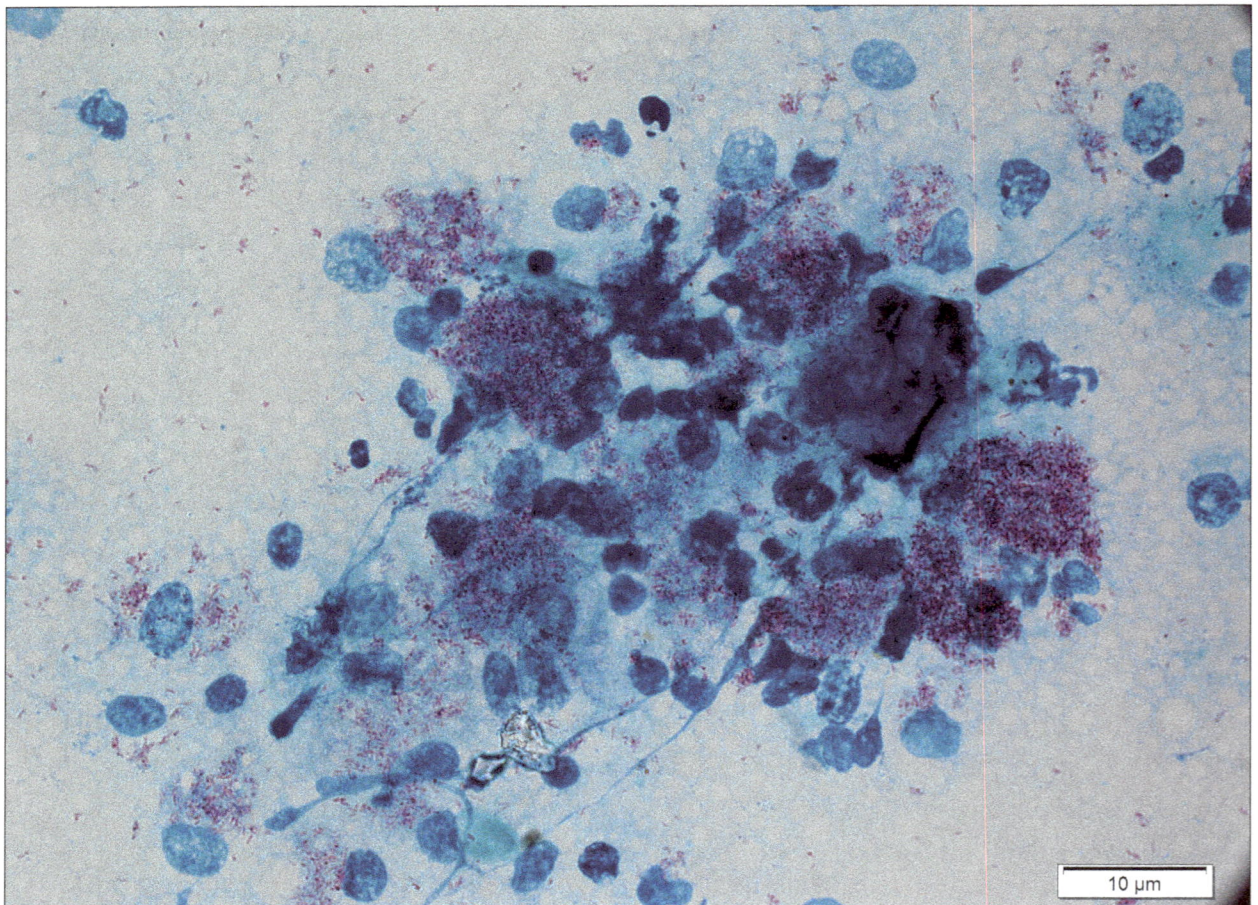

Figure 16.5: An impression smear of the intestinal mucosa of a sheep in the advanced stages of multibacilliary paratuberculosis (Ziehl-Neelsen stain). The mycobacteria are abundant and obvious but, as a diagnostic tool, the presence of mycobacteria is not specific for paratuberculosis. Source: KA Abbott.

Paratuberculosis infection can be confirmed by histopathology or bacteriology. Tests which rely on identification of the bacteria must distinguish between *M ptb* and other mycobacteria, several non-pathogenic species of which occur in the gut and faeces of healthy ruminants.

There are several liquid and solid media used for culture of *M ptb* from faeces or tissues. While all media may be suitable for the growth of C strains, only some are suitable for S strains. Growth of the cultures is very slow and routinely takes up to 16 weeks before the presence or absence of the organism in culture can be confirmed.

Isolation of acid-fast organisms on mycobactin-positive media when growth fails on mycobactin-negative media is considered consistent with *M ptb* even though other mycobactin-dependent species of mycobacteria are known to occur. (These organisms have been called wood pigeon bacteria or *M avium* subsp *silvaticum*[66]). In diagnostic laboratories, the identification of *M ptb* following the isolation of slow-growing, acid-fast, mycobactin-dependent organisms is confirmed with a PCR procedure specific for IS900. The presence of genetic elements similar to IS900 in mycobacteria other than *M ptb* is very uncommon and none identified to date has been mycobactin-dependent.[67]

This approach — culture on selective media, then PCR-based identification of IS900 — has been considered the gold standard assay for *M ptb* in faeces because of its high specificity and high sensitivity in animals which are shedding *M ptb* organisms. Its sensitivity is such that it can be used successfully to detect *M ptb* organisms even in pooled faecal specimens, where the faeces from one infected animal is mixed with faeces from many other non-infected sheep. Pooled faecal culture (PFC) of multiple groups, each 50 sheep, is a commonly used flock-level test in Australian control programmes.

All culture-based techniques suffer from the serious disadvantage of the extended time required to confirm or rule-out the presence of *M ptb*. Consequently, there has been much interest in developing reliable diagnostic techniques based on the identification of *M ptb* DNA.

DNA probe techniques based on PCR amplification of parts of IS900[68,69] can be applied to faecal samples and are highly specific tests because they detect only those mycobacteria containing the IS900 sequence. A real-time quantitative PCR test (qPCR) developed in 2007[70] has demonstrated a high level of sensitivity, even in paucibacilliary cases[71], and is the basis for an approved diagnostic test in Australia — the high throughput Johne's Disease PCR (HT-J PCR). This test can be used to rapidly identify flocks likely to be free from infection with *M ptb* when applied to pools of faeces from suitably selected adult sheep. Despite the relatively high specificity conferred by the presence of IS900, regulatory authorities still base confirmation of positive results on culture techniques, where mycobactin dependence can further limit the risk of a false positive finding.

PCR tests can also be applied to tissues removed at necropsy.[72]

Serology

Serological tests for paratuberculosis include a complement fixation test (CFT), an agar-gel immunodiffusion (AGID) test and an absorbed ELISA test. In cattle, the CFT has been used for many years and is often required by cattle-importing countries. In sheep, serum samples often show non-specific reactions in low dilutions with complement fixing antigens so the CFT lacks specificity in sheep.

The AGID test has a higher threshold for antibody detection than the CFT or ELISA and consequently has a lower sensitivity but higher specificity than either of those two tests. The high specificity makes the test useful for animals displaying clinical signs of OJD or for animals which have reacted positively to the ELISA as a screening test. In common with infections caused by other mycobacteria and by corynebacteria, detectable antibodies to *M ptb* generally only occur when the disease is advanced. In sheep, this usually coincides with the development of clinical signs and the appearance of grossly visible lesions in the intestines. The AGID test was, until recently, the preferred test in sheep. The available data suggest a 90% or greater sensitivity in clinically affected animals but the sensitivity is much lower in sheep without signs. Marshall et al.[73] estimated a sensitivity of 24% in an unbiased sample of six infected flocks in NSW. In small groups of prodromal cases, the AGID may fail to detect any cases.[74]

Early versions of the ELISA test appeared to lack specificity. The development of the technique of pre-absorption of serum samples with whole cells or antigen extracts of *Mycobacterium phlei* has been reported to increase the specificity of the ELISA for paratuberculosis[75], although it is still less specific than the AGID test. The ELISA has the advantage of being relatively inexpensive and rapid and is probably the most sensitive of the available serological tests. It may be more accurate than the CFT in cattle.[76,77] Nevertheless, in clinically normal infected sheep, the ELISA still has a very low sensitivity.

A comparison was carried out in New Zealand of the CFT, AGID test and two absorbed and unabsorbed ELISA tests in sheep. When applied to sheep from known OJD-free flocks, all tests had a similar specificity of at least 97%. In infected flocks, in sheep with histopathological lesions the sensitivity of all tests was similar and at least 98%. When applied to sera from sheep that were without histopathological lesions of OJD but were from infected flocks, the results were poorly correlated between the various tests. The CFT appeared to be the least sensitive, although true sensitivities of the tests were not calculated.[78]

Despite the low sensitivity of serological tests in sheep in the early stages of OJD infection, they can still be very useful when applied as flock-level tests. The likelihood of detecting an infected flock can be increased by testing multiple animals from the flock, selecting sheep of below-average condition score and including sheep from age groups which are most likely to contain infected animals.

Within an infected flock, condition score is strongly influenced by OJD lesion severity, so biasing the sample of the flock on the basis of condition score increases the apparent sensitivity of the test. In a serological and histopathological survey of six infected flocks, the sensitivity of an absorbed ELISA varied with the condition score of the sheep selected for testing. Depending on the ELISA optical density ratio used for declaring positivity, the ELISA detected around 80% of the infected sheep in condition score 1, but fewer than 40% or 20% of those in condition score 3 or 4 (Figure 16.6). Across all condition scores, the ELISA had a sensitivity of 22% (OD ratio 3.6) or 42% (OD ratio 2.4). The specificity was 95% (OD ratio 2.4) or 99% (OD ratio 3.6).[79]

Other diagnostic aids

The simultaneous presence of hypoalbuminaemia and normoglobulinaemia may be a useful screening test for OJD.[80] Albumin levels fall lower in OJD than is usually the case in

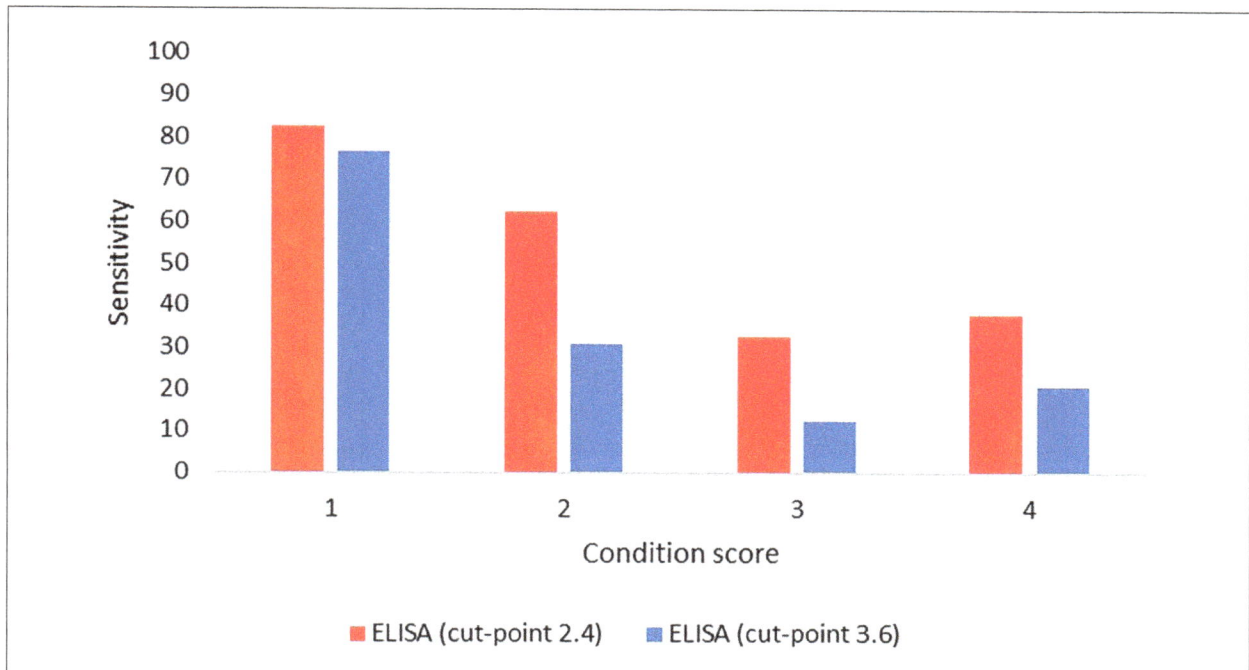

Figure 16.6: The sensitivity of the ELISA test for OJD varies with the condition score of the sheep selected for testing, and the cut-point (OD ratio) used for declaring positivity. Drawn by KA Abbott, with data from Sergeant et al. (2003).[79]

fascioliasis.[49] Chronic suppurative processes, which may also lead to hypoalbuminaemia, are usually accompanied by hypergamma-globulinaemia, which is not the case with OJD.

Diagnosis

OJD must be differentiated from other causes of wasting in adult sheep, including nematodiasis, fascioliasis and intestinal adenocarcinoma. In other countries, caseous lymphadenitis (CLA) is reported to be a cause of *thin-ewe syndrome*, but this presentation appears to be uncommon in Australia. With infestations of nematodes other than *Haemonchus contortus*, diarrhoea is more frequent than is the case with OJD. Faecal egg counts may be helpful in indicating a high worm burden, but egg counts are not always a reliable predictor of worm burden in adult sheep. Ultimately, a necropsy including a total worm count and histopathological examination of the ileum may be necessary for differentiation. With haemonchosis, anaemia is usually marked, in contrast to OJD. Dependent oedema can occur in haemonchosis, fascioliasis and OJD, although it is relatively uncommon in the latter.

Intestinal adenocarcinoma is uncommon in sheep 3 years of age or younger but in older sheep, necropsy may be necessary to differentiate between wasting caused by OJD or adenocarcinoma. Sheep with carcinoma often die from intestinal obstruction before they waste as severely as sheep with OJD.

National control programme

In Australia, the National Ovine Johne's Disease Control and Evaluation Program (NOJDP) was established in 1998 with the aims of limiting the spread of OJD from the main foci of infection in NSW and Victoria and evaluating the possibility of eradicating the disease from sheep flocks, properties and, ultimately, regions. The programme ran, as planned, until 2004. In that six-year period there were several notable developments in the history of OJD in Australia. First, the disease continued to spread and efforts to contain the disease in regions were largely unsuccessful, although it is likely that the rate of spread was substantially slowed by trading restrictions applied to infected flocks. Second, the trading restrictions themselves placed very significant hardships on a large number of flock owners and, in some cases, caused acrimony between regulatory bodies, producers whose flocks remained free of disease and producers whose livelihoods were negatively affected by the regulations. Third, there was a very strong and effective research effort applied to the epidemiology, diagnosis and control of OJD, resulting in a much better understanding of the disease in Australia and Australian sheep by the end of the programme.

In 2004 and 2007, two further management plans were established as a result of consultation between government bodies and sheep industry organisations. The plans were managed by Animal Health Australia. In 2013, the OJD Management Plan 2013-2018 commenced. This plan aimed to facilitate efforts by individual producers in Australia to reduce the likelihood (risk) of introducing OJD into their flocks with purchased sheep through two strategies:

- A Market Assurance Program (MAP) has been developed, in which flock owners can demonstrate the level of testing with negative results that the flock has undertaken. Flocks can achieve Monitored Negative levels 1, 2 or 3 (MN1, MN2, MN3) status and the highest status indicates the lowest risk that OJD is present in the flock.

- All sheep sold are accompanied by a Sheep Health Statement, a form which allows vendors to inform purchasers of the level of OJD testing that has been undertaken in the flock, if any.

Participation in the Sheep MAP is voluntary and the costs of testing are borne by the producer. Private veterinarians are responsible for the development, with the producer-client, of a flock management plan, identification of sheep to be tested and collection of the faecal or blood specimens which are to be tested at the laboratory. There are options and guidelines which are used for individual flocks, depending primarily on their size, but the most common form of testing performed is of bacteriological culture of faeces from 350 sheep. Faeces collected from groups of 50 sheep are pooled before culture so that the testing of 350 sheep is achieved by the culture of seven pools, each from 50 sheep — constituting the pooled faecal culture (PFC) test.

In order to advance from lower status to higher MN status, flocks are tested every 24 months. Flock owners who wish to maintain the MN status of their flocks, but not advance it, can opt for a less sensitive flock test (a smaller sample of animals). In the year when PFC testing is not performed, the veterinarian responsible for the flock in the MAP is required to examine and necropsy any sheep in low condition score as a component of the annual veterinary review.

Owners can also choose to vaccinate their flocks with Gudair® and the vaccination status of the flock is also recognised in the record of their flock's status.

The contribution of the MAP to the national or regional control of OJD is not yet clear but is probably not significant at a national level. A technical review of the programme in 2013 reported that the number of flocks in the MAP has declined from about 800 in 2005 to 432 in 2013.[81]

Elimination of infection from a farm

Elimination of the disease from individual properties is theoretically possible and destocking is the most realistic method. This could be successful provided that any animals which may harbour and transmit the infection are removed from the property for longer than the organism can survive on pasture. Previously recommended strategies for eradication of OJD included removal of all sheep and replacement with cattle, but the increasing number of cases of S strain infection in cattle makes this approach unreliable. If wildlife species are found to serve as a source of infection for sheep, attempts to eradicate the disease by destocking (of domestic livestock) are unlikely to succeed but, to date, this has not emerged as a problem in Australia.

The recommended length of time for destocking is 18 months, preferably including two summer periods.

A study in 1998-2002 on 40 farms which removed all sheep for a period of at least 15 months, including two summer periods, found that infection was detected in replacement stock by three years after restocking in 68% of cases.[3] Reasons for the failure of elimination attempts included reinfection from neighbours' flocks and inadvertent reintroduction of infection in sheep used for restocking. Not all causes of failure were determined. Persistence of infection in soil and pasture for the destocked period remains, therefore, a possibility.

In most states of Australia, there is no compulsory eradication or control scheme, nor is there any compensation scheme. Consequently, individual owners of infected flocks can decide whether to attempt to eradicate the disease from their flocks, or to attempt to manage their flock in ways which minimise the economic impact of it. In an assessment of the likely economic impact of different strategies to deal with OJD, one study found that the lowest-cost option for eradication varied between flocks, based on flock type and alternative land uses available to the producer. Options included cropping, destocking of sheep and replacing with cattle, or taking no steps to eradicate but adopting flock structures which minimised the production losses from OJD.[82] Vaccination has become the most realistic disease management strategy for most farms with OJD-infected flocks.

There is no effective treatment for individual cases of OJD.

Vaccination

Vaccination is a very useful disease management tool for OJD and is the preferred method to control the disease in sheep flocks in a number of countries.[61,83,63,84,62,85,86,87] The use of vaccination in sheep flocks contrasts with the approach preferred in cattle herds — which involves culling suspect clinical cases, cows shedding *M ptb* in faeces and seropositive cows. Partly, this is due to the higher individual value and higher intensity of management of dairy cattle compared to sheep, and partly it is due to the difficulties posed for tuberculosis control in cattle by the cross-reaction of paratuberculosis-vaccinated cattle to the tuberculin test.

Vaccines used widely in sheep in the field have included those based on heat-killed *M ptb*, such as those used in Iceland (Sigurdsson's vaccine), Spain and Australia (Gudair® vaccine), and live attenuated vaccines which have been used in Spain and New Zealand (Neoparasec®

vaccine) and Britain (Weybridge). The live vaccine was withdrawn in New Zealand in 2002. Both killed and live vaccines have similar levels of efficacy.[84,88]

Vaccination does not prevent the development of infection in all sheep but increases the rate and the intensity of the response to infection.[89,90,91,92,93] Consequently, compared to non-vaccinates, vaccinated animals tend to develop intestinal and lymph node lesions which are more granulomatous and contain fewer *M ptb* organisms, and they are more likely to clear the infection and to do so earlier.

In infected flocks, vaccination is effective in reducing the rate of clinical disease, the proportion of a flock which is shedding *M ptb* at detectable levels, the age of onset of faecal shedding and the mortality rate from *M ptb* infection — the latter by over 90%.[83,87,89,93] Because it is not 100% effective, vaccination is unlikely to lead to the elimination of *M ptb* from a flock unless its use is combined with other control measures. In Merino flocks, and possibly other breeds, a small number of vaccinated sheep will develop multibacillary forms of OJD and shed very large numbers of *M ptb* organisms into the environment before dying or culling.[93] Nevertheless it is possible that, if environmental factors or host factors are not favourable for transmission of paratuberculosis, vaccination could be the additional factor which leads to the ultimate disappearance of the disease from a flock.[94]

Vaccination has been performed on lambs from 2 weeks of age upwards. Cellular immune responses, as measured by IFN-δ, are stronger in sheep aged 5 months than those aged 15 months[95], but the significance of this result for protection in natural flock conditions is not clear. Excellent results have been obtained with vaccination at 2 weeks[89], 1 to 3 months[93] and at 6 months of age.[61] Revaccination is not necessary.[89]

It is generally considered desirable to vaccinate animals before exposure to the disease in question but, with paratuberculosis in sheep flocks, exposure can commence before, at or soon after, birth. In extensive management systems lambs are often not handled until they are 2 to 6 weeks of age, so, in infected flocks, vaccination before exposure to *M ptb* may be impossible or impractical. Several studies have shown that vaccination post-exposure is successful.[96,61,62] The only study presenting a contrary view was carried out on lambs aged 5 months which were killed and examined six weeks after vaccination. A low rate of infection in the control group and a relatively premature ending to the experiment may have led to a mistaken inference about the lack of success of post-exposure vaccination in that case.

The length of time between exposure and vaccination may also be a critical factor in determining how well post-exposure vaccination will work. In lambs aged 6 to 8 weeks, vaccination two weeks after artificial infection reduced incidence by 71% in the 44 weeks post-vaccination and markedly reduced the severity of intestinal lesions and the number of *M ptb* organisms detectable in intestinal tissues.[96] In field reports of infected flocks in which adult animals have been vaccinated, the incidence of paratuberculosis has been observed to decline three months[62], five months[84] or six months[97] post-vaccination. In cases where many adult sheep have relatively advanced paratuberculosis when vaccinated or when other disease control strategies are not simultaneously put in place, the effect of vaccination on the mortality rate may not be evident in the first year post-vaccination.[55]

Vaccination produces large granulomatous reactions at the site of injection in about 40% of animals, although the nodules tend to decrease in size over the following 10 months to an average of 26 mm in diameter.[93] The reactions are sufficiently large to discourage the use of vaccine in lambs which are destined for slaughter at young ages, and to encourage the use of the vaccine only in those sheep which will be retained for breeding.[83]

Gudair® (*Zoetis*) is the only vaccine registered for use in Australia and in New Zealand and is approved for use in both sheep and goats. In states where vaccine is generally available, each vaccinated sheep must be identified by a mark (usually the letter V in a circle) on the NLIS (National Livestock Information System) ear tag.

Vaccination interferes with serological testing for OJD for export and diagnostic testing[86] and animals intended for export should not be vaccinated.

Dangers to human health

Both killed and live vaccines present a potentially serious risk to human health following accidental self-inoculation, and such incidents may occur as frequently as once every 7400 sheep vaccinations.[98] Needle-stick injury of a finger can lead to incapacity lasting months and, possibly, permanent loss of function, or even amputation.[99] Risk factors for self-inoculation are likely to include inadequate restraint of the sheep, the age of the sheep, poor technique and the performance of multiple husbandry procedures at the time of vaccination.[98]

Crohn's disease

Crohn's disease is a chronic inflammatory bowel disease of humans. The histological appearance of the lesions is similar to ruminant paratuberculosis. Mycobacterial species have been isolated from Crohn's disease lesions.[100] PCR tests for IS900 were applied to DNA extracted from intestinal lesions surgically removed from patients with Crohn's disease and detected *M ptb* from 11 of 24 patients.[101] The relationship of the organism to the disease, however, is unclear and there remain uncertainties that *M ptb* has a causative role in Crohn's disease.

Intestinal carcinoma

In general, neoplasia of the gastrointestinal tract is uncommon in sheep and this tumour is the most likely form to be encountered. The only systematic study in Australia detected 17 cases in 6248 adult sheep, estimated to be 5 to 7 years of age, at export abattoirs in three centres of NSW.[102] It characteristically occurs in Merino sheep 5 years of age or older. Affected animals are dull, constipated and show abdominal distension. The clinical course is usually limited to three to six weeks before the sheep dies.[103]

At necropsy, the tumour appears as a dense, irregular ring of white fibrous tissue encircling the gut anywhere along the length of the jejunum and ileum. The lumen is usually markedly constricted and the intestine proximal to the lesion is grossly distended. In many cases, metastatic scirrhous plaques occur on the serosal surface of adjacent parts of the intestine.

The incidence of the tumour is likely to rise with the age of the sheep. An association has been noted with the use of herbicides of the phenoxy group (MCPA, for example) and pyridine group (Clopyralid, Picloram).[104]

SECTION B

GASTROINTESTINAL DISEASES OF SHEEP OF ALL AGES

Grain poisoning (grain overload)

The ingestion of large amounts of rapidly fermentable starch-containing foodstuffs in sheep which are not accustomed to the diet leads to the excessive production of lactic acid in the rumen. The consequent fall in ruminal pH may lead to systemic acidosis. Fermentation in the caecum can also occur and contribute to the syndrome. Cereal grains are the most common causes of ruminal acidosis — wheat being the most dangerous and oats the least, with triticale and barley being intermediate in risk. In sheep not accustomed to grain, amounts of 1 kg (weaners) or 2 kg (adults) are potentially fatal. Grain poisoning can be acute, subacute or chronic, largely depending on the magnitude and persistence of the depression in ruminal or large intestinal pH. Management of subacute ruminal acidosis (SARA) is particularly important in feedlot lamb production systems in order to maintain appetite and feed intake and to ensure that high rates of growth are achieved. Milder, more chronic forms of ruminal acidosis are accompanied by reduced feed intake and poor growth rate. Laminitis may develop in some animals, particularly with subacute or chronic grain overload.

Sheep are often fed cereal grains to supplement a pasture-based diet or are fed in feedlots with cereal grains as a dominant component of a ration. Overeating does not usually occur when animals are first offered grain because the novelty of the feed tends to limit their intake. Acute grain poisoning is more likely to occur when animals which have been previously habituated to the feed are

(1) offered large amounts of grain after a period of abstinence

(2) offered a sudden increase in amount or sudden change to a more 'dangerous' grain

(3) allowed access to their normal full ration after a short period of starvation or cold stress.

Pathogenesis

Bacteria, protozoa and other microbes in the rumen and caecum ferment dietary carbohydrates to volatile fatty acids (VFAs — principally acetate, butyrate and propionate) and lactate. Normally, lactate is only present in the digestive tract at low concentrations and there is a balance between lactate-producing bacteria and the lactate-utilising microflora in the rumen.

The pH of the rumen of roughage-fed sheep is normally maintained between 6.7 and 7.0 (pre-feeding) and 5.8 to 6.5 (post-feeding).[105] The acidifying effect of VFA production is ameliorated by the buffering action of phosphate and bicarbonate ions present in saliva and bicarbonate ions secreted into the rumen through the ruminal epithelium. Bicarbonate is particularly important when the rumen pH is below 6.3. In a more strongly acidic environment the VFAs themselves act as buffers, with their most effective buffering capacity in the region of their pKa (4.7-4.8).

When the supply of highly fermentable carbohydrate is suddenly increased the total production of all organic acids, including lactate, increases. The normal buffering systems are overwhelmed. Lactate is a much stronger acid (pKa of 3.8) than the VFAs and is chiefly responsible for the dramatic fall in pH. The lower pH and the ready supply of starch and glucose in the rumen favour the multiplication of gram-positive, lactate-producing bacteria, including *Streptococcus bovis* and *Lactobacillus* spp. These bacteria are normally present in relatively low numbers in the rumen and caecum. The increase in lactate production is not just a consequence of the increase in numbers of these bacteria, but also a change in the glycolytic pathways they employ for energy production. When the rumen pH is close to neutrality *S bovis* produces acetate, formate and ethanol from fermentation of glucose but, at a pH of 5.6 or lower, it switches to one which produces predominantly lactate.[106] When the pH falls below 5.5 the growth of lactate-utilising bacteria is inhibited, further exacerbating the accumulation of lactate.

The change in rumen microflora can occur very rapidly — the concentration of lactic-acid producing bacteria in rumen fluid can increase by a factor of 10^6 within 24-48 hours of a diet change from forage to wheat.[107]

If rumen pH continues to fall there is a cessation of normal ruminal flora activity and of ruminal motility. Below a pH of 5.0, there may be complete ruminal stasis. The high osmolality of the lactic acid in the rumen draws water in from the systemic circulation, leading to distension of the rumen, haemo-concentration and dehydration. Damage to the ruminal epithelium and the influx of water into the rumen interfere with the absorption of VFAs through the rumen wall and the secretion of bicarbonate from the ruminal epithelium. Salivation — normally stimulated by rumination, mastication and mechanical irritation of the ruminal epithelium with fibrous material in the rumen ingesta[108,109] — is reduced, further impeding the flow of bicarbonate into the rumen.

In severe cases, sufficient lactic acid is absorbed through the rumen wall and through the intestinal wall to challenge the systemic bicarbonate buffering system. If this system is overwhelmed, blood pH falls and peripheral perfusion declines as the circulatory system fails. Renal blood flow declines, the animal becomes anuric, hypovolemic and shocked, and may die within 24 to 48 hours of grain ingestion.

The low pH of the rumen favours the growth of fungi which invade the ruminal vessels causing thrombosis and infarction. Bacterial rumenitis with necrosis and sloughing of the wall can follow. Rarely, the abomasum can also be involved, possibly also due to mycotic infection. The damaged ruminal epithelium may allow the passage of bacteria including *Fusobacterium necrophorum* and *Arcanobacterium pyogenes* through the portal system to the liver, leading to liver abscessation. *F necrophorum* is a lactate-utilising normal inhabitant of the rumen which increases in number in grain-fed animals and is capable of invading the parakeratotic rumen wall.[110,111] The liver involvement may be the cause of mortalities days after the initial lactic acidosis is corrected.

Fermentation and acid accumulation also occur in the caecum and are responsible for the diarrhoea which is often observed. Diarrhoea due to hind-gut acidosis after ingestion of grain can sometimes occur even while rumen fermentation remains normal.

The pathogenesis of acute laminitis, which sometimes accompanies grain overload in sheep, is unclear. In dairy cattle, it is proposed that the fall in systemic pH is accompanied by seepage of serum, oedema and haemorrhage in the solar corium, expansion of the corium and severe pain, and that the vasoactive substances may be histamine, endotoxins absorbed from the rumen, or other substances as yet unidentified.[112]

Clinical signs

In the early stages of acute grain overload, affected sheep separate from the mob, stand or lie in sternal recumbency and are anorectic. Some abdominal distension may be evident; there is rumen stasis; and respiration is rapid and shallow. Laminitis may be present in some affected animals, causing a reluctance to rise and a tendency to stumble when first moving off. Diarrhoea develops after about two days, often in animals which have had mild or inapparent signs of ruminal stasis. The diarrhoea is usually brown in colour with a characteristic sweet-sour odour. Severely affected animals become extremely depressed and dehydrated and die quietly within three days of grain ingestion.

If ewes in late pregnancy are grain-fed, the few days off-feed while the ruminal acidosis runs its course, or a reluctance to graze due to laminitis, may precipitate pregnancy toxaemia.

Diagnosis

The clinical signs and a history of access to highly fermentable feedstuffs are usually sufficient to indicate a diagnosis of grain overload. On some occasions, if laminitis is the predominant presenting sign in a flock, the association with a gastrointestinal cause may not be immediately evident and a careful collection of the history is important. A ruminal pH below 5.0 is strong evidence of lactic acidosis but collection of reliable rumen fluid specimens in live sheep can be difficult. Rumen fluid collected by stomach tube may be contaminated with saliva and the pH found to be within a normal range. In cattle, rumenocentesis through the left paralumbar fossa is considered the most reliable method to collect a diagnostic specimen.[113]

Relying on a low pH of rumen fluid alone to confirm a diagnosis may be hazardous. The reflux of abomasal contents into the rumen can also reduce ruminal pH. Abomasal reflux can occur as a result of atony of the abomasum, or any of the fore-stomachs, independent of lactic acidosis. Cases of rumen acidosis caused by abomasal reflux can be differentiated from lactic acidosis by the high chloride concentrations in the rumen fluid[114], indicating abomasal hydrochloric acid as the source of acidity.

Elevated plasma D-lactate concentrations will confirm a diagnosis of grain overload. Blood concentrations are normally very low (<0.5 mM) but are raised substantially (>5 mM) in animals clinically affected or dead from lactic acidosis. Plasma, serum or whole blood from clinical cases or recently dead animals are suitable specimens.

Necropsy

Animals dead from the direct effects of acidosis are dehydrated and sunken-eyed. The rumen and reticulum are full of partly digested grain in a porridge-like 'soup' much thinner than normal rumen contents and with a characteristic odour. The rumen pH is much lower than its normal 6.8 to 7.0. Measurements of rumen pH in animals which have been dead for some time may be misleading because the pH will change after death and after exposure to air.[105] The rumen wall may show patches of haemorrhage and sloughing of the epithelium.

Treatment

Severe cases of acute ruminal acidosis benefit from treatment aimed at correcting the ruminal acidosis, the systemic acidosis and dehydration, and secondary effects resulting from damage to the ruminal epithelium.

The most effective method of correcting the ruminal acidosis is surgical removal of the rumen contents through a rumenotomy. The success rate is high if surgery is performed in the early stages of the disease — while the animal is depressed but before hypovolemic shock is advanced. Individual surgical treatment is only justified for valuable animals.

Non-surgical treatment is aimed at stopping the further production of lactic acid in the rumen and raising ruminal pH. Oral or intra-ruminal administration of antibiotics, such as virginiamycin, which are selectively effective against gram-positive organisms, may be useful by reducing the *Lactobacillus* spp and *S bovis* populations in the rumen.[115] Administration of sodium bicarbonate (baking soda) is rational but large amounts are required to provide sufficient buffering of the low pH induced by the lactic acid. For large groups of affected sheep, provision of sodium bicarbonate as a loose lick has been found to be effective — sheep reportedly ingesting amounts of around 100 g per head. Sheep which are too ill to move can be drenched with a sodium bicarbonate solution.[115] Both of these less intensive forms of treatment (ruminal antibiotics, oral sodium bicarbonate) are particularly useful when managing outbreaks of grain overload in flocks of sheep involving multiple affected animals.

More intensive medical treatment includes the daily administration by stomach tube of 0.5 to 1 L of ruminal fluid from a healthy ruminant, usually a cow for convenience of collection, and intravenous fluid therapy — isotonic sodium chloride or Ringer's solution to correct dehydration, and sodium bicarbonate to correct metabolic acidosis.[114] Intravenous thiamine and systemic antibiotics such as oxytetracycline are also advisable. The use of hypertonic solutions for rehydration of animals with grain overload should be avoided. It should be noted that both normal saline and Ringer's solution are acidifying[116], so sodium bicarbonate therapy is an important adjunct to rehydration with those solutions.

The amount of sodium bicarbonate solution which should be administered intravenously depends on the degree of acidosis. In the absence of laboratory aids to provide an objective measure (blood gas analysis, total CO_2 measurement), treatment can be provided on the assumption of a moderate degree of acidosis (a base deficit of 10 mmol/L).[117] In this case, a 75 kg sheep requires 19 g of $NaHCO_3$ — delivered as 375 mL of a 5% sodium bicarbonate solution or 1.5 L of an isotonic (13 g/L) solution. In the absence of laboratory support during treatment and in light of the risk of overtreatment with bicarbonate, Michel (1990)[118] recommends for cattle a maximum of 0.5 g/kg of $NaHCO_3$ initially (37.5 g to a 75 kg sheep), followed by the same amount, in isotonic solution, over the next 24 hours, if required. Sheep with metabolic acidosis are tachypnoeic, in order to blow off excess CO_2. As blood pH returns to normal the respiratory rate should approach a normal rate.

Management of an affected flock also involves changing the ration to avoid initiating new cases. The amount of grain ration should be reduced and roughage (hay) offered. If treatment is impractical and most cases are relatively mild, affected sheep are often left to recover on their own.

Prevention

Grain poisoning which occurs as a result of accidental access to dangerous grains can be prevented by ensuring that animals cannot gain access to bagged grains in sheds, spilled grain under silos or other grain stores or depots. The risk of grain poisoning which occurs as a consequence of deliberate feeding of grains can be reduced in a number of ways. Gradual introduction is very important. Animals can adapt to grain diets over a period of one to two weeks, principally by increasing saliva flow to increase the buffering of lactic acid in the rumen.

Sodium bicarbonate is often included in feedlot rations for sheep at the rate of 2% or 3% of the diet and there is evidence that it will improve the digestibility of cellulose and starch, increase ruminal pH and ruminal nitrogen concentrations, and increase lamb growth rates when added to rations predominantly composed of grain. The benefits of sodium bicarbonate are not, apparently, a simple consequence of buffering by acid neutralisation — indeed, the amount of acid produced by the fermentation of 1 kg of grain requires 300 to 500 g of sodium bicarbonate to neutralise it. The buffering effect is likely to be more related to the increase in osmotic pressure in the rumen and increased inflows of water and the rate of rumen outflow. Nevertheless, inclusion of sodium bicarbonate in grain-rich rations is not always successful in preventing ruminal acidosis and may predispose lambs to urolithiasis.[119] (See Chapter 18 for more detail.)

Sodium bentonite is sometimes added to the feedlot rations of sheep and cattle. It is not a buffer but is a clay with marked water-absorbing and toxin-absorbing properties. It may contribute to improvements in rumen digestion on high-grain diets but there is no evidence of a consistent benefit to prevention of ruminal acidosis.

Limestone, often added to grain diets as a source of calcium, is not a buffer and does not prevent ruminal acidosis.

Antibiotic compounds active against gram-positive bacteria are very effective against acidosis. These compounds include avoparcin and virginiamycin. Virginiamycin (Eskalin®, Philbro Animal Health) is available in several forms including a wettable powder which can be dissolved in water. The solution can be sprayed onto grain immediately before it is offered to sheep as it passes through an auger into a grain-feeding bin. The risk of lactic acidosis is substantially reduced by inclusion of virginiamycin when animals are introduced onto high-grain (80%) diets without a period of adaptation.[120,121,122] Virginiamycin may reduce feed costs and simplify management when switching sheep from all roughage to high-grain diets. It simplifies grain feeding in feedlots, during drought and during transhipment of livestock by sea.

The addition of antibiotics to livestock feeds, particularly for extended periods, increases the risk of selection for antibiotic resistance in bacteria. Virginiamycin is a streptogramin antibiotic — a class of antibiotics which may become increasingly important in human medicine as resistance to other human bacterial pathogens becomes more widespread. In Australia, virginiamycin is only available to producers or feed manufacturers with the approval of a veterinarian. The Australian Veterinary Association, through its published *Code of Practice for prescription and use of products which contain antimicrobial agents*[e], recommends

e   The full policy can be seen at https://www.ava.com.au/sites/default/files/documents/Other/Code%20of%20 Practice%20for%20Antimicrobial%20Agents.pdf.

that veterinarians explore non-antibiotic options to reduce the risk of ruminal acidosis before prescribing or approving the use of virginiamycin in feeds.

Ionophore rumen modifiers (monensin and lasalocid) have been shown to prevent lactic acidosis in cattle following infusion of readily fermentable carbohydrates.[123] They may be added to commercially prepared rations for sheep in order to assist in control of coccidiosis and to improve feed conversion efficiency.[119]

The selection of grains is also important in reducing the risk of acidosis. Cereal grains have a high starch content and a low fat content. The high starch content frequently leads to feeding problems; the safer grains are those with lower starch content or where the hull has not been removed, such as oats. Wheat is high in starch and its starch is very soluble — more soluble than barley starch. Oat starch is soluble but oats are relatively low in starch. Much more wheat starch is fermented in the reticulo-rumen than is the case for barley, oats, maize or sorghum. Of the alternative, non-cereal grains which can be fed to sheep, lupins are the most popular in Australia. These are a legume grain which contain very little starch and are relatively high in fat and protein. They can be introduced to sheep with little or no period of adaptation and present a very low risk of lactic acidosis, despite having a metabolisable energy content similar to barley.

When feeding cereal grains without antibiotics, feeding frequency is important to reduce the risk of grain overload. A lower feeding frequency can be employed with lupins. Sheep supplemented at pasture will grow at similar rates when fed barley or lupins on a daily basis. If fed once per week, however, animals fed barley perform less well due to a low level of ruminal acidosis. If the barley is treated with virginiamycin, animals will perform as well on barley once per week as on lupins at the same frequency.[124]

Red gut (haemorrhagic enteritis)

Red gut is a condition of sheep in which a catastrophic intestinal accident leads rapidly to death. At necropsy, affected animals have intense venous congestion of the intestines. In almost all cases there is evidence of a torsion of the intestinal mass with displacement of the rumeno-reticulum, small intestine and large intestine, with a torsion of the anterior mesentery.

Pathogenesis

While the condition has been reported in sheep grazing subterranean clover, red clover, white clover or lush ryegrass-white clover pastures, it is most strongly associated with the grazing of lucerne growing in swards with little or no grass or weed content and it is with lucerne pastures that most studies of red gut have been performed.

It is known that, compared to the grazing of pastures with significant grass content, the grazing of monoculture lucerne pastures leads to a change in the relative volumes of the rumeno-reticulum and the large intestine, as an apparent consequence of a shift in fermentative digestion from the proximal gastrointestinal tract to the hindgut.[125] Consequently, the rumeno-reticulum and the large intestine take up different positions in the abdomen, with the large intestine, particularly the caecum, moving anteriorly.

It is proposed that the new positions and altered proportions of the gastrointestinal organs make them rotationally unstable and a further twist of the intestinal mass can cause occlusion of the cranial mesenteric artery and vein.[126] Both the small intestine, from a point about 1 m distal to the pylorus, and the proximal large intestine are dependent on these vessels for blood supply and drainage. The occlusion of these vessels leads rapidly to profound shock and death.

In some cases of red gut, the massive reddening of the intestines and the changes in position and size of the parts of the gastrointestinal tract are clear, but the torsion is not evident at necropsy. This has led to conjecture that the torsion may not be essential to cause occlusion of the cranial mesenteric vessels and that, possibly, distension of the large intestine may be sufficient cause alone.

Clinical and necropsy findings

Affected sheep are usually recumbent with distended abdomens and signs of acute abdominal pain. Death occurs soon after signs first develop — possibly within two hours.[127]

In suspected cases, animals should be necropsied while laid on their backs so as to better view the position of the abdominal organs. The intense hyperaemia of the gut is striking. It commences about 1 m distal to the pylorus and, unlike the hyperaemia seen with infectious enteritis, there is a sudden transition from normal to abnormal intestinal tissue. Displacement of the gastrointestinal tract is typically a 180° to 360° clockwise rotation, viewed from the ventral surface.

Diagnosis

Diagnosis is generally made on the basis of post-mortem examination and a history of exposure, for two weeks at least, to lucerne or clover-dominant pastures. Other causes of sudden death on high-quality legume-rich pastures must be considered. Bloat does occur in sheep but is much less common than in cattle. Bloat will often occur within hours of introduction of sheep to the pasture and any affected animals can be successfully treated by insertion of a rumen trocar. The abdominal and ruminal distension in cases of bloat are much more pronounced than in cases of red gut. Enterotoxaemia, particularly cases which have been dead for some hours before necropsy, can appear similar to red gut. In the absence of a mesenteric torsion to confirm red gut, the diagnosis of enterotoxaemia can be confirmed or ruled out by necropsy including brain histopathology.

Prevention

Prevention involves reducing the continuity of access to the highly digestible forage which is leading to the changes in the gastrointestinal tract predisposing to the disease. There is little point in offering hay or lower-quality feeds while they are grazing lucerne-dominant or clover-dominant pastures — sheep are unlikely to eat significant quantities of the lower-quality feed. It is advisable to remove them from the high-risk forage for a period of time, placing them on grassy pasture, or offering food of lesser digestibility or higher fibre — to encourage ruminal fermentation and slower ruminal emptying. One strategy reported to be effective is grazing five days on and then two days off the high-risk pasture.[126] The changes in the gastrointestinal tract which appear to predispose to red gut take about three weeks to develop after sheep are introduced to the offending forage. One could expect, therefore, that strategies to reduce the risk of the condition should commence about two weeks after introduction.

Salmonellosis

In sheep, the main serovars involved in outbreaks are *S typhimurium* and *S bovis-morbificans*. Outbreaks of disease in sheep are usually associated with crowding and stress. It is known that transport and food deprivation lead to increased excretion by carrier sheep. It is no surprise, therefore, that salmonellosis has been found to be a significant cause of death of sheep transported live by sea for slaughter in the Middle East. Salmonellosis is the most common cause of death in the feedlots in which sheep are gathered before the voyage and the second most common cause aboard ship.[128,129] Failure to eat in the feedlot is a major risk factor for salmonellosis aboard ship.[130] The disease also occurs in sporadic outbreaks on farms.

Clinical signs

The disease generally occurs in outbreaks, with morbidity rates of 5% to 30% and case fatality rates of 25%. Some affected animals die within hours of the onset of signs; most die within one to five days. Those examined alive are depressed, anorectic, dehydrated and febrile, with severe, putrid, very fluid diarrhoea, sometimes with shreds of mucosa and blood. Recovery in survivors is slow and these sheep lose much body weight over the course of the illness.

Necropsy

Lesions are evident in the ileum, jejunum, caecum and colon. Gut contents are fluid, sometimes with blood, mucus and fibrin. Common additional lesions include abomasal congestion, hydropericardium and congestion of the liver and spleen.

Diagnosis

Confirmation is based upon clinical signs, histopathology of the gastrointestinal tract and isolation of the organism. The exotic disease *pestes des petits ruminants* should be considered in outbreaks of acute, severe diarrhoea with fever and severe depression. Diarrhoea is not a consistent sign in *bluetongue*, but affected sheep are severely depressed and febrile. Buccal lesions are a consistent finding in bluetongue.

Laboratory aids

Faecal specimens for bacteriology should be collected from a range of clinical cases. The organism can also be isolated from ileal and caecal contents, mesenteric lymph nodes, liver and spleen. Fixed sections[f] of the same organs plus lung and kidney should be submitted for histopathology.[131]

Control

Reduction of stocking density, removal of affected sheep and removal of the entire mob to another area may help limit an outbreak. Treatment with trimethoprin/sulphonamide combinations has been effective in some outbreaks.

---

f   Buffered formol-saline is satisfactory for routine fixation of tissues. Special attention should be given to the fixation of intestinal tract sections. A recommended method involves opening to a rectangle a length of intestine and placing it on a piece of cardboard, serosal surface down. Dry in air for a few minutes until the tissue adheres to the cardboard, then place both into the fixative.

Diarrhoea of unknown cause in adult sheep ('winter scours')

Diarrhoea in adult sheep is a serious problem for sheep producers because it predisposes sheep to flystrike of the breech area, reduces the value of wool from the breech area as a consequence of faecal contamination, and adds to the cost of sheep management when more frequent control measures must be put in place to reduce the risk of breech strike — chemical control or additional crutching.

Diarrhoeic faeces which accumulates around the breech in the perineal wool dries and forms hard masses known commonly as dags.

Adult sheep sometimes develop diarrhoea on high-quality, fast-growing, highly available improved pastures in winter and early spring but show no evidence of systemic illness or inappetence. In the absence of significant worm burdens or faecal egg counts, nematodiasis is usually discounted and it has been assumed that the high water content of the pasture and high soluble sugar content of the pasture causes diarrhoea by an osmotic effect in the caecum and colon.[132] *Arctotheca calendula* (capeweed) has long been suspected as a major contributor to this effect but attempts to reproduce it experimentally have failed.[133] In the absence of any infectious cause, the condition has been called *nutritional scours* or *winter scours*.

Other suggested causes for this form of diarrhoea include molybdenosis and hypocuprosis (a syndrome associated with diarrhoea in cattle) and high nitrate intake. While some pastures, fodder crops (oats and oaten hay) and weeds (including capeweed[134]) can be high in nitrate, there is no evidence which provides a clear link between high nitrate intake and diarrhoea in otherwise healthy, normal sheep. High levels of dietary nitrate are irritant to the gastrointestinal mucosa and, in cases of nitrate toxicity, diarrhoea may be present but is likely to be accompanied by other signs, including abdominal pain and excessive salivation.[135] Much more commonly, however, high levels of nitrate intake in ruminants lead to nitrite intoxication as a result of the ruminal reduction of nitrate. Rather than gastrointestinal signs, nitrite poisoning is characterised by respiratory distress and death from anoxia, as a result of the conversion of haemoglobin to methaemoglobin.[136]

In the absence of empirical evidence linking idiopathic diarrhoea in grazing adult sheep to dietary factors, Larsen (1997)[137] examined the likelihood that high intakes of water or soluble carbohydrates could cause winter scours, based on knowledge of ruminant nutritional physiology. He concluded that these were unlikely or insignificant contributors to the syndrome and that other, as yet unidentified, factors in green pasture may have a small role in predisposing some sheep to diarrhoea. The most important risk factor was the ingestion of larvae of *Trichostrongylus* spp and *Teladorsagia circumcincta* (2000 to 20 000 infective larvae per week) by sheep which had developed an acquired immunity to internal parasites. Some sheep develop a hypersensitivity response to the ingested larvae in the small intestinal mucosa, and the resultant hypersecretion and increased peristalsis leads to diarrhoea.[138,139] Even with sheep predisposed to the syndrome, the ingestion of larvae is not alone a sufficient cause for the development of diarrhoea. Other factors, presumably related to diet or the sheep's response to diet, are apparently also necessary.[140]

The hypersensitivity diarrhoea may be controlled to a large extent by administering anthelmintics through controlled-release capsules[141], but this strategy, too, has a significant cost and it is not always completely successful, particularly if given after the syndrome appears.[142]

Susceptibility to the condition has a genetic basis and it is possible to select for sheep with a low propensity for breech soiling (low dag score). Ultimately, the best way to prevent the condition may be through genetic selection, as discussed further in Chapter 9.

Alimentary tract diseases caused by toxic plants

Solanum elaeagnifolium (silverleaf nightshade) will cause gastroenteritis in sheep, with diarrhoea for a brief period before death. *Solanum pseudocapsicum* (winter cherry) and *S sturtianum* (Sturt's nightshade) have solanine-type alkaloids toxic to sheep which cause scouring. Cape tulips (*Homeria* spp), which contain cardiac glycosides, *Pimelea* spp and *Bryophyllum* spp will cause diarrhoea.

SECTION C

GASTROINTESTINAL DISEASES OF YOUNG SHEEP

Enteric diseases of neonatal lambs

In many countries other than Australia and New Zealand, it is common practice to confine ewes indoors for the lambing period. This results in a reduction in lamb mortality associated with exposure, mismothering or predation, but it increases the risks of infectious diseases, particularly those caused by agents passed from other neonates and those which survive well in warm bedding material. Neonatal diarrhoea is one of the more common health problems associated with this form of husbandry and the most common causes of enteric infection in lambs up to 3 weeks of age are rotavirus, enterotoxigenic *Escherichia coli* and *Cryptosporidium parvum*.[143,144] Lamb dysentery, caused by *Clostridium perfringens* type B, and salmonellosis are also occasional causes of neonatal diarrhoea and death in young lambs.

Viral diarrhoea

Rotavirus

Infection with ovine rotavirus can lead to enteritis and diarrhoea in neonatal lambs, sometimes with high morbidity and significant mortality. The virus is probably widely distributed in sheep flocks around the world, as it is in other domestic animals and humans, and occurrences of neonatal diarrhoea outbreaks caused by rotavirus are reported from the UK, USA, Japan, Egypt and India.[145] Rotavirus has been identified in neonatal lambs in Australia — both in lambs with diarrhoea and in clinically normal lambs.[146]

Lambs in their first two days of life are most susceptible to clinical infection and lambs over 4 days of age usually develop only subclinical infections following exposure.[147] Outbreaks of severe disease appear to be uncommon but have been described — including cases in shed-lambing operations involving lambs of 12-16 hours of age, a morbidity approaching 100% and mortality rates around 10%. Surviving lambs may remain ill-thrifty and particularly susceptible to other infectious diseases.[148]

Rotaviruses are a very important cause of diarrhoea in human infants globally and, in developing countries in particular, are one of the most significant causes of infant mortality.[149] Transmission of rotaviruses between host species can occur but does not appear to be important in the epidemiology of the disease in people or farm animals. While human rotaviruses have been shown experimentally to be capable of causing disease in calves, lambs and piglets, there is little evidence to suggest that animals contribute directly to the occurrence of rotavirus disease in humans.[150] There is, however, marked genetic diversity between isolates of rotaviruses from humans across the world and evidence that rotaviruses from animals are contributing to the genetic diversity and evolution of human rotavirus strains.[149,151]

There is no specific treatment for rotavirus infection in lambs. In the event of disease outbreaks, attention should be paid to decontamination of the environment, provision of clean bedding, isolation of clinical cases, and disinfection of pens and protective clothing; and it should be ensured that lambs receive adequate colostrum from their dams soon after birth, preferably receiving 100 to 200 mL within their first 30 minutes.[152] There is no commercial vaccine licensed for sheep in Australia but a rotavirus vaccine preparation, including antigens of coronavirus and E coli, is available for cattle. Newborn lambs can be given colostrum from rotavirus-vaccinated cows as a way of providing immediate protection in the face of an outbreak. Bovine colostrum can provide useful levels of immunoglobulins in lambs[153], although colostrum from a small minority of cows contains anti-sheep erythrocyte antibodies and will cause potentially fatal anaemia in lambs.[154] (Attention should also be given to the disease status of the cattle herd from which colostrum is sourced, to avoid transmission of diseases such as paratuberculosis.) If rotavirus vaccine is used in sheep, it should be administered to the ewes a few weeks before lambing to ensure that high levels of passive protection are provided in colostrum.

Coronavirus

Coronaviruses in sheep and goats are genetically similar to bovine coronavirus and may be host variants of the bovine strain. While there have been reports of coronavirus associated with diarrhoea in newborn lambs, their aetiological role remains unclear. Reports of coronavirus and other viruses which may be involved in enteric infections of lambs have been recently summarised.[155]

Enterotoxigenic E coli infection

Escherichia coli is a species of gram-negative bacteria that displays enormous diversity between strains with respect to the antigenicity of the cell wall lipopolysaccharides (the O serogroup), the flagella (the H serogroup) and a range of antigenic and non-antigenic virulence factors. Most strains of E coli are harmless, commensal organisms which inhabit the gastrointestinal tract of animals and do not cause disease. Strains which express virulence factors, however, may be highly pathogenic. Based on the type of virulence factors, pathogenic E coli are categorised into six pathotypes. Organisms from one of these pathotypes, the enterotoxigenic E coli (ETEC), are frequently associated with diarrhoea in young farm animals. The two important types of virulence factors in ETEC strains are colonisation factors (CFs) and enterotoxins.

Across all pathotypes, a very large number of O and H serotypes of E coli are recognised. While the O and H serotypes are commonly used in typing of pathogenic isolates, clinically important strains come from a range of serotypes and, to a certain extent, the presence of virulence factors is independent of serotype. Nevertheless, there are particular combinations of serotype, CF and enterotoxin type which occur more commonly together and these relationships tend to persist in clonal lines. Consequently, serotype can still be used to infer the likelihood that a strain of *E coli* is pathogenic and serotype analysis can be used in addition to genetic typing of virulence factors for epidemiological investigations.[156]

ETEC produce a heat-labile enterotoxin (LT) and a heat-stable enterotoxin (ST) — and any one strain of ETEC may produce either or both enterotoxins. The elaboration of enterotoxins is under the genetic control of transmissible plasmids. There are variants of ST and one of these, STa, is more important than others in farm animals. Both LT and ST enterotoxins disrupt the normal function of small intestinal epithelial cells, leading to increased chloride secretion, inhibition of sodium chloride absorption and a loss of water into the bowel. The hypersecretion arises from a functional interference and there is no inflammatory process involved. The enterotoxins are thus responsible for the diarrhoea and profound dehydration which occurs in clinical cases of ETEC infection.

Colonisation factors are protein structures on the surface of ETEC organisms that enable attachment of the bacteria to receptors on small intestinal mucosal cells. They may take the form of adhesins, pili, or fimbriae. These structures allow the organisms to colonise the small intestine — the site of action of the enterotoxins. Without CFs promoting the adherence of the organisms to the intestinal epithelium, the bacteria would not remain in the small intestine for sufficiently long to increase in number to a level capable of causing disease. In order to cause clinical disease, ETEC must express CFs in addition to enterotoxins.[157]

The epithelial receptors to which CFs attach in the small intestine differ between humans and animals. Consequently, a different range of CFs are found in ETEC infecting humans from those found in piglets, calves, lambs and kids. The most common CFs in animals have been designated F4, F5, F6 (formerly K88, K99, 987P), Fy/Att25, F18 and F41.[158] The different colonisation factors produced by ETEC strains are antigenically distinct.

Lambs are particularly susceptible to ETEC infection in their first 48 hours of life and much less susceptible once they over 4 days of age.[147] The age-dependency of susceptibility — also reported in piglets and calves, although at greater ages — is associated with a change in the intestinal cell surface structures bearing the receptors for CFs. There also appear to be differences unrelated to age between lambs in their susceptibility to the adhesion of CFs, associated with phenotypic differences in the brush border membrane of the intestinal epithelial cells and the intestinal mucus[159], suggesting that there are degrees of natural resistance to ETEC infection in lambs, as there are in piglets. While lambs older than 4 days are generally less susceptible to infection than younger lambs, they may succumb to ETEC infection in the presence of a mixed infection with other enteric pathogens.

Clinical findings

Affected lambs have a severe, watery diarrhoea. Dehydration is rapid and severe, and affected lambs are weak and collapsed. The very young age of the lambs and the severity of the dehydration lead to high mortality rates. In outbreaks in lambing sheds, morbidity may be high.

Diagnosis

Differentiation of the principal causes of neonatal diarrhoea in lambs on clinical grounds alone is very difficult and, in some cases, mixed infections (with rotavirus and *Cryptosporidium parvum*) may be present. Salmonellosis can be differentiated readily at necropsy by the profound inflammatory changes in the intestine.

Clinical pathology

In cases of ETEC colibacillosis, *E coli* can be readily cultured from faeces or intestinal contents but this alone is insufficient to confirm an aetiological diagnosis. Laboratory tests to confirm that the isolates are ETEC require the identification of fimbriae and STa enterotoxin. F5 and F41 fimbriae can be readily detected by slide agglutination tests. Generally, the confirmation of F5 or F41 fimbriae is taken as proof that *E coli* isolated from a case of neonatal diarrhoea is enterotoxigenic, but some *E coli* strains may possess fimbriae but not elaborate enterotoxin, or may have fimbriae which are not F5 or F41.[143,160]

A multiplex PCR test which can simultaneously detect the presence of F5 and F41 fimbriae and STa has been developed.[161] Some tests can be used without prior isolation of the infecting *E coli*. These include K99 (F5) enzyme immunoassay kits that can be applied to faeces and fluorescent antibody techniques for impression smears from intestinal tissue. These tests have value in that they are rapid and test the tissues which are directly involved in the pathology, but they may lack diagnostic accuracy if they do not test for F41 fimbriae and enterotoxin.

Pathology

In combination with microbiological tests, histopathology of the small intestine from an affected animal immediately after death, natural or euthanased, will generally confirm the diagnosis. There are diffuse or focal areas of colonisation of the distal jejunal and ileal mucosa, often forming a layer of adherent bacteria lining the brush border epithelium. In uncomplicated cases of ETEC colibacillosis there is no evidence of inflammatory change or extensive damage to the epithelium.[150]

Prevention

The risk of ETEC infection in lambs, like that of rotavirus, is reduced by attention to cleanliness, and disinfection of the lambing and lamb-raising areas. Vaccination of ewes against ETEC will ensure that strong colostral antibodies are present. Steps should be taken to ensure that lambs receive adequate protective colostrum as soon after birth as possible, and that they continue to receive colostrum over the subsequent 24 hours. The protection afforded by colostral antibodies is due to a local effect of the antibodies, rather than a systemic effect following antibody absorption. Within the gut lumen the antibodies coat the binding sites on ETEC fimbriae, thus preventing the attachment of the organisms to the mucosal epithelium.[162]

Human health risk

Because of the specificity of the CFs, animal ETEC strains normally do not infect humans and ETEC infections in animals are not generally considered zoonotic. This is in contrast to other pathotypes of *E coli*, such as those that produce shiga-like toxins (O157:H7) which are found in animals, mainly cattle but also sheep, and produce severe disease in humans. Most human cases in Australia result from foodborne infection. Cases of STEC transmission to children from animals in petting zoos and at Agricultural Shows are reported frequently from many countries, including Australia, highlighting a need for personal hygiene (hand-washing) when farm animals are handled.

Cryptosporidium infection

Cryptosporidium is a genus of unicellular parasites within the apicomplexan phylum, and therefore related to *Toxoplasma*, *Neospora* and the coccidia, with which it shares a number of similarities in life cycle and pathogenesis. The principal species affecting sheep and goats is *C parvum*.

In environments where ewes are housed or closely confined for lambing, infection with *Cryptosporidium* is often a major cause of diarrhoea in lambs and goat kids between the ages of 1 and 4 weeks. Reports of cryptosporidiosis in Australia in lambs and kids appear limited to situations where neonates were confined and raised together.[163,164] A periparturient increase in the excretion of *Cryptosporidium* oocysts may be an important source of infection for lambs maintained in close confinement.

The life cycle of *Cryptosporidium* spp is similar to that of other enteric coccidia.[165] Infection occurs most commonly by ingestion of oocysts passed in the faeces of asymptomatic adults or other infected neonates. Oocysts are fully sporulated when passed in the faeces and are therefore immediately infectious. On ingestion, oocysts excyst in the small intestine, releasing sporozoites which enter epithelial cells of the intestinal tract. These multiply asexually (merogony), producing type I meronts, some of which release merozoites which can invade more epithelial cells, while other merozoites produce type II meronts, releasing type II merozoites (gametogeny) to form macrogamonts and microgamonts. Microgamonts fertilise macrogamonts, producing zygotes. Most of these differentiate into thick-walled sporulated oocysts which are passed in the faeces, but a small proportion form thin-walled oocysts which rupture in the small intestine, releasing sporozoites and causing further autoinfection.[166,167]

In the intestine, sporozoites lodge and develop within the mucosal epithelial cell microvillous brush border. A membrane, probably of host cell origin, develops around the parasite, effectively making the parasite intracellular, although extra-cytoplasmic. (The location may be relevant to the inability of antimicrobial agents to kill the organism.) All development of the sporozoite leading up to the release of merozoites occurs within the membrane sac. The relationship of the parasite to the host cell is in contrast to that of other coccidia which develop and multiply within the cytoplasm of the host cell.[150,166,168]

The number of oocysts necessary to establish infection is low (five oocysts may be sufficient). *C parvum* is capable of such rapid colonisation and multiplication that an infective dose can lead to massive infection within a few days. The prepatent period, typically two to seven days, tends to be shortest when the initial dose is high and when the lambs are very young. Clinical signs may be evident between three and seven days after initial infection.

Clinical findings

Lambs are most frequently affected between the ages of 4 and 12 days. Infection leads to a profuse diarrhoea with dullness, abdominal pain and anorexia, which last for several days. Some affected lambs die, while some will recover after five to seven days.[169] Reported mortality rates in natural outbreaks vary from very low, following a mild illness, to 25%. Lambs become relatively resistant to infection at 15 to 20 days of age. The resistance develops with age and does not depend on any prior exposure or any immunological mechanism.[167]

Diagnosis

The age of lambs affected and the nature of the clinical signs mean that cryptosporidiosis can appear, clinically, very similar to rotavirus infection and ETEC colibacillosis. Differentiation by clinical pathology and, if available, necropsy, is important so that appropriate therapy and preventive actions can be instituted in the event of an outbreak of neonatal diarrhoea.

Clinical pathology

Identification of large numbers of cryptosporidial oocysts in the faeces of clinically affected animals is strong circumstantial evidence that the organism is responsible for, or contributing to, disease. Oocyst shedding is usually highest when diarrhoea is present and persists for several days after recovery begins. Oocysts can be demonstrated in faeces with a modified zinc sulphate centrifugal flotation or a Giemsa-stained smear, or by using a modified Ziehl-Neelson technique.[166]

Pathology

The primary site of infection is the distal jejunum and ileum but, after oocyst shedding commences, infection can spread to the caecum, colon and rectum. Grossly, the intestines are distended with gas and the contents are watery. The mucosa may appear inflamed or may be relatively normal on gross inspection but, on histopathological examination, villous atrophy, shortening of microvilli and the organisms themselves are evident.[168,169]

Treatment and prevention

There is no specific treatment for cryptosporidiosis that has any proven efficacy. The disease is ultimately self-limiting, so affected lambs should be given supportive treatment to correct dehydration while they recover. Antibacterial therapy may assist recovery by reducing the effect of any concurrent infections. Affected lambs should be separated from other, non-infected lambs to reduce the risk of spread. Prevention of infection is best achieved by maintaining low stocking densities and, in the case of housed animals, clean bedding.

Provided the ewe has antibodies against *C parvum*, colostrum appears to provide at least partial protection from *C parvum* infection in lambs, so, as for other enteric infections of neonates, ensuring adequate and early ingestion of colostrum is likely to be important in limiting the risk and severity of disease. As with ETEC infection, the colostral protection against *C parvum* occurs as a result of the presence of antibodies in the gut lumen. Absorbed and circulating colostral antibodies against *C parvum* are not protective.[170]

In the preparation of low-risk environments for lambs it must be remembered that infection can arise from contamination of feed or water sources by adult sheep, other lambs or other farm animals, such as cattle and humans. Oocysts are very resistant to exposure in the environment and are resistant to inactivation by commonly used drinking water disinfectants, including chlorine.

Coccidiosis

Coccidiosis is most common in lambs 1 to 6 months of age run under crowded conditions, such as feedlots, or at high densities on pastures. While infection at low levels is very common, clinical disease is uncommon in Australia, but only because lambs are not often managed at high stocking densities. When they are, however, there is a significant risk of coccidiosis.

Aetiology

While there are currently 11 species of *Eimeria* recognised in sheep, only two are considered strongly pathogenic (Table 16.2). Both of these infect the small and large intestine, while the less pathogenic species tend to be restricted to the small intestine. *Eimeria* spp are host-specific and there is no cross-infection between sheep, goats or other hosts.[171][172]

Pathogenesis

The pathogenesis of coccidiosis is related to the massive amplification of infection which occurs within the cells of the intestinal mucosa of each susceptible host and the cellular damage caused by the developing stages. Each ingested oocyst gives rise to eight sporozoites and each of these enters a mucosal cell. One or more cycles of asexual division, termed *merogony*, follow, releasing many merozoites (hundreds or thousands from each sporozoite, depending on the species), each of which in turn invades a cell and either continues another cycle of asexual multiplication or diverts into the sexual phase and forms a gamont. After fertilisation, oocysts are produced within mucosal cells, then released and passed in the faeces.

The damage done to the infected cell varies with the infecting species and its effects vary from functional disturbance to the loss of surface epithelial cells and villus atrophy. Acute inflammatory reactions are most commonly associated with destruction of cells by the sexual stages and oocysts rather than with a response to asexual stages.[180] The *Eimeria* spp which are the most pathogenic in sheep are those which invade the ileum, caecum and colon, rather than those in the jejunum. Virulence is also a function of the biotic potential of the parasite (the degree of asexual replication) and the degree of host response to infection. In any one episode

Table 16.2: *Eimeria* spp infecting sheep. Adapted from Taylor (2009).[172]

Most pathogenic	Low or unknown pathogenicity	
E ovinoidalis	E ahsata	E parva
E crandallis	E intricata	E pallida
	E faurei	E granulosa
	E bakuensis	E marsica
	E weybridgensis	

of exposure to coccidia, the severity of signs also depends on the size of the infecting dose and the susceptibility of the host. The untoward effects of infection mainly relate to malabsorption induced by villus atrophy, hypoproteinaemia and dehydration due to exudative enteritis, and colitis caused by epithelial erosion and ulceration.

Epidemiology

Coccidia are normally present in healthy sheep of all ages and small numbers of oocysts are usually present in the faeces. Disease outbreaks occur when susceptible animals are exposed to infection on the ground, bedding or pasture and left in the same environment for long enough to amplify the infection significantly and then exposed to the greatly increased levels of infectivity in a time too short for a useful degree of immunity to develop. This process is much more likely to occur when stocking rates are very high and when the environment is damp or muddy.

The pre-patent period of *E ovinoidalis* is 12 to 15 days; for *E crandallis* about three days longer. Typically, outbreaks of coccidiosis occur two to three weeks after young animals are confined on suitable pastures and oocyst output peaks four to five weeks after confinement, remains high for one to three weeks, then declines. In Britain, lambs are known to be infected in the first few days of life and commence excretion of oocysts at 2 to 3 weeks of age. The peak excretion of oocysts occurs when they are about 6 weeks of age and then falls rapidly.[173] Outbreaks have been reported in USA in weaned lambs following transport and confinement to a feedlot.[174]

The primary source of infection is usually the lambs or weaners themselves. Oocysts can survive prolonged periods on pasture, provided humidity is high. They do not survive prolonged dry periods or cold periods. Small amounts of contamination may survive from one season to the next, but the most significant source of infection is generally contemporaneous.

Infection does not persist in the host without constant reinfection. Consequently, moving animals away from heavily contaminated paddocks to 'clean' pastures at normal stocking rates will prevent the development of new cases after a week or two and allow chronic cases to recover.

The effect of host immunity appears to be a dramatic reduction of the biotic potential of the coccidia. Thus adult sheep may remain constantly exposed and constantly infected at low levels, providing a low level of contamination for young sheep.

Clinical signs

In acute coccidiosis diarrhoea is the predominant sign (dysentery may be present but this is a less common finding in sheep than calves), with depression, anorexia and abdominal pain. Diarrhoea may persist for only three days, by which time the animal has lost significant body weight, which is slow to return. The faeces are watery and vary in colour from yellow, brown, grey or dark tar. Mortality rates can be high in some cases.

The severity of signs is apparently influenced by the general nutritional and disease status of the lamb. Lambs which are less well nourished (twins and triplets) or have a concurrent parasitic or bacterial gastroenteritis are more likely to be seriously affected. In Britain, disease produced by *Nematodirus battus* may occur in lambs at the same time as coccidiosis, and the two conditions aggravate each other.[172]

Coccidiosis may also be subclinical or clinically mild. In such cases, at a flock level, some lambs may have mild clinical signs while others may suffer reduced growth rates with no clinical evidence of disease.

Necropsy findings

Lesions are most obvious in the terminal ileum, caecum and proximal colon, and include oedematous thickening of the wall, congestion and haemorrhage and sometimes ulceration in the mucosa. Small (1-2 mm diameter) grey-white nodules formed by the parasite may be visible on the mucosa. *E bakuensis* often produces larger nodules or polyps of 10 to 15 mm, but the pathological significance of these is unclear. In the case of the more pathogenic species, histopathological examination of the intestinal mucosa reveals a loss of surface epithelium and villous atrophy.

Diagnosis

The rapid and reasonably synchronous onset of diarrhoea in a number of young sheep within a few weeks of confinement is strongly suggestive of coccidiosis. Nematodiasis should also be considered but can often be ruled out quickly on history or faecal egg count. Faecal oocyst counts can become very high in infected animals; 5000 to 10 000 oocysts per gram of faeces is not unusual and counts 10 to 100 times higher do occur. High counts, however, do not necessarily indicate disease, nor do low counts necessarily mean that coccidia are not causing disease. A definite diagnosis should only be made when the history and clinical signs suggest coccidiosis and those findings are supported by either high oocyst counts or necropsy examination, or both. Identification of the oocysts to species level requires holding them at room temperature for two or three days until sporulation occurs, at which time the oocyst morphology will allow differentiation.

Treatment

For the treatment of clinical coccidiosis (in contrast to prophylaxis), sulphonamides have been widely used. Since 2000 there have been no sulphonamides registered in Australia specifically for treatment of coccidiosis in sheep. There are sulphonamide products registered for use in sheep but all are Schedule 4 poisons, available only on prescription from a veterinarian. Sulphonamide therapy for coccidiosis therefore requires a veterinarian to use registered injectable products or to use products registered for other food-producing animals under *off-label* conditions. The prescribing veterinarian is then responsible for establishing adequate withholding periods and for ensuring compliance with all label restraints which, if there are any, cannot be ignored under any circumstances.

Oral sodium sulphadimidine has been widely used and recommended for treatment for coccidiosis of sheep. Recommended dose rates vary but are typically around the rate of 1 g per 7 kg liveweight, orally, daily for three days[175,176] (6 mL of a 33.3% solution to a 14 kg lamb). Treatment for one day only, rather than three, may provide satisfactory clinical responses and may be a more feasible approach for treatment of large groups of lambs than repeated treatments. Currently, sulphadimidine is only registered for use in other food-producing species but can be prescribed for off-label use in sheep. In assessing the response to treatment it is important to remember that infection is self-limiting and animals may be entering a recovery phase at the time they are treated.

Toltrazuril (Baycox®, Bayer) is registered for treatment of coccidiosis in lambs in countries other than Australia, where it is registered for use in cattle and poultry. There is no information on withholding periods (WHP) or export slaughter intervals for sheep in Australian conditions. The WHP for cattle in Australia is 56 days. In the UK, the recommended treatment with the 5% solution (50 mg/L toltrazuril) is 1 mL per 2.5 kg body weight. The drug is effective against coccidia at all stages of the life cycle, so both lambs in the early stages of infection and those which are showing signs will benefit from treatment. If lambs are to be grazed on high-risk pastures they should be treated seven days after first exposure.

Diclazuril (Vecoxan®, Elanco) is effective in treatment and prevention and is a recommended medication for at-risk lambs 2 to 6 weeks of age in the UK.

If cases occur in a flock of lambs, it is generally advisable to treat all of the group exposed to the same environment, because other animals may be in the early stages of infection or may be excreting oocysts subclinically. Treatment should also be accompanied by a move to a clean environment. The disease is ultimately self-limiting and most therapies are ineffective against the late stages of the parasite.

Control

Prevention rests on avoidance of overcrowded conditions, particularly when the weather is warm and ground conditions are moist. If these conditions cannot be avoided, continuous medication can be attempted where animals at risk are in feedlots or offered supplementary feed. Monensin (2 mg/kg liveweight or 20 mg/kg of feed) and lasalocid (25-100 mg/kg of feed) can be used for preventive treatment.[176]

Decoquinate (Deccox®, Zoetis) is used in some countries as a prophylactic medication against coccidiosis in lambs.[171] It is registered in Australia only for use in poultry.

Yersiniosis

Yersiniosis is an enterocolitis, with mesenteric lymphadenitis, caused by infection with *Yersinia pseudotuberculosis* or, possibly, *Y enterocolitica*. Clinically, the disease is marked by persistent diarrhoea with a low to moderate prevalence, low mortality and a noticeable failure to thrive of affected animals. It occurs most commonly in animals 3-18 months of age. Microabscessation in the lamina propria of the distal half of the small intestine is a characteristic histological finding.

Both *Y pseudotuberculosis*[177] and *Y enterocolitica* can be isolated from apparently normal sheep. *Y pseudotuberculosis* is the species more frequently isolated from field cases of clinical yersiniosis and serotype III is more commonly isolated from sheep cases than types I and II. Experimentally, the disease has been reproduced both in terms of infection with microabscessation in the intestinal mucosa *and* clinical disease with this species of the bacterium.

Yersinia enterocolitica has also been isolated from affected sheep in Australia. The infection, with microabscesses, has been reproduced experimentally with this species, but clinical disease has not. The pathogenicity of this organism for sheep remains unclear.[178]

Epidemiology

The disease is usually seen in late winter and spring in winter rainfall areas or at all times of the year in other regions. The onset of cold, wet windy weather, particularly if associated with stressful events like weaning or shearing, appears to be an important predisposing factor. The events which lead to the development of the disease are poorly understood. The organism is present in apparently healthy sheep, so possibly, under conditions of stress, carrier animals commence or increase the excretion of the organism, providing the potential to infect susceptible sheep. Outbreaks of disease are preceded by a marked increase in the proportion of the flock which are shedding the organism in faeces.[179] As well as sheep, gregarious birds and rodents may serve as carriers of the organism and may be responsible for its spread. Infection is by the oral route. Disease may in part be due to the compromise of cell-mediated immunity, permitting establishment of invading organisms or the recrudescence of latent infection.[180]

Pathogenesis

Pathogenic strains of *Yersinia* spp have the capacity, conferred by surface proteins, to attach to and invade epithelial cells lining the gut and to resist phagocytosis and other host attempts to kill the bacteria. The bacteria invade as far as the lamina propria, where they form microcolonies, and host inflammatory responses lead to the production of microabscesses. An enterotoxin which causes hypersecretion enhances virulence but is not essential for the development of the disease.

Clinical findings

Clinically, there is a dark green to black diarrhoea, which tends to be mild and chronic and which leads to dagginess on the perineum and hocks. Tenesmus may be present. Affected sheep are mildly dehydrated and in poorer condition than expected from available feed. The mortality rate is generally lower than 2%, although intercurrent disease, particularly nematodiasis, is frequently present and may contribute to an increase in mortality rates.

Necropsy findings

Gross lesions are minimal. Intestinal contents are abnormally fluid and the wall of the small intestine, caecum and colon may show some thickening, congestion and oedema. Small foci of pallor, haemorrhage and erosions less than 0.5 cm in diameter may be evident. Mesenteric lymph nodes are congested and oedematous. If the liver is involved there may be a multifocal, necrotising hepatitis. Histopathology reveals an acute segmental erosive enteritis with microabscesses, more prevalent in the jejunum and ileum than in the large intestine.

Diagnosis

Nematodiasis should be ruled out by history, faecal egg count and total worm count if necessary. Definitive determination of the nature of the bacterial infection in the live animal is difficult. After necropsy, bacteriology with specimens of liver, spleen, mesenteric lymph nodes and intestinal contents of the jejunum, ileum, caecum and colon, using selective media, will confirm the diagnosis. Some fixed sections of gut should also be taken for histopathology.

Treatment and prevention

Tetracyclines should be effective but treatment is often of uncertain usefulness due to the persistent predisposition in most cases and the serial nature of infection through a mob of sheep. Sulphafurazole is also used frequently on farms to control outbreaks. Resistance to tetracyclines and sulphafurazoles has been found in some outbreaks, possibly related to use of the drugs in previous years.[179]

Steps should be taken to improve the environmental conditions for the sheep — reducing exposure, improving nutrition and attempting to reduce the stress associated with management practices as much as possible, particularly for young sheep. A major reduction in the risk of yersiniosis may be achieved if the periods that young sheep are held off-feed are kept as few and as short as possible. Internal parasite control is also important in reducing the susceptibility to yersiniosis. A vaccine against *Y pseudotuberculosis* is commercially available and widely used in the deer industry in New Zealand. In Australia, the usefulness of vaccination of sheep has not been established, partly due to the sporadic and unpredictable nature of outbreaks of disease.[181]

Campylobacteriosis

Campylobacter jejuni has been isolated from sheep and it is possible to produce enteritis experimentally with the organism. A number of other *Campylobacter* spp have been isolated from the intestinal tracts and faeces of both healthy and diseased sheep — including *C faecalis*, *C coli*, and *C fetus* subsp *fetus*. *Campylobacter* spp bacteria are difficult to isolate, a problem which has added to the difficulty of experimental reproduction of *Camplylobacter* enteritis. In Victoria, a clinical syndrome called *weaner colitis* has been associated with Campylobacter-like organisms[182,183,184], but firm proof of the aetiology of this disease is lacking.

Differential diagnosis of scouring in sheep

Nematodiasis, lactic acidosis, yersiniosis, campylobacteriosis and coccidiosis form the principal differential diagnoses of subacute diarrhoea in sheep beyond neonatal age. Diarrhoea is a prominent sign in salmonellosis and a minor clinical finding in enterotoxaemia, poisoning with copper, organophosphates, phosphorus, selenium and several plant intoxications.

Generally, the severe nature of the diarrhoea and presence of fever differentiate salmonellosis from nematodiasis, coccidiosis and yersiniosis. The protozoal and other bacterial enteritides are usually less dramatic epidemics than salmonellosis. Nematodiasis is by far the most common cause of scouring in sheep in Australia. Clinically, it may be difficult to differentiate from yersiniosis and campylobacteriosis. The preferred method to diagnose nematodiasis is to conduct worm counts of the abomasum and small intestine, preferably from three animals; faecal egg counts (from at least 10 animals) may be difficult to interpret if not supported by other information, but they should always be included as part of an investigation. If significant numbers of parasites are not present, the worm count can then be followed by a systematic collection of tissue specimens for laboratory examination (Table 16.3).

Yersiniosis, campylobacteriosis and coccidiosis are usually diseases of sheep under 2 years of age, but cases can occur in older animals. A presumptive diagnosis of coccidiosis can usually be made on history, epidemiology and clinical signs, and this can be confirmed by necropsy findings.

Grain poisoning may lead to diarrhoea, but generally the other signs characteristic of lactic acidosis — including laminitis, abdominal distension, an absence of rumen movements, and a history of access to a rapidly fermentable carbohydrate such as cereal grain — will differentiate this condition from infectious causes of diarrhoea. Diarrhoea occurs with clostridial enterotoxaemia but the disease is also characterised by neurological signs, a rapid course and almost pathognomonic necropsy findings. Diarrhoea has also been recorded in septicaemic listeriosis.

Diarrhoea caused by toxic plants will generally be diagnosed by the exclusion of the above possibilities, the presence of additional clinical signs and the offending plant. Diarrhoea due to nematode larval challenge and dietary factors is very difficult to diagnose definitively but a tentative diagnosis can be based on the absence of clinical findings other than diarrhoea and the spontaneous resolution as pastures dry off.

Table 16.3: Appropriate specimen selection for diagnosis of diarrhoea in sheep.

	Faeces	Fresh samples, chilled	Mucosal smears	Fixed tissues
Nematodiasis	Faecal egg count, minimum 10 sheep	Abomasum and intestinal tract for total worm count.		
Coccidiosis	Faecal oocyst count, minimum 10 sheep			Terminal ileum, caecum & colon
Bacterial infections	Culture, several sheep	Tied off sections, upper, middle, lower small int, colon, mesenteric nodes, liver & spleen	Caecum and colon for campylobacter, ileum for OJD	Abomasum, duodenum, jejunum, ileum, caecum, colon, liver, lung, kidney
Lactic acidosis		Plasma for D-lactate		Rumen wall

SECTION D

GASTROINTESTINAL DISEASES EXOTIC TO AUSTRALIA

Bluetongue

Bluetongue is a disease of sheep in the tropics and warm temperate regions of the world. It is characterised by stomatitis, rhinitis and enteritis. Lameness, due to coronitis and laminitis, may also be a prominent sign. Mild strains of the virus occur in Australia but, as yet, the restricted distribution and the vectors' preference for cattle has kept Australian sheep flocks free of clinical disease.

Aetiology

Bluetongue is caused by an arbovirus of the *Orbivirus* genus, an RNA virus. Orbiviruses are stable and resistant to decomposition. There are 24 recognised serotypes of bluetongue virus, which vary considerably in pathogenicity. Eight serotypes occur in Australia: BTV 1,3,9,15,16,20,21,23. These strains are only mildly to moderately pathogenic in sheep, and appear to be genetically distinct from the virulent strains which occur in Africa. Related orbiviruses also occur in Australia and, although not pathogenic for any Australian host, some (including the virus causing epizootic haemorrhagic disease (EHD) in deer) cause cross reactions in bluetongue serology. Infection with one serotype does not necessarily protect against infection with another, but some serotypes are partially cross-protective.

Transmission

The disease is transmitted by culicoides insects (variously called midges, gnats or sandflies), with the most efficient vectors varying between countries. There is a report of mechanical transmission by sheep ked. The virus multiplies in the culicoid insect and the insects are infective 10 to 15 days after ingesting infected blood. Transovarial transmission does not occur.

In countries where the disease is endemic, the virus of bluetongue persists in a number of ruminant species, including cattle, without producing any significant clinical illness. Outbreaks of disease occur when infection spills over into sheep. Neither infection nor disease occurs naturally in goats.

A serological survey in WA in 1978-79 indicated that 14% of Kimberley cattle (north of latitude 17°S), 1.5% of cattle in the northwest (north of 25°S) but no cattle south of latitude 25°S had positive titres for bluetongue or related orbiviruses.[187] The incidence of infection in cattle varies widely each year, consistent with movements in the vectors, which expand their range south in the wet season. The range of the vectors has not yet overlapped the major sheep-raising areas of Australia, and clinical field cases have not been recorded in sheep in Australia. The major bluetongue vectors in Australia are most commonly closely associated with cattle, and their range is limited to areas with pronounced summer rainfall.

Epidemiology

The morbidity rate in sheep flocks varies with the size of the insect population and the immune status of the sheep. When the disease first occurred as an epidemic in sheep flocks in Texas in

Table 16.4: *Culicoides* spp in Australia capable of transmitting bluetongue to sheep. Based on data from Standfast et al. (1978)[185] and Alexander (1990).[186]

Culicoides spp	Comment
C brevitarsis	Widespread; across northern Aust and down east coast to south of Sydney. Also spreads Akabane virus. Breeds in cattle dung.
C wadai	A more effective vector; apparently tolerant of colder and drier climate than *C brevitarsis*, range expanded south until 1985; considered a potential threat to sheep raising districts. Breeds in cattle dung.
C actoni	Efficient vector; range restricted to North
C fulvus	Efficient also, range restricted to North
C schultzei	Northern range only
C brevipalpis	Breeds in cattle dung

1952, the morbidity approached 100% in some flocks, the mortality rate 20% to 50%. After a flock has experienced infection, the morbidity rate may be as low as 1% to 2% with very few deaths.[188] Usually 2% to 5% of affected sheep die but, in some outbreaks, the mortality rate is 30% or even more. Young sheep, about 1 year old, are most susceptible and most severely affected, although this is probably a factor of immunity rather than age alone. British breeds and Merinos are more susceptible than indigenous African breeds. Recovery from infection is slow and losses of productivity are generally of greater economic significance than are deaths.[189]

Pathogenesis

After infection, viral replication occurs in regional lymph nodes and spleen. Viraemia occurs four to six days after infection. The virus infects the endothelium of arterioles, capillaries and venules, causing microvascular thrombosis and increased permeability. Gross lesions reflect the ischaemic necrosis, oedema and haemorrhage resulting from vascular damage. Leucopoenia and pyrexia occur within a day of the viraemia.

Clinical signs

A period of fluctuating high fever persists generally for five or six days. Two or three days after fever commences, lesions of the buccal and nasal mucosae appear. Initially, there is hyperaemia of those mucosae, salivation and nasal discharge which is at first watery, then muco-purulent or even bloodstained, and which dries out to form crusts.[190] There is oedema of the lips, ears, eyelids, conjunctiva and intermandibular space. Hyperaemia may occur over all skin areas, particularly axillary and inguinal areas, and focal haemorrhage may occur on lips and gums. Oedema, congestion and cyanosis of the tongue occur occasionally. Excoriation, erosions and ulcerations, consequent on the infarction of epithelium, occur on the tongue margins, buccal mucosa, hard palate and dental pad. Diarrhoea and dysentery may occur and, if they do, should be regarded as an unfavourable diagnostic sign. Respiration is obstructed and the respiratory rate increased.

The coronet, bulbs of the heel and interdigital skin may become hyperaemic. Lameness is not consistently present and, if it occurs, usually appears after oral lesions begin healing. Defects and brown lines, from haemorrhage, may grow down the hoof in sheep that have recovered. A wool break is likely. Pregnant ewes may abort.

Animals are often reluctant to move or are recumbent, due to the foot lesions or due to viral muscle damage, which can lead to torticollis and extreme weakness. Dysphagia may lead to apparent vomition and aspiration pneumonia. Affected sheep are inappetent and become rapidly emaciated. In fatal cases, death occurs about six days after signs appear.

Pathology

At necropsy, superficial lymph nodes are enlarged and moist. There is subcutaneous and intermuscular oedema. Necrosis may be present in the myocardium. The most consistent, almost pathognomonic lesion is focal haemorrhage in the tunica media at the base of the pulmonary artery, visible from both the internal and adventitial surfaces. Petechial haemorrhages may also occur at the base of the aorta and in the subendocardium and subepicardium.

In the respiratory tract, lesions may include oedema and ecchymotic haemorrhages in the pharyngeal or laryngeal area, purple discolouration and oedema of the lungs.

Lesions of the lower alimentary system may include mucosal ulceration in the rumen and reticulum, petechial haemorrhages of the abomasal mucosa and congestion and haemorrhage in the intestinal mucosa generally, particularly in the large intestine.[191]

Laboratory submissions

Laboratory diagnosis involves transmission to susceptible sheep with blood collected from field cases, serology, virus isolation and virus neutralisation tests, and cross-protection tests in susceptible sheep. Required specimens are 50 mL of jugular blood from each of six febrile sheep; 25 mL in EDTA, 25 mL clotted; mesenteric lymph nodes and spleen in buffered glycerol-saline. All samples should be packed in ice or refrigerated, but not frozen.[192]

Diagnosis

Although bluetongue is not a vesicular disease the oral lesions may resemble ruptured vesicles. In sheep, however, foot-and-mouth disease is more likely to cause lameness than signs referable to lesions of the buccal cavity. Vesicular stomatitis is rare in sheep. Sheep pox has more general and more typical pox lesions on other areas of the body. CPD is generally much milder and does not involve the oral mucosa, but severe outbreaks do occur and can resemble bluetongue lesions. Facial oedema occurs in photosensitisation but the skin lesions are restricted to non-woolled areas. Erosions of the buccal cavity also occur in peste des petits ruminants, also characterised by severe diarrhoea.

Prevention and control

Given the preference of the vector midges for cattle, outbreaks in sheep can be precipitated by the removal of cattle from an area which contains infective insects and sheep. Using cattle as *decoys* can therefore reduce the exposure of sheep to bluetongue virus.

Modified live viruses are used for vaccination in Africa and USA. The persistence of immunity is variable. Vaccination of pregnant ewes (four to eight weeks pregnant) may result in the birth of deformed *dummy* lambs with severe hypoplasia of the brain.

Foot-and-mouth disease (FMD)

FMD is the principal vesicular disease of sheep, but its worldwide impact on this species is moderate compared to the catastrophic effects of FMD on cattle production. Cattle-adapted strains of FMD virus often produce only mild disease in sheep, particularly in the initial phase of an outbreak, and the relatively minor lesions could easily be overlooked. FMD should be considered as a differential diagnosis in sheep when a high incidence of lameness with unusual lesions of the feet occur, accompanied by minor and short-lived lesions of the dental pad and posterior aspect of the dorsum of the tongue in some sheep.

Aetiology and epidemiology

There are seven antigenically distinct serotypes of FMD virus and a large number of subtypes of each serotype. Vaccination with one serotype does not protect against infection with another serotype. The virus is moderately resistant to environmental effects but is inactivated by high temperatures and strongly acid or alkaline conditions. It can survive several weeks outside the host if conditions are not extreme. Spread occurs principally between animals by direct contact but windborne spread can occur over several kilometres. The feeding of infected animal products to pigs is perceived as a major risk of introducing infection into clean areas or countries. The virus can be carried by humans, both on clothes and footwear and on hands or in the nostrils, so veterinarians or visitors returning to Australia from infected farms overseas are also potential sources of disease introduction.[193]

Clinical signs

Signs referable to mouth lesions are common in sheep but are often mild and are frequently missed. Lesions occur principally on the dental pad and also on the tongue, occasionally on the gums and cheeks. On the dental pad, vesicles appear and rupture within 24 hours, leaving a raw surface which heals rapidly. On the tongue, lesions usually occur on the posterior portion of the dorsum of the tongue, which is difficult to examine. The lesions differ from those of cattle and are not vesicular in character. Small necrotic patches slough leaving a raw painful surface but even the most severe lesions heal within in a few days.

Lameness, due to FMD lesions of the feet, is a much more prominent sign of FMD in sheep than oral lesions. Vesicles develop in the interdigital space, particularly on the skin-horn junction, on the coronet and on the bulbs of the heel. They vary in size from a few millimetres to extensive lesions involving extensive parts of the skin-horn junction or the coronet. They rupture and the raw wound is particularly prone to secondary infection. Healing of the large lesions may take up to two weeks.

Vesicles also may appear on the teats of ewes, vulva, prepuce and ruminal papillae.

In lambs up to 3 months and, occasionally, up to 8 months of age, the virus displays a tropism for the myocardium and skeletal muscle. If myocardial infection occurs, the clinical course is short and the lambs are usually found dead. Mortality is high, particularly in flocks where infection has not occurred for some years and ewes provide little if any colostral protection to their lambs.[194]

Diagnosis

The only other truly vesicular disease of sheep is vesicular stomatitis, but it is rare in sheep. Mouth lesions of FMD in sheep require differentiation from bluetongue, CPD and sheep pox. Pestes des petits ruminants causes erosions in the mouth which could superficially resemble ruptured vesicles, but the erosions are much more extensive throughout the alimentary tract and diarrhoea is characteristic. The lesions of CPD and sheep pox are usually more proliferative than ulcerative.

Foot lesions also occur in bluetongue and this disease may present a significant difficulty in clinical differentiation. CPD, footrot and strawberry footrot also require differentiation from the foot lesions of FMD. Secondary infections of the foot can substantially alter the appearance of the FMD lesions.

Laboratory confirmation

On suspicion of FMD, specimens should not be collected or any personnel allowed to leave the property until specialist diagnostic teams have been contacted. Fluid collected from freshly ruptured vesicles is used for CF and ELISA tests and for isolation in tissue culture to confirm FMD and to differentiate it from other vesicular diseases.

Rinderpest

Rinderpest was an acute, highly fatal, viral infection of ruminants and pigs which has now been eradicated from the world. Rinderpest virus is related to the measles virus, which is thought to have evolved from the virus of rinderpest about 1000 years ago. Natural infection occurred commonly in cattle and buffalo only, but sheep and goats did become affected. Introduction of the disease into Australia in 1923 followed the importation of an infected animal. The outbreak was quickly controlled by slaughter of all susceptible animals within a radius of one mile of the site of infection. The Pan-African Rinderpest Campaign began in 1987, using a vaccine and teams of veterinarians and lay staff to combat the disease and monitor for outbreaks. The vaccine had been developed by Walter Plowright — an English veterinary scientist who devoted his career to the eradication of rinderpest. By 1962 he and his colleagues had developed a safe and effective vaccine. On 28 June 2011, the Food and Agriculture Organization of the United Nations (FAO) announced that rinderpest had been eradicated from the world.

The rinderpest eradication effort is estimated to have cost $5bn and is considered to have had an economic benefit to the African region alone of at least $1bn per annum. Elimination of rinderpest through large-scale vaccination and surveillance campaigns stands as one of the greatest successes for veterinary science. It stands with smallpox as one of the two diseases to have ever been eradicated from the world.

The virus of rinderpest is closely related to that of pestes des petits ruminants (PPR), and the clinical signs of PPR are very similar to those of rinderpest.

Pestes des petits ruminants (PPR)

PPR occurs in sheep, goats and some wild ruminant species. The infectivity of the virus for cattle is not clear. The virus is shed in expired air, secretions, urine and faeces of infected animals and transmission occurs by direct contact of susceptible animals with infected animals. As with rinderpest, explosive outbreaks occur, with very high morbidity and mortality, when the disease is introduced into a susceptible population.

Clinical signs

Within five days of infection, there are fever, severe depression and anorexia. There is a nasal discharge, initially serous, then mucopurulent, which may encrust the nares. There is a heavy discharge from the eyes. Mouth lesions appear within a further three days as erosions on the gums, dental pad, hard palate, cheeks and tongue. Severe diarrhoea is usual, and secondary bacterial pneumonia is common. The mortality rate approaches 100%.

Diagnosis

The principle differential diagnoses in sheep are bluetongue and, to a lesser extent, sheep and goat pox. Bluetongue lesions differ in some characteristics from PPR, but both can be marked by high morbidity and mortality in susceptible populations, severe depression, nasal and ocular discharges and diarrhoea. The latter is less common in bluetongue, which also frequently presents foot lesions, unlike PPR.

RECOMMENDED READING

Juste RA and Perez V (2011) Control of paratuberculosis in sheep and goats. Veterinary Clinics of North America (Food Animal) **27** 127-38.
Windsor PA (2015) Paratuberculosis in sheep and goats. Vet Microbiol **181** 161-9.

REFERENCES

1 Seaman JT, Gardner IA and Dent CHR (1981) Johne's disease in sheep. Aust Vet J **57** 102-3.
2 Denholm LJ, Ottaway SJ, Corish JA et al. (1997) Control and eradication of ovine Johne's disease in Australia. In: Proceedings of the Fourth International Congress for Sheep Veterinarians, Australian Sheep Veterinary Society. 2-6 February, Armidale, Australia pp. 158-61.
3 Taylor PJ, Thornberry K and Florance L (2005) Efficacy of whole flock destocking and restocking strategies to eradicate ovine Johne's disease — the final results. In: Proceedings of the Australian Sheep Veterinarians 2005, Gold Coast Conference, Volume 15. 16-19 May. Australian Veterinary Association: St Leonards, Australia, pp. 140-3.
4 Anon (2005) Disease statistics. JD News, Official Newsletter of the National Johne's Disease Control Program **6(2)** 5. Animal Health Australia.
5 Thorel MF, Krichevsky M and Lévy-Frébault VV (1990) Numerical taxonomy of mycobactin-dependent mycobacteria, emended description of *Mycobacterium avium*, and description of *Mycobacterium avium* subsp. *avium* subsp. nov., *Mycobacterium avium* subsp. *paratuberculosis* subsp. nov., and *Mycobacterium avium* subsp. *silvaticum* subsp. nov. Int J Syst Bacteriol **40** 254-60. https://doi.org/10.1099/00207713-40-3-254.
6 Collins DM, Gabric DM and de Lisle GW (1989) Identification of a repetitive sequence specific to *Mycobacterium paratuberculosis*. FEMS Microbiol Lett **60** 175-8. https://doi.org/10.1111/j.1574-6968.1989.tb03440.x.
7 Green EP, Tizard MLV, Moss MT et al. (1989) Sequence and characteristics of IS900, an insertion element identified in a human Crohn's disease isolate of *Mycobacterium paratuberculosis*. Nucleic Acids Res **17** 9063-73. https://doi.org/10.1093/nar/17.22.9063.
8 Collins DM, Gabric DM and de Lisle GW (1990) Identification of two groups of *Mycobacterium paratuberculosis* strains by restriction endonuclease analysis and DNA hybridization. J Clin Microbiol **28** 1591-6.
9 Momotani E, Whipple E, Thierman A et al. (1988) Role of M cells and macrophages in the entrance of *Mycobacterium paratuberculosis* into domes of ileal Peyer's patches in calves. Vet Pathol **25** 131-7. https://doi.org/10.1177/030098588802500205.
10 Clarke CJ (1994) Host responses to *Mycobacterium paratuberculosis/M avium* infection. In: Proceedings of the Fourth International Colloquium on Paratuberculosis, eds RJ Chiodini, MT Collins and EOE Bassey. International Association for Paratuberculosis Inc: Madison, WI, pp. 345-8.
11 Larsen AB, Merkal RS and Cutlip RC (1975) Age of cattle as related to resistance to infection with *Mycobacterium paratuberculosis*. Am J Vet Res **35** 255-7.

12 Carrigan MJ and Seaman JT (1990) The pathology of Johne's disease in sheep. Aust Vet J **67** 47-50. https://doi.org/10.1111/j.1751-0813.1990.tb07693.x.

13 Clarke CJ and Little D (1996) The pathology of ovine paratuberculosis: Histopathological and morphometric changes and correlations in the intestine and other tissues. J Comp Pathol **114** 419-37. https://doi.org/10.1016/S0021-9975(96)80017-X.

14 Pérez V, Garcia Marin JF and Badiola JJ (1996) Description and classification of different types of lesion associated with natural paratuberculosis infection in sheep. J Comp Pathol **114** 107-22.

15 Marsh I, Whittington R and Cousins D (1999) PCR-restriction endonuclease analysis for identification and strain typing of *Mycobacterium avium* subsp. *paratuberculosis* and *Mycobacterium avium* subsp. *avium* based on polymorphisms in IS1311. Mol Cell Probe **13** 115-26. https://doi.org/10.1006/mcpr.1999.0227.

16 Whittington RJ, Marsh IB and Whitlock RH (2001) Typing of IS1311 polymorphisms confirms that bison (*Bison bison*) with paratuberculosis in Montana are infected with a strain of *Mycobacterium avium* subsp. *paratuberculosis* distinct from that occurring in cattle and other domesticated livestock. Mol Cell Probe **15** 139-45. https://doi.org/10.1006/mcpr.2001.0346.

17 Singh AV, Chauhan DS, Singh A et al. (2015) Application of IS1311 locus 2 PCR-REA assay for the specific detection of 'Bison type' *Mycobacterium avium* subspecies *paratuberculosis* isolates of Indian origin. Ind J Med Res **14** 55-64.

18 Stevenson K, Hughes VM, de Juan L et al. (2002) Molecular characterization of pigmented and non-pigmented isolates of *Mycobacteriun avium* subsp. *paratuberculosis*. J Clin Microbiol **40** 1798-1804. https://doi.org/10.1128/JCM.40.5.1798-1804.2002.

19 de Juan L, Álvarez J, Aranaz A et al. (2006) Molecular epidemiology of Types I/III strains of *Mycobacterium avium* subspecies *paratuberculosis* isolated from goats and cattle. Vet Microbiol **115** 102-10. https://doi.org/10.1016/j.vetmic.2006.01.008.

20 Biet F, Sevilla IA, Cochard T et al. (2012) Inter- and intra-subtype genotypic differences that differentiate *Mycobacterium avium* subspecies *paratuberculosis* strains. BMC Microbiol **12** 264. https://doi.org/10.1186/1471-2180-12-264.

21 Fernández M, Benavides J, Sevilla IA et al. (2014) Experimental infection of lambs with C and S-type strains of *Mycobacterium avium* subspecies *paratuberculosis*: Immunological and pathological findings. Vet Res **45** 5. https://doi.org/10.1186/1297-9716-45-5.

22 Cousins DV, Williams SN, Hope A et al. (2000) DNA fingerprinting of Australian isolates of *Mycobacterium avium* subsp *paratuberculosis* using IS900 RFLP. Aust Vet J **78** 184-90. https://doi.org/10.1111/j.1751-0813.2000.tb10590.x.

23 Whittington RJ, Hope AF, Marshall DJ et al. (2000) Molecular epidemiology of *Mycobacterium avium* subsp *paratuberculosis*: IS900 restriction fragment length polymorphism and IS1311 polymorphism analyses of isolates from animals and a human in Australia. J Clin Microbiol **38** 3240-8.

24 Moloney BJ and Whittington RJ (2008) Cross species transmission of ovine Johne's disease from sheep to cattle: an estimate of prevalence in exposed susceptible cattle. Aust Vet J **86** 117-23. https://doi.org/10.1111/j.1751-0813.2008.00272.x.

25 Sergeant E, Keatinge N, Allan D et al. (2014) Occurrence of sheep strain Johne's disease in the Australian beef industry. In: Proceedings of the 12th International Colloquium on Paratuberculosis. 22-26 June, Parma, Italy, p. 290.

26 Whittington RJ, Taragel CA, Ottaway S et al. (2001) Molecular epidemiological confirmation and circumstances of occurrence of sheep (S) strains of *Mycobacterium avium* subsp. *paratuberculosis* in cases of paratuberculosis in cattle in Iceland. Vet Microbiol **79** 311-22.

27 Albiston HE and Talbot RJdeC (1936) Johne's disease in Victoria. Aust Vet J **12** 125-38. https://doi.org/10.1111/j.1751-0813.1936.tb13608.x.

28 McCausland IP (1980) Apparent Johne's disease in a sheep. Aust Vet J **56** 564. https://doi.org/10.1111/j.1751-0813.1980.tb02599.x.

29 Whittington RJ, Marsh I, McAllister S et al. (1999) Evaluation of Modified BACTEC 12b

radiometric medium and solid media for culture of *Mycobacterium avium* subsp *paratuberculosis* from sheep. J Clin Microbiol **37** 1077-83.

30 Ris DR, Hamel KL and Ayling JM (1987) Can sheep become infected by grazing pasture contaminated by cattle with Johne's disease? NZ Vet J **35** 137. https://doi.org/10.1080/0048016 9.1987.35414.

31 Verdugo C, Jones G, Johnson W et al. (2014) Estimation of flock/herd true *Mycobacterium avium* subspecies *paratuberculosis* prevalence on sheep, beef cattle and deer farms in New Zealand using a novel Bayesian model. Prev Vet Med **117** 447-55. https://doi.org/10.1016/j. prevetmed.2014.10.004.

32 Verdugo C, Pleydell E, Price-Carter M et al. (2014) Molecular epidemiology of *Mycobacterium avium* subsp *paratuberculosis* from sheep, cattle and deer on New Zealand pastoral farms. Prev Vet Med **117** 436-46. https://doi.org/10.1016/j.prevetmed.2014.09.009.

33 Liapi M, Botsarisa G, Slana I et al. (2015) *Mycobacterium avium* subsp *paratuberculosis* sheep strains isolated from Cyprus sheep and goats. Transboundary and Emerging Diseases **62** 223-7. https://doi.org/10.1111/tbed.12107.

34 Ris DR, Hamel KL and Ayling JM (1988) Natural transmission of Johne's disease to feral goats. NZ Vet J **36** 98. https://doi.org/10.1080/00480169.1988.35496.

35 Sharp JM (1997) Johnes disease: Risks of interspecies transmission. In: Proceedings of Fourth International Congress for Sheep Veterinarians, ed MB Allworth. Australian Sheep Veterinary Society: Indooroopilly, Qld, pp. 155-7.

36 Grieg A, Stevenson K, Henderson D et al. (1999) Epidemiological study of paratuberculosis in wild rabbits in Scotland. J Clin Microbiol **37** 1746-51.

37 Daniels MJ, Henderson D, Greig A et al. (2003) The potential role of wild rabbits *Oryctolagus cuniculus* in the epidemiology of paratuberculosis in domestic ruminants. Epidemiol Infect **130** 553-59.

38 Beard PM, Stevenson K, Pirie A et al. (2001) Experimental paratuberculosis in calves following inoculation with a rabbit isolate of *Mycobacterium avium* subsp *paratuberculosis.* J Clin Microbiol **39** 3080-4. https://doi.org/10.1128/JCM.39.9.3080-3084.2001.

39 Raizman EA, Wells SJ, Jordan PA et al. (2005) *Mycobacterium avium* subsp *paratuberculosis* from free-ranging deer and rabbits surrounding Minnesota dairy herds. Can J Vet Res **69** 32-8.

40 Abbott KA (2000) Prevalence of Johne's disease in rabbits and kangaroos. Final report for Project TR050. Meat and Livestock Australia: North Sydney, Australia.

41 Kluver P, Hope A, Waldron B et al. (2000) A survey of potential wildlife reservoirs for *Mycobacterium paratuberculosis.* Final Report for Project TR054. Meat and Livestock Australia: North Sydney, Australia.

42 Cleland PC, Lehmann DR, Phillips PH et al. (2010) A survey to detect the presence of *Mycobacterium avium* subspecies *paratuberculosis* in Kangaroo Island macropods. Vet Microbiol **145** 339-46. https://doi.org/10.1016/j.vetmic.2010.03.021.

43 Adúriz JJ, Juste RA and Cortabarria N (1995) Lack of mycobactin dependence of mycobacteria isolated on Middlebrook 7H11 from clinical cases of ovine paratuberculosis. Vet Microbiol **45** 211-17. https://doi.org/10.1016/0378-1135(95)00037-B.

44 Collins MT (2003) Update on paratuberculosis: 1. Epidemiology of Johne's disease and the biology of *Mycobacterium paratuberculosis.* Ir Vet J **56** 565-74.

45 Eppleston J, Begg DJ, Dhand NK et al. (2014) Environmental survival of *Mycobacterium avium* subsp *paratuberculosis* in different climatic zones of eastern Australia. Appl Environ Microbiol **80** 2337-41. https://doi.org/10.1128/AEM.03630-13.

46 Whittington RJ, Marshall DJ, Nicholls PJ et al. (2004) Survival and dormancy of *Mycobacterium avium* subsp *paratuberculosis* in the environment. Appl Environ Microbiol **70** 2989-3004. https:// doi.org/10.1128/AEM.70.5.2989-3004.2004.

47 Whittington RJ, Marsh IB, Taylor PJ et al. (2003) Isolation of *Mycobacterium avium* subsp *paratuberculosis* from environmental samples collected from farms before and after destocking

sheep with paratuberculosis. Aust Vet J **81** 559-63. https://doi.org/10.1111/j.1751-0813.2003.tb12887.x.

48   Abbott KA, Whittington RJ and McGregor H (2004) Exposure factors leading to establishment of OJD infection and clinical disease. Final Report to MLA Project OJD.002A. Meat and Livestock Australia: North Sydney, Australia.

49   McGregor H, Abbott KA and Whittington RJ (2015) Effects of *Mycobacterium avium* subsp. *paratuberculosis* infection on serum biochemistry, body weight and wool growth in Merino sheep: A longitudinal study. Small Ruminant Res **125** 146-53. https://doi.org/10.1016/j.smallrumres.2015.02.004.

50   McGregor H, Dhand NK, Dhungyel O et al (2012) Transmission of *Mycobacterium avium* subsp. *paratuberculosis*: Dose-response and age-based susceptibility in a sheep model. Prev Vet Med **101** 76-84. https://doi.org/10.1016/j.prevetmed.2012.05.014.

51   Delgado L, Juste RA, Muñoz M et al. (2012) Differences in the peripheral immune response between lambs and adult ewes experimentally infected with *Mycobacterium avium* subspecies *paratuberculosis*. Vet Immunol Immunopathol **145** 23-31. https://doi.org/10.1016/j.prevetmed.2012.05.014.

52   Dennis MM, Reddacliff LA and Whittington RJ (2011) Longitudinal study of clinicopathological features of Johne's disease in sheep naturally exposed to *Mycobacterium avium* subspecies *paratuberculosis*. Vet Pathol **48** 565-75.

53   Gillan S, O'Brien R, Hughes AD et al. (2010) Identification of immune parameters to differentiate disease states among sheep infected with *Mycobacterium avium* subsp *paratuberculosis*. Clin Vaccine Immunol **17** 108-17. https://doi.org/10.1128/CVI.00359-09.

54   Rast L and Whittington RJ (2005) Longitudinal study of the spread of ovine Johne's disease in a sheep flock in southeastern New South Wales. Aust Vet J **83** 227-32. https://doi.org/10.1111/j.1751-0813.2005.tb11658.x.

55   Abbott KA, McGregor H, Windsor PA et al. (2002) The paratuberculosis-attributable mortality rate in a flock of Merino sheep in Australia. In: Proceedings of the Seventh International Colloquium on Paratuberculosis, eds RA Juste, MV Geijo and J Garrido. International Association for Paratuberculosis, Inc: Madison, WI, pp. 345-50.

56   Chaitaweesub P, Abbott KA, Whittington R et al. (1999) Shedding of organisms and subclinical effects on production in pre-clinical Merino sheep affected with paratuberculosis. In: Proceedings of the Sixth International Colloquium on Paratuberculosis, eds EJ Manning EJ and MT Collins. International Association for Paratuberculosis, Inc: Madison, WI, pp. 126-31.

57   Mainar-Jaime RC and Vásquez-Boland JA (1998) Factors associated with seroprevalence to *Mycobacterium paratuberculosis* in small-ruminant farms in the Madrid region (Spain). Prev Vet Med **34** 317-27. https://doi.org/10.1016/S0167-5877(97)00091-3.

58   Scott-Orr H, Everett RE, Ottoway SJ et al. (1988) Estimation of direct and indirect losses due to Johne's disease in New South Wales, Australia. Acta Vet Scand Suppl **84** 411-14.

59   Eppleston J, Simpson G, O'Neill S, et al. (2000) Reported levels of sheep mortalities in flocks infected with ovine Johne's disease in New South Wales. Asian-Aust J Anim Sci **13** 247.

60   McGregor H, Abbott KA and Whittington RJ (2004) Development of grazing management strategies for the control of ovine Johne's disease. In: Proceedings of the Australian Sheep Veterinarians 2004 — Canberra Conference, Volume 14. 3-6 May, Canberra, pp. 60-5.

61   Sigurdsson B (1960) A killed vaccine against paratuberculosis. Am J Vet Res **11** 54-67.

62   Crowther RW, Polydorou K, Nitti S et al. (1976) Johne's disease in sheep in Cyprus. Vet Rec **98** 463.

63   Cranwell MP (1993) Control of Johne's disease in a flock of sheep by vaccination. Vet Rec **133** 219-20. https://doi.org/10.1136/vr.133.9.219.

64   Bruere AN and West DM (1993) Johne's disease. In: The sheep: Health, disease and production. Foundation for Continuing Education of the NZ Veterinary Assoc: Massey University, Palmerston North, NZ, pp. 244-51.

65 Armstrong MC (1956) Johne's disease of sheep in the South Island of New Zealand. NZ Vet J **4** 56-9. https://doi.org/10.1080/00480169.1956.33220.

66 Collins MT (1994) Diagnosis and control of paratuberculosis. In: Proceedings of the Fourth International Colloquium on Paratuberculosis, eds RJ Chiodini, MT Collins and EOE Bassey. 17-21 July, Cambridge, UK, p. 325.

67 Gwozdz JM (2010) Paratuberculosis (Johne's disease). Available from: http://www.agriculture. gov.au/SiteCollectionDocuments/animal/ahl/ANZSDP-Johnes_disease.pdf. Accessed 16 August 2018.

68 Collins DM, Stephens DM and de Lisle GW (1993) Comparison of polymerase chain reaction tests and faecal culture for detecting *Mycobacterium paratuberculosis* in bovine faeces. Vet Microbiol **36** 289-99. https://doi.org/10.1016/0378-1135(93)90095-O.

69 Collins DM, Hilbink F, West DM et al. (1993) Investigation of *Mycobacterium paratuberculosis* in sheep by faecal culture, DNA characterisation and the polymerase chain reaction. Vet Rec **133** 599-600.

70 Kawaji S, Taylor DL, Mori Y et al. (2007) Detection of *Mycobacterium avium* subsp *paratuberculosis* in ovine faeces by direct quantitative PCR has similar or greater sensitivity compared to radiometric culture. Vet Microbiol **125** 36-48. https://doi.org/10.1016/j.vetmic.2007.05.002.

71 Sonawane GG and Tripathi BN (2013) Comparison of a quantitative real-time polymerase chain reaction (qPCR) with conventional PCR, bacterial culture and ELISA for detection of *Mycobacterium avium* subsp *paratuberculosis* infection in sheep showing pathology of Johne's disease. SpringerPlus **2(45)** 1-9.

72 Perez V, Bolea R, Chavez G et al. (1994) Efficiency of PCR and culture in the detection of *Mycobacterium avium* subsp *silvaticum* and *Mycobacterium avium* subsp *paratuberculosis* in tissue samples of sheep. In: Proceedings of the Fourth International Colloquium on Paratuberculosis, eds RJ Chiodini, MT Collins and EOE Bassey. 17-21 July, Cambridge. International Association for Paratuberculosis Inc: Madison, WI, p. 97.

73 Marshall DJ, Ottaway, SJ, Eamens GJ et al. (1996) Validation of a diagnostic strategy for detection of ovine Johne's disease in New South Wales sheep flocks. In: Proceedings of the Fifth International Colloquium on Paratuberculosis, 29 September-4 October, Madison, eds RJ Chiodini, ME Hines II and MT Collins. International Association for Paratuberculosis Inc: Madison, WI, p. 286.

74 Hillbink F (1990) Serology for paratuberculosis (Johne's disease). Surveillance **19(2)** 20.

75 Yokomizo Y, Yugi H and Merkal RS (1985) A method of avoiding false-positive reactions in an enzyme-linked immunosorbent assay (ELISA) for the diagnosis of bovine paratuberculosis. Jpn J Vet Sci **47** 111-19. https://doi.org/10.1292/jvms1939.47.111.

76 Collins MT, Sockett DC, Ridge S et al. (1991) Evaluation of a commercial enzyme-linked immunosorbent assay for Johne's disease. J Clin Microbiol **29** 272-6.

77 Ridge SE, Morgan IR, Sockett DC et al. (1991) Comparison of the Johne's absorbed EIA and the complement-fixation test for the diagnosis of Johne's disease in cattle. Aust Vet J **68** 253. https:// doi.org/10.1111/j.1751-0813.1991.tb03230.x.

78 Hilbink F, West DM, de Lisle GW et al. (1994) Comparison of a complement fixation test, a gel diffusion test and two absorbed and unabsorbed ELISAs for the diagnosis of paratuberculosis in sheep. Vet Microbiol **41** 107-16.

79 Sergeant ESG, Marshall DJ, Eamens GJ et al. (2003) Evaluation of an absorbed ELISA and an agar-gel immune-diffusion test for ovine paratuberculosis in sheep in Australia. Prev Vet Med **61** 235-48. https://doi.org/10.1016/j.prevetmed.2003.08.010

80 Scott PR, Clarke CJ and King TJ (1995) Serum protein concentrations in clinical cases of ovine paratuberculosis (Johne's disease). Vet Rec **137** 173.

81 Animal Health Australia (2014) Technical review of sheep MAP. Final Report. Available from: https://www.animalhealthaustralia.com.au/what-we-do/endemic-disease/market-assurance-programs-maps/sheepmap/. Accessed 16 August 2018.

82   Holmes, Sackett and Associates Pty Ltd (1996) Assessment of economic impact of options for eradication of ovine Johne's disease. Report M.835. Meat Research Corporation: Sydney.

83   Fridriksdottir V, Gunnarsson E, Sigurdarson S et al. (2000) Paratuberculosis in Iceland: Epidemiology and control measures, past and present. Vet Microbiol **77** 263. https://doi.org/10.1016/S0378-1135(00)00311-4.

84   Perez V, Garcia Marin JF, Bru R et al. (1995) Results of a paratuberculosis vaccination trial in adult sheep. (*Resultados obtenidos en la vacunacion de ovinos adultos frente a paratuberculosis*). Medicina Veterinaria **12** 196.

85   Sreenivasulu D, Krishnaswamy S and Janakiramasharma B (1986) Study on *in vitro* techniques to assess the cell-mediated immune responses to vaccination against Johne's disease in sheep. Indian Vet J **63** 519-22.

86   Hilbink F and West D (1990) The antibody response of sheep to vaccination against Johne's disease. NZ Vet J **38** 168-9. https://doi.org/10.1080/00480169.1990.35646.

87   Windsor P, Eppleston J, Whittington R et al. (2002) Efficacy of a killed *Mycobacterium paratuberculosis* vaccine for the control of OJD in Australian sheep flocks. In: Proceedings of the Seventh International Colloquium on Paratuberculosis, eds RA Juste, MV Geijo and J Garrido. International Association for Paratuberculosis, Inc: Bilbao, Spain, pp. 420-3.

88   Garcia Marin JF, Tellechea J, Gutierrez M et al. (1999) Evaluation of two vaccines (killed and attenuated) against small ruminant paratuberculosis. In: Proceedings of the Sixth International Colloquium on Paratuberculosis, eds EJ Manning and MT Collins. International Association for Paratuberculosis, Inc: Madison, WI, pp. 234-41.

89   Brotherston JG, Gilmour NJL and Samuel JMcA (1961) Quantitative studies of *Mycobacterium johnei* in the tissues of sheep II. Protection afforded by dead vaccines. J Comp Pathol **71** 300-10.

90   Larsen AB, Hawkins WW and Merkal RS (1964) Experimental vaccination of sheep against Johne's disease. Am J Vet Res **25** 974-6.

91   Gilmour NJL, Halhead WA and Brotherston JG (1965) Studies of immunity to *Mycobacterium johnei* in sheep. J Comp Path **75** 165-73.

92   Juste RA, Garcia Marin JF, Peris B et al. (1994) Experimental infection of vaccinated and non-vaccinated lambs with *Mycobacterium paratuberculosis*. J Comp Pathol **110** 185-94. https://doi.org/10.1016/S0021-9975(08)80189-2.

93   Reddacliff L, Eppleston J, Windsor P et al. (2006) Efficacy of a killed vaccine for the control of paratuberculosis in Australian sheep flocks. Vet Microbiol **115** 77-90. https://doi.org/10.1016/j.vetmic.2005.12.021.

94   Juste RA and Casal J (1993) An economic and epidemiologic simulation of different control strategies for ovine paratuberculosis. Prev Vet Med **15** 101-15. https://doi.org/10.1016/0167-5877(93)90106-4.

95   Corpa JM, Perez V and Garcia Marin JF (2000) Differences in the immune responses in lambs and kids vaccinated against paratuberculosis, according to the age of vaccination. Vet Microbiol **77** 475-85. https://doi.org/10.1016/S0378-1135(00)00332-1.

96   Gwozdz JM, Thompson KG, Manktelow BW et al. (2000) Vaccination against paratuberculosis of lambs already infected experimentally with *Mycobacterium avium* subspecies *paratuberculosis*. Aust Vet J **78** 560-6. https://doi.org/10.1111/j.1751-0813.2000.tb11902.x.

97   Garcia-Pariente C, Gonzalez J, Ferreras MC et al. (2002) Paratuberculosis vaccination of adult animals in two flocks of dairy sheep. In: Proceedings of the Seventh International Colloquium on Paratuberculosis, eds RA Juste, MV Geijo and J Garrido. International Association for Paratuberculosis, Inc: Bilbao, Spain, pp. 507-10.

98   Windsor PA, Bush R, Links I et al. (2005) Injury caused by self-inoculation with a vaccine of a Freund's complete adjuvant nature (Gudair™) used for control of ovine paratuberculosis. Aust Vet J **83** 216-20. https://doi.org/10.1111/j.1751-0813.2005.tb11654.x.

99   Richardson GD, Links II and Windsor PA (2005) Gudair (OJD) vaccine self-inoculation: A case for early debridement. Med J Aust **183** 151-2.

100 Chiodini RJ, van Kruiningen HJ, Merkel RS et al. (1984) Characteristics of an unclassified mycobacterium species isolated from patients with Crohn's disease. J Clin Microbiol **20** 966-71.

101 Lisby G, Anderson J, Engbaek K et al. (1994) *Mycobacterium paratuberculosis* in intestinal tissue from patients with Crohn's disease demonstrated by a nested primer polymerase chain reaction. Scand J Gastroenterol **29** 923-9. https://doi.org/10.3109/00365529409094864.

102 Ross AD (1980) Small intestinal carcinoma in sheep. Aust Vet J **56** 25. https://doi.org/10.1111/j.1751-0813.1980.tb02536.x.

103 McDonald JW and Leaver DD (1965) Adenocarcinoma of the small intestine of Merino sheep. Aust Vet J **41** 269. https://doi.org/10.1111/j.1751-0813.1965.tb06558.x.

104 Newell KW, Ross AD and Renner RM (1984) Phenoxy and picolinic acid herbicides and small-intestinal adenocarcinoma in sheep. Lancet **2** 1301. https://doi.org/10.1016/S0140-6736(84)90821-3.

105 Briggs PK, Hogan JP and Reid RL (1957) The effect of volatile fatty acids, lactic acid, and ammonia on rumen pH in sheep. Aust J Ag Res **8** 674-90. https://doi.org/10.1071/AR9570674.

106 Russell JB and Hino T (1985) Regulation of lactate production in *Streptococcus bovis*: A spiralling effect that contributes to rumen acidosis. J Dairy Sci **68** 1712-21. https://doi.org/10.3168/jds.S0022-0302(85)81017-1.

107 Allison MJ, Robinson IM, Dougherty RW et al. (1975) Grain overload in cattle and sheep; changes in microbial populations in the caecum and rumen. Am J Vet Res **36** 181-5.

108 Clark R and Weiss KE (1957) Reflex salivation in sheep and goats initiated by mechanical stimulation of the cardiac area of the fore stomachs. J Sth Afr Vet Med Assoc **23** 163-5.

109 Beauchemin KA (1991) Ingestion and mastication of feed by dairy cattle. Veterinary Clinics of North America Food Animal Practice **7** 439-63. https://doi.org/10.1016/S0749-0720(15)30794-5.

110 Tadepalli S, Narayanan SK, Stewart GC et al. (2009) *Fusobacterium necrophorum*: A ruminal bacterium that invades liver to cause abscesses in cattle. Anaerobe **15** 36-43. https://doi.org/10.1016/j.anaerobe.2008.05.005.

111 Al-Qudah K and Al-Majali A (2003) Bacteriologic studies of liver abscesses of Awassi sheep in Jordan. Small Ruminant Res **47** 249-53. https://doi.org/10.1016/S0921-4488(02)00260-2.

112 Nocek JE (1997) Bovine acidosis: Implications on laminitis. J Dairy Sci **80** 1005-28. https://doi.org/10.3168/jds.S0022-0302(97)76026-0.

113 Reference Advisory Group on Fermentative Acidosis of Ruminantts (RAFGAR) (2007) Ruminal acidosis — aetiopathogenesis, prevention and treatment. A review for veterinarians and nutritional professionals. Australian Veterinary Association: Australia.

114 Braun U, Rihs, T and Schefer U (1992) Ruminal lactic acidosis in sheep and goats. Vet Rec **130** 343-9. http://dx.doi.org/10.1136/vr.130.16.343.

115 Morton A (2009) Virginiamycin and *ad lib* sodium bicarbonate in the treatment of grain poisoning in sheep. Flock and Herd Case notes. Available from: http://www.flockandherd.net.au/sheep/reader/grain-poisoning-II.html. Accessed 18 July 2018.

116 Jones M and Navarre C (2014) Fluid therapy in small ruminants and camelids. Veterinary Clinics of North America Food Animal Practice **30** 441-53. https://doi.org/10.1016/j.cvfa.2014.04.006.

117 Roussel AJ (2014) Fluid therapy in mature cattle. Veterinary Clinics of North America Food Animal Practice **30** 429-39. https://doi.org/10.1016/j.cvfa.2014.04.005.

118 Michel AR (1990) Ruminant acidosis. Practice **12** 245-9. https://doi.org/10.1136/inpract.12.6.245.

119 Jolly S and Wallace A (2007) Best practice for production feeding of lambs: A review of the literature. Project SCSB.091. Meat and Livestock Australia: North Sydney, Australia.

120 Rowe JB and Zorrilla-Rios J (1993) Simplified systems for feeding grain to cattle in feed lot and under grazing conditions. In: Recent advances in animal nutrition in Australia — 1993, ed DJ Farrell. University of New England: Armidale, pp. 89-96.

121 Zorrilla-Rios J, May PJ and Rowe JB (1991) Rapid introduction of cattle to grain diets using virginiamycin. In: Recent advances in animal nutrition in Australia — 1991, ed DJ Farrell. University of New England: Armidale, p. 10A.

122 Godfrey SI, Thorniley GR, Boyce MD et al. (1994) Virginiamycin reduces acidosis in hungry sheep fed wheat grain. Proc Aust Soc Anim Prod **20** 447.

123 Nagaragja TG, Avery TB, Bartley EE et al. (1981) Prevention of lactic acidosis in cattle by lasalocid or monensin. J Anim Sci **53** 206. https://doi.org/10.2527/jas1981.531206x.

124 Godfrey SI, Rowe JB, Speijers EJ et al. (1993) Lupins, barley, or barley plus virginiamycin as supplements for sheep at different feeding intervals. Aust J Exp Agric **33** 135-40. https://doi.org/10.1071/EA9930135.

125 Jagusch KT, Gumbrell RC and Dellow DW (1976) Red gut in lamb lucerne grazing trials at Lincoln. Proc NZ Soc Anim Prod **36** 190-7.

126 Gumbrell RC (1997) Redgut in sheep: A disease with a twist. NZ Vet J **45** 217-21. https://doi.org/10.1080/00480169.1997.36033.

127 Gumbrell RC (1973) 'Red gut' syndrome in lambs grazing Lucerne. NZ Vet J **21** 178-9. https://doi.org/10.1080/00480169.1973.34104.

128 Norris RT and Kelly AP (1988) The export of live sheep from Australia to the Middle East. In: Sheep health and production. Proceedings No 110. University of Sydney Postgraduate Committee in Veterinary Science: Australia, p. 37.

129 Richards RB, Norris RT, Dunlop RH et al. (1989) Causes of death in sheep exported live by sea. Aust Vet J **66** 33-8. https://doi.org/10.1111/j.1751-0813.1989.tb03011.x.

130 Norris RT, Richards RB and Dunlop RH (1989) Pre-embarkation risk factors for sheep deaths during export by sea from Western Australia. Aust Vet J **66** 309-14. https://doi.org/10.1111/j.1751-0813.1989.tb09713.x.

131 Glastonbury JRW (1990) Non-parasitic scours. In: Sheep medicine. Proceedings No 141. University of Sydney Postgraduate Committee in Veterinary Science: Sydney, pp. 459-81.

132 Oddy H, cited by Glastonbury JRW (1990) Non-parasitic scours. In: Sheep medicine. Proceedings No 141. University of Sydney Postgraduate Committee in Veterinary Science: Sydney, p. 472.

133 Pethick DW and Chapman HM (1991) The effects of *Arctotheca calendula* (capeweed) on digestive function of sheep. Aust Vet J **68** 361-3. https://doi.org/10.1111/j.1751-0813.1991.tb00737.x.

134 Fairnie IJ (1969) Nitrite poisoning in sheep due to capeweed (*Arctotheca calendula*). Aust Vet J **45** 78-9. https://doi.org/10.1111/j.1751-0813.1969.tb13697.x.

135 Radostits OM, Blood DC and Gay CC (1994) Diseases caused by major phytotoxins. In: Veterinary medicine — a textbook of diseases of cattle, sheep, pigs, goats and horses. 8th ed. Baillière Tindall: London, p. 1538.

136 Setchell BP amd Williams AJ (1962) Plasma nitrate and nitrite concentration in chronic and acute nitrate poisoning in sheep. Aust Vet J **38** 58-61. https://doi.org/10.1111/j.1751-0813.1962.tb08721.x.

137 Larsen JWA (1997) The pathogenesis and control of diarrhoea and breech soiling ('winter scours') in adult merino sheep. PhD thesis, School of Veterinary Science, The University of Melbourne. https://doi.org/10.1016/S0020-7519(99)00050-8.

138 Larsen JWA, Anderson N and Vizard AL (1999) The pathogenesis and control of diarrhoea and breech soiling in adult Merino sheep. Int J Parasitol **29** 893-902.

139 Williams AR and Palmer DG (2012) Interactions between gastrointestinal parasites and diarrhoea in sheep: Pathogenesis and control. Vet J **192** 279-85. https://doi.org/10.1016/j.tvjl.2011.10.009.

140 Larsen JWA and Anderson N (2000) The relationship between the rate of intake of trichostrongylid larvae and the occurrence of diarrhoea and breech soiling in adult sheep. Aust Vet J **78** 112-16. https://doi.org/10.1111/j.1751-0813.2000.tb10537.x.

141 Vizard A (1993) Capsules in ewes — good and bad news. The Mackinnon Project Newsletter, April, University of Melbourne, Werribee.

142 Jacobson C, Bell K, Forshaw D et al. (2009) Association between nematode larvae and 'low worm egg count diarrhoea' in sheep in Western Australia. Vet Parasitol **165** 66-73. https://doi.org/10.1016/j.vetpar.2009.07.018.

143 Muñoz M, Álvarez M, Lanza I et al. (1996) Role of enteric pathogens in the aetiology of neonatal diarrhoea in lambs and goat kids in Spain. Epidemiol Infect **117** 203-11. https://doi.org/10.1017/S0950268800001321.

144 Sargison N (2004) Differential diagnosis of diarrhoea in lambs. Practice **26** 20-7. https://doi.org/10.1136/inpract.26.1.20.

145 Gazal S, Mir IA, Iqbal A et al. (2011) Ovine rotaviruses. Open Veterinary Journal **1** 50-4.

146 Ellis GR and Daniels E (1988) Comparison of direct electron microscopy and enzyme immunoassay for the detection of rotaviruses in calves, lambs, piglets and foals. Aust Vet J **65** 133-5. https://doi.org/10.1111/j.1751-0813.1988.tb14439.x.

147 Tzipori S, Sherwood D, Angus KW et al. (1981) Diarrhea in lambs: Experimental infections with enterotoxigenic *Escherichia coli*, rotavirus and *Cryptosporidium* sp. Infect and Immun **33** 401-6.

148 Theil KW, Lance SE and McCloskey CM (1996) Rotavirus associated with neonatal lamb diarrhea in two Wyoming shed-lambing operations. J Vet Diagn Invest **8** 245-8. https://doi.org/10.1177/104063879600800217.

149 Martella V, Bányai K, Matthijnssens J et al. (2010) Zoonotic aspects of rotaviruses. Vet Microbiol **140** 246-55. https://doi.org/10.1016/j.vetmic.2009.08.028.

150 Holland RE (1990) Some infectious causes of diarrhea in young farm animals. Clin Microbiol Rev **3** 345-75. https://doi.org/10.1128/CMR.3.4.345.

151 Midgley SE, Bányai K, Buesa J et al. (2012) Diversity and zoonotic potential of rotaviruses in swine and cattle across Europe. Vet Microbiol **156** 238-45. https://doi.org/10.1016/j.vetmic.2011.10.027.

152 Hindson JC and Winter AC (2002) Diarrhoea. In: Manual of sheep diseases. Blackwell Publishing: USA, pp. 90-102. https://doi.org/10.1002/9780470752449.ch10.

153 Tsiligianni Th, Dovolou E and Amiridis GS (2012) Efficacy of feeding cow colostrum to newborn lambs. Livestock Science **149** 305-9. https://doi.org/10.1016/j.livsci.2012.07.016.

154 Ruby RE, Balcomb CC, Hunter SA et al. (2012) Bovine colostrum-induced anaemia in a 2-week old-lamb. NZ Vet J **60** 82-3. https://doi.org/10.1080/00480169.2011.628634.

155 Martella V, Decaro N and Buonavoglia C (2015) Enteric viral infections in lambs and kids. Vet Microbiol **181** 154-60. https://doi.org/10.1016/j.vetmic.2015.08.006.

156 Qadri F, Svennerholm A-M, Faruque ASG et al. (2005) Enterotoxigenic *Escherichia coli* in developing countries: Epidemiology, microbiology, clinical features and prevention. Clin Microbiol Rev **18** 465-83. https://doi.org/10.1128/CMR.18.3.465-483.2005.

157 Turner SM, Scott-Tucker A, Cooper LM et al. (2006) Weapons of mass destruction: Virulence factors of the global killer enterotoxigenic *Escherichia coli*. FEMS Microbiol Lett **263** 10-20. https://doi.org/10.1111/j.1574-6968.2006.00401.x.

158 Nagy B and Fekete PZ (2005) Enterotoxigenic *Escherichia coli* in veterinary medicine. Int J Med Microbiol **295** 443-54. https://doi.org/10.1016/j.ijmm.2005.07.003.

159 Ouadia A, El Khalil H and Maarouf A (2007) Lamb sensitivity and resistance to K99-fimbriated enterotoxigenic *Escherichia coli* are correlated to the presence of intestinal mucin sialoglycoprotein receptors. Revue Med Vet **12** 618-26.

160 Orden JA, Ruiz-Santa-Quiteria JA, Cid D et al. (2002) Presence and enterotoxigenicity of F5 and F41 *Escherichia coli* strains isolated from diarrhoeic small ruminants in Spain. Small Ruminant Res **44** 159-61. https://doi.org/10.1016/S0921-4488(02)00046-9.

161 Franck SM, Bosworth BT and Moon HW (1998) Multiplex PCR for enterotoxigenic, attaching and effacing, and shiga toxin-producing *Escherichia coli* strains from calves. J Clin Microbiol **36** 1795-7.

162 Morris JA, Wray C and Sojka WJ (1980) Passive protection of lambs against enteropathogenic *Escherichia coli*: role of antibodies in serum and colostrum of dams vaccinated with K99 antigen. J Med Microbiol **13** 265-71. https://doi.org/10.1099/00222615-13-2-265.

163 Barker IK and Carbonell PL (1974) Cryptosporidium agni sp.n. from lambs, and *Cryptosporidium bovis* sp.n from a calf, with observations on the oocyst. Z Parasitenk **44** 289-98.

164 Mason RW, Hartley WJ and Tilt L (1981) Intestinal cryptosporidiosis in a kid goat. Aust Vet J **57** 386-8. https://doi.org/10.1111/j.1751-0813.1981.tb00529.x.

165 O'Donoghue PJ (1985) *Cryptosporidium* infections in man, animals, birds and fish. Aust Vet J **62** 253-8. https://doi.org/10.1111/j.1751-0813.1985.tb14245.x.

166 Tzipori S (1983) Cryptosporidiosis in animals and humans. Microbiol Rev **47** 84-96.

167 Tzipori S (1988) Cryptosporidiosis in perspective. Adv Parasitol **27** 63-121.

168 De Graaf DC, Vanopdenbosch E, Ortega-Mora LM et al. (1999) A review of the importance of cryptosporidiosis in farm animals. Int J Parasitol **29** 1269-87. https://doi.org/10.1016/S0020-7519(99)00076-4.

169 Coop RL and Wright SE (2000) Cryptosporidiosis and coccidiosis. In Diseases of sheep, eds WB Martin and ID Aitken. 3rd ed. Blackwell Scientific Publications: London, p. 153.

170 Naciri M, Mancassola R, Réperant JM et al. (1994) Treatment of experimental ovine cryptosporidiosis with ovine or bovine hyperimmune colostrum. Vet Parasitol **53** 173-90. https://doi.org/10.1016/0304-4017(94)90181-3.

171 Andrews AH (2013) Some aspects of coccidiosis in sheep and goats. Small Ruminant Res **110** 93-5. https://doi.org/10.1016/j.smallrumres.2012.11.011.

172 Taylor MA (2009) Changing patterns of parasitism in sheep. Practice **31** 474-83. https://doi.org/10.1136/inpract.31.10.474.

173 Chartier C and Paraud C (2012) Coccidiosis due to *Eimeria* in sheep and goats, a review. Small Ruminant Res **103** 84-92. https://doi.org/10.1016/j.smallrumres.2011.10.022.

174 Mahrt JL and Sherrick GW (1965) Coccidiosis due to *Eimeria ahsata* in feedlot lambs in Illinois. J Am Vet Med Assoc **146** 1415-6.

175 Irvine JH (1962) Ovine coccidiosis. Veterinary Inspector 61-5. NSW Department of Agriculture: New South Wales.

176 Gregory MW, Joyner LP and Catchpole J (1982) Medication against ovine coccidiosis — a review. Vet Res Comm **5** 307-25.

177 Slee KJ and Button C (1990) Enteritis in sheep, goats and pigs due to *Yersinia pseudotuberculosis* infection. Aust Vet J **67** 320-2. https://doi.org/10.1111/j.1751-0813.1990.tb07814.x.

178 Slee KJ and Button C (1990) Enteritis in sheep and goats due to *Yersinia enterocolitica* infection. Aust Vet J **67** 396-8. https://doi.org/10.1111/j.1751-0813.1990.tb03024.x.

179 McGregor H, Stanger K and Larsen J (2015) Insights into the epidemiology of a weaner scours syndrome in Merino lambs in SE Australia. Proceedings of the combined ACV/ASV Annual Conference, eds DS Beggs and RG Batey. Australian Cattle Veterinarians: Brisbane, Qld, Australia, pp. 234-40.

180 Barker IK, van Dreumel AA and Palmer N (1993) The alimentary system. In: Pathology of domestic animals, eds KVF Jubb, PC Kennedy and N Palmer. 4th ed. Academic Press: Canada.

181 Stanger K, McGregor H and Larsen J (2015) Preliminary evaluation of an autogenous vaccine containing *Yersinia pseudotuberculosis* for the control and prevention of Yersiniosis in weaned Merino lambs. Proceedings of the combined ACV/ASV Annual Conference, eds DS Beggs and RG Batey. Australian Cattle Veterinarians: Brisbane, Qld, Australia, pp. 293-8.

182 Stephens LR (1983) Weaner colitis in sheep due to *Campylobacter*-like bacteria. In: Sheep production and preventive medicine. Proceedings No 67. University of Sydney Post-graduate Committee in Veterinary Science, Australia: p. 53.

183 Stephens LR, Browning JW, Slee KJ et al. (1984) Colitis in sheep due to a *Campylobacter*-like bacterium. Aust Vet J **61** 183-7. https://doi.org/10.1111/j.1751-0813.1984.tb07237.x.

184 McOrist S, Stephens LR and Skilbeck N (1987) Experimental reproduction of ovine weaner colitis with a *Campylobacter*-like organism. Aust Vet J **64** 29-31. https://doi.org/10.1111/j.1751-0813.1987.tb06056.x.

185 Standfast HA, St George TD and Cybinski DH (1978) Experimental infection of *Culicoides* with a Bluetongue virus isolated in Australia. Aust Vet J **54** 457-8. https://doi.org/10.1111/j.1751-0813.1978.tb05586.x.

186 Alexander GI (1990) Bluetongue in Australia. Aust Vet J **67** 277. https://doi.org/10.1111/j.1751-0813.1990.tb07795.x.

187 Coackley W, Smith VW and Maker D (1980) A serological survey for bluetongue virus antibody in Western Australia. Aust Vet J **56** 487-91. https://doi.org/10.1111/j.1751-0813.1980.tb02562.x.

188 Hourrigan JL and Klingsporn AL (1975) Epizootiology of Bluetongue: The situation in the United States of America. Aust Vet J **51** 203-8. https://doi.org/10.1111/j.1751-0813.1975.tb00056.x.

189 Radostits OM, Blood DC, and Gay CC (1994) Viral diseases characterised by alimentary tract signs In Veterinary medicine — a textbook of diseases of cattle, sheep, pigs, goats and horses. 8th ed. Baillière Tindall: London, p. 1031.

190 Erasmus BJ (1975) Bluetongue in sheep and goats. Aust Vet J **51** 165-70. https://doi.org/10.1111/j.1751-0813.1975.tb00048.x.

191 Barker IK, van Dreumel AA and Palmer N (1993) Bluetongue and related diseases. In: Pathology of domestic animals, eds KVF Jubb, PC Kennedy and N Palmer. 4th ed. Vol 2. Academic Press: Canada, p. 174.

192 French El (1978) Bluetongue. In: Exotic diseases of animals: A manual for diagnosis, eds EL French and WA Geering. Aust Govt Publ Service: Canberra, p. 54.

193 Geering WA and Foreman AJ (1987) Foot-and-mouth disease. In: Exotic diseases, animal health in Australia. Vol 9. Bureau of Rural Science, Department of Primary Industries and Energy: Canberra.

194 Geering WA (1967) Foot-and-mouth disease of sheep. Aust Vet J **43** 485-9. https://doi.org/10.1111/j.1751-0813.1967.tb04774.x.

17

Diseases of the liver

This chapter discusses
- liver damage caused by the mycotoxins phomopsin and aflatoxin
- liver damage caused by pyrrolizidine alkaloids
- chronic copper toxicity.

Other diseases of the liver are also briefly mentioned in this chapter, and some are covered in greater detail in other chapters, including liver fluke, black disease, hepatic abscesses, lipidosis and cobalt deficiency.

There are other important disease conditions affecting the liver caused by plant, bacterial and fungal toxins but, because photosensitisation is the principal presenting sign, these are discussed in Chapter 10. Specifically, these conditions are facial eczema, caused by the mycotoxin sporidesmin, and hepatic injury with photosensitisation caused by cyanobacterial toxins, and phytotoxins including the saponins (in *Tribulus* spp and *Panicum* spp) and lantadenes (in *Lantana camara*).

PHOMOPSIN INTOXICATION (LUPINOSIS)

Lupins (*Lupinus* spp) are a grain legume crop planted extensively in parts of southern Australia. Several species of lupins are grown commercially in Australia. The narrow-leafed lupin (*Lupinus angustifolius*, also known as the Australian sweet lupin) is widely planted in Western Australia with some grown in South Australia and small areas grown in Victoria. The crop is well suited to the acid and sandy soils common in much of the Wheatbelt region of WA and is frequently grown in rotation with cereal crops. The seed (lupin grain) is high in protein and energy but low in starch, and is used widely as animal feed in Australia. There is a strong export market from WA to countries where it is used in both stockfeed and in human foods.

Much smaller areas of the European white lupin or albus lupin (*L albus*) are grown — principally for export for human consumption. The plant has a broader leaf and a larger seed than *L angustifolius*. Yellow lupin (*Lupinus luteus*) is also grown in WA but to a very limited extent. The grain has a higher protein content and better amino acid balance than narrow-leafed lupin grain and is used particularly as stockfeed in the pig and poultry industries.

Sandplain lupins (*L cosentenii*, also known as WA blue lupin) are grown for soil improvement and to provide summer feed, but are not harvested for grain. The plant will grow successfully on soils of very low fertility. It is hard-seeded and will self-regenerate from seed each year. On some farms it is valued as a source of stockfeed but, on other farms, it is considered undesirable because it can grow with, and contaminate, narrow-leafed lupin crops.

Stubbles of narrow-leafed or white lupin are available in summer and are frequently grazed by sheep, which benefit from the nutrition supplied by the dead plant tissue and, particularly, any spilled seed.

The fungus *Diaporthe toxica* (anamorph = *Phomopsis leptostromiformis*[a])[1] grows on the stems of lupin plants and causes the plant disease *phomopsis stem blight*. The fungus persists on the plant stems after harvest and increases in density after rainfall. The fungus produces metabolites — phomopsins — which are toxic to sheep, causing the condition *lupinosis*. Sheep are particularly susceptible to lupinosis. Sheep grazing affected lupin stubbles may ingest sufficient mycotoxin to suffer liver injury which, if severe or prolonged, can lead to fibrosis and irreversible hepatic damage and fibrosis. The degree of insult is related to the amount of toxin ingested. Lower doses of the mycotoxin cause changes which can be detected histologically but are not associated with clinical signs. Higher doses, causing more severe damage to the liver, lead to clinical signs and mortalities.

The condition most often occurs in summer and autumn when sheep have access to stubbles or are grazing sandplain lupins as a summer fodder. In stubbles, weaners are more commonly affected than adult sheep because they are unaccustomed to the grain and graze more heavily on the stems than adult sheep, which prefer to eat the grain. Lupin stubbles typically contain over 200 kg of lupin grain per hectare after harvest, an amount which, in conjunction with plant stems and weeds, is sufficient to nourish sheep at high stocking densities for several weeks. As the grain is consumed the sheep will consume more of the dead lupin plant stem material which may have toxic levels of phomopsin. Particular care should be taken if the amount of grain in the stubble falls below 50 kg per hectare.

Lupin grain can also contain toxic levels of phomopsin[2] but this is relatively uncommon and lupin grain is generally considered safe. The grain is more likely to be infected if there has been heavy rain during the period of seed and pod maturation. Infected, toxic grain is discoloured, ranging in colour from pale yellow to dark purple-brown. If more than 10% of the lupin seed is discoloured, it should not be fed. If infection is present at lower rates, the grain should be fed only in small amounts.

Photosensitisation is sometimes seen in sheep which have chronic lupinosis-induced liver damage and gain sudden access to green feed[3,4] (see Chapter 11). Lupinosis, although primarily a hepatotoxicity, has also been associated with a myopathy in sheep in Western Australia (lupinosis-associated myopathy, LAM).[5] In Chapter 11, LAM is discussed further as a differential diagnosis of nutritional myopathies.

Clinical findings

Both acute and chronic syndromes are recognised. Acute disease is most common in livestock on sandplain lupins following summer rains, while the chronic syndrome is more common in sheep grazing narrow-leafed lupin stubbles or being fed lupin grain.

Sheep with acute lupinosis are severely depressed and likely to be isolated from the flock. Close examination shows marked icterus.

a   The asexual reproductive form of a fungus is the anamorph. The sexual reproductive form is the teleomorph.

With the chronic condition there is inappetence, a gradual loss of condition and lethargy, unresponsiveness to stimuli and a staggering gait. If the condition progresses there are signs of an hepatic encephalopathy with aimless wandering, apparent blindness, sometimes with head pressing.[6]

Animals with subclinical liver damage may be more likely to succumb to other conditions which place further demands on the already-compromised liver. Exposure of ewes to phomopsin around the time of mating will lead to a reduction in ovulation rates and pregnancy rates.[7]

Pathology

In animals which have died from lupinosis the generalised jaundice is the most prominent post-mortem finding in the field. The more subtle effects of the phomopsin toxin on hepatic tissue were described by Gardiner (1965).[3] He classified the changes which were apparent grossly as either early, subacute or chronic. The subacute and chronic conditions were associated with accumulation of granular degenerated hepatocytes in centrilobular areas with proximate fibrosis, developing into portal tract fibrosis, bile duct cell and Kupffer cell proliferations, and cirrhosis. A mechanism for quantifying the degree of liver injury based on the extent of these histological findings has been described by Allen et al. (1978).[8] These workers found a high correlation between the injury score and the clinical presentation — low scores associated with no or subclinical expression only, medium scores with moderate clinical expression, and high scores with severe and very severe clinical lupinosis. While this technique has most relevance to research studies rather than field investigations it does support the strong association between phomopsin exposure and the impact on the health and productivity of a flock.

Treatment

Following recognition of cases of lupinosis all exposed sheep should be removed from the stubble or lupin crop; or, if lupin grain is likely to be the source, feeding of the grain should cease. The exposed sheep should then be subject to management which avoids placing further stress on their damaged livers. They should be placed in paddocks with adequate shade and easy access to good-quality water. Access to more than small amounts of green feed is contraindicated because of the risk of photosensitisation. Supplementation with cereal grains, cereal hay or pasture hay is advised but higher protein feeds, such as legume hays, should be avoided.

Sheep with mild to moderate liver damage can be expected to recover over the following few weeks or months. It is advisable to check the affected flock two to three days after removing them from the phomopsin source and to identify those animals which are still clinically ill and, particularly, still not eating. These animals should be euthanased.

Prevention

Plant-breeding programmes in WA have led to the development of varieties of narrow-leafed lupins which have some resistance to phomopsis stem blight. The resistance is not, however, complete and most lupin crops and stubbles will be infected to some degree by the fungus. The Western Australian Department of Primary Industries and Regional Development (Agriculture and Food) offers the following advice to producers to minimise the risk of lupinosis:

- Graze lupin stubbles early and before grazing cereal stubbles, as toxicity slowly increases with each summer rain event.

- Provide multiple watering points in a paddock to promote even grazing of the stubble. Weaner sheep concentrate their grazing within 600-800 m of watering points.

- Pre-feed lupin seed to train stock to seek out lupin seed in stubbles.

- Check stock regularly for signs of lethargy, reduced appetite and hollow flanks.

- Remove stock from stubble paddocks before the lupin seed count gets below 40 seeds per m² (50 kg/ha of lupin seed).

- Keep the flock size less than 600 for weaner sheep.

- Sheep should be removed from sandplain lupin paddocks after summer rainfall and not returned until the stalks have dried out. Check the sheep daily for the first week after reintroduction.

- It is not necessary to remove sheep from white lupins after summer rain but sheep should be checked daily for a week after significant rainfall.

AFLATOXIN INTOXICATION

Aflatoxins are the products of some species of fungi, particularly *Aspergillus* spp and *Penicillium* spp, which grow on feedstuffs which have been kept under moist conditions or, occasionally, on crops when harvested or fed as standing fodder. High doses of the toxins cause an acute and severe hepatic injury leading to liver failure, while lower doses over prolonged periods cause less severe damage with subclinical effects on health and productivity.

In Australia, *Aspergillus flavus* has been detected in Queensland in peanuts, peanut meals, sorghums, pig feeds and poultry feeds.[9] An outbreak of mortalities in ewes in NSW was attributed to aflatoxicosis from mouldy maize. The ewes were malnourished as a result of drought, which may have explained their readiness to eat mouldy feeds.[10]

PYRROLIZIDINE ALKALOID POISONING

Pyrrolizidine alkaloids (PA) are chemical substances which occur naturally in many plant species and probably have a role in protecting plants from insect attacks. There are many different PAs, all of similar chemical structure based on the pyrrolizidine nucleus, and many of them are hepatotoxic to mammals. Ruminants are generally less susceptible to PA poisoning than monogastric animals, including pigs and horses.

PAs are present at high levels in several species of plants commonly grazed by sheep in Australia. The most important two species in the sheep industry are naturalised weeds which have been introduced into Australia from southern Europe and north Africa — *Echium plantagineum* (Paterson's curse or Salvation Jane) and *Heliotropium europaeum* (common heliotrope). Both of these plants are widespread in western and southern New South Wales, northern and central Victoria, Western Australia and South Australia, and in some regions the distribution overlaps. *E plantagineum* also occurs in Tasmania and southeast Queensland and is considered the most costly weed of annual pastures in southern Australia.[11] It is a winter-active annual weed, typically growing from germination with autumn rains (April-May), flowering in late spring and dying in December-January. Heliotrope is a summer-growing annual weed.

Generally, heliotrope is considered more likely to cause PA poisoning of sheep than *E plantagineum*. Although PA levels in the plants vary within and between seasons and, possibly, between environments, the concentrations of PAs are usually higher in heliotrope (around 1.2% of dry weight) than in *E plantagineum* (around 0.3%).[12,13] Nevertheless, there are numerous reports implicating *E plantagineum* in cases of PA poisoning, indicating that its relative safety for sheep is unpredictable.[14,15] If the stand of *E plantagineum* is relatively high in PAs, and the diet consists principally or wholly of *E plantagineum*, high mortality rates due to PA poisoning can be expected within a few months.[16]

Two species of *Senecio* — *S madagascariensis* (fireweed) and *S jacobaea* (ragwort) — also contain high levels of PAs and may cause poisoning of grazing animals.[15] *Crotalaria* spp contain PAs and several species growing in northern Australia are capable of causing *Walkabout disease (Kimberley Horse disease)* of horses. *Crotalaria eremea* (bluebush pea) has caused PA poisoning of sheep in Queensland at a time when there was little other fodder available.

Syndromes caused by PA ingestion

Compared to non-ruminants and cattle, sheep are relatively resistant to PA poisoning but chronic ingestion — over months or years — leads to liver damage which is expressed in two distinct clinical forms. One — *primary pyrrolizidine alkaloid poisoning* — is the result of chronic liver failure caused by the direct effect of ingested PAs on liver cells. The other is an acute haemolytic syndrome caused by massive copper release from the liver, following an extended period during which copper accumulates to very high levels in the liver, as a result of the PA damage. This second form is known as *hepatogenous chronic copper poisoning* and is discussed further below, with other causes of chronic copper poisoning.

Pathophysiology

Both conditions start with chronic hepatic damage which is irreversible. Ingested PAs are absorbed from the intestinal tract and metabolised by hepatocytes, especially those cells in the centrilobular region of the liver lobules which are most important in detoxification of drugs and toxins. The pyrrole derivatives of the PA are reactive with DNA and therefore interfere with cell division and protein synthesis.[17] This effect is presumably responsible for the marked increase in size of the hepatocytes (megalocytosis) which is characteristic of PA intoxication.[18] Histologically, in addition to the megalocytosis and karyomegaly, there is evidence of death of some cells and variable degrees of regeneration, fibrosis and bile duct proliferation. Grossly, the liver becomes smaller and firmer because of the progressive fibrosis. Liver function declines, leading to the reduced production of proteins, including albumin, and reduced capacity to detoxify and excrete other metabolites. Hepatogenous photosensitisation, which occurs in some cases, reflects the impaired metabolism of phylloerythrin. Affected animals lose condition as a result of reduced appetite and impaired metabolism. Death, when it occurs, is a result of liver failure.

The relative tolerance of sheep to PA ingestion may be the result of enhanced ruminal detoxification of PAs — an effect which is more pronounced in sheep exposed constantly to PA-containing plants[13] — and differences in the way in which PAs are metabolised in the liver in different species. Compared to cattle, sheep liver cells *in vitro* have a low rate of pyrrole production[19], which may reduce the relative intracellular toxicity of PAs for sheep. There is some evidence that the several PAs of *E plantagineum* are less toxic to sheep than those of *H europaeum*, and some evidence that Merino sheep are better able to safely detoxify PAs in the liver than British breed and British breed-Merino crossbred sheep.[13]

Predisposing plant and grazing factors

Sheep can usually graze heliotrope with relative safety for one summer season, although losses of lambs following one season of grazing have been observed.[20] Disease occurs generally in sheep which graze the plant for a second or subsequent season.[21,22] When first exposed to heliotrope sheep do not readily consume it unless there is little alternative feed source but, once they have become accustomed to eating the plant, the initial aversion is overcome. The strength of the initial aversion is much greater in Merino sheep than crossbred sheep or non-Merino breeds. Straight-bred Merino sheep consume less heliotrope than crossbred or British breed sheep when offered similar availability of the plants and are less often affected by PA poisoning caused by that plant.

Heliotrope thrives in fallow paddocks and in crop stubbles following late spring or summer rains, where the seedlings can germinate without competition or shading from perennial pasture plants. The environment and farming practices of the Victorian Mallee district are particularly suitable for heliotrope, and PA poisoning is considered to be a major contributor to the mortality of ewes in that region, particularly in ewes 3 years of age or older, and of non-Merino breed.[23]

Sheep can graze pastures in which *E plantagineum* is subdominant for several years without suffering any ill effects, and the plant is reasonably nutritious while green and vegetative[12,24] (Figure 17.1). In parts of north-central South Australia in particular, the plant was viewed by pastoralists as an important fodder plant, well deserving of its local name, Salvation Jane. In higher rainfall areas the plant is viewed as a weed which competes with more productive pasture plants. On infested, ungrazed land the plant can be dominant, forming a purple carpet as it flowers (Figure 17.2). When sheep graze the pasture, however, few flowers are seen — such is the sheep's fondness for the plant. The levels of PA and, therefore, the relative toxicity of the plant vary from season to season. That fact, plus variation in the plant density to which sheep are exposed while grazing over one or more years, lead to difficulties in predicting the outcome of exposure of sheep to the plant.

There is no apparent breed predisposition for PA intoxication from *E plantagineum* and Merino sheep and British breed sheep are similarly affected by PA poisoning induced by that plant.[14]

Biological control strategies, based on introduced species of root weevils and flea beetles which can attack and kill *E plantagineum*, commenced in the 1980s, despite resistance from some graziers and apiarists who valued the plant as a feed source or pollen source. Biological control has been partially effective but the weed is still abundant in most parts of its territory.

Heavy grazing by sheep of *E plantagineum*, particularly through the flowering season, will lead to a marked reduction in the seed bank in the soil, although it may take four years to achieve significant benefits to the pasture composition. A widely used control strategy for *E plantagineum* is the *spray-graze* technique, in which a low application rate of a selective

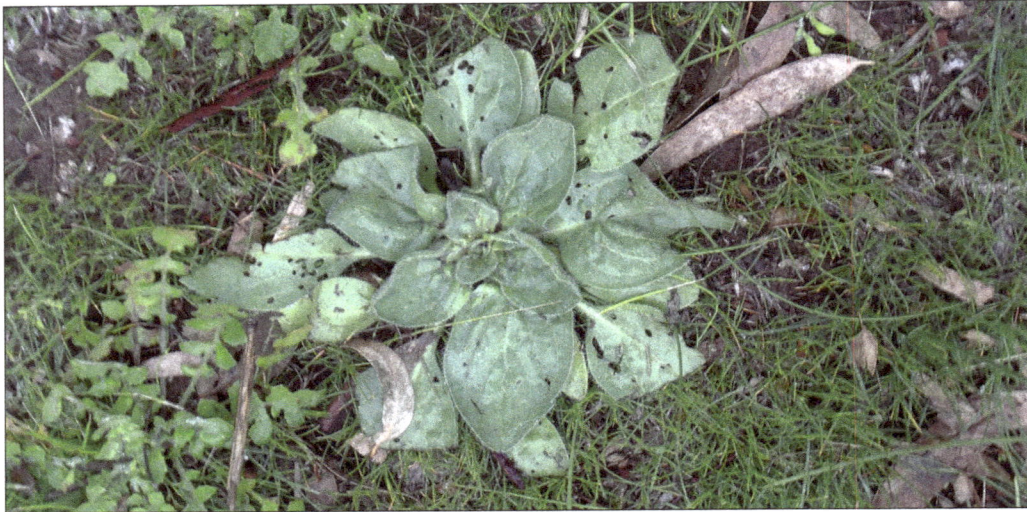

Figure 17.1: *Echium plantagineum* (Paterson's curse, Salvation Jane), in the rosette stage soon after germination. In spring and early summer the plant develops upright stems and flowers. Source: KA Abbott.

Figure 17.2: A dense carpet of *Echium plantagineum* in flower. Source: KA Abbott.

herbicide is applied to the pasture, followed by a period of very high stocking density. Adult wethers are best suited for this strategy and two seasons of herbicide treatment and grazing are expected to reduce the plant density by more than 50% without clinically apparent ill effects on the health of the sheep as a consequence of the PA ingestion.[11]

Clinical and necropsy findings

The main clinical signs of primary PA poisoning are weight loss and progressive emaciation.[15] Photosensitisation of hepatogenous origin occurs in some cases.[23] Jaundice may occur but is uncommon with primary PA poisoning. (By contrast, it is a characteristic of hepatogenous chronic copper poisoning.) Primary PA poisoning can lead to clinical signs and deaths in sheep grazing heliotrope, *E plantagineum,* or both and, less commonly, *Senecio* spp. In a review of field cases over seven years in NSW, Seaman (1987)[15] found that the fatality rate in 10 outbreaks caused by heliotrope was 9% while in 19 cases in sheep exposed to *E plantagineum,* alone or in combination with heliotrope, the fatality rate was less than 2.5%.

At necropsy, the liver of affected sheep is shrunken, grey-yellow, smooth and fibrotic. Histologically, the presence of megalocytosis of liver cells is characteristic, even pathognomonic, for PA poisoning. Regenerative nodules of new hepatic cells are common. There is usually evidence of proliferation of the bile duct epithelium within the portal triads, sometimes accompanied by biliary fibrosis.[15,21] Other tissues, especially kidneys, may also be affected.

Treatment and prevention

There is no treatment which will reverse the liver damage done by PA ingestion. Because PA poisoning is cumulative, sheep which have been exposed to PA-containing plants, particularly heliotrope, should be prevented from further grazing of the weeds. This is particularly important for ewes because the physiological stresses associated with pregnancy, lambing and lactation may precipitate liver failure.

Prevention requires control of the weeds in pastures or crop residues or fallows, by cultivation, herbicide treatment, introduction of competitive pasture species or grazing management. Access to heliotrope should be restricted such that sheep only graze the plant for one season and steps should be taken to provide additional alternative feed materials while they graze the heliotrope. If grazing by sheep of heliotrope is unavoidable, consideration should be given to choosing Merino sheep rather than British-breeds or their crosses.

Sheep that have incurred some liver damage due to PA ingestion are predisposed to chronic copper toxicity. Steps which can be taken to reduce the risk of a subsequent haemolytic crisis are described below.

CHRONIC COPPER TOXICITY (TOXAEMIC JAUNDICE)

Sheep grazing in Australia and some other countries are at risk of syndromes associated with too little copper (a primary dietary deficiency or one induced by a high intake of molybdenum) or with excessive copper. Disease conditions associated with deficiencies of copper and excesses of molybdenum are discussed in Chapter 11.

Acute copper toxicity usually presents as a condition of gastroenteritis and abdominal pain following the oral administration of excessive amounts of copper salts. The liver is not the principal organ involved and the syndrome will not be discussed further in this chapter.

The liver is, however, centrally involved in the development of chronic copper toxicity because it is the site in which absorbed copper is stored. In all animal species copper storage is moderated in the healthy liver by absorption and storage on one hand and, on the other, the secretion of excess copper in the bile. Sheep are generally well adapted to low-copper diets and readily accumulate the element in the liver. Sheep hepatocytes are very efficient at capturing copper from the portal circulation but biliary excretion of the stored element is less efficient in sheep than in other domestic animal species.[25] Consequently, under some conditions, the levels of stored copper can become too high, and a syndrome of chronic copper toxicity develops.

The conditions which lead to the excessive accumulation of copper can be categorised under three headings:

(1) high levels of dietary copper, often following its excessive inclusion in prepared feeds

(2) low levels of dietary molybdenum with moderate (or higher) levels of dietary copper

(3) exposure to toxins which compromise the capacity of the liver to moderate copper storage.

Stored liver copper concentrations up to 350 mg/kg (liver dry weight) present no risk of cellular damage. At higher concentrations, however, the rate of hepatic cellular apoptosis increases. Initially, this cellular damage is without clinical signs but is evident biochemically through the increased plasma concentrations of hepatic cellular enzymes[26] and, histologically, through an increase in the number of apoptotic cells, swelling of cells, changes to the appearance of nuclei and the accumulation in extracellular spaces of macrophages containing copper-rich lipofuscins.

The rate of formation of new hepatocytes increases in response to the increased rate of cell loss through apoptosis. Dying cells release copper which is taken up by the new cells but, if the rate of regeneration is insufficient, plasma copper levels rise. The copper is taken up by erythrocytes, achieving very high levels and causing two critical disturbances of erythrocyte function. Some haemoglobin is converted to methaemoglobin, thus reducing the oxygen-carrying capacity of the blood. In addition, copper-induced oxidative damage to the cell membrane leads to lysis and the release of haemoglobin.

The products of haemolysis, coupled with the anaemia, further compromise the capacity of hepatocytes to store copper and the rate at which copper enters the circulation accelerates. The sudden release of massive amounts of liver copper cause blood copper concentrations to rise very quickly, reaching levels 10 times higher than normal within a 48-hour period. The acute syndrome is characterised by massive intravascular haemolysis, liver failure and renal failure.

During the period of copper accumulation in the liver, associated with subclinical changes to liver tissue, sheep are considered to be in the pre-haemolytic phase of copper poisoning. Following the sudden increase in blood copper concentrations and consequent haemolytic episode, sheep are considered to be in the post-haemolytic phase. In most field cases in which a haemolytic crisis occurs, death occurs within one to three days of the event.

The sudden elevation of blood copper concentration and the acute fatal haemolytic syndrome is the best recognised clinical expression of the chronic intoxication, but cases have also been reported in which copper concentrations in the blood are elevated for several weeks before the acute haemolytic episode. In some cases, milder forms occur from which animals recover.

Chronic (cumulative) copper poisoning following high dietary or parenteral intake of copper

Copper accumulates to elevated and dangerous levels in the otherwise healthy livers of sheep as a consequence of sustained high dietary intake of the element. Pastures in southern Australia typically contain 3-20 mg/kg of copper (on a dry matter basis). Deficiency syndromes are only likely to occur on pastures containing less than 7 mg/kg and cumulative copper poisoning is only likely if pastures contain more than 20 mg/kg. Both syndromes (deficiency and excessive accumulation) are strongly influenced by the dietary intake of other elements, particularly molybdenum, and both are only likely to occur when the predisposing conditions (low copper/high molybdenum or high copper/low molybdenum) are sustained for several months.

Copper poisoning has been reported in sheep grazing pastures growing on soils which are high in copper. Such soils occur naturally in some parts of Australia — such as the Lachlan River valley in New South Wales — or alternatively the soils may be contaminated by copper-rich dust from nearby copper mines.[27] The high intakes of copper result from high levels in the pasture plants, dust and soil ingested with the plants, or both.

Copper is frequently added to superphosphate fertiliser in regions of Australia where soils are deficient in the element. Generally, this is a safe procedure and the copper is readily incorporated into the soil profile by rain after application. Nevertheless, cumulative copper poisoning of sheep may occur if sheep graze the treated pastures after application and before rainfall has washed the copper from the pasture into the soil.[28] Copper-based fungicides applied in orchards or vineyards may contaminate pasture grazed by sheep, and lead to toxicity.[29]

Sheep which are housed and fed prepared dry feeds are prone to chronic copper poisoning. Feed mills preparing sheep feeds often have difficulty keeping copper levels below 15 mg/kg because some constituent feed sources may be high in copper.[26] If molybdenum levels in the feed are low, cumulative poisoning can occur.[30] Toxicity sometimes occurs as the result of feeding sheep with feed-mixes or supplements prepared for other livestock species. Cattle, pigs, poultry and horses can safely consume feeds containing copper at much higher levels than sheep can tolerate and copper is deliberately added to pig feeds at levels of 100 to 250 mg/kg DM because it acts as a growth promotant in that species. Pig feeds with elevated copper levels are therefore likely to cause chronic copper poisoning if given to sheep, and faeces from pigs fed on high-copper feeds also present a danger of intoxication to sheep. Much of the copper (about 95%) included in pig feeds is not absorbed by the pig and passes out in the faeces. Pastures which have been fertilised with slurry from pig sheds — particularly if the application is repeated over extended periods — may lead to cumulative copper poisoning in sheep grazing the pasture or consuming conserved fodder prepared from it.[31] As poultry are also often fed high-copper diets, poultry manure may pose a similar risk.

Usually, the syndrome of chronic copper poisoning occurs when sheep ingest and accumulate excessive amounts of copper over a period of weeks or months, ultimately resulting in a relatively acute manifestation, marked by haemolysis and jaundice, when large amounts of copper are released by the damaged liver over a short time period. The syndrome may also be precipitated by copper intake over a much shorter period when sheep receive copper supplementation, orally or parenterally, at excessive levels. The intake of copper in such cases is not so high that acute poisoning occurs as a direct result of the administration of the metal, but it is sufficient to overload the hepatic storage capacity, and it leads to the death of liver cells, copper release into the blood and a haemolytic crisis in a way similar to the more chronic form. Clinical signs may be evident within 48 hours of dosing.[32]

Chronic (cumulative) copper poisoning as a consequence of low molybdenum intake

Legumes have a higher affinity for copper than do grasses, and cumulative copper poisoning of sheep grazing subterranean clover (*Trifolium subterraneum*) has been reported on numerous occasions in Victoria and New South Wales. The condition is referred to as *phytogenous copper toxicity*, to distinguish it from chronic toxicity associated with prior liver damage. Generally, copper levels in the pasture are in the range of 10-20 mg/kg and molybdenum intakes are very low, typically less than 0.2 mg/kg.[27,30] Seasonal conditions which favour clover growth, such as warm autumns with adequate soil moisture, can lead to clover-dominance of the pastures. Copper (Cu) accumulates in the plant during winter at the same time as molybdenum (Mo) levels in the plant decline, leading to rapid development of Cu:Mo ratios of 100 or more. The copper-molybdenum imbalance rectifies itself as pastures dry off in spring and summer. Outbreaks of phytogenous copper toxicity occur sporadically in southern Australia, typically on acid soils, with a very strong seasonal variation, and are more common in British breed and British-breed Merino crossbreds than in pure Merino sheep.[27,33]

Chronic (cumulative) copper poisoning following phytotoxic liver damage

The chronic ingestion of pyrrolizidine alkaloids causes liver damage, as described above, which leads to primary PA poisoning, but also predisposes sheep to *hepatogenous chronic copper poisoning*. The mechanism which leads to this syndrome is the increased avidity of the PA-damaged hepatocytes for copper.[21] Sheep consuming PA at toxic levels for sustained periods accumulate copper to very high levels (over 1000 mg/kg) despite a normal or moderate intake of dietary copper. The enhanced storage of copper continues after the PA is removed from the diet, if dietary copper is available. The increased storage is associated with a reduced rate of loss of copper from the liver during periods of low copper intake but the mechanism for this unclear. There is evidence that bile flow is not reduced in PA-damaged livers and that the bile duct proliferation which occurs in PA poisoning consists of functional new ductule cells.[34]

The development of the acute haemolytic crisis follows the same pattern as described above for phytogenous chronic copper poisoning or for chronic copper poisoning following excessive parenteral or dietary intakes.

Clinical signs

There are virtually no clinical signs of pre-haemolytic chronic copper poisoning until the last few days preceding the haemolytic crisis, at which time there is a progressive loss of appetite which is only likely to be observed in housed sheep. The appearance of cases most commonly follows husbandry events which apply some stress to the flock, such as shearing or lambing. Affected animals are depressed; they separate from the flock and are reluctant to move. Jaundice is marked and urine is coloured brown to black by the haemoglobinuria.[15] The clinical course is short, with death often occurring within 48 hours of the first clinical signs. Sheep which die from hepatogenous copper poisoning are usually in good condition.

Clinical pathology

During the pre-haemolytic phase of chronic copper poisoning, the liver copper concentration rises from normal levels to levels of 1000 mg/kg or higher. The varying concentrations at which a haemolytic crisis occurs may be due to differences between individual sheep and, possibly, to the degree of physiological stress (late pregnancy, for example) being applied to liver function, but the severity of liver damage can be expected to increase progressively as the levels rise above 350 mg/kg.[26]

Plasma copper levels do not increase until the haemolytic crisis is imminent. In the late stages of the pre-haemolytic phase, hepatic enzymes can be detected at abnormal levels in plasma. Of these, sorbitol dehydrogenase (SDH), gamma glutamyl transferase (γGT) and aspartate aminotransferase (AST, formerly SGOT) provide early evidence of liver damage, becoming increasingly elevated for the two to four weeks preceding the haemolytic crisis. SDH may be the first to rise, some weeks before γGT[35], but the levels in plasma during the pre-haemolytic phase may be more variable than those of γGT.[36]

Necropsy findings

At necropsy, jaundice is marked and obvious. There are ecchymotic haemorrhages on serosal surfaces and increased fluid in cavities, including the peritoneal. The liver may be enlarged and friable or reduced in size and firmer than normal. Kidneys are swollen, soft and black — often described as gun-metal blue — as a result of haemoglobinuric nephrosis.

Diagnosis

Confirmation of chronic copper poisoning rests on the presence of haemolytic jaundice and high concentrations of copper in both liver and kidney. Histopathology of the liver will be necessary to determine whether the excessive copper storage leading to the acute haemolytic crisis was precipitated by chronic pyrrolizidine alkaloid ingestion.

Treatment

Usually, the relatively sudden death of one or more sheep in the flock is the first indication that an outbreak of chronic copper poisoning is underway. Sheep showing clinical signs of the acute crisis of chronic copper poisoning (jaundice, anaemia and inappetence) can be administered treatments aimed at reducing the copper load in the liver. Surviving sheep in the flock which have the same grazing and feeding history as clinical cases should be considered to be in the pre-haemolytic phase and treated similarly. Prompt and effective treatment may prevent further deaths.

Effective therapy is based on the provision of salts of molybdenum and sulphur. Molybdate reacts with sulphide in the rumen to form thiomolybdate, which combines with copper to form complexes which do not release copper. In the alimentary tract insoluble thiomolybdate-copper complexes are passed in the faeces. In the bloodstream and tissues, including the liver, thiomolybdate also forms complexes from which copper is not available. Circulating thiomolybdate leads to rapid falls in liver copper concentrations.

Groups of ewes with chronic copper poisoning have been successfully treated with ammonium molybdate (40 g) and sodium sulphate (1200 g) mixed in water (10 L) and sprayed on hay such that the sheep each received 40 to 50 mg ammonium molybdate per head each day. Treatment was continued for three weeks but deaths in the flock ceased three days after treatment began.[37]

There are limited, if any, products for treatment of copper poisoning registered in Australia. Provision of a home-made loose lick consisting of salt, finely ground gypsum (calcium sulphate) and sodium molybdate in the ratio by weight of 187:140:1 has been recommended[33] and reported to be successful when used in Australia.[38] The salt (75 kg) and gypsum (56 kg) are mixed together on a flat surface and then the sodium molybdate in solution (400 g in 20 L) is sprayed onto the mixture and thoroughly mixed. The product should then be placed in open containers and offered to the sheep in the field, ensuring that all sheep have uninhibited access.

For animals receiving individual attention, parenteral administration of ammonium tetrathiomolybdate (ATTM) is recommended for sheep in both the pre-haemolytic or post-haemolytic phases of chronic copper poisoning. (ATTM is very expensive and its use would only be considered in valuable animals.) Affected sheep should be treated with 1.7-3.4 mg ATTM per kg body weight, on alternate days, for three treatments. The lower dose should be used when treatment is administered intravenously; the higher dose for subcutaneous administration. TTM has the side effect of inactivating cupro-enzymes, including erythrocyte superoxide dismutase, for two to three weeks after treatment, so Suttle (2012)[39] has recommended that a single, low intravenous dose (1 mg/kg) be used for treatment of sheep in the pre-haemolytic phase, with repeated treatment only in the event of recrudescence.

Prevention

Where high copper intake in the diet is suspected, molybdenum can be added to ensure that the Cu:Mo ratio in the diet is around 10:1 or lower. Molybdenum can be added to feed, water, lick blocks or fertiliser. Care must be taken to ensure that excessive molybdenum intakes do not lead to copper deficiency.

LIVER FLUKE (FASCIOLA HEPATICA)

Liver fluke can cause clinical disease in sheep ranging from acute disease with high mortality, as a result of the ingestion of large numbers of metacercariae and their subsequent migration through the liver, to chronic disease as a result of the adult flukes in the bile ducts causing cholangitis, biliary obstruction and anaemia. Fascioliasis is discussed in detail in Chapter 9.

BLACK DISEASE (NECROTIC HEPATITIS)

Black disease is a peracute disease of sheep and other animals, caused by the toxin produced by *Clostridium novyi* type B, when it proliferates in the anaerobic conditions of the necrotic tracts in the liver produced by migrating *F hepatica* larvae or, rarely, other migrating parasitic larvae. The infection and intoxication are a cause of sudden death and are discussed further in Chapter 14.

HEPATIC ABSCESSES

Abscessation of the liver of lambs can occur following an infection — usually with *Fusobacterium necrophorum* — emanating from the rumen wall or instigated by liver damage caused by a migrating parasite.[40] If there are only a few abscesses the animal's apparent health may be unaffected and the lesions only discovered at necropsy or slaughter.

COBALT DEFICIENCY (WHITE LIVER DISEASE)

Ovine white liver disease has been reported from Australia and New Zealand. The disease appears to be a vitamin B_{12} deficiency complicated by liver damage. It principally affects lambs 2 to 6 months of age causing severe ill-thrift, hepatopathy, depression, serous ocular discharge and crusty lesions on the ears probably as a result of photosensitisation. There is a high morbidity with significant mortality rates. Affected sheep lose weight rapidly. Clinical signs are present for up to 14 days before death. At necropsy, the liver is pale and, in some cases, almost white, and very swollen. Lipidosis is usually present. Cobalt and vitamin B_{12} deficiency are discussed further in Chapter 11.

HEPATIC LIPIDOSIS (FATTY LIVER)

While hepatic lipidosis in sheep can result from cytotoxic injury, it is seen most commonly in ewes which have died with pregnancy toxaemia because of the impaired metabolism of fatty acids which occurs with that condition (Chapter 7). It is also an occasional finding in white liver disease caused by cobalt deficiency.

REFERENCES

1   Williamson PM, Highet AS, Gams W et al. (1994) *Diaporthe toxica* sp nov, the cause of lupinosis in sheep. Mycol Res **98** 1364-8. https://doi.org/10.1016/S0953-7562(09)81064-2.
2   Allen JG, Moir RJ and Mackintosh JB (1983) Ovine lupinosis resulting from the ingestion of lupin seed naturally infected with *Phomopsis leptostromiformis*. Aust Vet J **60** 206-8. https://doi.org/10.1111/j.1751-0813.1983.tb09584.x.
3   Gardiner MR (1965) The pathology of lupinosis of sheep. Vet Pathol **2** 417-45.
4   Allen JG, Moir RJ and Mackintosh JB (1983) Ovine lupinosis resulting from the ingestion of lupin seed naturally infected with *Phomopsis leptostromiformis*. Aust Vet J **60** 206-8. https://doi.org/10.1111/j.1751-0813.1983.tb09584.x.
5   Allen JG (1978) The emergence of a lupinosis-associated myopathy in sheep in Western Australia. Aust Vet J **54** 548-9.
6   Braddon E and Lachlan SDV (2012) A case of lupinosis: Flock and herd case notes. Available at: http://www.flockandherd.net.au/sheep/reader/lupinosis.html. Accessed 15 August 2017.
7   Barnes AL, Croker KP, Allen JG et al. (1996) Lupinosis of ewes around the time of mating reduces reproductive performance. Aust J Agric Res **47** 1305-14. https://doi.org/10.1071/AR9961305.
8   Allen JG, Wood PMc and O'Donnell FM (1978) Control of ovine lupinosis: Experiments on the making of lupin hay. Aust Vet J **54** 19-22. https://doi.org/10.1111/j.1751-0813.1978.tb00263.x.
9   Connole MD and Hill MWM (1970) Aspergillus flavus contaminated sorghum grain as a possible cause of aflatoxicosis in pigs. Aust Vet J **46** 503-5. https://doi.org/10.1111/j.1751-0813.1970.tb09175.x.
10 Baird JD (1972) cited by Culvenor CCJ (1974) The hazard from toxic fungi in Australia. Aust Vet J **50** 69-78.

11 Smyth MJ, Sheppard AW and Swirepik A (1997) The effect of grazing on seed production in *Echium plantagineum*. Weed Res **37** 63-70. https://doi.org/10.1046/j.1365-3180.1996.d01-2.x.

12 Seaman JT, Turvey WS, Ottaway SJ et al. (1989) Investigations into the toxicity of *Echium plantagineum* in sheep. 1. Field grazing experiments. Aust Vet J **66** 279-85. https://doi.org/10.1111/j.1751-0813.1989.tb13952.x.

13 Culvenor CCJ, Jago MV, Peterson JE et al. (1984) Toxicity of *Echium plantagineum* (Paterson's curse). 1. Marginal toxic effects in Merino wethers from long-term feeding. Aust J Agric Res **35** 293-304. https://doi.org/10.1071/AR9840293.

14 St. George-Grambauer TD and Rac R (1962) Hepatogenous chronic copper poisoning in sheep in South Australia due to the consumption of *Echium plantagineum* L. (Salvation Jane). Aust Vet J **38** 288-93. https://doi.org/10.1111/j.1751-0813.1962.tb04033.x.

15 Seaman JT (1987) Pyrrolizidine alkaloid poisoning of sheep in New South Wales. Aust Vet J **64** 164-7.

16 Seaman JT and Dixon RJ (1989) Investigations into the toxicity of *Echium plantagineum* in sheep. 2. Pen feeding experiments. Aust Vet J **66** 286-92. https://doi.org/10.1111/j.1751-0813.1989.tb13953.x.

17 Cheeke PR (1988) Toxicity and metabolism of pyrrolizidine alkaloids. J Anim Sci **66** 2343-50. https://doi.org/10.2527/jas1988.6692343x.

18 Bull LB (1955) The histological evidence of liver damage from pyrrolozidine alkaloids: Megalocytosis of the liver cells and inclusion granules. Aust Vet J **31** 33-40. https://doi.org/10.1111/j.1751-0813.1955.tb05488.x.

19 Shull LR, Buckmaster GW and Cheeke PR (1976) Factors influencing pyrrolizidine (Senecio) alkaloid metabolism: Species, liver sulfhydryls and rumen fermentation. J Anim Sci **43** 1247-53. https://doi.org/10.2527/jas1976.4361247x.

20 Salmon D (2011) Pyrrolizidine alkaloid poisoning of sheep. In: Proceedings of the Australian Sheep Veterinarians 2011 Conferences (Barossa Valley SA and AVA Adelaide). Australian Veterinary Association: St Leonards, pp. 82-4.

21 Bull LB, Dick AT, Keast JC et al. (1956) An experimental investigation of the hepatotoxic and other effects on sheep of consumption of *Heliotropium europaeum* L: Heliotrope poisoning of sheep. Aust J Agric Res **7** 281-332. https://doi.org/10.1071/AR9560281.

22 Bull LB (1961) Liver diseases in livestock from intake of hepatotoxic substances. Aust Vet J **37** 126-30. https://doi.org/10.1111/j.1751-0813.1961.tb03877.x.

23 Harris DJ and Nowara G (1995) The characteristics and causes of sheep losses in the Victorian Mallee. Aust Vet J **72** 331-40.

24 Piggin CM (1977) The nutritive value of *Echium plantagineum* L. and *Trifolium subterraneum* L. Weed Res **17** 361-5. https://doi.org/10.1111/j.1365-3180.1977.tb00494.x.

25 Corbett WS, Saylor WW, Long TA et al. (1978) Intracellular distribution of hepatic copper in normal and copper-loaded sheep. J Anim Sci **47** 1174-9. https://doi.org/10.2527/jas1978.4751174x.

26 Suttle N (2010) Copper. In: Mineral nutrition of livestock. 4th ed. CABI Oxfordshire: UK, pp. 255-305. https://doi.org/10.1079/9781845934729.0255.

27 Bull LB (1949) The occurrence of chronic copper poisoning in grazing sheep in Australia. Paper presented to British Commonwealth Scientific Official Conference, Specialist Conference in Agriculture — Australia, February, Melbourne.

28 Pryor WJ (1959) An outbreak of copper poisoning in sheep following copper top-dressing of pastures. Aust Vet J **35** 366-9. https://doi.org/10.1111/j.1751-0813.1959.tb03697.x.

29 Oruc HH, Cengiz M and Beskaya A (2009) Chronic copper toxicosis in sheep following the use of copper sulfate as a fungicide on fruit trees. J Vet Diagn Invest **21** 540-3. https://doi.org/10.1177/104063870902100420.

30 Hosking, WJ, Caple IW, Halpin CG et al. (1986) Trace elements for pastures and animals in Victoria. Victorian Government Printer on behalf of the Department of Agriculture and Rural Affairs: Melbourne.

31 Gracey HI, Stewart TA and Woodside JD (1976) The effect of disposing high rates of copper-rich pig slurry on grassland on the health of grazing sheep. J Agric Sci Camb **87** 617-23.

32 Ishmael J, Gopinath C and Howell J McC (1971) Studies with copper calcium EDTA. J Comp Pathol **81** 279-90.

33 Toxaemic Jaundice Investigation Committee (1956) Toxaemic jaundice of sheep: Phytogenous chronic copper poisoning, heliotrope poisoning, and hepatogenous chronic copper poisoning. Final Report. Aust Vet J **32** 229-36.

34 Caple IW and Heath TJ (1979) Effect of chronic liver damage caused by ingestion of *Heliotropium europaeum* on bile formation in sheep. J Comp Pathol **89** 83-8.

35 Farquharson BC (1984) Some aspects of copper toxicity in sheep grazing New Zealand pastures. Thesis prepared in partial fulfilment of the requirements for the degree of Doctor of Philosophy in Veterinary Science at Massey University, New Zealand. Available at: http://mro.massey.ac.nz/handle/10179/3577. Accessed 2 August 2018.

36 Ortolani EL, Machado CH and Sucupira MC (2003) Assessment of some clinical and laboratory variables for early diagnosis of cumulative copper poisoning in sheep. Vet Hum Toxicol **45** 289-93.

37 Pierson RE and Aanes WA (1958) Treatment of chronic copper poisoning in sheep. J Amer Vet Med Assoc **133** 312-15.

38 Watt B and Maugauret T (2009) Primary copper poisoning in two flocks of first cross ewes: Flock and herd case notes. Available at: http://flockandherd.net.au/sheep/reader/copper-poisoning-primary.html. Accessed 2 August 2018.

39 Suttle NF (2012) Responsiveness of prehaemolytic copper poisoning in sheep from a specific pathogen-free environment to a relatively high dose of tetrathiomolybdate. Vet Rec **171** 246-10. http://dx.doi.org/10.1136/vr.100722.

40 Scanlan CM and Edwards JF (1990) Bacteriologic and pathologic studies of hepatic lesions in sheep. Am J Vet Res **51** 363-6.

18

Diseases of the urinary system

UROLITHIASIS

The obstruction of the urethra or, occasionally, the ureter of male sheep by *calculi* or *uroliths* produces the disease of *urolithiasis* or *calculosis*. When bladder rupture occurs the resultant uroperitoneum gives the condition its common name of *water belly*. The formation of calculi occurs also in ewes but, because of their short, wide urethra, obstruction rarely occurs.

The disease can cause significant economic losses of wethers and rams as a result of urethral obstruction, hydronephrosis, rupture of the urethra or bladder and death from uraemia. The morbidity in wether flocks is generally low but can be as high as 10% annually in exceptional circumstances.[1] The disease in valuable rams or pet sheep may warrant attempts at intensive medical or surgical treatment, but generally intervention is aimed at salvage for slaughter, minimising the number of further cases in the affected group and preventing reoccurrences in the future.

Urolithiasis in Australia occurs usually in one of three circumstances. The first is in association with the feeding of cereal grains at high levels — either as supplementary feeding, feeding for growth (feedlots) or in stubbles — and the consequent imbalance in the dietary ratio of calcium (Ca) and phosphorous (P). Calculi in these cases usually contain phosphates. The second is in sheep grazing grass-dominant or unimproved pastures or cereal stubbles which predispose them to the formation of urinary calculi composed of silicates, oxalates or carbonates, or combinations of these. The third circumstance is in wethers grazing phyto-oestrogenic clovers, in which case the obstructing material is soft and not heavily mineralised, and the underlying pathophysiology is related primarily to hormonal effects on the renal tissue, rather than to chemical instability of the urinary solution.

The development of uroliths

Uroliths are usually mineral concretions formed by the precipitation or crystallisation of mineral salts on an organic matrix. The chemistry of their formation is not well understood but, in nearly all cases where the stones are composed principally of minerals, the instigating factors are the presence of organic material, largely mucoproteins and mucopolysaccharides, formed from urinary tract debris and the oversaturation of the urine with excreted products in solution or colloidal suspension. Urinary concentration has a role in determining the presence of the organic matrix because urinary mucoproteins are increasingly likely to precipitate as the concentration of sodium, potassium, calcium and magnesium increases — as happens when sheep are on dry feed.[2] When the urinary solution becomes chemically unstable the organic material presents a matrix for the precipitation of minerals and the formation of the urolith. The particular chemical composition in each case is influenced by the pH of the urine, the minerals which are being excreted in the urine and their solubility in urine of particular pH — factors which are principally determined by the diet.

Frequently uroliths are a combination of two or more different chemical compounds. Calcium and magnesium are the cations involved in most cases.

Silica, oxalates and carbonate calculi

Several investigations of outbreaks of urolithiasis in grazing sheep in different regions of Australia were carried out and reported in the latter half of last century. In Queensland, the condition was found to be caused most often by calculi of calcium and magnesium carbonate, with silica calculi occurring occasionally. Outbreaks occurred mostly between May and August and involved adult wethers that were in good condition and grazing lush natural pastures.[3] In Western Australia, urolithiasis has been a significant problem for sheep producers in the eastern (drier) parts of the Wheatbelt — a roughly square region running from the west coast, centred approximately on Perth, to about 400 km inland. Calculi composed of silica are more common than those composed of calcium carbonate, although both types occur, sometimes on the same property. The affected animals are most commonly wether weaners grazing oat stubbles or receiving oat grain supplements during the dry season.[4] In a study of urolithiasis in wethers and steers grazing cereal stubbles and native grass pastures in South Australia, silica and calcium oxalate uroliths were the most common types of calculi identified in legume-free pastures, while calcium carbonate uroliths, alone or with silica, were found in sheep grazing clover- or medic-dominant pastures.[1]

Cereal grains, cereal hays and grasses generally, including many Australian native grasses, are high in monosilicic acid and in solid silica. Legumes contain much lower amounts of silica of either form.[5] Sheep on diets based principally on the grass family (*Gramineae*) are at risk of silica urolithiasis — a fact aligned with the observations of the incidence of the condition from both South Australia and Western Australia. Over 96% of ingested silica is excreted in faeces but some is absorbed and excreted in urine as silicic acid. There is an upper limit on silicon absorption, probably related to the concentration of silica when at saturation in the rumen liquor, and therefore an upper limit on the amount of silicon excreted in urine each day.[2,5] The maximum concentration of silicon in the urine in sheep on silicon-rich diets is therefore more related to urinary volumes, as a consequence of water intake, than it is related to dietary silicon ingestion, once a certain upper limit is reached. Under certain conditions, possibly related to water intake and the excretion of other salts, silica can precipitate in the urine, forming calculi composed either mainly of silica or of silica in combination with other minerals.

Oxalate is normally present in urine and is commonly a component of uroliths. Oxalate is present in most forages eaten by sheep and, while some plants can contain a high concentration of oxalate, the reason for the formation of oxalate-containing uroliths is not necessarily a consequence of high dietary intake of the compound. Oxalate is an end-product of the metabolism of a number of compounds — including the amino acid glycine — and most urinary oxalate arises from endogenous sources. Once adapted to high-oxalate diets, ruminants can ingest relatively large amounts of plant oxalate without any increase in urinary oxalate excretion, or other ill effects, because of the microbial breakdown of the compound in the rumen. Carbonate is an end-product of the ruminal degradation of oxalate and it seems likely that plants high in oxalates may predispose sheep to the development of carbonate uroliths, with or without oxalate.[3,6] Calcium carbonate is less soluble in alkaline urine than acidic urine, and carbonate uroliths — most commonly calcium carbonate — tend to occur in alkaline urine.

In sheep which are not adapted to a high-oxalate diet, high levels of calcium oxalate may be excreted in the kidneys. The compound is highly insoluble in urine, regardless of pH, and crystals may form in the renal tubules, causing nephrosis (discussed later in this chapter). Acute oxalate nephrosis is not therefore associated with the chronic ingestion of high oxalate diets, nor is it an indicator of the likelihood of oxalate urolithiasis.

Phosphate calculi

Sheep on diets high in cereal grain and low in roughage are particularly disposed to the development of uroliths composed of salts of phosphate — particularly magnesium ammonium phosphate (struvite) and calcium phosphate (apatite). Amorphous magnesium calcium phosphate (AMCP) has recently been identified as a type of phosphate urolith which may occur as a majority component with struvite or alone.[7]

Cereal grain is relatively high in phosphorus and low in calcium but the development of phosphate calculi in sheep on diets comprising mainly cereal grains is not simply a result of the ingestion of high levels of phosphorus. There are at least three factors which lead to the precipitation of the compounds.

The first of these is the low dietary calcium intake and the low Ca:P ratio in the diet. When the intake of calcium is low, the urinary excretion of phosphorus is raised — an effect mediated by parathyroid hormone and other components of the phosphatonin system which act to decrease the renal tubular reabsorption of phosphorus, increase absorption of phosphorus from the gastrointestinal tract, and increase osteoclastic resorption from bone.[8] Consequently, urinary excretion of phosphate is high (compared to non-cereal diets) and, depending on the urine volume, concentrations of urinary phosphate can become very high.[9] Evidence that the high urinary phosphate is mediated by parathyroid hormone and phosphatonins in response to low dietary calcium, rather than high dietary phosphorus, is provided by the response to the addition of calcium carbonate to the diet. Without changing dietary phosphorus or plasma phosphate levels, urinary excretion of phosphate falls markedly in response to the added calcium.[10]

Second, with diets that are rich in highly digestible feedstuffs, like cereal grains, and low in roughage, sheep ruminate less and produce less saliva. Saliva is rich in phosphate, and the recycled salivary phosphate, added to dietary phosphorus, is available for absorption from the gastrointestinal tract or, if not absorbed, for excretion in faeces. When saliva production declines, faecal excretion of phosphorus also falls. If this results in an increase in plasma phosphorus, urinary excretion becomes a more important mechanism for phosphorus homeostasis. The addition of roughage to the diets of sheep on diets composed principally of cereal grains increases the faecal excretion of phosphorus and reduces the urinary phosphate excretion.[10]

Third, the urine of sheep on roughage diets is usually alkaline and the addition of cereal grains to the diet has the effect of reducing urinary pH, sometimes to levels below 7.0.[9]

If, however, the diet which leads to the excretion of high levels of phosphate is accompanied by an alkaline urine, calcium and magnesium salts of phosphate are more likely to precipitate in the urine. In such cases, reducing the urinary pH alone can reduce the risk of formation of calcium and magnesium phosphate calculi.

Phyto-oestrogens

Grazing of phyto-oestrogenic clovers can lead to urinary tract obstruction in wethers. The obstruction may be due to one or a combination of three materials:

(1) desquamated cells and secretions of accessory glands under oestrogen influence

(2) benzcoumarin calculi in the renal pelvis (*clover stone*) which cause renal fibrosis in ewes and wethers

(3) calcium carbonate in a white organic paste.

Cases of the latter have been reported in outbreak form, with losses up to 20% or more of wethers 10 days after introduction to phyto-oestrogenic pasture in the late spring.[11]

Clinical findings

The presenting clinical signs vary with the degree of development of the condition, as follows:

- *partial obstruction*: inappetence, stretching, dribbling of blood-stained urine, stranguria. Fine calculi particles may be detectable (between the fingers) on the preputial hairs

- *total obstruction*: anorexia, *stretched-out* stance or recumbency, and straining to urinate

- *rupture of urethra*: swelling of subcutaneous tissues of ventral abdominal wall as urine escapes from the ruptured urethra

- *rupture of bladder*: or, more commonly, leakage through the stretched bladder wall, leading to uroperitoneum and abdominal distension, giving the condition its common name of *water belly*. Initially a rupture of the bladder may give temporary relief but, as the uraemia progresses, the animal becomes increasingly depressed, standing or lying isolated from the flock and often still straining.

Diagnostic radiography may be useful in the evaluation of the condition in valuable animals and provide a means to identify the position of the most distal obstructing stone and the number of calculi still present proximal to the obstruction and in the bladder. This information is useful for determining the best surgical approach and the prognosis. Calcium carbonate, calcium oxalate and silica uroliths are radiopaque while AMCP and struvite are often small and difficult or impossible to detect with plain radiographs.[7,12] Contrast media can be used with radiolucent stones — the medium being introduced after amputation of the urethral process.

Ultrasonography is also useful and can identify the distended bladder, hydronephrosis and in many cases, the calculi in the urethra as well as the bladder.[13,14]

Diagnosis

The usual presenting signs of dysuria and stranguria, with the passage of a few drops or trickle of bloodstained urine, are usually sufficient to diagnose the condition. Urinary tract infections and cystitis — which may present some signs similar to urolithiasis — are uncommon in sheep and are not accompanied by bladder distension or urine accumulation in the abdomen or subcutaneous tissues. In those animals with urethral rupture the ventral accumulation of fluid must be differentiated from other causes of ventral oedema, including clostridial infections. In the event that blood tests reveal a uraemia, the presence of uroperitoneum differentiates the uraemia from that of renal failure.

Necropsy findings

At necropsy, urine will be present in large quantities in the ventral subcutaneous tissues, abdominal cavity or both. The bladder will be very distended or, possibly, ruptured. The kidneys show changes of hydronephrosis which occurs as a result of the back-pressure to the kidneys.[13] Apart from the calculus causing the obstruction there are usually numerous further stones in the bladder, urethra, kidneys or ureters.

Treatment

Broadly, cases of urolithiasis in sheep present as either an outbreak involving groups of sheep — usually wethers of commercial value — or involving one or a few valuable animals such as rams or pet sheep. After an initial assessment, the approach to treatment differs substantially between the two scenarios.

In the case of an outbreak involving sheep of (relatively) low value, each affected sheep should be examined and, if the diagnosis of urinary obstruction is confirmed, the animal should be restrained so that the penis can be exteriorised and examined and the urethral process amputated with scissors at its junction with the glans. The urethral process is the narrowest part of the lower urinary tract and therefore is a common site for uroliths to lodge. It may be that a calculus can be seen and palpated in the urethral process but, even if the obstruction is more proximal, this simple and harmless procedure removes the risk of a descending calculus later lodging in this part of the urethra.

Individual assessment of each animal also provides the opportunity to identify those animals which have uroperitoneum, those with urethral rupture and those where immediate euthanasia is the best course of treatment. Euthanasia and necropsy also allow the collection of calculi for later determination of their chemical content — information which is helpful in understanding the cause of the outbreak and in determining the best way to manage the remainder of the flock.

Generally, if some sheep in the flock are obstructed, others in the same management group are at high risk of developing an obstruction. Attention should then be given next to reducing the risk, or the number, of further cases.

Non-surgical treatments

Whatever the particular dietary conditions are that have led to this outbreak, and whatever the chemistry leading to calculus formation, there is one preventive strategy which is very unlikely to do harm and which will assist in many cases. That strategy is to attempt to increase the water intake of the remaining sheep in the flock. Both the availability — adequate trough space, to reduce competition for space when drinking, and distance from grazing areas to water — and water quality should be assessed. If either is lacking, efforts should be made immediately to improve the flock's access to good-quality drinking water. The addition of salt (NaCl) to the diet such that they ingest 50 to 100 g of NaCl daily (for an adult sheep) may stimulate water intake. Allowance should be made for the salt content of the drinking water. Excess salt may reduce feed intake in sheep unaccustomed to high salt levels.

The second and equally urgent step is to adjust the diet in an attempt to prevent the formation of more calculi — or the enlargement of existing calculi — in other sheep in the flock which are likely to have formed uroliths in their bladders but have not yet obstructed.

For phosphate calculi which have occurred in sheep on cereal grain diets, adding calcium carbonate or legume hay to the diet, changing the diet completely, acidifying the urine or any successful strategies to increase urine flow may reduce the continuing incidence of obstruction in the flock at risk.

Calcium carbonate, in the form of ground limestone and at the rate of 1.5% to 2% of the diet (20 kg per tonne of grain), should be included in all cereal-grain-based diets to reduce the risk of phosphate calculi as well as to ensure adequate calcium for other physiological functions (see Chapter 13 for a discussion of osteodystrophies). The diet should contain calcium and phosphorus in the ratio of 2:1 or higher (at least twice as much calcium by weight as phosphorus)[15] and cereal grains often have Ca:P ratios much lower in calcium than 1:1.

On the basis that phosphate calculi, such as struvite, form in alkaline urine, urine acidifiers can be included in the diet or administered orally. Ammonium chloride is a potent urine-acidifying agent and will dramatically reduce the risk of formation of phosphate uroliths.[15,16] The addition of ammonium chloride to the diet of the flock in which some cases of obstruction have occurred may be helpful but cases can be expected to continue for some weeks caused by uroliths already formed in the urinary tract before the urine-acidifying strategy commenced.

As a preventive treatment ammonium chloride should be included in the diet at the rate of 0.5-1.5% of the feed — for example, 5-15 g per kg of feed on a dry matter basis. The higher rate has been more effective than the lower rate in some studies but caution should be exercised in using rates above 1.5% or using rates of 1% for extended periods (three months or more). With very high rates the feed becomes relatively unpalatable[15] and there is also a risk of inducing metabolic acidosis[17] and decreasing bone density.

Ammonium chloride can be administered as a drench, with 10 g dissolved in 40 mL of water as a dose for an adult sheep, but treatment needs to be given daily.

For calcium carbonate, silica, oxalate and phyto-oestrogenic calculi, after ensuring adequate water intake, changing the diet may be the only nonsurgical strategy which may help. The provision of access to legume-dominant pastures or legume hay may be helpful.

Unfortunately, in a flock which has been exposed to the conditions leading to urolithiasis in some animals, deaths may continue despite control measures, and the owner may choose to sell the unaffected sheep to slaughter before further cases occur. Because of the risks of urolith formation in sheep grazing cereal stubbles, many producers prefer to use ewes rather than wethers (or rams) to 'clean up' after harvest.

Surgical treatments

When dealing with animals of relatively high value, if amputation of the urethral process has not relieved the obstruction then more complex medical or surgical procedures can be considered. These have been recently reviewed and described by Videla and van Amstel (2016).[17] The technique for tube cystotomy appears to offer a reliable surgical solution for valuable animals.[18]

Urethrostomy alone may provide temporary relief but without effective efforts to remove further calculi from the bladder, closure of the urethral stoma and recurrence of obstruction is likely.[19] An ascending pyelonephritis can develop and further reduce the chances of recovery. In all cases where surgery (beyond urethral process amputation or urethrostomy) is contemplated, attempts should be made to evaluate the state of the kidneys because hydronephrosis may lead to mortality even after the calculi are successfully removed. When continuing veterinary supervision of surgical interventions is not economically justified, euthanasia is often the best solution.

POSTHITIS

Also called balanoposthitis, enzootic posthitis or pizzle rot, the disease is an inflammation of the prepuce which extends in some cases to the penis, caused by *Corynebacterium renale* and, probably, other *Corynebacterium* spp under particular dietary conditions. The disease affects wethers, particularly mature wethers, and young rams. The lower incidence in mature rams is probably associated with the separation of the penile and preputial epithelial surfaces as rams mature. (A balanitis (knob rot) of unknown aetiology has also been reported in rams.) The organism has also been isolated from vulvitis of ewes and posthitis in goat wethers.

Development of lesions

C renale is capable of hydrolising urea and proliferates more rapidly when higher concentrations of urea are present. The tissue necrosis may be caused by ammonia rather than the organism itself. Urinary concentration of urea is raised when sheep graze high-protein pastures such as lush, legume-dominated pastures. Under conditions favourable for the development of the disease, up to 60% of wethers may be affected with clinically significant posthitis.[20]

The disease has two distinct stages. The mild, early stage is an *external ulceration* of the skin near the preputial orifice. Scabs develop over these ulcers. After a period of weeks or months, extension of the lesion within the prepuce may occur, leading to the second stage of *internal ulceration*. In animals with internal ulceration, the prepuce swells and the passage of urine may be difficult. Necrotic material and pus accumulate inside the prepuce and discharges accumulate in the belly wool. The area becomes attractive to flystrike. The serious losses of production which make the disease of economic significance are associated with the development of internal lesions.

Clinical signs

Sheep that only have external lesions show no clinical signs other than a mild, superficial necrosis of the skin of the preputial orifice which is usually covered in a brown scab. Clinical signs of the internal lesions include swelling of the sheath, dribbling of urine, restlessness, irritation or pain, humpbacked stance and frequent recumbency. The wool around the prepuce is heavily stained and the sheep may be flystruck on the belly. In advanced cases, the penis becomes involved (balanitis) and the urethra may become obstructed. The major economic consequences of posthitis are from deaths from severe lesions and from flystrike. Losses of wool production and liveweight gain are probably small except in the most severe cases.[21]

Prevention

Testosterone injections are the preferred prophylactic treatment. A number of testosterone products are available. For best effect they should be given immediately before a high-risk period. In one experiment, 75 mg of testosterone enanthate liquid (Tesgro, MSD Agvet) and 75 mg of testosterone cypionate liquid (Banrot, Pitman Moore) produced higher blood levels of a shorter duration than 70 mg of testosterone propionate pellets (Ropel, Schering Pty Ltd) implanted subcutaneously which provided adequate preventive levels for at least 10 weeks.[22] Testosterone treatment will only slightly reduce the incidence of external lesions but will markedly reduce the incidence of internal ulceration. There are side effects to treatment with testosterone. Two doses of 70 mg of testosterone propionate (autumn and spring) will increase greasy fleece weight by up to 7%, decrease yield, increase clean fleece weight by up to 4%, increase mean fibre diameter by about 1% and increase liveweight by 2 kg or more[23] independent of effects on production which occur as a result of controlling pizzle rot.

Prophylactic testosterone should be given to wethers grazing clover-dominant pastures in spring and possibly in autumn, at the start of the seasonal pasture growth.

The removal of wool from the entire belly by shearing is known to reduce the incidence of posthitis but *ringing* — the shearing of wool just around the preputial orifice which is commonly performed when wethers are crutched to prevent the contamination of wool near the orifice and subsequent flystrike — has little effect on the incidence of pizzle rot.[24]

Pizzle dropping — the surgical lowering of the preputial orifice by incising the skin fold between the abdominal skin and the prepuce — has also been claimed to reduce the incidence of pizzle rot. Pizzle dropping, if carried out properly, will dramatically reduce urine staining of wool and the incidence of flystrike of the belly.[25] There may be an additional benefit of reducing the incidence of severe pizzle rot.[26]

Treatment

Dietary change

For cases of external ulceration, a change to a low-protein diet will reduce the prevalence of lesions.

Local treatment

The wool around the prepuce should be removed, the scabs removed, and the prepuce irrigated with antiseptics. The installation of 2-3 mL of antiseptic solution into the sheath from a plastic squeeze bottle and then the smearing of the antiseptic around the orifice as it drains have been shown to be effective. Treatment should be repeated every two or three days until the lesion has resolved. Originally, an alcoholic solution of 20% cetrimide was recommended but this product is no longer readily available. Chlorhexidine digluconate in 0.8% solution (undiluted Hibitane, Coopers) is also effective. It may be less effective than cetrimide but is also less irritant to tissues. Cetrimide should not be used in rams and should not be applied to the vagina of ewes.[27]

Parenteral treatment

The causative organism is sensitive *in vitro* to penicillin and slightly sensitive to oxytetracycline and nitrofurantoin. It is less sensitive to erythromycin, neomycin and streptomycin[28]: 500 mg of oxytetracycline *per os* daily for four days is ineffective *in vivo*.[27] Procaine penicillin parenterally is reportedly effective in resolving lesions[29] but should be given in conjunction with local therapy, if at all.

Testosterone, at double the dose recommended for prophylaxis, is recommended for treatment of internal ulceration. Lower doses have shown marginal efficacy when given alone. In one experiment, 90 mg of testosterone propionate was 49% effective in treatment and prevention, while 90 mg testosterone plus local treatment with cetrimide were 88% effective.[20]

Surgical drainage

Slitting the prepuce to provide drainage is an effective treatment of internal ulceration even in advanced cases. The incision is either a single ventral slit posteriorly from and including the preputial orifice or the removal of a V-shaped section of the ventral prepuce. If the single slit is performed, the incision should extend to the tip of the urethral process but not too far posteriorly to avoid exposing the penis to trauma and drying conditions.[29] Wool adjacent to the incision may become urine-stained and attractive to strike flies.

ENLARGEMENT OF BULBO-URETHRAL GLANDS IN WETHERS

This occurs in wethers, but not rams, grazing oestrogenic pastures. The glands undergo hyperplasia and the wethers suffer a loss in condition and, in some cases, death. Testosterone is protective of the oestrogenic effect.[30,31]

DISEASES OF THE KIDNEY

Renal diseases of the sheep are not common and, with the exception of some plant and chemical intoxications, tend to occur very sporadically. Ovine renal disease can be classified, on the basis of aetiology, into four major classifications (Table 18.1).

Congenital malformations

These are uncommon in sheep. The most common urinary tract defect in a Western Australian survey of dead lambs was agenesis of one or both kidneys.[32]

Immunologically mediated glomerulonephritides

In sheep, glomerulonephritis is an inflammatory disease resulting from the deposition of antigen-antibody complexes along the basement membrane of the glomerular capillaries. The most common form in sheep is membranous glomerulonephritis[34] or membranoproliferative glomerulonephritis.[33] The disease occurs only sporadically, usually in adult sheep, but has been reported in lambs with coccidiosis.[33] A genetically controlled congenital glomerulonephritis (mesangiocapillary glomerulonephritis or MCGN) has been reported in Finnish Landrace sheep and their crosses. Lambs with MCGN are clinically normal at birth but within a few weeks stop sucking, often demonstrating signs of a secondary encephalopathy. At examination either pre- or post-mortem, the gross renal enlargement is the outstanding sign; the kidneys may be up to six times normal size.[34]

Infectious nephropathies

A number of bacterial species can cause infections, particularly abscessation, of the kidney following haematogenous spread. These include *Corynebacterium pseudotuberculosis*, *Streptococcus* spp, *Staphylococcus* spp and other pyogenic bacteria which may enter the body through wounds or by navel infection and cause simultaneous infections in other organs and joints. Renal lesions may also occur in lambs infected with pestivirus (hairy shaker lambs).

Ascending infections of the kidney are uncommon in sheep but may occur in animals with urolithiasis and in lambs with navel infections.

Toxic nephropathies

Toxins exert their effects on the tissues of the nephron in two ways. First, they may do so by inducing ischaemia, as many bacterial endotoxins do as part of a generalised process of capillary endothelial damage; and, second, by direct effect on the sensitive tissues of the nephron. The proximal tubule is particularly susceptible to this form of damage, possibly because it is highly specialised and it is exposed to toxins which are reabsorbed after glomerular filtration. The primary lesion in most forms of toxic nephrosis is tubular; glomerular dysfunction may follow.

The epsilon toxin of *Clostridium perfringens* type D is particularly lethal to the renal tubule. The destruction of the renal tubule is an important component of the pathogenesis and post-mortem diagnosis of enterotoxaemia in sheep.

Table 18.1: Classification of ovine renal disease. Adapted from Angus (2000).[34]

Classification	Examples
Congenital malformations	Renal agenesis Hydronephrosis Cystic and polycystic kidneys
Immunologically mediated glomerulonephritides	Membranous glomerulonephritis Mesangiocapillary glomerulonephritis of Finnish Landrace sheep
Infectious nephropathies	Embolic nephritis and renal abscessation
Toxic nephropathies	Enterotoxaemia Plant intoxications Chemical intoxications

Isotropis spp poisoning (lamb poisons)

There are nine recorded species of this plant genus; six of them occur in WA where most cases of intoxication have been recorded. Of the nine species, five have been associated with toxicity. *Isotropis atropurpurea* occurs in all states except Queensland and has caused toxicity in WA, NT and NSW. In WA, poisoning has been recorded in the northwest desert regions, the southwest agricultural region and the pastoral zones.[35] Poisoning occurs most commonly in spring when green suckers occur. The plants appear to be very palatable.[36]

The experimental intoxication of sheep with a bolus dose of 300 g of dried plant material of *Isotropis forrestii* has been described.[37] Clinical signs included weakness, depression, anorexia, diarrhoea and oliguria. Some animals showed clinical signs on day one; all were anorexic by day five. At necropsy, the kidneys were pale, surrounded by gelatinous oedema extending into the mesenteries; some had ascites and hydrothorax. Histologically there was severe renal tubular damage and mild to moderate gastroenteritis (possibly through uraemia rather than a direct effect of the toxin). The tubular damage appeared to be a result of a direct effect of the toxin rather than one mediated by ischaemia. Peracute deaths have also been reported from ingestion of *Isotropis* spp. The toxic principle has not been identified.

Oxalate nephrosis

Some weeds and cultivated plants which contain oxalates and which are occasionally associated with cases of oxalate nephrosis in sheep are listed in Tables 18.3 and 18.4.

Table 18.2: Toxic nephropathies of sheep.

Source of toxin	Form of damage	Comment
Metallic salts (As, Hg, Pb, Cd and organic copper compounds)	Generally a direct toxic effect on renal tubular cells.	Most cause damage to alimentary tract and other organs in addition to nephrosis.
Paraquat and diquat	Acute renal tubular necrosis, plus injury to alimentary tract and lung.	Concentrate particularly dangerous. When sprayed on pasture (*Gramoxone*, ICI) generally too dilute to be toxic.
Aminoglycoside and tetracycline antibiotics	Aminoglycosides potentially nephrotoxic; overdosage of tetracyclines in lamb may cause kidney damage.	Aminoglycosides also potentially ototoxic and can cause neuromuscular blockade and hypocalcaemia.
Oxalate-containing plants	Oxalate ions damage numerous tissues; crystal precipitate in renal tubules.	Soursob is the principal plant in medium rainfall zones; chenopodiaceous shrubs in inland regions.
Tannin-containing plants	Unknown pathogenesis.	Yellowwood and other plants.
Anagallis arvensis ingestion	Coagulative necrosis of proximal renal tubular epithelium.	Commonly called scarlet pimpernel. Suspected case of poisoning in NSW Riverina in 1983-84 but feeding trial failed to reproduce disease.[38]
Ipomoea spp plants	Chronic ingestion over five weeks leads to neurological signs and kidney lesions.	Commonly called weir vine (Qld) or morning glory (WA).
Isotropis spp plants	Unknown toxic principle strongly nephrotoxic.	Plants commonly called *lamb poisons*. Poisoning occurs in pastoral areas of Aust.

Table 18.3: Weeds with high levels of oxalates.

Common name	Scientific name
Family *Oxalidaceae*	
Soursob	*Oxalis pes-caprae*
Family *Chenopodiaceae*	
Soft roly poly	*Salsola kali*
Black pigweed	*Trianthema portulacastrum*
Soda bush	*Threlkeldia proceriflora*
Family *Polygonaceae*	
Sheep sorrel	*Acetosella vulgaris*
Dock	*Rumex* spp
Spiny emex	*Emex australis*
Family *Grystomaceae*	
Pigweeds	*Portulaca* spp
Family *Aizoaceae*	
Slender iceplant	*Mesembryanthemum nodiflorum*

Table 18.4: Cultivated grasses which generally have only moderate levels of oxalate.

Common name	Scientific name
Setaria	*Setaria sphacelata*[39]
Panic	*Panicum antidotale, P maximum*
Buffel grass	*Cenchrus ciliaris*
Kikuyu grass	*Pennisetum clandestinum*

Soursob occurs extensively in southern NSW, northern Victoria, SA and southwest WA. Many of these areas are of medium rainfall (around 480 mm per annum) and are populated by many sheep. The chenopodiaceous plants are saltbushes and succulents, which are prevalent mainly in the inland areas of Australia and are grazed extensively by sheep. The cultivated grasses occur chiefly in Queensland.

Seawright (1982)[40] describes the pathogenesis of plant oxalate poisoning as follows: ruminants are able to graze these plants in most circumstances without oxalate intoxication. Soluble oxalate is rapidly broken down to carbonate and formate in the rumen through microbial action, or else precipitated as insoluble calcium oxalate in the rumen contents. Both processes prevent or reduce the uptake of oxalate into the ruminal mucosa and into the bloodstream. If substantial amounts of oxalate are absorbed, calcium oxalate will precipitate in submucosal arterioles and damage the rumen mucosa. Similar damage may occur in lung capillaries, causing pulmonary oedema. In the kidneys, oxalate is secreted by the tubular epithelial cells, causing them damage, and calcium oxalate crystals are precipitated in the tubular lumina.

The plasma calcium concentration falls in the presence of high plasma oxalate levels and clinical signs of hypocalcaemia may occur. Animals dying from acute oxalate poisoning often have terminal pulmonary oedema, with copious amounts of stable froth in the trachea and bronchi and marked hyperaemia of the mucosa of the fore-stomachs and of the lower alimentary tract. Kidneys are usually swollen and the cortex pale.

Acute oxalate poisoning usually occurs when hungry stock are placed on pasture dominated by oxalate-containing plants which are, at the time, high in oxalate levels. Fasting or low-calcium diets prior to exposure predispose the sheep to intoxication. Fasting reduces the numbers of methanogenic bacteria which utilise formate, and if formate is not metabolised oxalate breakdown itself is inhibited. Damage to the ruminal mucosa can be followed by fungal or bacterial invasion, leading to acute rumenitis and intractable indigestion. Animals acutely or subacutely intoxicated with oxalate respond only transiently or not at all to calcium borogluconate therapy and death occurs due to severe metabolic disturbance complicated by renal, pulmonary and gastrointestinal injury. Three outbreaks of acute oxalate poisoning by buffel grass (*Cenchrus ciliaris*) have been described in which 5%, 1% and 18% of the mob died within 12 to 120 hours of exposure to the pasture, following two days of starvation. Clinical signs included dyspnoea, collapse, loss of consciousness and death without struggling.[41] Slender iceplant (*Mesembryanthemum nodiflorum*) has also been associated with acute poisoning in WA.[42]

Chronic oxalate toxicity usually occurs in animals grazing soursob (*Oxalis pes-caprae*). Animals grazing soursob develop a high degree of tolerance to ingested oxalate. There is, however, usually a low level of oxalate absorption with progressive renal injury and deposition in the kidneys of calcium oxalate crystals. Over a period of 2 to 12 months of grazing on such pastures, animals gradually develop renal failure and the occasional death occurs. At necropsy, the kidneys are small, white and fibrotic with large amounts of crystals.

REFERENCES

1   McIntosh GH, Pulsford MF, Spencer WG et al. (1978) A study of urolithiasis in grazing ruminants in South Australia. Aust Vet J **50** 345-50.
2   Nottle MC and Armstrong JM (1966) Urinary excretion of silica by grazing sheep. Aust J Agric Res **17** 165-73. https://doi.org/10.1071/AR9660165.
3   Sutherland AK (1958) Urinary calculi in sheep. Aust Vet J **34** 44-6. https://doi.org/10.1111/j.1751-0813.1958.tb05834.x.
4   Bennets HW (1956) Urinary calculi of sheep in Western Australia. J Dept Agric West Aust **5** 421-33.
5   Jones LHP and Handreck KA (1965) The relation between the silica content of the diet and the excretion of silica by sheep. J Agric Sci **65** 129-34. https://doi.org/10.1017/S0021859600085439.
6   Manning RA and Blaney BJ (1986) Epidemiological aspects of urolithiasis in domestic animals in Queensland. Aust Vet J **63** 423-4.
7   Jones ML, Gibbons PM, Roussel AJ et al. (2017) Mineral composition of uroliths obtained from sheep and goats with obstructive urolithiasis. J Vet Intern Med **31** 1202-8. https://doi.org/10.1111/jvim.14743.
8   Hardcastle MR and Dittmer KE (2015) Fibroblast growth factor 23: A new dimension to diseases of calcium-phosphorus metabolism. Vet Pathol **52** 770-84. https://doi.org/10.1177/0300985815586222.
9   Scott D (1972) Excretion of phosphorus and acid in the urine of sheep and calves fed either roughage or concentrate diets. Quarterly J Exper Physiol **57** 379-92. https://doi.org/10.1113/expphysiol.1972.sp002174.

10 Godwin IR and Williams VJ (1982) Urinary calculi formation in sheep on high wheat grain diets. Aust J Agric Res **33** 843-55. https://doi.org/10.1071/AR9820843.

11 Gardiner MR, Nairn ME and Meyer EP (1966) Urinary calculi associated with oestrogenic subterranean clover. Aust Vet J **42** 315-20. https://doi.org/10.1111/j.1751-0813.1966.tb04728.x.

12 Kinsley MA, Semevolos S, Parker JE et al. (2013) Use of plain radiography in the diagnosis, surgical management and postoperative treatment of obstructive urolithiasis in 25 goats and 2 sheep. Vet Surg **42** 663-8. https://doi.org/10.1111/j.1532-950X.2013.12021.x.

13 AlLugami A, von Pückler K, Wehrend A et al. (2017) Sonography of the distal urethra in lambs. Acta Vet Scand **59** 16. https://doi.org/10.1186/s13028-017-0283-2.

14 Scott P (2017) Abdominal ultrasonography as an adjunct to clinical examination in sheep. Small Ruminant Res **152** 132-43.

15 Bushman DH, Embry LB and Emerick RJ (1967) Efficacy of various chlorides and calcium carbonate in the prevention of urinary calculi. J Anim Sci **26** 1199-1204. https://doi.org/10.2527/jas1967.2651199x.

16 Crookshank HR (1970) Effect of ammonium salts on the production of ovine urinary calculi. J Anim Sci **30** 1002-4. https://doi.org/10.2527/jas1970.3061002x.

17 Videla R and van Amstel S (2016) Urolithiasis. Vet Clin Food Anim **32** 687-900. https://doi.org/10.1016/j.cvfa.2016.05.010.

18 Ewoldt JM, Jones ML and Miesner MD (2008) Surgery of obstructive urolithiasis in ruminants. Vet Clin Food Anim **24** 455-65.

19 Van Weeren PR, Klein WR and Voorhout G (1987) Urolithiasis in small ruminants. I. A retrospective evaluation of urethrostomy. Vet Quarterly **9** 76-9. https://doi.org/10.1080/01652176.1987.9694078.

20 Southcott WH (1962) The prevention and treatment of ovine posthitis with testosterone propionate. Aust Vet J **38** 33-41. https://doi.org/10.1111/j.1751-0813.1962.tb08715.x.

21 Osborne HG, Bentley JB and Pearson IG (1989) Economic aspects of balanoposthitis in Merino wethers. In: Proceedings of the annual conference of the Australian Veterinary Association (Australian Advances in Veterinary Science). Australian Veterinary Association: Artarmon, Sydney.

22 Sacket DM, Wright PJ and Darvill FM (1987) Plasma testosterone concentrations in wethers treated with commercial preparations of testosterone. Aust Vet J **64** 386-7. https://doi.org/10.1111/j.1751-0813.1987.tb09613.x.

23 Osborne WB (1968) The effect of testosterone on the components of fleece weight in Merino wethers. Proc Aust Soc Anim Prod **7** 407-12.

24 Southcott WH (1965) Epidemiology and control of ovine posthitis and vulvitis. Aust Vet J **41** 225-34. https://doi.org/10.1111/j.1751-0813.1965.tb04571.x.

25 Marchant RS and Burbidge SG (1986) Effect of pizzle dropping and testosterone propionate on production of Merino wethers. Proc Aust Soc Anim Prod **16** 419-20.

26 Watts TJ (1986) Pizzle dropping. In: Mackinnon Project newsletter. May, University Melbourne, p. 4.

27 Swan RA (1971) Treatment of posthitis in sheep. Vet Rec **88** 304-5. https://doi.org/10.1136/vr.88.12.304.

28 Southcott WH (1965) Etiology of ovine posthitis: Description of a causal organism. Aust Vet J **41** 193-200. https://doi.org/10.1111/j.1751-0813.1965.tb01830.x.

29 Dent CHR (1971) Ulcerative vulvitis and posthitis in Australian sheep and cattle. Vet Bull **41** 719-23.

30 Gardner JJ and Adams NR (1986) The effect of zeranol and testosterone on Merino wethers exposed to highly oestrogenic clover pasture. Aust Vet J **63** 188-90. https://doi.org/10.1111/j.1751-0813.1986.tb02972.x.

31 Chamley WA, Findlay JK and Nairn ME (1977) The effect of testosterone and an anti-oestrogen on hypertrophy of bulbourethral glands of wethers grazing oestrogenic pastures. Aust Vet J **53** 476-7. https://doi.org/10.1111/j.1751-0813.1977.tb05466.x.

32   Dennis SM (1979) Urogenital defects in sheep. Vet Rec **105** 344-7. https://doi.org/10.1136/vr.105.15.344.

33   Majid HN and Winter H (1986) Glomerulonephritis in lambs with coccidiosis. Aust Vet J **63** 314-16. https://doi.org/10.1111/j.1751-0813.1986.tb08081.x.

34   Angus KW (2000) Diseases of the urinary system. In: Diseases of sheep, ed WB Martin and ID Aitken. 3rd ed. Blackwell Scientific Publications: London.

35   Gardiner MR and Royce RD (1967) Poisoning of sheep and cattle in Western Australia due to species of *Isotropis* (Papilionaceae). Aust J Agric Res **18** 505-13. https://doi.org/10.1071/AR9670505.

36   Seawright AA (1982) *Isotropis* spp in chemical and plant poisons. In: Animal health in Australia. Vol 2. Australian Bureau of Animal Health: Canberra, p. 53.

37   Cooper TB, Huxtable CR and Vogel P (1986) The nephrotoxicity of *Isotropis forrestii* in sheep. Aust Vet J **63** 178-82. https://doi.org/10.1111/j.1751-0813.1986.tb02968.x.

38   Rothwell JT and Marshall DJ (1986) Suspected poisoning of sheep by Anagallis arvensis (scarlet pimpernel). Aust Vet J **63** 316. https://doi.org/10.1111/j.1751-0813.1986.tb08082.x.

39   Seawright AA (1970) An outbreak of oxalate poisoning in cattle grazing Setaria sphac*elate*. Aust Vet J **46** 293-6. https://doi.org/10.1111/j.1751-0813.1970.tb07900.x.

40   Seawright AA (1982) Oxalates in chemical and plant poisons. In: Animal health in Australia. Vol 2. Australian Bureau of Animal Health: Canberra, p. 74.

41   McKenzie RA, Bell AM, Storie GJ et al. (1988) Acute oxalate poisoning of sheep by buffel grass (*Cenchrus ciliaris*). Aust Vet J **65** 26. https://doi.org/10.1111/j.1751-0813.1988.tb14926.x.

42   Jacob RH and Peet RL (1989) Acute oxalate toxicity of sheep associated with slender iceplant (*Mesembryanthemum nodiflorum*). Aust Vet J **66** 91-2. https://doi.org/10.1111/j.1751-0813.1989.tb09752.x.

19

DISEASES OF THE BLOOD AND LYMPHATIC SYSTEM

CASEOUS LYMPHADENITIS (CLA)

This disease is a chronic bacterial infection, predominantly of sheep and goats, characterised by abscessation of lymph nodes, lungs and other visceral organs. It is caused by *Corynebacterium pseudotuberculosis* (formerly *Corynebacterium ovis*). The disease is very common in the small ruminant populations of Australia, New Zealand, South Africa, Canada, USA, South America and some European countries. It was introduced into the UK, presumably with imported goats, in the late 1980s.[1]

In Australia, clinical signs of CLA are rarely seen unless there is detectable enlargement of the superficial lymph nodes, a superficial abscess not associated with a lymph node or a discharging fistula from a CLA abscess. Because of its largely subclinical nature, most sheep producers are unaware of or relatively unconcerned about the condition despite the often very high within-flock prevalence. Subclinical losses of productivity are presently the major form of direct loss to individual producers but the more important source of economic loss to the Australian sheep industry is the condemnation of carcases at abattoirs (Figure 19.1), for which CLA is the most common reason. The cost of the disease in Australia is estimated to be around $18m.[2,3,4] The prevalence of infection in adult sheep sent to Australian abattoirs is between 26% and 50%.[5,6]

In North America, the prevalence of infection is also high — around 40%.[7] The condition has previously been considered a common cause of the condition described as *thin ewe syndrome* there[8] but a more recent Canadian study has found no relationship between infection and body condition score in sheep examined at slaughter.[9] Emaciation is not a typical manifestation of the disease in Australia but occurs in some cases, particularly when visceral organs are infected.

C pseudotuberculosis exists in two biotypes, each with different biochemical properties. With occasional exceptions, the nitrate-reducing biotype infects cattle and horses, while the nitrate-negative biotype infects sheep and goats.[10] Infections have been recorded less commonly in other animal species. Infection of humans — causing a chronic lymphadenitis — also occurs and is most commonly reported in people who work with or handle sheep or sheep carcases.[11] Presumably, the disease transmission from sheep to humans occurs most readily when any wounds on the hands or arms are exposed to abscess material from sheep. Ingestion of raw sheep or goats' milk has also been suggested as an explanation for cervical lymphadenitis in a human.

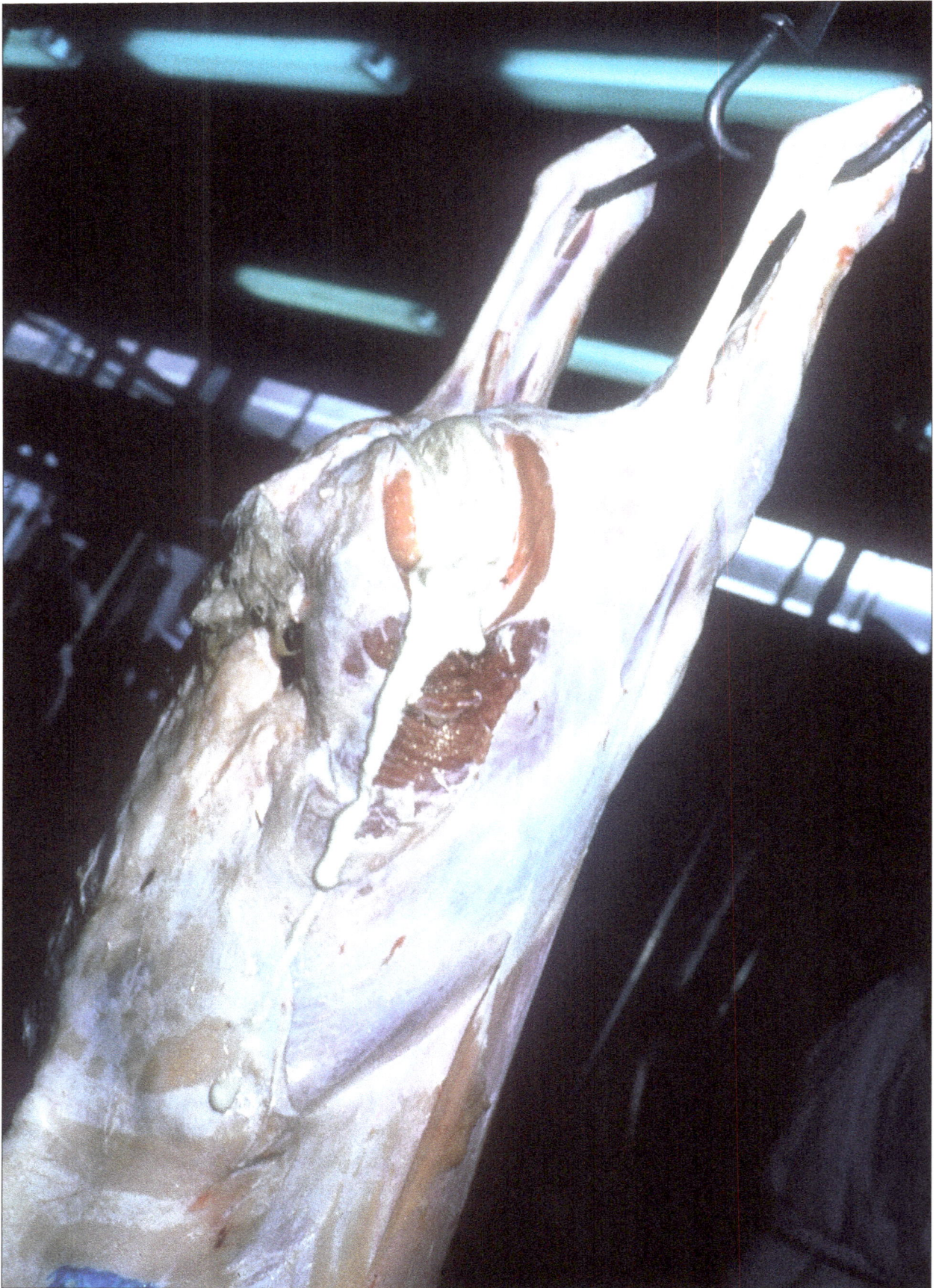

Figure 19.1: CLA abscessation is a major cause of condemnation of sheep carcases. Photograph courtesy of Michael Paton.

Pathogenesis

C pseudotuberculosis can cause CLA lesions in peripheral lymph nodes and in the lungs of sheep when applied to skin wounds or to intact, recently shorn or, possibly, non-woolled skin. Skin lesions develop first, and then spread occurs either through lymphatic drainage or haematogenously.

The pathogenicity of *C pseudotuberculosis* is related to an exotoxin and a surface lipid which is leucotoxic. The local effect of the exotoxin is to cause intense dermal inflammation and necrosis and to increase vascular permeability.[12] These effects presumably promote the invasiveness of the organism and its movement to regional lymph nodes from a primary skin infection. The organisms are transported to regional lymph nodes in phagocytes in which they multiply, and which they ultimately destroy. The surface lipid possibly provides the bacterium with the ability to resist the antibacterial activity of the phagocytes.

In some cases (perhaps 25% to 50%) the inflammatory responses successfully prevent the development of infection beyond the cutaneous lesion. The ability to successfully resist infection is evidenced by observations that serological prevalence exceeds lesion prevalence within flocks.[5] In most challenged sheep, however, a local lymphadenitis occurs, followed by suppuration and destruction of the lymph node. Natural resolution of infection from this stage is unusual. The abscesses become encapsulated and the inflammatory reaction in surrounding tissues subsides. The abscesses usually continue to enlarge with progressive necrosis and reformation of the capsule — producing an *onion-ring* effect of concentric layers of necrotic material. In the early stages of abscessation the pus is a green-tinged yellow colour but, with age, it loses the green colour and becomes inspissated and *cheesy*.[13] Infected lymph nodes are considerably enlarged, commonly becoming 5-15 cm in diameter. The lymph nodes most commonly affected are the prefemoral (precrural), prescapular, mediastinal and bronchial (Figure 19.2).

Continued enlargement of the abscesses may lead to their rupturing externally if superficial nodes are involved. Extension from mediastinal and bronchial lymph node lesions may lead to lung involvement with formation of abscesses in the lung tissue. Pulmonary lesions can also be in the form of extensive bronchopneumonia with numerous foci of caseation and pleuritis with adhesions. Lung infections presumably originate from haematogenous spread and local extension rather than from inhalation; most early lung lesions have no obvious connection to the air passages. Some 15% to 20% of affected sheep have lesions only in the lungs; the mechanism by which lung infection develops without peripheral lymphadenitis remains unclear.

CLA of the inguinal and other scrotal lymph nodes of rams is a frequent finding when rams are examined. The lesions are readily palpable but are not connected to the testes or epididymis, usually lying in the fascia adjacent to the spermatic cord high in the neck of the scrotum. Semen quality is normal and the organism is not excreted in the semen.[14]

Extension of infection to other organs occasionally occurs, involving liver, spleen and kidney. The presence of visceral lesions of CLA (including thoracic lesions) has been associated with thin ewe syndrome in the United States but, in these reports, lesions in the bronchial nodes and lungs still predominate over lesions in other organs and CLA was not invariably associated with the syndrome.[7,15] CLA has not been associated with such a syndrome in Australia.

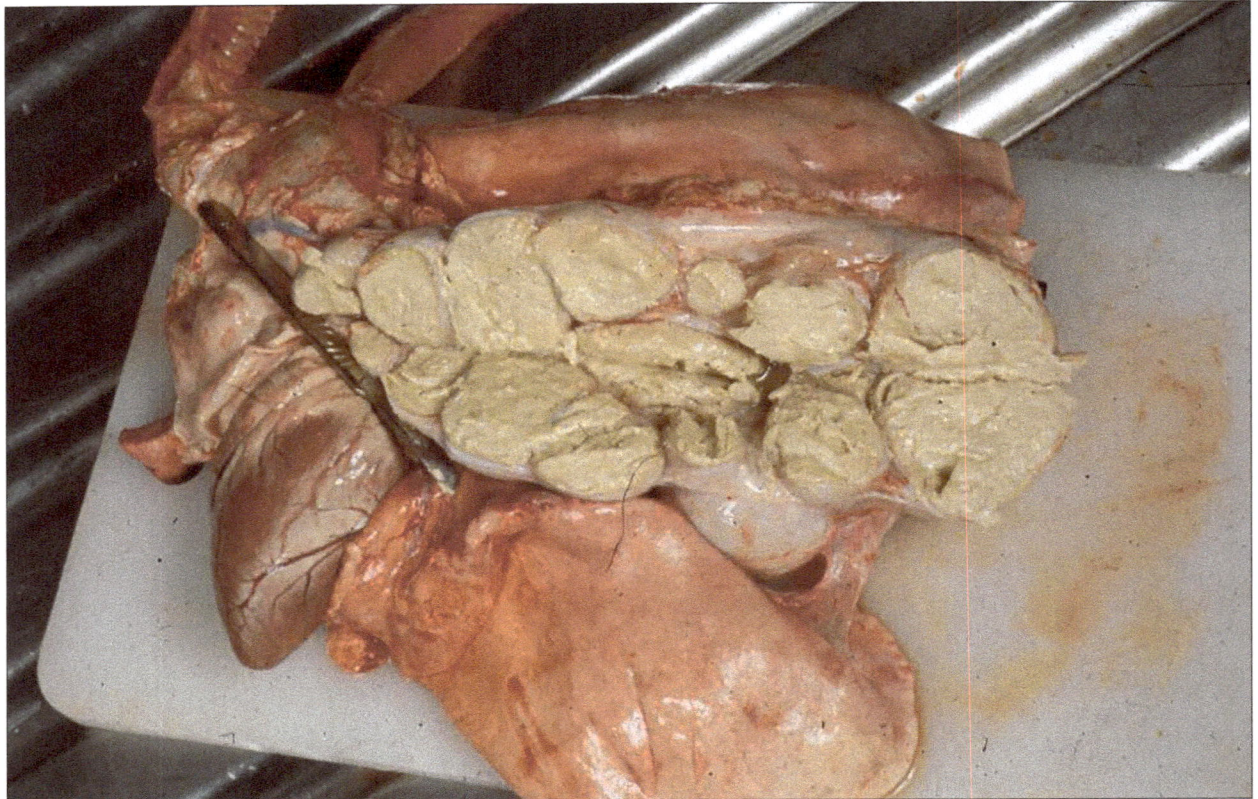

Figure 19.2: CLA abscesses in the mediastinal lymph nodes. Photograph courtesy of Michael Paton.

Epidemiology

The organism can survive for several weeks in faeces and on fomites. The major source of infection for uninfected sheep, however, is infected sheep with lung lesions or with lesions discharging externally. Under Australian farm conditions, lung-infected sheep are probably the most important source of new infection. The ability of infected sheep with no externally discharging CLA abscesses to infect freshly shorn uninfected sheep when confined together has been clearly demonstrated.[16,17] Lung abscesses examined at necropsy are frequently observed to have fistulae connecting the abscess to an air passage and the organisms have been isolated from the tracheas of infected sheep (Figure 19.3). The organism can survive in dipping fluids under field conditions for at least two hours, so dips contaminated by *C pseudotuberculosis* in purulent material from an open abscess should also be considered a potential source of infection.[18]

In Australia it is likely that all or nearly all sheep flocks are infected, although the prevalence of infection within flocks varies considerably — from near 0% to over 90%[4,5,6,19] (Figure 19.4). There are a number of management factors — including the use of vaccine — which strongly influence the within-flock prevalence.

On most farms, the prevalence of CLA infection in sheep up to 1 year of age is very low. In flocks where the disease is uncontrolled the prevalence of infection rises dramatically after the first and second adult shearings[20], probably as a result of the close contact between susceptible, fleece-free sheep and adult sheep with lung lesions excreting organisms by coughing. Normally,

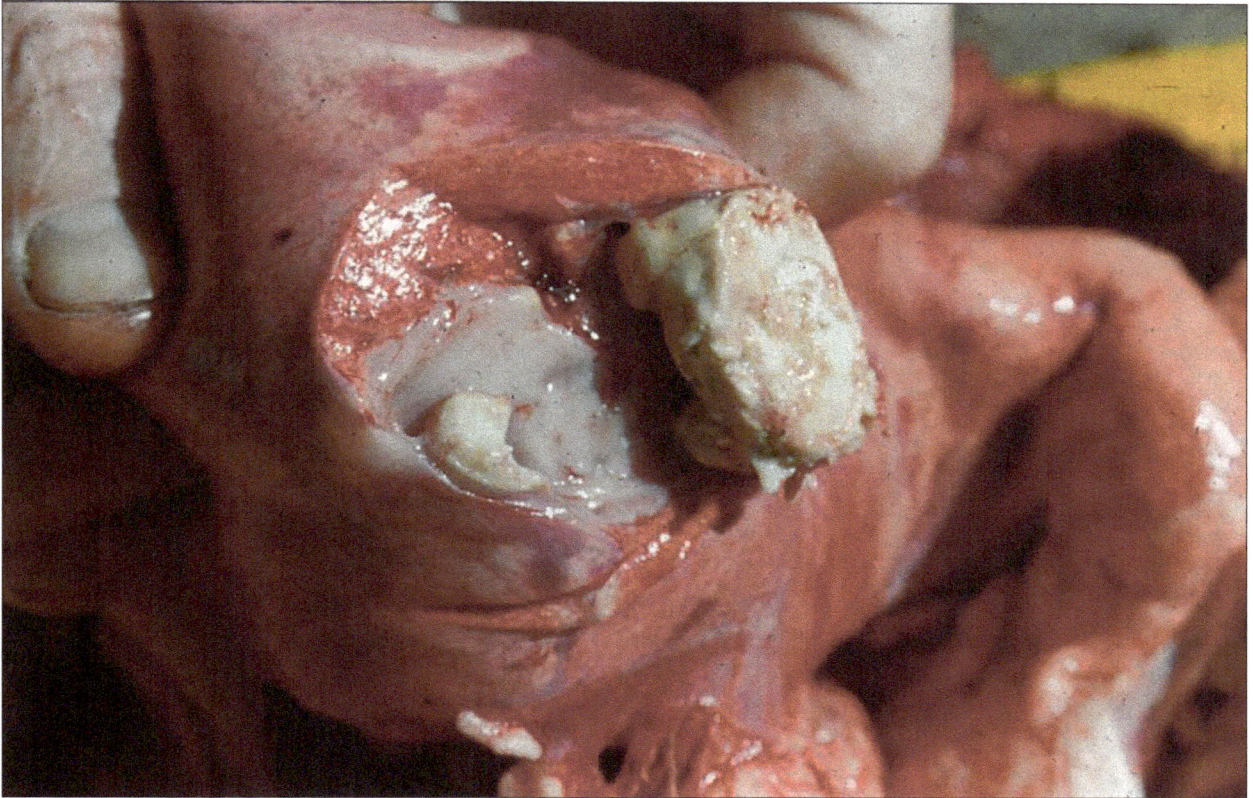

Figure 19.3: The connection of CLA lung abscesses to the airways provides a portal for the excretion of organisms by coughing, which is a probable source of transmission of the disease. Photograph courtesy of Michael Paton.

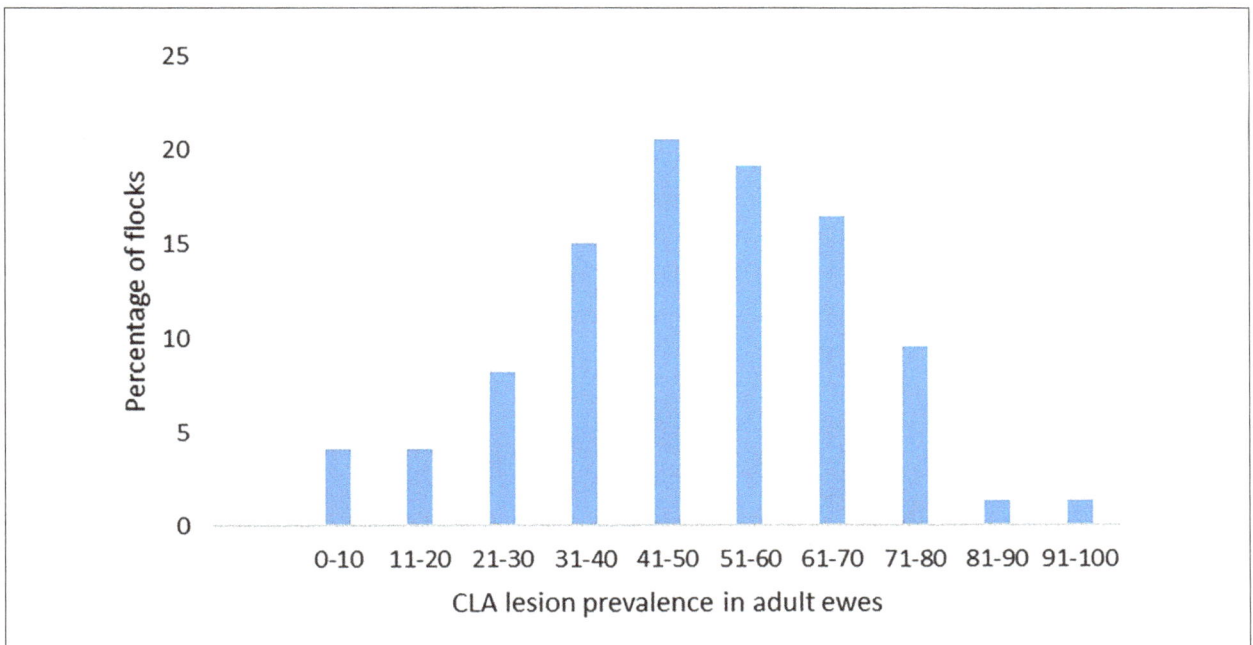

Figure 19.4: The prevalence of CLA lesions in unvaccinated mature-age Merino sheep varies markedly between flocks. Adapted from Paton (1990)[4]; Middleton et al. (1991)[5]; Paton et al. (1996).[19]

in Australia, sheep are released into *counting-out* pens immediately after shearing and held there for the duration of a two-hour *run* of shearing, before being counted and released into larger pens or paddocks (Figure 19.5). During this period the sheep are in close confinement and those with fresh shearing wounds are exposed to sources of infection from other sheep — sources which may include respiratory discharges or discharging infected cutaneous lesions.

One Australian study[19] has identified particular risk factors for a high incidence of new infections of young sheep. (For this study, a high incidence was one exceeding 5.5%, and a low incidence was one of 5.5% or less.) Four of the most significant factors identified were

- serological prevalence before the period during which the incidence is recorded (i.e. the more infected sheep in the group, the higher the challenge)

- high prevalence of lesions in aged ewes detected at the abattoirs (another indicator of the level of challenge in the flock)

- the use of a shower dip (presumably a plunge dip would also be a risk factor, but none of the farms under study had one). Shower dipping increased the risk of having a high CLA incidence by more than five times

- the time under cover after shearing. Flocks in which sheep were held together after shearing for more than one hour had nearly three times the risk of being in the high-incidence group compared to those in which sheep were held for less than one hour.

Figure 19.5: Sheep collected in a counting-out pen immediately after shearing. The risk of transmission of CLA increases as the duration of time in the counting-out pen increases. Source: KA Abbott.

The level of shearing cuts was not found to be a risk factor, possibly because the frequency of shearing cuts even in the cleanest-shorn flocks was not a limiting factor on the incidence of infection.

The first two risk factors above merely state what one would expect: the more infected sheep there are, the greater the chance of exposure. Reducing the number of infected sheep will also reduce the challenge, the proportion of the flock which seroconvert and the proportion which become chronically infected.

The third risk factor (dipping) suggests there is a high probability of disease transmission from contaminated dipping fluid through unhealed shearing cuts, intact skin or other sites of entry to the body. The fourth risk factor supports the importance of the counting-out pens as a place where disease transmission occurs.

Effect on productivity

In one experiment, artificial challenge with *C pseudotuberculosis* resulted in 7% lower clean wool production compared to control sheep, but challenge did not significantly affect body weight.[21] Challenged sheep which successfully resisted the development of lymphadenitis (and therefore had no lesions) had the same fleece weights as challenged sheep which developed CLA lesions. These findings suggest that it is the challenge of infection and the development of a skin lesion without extension to lymph nodes which reduce production, rather than the presence of abscessed lymph nodes. Vaccinated sheep which were challenged produced less wool than controls, and more wool than unvaccinated challenged sheep, but the differences were not statistically significantly. It may be that the objective of control measures — both immunological control measures and those using husbandry strategies — should be the reduction of challenge in a flock rather than just protection of some age groups or sex groups.

The thin ewe syndrome is associated with low reproductive rates in the United States[22] but the syndrome is not recognised as such in Australia. An outbreak of serious visceral infection in 11 of 89 sheep and goats imported from USA in a South Australian quarantine station resembled the North American description of CLA much more than the usual Australian one.[23] Subsequent comparative genetic analysis of the USA strains and the Australian ones failed to detect any differences between the two isolates.[24]

Control measures

A vaccine prepared from formalised *C pseudotuberculosis* toxin was shown in 1977 to protect sheep from the development of CLA lesions, although it did not protect them from skin infections.[25] It is likely that protection comes from neutralisation of the permeability-increasing effect of the exotoxin, thus protecting sheep by reducing the spread of bacteria from the local site of primary infection.

In 1984 the vaccine was produced commercially as Glanvac and is now available in combination with clostridial vaccines from three manufacturers. It has recently been shown that the addition of bacterial cells to the toxoid will not improve vaccine efficacy and that the vaccine is still protective when given with clostridial vaccine components.[26,27] It is currently only available in combination with clostridial vaccine.

The vaccine is not 100% protective but, in field trials, has been 25% to 90% protective.[28] In one study, sheep vaccinated against CLA and naturally exposed to infection had a 74% lower infection rate than unvaccinated sheep and those vaccinates that did become infected had 96% fewer lung abscesses than infected non-vaccinates.[17] The response to vaccination in a previously unvaccinated and infected flock, therefore, can be expected to occur in two stages. Initially, vaccinated sheep are exposed to high levels of challenge, many becoming infected but relatively few developing lung abscesses because of the protection afforded by vaccination. Subsequently, older sheep are sold from the flock and the challenge of infection decreases because of the small number of sheep in the flock with lung abscesses. From this point, if effective vaccination continues, very few sheep will develop infection.

Vaccination must be repeated annually after two initial vaccinations and boosters should be given in the three to six months before shearing, but no less than two weeks before shearing. To provide significant reduction in the prevalence of CLA within a flock, it is necessary to vaccinate all sheep in the flock according to the recommended protocol (two doses as lambs, annual boosters before shearing). Less frequent vaccination has a much less significant effect on the prevalence of CLA within the flock — perhaps no better than no vaccination at all.[29] Maternal antibody does interfere with the immunological response of lambs and, in flocks with a high prevalence of CLA, vaccinating lambs under 10 weeks of age will not give protection as strong as that produced by later vaccination.[30] Nevertheless, practicability may necessitate vaccination at ages below 10 weeks and the resultant protection is probably satisfactory, particularly because the lambs are not at high risk of exposure to CLA until later in life, by which time there have been further opportunities for vaccination.

The usefulness of other techniques to reduce flock prevalence is untested. It may be useful to attempt control by isolating young, recently shorn sheep from older sheep, by shearing young sheep first and by reducing the time during which sheep are held together under cover off-shears. The evidence of the importance of shower dips as a risk factor for disease spread and of the survival of organisms in plunge dips provides additional reasons for the need to reduce the frequency of lice infestations in flocks, a topic that is discussed further in Chapter 10.

Economics of vaccination

Estimates of the loss in wool production associated with challenge from *C pseudotuberculosis* lie between 4% and 7%. A loss appears to occur on the first occasion that a sheep is challenged, whether it is vaccinated or not. Consequently, the full benefit from a reduction in flock prevalence will not occur until all sheep in the flock have been vaccinated for several years and the prevalence of chronically infected sheep is low.

Lambs should be vaccinated twice at an interval of approximately six weeks and then vaccination must be repeated annually to maintain effective immunity. Commonly, wool-producing flocks in Australia have six age groups of sheep, so the number of vaccinations required to be given per year is between 1.5 and 2.0 times the number of adult sheep in the flock, depending on the reproductive rates and selling policies within the flock.

For the purposes of illustration, assume that vaccination reduces the number of sheep which are challenged by 90%. If the current infection rate is high, approximately 10 to 15 times as many sheep will be vaccinated annually as will be saved from challenge, depending on the flock age and sex structure. If the infection rate is low, 20 to 30 times as many sheep will be vaccinated annually as are saved from challenge.

For individual producers, the economics of vaccination against CLA will therefore vary with

- the prevalence of CLA in the flock
- the cost of the vaccine
- the value of the wool of the sheep subjected to CLA challenge
- the flock structure, particularly the ratio of lambs to adult sheep
- the incidence of clostridial diseases in the flock, which could be simultaneously reduced by a combined CLA-clostridial vaccine
- the additional price received for selling aged sheep to the abattoirs which are relatively free of CLA.

For the Australian flock as a whole, there could be significant economic benefits to the sheep industry if the rate of carcase condemnation due to CLA was reduced. Currently, fewer than 10% of producers use CLA vaccine effectively, although 40% to 50% of producers use it to some degree[29], with little benefit to CLA prevalence (although there may be benefits for control of clostridial diseases).

Treatment

Penicillin and other antibiotics active against gram-positive bacteria are effective against the organism *in vitro* but have little or no effect against chronic lesions. It is relevant to the prognosis for treatment of affected sheep that, in human cases where therapy is continued until resolution, ultimately it is sometimes necessary to surgically remove the abscessed lymph node.[11]

Diagnosis

CLA is by far the most common condition causing abscessation of lymph nodes in sheep in Australia. Vaccination reactions to ovine Johne's disease vaccine (Gudair®) may resemble a CLA abscess (see Chapter 16). In tropical Australia, melioidosis can cause fatal multiple abscessation in sheep and, in that region, may need to be differentiated from CLA.

INFECTION WITH *MYCOPLASMA OVIS*

Mycoplasma ovis was formerly named *Eperythrozoon ovis*. It is now known to be a mycoplasma of the haemotrophic group — organisms which parasitise the surface of erythrocytes in numerous different host species and are transmitted mainly by arthropod vectors.[31] The disease caused by this organism is still commonly referred to as eperythrozoonosis. It is characteristically a disease of weaner sheep but in fact sheep of all ages are susceptible. Mature sheep with no previous experience of infection are as badly affected as weaners.[32] Merino sheep are particularly susceptible. Epizootics occur when three particular conditions come together — susceptible animals, a source of infection and a vector. Recovered animals remain infected for long periods, probably for life, and are *premune*.[33] The flock prevalence in southern Australia is very high: at least 90%.[34]

Transmission and infection

The incubation period in cases of natural infection is one to six weeks. Transmission occurs by flying insects, such as mosquitoes. There is no transplacental or intrauterine transmission. Episodes of infection last 14 to 28 days. These episodes are sometimes followed by a series of cycles at intervals of two to four months. The Coombs test is positive from the time that parasitaemia is maximal until up to six months after the parasites are no longer detectable.

Clinical and necropsy signs

When first infected with *M ovis*, sheep are anaemic, leading to weakness and a reduction in exercise tolerance. Slight to moderate icterus is sometimes observable, particularly three to four days after stress has exacerbated an already existing infection. *M ovis* is probably a frequent contributor to the development of the *tail* that occurs often in a flock of weaner sheep. Although not observable clinically, infection may also reduce wool production by 5% to 10% in the year of first infection, and may also reduce body weight gain during the initial period of infection.[35]

At necropsy, splenomegaly is a prominent feature and the haemal nodes are enlarged.

Laboratory aids to diagnosis

A blood profile reveals anaemia, often severe, with a macrocytic, hypo- or normochromic regenerative response. Parasites may appear in smears, but often they disappear about the same time as anaemia develops. The Coombs (anti-globulin) test is not specific for *M ovis* but, because other causes of a positive test are unusual, it is a useful diagnostic aid. A complement fixation test (CFT)[36] and a modified Indirect Immunofluorescent Antibody Assay (IFAA) are also used.[37]

Treatment and control

Infection and spread of *M ovis* are inevitable if the vector and infected sheep are present. In attempting to control the disease, weaners should be maintained in good nutritional state and free from intercurrent diseases. Tetracycline and oxytetracycline are effective therapies for *Mycoplasma suis* infection in pigs and so are rational treatments for sheep.

RECOMMENDED READING

Paton M (1990) Caseous lymphadenitis. Proceedings No 141. University of Sydney Postgraduate Committee in Veterinary Science: Sydney, pp. 149-55.

REFERENCES

1   Meldrum KC (1990) Caseous lymphadenitis outbreak. Vet Rec **126** 369.
2   Paton MW (1989) The impact of caseous lymphadenitis on the wool and sheep meat industries. Proceedings of the annual conference of the Australian Veterinary Association (Australian Advances in Veterinary Science). Australian Veterinary Association: Artarmon, Sydney, p. 147.
3   GHD Pty Ltd (2015) Priority list of endemic diseases for the red meat industries. Final Report for Project B.AHE.0010. Meat and Livestock Australia: North Sydney.

4   Paton M (1990) Caseous lymphadenitis. Proceedings No 141. University of Sydney Postgraduate Committee in Veterinary Science: Sydney, pp. 149-55.

5   Middleton MJ, Epstein VM and Gregory GG (1991) Caseous lymphadenitis on Flinders Island: Prevalence and management surveys. Aust Vet J **68** 311-12. https://doi.org/10.1111/j.1751-0813.1991.tb03272.x.

6   Batey RG (1986) Frequency and consequence of caseous lymphadenitis in sheep and lambs slaughtered at a Western Australian abattoir. Am J Vet Res **47** 482-5.

7   Stoops SG, Renshaw HW and Thilsted JP (1984) Ovine caseous lymphadenitis: Disease prevalence, lesion distribution, and thoracic manifestations in a population of mature culled sheep from western United States. Am J Vet Res **45** 557-61.

8   Williamson LH (2001) Caseous lymphadenitis in small ruminants. Vet Clin North Am Food Animal Practice **17** 359-71.

9   Girard C, Dubreuil P, Daignault D et al. (2003) Prevalence of and carcass condemnation from maedi-visna, paratuberculosis and caseous lymphadenitis in culled sheep from Quebec, Canada. Prev Vet Med **59** 67-81. https://doi.org/10.1016/S0167-5877(03)00060-6.

10   Biberstein EL, Knight HD and Jang S (1971) Two biotypes of *Corynebacterium pseudotuberculosis*. Vet Rec **89** 691. https://doi.org/10.1136/vr.89.26.691.

11   Peel MM, Palmer GG, Stacpoole AM et al. (1997) Human lymphadenitis due to *Corynebacterium pseudotuberculosis*: Report of ten cases from Australia and review. Clin Infect Dis **24** 185-91. https://doi.org/10.1093/clinids/24.2.185.

12   Jolly RD (1965) The pathogenic action of the exotoxin of *Corynebacterium ovis*. J Comp Pathol **75** 417-31.

13   Jubb KVF and Kennedy PC (1970) Caseous lymphadenitis. In: Pathology of domestic animals. 2nd ed. Vol 1. Academic Press: New York, p. 373.

14   Williamson P and Nairn ME (1980) Lesions caused by *Corynebacterium pseudotuberculosis* in the scrotum of rams. Aust Vet J **56** 496-8. https://doi.org/10.1111/j.1751-0813.1980.tb02565.x.

15   Renshaw HW, Graff VP and Gates NL (1979) Visceral caseous lymphadenitis in thin ewe syndrome: Isolation of *Corynebacterium, Staphylococcus,* and *Moraxella* spp from internal abscesses in emaciated ewes. Am J Vet Res **40** 1110-14.

16   Ellis TM, Sutherland SS, Wilkinson FC, Mercy AR and Paton MW (1987) The role of *Corynebacterium pseudotuberculosis* lung lesions in the transmission of this bacterium to other sheep. Aust Vet J **64** 261-3. https://doi.org/10.1111/j.1751-0813.1987.tb15952.x.

17   Paton MW, Sutherland SS, Rose IR, Hart RA, Mercy AR and Ellis TM (1995) The spread of *Corynebacterium pseudotuberculosis* infection to unvaccinated and vaccinated sheep. Aust Vet J **72** 266-9. https://doi.org/10.1111/j.1751-0813.1995.tb03542.x.

18   Nairne ME and Robertson JP (1974) *Corynebacterium pseudotuberculosis* infection of sheep: Role of skin lesions and dipping fluids. Aust Vet J **50** 537-42. https://doi.org/10.1111/j.1751-0813.1974.tb14072.x.

19   Paton M, Rose I, Hart R, Sutherland S et al. (1996) Post-shearing management affects the seroincidence of *Corynebacterium pseudotuberculosis* infection in sheep flocks. Prev Vet Med **26** 275-84.

20   Paton MW, Mercy AR, Sutherlands SS et al. (1988) The influence of shearing and age on the incidence of caseous lymphadenitis in Australian sheep flocks. Acta Vet Scand **84** (suppl) 101-3. https://doi.org/10.1016/0167-5877(95)00544-7.

21   Paton MW, Mercy AR, Wilkinson FC et al. (1987) The effects of caseous lymphadenitis on wool production and bodyweight in young sheep. Aust Vet J **65** 117-19. https://doi.org/10.1111/j.1751-0813.1988.tb14429.x.

22   Gates NL, Everson DO and Hulet CV (1977) Effects of thin ewe syndrome on reproductive efficiency. J Am Vet Med Ass **171** 1266-7.

23   Bunn CM (1991) An outbreak of lethal *C ovis* infection in imported sheep and goats. Proceedings

of the Australian Sheep Veterinary Society Conference. 13-17 May, Darling Harbour, Australia, pp. 111-12.

24   Sutherland SS, Hart RA and Buller NB (1993) Ribotype analysis of *Corynebacterium pseudotuberculosis* isolates from sheep and goats. Aust Vet J **70** 454-6.

25   Nairne ME, Robertson JP and McQuade NC (1977) The control of caseous lymphadenitis in sheep by vaccination. Proc Ann Conference of the Aust Vet Assoc. Australian Veterinary Association: Artarmon, Sydney, p. 159.

26   Eggleton DG, Middleton HD, Doidge CV et al. (1991) Immunisation against ovine caseous lymphadenitis: comparison of *Corynebacterium pseudotuberculosis* vaccines with and without bacterial cells. Aust Vet J **68** 317-19. https://doi.org/10.1111/j.1751-0813.1991.tb03085.x.

27   Eggleton DG, Doidge CV, Middleton HD et al. (1991) Immunisation against ovine caseous lymphadenitis: Efficacy of monocomponent *Corynebacterium pseudotuberculosis* toxoid vaccine and combined clostridial-corynebacterial vaccines, Aust Vet J **68** 320-1. https://doi.org/10.1111/j.1751-0813.1991.tb03087.x.

28   Eggleton DG, Haynes JA, Middleton HD et al. (1991) Immunisation against ovine caseous lymphadenitis: Correlation between *Corynebacterium pseudotuberculosis* toxoid content and protective efficacy in combined clostridial-corynebacterial vaccines. Aust Vet J **68** 322-5. https://doi.org/10.1111/j.1751-0813.1991.tb03088.x.

29   Paton MW, Walker SB, Rose IR et al. (2003) Prevalence of caseous lymphadenitis and usage of caseous lymphadenitis vaccines in sheep flocks. Aust Vet J **81** 91-5. https://doi.org/10.1111/j.1751-0813.2003.tb11443.x.

30   Paton MW, Mercy AR, Sutherland SS et al. (1991) The effect of antibody to caseous lymphadenitis in ewes on the efficacy of vaccination in lambs. Aust Vet J **68** 143-6. https://doi.org/10.1111/j.1751-0813.1991.tb03158.x.

31   Neimark H, Hoff B and Ganter M (2004) Mycoplasma ovis comb. nov. (formerly Eperythrozoon ovis) an epierythrocytic agent of haemolytic anaemia in sheep and goats. Int J Syst Evol Microbiol **54** 365-71. https://doi.org/10.1099/ijs.0.02858-0.

32   Sheriff D (1976) Infections with *Eperythrozoa* and *Haemobartonellae*. Proceedings No 27. University of Sydney Postgraduate Committee in Veterinary Science: Sydney, pp. 5-11.

33   Sutton RH (1990) Eperythrozoonosis. In: Sheep Medicine. Proceedings No 141. University of Sydney Postgraduate Committee in Veterinary Science: Sydney, pp. 133-47.

34   Nicholls TJ and Veale PI (1986) The prevalence of *Eperythrozoon ovis* infection in weaner and adult sheep in north eastern Victoria. Aust Vet J **63** 118-20. https://doi.org/10.1111/j.1751-0813.1986.tb07678.x.

35   Daddow KN (1979) *Eperythrozoon ovis* — a cause of anaemia, reduced production and decreased exercise tolerance in sheep. Aust Vet J **55** 433-4. https://doi.org/10.1111/j.1751-0813.1979.tb05600.x.

36   Daddow KN (1977) A complement fixation test for the detection of Eperythrozoon infection in sheep. Aust Vet J **53** 139-43. https://doi.org/10.1111/j.1751-0813.1977.tb00140.x.

37   Nicholls TJ and Veale PI (1986) A modified indirect immunofluorescent assay for the detection of antibody to *Eperythrozoon ovis* in sheep. Aust Vet J **63** 157-9. https://doi.org/10.1111/j.1751-0813.1986.tb02956.x.

20

DISEASES OF THE RESPIRATORY SYSTEM

PNEUMONIA

Diseases of the lower respiratory tract of sheep are not generally considered important causes of lost productivity in Australian flocks because of the sporadic nature of serious respiratory disease and the subclinical nature of most cases of respiratory infection. The extensive conditions under which Australian sheep are usually managed do not favour transmission of pneumonic infectious agents, and they also make detection of the disease less likely.

The evidence from abattoir surveys in Australia and more detailed investigations in New Zealand and European countries suggest that we may be underestimating the effect that pneumonic diseases have on Australian sheep flocks. Pneumonic lesions are frequently seen at necropsy in sheep of all ages and clinical signs are frequently evident when flocks of sheep are driven. It is likely that, either alone or in combination with other disease conditions, respiratory diseases are a significant cause of loss to the Australian sheep industry. Neither the extent of the problem in Australia, nor its economic impact, nor the success of any control measures have been investigated since the 1960s.

The important pneumonic conditions of sheep in Australia can be first classified as *parasitic* or *microbiological*. The latter classification is nonspecific and encompasses conditions which vary between mild and severe, chronic and acute; it includes a number of aetiological agents which can be viruses, mycoplasmas or other bacteria or any combination of these.

Of lesser importance are *aspiration pneumonia* and some bacterial infections which may or may not involve the lungs (caseous lymphadenitis (CLA), tuberculosis, melioidosis).

The microbiological agents associated with pneumonia in sheep include

- *Mycoplasma ovipneumoniae, M arginini*
- viruses: *Parainfluenza 3 (PI3)* virus principally, but also *ovine adenovirus type 6, bovine adenovirus type 7* and *respiratory syncytial virus, type 3 reovirus, maedi, pulmonary adenomatosis*
- *Bordetella parapertussis*[1]
- *Mannheimia haemolytica* (formerly *Pasteurella haemolytica* biotype A) and *Pasteurella multocida* (infections caused by these agents are referred to as pasteurellosis, reflecting their previous common genus nomenclature)
- a number of other bacteria including *Actinobacillus lignieresii, Trueperella pyogenes, Klebsiella* spp, *Escherichia coli, Histophilus somni, Fusobacterium necrophorum* and *Neisseria* spp.[2,3]

These and possibly other, as yet unidentified, agents cause a spectrum of pneumonic diseases which vary from mild to severe, acute to chronic, proliferative interstitial to exudative. The most severe syndromes, which are accompanied by obvious clinical signs and significant mortality rates, are associated with bacterial infections, particularly *M haemolytica*. The mildest forms, in which clinical signs are absent or develop only after extended periods, are usually associated with mycoplasma infection, with or without concurrent viral infection.

The spectrum of disease is such that pneumonic lesions and clinical signs may fall anywhere between the two extremes. It is evident, however, that the mycoplasmas, with or without viral assistance or induction, are the usual initiating infectious agents and other bacteria the secondary invaders. The outcome of the initial infection is determined by a number of complex environmental, genetic and immunological factors. A number of classifications, which are in common usage but do not make the aetiology or severity of infection any clearer, are applied to different types of pneumonia in different types of sheep and in different countries.

The term *enzootic pneumonia* is applied by some authors to the whole spectrum of lower respiratory tract infections, from the mild, chronic proliferative interstitial pneumonia caused by mycoplasmas and PI3; to the subacute pneumonia associated with mycoplasma infection and secondary bacterial infection (commonly called *summer pneumonia*), involving a combination of proliferative and exudative reactions in the lungs; to the acute exudative pneumonia of pasteurellosis, which has much in common, pathologically, with shipping fever of cattle.[2]

Other authors use the term *enzootic pneumonia* to describe only the most severe, acute and exudative forms of pneumonia which are associated with bacterial (usually *Mannheimia haemolytica*) infection.[4] This restricted use of the term denies the almost certain connection to milder forms of pneumonia through the common mycoplasmal initiation and also loses the hint of consistency in terminology with the porcine disease associated with *M hyopneumoniae*.

In New Zealand, the term *chronic non-progressive pneumonia* or CNP is used to describe a subacute or chronic pneumonia which is associated with *Mycoplasma ovipneumoniae* and is enzootic in lambs of slaughter age.[5,6,7] This disease is associated with reduced growth rates of lambs[8] and with increased trimming and condemnation of carcases at slaughter.[9] The disease appears to have much in common with *summer pneumonia* of Australian flocks, including the seasonal rise in incidence, but its impact is recognised in a different way because of the different management systems and production objectives of sheep flocks in the two countries. *Acute fibrinous pneumonia* may be a progression from the milder form and associated with secondary infection with *Mannheimia haemolytica* and other bacteria[10] or it may be caused by *Mannheimia haemolytica* alone.

Mycoplasma pneumonia

In Australia, outbreaks of pneumonia of lambs with high morbidity, low mortality and apparently significant deleterious effects on growth rate and exercise tolerance have been recorded from Queensland[11] and southeastern Australia.[12,13] The condition in southeastern Australia became known as *summer pneumonia* because of an increase in the prevalence of the disease in the summer and early autumn months. Two types of *Mycoplasma* spp were frequently associated with the disease and these were subsequently identified as *Mycoplasma ovipneumoniae* and *M arginini*.[14]

Experimental reproduction of the disease with pure cultures of *M ovipneumoniae* have shown that it is capable of causing a mild interstitial pneumonia with few clinical signs, generally only detectable by auscultation.[15,16,5] Other bacteria, which are incapable of causing pneumonia alone, subsequently infect the lungs and cause a more severe inflammatory response and clinical signs, including coughing and nasal discharge.

The natural history of infection with mycoplasmas appears to be as follows: in flocks in which *Mycoplasma ovipneumoniae* is endemic, lambs become infected within a few days of birth from ewes which are carriers. The disease is slowly progressive over a period of weeks and, at ages over 5 to 10 weeks, secondary bacterial infection occurs. The pneumonia then becomes more severe. Some lambs recover over the following few weeks but clinical signs in some lambs persist until they are up to the age of 6 months or more. The extra demand placed on the lungs for heat exchange by high atmospheric temperatures may be a precipitating factor in the exacerbation of an otherwise mild condition and the appearance of obvious clinical signs in summer.[11,15]

Viral pneumonia

Parainfluenza 3 virus is the only virus which has been isolated from the lungs of sheep with pneumonia in Australia.[17] In 1968 the prevalence of flocks in Australia with antibody to PI3 was estimated to be 87%.[18]

The virus is capable of producing mild pneumonic lesions but, without secondary bacterial infection, is unlikely to produce pneumonia with observed clinical signs. It has been linked to outbreaks of pneumonia in Australian sheep, either by viral isolation or by serological evidence, but that association does not imply causation.[17] It is clear that prior PI3 infection is not necessary for the development of pneumonia because cases of pneumonia associated with *Mycoplasma* spp occur in animals serologically negative for PI3 or in animals that become infected with PI3 months after pneumonia commenced in the flock.[13]

PI3 virus can have a role in the development of pneumonic pasteurellosis. *M haemolytica* will proliferate in lungs previously infected with PI3, leading to an acute exudative bronchopneumonia and pleurisy, whereas it is usually cleared rapidly from lungs not infected with virus.[19,20] A vaccine specific for PI3 was used in field trials in New Zealand and reduced the prevalence of pneumonia in naturally exposed lambs but did not prevent all cases of pneumonia.[21] This indicates that agents other than PI3 are responsible for initiating at least some natural cases of pneumonia in lambs.

Bacterial pneumonia

In one small study in New Zealand, *Bordetella parapertussis* was found to be present in the nasal cavity and bronchial washings of a high proportion of 6- to 10-month-old lambs with lesions of CNP, but a low proportion of lambs without lung lesions.[1] On experimental intratracheal inoculation of colostrum-deprived, 1-week-old lambs, lesions resembling those of early naturally occurring CNP were observed. This organism was considered to be capable of initiating a mild and short-lived respiratory infection which, with subsequent secondary bacterial infection, could develop to CNP.[22]

Bacteria such as *Mannheimia haemolytica, Pasteurella multocida, Actinobacillus lignierisii, Trueperella pyogenes*[a], *Corynebacterium equi, Klebsiella* spp, *Escherichia coli, Fusobacterium necrophorum* and *Neisseria* spp have all been recorded from the lungs of sheep with pneumonia.[2] These bacteria act as secondary invaders and are important in increasing the severity of the clinical and pathological signs of pneumonia. They effectively render subclinical pneumonia, induced by *Mycoplasma* spp or PI3, a clinical entity such as *summer pneumonia* or CNP. Without them the effect on productivity or mortality rates of the milder pneumonias would be much less significant.

M haemolytica is the primary bacterial species responsible for the most severe and acute forms of ovine pneumonia. The disease is an acute exudative pneumonia with septicaemia and very high mortality rates. Difficulties in reproducing the disease experimentally with pure cultures of *M haemolytica* suggest that a previous infection with a virus or mycoplasma is usual in field cases.

Histophilus somni (previously *H ovis*) has been associated with embolic pneumonic infection, characterised histologically by a necrotising vasculitis and grossly by abscessation.

Chlamydial pneumonia

Chlamydia spp have been considered a cause of pneumonia in sheep in Europe and North America and were initially suspected of contributing to ovine pneumonia in Australia because the organisms were identified in the faeces of pneumonic and healthy sheep.[2] Subsequent investigations have failed to identify them in cases of pneumonia in either Australia or New Zealand and they are largely discounted as lung pathogens in these countries now.

Effects on productivity

Graziers and research workers observing flocks with enzootic pneumonia have noted the inferior growth rate of lambs affected by pneumonia.[13] There has been no quantification of this loss in Australia nor have there been any studies of the effect of pneumonia on wool growth rates in any country. One New Zealand study has estimated that moderate to severe pneumonia is responsible for losses in carcase weight of 1.5 kg and a reduction in liveweight gain of 0.8 to 0.9 kg per month for every 10% of the lung surface grossly affected by pneumonia.[23] An earlier and extensive study of lambs slaughtered from a flock with no evidence of clinical pneumonia found that 60% of the lambs slaughtered had lesions of pneumonia. Only 6.5% of lambs had moderate to severe lesions and the deleterious effect on carcase weight of such lesions was estimated to be 0.45 kg.[8,24] The economic consequence in the whole flock of this loss of carcase weight of lambs is approximately 0.2% only and, therefore, quite insignificant. This study has been criticised for underestimating the degree of weight loss and the prevalence of moderate to severe lesions[25] but it remains the only attempt to put an economic cost on CNP of lambs in Australasia.

The disease in most flocks appears to be clinically mild or subclinical and self-limiting. Most lambs recover within the first year of life. In line with many other interruptions to weight gain in early life, compensatory gain in liveweight may afford a *catch-up* within a few months of recovery. The loss of wool production, a more important source of economic loss in Merinos than weight gain, is probably not recovered but has not been measured. Increased mortalities,

---

a   Formerly *Corynebacterium pyogenes, Actinomyces pyogenes* and *Arcanobacterium pyogenes*.

both from pneumonia and from increased susceptibility to other diseases, are likely to be additional and significant sources of economic loss in some flocks.

ACUTE PASTEURELLOSIS

Pasteurellosis is regarded as the most important bacterial disease of sheep in many countries, particularly those in which sheep rearing is practised intensively. It is of much less importance in countries where grazing conditions are extensive, as in Australia and New Zealand. In these countries, outbreaks are uncommon and sporadic and usually have a low morbidity.

Mannheimia haemolytica exists in 12 different serotypes (1-17 except 3, 4, 10, 11 and 15), while *Bibersteinia trehalosi* (formerly *Pasteurella trehalosi* and, before that, *Pasteurella haemolytica* type T) exists in four serotypes (3, 4, 10 and 15). The differences are important in vaccine considerations. In the UK, *M haemolytica* causes pneumonic pasteurellosis in cattle (mainly serotype A_1) and pneumonia in goats and sheep (mainly A_2 but, in decreasing order of frequency, A_6, A_7, A_9 and A_1). *B trehalosi* causes systemic pasteurellosis in sheep, rather than a primarily pulmonary infection. Sheep strains of *M haemolytica* appear to be different from cattle strains even within serotypes, principally in the way the organism is able to use host haemopexin as a source of iron. *M haemolytica* and *B trehalosi* both occur in the nasopharynx of apparently normal sheep.

Pasteurella multocida has been considered an uncommon pathogen in sheep except in tropical countries.[26] While it is commonly isolated from pneumonic ovine lungs in NZ it is reportedly a much less virulent pathogen than *M haemolytica*.[7]

There are two important epidemiological associations with acute pasteurella pneumonias. One is the probable initiating role of PI3 virus in most cases; the other is that with pasteurella mastitis in ewes. Ewes with pasteurella mastitis commonly infect their lambs, and infected lambs may induce pasteurella mastitis in their dams.

Clinical signs

Outbreaks of acute pneumonic pasteurellosis often commence with sudden deaths before clinical signs are observed. As an outbreak proceeds, respiratory signs become more apparent, particularly in older sheep rather than in lambs. Signs then include dullness, anorexia, fever, dyspnoea or hyperpnoea. On auscultation, respiratory sounds are loud and prolonged. Affected sheep froth at the mouth, cough and have a serous nasal discharge. In acute cases, death occurs in one to three days.

Necropsy findings

The findings at necropsy vary with the chronicity of the condition before death. In acute cases, the lungs are swollen and heavy with bright purple-red patches which are solid and exude a frothy haemorrhagic fluid when incised. Consolidation is evident and some areas may contain greenish-brown areas of necrosis surrounded by dark, haemorrhagic zones. In less acute cases, areas of the cranial lobes are greyish-pink, raised and consolidated. On cutting, the tissue is dense and the septa thickened and prominent. The bronchial lymph nodes are enlarged and may have petechial haemorrhages.[26]

Diagnosis

Aspiration pneumonia, following careless oral dosing or plunge-dipping, can sometimes occur in outbreak form and may resemble pasteurellosis clinically. A careful collection of history may reveal that an affected flock has been recently drenched or dipped, suggesting that aspiration may be the cause of the outbreak, but it should be remembered that pasteurellosis is often precipitated by stressful events, particularly in young sheep. The post-mortem diagnosis of acute pneumonia must be made with caution in animals which have died naturally because post-mortem change in the lungs and following some clostridial diseases can grossly resemble acute pasteurella pneumonia. Histopathology and bacteriology of freshly dead specimens are necessary to confirm the diagnosis.

Treatment and control

Vaccination is practised successfully in a number of countries, including the UK. One treatment with long-acting oxytetracycline intramuscularly at 20 mg/kg is effective in controlling the development of pneumonic pasteurellosis for four days in animals with early clinical signs or in animals which are post-exposure but preclinical.[27] Retreatment after three to four days is advisable because relapses occur. Feeding of broad-spectrum antibiotics, especially tetracyclines, to feedlot lambs is often done for recently weaned lambs, in an effort to reduce the incidence of acute pasteurellosis.

TUBERCULOSIS

The disease is rare in sheep in Australia. It can be caused by *M bovis* and *M avium*.[28,29]

MELIOIDOSIS

Melioidosis is caused by *Burkholderia* (formerly *Pseudomonas*) *pseudomallei*. It occurs in Australia but outbreaks are restricted to tropical zones. The chief sources of infection are rodents, which develop protracted infections during which organisms are excreted in the faeces. After the wet season in northern Australia, climatic conditions are conducive to the growth of the organism and it can be found in surface soil and water. Outbreaks occur in all farm animals, particularly sheep and goats. Infection occurs through ingestion, insect bites, abrasions and inhalation. The disease is highly fatal in humans.

Clinical signs in sheep include lameness, respiratory distress, weakness and recumbency and death in one to seven days. Chronic forms of the disease are more common in goats, which can be infected without showing clinical signs. At necropsy, multiple abscesses are found in most organs, including lungs, spleen, liver, subcutaneous tissues and lymph nodes, particularly nodes of the thoracic cavity. The pus in the abscesses resembles that of CLA.

Control of the disease involves the elimination of infected farm animals, reduction of rodent infestations and disinfection of premises.

PARASITIC PNEUMONIA

Dictyocaulus filaria is the only significant lungworm of sheep in Australia. Adult parasites live freely in the bronchi, and the exudative bronchitis that they induce can lead to bronchiolar obstruction and areas of atelectasis. Rarely are the affected areas extensive, so dyspnoea is not evident, and coughing is the predominant clinical sign. Occasionally, very heavy infestations occur and clinical signs of dyspnoea, moist râles on auscultation and even death result. Additionally, secondary bacterial infections can exacerbate the verminous pulmonary damage and the clinical signs.

Protostrongylus rufescens causes similar clinical signs but infestations are usually light and lung lesions only ever significant in lambs.

Muellerius capillaris adults live in fibrous nodules in the interstitial tissue of the lungs and rarely cause any clinical signs or deleterious effects. (In goats, nodule formation is not a characteristic and an interstitial pneumonia can develop.[30])

ASPIRATION PNEUMONIA

In sheep, most cases of aspiration pneumonia occur from careless or improper drenching technique or from prolonged submersion in plunge dips — either through weakness of the sheep, crowding or overzealous dunking by the operator. Isolated cases of aspiration pneumonia also occur following vomition and inhalation of rumen contents, paralysis of larynx, pharynx or oesophagus, or rupture of a pulmonary or pharyngeal abscess.

Clinical signs vary with the dose and nature of the aspirant. Significant amounts of infective material can lead to acute, severe pneumonia and toxaemia, with death in one or two days. Lesser amounts or less noxious material may lead to chronic pneumonic lesions, pulmonary abscessation, pleurisy and a prolonged period of ill health. Gangrenous pneumonia occurs sometimes following aspiration and is marked by a putrid odour on the breath.

SHEEP PULMONARY ADENOMATOSIS (OVINE PULMONARY ADENOMATOSIS OR OPA, JAAGSIEKTE)

This is a chronic progressive pneumonia of sheep marked histologically by adenomatous ingrowths of the alveolar walls and clinically by the production of a profuse watery mucus from the lungs which is discharged from the nose. It is caused by a retrovirus and is one of the three *slow virus* infections of sheep with maedi-visna and scrapie. The disease occurs in Britain, continental Europe, South Africa, Israel, Asia and Iceland. There is no treatment.

Epidemiology

Only sheep are infected in natural cases. Transmission is presumed to occur by inhalation of infected droplets and by vertical transmission to the foetus. Close housing in winter, such as occurs routinely in Iceland, promotes transmission but is not essential for flock outbreaks. The incubation period is one to three years, but many infected sheep never show clinical signs within their lifetime.

Clinical signs

The adenomatous ingrowths encroach on the alveolar space and lead to anoxic anoxia. Coughing occurs but is not a prominent sign. Emaciation, dyspnoea and panting after exercise, and profuse watery discharge from the nose are characteristic signs. Moist râles are audible over affected areas of lung. The disease is inevitably fatal.

Necropsy

The lungs are enlarged, heavy and consolidated and there is frothy fluid in the bronchi. The histopathology is characteristic and diagnostic.

MAEDI

Maedi-visna is a chronic, progressive viral infection characterised by a prolonged incubation period and predominantly two clinical manifestations: pneumonia (*maedi* means dyspnoea) and encephalomyelitis (*visna* means wasting in Icelandic). The virus also infects the udder, causing a chronic mastitis and reduced milk yield. Generally, only one form of the disease occurs in one animal and often one form predominates in any one flock. The disease does occur in goats but transmission between sheep and goats does not usually occur in field cases.

The disease occurs widely throughout the world but does not occur in Australia or New Zealand. It was inadvertently introduced into Iceland in 1933 and, before the disease was eradicated in 1965 by a slaughter and restocking programme, 650 000 sheep had been slaughtered and over 100 000 sheep had died of the disease.

Epidemiology

Vertical transmission, by the excretion of virus-infected leucocytes in colostrum and milk, is the main form of spread of maedi-visna in flocks. Transmission to the foetus *in utero* occurs but is rare. It is believed that transmission via ova or sperm does not occur. Horizontal transmission occurs, chiefly by the inhalation of infected respiratory secretions from infected sheep. Infection is effectively always introduced into clean flocks by horizontal transmission from an introduced, infected sheep.[31] The incubation period is very long — often two, three or four years or more. The virus does not survive for more than two weeks outside the host.

Pathogenesis

Following infection, viral replication remains very limited for a prolonged period, during which the viral genetic material resides in infected cells as proviral DNA. As the infection progresses, clinical signs are associated with the chronic progressive proliferation of lymphoid tissue in lungs, brain, udder and joints. The pulmonary lesions are effectively an interstitial pneumonia with a restriction of the alveolar space leading to anoxia. Many sheep remain in a subclinical phase of the disease. The variation in expression of the disease as primarily respiratory, nervous or mastitic is presumed to be due to different tissue tropism of the strains of the virus, differences between breeds and flocks of sheep in their susceptibility, and the effects of different management strategies.

Two characteristics of the disease are particularly important in the diagnosis and the epidemiology of the disease. First, infected animals may not develop any detectable antibodies against the infection, or they may take months or years to do so. These animals, which may or may not be presenting clinical abnormalities, are still capable of infecting other animals. Second, even in cases where the animals do mount an immunological response, the disease is progressive and virus multiplication continues.

Clinical signs

Animals with maedi are listless, emaciated and dyspnoeic. Respirations are laboured and rapid: 80 per minute or higher. There is coughing and nasal discharge, but most affected sheep retain their appetite. Udder induration, hind limb paralysis and, in some cases, swollen joints with or without lameness may also be present in the flock. Clinical signs last for 3 to 12 months but the disease is inevitably fatal.

Diagnosis

Sheep pulmonary adenomatosis can produce similar clinical signs with similar flock history and, in some countries, can be simultaneously present in the same flock and the same sheep. Pulmonary adenomatosis is characterised by a profuse nasal discharge and a shorter clinical course. The two diseases can be readily differentiated histopathologically. Parasitic pneumonia and melioidosis also have signs of chronic respiratory disease.

Clinical pathology can aid in diagnosis, either by one of a number of techniques for virus identification or by serology. Serology is used as a flock diagnosis but negative serology in individual cases is not reliable evidence of freedom from infection. The time between infection and seroconversion is variable and may be as long as one or more years. Some infected animals remain seronegative.

Control

There is no treatment for maedi-visna. National control programmes in endemically infected countries vary in their approach but are based largely on the identification of clean flocks by serological testing; separation of lambs from ewes along with artificial rearing of lambs in flocks with low prevalence of infection; or complete destocking and replacement in flocks with moderate or higher levels of infection.

REFERENCES

1   Cullinane LC, Alley MR, Marshall RB et al. (1987) *Bordetella parapertussis* from lambs. NZ Vet J **35** 175. https://doi.org/10.1080/00480169.1987.35433.

2   St George TD and Sullivan ND (1973) Pneumonias of sheep in Australia. Veterinary Review No 13. The University of Sydney Post-Graduate Foundation in Veterinary Science: Sydney.

3   Davies DH (1985) Aetiology of pneumonias of young sheep. Prog Vet Microbiol Immun **1** 229.

4   Hore DE (1976) A review of respiratory agents associated with disease of sheep, cattle and pigs in Australia and overseas. Aust Vet J **52** 502-9. https://doi.org/10.1111/j.1751-0813.1976.tb06985.x.

5   Alley MR and Clarke JK (1979) The experimental transmission of ovine chronic non-progressive pneumonia. NZ Vet J **27** 217-20. https://doi.org/10.1080/00480169.1979.34653.

6   Alley MR, Ionas G and Clarke JK (1999) Chronic non-progressive pneumonia of sheep in New Zealand — a review of the role of *Mycoplasma ovipneumoniae*. NZ Vet J **47** 155-60. https://doi.org/10.1080/00480169.1999.36135.

7   Alley MR (2002) Pneumonia in sheep in New Zealand: An overview. NZ Vet J **50(3)** Supplement 99-101.

8   Kirton AH, O'Hara PJ, Shortridge EA et al. (1976) Seasonal incidence of enzootic pneumonia and its effect on the growth of lambs. NZ Vet J **24** 59-64. https://doi.org/10.1080/00480169. 1976.34284.

9   Hathaway SC and McKenzie AI (1987) Pleurisy in slaughter age lambs. In: Proceedings of the Sheep and Beef Cattle Society of the New Zealand Veterinary Association 17th Seminar. Waikato University, Hamilton, NZ, p. 156.

10   Alley MR (1987) Effects of pneumonia on lamb production. Proceedings of the Sheep and Beef Cattle Society of the New Zealand Veterinary Association 17th Seminar. Waikato University, Hamilton, NZ, p. 163.

11   St George TD, Sullivan ND, Love JA et al. (1971) Experimental transmission of pneumonia in sheep with a mycoplasma isolated from pneumonic sheep lung. Aust Vet J **47** 282-3. https://doi. org/10.1111/j.1751-0813.1971.tb02159.x.

12   Cottew GS (1971) Characterisation of mycoplasmas isolated from sheep with pneumonia. Aust Vet J **47** 591-6. https://doi.org/10.1111/j.1751-0813.1971.tb02078.x.

13   St George TD (1972) Investigations of respiratory disease of sheep in Australia. Aust Vet J **48** 318-22. https://doi.org/10.1111/j.1751-0813.1972.tb02259.x.

14   Carmichael LE, St George TD, Sullivan ND et al. (1972) Isolation, propagation and characterization studies on an ovine mycoplasma responsible for interstitial pneumonia. Cornell Vet **62** 654-79.

15   Sullivan ND, St George TD and Horsfall N (1973) A proliferative interstitial pneumonia of sheep associated with mycoplasma infection. 1. Natural history of the disease in a flock. Aust Vet J **49** 57-62. https://doi.org/10.1111/j.1751-0813.1973.tb09314.x.

16   Sullivan ND, St George TD and Horsfall N (1973) A proliferative interstitial pneumonia of sheep associated with mycoplasma infection. 2. The experimental exposure of young lambs to infection. Aust Vet J **49** 63-8. https://doi.org/10.1111/j.1751-0813.1973.tb09316.x.

17   St George TD (1969) The isolation of *Myxovirus parainfluenza* type 3 from sheep in Australia. Aust Vet J **45** 321-5. https://doi.org/10.1111/j.1751-0813.1969.tb05007.x.

18   St George TD (1971) A survey of sheep throughout Australia for antibody to Parainfluenza type 3 virus and to mucosal disease virus. Aust Vet J **47** 370-4. https://doi.org/10.1111/j.1751-0813.1971. tb09214.x.

19   Davies DH, Dungworth DL, Humphreys S et al. (1977) Concurrent infection of lambs with parainfluenza virus type 3 and *Pasteurella haemolytica*. NZ Vet J **25** 263-5. https://doi.org/10.108 0/00480169.1977.34425.

20   Davies DH, Herceg M, Jones BAH et al. (1981) The pathogenesis of sequential infection with parainfluenza virus type 3 and *Pasteurella haemolytica* in sheep. Vet Microbiol **6** 173-82. https:// doi.org/10.1016/0378-1135(81)90010-9.

21   Davies DH, Davis GB, McSporran KD et al. (1983) Vaccination against ovine pneumonia: A progress report. NZ Vet J **31** 87. https://doi.org/10.1080/00480169.1983.34977.

22   Chen W, Alley MR and Manktelow BW (1988) Pneumonia in lambs inoculated with *Bordetella parapertussis*: Clinical and pathological studies. NZ Vet J **36** 138-42. https://doi.org/10.1080/ 00480169.1988.35509.

23   Alley MR (1987) The effect of chronic non-progressive pneumina on weight gain of pasture-fed lambs. NZ Vet J **35** 163-6. https://doi.org/10.1080/00480169.1987.35429.

24   Kirton AH, O'Hara PJ, Shortridge EH et al. (1977) Pneumonia in sheep: Does it affect weight gain? (a rejoinder). NZ Vet J **25** 195. https://doi.org/10.1080/00480169.1977.34404.

25   Harris RE and Alley MR (1977) Pneumonia in sheep: Does it affect weight gain? NZ Vet J **25** 108. https://doi.org/10.1080/00480169.1977.34374.

26   Gilmour NJL, Angus KW and Gilmour JS (1991) Pasteurellosis. In: Diseases of sheep, eds WB Martin and ID Aitken. 2nd ed. Blackwell Scientific: London, p. 133.

27 Gilmour NJL and Appleyard WT (1989) Long acting oxytetracycline in pneumonic pasteurellosis in lambs. In: Second International Congress for sheep veterinarians, Sheep and Beef Cattle Society of New Zealand. Massey University, Palmerston North, New Zealand, p. 308.

28 Barton and Acland HM (1973) *Mycobacterium avium* serotype 2 infection in a sheep. Aust Vet J **49** 212-13. https://doi.org/10.1111/j.1751-0813.1973.tb06794.x.

29 Cordes DO, Bullians JA, Lake DE et al. (1981) Observations on tuberculosis caused by *Mycobacterium bovis* in sheep. NZ Vet J **29** 60-2. https://doi.org/10.1080/00480169.1981.34798.

30 Arundel JH (1994) Diseases caused by helminth parasites. In: Veterinary medicine, eds DC Blood, OM Radostits and CC Gay. 8th ed. Baillière Tindall: London, pp. 1223-79.

31 Hoff-Jorgensen R (1989) Maedi-Visna. In: Second international congress for sheep veterinarians, Sheep and Beef Cattle Society of New Zealand. Massey University, Palmerston North, New Zealand, p. 333.

www.ingramcontent.com/pod-product-compliance
Lightning Source LLC
Chambersburg PA
CBHW041620220326
41597CB00035BA/6178